T0181847

Lecture Notes in Computer Science   10895

Commenced Publication in 1973
Founding and Former Series Editors:
Gerhard Goos, Juris Hartmanis, and Jan van Leeuwen

More information about this series at http://www.springer.com/series/7407

Jeremy Avigad · Assia Mahboubi (Eds.)

Interactive
Theorem Proving

9th International Conference, ITP 2018
Held as Part of the Federated Logic Conference, FloC 2018
Oxford, UK, July 9–12, 2018
Proceedings

 Springer

Editors
Jeremy Avigad
Carnegie Mellon University
Pittsburgh, PA
USA

Assia Mahboubi
Inria
Nantes
France

ISSN 0302-9743   ISSN 1611-3349 (electronic)
Lecture Notes in Computer Science
ISBN 978-3-319-94820-1   ISBN 978-3-319-94821-8 (eBook)
https://doi.org/10.1007/978-3-319-94821-8

Library of Congress Control Number: 2018947441

LNCS Sublibrary: SL1 – Theoretical Computer Science and General Issues

Printed on acid-free paper

This Springer imprint is published by the registered company Springer Nature Switzerland AG
The registered company address is: Gewerbestrasse 11, 6330 Cham, Switzerland

Preface

The International Conference on Interactive Theorem Proving (ITP) is a premier venue for publishing research in the area of logical frameworks and interactive proof assistants. Its topics include both theoretical foundations and implementation aspects of the technology, as well as applications to verifying hardware and software systems to ensure their safety and security, and applications to the formal verification of mathematical results.

ITP grew out of TPHOLs conferences and ACL2 workshops that began in the early 1990s. Previous editions of ITP have taken place in Brasília (Brazil), Nancy (France), Nanjing (China), Vienna (Austria), Rennes (France), Princeton (USA), Berg en Dal (The Netherlands), and Edinburgh (UK).

This ninth edition, ITP 2018, was part of the Federated Logic Conference (FLoC) 2018, and took place in Oxford, UK, during July 9–12, 2018. We thank the FLoC Organizing Committee for undertaking the Herculean task of planning and organizing the event.

In all, 55 regular papers and ten short papers were submitted to the conference. Each paper was reviewed by at least three people, either members of the Program Committee or external reviewers. The committee ultimately accepted 32 regular papers and five short papers.

ITP 2018 included memorial lectures in honor of Mike Gordon and Vladimir Voevodsky, two influential figures in interactive theorem proving who had passed away over the course of the previous year. John Harrison was invited to present the lecture for Gordon, and Daniel Grayson was invited to present the lecture for Voevodsky. In addition, Jean-Christophe Filliâtre was invited to present a third keynote talk.

The present volume collects all the scientific contributions to the conference as well as abstracts of the three keynote presentations. We are grateful to the members of the ITP Steering Committee for their guidance and advice, and especially grateful to our colleagues on the Program Committee and the external reviewers, whose careful reviews and thoughtful deliberations served to maintain the high quality of the conference. We extend our thanks to the authors of all submitted papers and the ITP community at large, without whom the conference would not exist.

Finally, we are grateful for Springer for once again publishing these proceedings as a volume in the LNCS series, and we thank the editorial team for the smooth interactions.

June 2018

Jeremy Avigad
Assia Mahboubi

Organization

Program Committee

Andreas Abel	Gothenburg University, Sweden
Benedikt Ahrens	University of Birmingham, UK
June Andronick	CSIRO's Data 61 and UNSW, Australia
Jeremy Avigad	Carnegie Mellon University, USA
Jasmin Christian Blanchette	Vrije Universiteit Amsterdam, The Netherlands
Adam Chlipala	Massachusetts Institute of Technology, USA
Thierry Coquand	Chalmers University of Technology, Sweden
Karl Crary	Carnegie Mellon University, USA
Leonardo de Moura	Microsoft, USA
Delphine Demange	University of Rennes 1/IRISA, France
Timothy Griffin	University of Cambridge, UK
Thomas Hales	University of Pittsburgh, USA
John Harrison	Amazon Web Services, USA
Chung-Kil Hur	Seoul National University, South Korea
Johannes Hölzl	Vrije Universiteit Amsterdam, The Netherlands
Jacques-Henri Jourdan	MPI-SWS, Germany
Cezary Kaliszyk	University of Innsbruck, Austria
Ambrus Kaposi	Eötvös Loránd University, Hungary
Chantal Keller	LRI, Université Paris-Sud, France
Assia Mahboubi	Inria, France
Panagiotis Manolios	Northeastern University, USA
Mariano Moscato	National Institute of Aerospace, USA
Magnus O. Myreen	Chalmers University of Technology, Sweden
Tobias Nipkow	Technical University of Munich, Germany
Lawrence Paulson	University of Cambridge, UK
André Platzer	Carnegie Mellon University, USA
Andrei Popescu	Middlesex University London, UK
Matthieu Sozeau	Inria, France
Pierre-Yves Strub	École Polytechnique, France
Enrico Tassi	Inria, France
Zachary Tatlock	University of Washington, USA
Laurent Théry	Inria, France
Cesare Tinelli	The University of Iowa, USA
Alwen Tiu	The Australian National University, Australia
Makarius Wenzel	sketis.net, Germany
Freek Wiedijk	Radboud University, The Netherlands

Additional Reviewers

Åman Pohjola, Johannes
Ahrendt, Wolfgang
Anguili, Carlo
Becker, Heiko
Booij, Auke
Bourke, Timothy
Brecknell, Matthew
Brunner, Julian
Chen, Zilin
Cordwell, Katherine
Czajka, Łukasz
Dawson, Jeremy
Eberl, Manuel
Filliâtre, Jean-Christophe
Fleury, Mathias
Fulton, Nathan
Gammie, Peter
Geuvers, Herman
Hou, Zhe
Immler, Fabian
Jung, Ralf
Komendantskaya, Ekaterina
Kovács, András Kovács
Kraus, Nicolai
Larchey-Wendling, Dominique
Le Roux, Stephane

Lewis, Robert
Martins, João G.
Mitsch, Stefan
Murray, Toby
Mörtberg, Anders
Nagashima, Yutaka
Naumowicz, Adam
Ringer, Talia
Rot, Jurriaan
Sanan, David
Scapin, Enrico
Schmaltz, Julien
Schürmann, Carsten
Sewell, Thomas
Sickert, Salomon
Sison, Robert
Sternagel, Christian
Tanaka, Miki
Tassarotti, Joseph
Thiemann, René
Traytel, Dmitriy
Turaga, Prathamesh
Verbeek, Freek
Villadsen, Jørgen

Abstracts of Invited Talks

Deductive Program Verification

Jean-Christophe Filliâtre[1,2]

[1] Lab. de Recherche en Informatique, Univ. Paris-Sud, CNRS, Orsay, F-91405
[2] Inria Saclay – Île-de-France, Orsay, F-91893
Jean-Christophe.Filliatre@lri.fr

Abstract. Among formal methods, the deductive verification approach consists in first building verification conditions and then resorting to traditional theorem proving. Most deductive verification tools involve a high degree of proof automation through the use of SMT solvers. Yet there may be a substantial part of interactive theorem proving in program verification, such as inserting logical cuts, ghost code, or inductive proofs via lemma functions. In this talk, I will show how the Why3 tool for deductive verification resembles more and more a traditional ITP, while stressing key differences between the two.

Keywords: Deductive verification · Theorem proving

Voevodsky's Work on Formalization of Proofs and the Foundations of Mathematics

Daniel R. Grayson

Abstract. A consistent thread running through the three decades of Voevodsky's work is the application of the ideas of homotopy theory in new and surprising ways, first to motives, and then to formalization of proofs and the foundations of mathematics. I will present the story of the latter development, focusing on the points of interest to mathematicians.

Mike Gordon: Tribute to a Pioneer
in Theorem Proving and Formal Verification

John Harrison

Amazon Web Services
jrh013@gmail.com

Abstract. Prof. Michael J. C. Gordon, FRS was a great pioneer in both computer-aided formal verification and interactive theorem proving. His own work and that of his students helped to explore and map out these new fields and in particular the fruitful connections between them. His seminal HOL theorem prover not only gave rise to many successors and relatives, but was also the framework in which many new ideas and techniques in theorem proving and verification were explored for the first time. Mike's untimely death in August 2017 was a tragedy first and foremost for his family, but was felt as a shocking loss too by many of us who felt part of his extended family of friends, former students and colleagues throughout the world. Mike's intellectual example as well as his unassuming nature and personal kindness will always be something we treasure. In my talk here I will present an overall perspective on Mike's life and the whole arc of his intellectual career. I will also spend time looking ahead, for the research themes he helped to establish are still vital and exciting today in both academia and industry.

Contents

Physical Addressing on Real Hardware in Isabelle/HOL

Reto Achermann[✉], Lukas Humbel, David Cock, and Timothy Roscoe

Department of Computer Science, ETH Zurich, Zürich, Switzerland
{reto.achermann,humbell,david.cock,troscoe}inf.ethz.ch

Abstract. Modern computing platforms are inherently complex and diverse: a heterogeneous collection of cores, interconnects, programmable memory translation units, and devices means that there is no single physical address space, and each core or DMA device may see other devices at different physical addresses. This is a problem because correct operation of system software relies on correct configuration of these interconnects, and current operating systems (and associated formal specifications) make assumptions about global physical addresses which do not hold. We present a formal model in Isabelle/HOL to express this complex addressing hardware that captures the intricacies of different real platforms or Systems-on-Chip (SoCs), and demonstrate its expressivity by showing, as an example, the impossibility of correctly configuring a MIPS R4600 TLB as specified in its documentation. Such a model not only facilitates proofs about hardware, but is used to generate correct code at compile time and device configuration at runtime in the Barrelfish research OS.

1 Introduction

The underlying models of system hardware used by both widely-used operating systems like Linux and verified kernels like seL4 [15] or CertiKOS [12] are highly over-simplified. This leads to both sub-optimal design choices and flawed assumptions on which correctness proofs are then based. Both of these systems treat memory as a flat array of bytes, and model translation units (MMUs) in a limited fashion, if at all. This model of the machine dates to the earliest verified-systems projects (and earlier), and does not reflect the reality of modern hardware, in particular systems-on-chip (SoCs) and expansion busses such as PCI.

Early verified CPUs such as CLI's FM9001 [7] do not include anything beyond what would today be described as the CPU core. The later Verisoft VAMP [6] added a cache, but was still extremely simple, even compared to a mobile phone processor of the same era. None of these models attempted to capture the complexity of, for example, the PCI bus, or a multiprocessor NUMA interconnect: both already commonplace by that time. Modern instruction-set models, such as

© Springer International Publishing AG, part of Springer Nature 2018
J. Avigad and A. Mahboubi (Eds.): ITP 2018, LNCS 10895, pp. 1–19, 2018.
https://doi.org/10.1007/978-3-319-94821-8_1

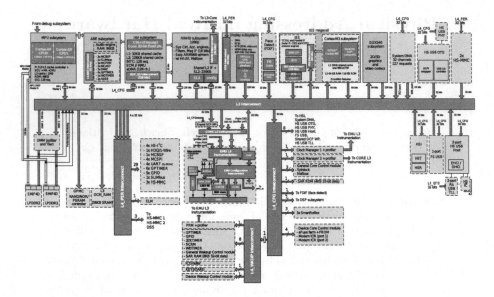

Fig. 1. The OMAP4460—A 'Simple' SoC (OMAP4460 TRM [21])

the HOL4 ARM model [10] or the ARM machine-readable specification [19] provide an excellent reference for reasoning about the behaviour of software, but say nothing about the complex interconnects in modern SoCs (which now include essentially all processor chips). No industrial projects [13] appear to claim to have tackled this area.

The weak memory modeling work of Sewell et al. [4,9], goes deepest, defining the software-visible semantics of memory operations including the effects of pipelining and reordering (e.g. write buffers), but nevertheless only gets us as far as the last-level cache: once we go beyond that, we're really in the Wild West and, as we demonstrate, the path an address takes from the CPU core to its final destination can be extremely complex (if it ever gets there at all)!

Addressing in a system is semantically far more complex than it first appears. Both Linux and seL4 assume a per-core virtual address space translated, at page granularity via a memory management unit (MMU), to a single global physical address space containing all the random access memory (RAM) and memory-mapped devices in the system. This model, found in many undergraduate textbooks, has been hopelessly inaccurate for some time.

Figure 1 shows the manufacturer's *simplified* block diagram for a 10-year-old mobile phone SoC, the Texas Instruments OMAP4460. Already on this chip we can identify at least 13 distinct interconnects, implementing complex address remapping, and at least 7 cores (not counting DMA-capable devices) each with a different view of physical addresses in the system. In addition, the SoC can be configured such that a memory access enters the same interconnect twice, effectively creating a loop.

Correct OS operation requires software to configure all the address translation and protection elements in this (or any other) platform correctly, and hence formal reasoning about the system requires a model which captures the complexity of addressing. Such a model does not fully exist, but the need is recognized even in the Linux community. The state of the art is DeviceTree [8], essentially a binary file format encoding how a booting OS can configure platform hardware in the absence of device discovery. However, DeviceTree's lack of semantics and narrow focus prevent both reasoning about correctness and runtime use beyond initialization.

As we have shown [1,3,11], systems of various architectures and sizes have no single physical address space, which may have been an illusion since early on. Thus, those systems are better modeled as a network of address spaces. We therefore introduced a "decoding net" model and demonstrated how it captures a wide variety of complex modern hardware, from the OMAP SoC, to multi-socket Intel Xeon systems with peripheral component interconnect (PCI)-connected accelerators containing general-purpose cores (e.g. a Xeon Phi).

The contribution of this paper is our formal decoding-net model, mechanised in Isabelle/HOL and expanded relative to our previously-published descriptions, particularly in the treatment of possibly-non-terminating decoding loops. We show its utility in seL4-style refinement proofs by modeling the MIPS4600 TLB [14] and demonstrating that the imprecision of its specification prevents any proof of correct initialization.

2   Model

Our goals in formally specifying the addressing behavior of hardware include the highly practical aim of more easily engineering code for a real OS (Barrelfish [5]) which we are confident operates correctly on a diverse range of hardware platforms. Our model (accessible on Github [2]) is therefore a compromise between the simplicity required to provide meaningful abstractions of the system, and the detail needed to capture features of interest and make the model usable in the OS at compile time and run time.

At the same time, the characteristics of the underlying formalism (here Higher-Order Logic), and the kinds of reasoning efficiently supported by the existing tools and libraries (Isabelle/HOL) also influence the choice of model. Specifically, we make limited use of HOL's relatively simple type system (a formalization in Coq would look very different), but exploit Isabelle's extensive automation for reasoning with relations and flexible function definitions.

Our core abstraction is the *qualified name*: An address is a name, defined in the context of some *namespace*, identified by a natural number. As we have previously shown [3], this suffices to model a large number of interesting real-world examples.

In this view a processor's page tables, for example, define a namespace n by mapping names (addresses) *qualified* by the identifier n into another address space n', the "physical" address of the processor. In general, a name may be mapped to any name in any address space (even itself) or to no name at all. As addresses are discrete we also label them with natural numbers, and the translation behavior of an address space is a function:

$$\texttt{translate} : \mathbb{N} \times \mathbb{N} \to \{\mathbb{N} \times \mathbb{N}\}$$

mapping a fully-qualified name (n, a) (address a in address space n) to some set of names $\{(n', a')\}$ (address a' in space n'). That $\texttt{translate}$ returns a *set*, not just an address, allows for nondeterminism and refinement e.g. the possible configurations of a translation unit can be modeled as "any input may map to any output", of which any particular configuration is a refinement. We do not yet use this feature of the model.

This process should end somewhere: any address should (hopefully) eventually refer to some device (e.g. RAM). To distinguish between this and the case where the translation of an address is simply undefined, we add a per-address space *accept set*:

$$\texttt{accept} : \mathbb{N} \to \{\mathbb{N}\}$$

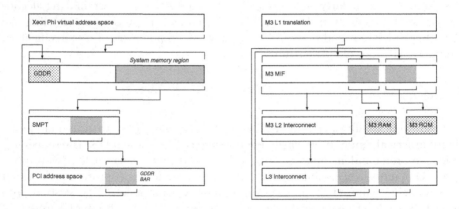

Fig. 2. Existing loops in hardware. Xeon Phi left and OMAP 4460 on the right.

Address $a \in \texttt{accept}\ n$ if a is accepted in address space n and thus address resolution *terminates* in address space n. We deliberately allow an address space to have both a non-empty accept set and non-empty translate sets to cover the behavior of e.g. a cache: some addresses may hit in the cache (and thus be accepted locally), while others miss and are passed through.

These two primitives define the entire semantics of the model: everything else is derived from these. The combination implicitly defines a directed graph on qualified names, where every well-defined translation path ends in an accepting set. We explicitly construct the associated (one-step) relation as follows:

$$\texttt{decodes\_to} = \{((n, a), (n', a')).\ (n', a') \in \texttt{translate}\ (n, a)\}$$

Likewise the set of all fully-qualified names accepted anywhere in the network:

$$\texttt{accepted\_names} = \{(n, a).\ a \in \texttt{accept}\ n\}$$

Finally we define the *net* to be a function from nodeid to node:

$$\texttt{net} : \mathbb{N} \to node$$

2.1 Views and Termination

One might think the model we have just described is overkill in generality for modeling address spaces. To motivate it, we show some examples of the complexity inherent in address resolution in modern hardware.

On the right of Fig. 2 is a subgraph of the interconnect of the OMAP4460 from Fig. 1, showing that it is not a tree. In fact, it is not even acyclic: For example, there is both an edge from the Cortex M3 cores (the ISS megacell) to the L3 interconnect, and another from the L3 back to the M3. The system can be configured so that an address issued by the M3 passes through its local address space *twice* before continuing to one of the L4 interconnects. There's no sensible reason to configure the system like this, but we must be able to express the possibility, to verify that initialisation code *doesn't*. The left of Fig. 2 shows a similar situation arising with a PCI-attached Intel Xeon Phi accelerator. Both examples are from our previous work [3], in which much more complex examples are modeled using the formalism presented here.

Thus the absence of true loops is a property we must formally prove from a description of the system, rather than an *a priori* assumption. Importantly, this proof obligation is not manually appended to the model, but (as we will see) arises naturally when attempting to define a functional representation of an address space.

The possibility of loops, and thus undefined translations, is captured by the general `decodes_to` relation above. While faithful, this relational model is not particularly usable: for practical purposes we are more interested in deriving the complete view of the system from a given processor: the eventual accepting set (if any) for each unqualified name in the processor's local name space.

This is expressible via the reflexive, transitive closure of the decoding relation[1] (R ' S is here the image of the set S under the relation R).

$$\lambda(n, a).\ \texttt{accepted\_names} \cap (\texttt{decodes\_to}^* {}' \{(n, a)\})$$

This is the set of names reachable in 0 or more steps from the root which are accepted by the network. The view from a particular node (the local address space) is then simply the curried function obtained by fixing a particular n.

---

[1] Note that this defines only the decoding *relation* i.e. the set of (name, address) pairs. We only need to show termination once we reformulate it as a recursive function: relations in Isabelle/HOL need only be well-founded if used in a recursive definition (or equivalent).

The model so far is still not quite what we want: we'd like to express the resolution process as a function, preferably with an operational interpretation corresponding (hopefully) meaningfully to the actual hardware behavior. For this we exploit the flexibility of Isabelle's function definition mechanism to separate the simple operational definition of resolution from the more difficult proof of termination:

resolve $(n, a) =$

$$(\{(n, a)\} \cap \texttt{accepted\_names}) \ \cup \ \bigcup \texttt{resolve}\,`(\texttt{decodes\_to}\,`\{(n, a)\})$$

The resolution of a name is the set containing that name (if it's accepted here), together with the resolutions of all names reachable in one step via the decode relation. With this carefully-chosen definition, the correspondence with the relational model is trivial:

assumes resolve_dom (n, a)

shows resolve $(n, a) =$ accepted_names $\cap$ (decodes_to* $`(n, a)$)

The resolve_dom predicate is produced by the Isabelle function definition mechanism thanks to our incomplete definition of the resolve function. It asserts that the name n is in the *domain* of the function resolve i.e. that the function terminates for this argument (or equivalently that it lies in the reachable part of the recurrence relation). Establishing a sufficient, and significantly a *necessary*, condition for the domain predicate comprises the bulk of the proof effort.

The general termination proof for resolve (i.e. establishing the size of resolve_dom) is roughly 500 lines of Isabelle, and consist of establishing a *variant*, or a well-formed ranking of addresses:

wf_rank f $(n, a) =$

$$\forall x, y. \ (x, y) \in \texttt{decodes\_to} \wedge ((n, a), x) \in \texttt{decodes\_to}^* \ \longrightarrow \ f(y) < f(x)$$

From the existence of a well-formed ranking it follows by a straightforward inductive argument that resolve terminates. A rather more complex argument shows that if each decoding step produces at most *finitely many* translations of a name (trivially true for any actual hardware), then the converse also holds i.e. we can find a well-formed ranking of names for any terminating resolution. This establishes a precise equivalence between relational and recursive-functional models:

$$\exists f. \ \texttt{wf\_rank} \ f \ (n, a) \ \longleftrightarrow \ \texttt{resolve\_dom} \ (n, a)$$

The argument proceeds by induction over the structure of the decode relation: For any leaf node, finding a well-formed ranking is trivial; if a well-formed ranking exists for all successors, then take the greatest rank assigned to any successor by any of these rank functions (here is where the finite branching condition is required), add one, and assign it to the current node.

2.2   Concrete Syntax, Prolog, and Sockeye

The goal of our work is to model real hardware and verify the algorithms used to configure it in the context of a real operating system. We therefore define a simple concrete syntax for expressing decoding nets:

$$net_s = \Big\{ \text{N is } node_s \mid \text{N..N are } node_s \Big\}$$

$$node_s = \Big[\textbf{accept } [\, \big\{ block_s \big\} \,] \Big] \Big[\textbf{map } [\, \big\{ map_s \big\} \,] \Big] \Big[\textbf{over N} \Big]$$

$$map_s := block_s \ \textbf{ to N} \Big[\textbf{ at N} \Big] \Big\{ , \text{N} \Big[\textbf{ at N} \Big] \Big\}$$

$$block_s := \text{N} - \text{N}$$

The interpretation of a decoding net expressed in this syntax is given by the parse function, in the accompanying theory files [2].

We use this syntax in the next section, in showing that important operations on the abstract model can be expressed as simply syntactic translations. It is also the basis for the (much more expressive) Sockeye language [18,20], now used for hardware description and configuration in Barrelfish. Programmers write descriptions of a hardware platform, and the Sockeye compiler generates both HOL and a first-order representation of the decode relation as Prolog assertions.

A set of Prolog inference rules is used to query or transform the model (for example implementing the flattening described above), both offline (for example, to preinitialize kernel page tables for bootstrap) and online (for device driver resource allocation) in the Barrelfish OS. Representing the model in Prolog allows it to be dynamically populated at run time in response to core and device discovery, while retaining a formal representation. Establishing equivalence with the HOL model (i.e. verifying the Sockeye compiler) should be straightforward, and is an anticipated extension of this work.

2.3   View-Equivalence and Refinement

To use the model we need efficient algorithms to manipulate it. For example, to preinitialize kernel page tables (as now occurs in Barrelfish) we need to know where in the "physical" address space of a particular processor each device of interest appears. This information is implicit in the decoding net model, but not easily accessible.

To build this view, we transform the network in a way that preserves the processor's views while constructing an explicit physical address space for each. We first split every node such that it either accepts addresses (is a resource), or remaps them (is an address space), but not both. Next, we flatten the address space nodes by mapping each input address directly to all names at which it is accepted, i.e. construct the 1-step transitive closure of the decode relation. Eventually, we terminate with a single address space whose translate function maps directly to the resource of interest.

We say that two decoding networks are *view-equivalent*[2], written $(f, net) \sim_S (g, net')$ if all observers (nodes) in S have the same view (i.e. the results of resolve are the same), modulo renaming (f and g) of the accepting nodes. Given some c greater than the label of any extant node, define the accept and translate functions of the split net, ($\texttt{accept}'_n$ and $\texttt{translate}'_n$, for node n) as:

$$\texttt{accept}'_n = \emptyset$$
$$\texttt{accept}'_{(n+c)} = \texttt{accept}_n$$
$$\texttt{translate}'_n \, a = \{(n+c, a) : a \in \texttt{accept}_n\} \cup \texttt{translate}_n \, a$$
$$\texttt{translate}'_{(n+c)} \, a = \emptyset$$

This new net is view-equivalent to the original, with names that were accepted at n now accepted at $n + c$, and no node both accepting and translating addresses:

$$(n \mapsto n + c, net) \sim_S (\emptyset, \texttt{split}(net)) \tag{1}$$

Splitting on the concrete representation ($\texttt{split}_C$) is a simple syntactic operation:

$$n \text{ is accept } A \text{ map } M \;\mapsto\; [n + c \text{ is accept } A, \; n \text{ is map } M(n \mapsto n + c)]$$

Refinement (in fact equivalence) is expressed as the commutativity of the operations (here $\texttt{split}$ and $\texttt{split}_C$) with the lifting function ($\textbf{parse}$):

$$\texttt{split} \, (\textbf{parse} \, s) = \textbf{parse} \, (\texttt{split}_C \, s) \tag{2}$$

Combining Eq. 1 with Eq. 2 we have the desired result, that the concrete implementation preserves the equivalence of the nets constructed by parsing:

$$(n \mapsto n + c, \textbf{parse} \, s) \sim_S (\emptyset, \textbf{parse} \, (\texttt{split}_C \, s)) \tag{3}$$

Together with the equivalent result for flattening, we can verify that physical address spaces read directly from the transformed model are exactly those that we would have found by traversing the original hardware-derived model for all addresses.

3   Refinement: Example of the MIPS R4600 TLB

Probably the most complex single translation element in a typical system is the processor's TLB, used to implement the abstraction of Virtual Memory. Translation hardware, such as an MMU, intercepts any load and store to a virtual address and maps it to a physical address, or triggers an exception (page fault). The translation implemented by the MMU is generally under the control of the operating system.

---

[2] See definition `view_eq` in `Equivalence.thy` in the attached sources.

As a demonstration of our decoding net model's ability to capture real hardware, and to support reasoning about it, we present a model of the MIPS R4600 TLB [14]; a well-understood and clearly documented, but still comparatively simple such device.

In this section, we show that the behavior of the TLB can be captured by the decoding net model, that we can use refinement to abstract the behavior of a correctly-configured TLB, and prove that the manufacturer's specification is too vague to allow provably-correct initialization.

3.1   The TLB Model

The MIPS TLB is software-loaded: It does not walk page tables in memory for itself, but rather generates a fault whenever a virtual address lookup fails, and relies on the operating system to replace an existing entry with a mapping for the faulting address.

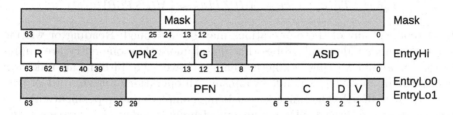

Fig. 3. A MIPS R4600 TLB entry with non-zero fields labelled.

Figure 3 gives the layout of a TLB entry. There are 48 entries, each of which maps two adjacent virtual pages (identified by their virtual page number, or VPN) specified in `EntryHi`, to the physical frames (identified by their physical frame number, or PFN) specified by `EntryLo0` and `EntryLo1`. The TLB (and our model) supports up to seven pre-defined page sizes, but here we consider only the 4kiB case. Physical addresses are 36 bit, while virtual addresses are 40 bit. Addresses are matched against `EntryHi` i.e. on the VPN and address-space identifier (ASID). An entry with the global bit set matches any ASID. We represent a TLB entry with the following Isabelle record type:

$$TLBEntryHi \ = \ (\text{region} : \mathbb{N}, \ \text{vpn2} : \mathbb{N}, \ \text{asid} : \mathbb{N})$$
$$TLBEntryLo \ = \ (\text{pfn} : \mathbb{N}, \ \text{v} : bool, \ \text{d} : bool, \ \text{g} : bool)$$
$$TLBEntry \ = \ (\text{hi} : TLBEntryHi, \ \text{lo0} : TLBEntryLo, \ \text{lo1} : TLBEntryLo)$$

The TLB consists of an indexed set of entries, and two state terms (wired and random) which we will describe shortly. The capacity of the TLB is static, and is only included here to support our refinement proof.

$$MIPSTLB = (\text{wired} : \mathbb{N}, \quad \text{capacity} : \mathbb{N}, \quad \text{random} : \mathbb{N}, \quad \text{entries} : \mathbb{N} \rightarrow TLBEntry)$$

All TLB state changes are made by the OS via 4 special instructions:

tlbp *TLB Probe* performs an associative lookup using the contents of the `EntryHi` register, and either returns the matching index or indicates a miss. The result of a multiple match is undefined (this is important).

tlbr *TLB Read* returns any requested TLB entry.

tlbwi *TLB Write Indexed* updates the entry at a user-specified index.

tlbwr *TLB Write Random* updates the entry at the index specified by the Random register, which is updated nondeterministically to some value in [wired, capacity).

In HOL these are state updates with the following types:

$$\text{tlbp} : TLBENTRYHI \rightarrow MIPSTLB \rightarrow \{\mathbb{N}\}$$
$$\text{tlbr} : \mathbb{N} \rightarrow MIPSTLB \rightarrow \{TLBENTRY\}$$
$$\text{tlbwi} : \mathbb{N} \rightarrow TLBENTRY \rightarrow MIPSTLB \rightarrow \{MIPSTLB\}$$
$$\text{tlbwr} : TLBENTRY \rightarrow MIPSTLB \rightarrow \{MIPSTLB\}$$

The outcome of any of these operations may be undefined: Reading or writing an out-of-bounds index and probes that match more than one entry are unpredictable; moreover writing conflicting entries leaves the TLB in an unknown state. Both are modeled as nondeterminism: All operations return a *set* of possible outcomes (UNIV, the universal set being complete underspecification). For example, `tlbwi` returns UNIV for an out-of-bounds index ($i \geq$ capacity tlb), and otherwise updates the specified entry[3], the singleton set indicating that the result is deterministic:

$$\text{tlbwi } i \ e \ tlb \ = \textbf{if } i < \text{capacity } tlb$$
$$\textbf{then } \{tlb \, (\text{entries} := (\text{entries } tlb)(i := e))\} \textbf{ else UNIV}$$

A TLB random write is then the nondeterministic choice of some indexed write:

$$\text{tlbwr } e \ tlb = \bigcup_{i=\text{wired } tlb}^{(\text{capacity } tlb)-1} \text{tlbwi } i \ e \ tlb$$

3.2 The Validity Invariant

The MIPS TLB famously permits the programmer to configure the TLB in an unsafe manner, such that the future behavior of the processor is undefined. Indeed in early versions of the chip it was possible to permanently damage the hardware in this way. The source of the problem is in the virtual-address match process: this is implemented using a parallel comparison against all 48 entries. The hardware to implement such an associative lookup is very expensive, and

---

[3] $f(x := y)$ is Isabelle/HOL syntax for the function f updated at x with value y.

moreover is on the critical path of any memory access. It is therefore highly optimized, taking advantage of the assumption that there will never be more than one match. Violating this assumption leads to two buffers attempting to drive the same wire to different voltages and, eventually, to smoke.

This assumption is exposed as a requirement that the programmer never configure the TLB such that two entries match the same virtual address. Note, the requirement is not just that two entries never *do* actually match a load or store, but that they never *can*[4]. Also note, that a match occurs independently of the value of the valid bit (V) and therefore even invalid entries must not overlap. This will shortly become important.

We model the above condition with the following invariant on TLB state:

$$\texttt{TLBValid } tlb = \texttt{wired } tlb \leq \texttt{capacity } tlb \wedge$$

$$(\forall i < \texttt{capacity } tlb. \texttt{ TLBEntryWellFormed } (tlb \ i) \wedge \qquad (4)$$

$$\texttt{TLBEntryConflictSet } (\texttt{entries } (tlb \ i)) \ tlb \subseteq \{i\})$$

This predicate states that all entries of the TLB are well formed and do not conflict (overlap) with each other. An entry is well formed if its fields are within valid ranges. We further define the `TLBEntryConflictSet` function:

$$\texttt{TLBEntryConflictSet } :: \ TLBEntry \Rightarrow MIPSTLB \Rightarrow \{\mathbb{N}\}$$

This returns the indices of TLB entries that overlap the provided entry. The correctness invariant is thus that either this set is empty, or contains just the entry under consideration (e.g. the one being replaced). The TLB validity invariant is preserved by all 4 primitives e.g.

> **assumes** TLBValid *tlb*   **and** TLBENTRYWellFormed *e*
>   **and** *i* < capacity *tlb*   **and** TLBEntryConflictSet *e tlb* $\subseteq \{i\}$
>   **shows** $\forall t \in$ tlbwi *i e tlb*. TLBValid *t*

3.3 Invariant Violation at Power On

After reset (e.g. after power on), it is software's responsibility to ensure that the TLB validity invariant is established. However, the specification of the power-on state of the TLB is sufficiently loose to render this impossible!

The MIPS R4600 manual [14] describes the reset state as follows: "*The Wired register is set to 0 upon system reset.*" The `random` register is set to `capacity`−1. The state of the TLB entries is undefined: "*The TLB may be in a random state and must not be accessed or referenced until initialized*". As the MIPS TLB is always on (the kernel is provided with a special untranslated region of

---

[4] We can only speculate as to the writer's intent here. One reason for such a restriction would be speculative execution: The CPU might *speculatively* cause a TLB lookup on an address that it never actually computes. The results would be discarded, but the damage would be done.

virtual addresses to solve the bootstrapping problem), and a strict reading of the invariant requires that there are never two matching entries for any address, even if invalid, the unpredictable initial state cannot be guaranteed to satisfy the invariant. We prove this formally by constructing a TLB state that satisfies the reset condition but not the invariant. A plausible initial state is one where all bits are zero (the null_entry):

$$(\text{wired} = 0, \ \text{random} = 47, \text{capacity} = 48, \text{entries} = \lambda\_. \ \text{null\_entry})$$

While this TLB does not actually translate anything as the valid bits of all entries are zero, addresses from the first page in memory will match all entries of the TLB. The straightforward reading of the manufacturer's specification requires that such a situation is impossible *even if it doesn't actually occur*.

Of course, in practice, operating systems demonstrably *do* successfully initialize MIPS processors. This indicates that the obvious solution is likely also the correct one: as long as two entries never actually match i.e. no translatable access is issued before the TLB is configured, there's no actual problem. In practice the operating system will execute in the non-translated physical window (KSEG0) until the TLB is configured.

This is an example of a *specification bug*, specifically an excessively cautious abstraction that inadvertently hides a correctness-critical detail. While this case is most likely harmless (and this hardware obsolete), recent experience (such as the Meltdown and Spectre attacks [16,17]) demonstrate that supposedly-invisible behavior hidden by an abstraction can unexpectedly become correctness- or security-critical. Indeed, the mechanism exploited by these attacks (speculative execution), could expose this invariant violation: even if no kernel code actually accesses a translatable address, the processor is free to *speculate* such an access, thus triggering the failure.

3.4   What Does a Fully-Wired TLB Do?

The entries of the MIPS TLB can be *wired* (see Fig. 4). The lower w entries are protected from being overwritten by the random write operation (tlbwr). The manual states that *"Wired entries are non-replaceable entries, which cannot be overwritten by a TLB write random operation."* The number of wired entries w can be configured.

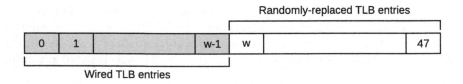

Fig. 4. Wired TLB entries

The random write operation uses the **random** register to select the entry to be overwritten. This register is initialized to **capacity** -1 at reset and is decremented whenever the processor retires an instruction, skipping entries $w-1$ down to 0. As its reset value is **capacity** -1, the random write operation will always succeed regardless of the current value of value w. We can express the bounds as:

$$\text{RandomRange } tlb = \{x.\ \text{wired } tlb \leq x \wedge x < \text{capacity } tlb\}$$

This definition is problematic when we wire all entries of the TLB i.e. by setting w to **capacity**. Note, hardware does not prevent wiring more entries than the capacity. The manual does not mention this case at all. Assuming that $w = \text{capacity}$ we obtain $\text{RandomRange } tlb = \{x.\ \text{capacity } tlb \leq x \wedge x < \text{capacity } tlb\} = \{\}$. This suggests, that no entries will be replaced randomly as intended. However, we know that upon reset the random register is set to **capacity** -1 which is not in the (empty) **RandomRange** set. This contradicts the specification of either the semantics of the wired entries or the random write instruction. We therefore express the random range as follows:

$$\text{RandomRange } tlb = \{x.\ \text{wired } tlb \leq x \wedge x < \text{capacity } tlb\} \cup \{\text{capacity } tlb - 1\}$$

3.5 The TLB Refines a Decoding Net

The preceding specification bugs notwithstanding, we can nevertheless use the TLB model to do useful work. In the remainder of this section we first show that with an appropriate lifting function, our operational model of the TLB refines a decoding-net model of a translate-only node, such that the **tlbwi** operation corresponds to simply updating the appropriate translation. Finally, in Sect. 3.6 through Sect. 3.9 we model the action of TLB refill handler and show that, combined with a valid page table and the operational TLB model, that its action is indistinguishable (again under decoding-net semantics) from that of a TLB large enough to hold all translation entries at once (i.e. no TLB miss exceptions). We lift a single TLB entry to a pair of address-range mappings as follows:

$\text{EntryToMap} : nodeid \Rightarrow \text{TLBENTRY} \Rightarrow addr \Rightarrow \{name\}$

$\text{EntryToMap } n\ e\ va =$

    (**if** $\text{EntryIsValid0 } e \wedge va \in \text{EntryExtendedRange0 } e$

        then $\{(n, \text{EntryPA0 } e + (va \bmod VASize) - \text{EntryMinVA0 } e)\}$ **else** $\{\}$) $\cup$

    (**if** $\text{EntryIsValid1 } e \wedge\ va \in \text{EntryExtendedRange1 } e$

        then $\{(n, \text{EntryPA1 } e + (va \bmod VASize) - \text{EntryMinVA1 } e)\}$ **else** $\{\}$)

The $\text{EntryExtendedRange(0,1)}$ functions consider the virtual address, the address-space identifier and the global bit, by extending the virtual address with the ASID bits such that the *extended virtual address* space contains all virtual addresses for ASID 0, followed by those for ASID 1, and so forth.

The TLB's representation is then the union of these translations:

ConvertToNode n tlb =

$$\left(accept = \{\}, \; translate = \lambda a. \bigcup \text{EntryToMap } n \; (\text{entries } tlb \; i) \; a\right)$$

The abstract equivalent of tlbwi is the replace_entry function, which replaces entry $e1$ with $e2$ by updating the translation as follows:

translate n $a \mapsto$ (translate n a − EntryToMap n $e1$ a) $\cup$ EntryToMap n $e2$ a

The following lemma shows the equivalence of the tlbwi instruction and the TLB indexed write (replace_entry) function, i.e. that commuting with the lifting function maps one to the other.

assumes $i <$ capacity tlb and TLBValid tlb and TLBEntryWriteable i e tlb

shows (ConvertToNode n)$'$(tlbwi i e tlb) =

replace_entry n (entries tlb i) e (ConvertToNode n tlb)

3.6 Modeling TLB Lookups and Exceptions

An MMU provides the illusion of a large virtual address space, using a small TLB, by loading entries on demand from a large in-memory translation table. On the MIPS, this is handled in software, according to the exception flowchart in Figure 4–19 of the MIPS manual [14]. The following three exceptions are defined:

TLB Refill: No entry matched the given virtual address.
TLB Invalid: An entry matched, but was invalid.
TLB Modified: Access violation e.g. a write to a read-only page.

Table 1. The outcome of the translate function.

Match	Valid	Entry writable/Memory write	VPN even	Result
No	*	*	*	TLB Refill Exception
Yes	No	*	*	TLB Invalid Exception
Yes	No	No and memory write	*	TLB Modified Exception
Yes	Yes	Yes or memory read	Yes	Translate using EntryLo0
Yes	Yes	Yes or memory read	No	Translate using EntryLo1

The possible outcomes of a TLB lookup are summarized in Table 1, and modeled (for a particular entry) in our nondeterministic operational style as follows:

TLBENTRY_translate e as vpn =

if EntryMatchVPNASID vpn as e then

if even vpn $\wedge$ EntryIsValid0 e

then $\{(\text{pfn } (lo0 \; e)) + (vpn - \text{EntryMin4KVPN } e)\}$

else if odd vpn $\wedge$ EntryIsValid1 e

then $\{(\text{pfn } (lo1 \; e)) + (vpn - \text{EntryMin4KVPN1 } e)\}$ else $\{\}$

else $\{\}$

Again exploiting nondeterminism, we define `MIPSTLB_translate` as follows:

$$\text{MIPSTLB\_translate } tlb \; vpn \; as =$$
$$\bigcup_{i<\texttt{capacity } tlb} \text{TLBENTRY\_translate } ((\text{entries } tlb) \; i) \; as \; vpn$$

The TLB invariant (Eq. 4) implies that at most one entry will match, and thus the union is trivial.

3.7 Adding a Page Table

With a software-loaded TLB, the OS programmer is free to select any data structure for the page tables. The simplest, and a very common, choice is an array of `TLBEntryLo` values, indexed by address space and virtual page number:

$$MIPSPT : \mathbb{N} \; \rightarrow \; \mathbb{N} \; \rightarrow \; TLBENTRYLO$$

The replacement handler must then simply load the entry corresponding to the faulting address (if any) and restart the faulting process. In order to guarantee that the TLB invariant is maintained, we show that the following invariant holds of the page table, which simply applies the invariant to all elements of the in-memory representation:

assumes `MIPSPT_valid` pt **and** `ASIDValid` as **and** $vpn < MIPSPT\_EntriesMax$
shows `TLBENTRYWellFormed` (`MIPSPT_mk_tlbentry` $pt \; as \; vpn$)

3.8 Modeling Replacement Handlers

Combining the page table representation above with the TLB, we can model a replacement handler that "caches" translations from the page table in the TLB:

$$MipsTLBPT = \big(tlb : MIPSTLB, \quad pte : MIPSPT\big)$$

The replacement handler writes entries constructed from the page table into the TLB. The TLB should thus always be an "instance" of (hold a subset of entries from) the page table:

`MipsTLBPT_is_instance` $mt = \forall i < $ `capacity` (`tlb` mt).
$\qquad$ `entries` (`tlb` mt) $i = $ `MIPSPT_mk_tlbentry` (`pte` mt)
$\qquad\qquad\qquad\qquad$ (`asid` (`hi` (`entries` (`tlb` mt) i)))
$\qquad\qquad\qquad\qquad$ (`vpn2` (`hi` (`entries` (`tlb` mt) i)))

This predicate ensures there are no other entries in the TLB than those constructed from the page table—a property we will use later when we show equivalence to a large TLB.

The replacement handler can either replace an entry deterministically by choosing the index as a function of the entry's VPN:

MIPSTLBIndex tlb $entry$ $=$ (vpn2 (hi $entry$)) mod (capacity tlb)

or make use of the nondeterministic random write function. The deterministic update function implements a direct mapped replacement policy i.e. an entry can only ever be present in a well defined slot which simplifies reasoning about the TLB invariant.

However, this placement policy is not applicable in general e.g. when the OS wants to divide the entries into wired and random (Sect. 3.4) or in the presence of hardware table walkers and associative TLBs that non-deterministically replace an entry. Hence, the location of the entry in the TLB is no longer fixed.

In the non-deterministic model of the replacement handler, we make sure that we only ever change the state of the TLB when a translation attempt would trigger a *refill* exception:

MipsTLBPT_fault $mtlb$ as vpn =

if MIPSTLB_try_translate (tlb $mtlb$) as vpn $=$ $EXNREFILL$

then   MipsTLBPT_update_tlb $mtlb$ as vpn else $\{mtlb\}$

Therefore, when we construct the new entry from the page table and update the TLB by replacing an existing entry with the new one, we are guaranteed not to cause and conflicts. Hence, we prove that MipsTLBPT_fault preserves the TLB invariance:

assumes MipsTLBPT_valid mpt and   ASIDValid as

and vpn < MIPSPT_EntriesMax

shows $\forall m \in$ MipsTLBPT_fault mpt as vpn. MipsTLBPT_valid m

Note, we use a stricter definition of validity in this case requiring also the *is_instance* predicate and that the page tables are well formed.

MipsTLBPT_valid mt =MIPSPT_valid (pte mt) $\land$ TLBValid (tlb mt)

$\land$ MipsTLBPT_is_instance mt

3.9   Equivalence to Infinitely Large TLB

The final result regarding the TLB is to show that, together with the refill handler, it implements the expected abstraction: a single decoding-net node, that maps the virtual address space to the physical. We do this by showing that the TLB plus refill handler is equivalent, in the decoding-net semantics, to a hypothetical giant TLB, large enough to hold all mappings at once, and that therefore never faults.

We construct the large TLB by pre-loading all entries from the page table, according to our "extended virtual address" scheme, giving a unique, deterministic location for each entry:

MipsTLBLarge_create $pt =$

$\quad (capacity = \text{MaxEntries}, \quad wired = \text{MaxEntries},$

$\quad entries = \lambda n.\ \text{MIPSPT\_mk\_tlbentry}\ pt\ (\text{idx2asid}\ n)\ (\text{idx2vpn}\ n))$

We first define a translate function $TLB \Rightarrow ASID \Rightarrow VPN \Rightarrow PFN$ for both, the large TLB and the TLB with replacement handler, and we show that the two are equivalent for any sensible VPN and ASID:

assumes $vpn < MIPSPT\_EntriesMax$ **and** $as < ASIDMax$

$\quad$ **and** capacity (tlb mpt) > 0 **and** MipsTLBPT_valid mpt

$\quad$ **shows** MipsTLBPT_translate mpt as $vpn =$

$\quad\quad$ MIPSTLB_translate (MipsTLBLarge_create(pte mpt)) as vpn(5)

We use the translate functions and the equivalence result above when lifting the large TLB and the TLB with replacement handler to the decoding net node. Here we show the variant for the real TLB:

MipsTLBPT_to_node nid $mpt =$

$\quad (accept = \{\}, translate = (\lambda a.(\textbf{if}\ \text{AddrValid}\ a\ \textbf{then}$

$\quad\quad (\bigcup x \in (\text{MipsTLBPT\_translate}\ mpt\ (\text{addr2asid}\ a)(\text{addr2vpn}\ a)).$

$\quad\quad \{(nid, \text{pfn2addr}\ x\ a)\})\ \textbf{else}\ \{\})))$

The accept set is empty. The node's translate function checks whether the address falls within defined range (**AddrValid**) and then either return an empty set or the result of the TLB's translate function. Note, we need to convert between addresses and VPN/ASID and PFN.

Lastly, we prove the equivalence of the lifting functions and their result when applied to the large TLB and the TLB with replacement handler respectively:

assumes capacity (tlb mpt) > 0 **and**   MipsTLBPT_valid mpt

$\quad$ **shows** MipsTLBPT_to_node nid $mpt =$

$\quad\quad$ MIPSLARGE_to_node nid (MipsTLBLarge_create (pte mpt))

Recall, we have shown that the translate function of the two TLB's have identical behavior (Eq. 5) if the large TLB was pre-populated with the same page tables. Therefore, applying the lifting functions produces equivalent nodes in decoding net semantics.

4   Conclusion

In this paper, we have demonstrated the use of Isabelle/HOL to formally model and reason about the increasingly complex process of address resolution and mapping in modern processors. The traditional model of a single virtual address space, mapped onto a global physical address space has been a gross oversimplification for a long time, and this is becoming more and more visible.

Our decoding-net model, and the Sockeye language that we have developed from it, present an semantically rigorous and formally-verified alternative to Device Trees. In our prior work, we have demonstrated that this model can be applied to a wide range of very complex real hardware, and in this paper we further demonstrate its application to modelling the MIPS TLB. That we were able to prove that this correctly implements the virtual-memory abstraction shows not just that this particular hardware is indeed correct, but that our model is tractable for such proofs. That we discovered a number of specification bugs demonstrates clearly the benefit of a rigorous formal proof, and is further evidence in favor of the formal specification of the semantics of hardware.

The Sockeye language is already in use in the Barrelfish operating system, and we anticipate verifying the Sockeye compiler, in particular that the Prolog assertions generated are equivalent to the HOL model. Further, the model as it stands is principally a static one: expressing the configuration space of the system in a more systematic manner than simply allowing general functions (for example to model region-based remapping units, as used in PCI), and reasoning about the dynamic behavior of requests as translations are updated (for example that a series of updates to different translation units never leaves the system in an intermediate state that violates invariants) is an exciting future direction that we intend to explore. Likewise modelling request properties (read, write, cacheable, etc.), and their interaction with existing weak memory models, presents a challenge.

The ultimate prize is to model the memory system in sufficient detail to be able to specify the behavior of a system including partly-coherent caches (such as ARM) and table-walking MMUs that themselves load page table entries via the cache and/or second-level translations (as in two-level paging for virtualization). This goal is still a long way off, but the increasing quality and availability of formal hardware models leaves us hope that it is indeed attainable.

References

1. Achermann, R.: Message passing and bulk transport on heterogenous multiprocessors. Master's thesis, Department of Computer Science, ETH Zurich, Switzerland (2017)
2. Achermann, R., Cock, D., Humebl, L.: Hardware Models in Isabelle/HOL, January 2018. https://github.com/BarrelfishOS/Isabelle-hardware-models
3. Achermann, R., Humbel, L., Cock, D., Roscoe, T.: Formalizing memory accesses and interrupts. In: Proceedings of the 2nd Workshop on Models for Formal Analysis of Real Systems, MARS 2017, pp. 66–116 (2017)
4. Alglave, J.: A formal hierarchy of weak memory models. Form. Methods Syst. Des. **41**(2), 178–210 (2012)
5. The Barrelfish Operating System. https://www.barrelfish.org

6. Beyer, S., Jacobi, C., Kröning, D., Leinenbach, D., Paul, W.J.: Putting it all together – formal verification of the VAMP. Int. J. Softw. Tools Technol. Transf. **8**(4), 411–430 (2006)
7. Bishop, M.K., Brock, C., Hunt, W.A.: The FM9001 Microprocessor Proof. Technical report 86, Computational Logic Inc. (1994)
8. devicetree.org: Devicetree Specification, May 2016. Release 0.1. http://www.devicetree.org/specifications-pdf
9. Flur, S., Gray, K.E., Pulte, C., Sarkar, S., Sezgin, A., Maranget, L., Deacon, W., Sewell, P.: Modelling the ARMv8 architecture, operationally: concurrency and ISA. In: Proceedings of the 43rd Annual ACM SIGPLAN-SIGACT Symposium on Principles of Programming Languages, POPL 2016, pp. 608–621. ACM, St. Petersburg (2016)
10. Fox, A., Myreen, M.O.: A trustworthy monadic formalization of the ARMv7 instruction set architecture. In: Kaufmann, M., Paulson, L.C. (eds.) ITP 2010. LNCS, vol. 6172, pp. 243–258. Springer, Heidelberg (2010). https://doi.org/10.1007/978-3-642-14052-5_18
11. Gerber, S., Zellweger, G., Achermann, R., Kourtis, K., Roscoe, T., Milojicic, D.: Not your parents' physical address space. In: Proceedings of the 15th USENIX Conference on Hot Topics in Operating Systems, HOTOS 2015, p. 16 (2015)
12. Gu, R., Shao, Z., Chen, H., Wu, X., Kim, J., Sjöberg, V., Costanzo, D.: CertiKOS: an extensible architecture for building certified concurrent OS kernels. In: Proceedings of the 12th USENIX Conference on Operating Systems Design and Implementation, OSDI 2016, pp. 653–669. USENIX Association, Savannah (2016)
13. Hunt, W.A., Kaufmann, M., Moore, J.S., Slobodova, A.: Industrial hardware and software verification with ACL2. Phil. Trans. R. Soc. A **375**(2104), 20150399 (2017)
14. Integrated Device Technology, Inc.: IDT79R4600 TM and IDT79R4700 TM RISC Processor Hardware User's Manual, revision 2.0 edition, April 1995
15. Klein, G., Elphinstone, K., Heiser, G., Andronick, J., Cock, D., Derrin, P., Elkaduwe, D., Engelhardt, K., Kolanski, R., Norrish, M., Sewell, T., Tuch, H., Winwood, S.: seL4: formal verification of an OS kernel. In: Proceedings of the ACM SIGOPS 22nd Symposium on Operating Systems Principles, SOSP 2009, pp. 207–220. ACM, Big Sky (2009)
16. Kocher, P., Genkin, D., Gruss, D., Haas, W., Hamburg, M., Lipp, M., Mangard, S., Prescher, T., Schwarz, M., Yarom, Y.: Spectre Attacks: Exploiting Speculative Execution. ArXiv e-prints, January 2018
17. Lipp, M., Schwarz, M., Gruss, D., Prescher, T., Haas, W., Mangard, S., Kocher, P., Genkin, D., Yarom, Y., Hamburg, M.: Meltdown. ArXiv e-prints, January 2018
18. T.B. Project: Sockeye in Barrelfish
19. Reid, A.: Trustworthy specifications of ARM V8-A and V8-M system level architecture. In: FMCAD 2016, pp. 161–168. FMCAD Inc., Austin (2016)
20. Schwyn, D.: Hardware configuration with dynamically-queried formal models. Master's thesis, Department of Computer Science, ETH Zurich, Switzerland (2017)
21. Texas Instruments: OMAP44xx Multimedia Device Technical Reference Manual, April 2014. Version AB. www.ti.com/lit/ug/swpu235ab/swpu235ab.pdf

Towards Certified Meta-Programming with Typed TEMPLATE-COQ

Abhishek Anand[1]([⊠]), Simon Boulier[2], Cyril Cohen[3], Matthieu Sozeau[4], and Nicolas Tabareau[2]

[1] Cornell University, Ithaca, NY, USA
aa755@cs.cornell.edu
[2] Gallinette Project-Team, Inria Nantes, Rennes, France
[3] Université Côte d'Azur, Inria, Nice, France
[4] Pi.R2 Project-Team, Inria Paris and IRIF, Paris, France

Abstract. TEMPLATE-COQ (https://template-coq.github.io/template-coq) is a plugin for COQ, originally implemented by Malecha [18], which provides a reifier for COQ terms and global declarations, as represented in the COQ kernel, as well as a denotation command. Initially, it was developed for the purpose of writing functions on COQ's AST in GALLINA. Recently, it was used in the CERTICOQ certified compiler project [4], as its front-end language, to derive parametricity properties [3], and to extract COQ terms to a CBV λ-calculus [13]. However, the syntax lacked semantics, be it typing semantics or operational semantics, which should reflect, as formal specifications in COQ, the semantics of COQ's type theory itself. The tool was also rather bare bones, providing only rudimentary quoting and unquoting commands. We generalize it to handle the entire Calculus of Inductive Constructions (CIC), as implemented by COQ, including the kernel's declaration structures for definitions and inductives, and implement a monad for general manipulation of COQ's logical environment. We demonstrate how this setup allows COQ users to define many kinds of general purpose plugins, whose correctness can be readily proved in the system itself, and that can be run efficiently after extraction. We give a few examples of implemented plugins, including a parametricity translation. We also advocate the use of TEMPLATE-COQ as a foundation for higher-level tools.

1 Introduction

Meta-programming is the art of writing programs (in a *meta-language*) that produce or manipulate programs (written in an *object language*). In the setting of dependent type theory, the expressivity of the language permits to consider the case were the meta and object languages are actually the same, *accounting for well-typedness*. This idea has been pursued in the work on inductive-recursive (IR) and quotient inductive-inductive types (QIIT) in Agda to reflect a syntactic model of a dependently-typed language within another one [2,9]. These term encodings include type-correctness internally by considering only well-typed terms of the syntax, i.e. derivations. However, the use of IR or QIITs complicates considerably the

J. Avigad and A. Mahboubi (Eds.): ITP 2018, LNCS 10895, pp. 20–39, 2018.
https://doi.org/10.1007/978-3-319-94821-8_2

meta-theory of the meta-language which makes it difficult to coincide with the object language represented by an inductive type. More problematically in practice, the concision and encapsulation of the syntactic encoding has the drawback that it is very difficult to use because any function from the syntax can be built only at the price of a proof that it respects typing, conversion or any other features described by the intrinsically typed syntax right away.

Other works have taken advantage of the power of dependent types to do meta-programming in a more progressive manner, by first defining the syntax of terms and types; and then defining out of it the notions of reduction, conversion and typing derivation [11,26] (the introduction of [11] provides a comprehensive review of related work in this area). This can be seen as a type-theoretic version of the functional programming language designs such as TEMPLATE HASKELL [22] or METAML [24]. This is also the approach taken by Malecha in his thesis [18], where he defined TEMPLATE-COQ, a plugin which defines a correspondence—using quoting and unquoting functions—between COQ kernel terms and inhabitants of an inductive type representing internally the syntax of the calculus of inductive constructions (CIC), as implemented in COQ. It becomes thus possible to define programs in COQ that manipulate the representation of COQ terms and reify them as functions on COQ terms. Recently, its use was extended for the needs of the CERTICOQ certified compiler project [4], which uses it as its front-end language. It was also used by Anand and Morissett [3] to formalize a modified parametricity translation, and to extract COQ terms to a CBV λ-calculus [13]. All of these translations however lacked any means to talk about the semantics of the reified programs, only syntax was provided by TEMPLATE-COQ. This is an issue for CERTICOQ for example where both a non-deterministic small step semantics and a deterministic call-by-value big step semantics for CIC terms had to be defined and preserved by the compiler, without an "official" specification to refer to.

This paper proposes to remedy this situation and provides a formal semantics of COQ's implemented type theory, that can independently be refined and studied. The advantage of having a very concrete untyped description of COQ terms (as opposed to IR or QIITs definitions) together with an explicit type checker is that the extracted type-checking algorithm gives rise to an OCAML program that can directly be used to type-check COQ kernel terms. This opens a way to a concrete solution to bootstrap COQ by implementing the COQ kernel in COQ. However, a complete reification of CIC terms and a definition of the checker are not enough to provide a meta-programming framework in which COQ plugins could be implemented. One needs to get access to COQ logical environments. This is achieved using a monad that reifies COQ general commands, such as lookups and declarations of constants and inductive types.

As far as we know this is the only reflection framework in a dependently-typed language allowing such manipulations of terms and datatypes, thanks to the relatively concise representation of terms and inductive families in CIC. Compared to the METACOQ project [27], LEAN's tactic monad [12], or AGDA's reflection framework [26], our ultimate goal is not to interface with COQ's

```
Inductive term : Set :=
| tRel       : N → term
| tVar       : ident → term
| tEvar      : N → list term → term
| tSort      : universe → term
| tCast      : term → cast_kind → term → term
| tProd      : name → term → term → term
| tLambda    : name → term → term → term
| tLetIn     : name → term → term → term → term
| tApp       : term → list term → term
| tConst     : kername → universe_instance → term
| tInd       : inductive → universe_instance → term
| tConstruct: inductive → N → universe_instance → term
| tCase      : inductive * N → term → term → list (N * term) → term
| tProj      : projection → term → term
| tFix       : mfixpoint term → N → term
| tCoFix     : mfixpoint term → N → term.
```

Fig. 1. Representation of the syntax in TEMPLATE-COQ

unification and type-checking algorithms, but to provide a self-hosted, bootstrappable and verifiable implementation of these algorithms. On one hand, this opens the possibility to verify the kernel's implementation, a problem tackled by BARRAS [6] using set-theoretic models. On the other hand we also advocate for the use of TEMPLATE-COQ as a foundation on which higher-level tools can be built: meta-programs implementing translations, boilerplate-generating tools, domain-specific proof languages, or even general purpose tactic languages.

Plan of the Paper. In Sect. 2, we present the complete reification of COQ terms, covering the entire CIC and define in Sect. 3 the type-checking algorithm of COQ reified terms in COQ. In Sect. 4, we show the definition of a monad for general manipulation of COQ's logical environment and use it to define plugins for various translations from COQ to COQ (Sect. 5). Finally, we discuss related and future work in Sect. 6.

2   Reification of COQ Terms

Reification of Syntax. The central piece of TEMPLATE-COQ is the inductive type **term** which represents the syntax of COQ terms, as defined in Fig. 1. This inductive follows directly the **constr** datatype of COQ terms in the OCAML code of COQ, except for the use of OCAML's native arrays and strings; an upcoming extension of COQ [5] with such features should solve this mismatch.

Constructor **tRel** represents variables bound by abstractions (introduced by **tLambda**), dependent products (introduced by **tProd**) and local definitions (introduced by **tLetIn**), the natural number is a De Bruijn index. The **name** is a printing annotation.

Sorts are represented with tSort, which takes a universe as argument. A universe is the supremum of a (non-empty) list of level expressions, and a level is either Prop, Set, a global level or a De Bruijn polymorphic level variable.

```
Inductive level := lProp | lSet | Level (_ : string) | Var (_ : N).
Definition universe := list (level * bool). (* level+1 if true *)
```

The application (introduced by tApp) is n-ary. The tConst, tInd and tConstruct constructors represent references to constants (definitions or axioms), inductives, or constructors of an inductive type. The universe_instances are non-empty only for polymorphic constants. Finally, tCase represents pattern-matchings, tProj primitive projections, tFix fixpoints and tCoFix cofixpoints.

Quoting and Unquoting of Terms. TEMPLATE-COQ provides a lifting from concrete syntax to reified syntax (quoting) and the converse (unquoting). It can reify and reflect all kernel COQ terms.

The command Quote Definition reifies the syntax of a term. For instance,

```
Quote Definition f := (fun x ⇒ x + 0).
```

generates the term f defined as

```
f = tLambda (nNamed "x") (tInd {| inductive_mind := "Coq.Init.Datatypes.
    nat"; inductive_ind := 0 |} []) (tApp (tConst "Coq.Init.Nat.add" [])
    [tRel 0; tConstruct {| inductive_mind := "Coq.Init.Datatypes.nat";
    inductive_ind := 0 |} 0 []]) : term
```

On the converse, the command Make Definition constructs a term from its syntax. This example below defines zero to be 0 of type N.

```
Make Definition zero := tConstruct (mkInd "Coq.Init.Datatypes.nat" 0)
                        0 [].
```

where mkInd n k is the kth inductive of the mutual block of the name n.

Reification of Environment. In COQ, the meaning of a term is relative to an environment, which must be reified as well. Environments consist of three parts: (i) a graph of universes (ii) declarations of definitions, axioms and inductives (iii) a local context registering types of De Bruijn indexes.

As we have seen in the syntax of terms, universe levels are not given explicitly in COQ. Instead, level variables are introduced and constraints between them are registered in a *graph of universes*. This is the way typical ambiguity is implemented in COQ. A constraint is given by two levels and a constraint_type (Lt, Le or Eq):

```
Definition univ_constraint := Level * constraint_type * Level.
```

Then the graph is given by a set of level variables and one of constraints. Sets, coming from the COQ standard library, are implemented using lists without duplicates. LevelSet.t means the type t of the module LevelSet.

```
Definition uGraph := LevelSet.t * ConstraintSet.t.
```

Functions to query the graph are provided, for the moment they rely on a naive implementation of the Bellman-Ford algorithm. `check_leq u1 u2` checks if the graph enforces $u1 \leq u2$ and `no_universe_inconsistency` checks that the graph has no negative cycle.

Constant and inductive declarations are grouped together, properly ordered according to dependencies, in a global context (`global_ctx`), which is a list of global declarations (`global_decl`).

```
Inductive global_decl :=
| ConstantDecl  : ident → constant_decl  → global_decl
| InductiveDecl : ident → minductive_decl → global_decl.
```

Definitions and axioms just associate a name to a universe context, and two terms for the optional body and type. Inductives are more involved:

```
(* Declaration of one inductive type *)
Record inductive_body := { ind_name : ident;
   ind_type  : term; (* closed arity *)
   ind_kelim : list sort_family; (* allowed elimination sorts *)
   (* names, types, number of arguments of constructors *)
   ind_ctors : list (ident * term * nat);
   ind_projs : list (ident * term) (* names and types of projections *)}.
```

```
(* Declaration of a block of mutual inductive types *)
Record minductive_decl := { ind_npars : nat; (* number of parameters *)
   ind_bodies : list inductive_body; (* inductives of the mutual block *)
   ind_universes : universe_context (* universe constraints *) }.
```

In COQ internals, there are in fact two ways of representing a declaration: either as a "declaration" or as an "entry". The kernel takes entries as input, type-check them and elaborate them to declarations. In TEMPLATE-COQ, we provide both, and provide an erasing function `mind_decl_to_entry` from declarations to entries for inductive types.

Finally, local contexts are just list of local declarations: a type for lambda bindings and a type and a body for let bindings.

Quoting and Unquoting the Environment. TEMPLATE-COQ provides the command `Quote Recursively Definition` to quote an environment. This command crawls the environment and quote all declarations needed to typecheck a given term.

The other way, the commands `Make Inductive` allows declaring an inductive type from its entry. For instance the following redefines a copy of $\mathbb{N}$:

```
Make Inductive (mind_decl_to_entry
    {| ind_npars := 0;  ind_universes := [];
       ind_bodies := [{|
         ind_name := "nat";
         ind_type := tSort [(lSet, false)];
         ind_kelim := [InProp; InSet; InType];
         ind_ctors := [("O", tRel 0, 0);
```

```
                    ("S", tProd nAnon (tRel 0) (tRel 1), 1)];
        ind_projs := [] |}] |} ).


Make Inductive (mind_decl_to_entry
    {| ind_npars := 0;  ind_universes := [];
       ind_bodies := [{|
         ind_name := "nat";
         ind_type := tSort [(lSet, false)];
         ind_kelim := [InProp; InSet; InType];
         ind_ctors := [("0", tRel 0, 0);
                       ("S", tProd nAnon (tRel 0) (tRel 1), 1)];
         ind_projs := [] |}] |} ).
```

Inductive cumul (Σ : global_ctx) (Γ : context) : term $\rightarrow$ term $\rightarrow$ Prop
:= | cumul_refl t u : leq_term (snd Σ) t u = true $\rightarrow$ Σ ; $\Gamma \vdash$ t $\leq$ u
| cumul_red_l t u v : red1 Σ Γ t v $\rightarrow$ Σ ; $\Gamma \vdash$ v $\leq$ u $\rightarrow$ Σ ; $\Gamma \vdash$ t $\leq$ u
| cumul_red_r t u v : Σ ; $\Gamma \vdash$ t $\leq$ v $\rightarrow$ red1 Σ Γ u v $\rightarrow$ Σ ; $\Gamma \vdash$ t $\leq$ u
where " Σ ; $\Gamma \vdash$ t $\leq$ u " := (cumul Σ Γ t u).

Inductive typing (Σ : global_ctx) (Γ : context) : term $\rightarrow$ term $\rightarrow$ Set
:= | type_Rel n : $\forall$ (H : n < List.length Γ),
      Σ ; $\Gamma \vdash$ tRel n : lift$_0$ (S n) (safe_nth Γ (exist _ n H)).(decl_type)
| type_Sort (l : level) :
      Σ ; $\Gamma \vdash$ tSort (Universe.make l) : tSort (Universe.super l)
| type_Prod n t b s1 s2 :
      Σ ; $\Gamma \vdash$ t : tSort s1 $\rightarrow$ Σ ; Γ , vass n t $\vdash$ b : tSort s2 $\rightarrow$
      Σ ; $\Gamma \vdash$ tProd n t b : tSort (max_universe s1 s2)
| type_App t l t_ty t' :
      Σ ; $\Gamma \vdash$ t : t_ty $\rightarrow$ typing_spine Σ Γ t_ty l t' $\rightarrow$
      Σ ; $\Gamma \vdash$ tApp t l : t'
| ...
where " Σ ; $\Gamma \vdash$ t : T " := (typing Σ Γ t T)

with typing_spine Σ Γ : term $\rightarrow$ list term $\rightarrow$ term $\rightarrow$ Prop :=
| type_spine_nil ty : typing_spine Σ Γ ty [] ty
| type_spine_const hd tl na A B B' T :
      Σ ; $\Gamma \vdash$ T $\leq$ tProd na A B $\rightarrow$ Σ ; $\Gamma \vdash$ hd : A $\rightarrow$
      typing_spine Σ Γ (subst$_0$ hd B) tl B' $\rightarrow$
      typing_spine Σ Γ T (cons hd tl) B'

Fig. 2. Typing judgment for terms, excerpt

More examples of use of quoting/unquoting commands can be found in the
file test-suite/demo.v.

3   Type Checking CoQ in CoQ

In Fig. 2, we present (an excerpt of) the specification of the typing judgment of the kernel of CoQ using the inductive type **typing**. It represents all the typing rules of CoQ[1]. This includes the basic dependent lambda calculus with lets, global references to inductives and constants, the match construct and primitive projections. Universe polymorphic definitions and the well-formedness judgment for global declarations are dealt with as well.

The only ingredients missing are the guard check for fixpoint and productivity of cofixpoints and the positivity condition of mutual (co-) inductive types. They are work-in progress.

The typing judgment **typing** is mutually defined with **typing_spine** to account for n-ary applications. Untyped reduction **red1** and cumulativity **cumul** can be defined separately.

Implementation. To test this specification, we have implemented the basic algorithms for type-checking in CoQ, that is, we implement type inference: given a context and a term, output its type or produce a type error. All the rules of type inference are straightforward except for cumulativity. The cumulativity test is implemented by comparing head normal forms for a fast-path failure and potentially calling itself recursively, unfolding definitions at the head in CoQ's kernel in case the heads are equal. We implemented weak-head reduction by mimicking CoQ's kernel implementation, which is based on an abstract machine inspired by the KAM. CoQ's machine optionally implements a variant of lazy, memoizing evaluation (which can have mixed results, see CoQ's PR #555 for example), that feature has not been implemented yet.

The main difference with the OCAML implementation is that all of the functions are required to be shown terminating in CoQ. One possibility could be to prove the termination of type-checking separately but this amounts to prove in particular the normalization of CIC which is a complex task. Instead, we simply add a fuel parameter to make them syntactically recursive and make **OutOfFuel** a type error, *i.e.*, we are working in a variant of the option monad.

Bootstrapping It. We can extract this checker to OCAML and reuse the setup described in Sect. 2 to connect it with the reifier and easily derive a (partialy verified) alternative checker for CoQ's .vo object files. Our plugin provides a new command **Template Check** for typechecking definitions using the alternative checker, that can be used as follows:

```
Require Import Template.TemplateCoqChecker List. Import ListNotations.
Definition foo := List.map (fun x ⇒ x + 3) [0; 1].
Template Check foo.
```

Our initial tests indicate that its running time is comparable to the coqchk checker of CoQ, as expected.

---

[1] We do not treat metavariables which are absent from kernel terms and require a separate environment for their declarations.

4   Reification of COQ Commands

COQ plugins need to interact with the environment, for example by repeatedly looking up definitions by name, declaring new constants using fresh names, or performing computations. It is desirable to allow such programs to be written in COQ (GALLINA) because of the two following advantages. Plugin-writers no longer need to understand the OCAML implementation of COQ and plugins are no longer sensitive to changes made in the OCAML implementation. Also, when plugins implementing syntactic models are proven correct in COQ, they provide a mechanism to add axioms to COQ without compromising consistency (Sect. 5.3).

In general, interactions with the environment have side effects, e.g. the declaration of new constants, which must be described in COQ's pure setting. To overcome this difficulty, we use the standard "free" monadic setting to represent the operations involved in interacting with the environment, as done for instance in Mtac [27].

```
Inductive TemplateMonad : Type → Type :=
(* Monadic operations *)
| tmReturn : ∀ {A}, A → TemplateMonad A
| tmBind   : ∀ {A B},
    TemplateMonad A → (A → TemplateMonad B) → TemplateMonad B

(* General operations *)
| tmPrint       : ∀ {A}, A → TemplateMonad unit
| tmFail        : ∀ {A}, string → TemplateMonad A
| tmEval        : reductionStrategy → ∀ {A}, A → TemplateMonad A
| tmDefinition  : ident → ∀ {A}, A → TemplateMonad A
| tmAxiom       : ident → ∀ A, TemplateMonad A
| tmLemma       : ident → ∀ A, TemplateMonad A
| tmFreshName   : ident → TemplateMonad ident
| tmAbout       : ident → TemplateMonad (option global_reference)
| tmCurrentModPath : unit → TemplateMonad string

(* Quoting and unquoting operations *)
| tmQuote         : ∀ {A}, A  → TemplateMonad term
| tmQuoteRec      : ∀ {A}, A  → TemplateMonad program
| tmQuoteInductive : kername → TemplateMonad mutual_inductive_entry
| tmQuoteConstant : kername → bool → TemplateMonad constant_entry
| tmMkDefinition  : ident → term → TemplateMonad unit
| tmMkInductive   : mutual_inductive_entry → TemplateMonad unit
| tmUnquote       : term  → TemplateMonad {A : Type & A}.
| tmUnquoteTyped  : ∀ A, term → TemplateMonad A
```

Fig. 3. The monad of commands

Table 1. Main TEMPLATE-COQ commands

Vernacular command	Reified command with its arguments	Description
Eval	tmEval red t	Returns the evaluation of t following the evaluation strategy red (cbv, cbn, hnf, all or lazy)
Definition	tmDefinition id t	Makes the definition id := t and returns the created constant id
Axiom	tmAxiom id A	Adds the axiom id of type A and returns the created constant id
Lemma	tmLemma id A	Generates an obligation of type A, returns the created constant id after all obligations close
About or Locate	tmAbout id	Returns Some gr if id is a constant in the current environment and gr is the corresponding global reference. Returns None otherwise.
	tmQuote t	Returns the syntax of t (of type term)
	tmQuoteRec t	Returns the syntax of t and all the declarations on which it depends
	tmQuoteInductive kn	Returns the declaration of the inductive kn
	tmQuoteConstant kn b	Returns the declaration of the constant kn, if b is true the implementation bypass opacity to get the body of the constant
Make Definition	tmMkDefinition id tm	Adds the definition id := t where t is denoted by tm
Make Inductive	tmMkInductive d	Declares the inductive denoted by the declaration d
	tmUnquote tm	Returns the pair (A;t) where t is the term whose syntax is tm and A it's type
	tmUnquoteTyped A tm	Returns the term whose syntax is tm and checks that it is indeed of type A

TemplateMonad is an inductive family (Fig. 3) such that TemplateMonad A represents a program which will finally output a term of type A. There are special constructor `tmReturn` and `tmBind` to provide (freely) the basic monadic operations. We use the monadic syntactic sugar x ← t ; u for `tmBind` t (fun x ⇒ u).

The other operations of the monad can be classified in two categories: the traditional COQ operations (`tmDefinition` to declare a new definition, ...) and the quoting and unquoting operations to move between COQ term and their syntax or to work directly on the syntax (`tmMkInductive` to declare a new inductive from its syntax for instance). An overview is given in Table 1.

A program `prog` of type `TemplateMonad A` can be executed with the command `Run TemplateProgram prog`. This command is thus an interpreter for `TemplateMonad`programs, implemented in OCAML as a traditional COQ plugin. The term produced by the program is discarded but, and it is the point, a program can have many side effects like declaring a new definition or a new inductive type, printing something, ...

Let's look at some examples. The following program adds the definitions `foo := 12` and `bar := foo + 1` to the current context.

```
Run TemplateProgram (foo ← tmDefinition "foo" 12 ;
                     tmDefinition "bar" (foo +1)).
```

The program below asks the user to provide an inhabitant of ℕ (here we provide 3 * 3) and records it in the lemma `foo` ; prints its normal form ; and records the syntax of its normal form in `foo_nf_syntax` (hence of type `term`). We use PROGRAM's obligation mechanism[2] to ask for missing proofs, running the rest of the program when the user finishes providing it. This enables the implementation of *interactive* plugins.

```
Run TemplateProgram (foo ← tmLemma "foo" ℕ ;
                     nf  ← tmEval all foo ;
                     tmPrint "normal form: " ; tmPrint nf ;
                     nf_ ← tmQuote nf ;
                     tmDefinition "foo_nf_syntax"  nf_).
Next Obligation. exact (3 * 3). Defined.
```

5   Writing COQ Plugins in COQ

The reification of syntax, typing and commands of COQ allow writing a COQ plugin directly inside COQ, without requiring another language like OCAML and an external compilation phase.

In this section, we describe three examples of such plugins: (i) a plugin that adds a constructor to an inductive type, (ii) a re-implementation of LASSON's parametricity plugin[3], and (iii) an implementation of a plugin that provides an extension of CIC—using a syntactic translation—in which it is possible to prove the negation of functional extensionality [8].

---

[2] In COQ, a proof obligation is a goal which has to be solved to complete a definition. Obligations were introduced by SOZEAU [23] in the PROGRAM mode.

[3] https://github.com/parametricity-coq/paramcoq.

5.1   A Plugin to Add a Constructor

Our first example is a toy example to show the methodology of writing plugins in
TEMPLATE-COQ. Given an inductive type I, we want to declare a new inductive
type I' which corresponds to I plus one more constructor.

For instance, let's say we have a syntax for lambda calculus:

```
Inductive tm : Set :=
    | var : nat → tm  | lam : tm → tm  | app : tm → tm → tm.
```

And that in some part of our development, we want to consider a variation of tm
with a new constructor, e.g., let in. Then we declare tm' with the plugin by:

```
Run TemplateProgram
    (add_constructor tm "letin" (fun tm' ⇒ tm' → tm' → tm')).
```

This command has the same effect as declaring the inductive tm' by hand:

```
Inductive tm' : Set :=
    | var' : nat → tm'          | lam' : tm' → tm'
    | app' : tm' → tm' → tm'   | letin : tm' → tm' → tm'.
```

but with the benefit that if tm is changed, for instance by adding one new constructor, then tm' is automatically changed accordingly. We provide other examples in the file test-suite/add_constructor.v, e.g. with mutual inductives.

We will see that it is fairly easy to define this plugin using TEMPLATE-COQ.
The main function is add_constructor which takes an inductive type ind (whose
type is not necessarily Type if it is an inductive family), a name idc for the new
constructor and the type ctor of the new constructor, abstracted with respect
to the new inductive.

```
Definition add_constructor {A} (ind : A) (idc : ident) {B} (ctor : B)
    : TemplateMonad unit
    := tm ← tmQuote ind ;
        match tm with
        | tInd ind₀ _ ⇒
          decl ← tmQuoteInductive (inductive_mind ind₀) ;
          ctor ← tmQuote ctor ;
          d' ← tmEval lazy (add_ctor decl ind₀ idc ctor) ;
          tmMkInductive d'
        | _ ⇒ tmFail "The provided term is not an inductive"
        end.
```

It works in the following way. First the inductive type ind is quoted, the
obtained term tm is expected to be a tInd constructor otherwise the function fails.
Then the declaration of this inductive is obtained by calling tmQuoteInductive,
the constructor is reified too, and an auxiliary function is called to add the
constructor to the declaration. After evaluation, the new inductive type is added
to the current context with tmMkInductive.

It remains to define the add_ctor auxiliary function to complete the definition
of the plugin. It takes a minductive_decl which is the declaration of a block of
mutual inductive types and returns a minductive_decl.

```
Definition add_ctor (mind : minductive_decl) (ind₀ : inductive)
                    (idc : ident) (ctor : term) : minductive_decl
  := let i₀ := inductive_ind ind₀ in
     {| ind_npars := mind.(ind_npars) ;
        ind_bodies := map_i (fun (i : nat) (ind : inductive_body) ⇒
          {| ind_name := tsl_ident ind.(ind_name) ;
             ind_type  := ind.(ind_type) ;
             ind_kelim := ind.(ind_kelim) ;
             ind_ctors :=
               let ctors := map (fun '(id, t, k) ⇒ (tsl_ident id, t, k)
     )
                                ind.(ind_ctors) in
               if Nat.eqb i i₀ then
                 let n := length mind.(ind_bodies) in
                 let typ := try_remove_n_lambdas n ctor in
                 ctors ++ [(idc, typ, 0)]
               else ctors;
             ind_projs := ind.(ind_projs) |})
        mind.(ind_bodies) |}.
```

$$[t]_0 = t$$

$$[x]_1 = x^t$$

$$[\forall(x:A).B]_1 = \lambda f.\forall(x:[A]_0)(x^t:[A]_1 x).[B]_1(f\ x)$$

$$[\lambda(x:A).t]_1 = \lambda(x:[A]_0)(x^t:[A]_1 x).[t]_1$$

$$[\![\Gamma, x:A]\!] = [\![\Gamma]\!], x:[A]_0, x^t:[A]_1 x$$

$$\Gamma \vdash t : A$$

$$\overline{\rule{0pt}{0pt}\hspace{3cm}}$$

$$[\![\Gamma]\!] \vdash [t]_0 : [A]_0$$

$$[\![\Gamma]\!] \vdash [t]_1 : [A]_1 [t]_0$$

Fig. 4. Unary parametricity translation and soundness theorem, excerpt (from [7])

The declaration of the block of mutual inductive types is a record. The field `ind_bodies` contains the list of declarations of each inductive of the block. We see that most of the fields of the records are propagated, except for the names which are translated to add some primes and `ind_ctors`, the list of types of constructors, for which, in the case of the relevant inductive (i_0 is its number), the new constructor is added.

5.2   Parametricity Plugin

We now show how TEMPLATE-COQ permits to define a parametricity plugin that computes the translation of a term following Reynolds' parametricity [21,25]. We follow the already known approaches of parametricity for dependent type theories [7,15], and provide an alternative to Keller and Lasson's plugin.

The definition in the unary case is described in Fig. 4. The soundness theorem ensures that, for a term t of type A, $[t]_1$ computes a proof of parametricity of $[t]_0$ in the sense that it has type $[A]_1$ $[t]_0$. The definition of the plugin goes in two steps: first the definition of the translation on the syntax of term in TEMPLATE-COQ and then the instrumentation to connect it with terms of COQ using the TemplateMonad. It can be found in the file translations/tsl_param.v.

The parametricity translation of Fig. 4 is total and syntax directed, the two components of the translation $[\ ]_0$ and $[\ ]_1$ are implemented by two recursive functions tsl_param$_0$ and tsl_param$_1$.

```
Fixpoint tsl_param₀ (n : nat) (t : term) {struct t} : term :=
match t with
| tRel k ⟹ if k >= n then (* global variable *) tRel (2*k-n+1)
                      else (* local  variable *) tRel k
| tProd na A B ⟹ tProd na (tsl_param₀ n A) (tsl_param₀ (n+1) B)
| _ ⟹ ...
end.


Fixpoint tsl_param₁ (E : tsl_table) (t : term) : term :=
match t with
| tRel k  ⟹ tRel (2 * k)
| tSort s ⟹ tLambda (nNamed "A") (tSort s)
                    (tProd nAnon (tRel 0) (tSort s))
| tProd na A B ⟹
    let A0 := tsl_param₀ 0 A in let A1 := tsl_param₁ E A in
    let B0 := tsl_param₀ 1 B in let B1 := tsl_param₁ E B in
    tLambda (nNamed "f") (tProd na A0 B0)
    (tProd na (lift₀ 1 A0)
          (tProd (tsl_name na) (subst_app (lift₀ 2 A1) [tRel 0])
                (subst_app (lift 1 2 B1) [tApp (tRel 2) [tRel 1] ])))
| tConst s univs ⟹ match lookup_tsl_table E (ConstRef s) with
                    | Some t ⟹ t
                    | None ⟹ default_term
                    end
| _ ⟹ ...
end.
```

On Fig. 4, the translation is presented in a named setting, so the introduction of new variables does not change references to existing ones. That's why, $[\ ]_0$ is the identity. In the De Bruijn setting of TEMPLATE-COQ, the translation has to take into account the shift induced by the duplication of the context. Therefore, the implementation tsl_param$_0$ of $[\ ]_0$ is not the identity anymore. The argument n of tsl_param$_0$ represents the De Bruijn level from which the variables have been duplicated. There is no need for such an argument in tsl_param$_1$, the implementation of $[\ ]_1$, because in this function all variables are duplicated.

The parametricity plugin not only has to be defined on terms of CIC but also on additional terms dealing with the global context. In particular, constants are translated using a translation table which records the translations of previously processed constants.

```
Definition tsl_table := list (global_reference * term).
```

If a constant is not in the translation table we return a dummy `default_term`, considered as an error (this could also be handled by an option monad).

We have also implemented the translation of inductives and pattern matching. For instance the translation of the equality type `eq` produces the inductive type:

```
Inductive eqᵗ A (Aᵗ : A → Type) (x : A) (xᵗ : Aᵗ x)
              : ∀ H, Aᵗ H → x = H → Prop :=
          | eq_reflᵗ : eqᵗ A Aᵗ x xᵗ x xᵗ eq_refl.
```

Then $[eq]_1$ is given by `eqᵗ` and $[eq\_refl]_1$ by `eq_reflᵗ`.

Given `tsl_param₀` and `tsl_param₁` the translation of the declaration of a block of mutual inductive types is not so hard to get. Indeed, such a declaration mainly consists of the arities of the inductives and the types of constructors; and the one of the translated inductive are produced by translation of the original ones.

```
Definition tsl_mind_decl (E : tsl_table) (kn : kername)
              (mind : minductive_decl) : minductive_decl.
```

In a similar manner, we can translate pattern-matching. Note that the plugin does not support fixpoints and cofixpoints for the moment.

Now, it remains to connect this translation defined on reified syntax `term` to terms of COQ. For this, we define the new command `tTranslate` in the `TemplateMonad`.

```
Definition tTranslate (E : tsl_table) (id : ident)
              : TemplateMonad tsl_table.
```

When `id` is a definition, the command recovers the body of `id` (as a `term`) using `tmQuoteConstant` and then translates it and records it in a new definition `idᵗ`. The command returns the translation table `E` extended by `(id, idᵗ)`. In the case `id` is an inductive type or a constructor then the command does basically the same but extends the translation table with both the inductive and the constructors. If `id` is an axiom or not a constant the command fails.

Here is an illustration coming from the work of Lasson [16] on the automatic proofs of $(\omega\text{-})$groupoid laws using parametricity. We show that all function of type `ID := ∀ A x y, x = y → x = y` are identity functions. First we need to record the translations of `eq` and `ID` in a term `table` of type `tsl_table`.

```
Run TemplateProgram (table ← tTranslate [] "eq" ;
                     table ← tTranslate table "ID" ;
                     tmDefinition "table" table).
```

Then we show that every parametric function on `ID` is pointwise equal to the identity using the predicate `fun y ⇒ x = y`.

```
Lemma param_ID (f : ID) : IDᵗ f → ∀ A x y p, f A x y p = p.
Proof.
  intros H A x y p. destruct p.
  destruct (H A (fun y ⇒ x = y) x eq_refl
```

x eq_refl eq_refl (eq_refl^t _ _)).
  reflexivity.
Qed.

Then we define a function myf $:= p \mapsto p . p^{-1} . p$ and get its parametricity proof using the plugin.

```
Definition myf : ID := fun A x y p ⇒ eq_trans (eq_trans p(eq_sym p)) p.
```

```
Run TemplateProgram (table ← tTranslate table "eq_sym" ;
                     table ← tTranslate table "eq_trans" ;
                     tTranslate table "myf").
```

It is then possible to deduce automatically that $p . p^{-1} . p = p$ for all $p : x = y$.

```
Definition free_thm_myf: ∀ A x y p, myf A x y p = p := param_ID myf myf^t.
```

5.3 Intensional Function Plugin

Our last illustration is a plugin that provides an intensional flavour to functions and thus allows negating functional extensionality (FunExt). This is a simple example of syntactical translation which enriches the logical power of CoQ, in the sense that new theorems can be proven (as opposed to the parametricity translation which is conservative over CIC). See [8] for an introduction to syntactical translations and a complete description of the intensional function translation.

$$[x] := x \qquad\qquad [\lambda(x : A). t] := (\lambda(x : [A]). [t], \text{true})$$
$$[t\ u] := (\pi_1\ [t])\ [u] \qquad\qquad [\forall(x : A). B] := (\forall(x : [A]). [B]) \times \mathbb{B}$$

Fig. 5. Intensional function translation, excerpt (from [8])

Even if the translation is very simple as it just adds a boolean to every function (Fig. 5), this time, it is not fully syntax directed. Indeed the notation for pairs hide some types:

```
[fun (x:A) ⇒ t] := pair (∀ x:[A]. ?T) bool (fun (x:[A]) ⇒ [t]) true
```

and we can not recover the type ?T from the source term. There is thus a mismatch between the lambdas which are not fully annotated and the pairs which are.[4]

However we can use the type inference algorithm of Sect. 3 implemented on TEMPLATE-CoQ terms to recover the missing information.

```
[fun (x:A) ⇒ t] := let B := infer Σ (Γ, x:[A]) t in
                   pair (∀ (x:[A]). B) bool (fun (x:[A]) ⇒ [t]) true
```

---

[4] Note that there is a similar issue with applications and projections, but which can be circumvented this time using (untyped) primitive projections.

Compared to the parametricity plugin, the translation function has a more complex type as it requires the global and local contexts. However, we can generalize the tTranslate command so that it can be used for both the parametricity and the intensional function plugins. The implementation is in the files translations/translation_utils.v and translations/tsl_fun.v.

Extending COQ *Using Plugins.* The intensional translation extends the logical power of COQ as it is possible for instance to negate FunExt. In this perspective, we defined a new command:

```
Definition tImplement (Σ : global_ctx * tsl_table)
                       (id : ident) (A : Type)
    : TemplateMonad (global_ctx * tsl_table).
```

which computes the translation A' of A, then asks the user to inhabit the type A' by generating a proof obligation and then safely adds the axiom id of type A to the current context. By safely, we mean that the correction of the translation ensures that no inconsistencies are introduced.

For instance, here is how to negate FunExt. We use for that two pairs (fun x ⇒ x; true) and (fun x ⇒ x; false) in the interpretation of functions from unit to unit, which are extensionally both the identity, but differ intensionally on their boolean.

```
Run TemplateProgram (TC ← tTranslate ([],[]) "eq" ;
                     TC ← tTranslate TC "False" ;
                     tImplement TC "notFunext"
   ((∀ A B (f g : A → B), (∀ x:A, f x = g x) → f = g) → False)).
Next Obligation.
  tIntro H. tSpecialize H unit. tSpecialize H unit.
  tSpecialize H (fun x ⇒ x; true). tSpecialize H (fun x ⇒ x; false).
  tSpecialize H (fun x ⇒ eq_refl^t _ _; true).
  apply eq^t_eq in H; discriminate.
Defined.
```

where tIntro and tSpecialize are special versions of the corresponding intro and specialize tactics of COQ to deal with extra booleans appearing in the translated terms. After this command, the axiom notFunext belongs to the environment, as if it where added with the Axiom command. But as we have inhabited the translation of its type, the correctness of the translation ensures that no inconsistency were introduced.

Note that we could also define another translation, e.g. the setoid translation, in which FunExt is inhabited. This is not contradictory as the two translations induce two different logical extensions of COQ, which can not be combined.

6   Related Work and Future Work

Meta-Programming is a whole field of research in the programming languages community, we will not attempt to give a detailed review of related work here. In contrast to most work on meta-programming, we provide a very rough interface to the object language: one can easily build ill-scoped and ill-typed terms in our framework, and staging is basic. However, with typing derivations we provide a way to verify meta-programs and ensure that they do make sense.

The closest cousin of our work is the Typed Syntactic Meta-Programming [11] proposal in AGDA, which provides a well-scoped and well-typed interface to a denotation function, that can be used to implement tactics by reflection. We could also implement such an interface, asking for a proof of well-typedness on top of the tmUnquoteTyped primitive of our monad.

Intrinsically typed representations of terms in dependent type-theory is an area of active research. Most solutions are based on extensions of Martin-Löf Intensional Type Theory with inductive-recursive or quotient inductive-inductive types [2,9], therefore extending the meta-theory. Recent work on verifying soundness and completeness of the conversion algorithm of a dependent type theory (with natural numbers, dependent products and a universe) in a type theory with IR types [1] gives us hope that this path can nonetheless be taken to provide the strongest guarantees on our conversion algorithm. The intrinsically-typed syntax used there is quite close to our typing derivations.

Another direction is taken by the Œuf certified compiler [19], which restricts itself to a fragment of COQ for which a total denotation function can be defined, in the tradition of definitional interpreters advocated by Chlipala [10]. This setup should be readily accommodated by TEMPLATE-COQ.

The translation + plugin technique paves the way for certified translations and the last piece will be to prove correctness of such translations. By correctness we mean computational soundness and typing soundness (see [8]), and both can be stated in TEMPLATE-COQ. Anand has made substantial attempts in this direction to prove the computational soundness, in TEMPLATE-COQ, of a variant of parametricity providing stronger theorems for free on propositions [3]. This included as a first step a move to named syntax that could be reused in other translations.

Our long term goal is to leverage this technique to extend the logical and computational power of COQ using, for instance, the forcing translation [14] or the weaning translation [20].

When performance matters, we can extract the translation to OCAML and use it like any ordinary COQ plugin. This relies on the correctness of extraction, but in the untyped syntax + typing judgment setting, extraction of translations is almost an identity pretty-printing phase, so we do not lose much confidence. We can also implement a template monad runner in OCAML to run the plugins outside COQ. Our first experiments show that we could gain a factor 10 for the time needed to compute the translation of a term. Another solution would be to use the certified CERTICOQ compiler, once it supports a kind of foreign function interface, to implement the TemplateMonad evaluation.

The last direction of extension is to build higher-level tools on top of the syntax: the unification algorithm described in [28] is our first candidate. Once unification is implemented, we can look at even higher-level tools: elaboration from concrete syntax trees, unification hints like canonical structures and type class resolution, domain-specific and general purpose tactic languages. A key inspiration in this regard is the work of Malecha and Bengston [17] which implemented this idea on a restricted fragment of CIC.

Acknowledgments. This work is supported by the CoqHoTT ERC Grant 64399 and the NSF grants CCF-1407794, CCF-1521602, and CCF-1646417.

References

1. Abel, A., Öhman, J., Vezzosi, A.: Decidability of conversion for type theory in type theory. PACMPL **2**(POPL), 23:1–23:29 (2018). http://doi.acm.org/10.1145/3158111
2. Altenkirch, T., Kaposi, A.: Type theory in type theory using quotient inductive types. In: POPL 2016, pp. 18–29. ACM, New York (2016). http://doi.acm.org/10.1145/2837614.2837638
3. Anand, A., Morrisett, G.: Revisiting parametricity: inductives and uniformity of propositions. In: CoqPL 2018, Los Angeles, CA, USA (2018)
4. Anand, A., Appel, A., Morrisett, G., Paraskevopoulou, Z., Pollack, R.,Belanger, O.S., Sozeau, M., Weaver, M.: CertiCoq: a verified compiler for Coq. In: CoqPL, Paris, France (2017). http://conf.researchr.org/event/CoqPL-2017/main-certicoq-a-verified-compiler-for-coq
5. Armand, M., Grégoire, B., Spiwack, A., Théry, L.: Extending COQ with imperative features and its application to SAT verification. In: Kaufmann, M., Paulson, L.C. (eds.) ITP 2010. LNCS, vol. 6172, pp. 83–98. Springer, Heidelberg (2010). https://doi.org/10.1007/978-3-642-14052-5_8
6. Barras, B.: Auto-validation d'un système de preuves avec familles inductives. Thèse de doctorat, Université Paris 7, November 1999
7. Bernardy, J.P., Jansson, P., Paterson, R.: Proofs for free: parametricity for dependent types. J. Funct. Program. **22**(2), 107–152 (2012)
8. Boulier, S., Pédrot, P.M., Tabareau, N.: The next 700 syntactical models of type theory. In: CPP 2017, pp. 182–194. ACM, Paris (2017)
9. Chapman, J.: Type theory should eat itself. Electron. Notes Theor. Comput. Sci. **228**, 21–36 (2009). Proceedings of LFMTP 2008. http://www.sciencedirect.com/science/article/pii/S157106610800577X
10. Chlipala, A.: Certified Programming with Dependent Types, vol. 20. MIT Press, Cambridge (2011)
11. Devriese, D., Piessens, F.: Typed syntactic meta-programming. In: ICFP 2013, vol. 48, pp. 73–86. ACM (2013). http://doi.acm.org/10.1145/2500365.2500575
12. Ebner, G., Ullrich, S., Roesch, J., Avigad, J., de Moura, L.: A metaprogramming framework for formal verification, pp. 34:1–34:29, September 2017
13. Forster, Y., Kunze, F.: Verified extraction from Coq to a lambda-calculus. In: Coq Workshop 2016 (2016). https://www.ps.uni-saarland.de/forster/coq-workshop-16/abstract-coq-ws-16.pdf

14. Jaber, G., Lewertowski, G., Pédrot, P.M., Sozeau, M., Tabareau, N.: The definitional side of the forcing. In: LICS 2016, New York, NY, USA, pp. 367–376 (2016). http://doi.acm.org/10.1145/2933575.2935320
15. Keller, C., Lasson, M.: Parametricity in an impredicative sort. CoRR abs/1209.6336 (2012). http://arxiv.org/abs/1209.6336
16. Lasson, M.: Canonicity of weak ω-groupoid laws using parametricity theory. Electron. Notes Theor. Comput. Sci. **308**, 229–244 (2014)
17. Malecha, G., Bengtson, J.: Extensible and efficient automation through reflective tactics. In: Thiemann, P. (ed.) ESOP 2016. LNCS, vol. 9632, pp. 532–559. Springer, Heidelberg (2016). https://doi.org/10.1007/978-3-662-49498-1_21
18. Malecha, G.M.: Extensible proof engineering in intensional type theory. Ph.D. thesis, Harvard University (2014)
19. Mullen, E., Pernsteiner, S., Wilcox, J.R., Tatlock, Z., Grossman, D.: Œuf: minimizing the Coq extraction TCB. In: Proceedings of CPP 2018, pp. 172–185 (2018). http://doi.acm.org/10.1145/3167089
20. Pédrot, P., Tabareau, N.: An effectful way to eliminate addiction to dependence. In: LICS 2017, Reykjavik, Iceland, pp. 1–12 (2017). https://doi.org/10.1109/LICS.2017.8005113
21. Reynolds, J.C.: Types, abstraction and parametric polymorphism. In: IFIP Congress, pp. 513–523 (1983)
22. Sheard, T., Jones, S.P.: Template meta-programming for Haskell. SIGPLAN Not. **37**(12), 60–75 (2002). http://doi.acm.org/10.1145/636517.636528
23. Sozeau, M.: Programming finger trees in Coq. In: ICFP 2007, pp. 13–24. ACM, New York (2007). http://doi.acm.org/10.1145/1291151.1291156
24. Taha, W., Sheard, T.: Multi-stage programming with explicit annotations. In: PEPM 1997, pp. 203–217. ACM, New York (1997). http://doi.acm.org/10.1145/258993.259019
25. Wadler, P.: Theorems for free! In: Functional Programming Languages and Computer Architecture, pp. 347–359. ACM Press (1989)
26. van der Walt, P., Swierstra, W.: Engineering proof by reflection in Agda. In: Hinze, R. (ed.) IFL 2012. LNCS, vol. 8241, pp. 157–173. Springer, Heidelberg (2013). https://doi.org/10.1007/978-3-642-41582-1_10
27. Ziliani, B., Dreyer, D., Krishnaswami, N.R., Nanevski, A., Vafeiadis, V.: Mtac: a monad for typed tactic programming in Coq. J. Funct. Program. **25** (2015). https://doi.org/10.1017/S0956796815000118
28. Ziliani, B., Sozeau, M.: A comprehensible guide to a new unifier for CIC including universe polymorphism and overloading. J. Funct. Program. **27**, e10 (2017). http://www.irif.univ-paris-diderot.fr/sozeau/research/publications/drafts/unification-jfp.pdf

Formalizing Ring Theory in PVS

Andréia B. Avelar da Silva[1(✉)], Thaynara Arielly de Lima[2],
and André Luiz Galdino[3]

[1] Faculdade de Planaltina, Universidade de Brasília, Brasília D.F., Brazil
andreiaavelar@unb.br
[2] Instituto de Matemática e Estatística, Universidade Federal de Goiás,
Goiânia, Brazil
thaynaradelima@ufg.br
[3] Unidade Acadêmica Especial de Matemática e Tecnologia,
Universidade Federal de Goiás, Catalão, Brazil
andre_galdino@ufg.br

Abstract. This work describes the ongoing specification and formalization in the PVS proof assistant of some definitions and theorems of ring theory in abstract algebra, and briefly presents some of the results intended to be formalized. So far, some important theorems from ring theory were specified and formally proved, like the First Isomorphism Theorem, the Binomial Theorem and the lemma establishing that every finite integral domain with cardinality greater than one is a field. The goal of the project in progress is to specify and formalize in PVS the main theorems from ring theory presented in undergraduate textbooks of abstract algebra, but in the short term the authors intended to formalize: (i) the Second and the Third Isomorphism Theorems for rings; (ii) the primality of the characteristic of a ring without zero divisors; (iii) definitions of prime and maximal ideals and theorems related with those concepts. The developed formalization applies mainly a part of the NASA PVS library for abstract algebra specified in the *theory* **algebra**.

1 Introduction

Ring theory has a wide range of applications in the most varied fields of knowledge. According to [18], the segmentation of digital images becomes more efficiently automated by applying the $\mathbb{Z}_n$ ring to obtain index of similarity between images. Furthermore, according to [3] finite commutative rings has an important role in areas like combinatorics, analysis of algorithms, algebraic cryptography and coding theory. In particular in coding theory, finite fields (which are commutative rings with unity) and polynomials over finite fields has been widely applied in description of redundant codes [16].

Andréia B. Avelar da Silva was partially supported by District Federal Research Foundation - FAPDF.

© Springer International Publishing AG, part of Springer Nature 2018
J. Avigad and A. Mahboubi (Eds.): ITP 2018, LNCS 10895, pp. 40–47, 2018.
https://doi.org/10.1007/978-3-319-94821-8_3

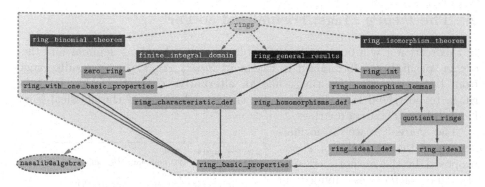

Fig. 1. Hierarchy of the PVS *theory* `rings`, which imports the *theory* `algebra` from *nasalib*. The four main branches developed so far are highlighted.

The authors has the project that consists in to formalize in the PVS proof assistant the basic theory for rings presented in undergraduate textbooks of abstract algebra. This formalization would make possible the formal verification of more complex theories involving rings in their scope. This is an ongoing formalization and the lemmas already verified constitute the *theory* `rings`, which is a collection of *subtheories* that will be described in Sect. 2. The PVS is a specification and verification system which provides an integrated environment for development and analysis of formal specifications. An important and well-known library for PVS is the NASA PVS Library[1] (*nasalib*) that contains many *theories* in several subjects, like analysis [5], topology [15], term rewriting systems [8], among others. In particular, a formal verification for basic abstract algebra is part of *nasalib*, in the *theory* `algebra` [4], where basic concepts about groups, rings and fields were specified. However the content of the *theory* `algebra`, for instance about rings, is essentially definitions and basic results obtained from such definitions. To the best knowledge of the authors, the only formalization involving rings in PVS is the *theory* `algebra`. The project proposed by the authors was motivated by the wish to contribute with the enrichment of mathematics formalizations in the available PVS libraries, by formalizing non basic results about rings that are not in *nasalib*.

The main contributions presented in this paper consist in the formalization of important theorems such that the First Isomorphism Theorem, the Binomial Theorem for rings and the result establishing that every finite integral domain with cardinality greater than one is a field. Furthermore, important concepts and lemmas from *nasalib theories* and `prelude` (the native library of PVS which contains a collection of theories about functions, sets, predicates, logic, numbers, among others) were generalized in order to build the ongoing *theory* `rings`. The present formalization follows the approach of the textbooks [2,7,11,12], but mainly the Hungerford textbook [12].

---

[1] Available at https://shemesh.larc.nasa.gov/fm/ftp/larc/PVS-library/.

2   The *theory* rings: Formalized so Far

In this section it will be described the collection of definitions, lemmas and theorems specified and formalized in the main *theory* rings. These results range from basic properties for rings, like an alternative characterization for subrings, to nontrivial formalizations, like the formalization of the classical First Isomorphism Theorem for rings.

The current state of formalization of the PVS *theory* rings, proposed by the authors, consists of some *subtheories* divided in four main branches (Fig. 1), each one dedicated to formalize lemmas involving: (i) characteristic of rings and rings with one; (ii) finite integral domain; (iii) Binomial Theorem for rings; (iv) homomorphism of rings. Those branches will be described in the following subsections. The basis of the development is constituted by some *subtheories* for fundamental definitions and results regarding ring theory, namely:

ring_basic_properties: This *subtheory* contains basic results about rings not specified in the *theory* algebra. The main contributions of this *subtheory* are: (i) An alternative characterization for subrings, Lemma **subring_equiv** (Fig. 2); (ii) The formalization of the recursive Function **R_sigma** that performs a summation of elements of arbitrary types and its properties (Fig. 2). In order to ensure its totality it was necessary to provide a decreasing measure applied to prove the TCC's (type correctness conditions - lemmas automatically generated by the prover during the process of type checking) for termination. Such function generalizes the summation of reals defined in the *nasalib theory* reals; and

```
subring_equiv: LEMMA
  subring?(S,R) IFF nonempty?(S) AND subset?(S,R)
  AND (FORALL (x,y:(S)):
       member(x-y,S) AND member(x*y,S))

R_sigma(low,high,F): Recursive T =
  IF low > high THEN zero ELSIF high = low
  THEN F(low) ELSE R_sigma(low,high-1,F)+F(high)
  ENDIF MEASURE abs(high+1-low)

left_zd?(x: nz_T): bool =
  EXISTS (y:nz_T): x*y = zero

nlzd: TYPE = {x:T | x = zero OR NOT left_zd?(x)}

nzd_cancel_left: LEMMA FORALL (a:nlzd, b,c:T):
  a*b = a*c IMPLIES (a = zero OR b = c)
----------------------------------------------------
R_homomorphism?(R1,R2)(phi:[(R1)->(R2)]): bool =
  FORALL(a,b:(R1)):phi(s1(a,b))=s2(phi(a),phi(b))
             AND phi(p1(a,b))=p2(phi(a),phi(b))

R_monomorphism?(R1,R2)(phi:[(R1)->(R2)]): bool =
  injective?(phi) AND R_homomorphism?(R1,R2)(phi)

R_epimorphism?(R1,R2)(phi:[(R1)->(R2)]): bool =
  surjective?(phi) AND R_homomorphism?(R1,R2)(phi)

R_isomorphism?(R1,R2)(phi:[(R1)->(R2)]): bool =
  R_monomorphism?(R1,R2)(phi)
  AND R_epimorphism?(R1,R2)(phi)

R_kernel(R1,R2)(phi: R_homomorphism(R1,R2)):
  subgroup[T1,s1,zero1](R1) = kernel(R1,R2)(phi)
----------------------------------------------------
multiple_char: LEMMA
  (EXISTS (m:int): k = m * charac(R))
  IFF (FORALL (x:(R)): times(x, k) = zero)

char_1_zero_ring: LEMMA
  charac(R) = 1 IFF R = singleton(zero)
----------------------------------------------------
power_commute: LEMMA x*y = y*x IMPLIES
  power(x,m)*power(y,i) = power(y,i)*power(x,m)

gen_times_int_one: LEMMA times(one,k) = zero
  IMPLIES times(x, k) = zero
```

Fig. 2.   Highlighted specifications in the *subtheories* **ring_basic_properties**, **ring_homomorphisms_def**, **ring_characteristic_def** and **ring_with_one_basic_properties**.

(iii) The definition of a non zero divisor element type, necessary in the formalization of a more general cancellation law that holds in an arbitrary ring since the cancelled element has the non zero divisor type (Fig. 2).

ring_ideal_def: The concepts of left and right ideal, as well as the type ideal of a ring were established.

ring_homomorphisms_def: Such *subtheory* contains the definition of homomorphism of rings and its variants: injective, surjective and bijective homomorphism. In addition, the kernel of a homomorphism of rings (Fig. 2) is defined from the kernel of a homomorphism of groups specified in the *theory* algebra.

ring_characteristic_def: The specification of the notion of characteristic of a ring and basic results were established. Two lemmas deserve to be highlighted: multiple_char and char_1_zero_ring (Fig. 2). The former is a characterization of multiples of the characteristic of a ring, and the latter states the characteristic of the zero ring as being the integer 1.

ring_with_one_basic_properties: In this *subtheory* one has two important results, power_commute and gen_times_int_one (Fig. 2), to formalize a version of the Binomial Theorem for rings and properties involving characteristic of a ring.

Note that, in some specified lemmas in Fig. 2, the universal quantifier on free variables is implicit. This is possible because the PVS syntax allows one to declare free variables anywhere in the specification file before lemmas, functions and definitions that use those variables. Furthermore, it is possible to use a set inside parentheses to denote the type of its elements.

2.1 The *subtheory* ring_general_results

The main result in this branch consists in to determine the kernel of the homomorphism from the ring of integers to a ring R, illustrated in the Lemma homomorphism_Z_to_R (Fig. 3), as the set of multiples of the characteristic of the ring R. Its proof follows from the Lemmas gen_times_int_one and multiple_char, respectively.

It is intended to extend this *subtheory* establishing results about, for instance, the characteristic of non zero divisor rings and, in particular, of integral domains.

```
homomorphism_Z_to_R: LEMMA
  charac(R) > 0 IMPLIES
  (LET phi:[(fullset[int])->(R)] =
      (LAMBDA (m:int): times(one, m)) IN
    R_homomorphism?(fullset[int],R)(phi) AND
    R_kernel(fullset[int],R)(phi)
    ={x:int | EXISTS (k:int): x = k*charac(R)})
-------------------------------------------------
R_bino_theo: LEMMA
  FORALL(x,y:(R)): x*y = y*x IMPLIES
    power(x+y,n) = R_sigma(0,n,F_bino(n,x,y))

F_bino(n,x,y): [nat -> T] = LAMBDA k:
  IF k > n THEN zero ELSE
    times(power(x,k)*power(y,n-k),C(n,k)) ENDIF
```

Fig. 3. Highlighted specifications in the *subtheories* ring_general_results and ring_binomial_theorem.

2.2 The *subtheory* finite_integral_domain

The *subtheory* finite_integral_domain extends the *subtheory* integral_domain from algebra. The most important theorem states that every finite integral domain with cardinality greater than 1 is a field. The formalization follows the approach in [11]. However, it is important to remark that in [11] a necessary hypothesis is omitted, since the author does not require that the cardinality of the finite integral domain is greater than 1, and the lack of this requirement makes the formal proof unachievable, once in this case the zero ring must be consider and obviously such integral domain is not a field.

Also, in this *subtheory* it was necessary to formalize a result generalizing the pigeonhole principle for an arbitrary set with elements of an arbitrary type, since the pigeonhole principle in the prelude is restricted to subsets of $\mathbb{N}$.

2.3   The *subtheory* ring_binomial_theorem

From the recursive Function R_sigma in ring_basic_properties and its properties and the Lemma power_commute in ring_with_one_basic_properties one can formally prove the Binomial Theorem for rings R_bino_theo (Fig. 3), where $\text{F\_bino}(n, x, y) = \binom{n}{k} \cdot x^k y^{n-k}$.

2.4   The *subtheory* ring_isomorphism_theorems

The *subtheory* ring_isomorphism_theorems is the more elaborated one among the four highlighted *subtheories* in Fig. 1. At this point, the most important lemma of such *subtheory* is the First Isomorphism Theorem for rings. In order to formalize the results in ring_isomorphism_theorems relevant notions related with ideals, quotient rings and homomorphisms of rings were specified in the *subtheories*:

ring_ideal: The main lemma formalized in this *subtheory* states that the ideal of ring is a normal subgroup (Fig. 4). This result was strongly applied to verify the TCC's in the *subtheory* ring_isomorphism_theorems, generated from the specification of quotient rings, in the *subtheory* quotient_ring, which in turn imports the *subtheory* factor_groups

```
ideal_is_normal_subgroup: LEMMA
  ideal?(I,R) IMPLIES normal_subgroup?(I,R)
-------------------------------------------------
cosets(R:ring,I:ideal(R)):TYPE
  = left_cosets(R,I)

add(R:ring,I:ideal(R)):
[cosets(R,I),cosets(R,I)->cosets(R,I)]
  = mult(R,I)

product(R:ring, I:ideal(R))
(A,B: cosets(R,I)):cosets(R,I) =
      lc_gen(R,I,A)*lc_gen(R,I,B) + I

ring_cosets: LEMMA FORALL(R:ring,I:ideal(R)):
  ring?[cosets(R,I),add(R,I),product(R,I),I]
  ({s:cosets(R,I) | EXISTS (a:(R)):s = a+I})
-------------------------------------------------
image_homo_is_subring: LEMMA
  FORALL (phi: R_homomorphism(R1,R2)):
     subring?(image(phi)(R1),R2)

monomorphism_charac: LEMMA
  FORALL (phi: R_homomorphism(R1,R2)):
     R_monomorphism?(R1,R2)(phi) IFF
     R_kernel(R1,R2)(phi) = singleton(zero1))

kernel_homo_is_ideal: LEMMA
  FORALL (phi: R_homomorphism(R1,R2)):
     ideal?(R_kernel(R1,R2)(phi),R1)
-------------------------------------------------
first_isomorphism_th: THEOREM
FORALL(phi: R_homomorphism(R,S)):
  R_isomorphic?[cosets(R, R_kernel(R,S)(phi)),
                 add(R,R_kernel(R,S)(phi)),
                 product(R,R_kernel(R,S)(phi)),
                 R_kernel(R,S)(phi),D,s,p,zerod]
  (/[T,+,*,zero]
   (R,R_kernel(R,S)(phi)),image(phi)(R))
```

Fig. 4. Highlighted specifications in the *subtheories* ring_ideal, quotient_ring, ring_homomorphism_lemmas and ring_isomorphism_theorems.

from the *theory* `algebra`, where it is required that the type of the parameters in the quotient of groups has to be a group G and a normal subgroup of G.

`quotient_rings`: The algebra of quotient rings is builded by specifying the type `cosets` and defining the operations of addition, `add`, and multiplication, `product`, between two cosets (Fig. 4). From that it was formalized that the structure `(cosets(R,I),add(R,I),product(R,I),I)` (Fig. 4) is a ring, where `R` is a ring and `I` is an ideal of `R`.

`ring_homomorphism_lemmas`: Classical results were formalized, such as, given a function $\phi : R \to S$ from a ring $(R, +_R, *_R, e_R)$ to a ring $(S, +_S, *_S, e_S)$, if ϕ is a homomorphism then: (i) the kernel of ϕ, denoted as $ker(\phi)$, is an ideal of R; (ii) the image of ϕ is a subring of S; and (iii) ϕ is a monomorphism iff the kernel of ϕ is the set $ker(\phi) = \{e_R\}$ (Fig. 4).

Additionally, in order to formalize the First Isomorphism Theorem (Theorem 1), whose specification is in Fig. 4 (Theorem `first_isomorphism_th`), it was necessary to specify and prove, in the *subtheory* `ring_isomorphism_theorems`, other six auxiliary lemmas corresponding to the Lemma 1.

Lemma 1. *If $\phi : R \to S$ is a homomorphism of rings and I is an ideal of R which is contained in the kernel of ϕ, then there is a unique homomorphism of rings $f : R/I \to S$ such that $f(a + I) = \phi(a)$ for all $a \in R$. The image of f is equal to the image of ϕ and $ker(f) = ker(\phi)/I$. f is an isomorphism if and only if ϕ is an epimorphism and $ker(\phi) = I$.*

Theorem 1 (First Isomorphism Theorem). *If $\phi : R \to S$ is a homomorphism of rings then ϕ induces an isomorphism of rings from $R/ker(\phi)$ to the image of ϕ.*

3   Related Work

In the literature, abstract algebra formalizations are available. In Coq results about groups, rings and ordered fields were formalized as part of the FTA project [9]. Also in Coq, [6] presents a formalization of rings with explicit divisibility. In Nuprl and in Mizar it is provided a formal proof of the Binomial Theorem for rings, [13,17] respectively. In ACL2 it is built a hierarchy of algebraic structures ranging from setoids to vector spaces focused on the verification of computer algebra systems [10]. The Algebra Library of Isabelle/HOL [1] presents an interesting collection of results in the algebraic hierarchy of rings, mainly about groups, factorization over ideals, ring of integers and polynomial ring. To the best of the authors knowledge, only in Mizar it was formalized the First Isomorphism Theorem for rings [14]. However, the Mizar formalization differs from the one presented in this paper in the sense that Mizar is a system of first order set theory whereas PVS is a higher order logic system.

4   Conclusions and Future Work

The formalization presented in this paper shows the beginning of a project where it is planned to develop in PVS the specification and formal verification of the main theorems from ring theory. Some important theorems were formalized, as well as several auxiliary results necessary to complete the current formalization (Sect. 2). In numbers the *theory* rings consists of 141 proved formulas, from which 68 are TCC's. The specification files have together 1134 lines and their size is 64 KB; the proof files have 17503 lines and 1.2 MB.

The next step would be the formalization of: (i) the Second and the Third Isomorphism Theorems; (ii) the Correspondence Theorem for rings; (iii) a theorem establishing the primality of the characteristic of a ring without zero divisors, in particular of a integral domain; (iv) definitions of prime and maximal ideals and theorems related with those concepts, as for example the equivalence between fields and the non existence of a proper ideal in commutative rings with one.

Ring theory has a number of applications, for example, coding theory, segmentation of digital images, cryptography, among others. In this sense, this formalization forms a basis for future formal verifications of more elaborated *theories* involving rings and their properties.

References

1. Aransay, J., Ballarin, C., Hohe, S., Kammüller, F., Paulson, L.C.: The Isabelle/HOL Algebra Library. Technical report, University of Cambridge - Computer Laboratory, October 2017. http://isabelle.in.tum.de/library/HOL/HOL-Algebra/document.pdf
2. Artin, M.: Algebra, 2nd edn. Pearson, Upper Saddle River (2010)
3. Bini, G., Flamini, F.: Finite commutative rings and their applications, vol. 680. Springer Science & Business Media (2012)
4. Butler, R., Lester, D.: A PVS Theory for Abstract Algebra (2007). http://shemesh.larc.nasa.gov/fm/ftp/larc/PVS-library/pvslib.html. Accessed 22 Jan 2018
5. Butler, R.W.: Formalization of the integral calculus in the PVS theorem prover. J. Formalized Reasoning 2(1), 1–26 (2009)
6. Cano, G., Cohen, C., Dénès, M., Mörtberg, A., Siles, V.: Formalized linear algebra over elementary divisor rings in coq. Logical Meth. Comput. Sci. 12(2), Jun 2016
7. Dummit, D.S., Foote, R.M.: Abstract Algebra, 3rd edn. Wiley, New York (2003)
8. Galdino, A.L., Ayala-Rincón, M.: A PVS theory for term rewriting systems. Electron. Notes Theoret. Comput. Sci. 247, 67–83 (2009)
9. Geuvers, H., Pollack, R., Wiedijk, F., Zwanenburg, J.: A constructive algebraic hierarchy in coq. J. Symbolic Comput. 34(4), 271–286 (2002)
10. Heras, J., Martín-Mateos, F.J., Pascual, V.: Modelling algebraic structures and morphisms in acl2. Appl. Algebra Eng. Commun. Comput. 26(3), 277–303 (2015)
11. Herstein, I.N.: Topics in Algebra, 2nd edn. Xerox College Publishing, Lexington (1975)
12. Hungerford, T.W.: Algebra, Graduate Texts in Mathematics, vol. 73. Springer-Verlag, New York-Berlin (1980)
13. Jackson, P.B.: Enhancing the Nuprl Proof Development System and Applying it to Computational Abstract Algebra. Ph.D. thesis, Cornell University (1995)

14. Kornilowicz, A., Schwarzweller, C.: The first isomorphism theorem and other properties of rings. Formalized Math. **22**(4), 291–301 (2014)
15. Lester, D.: A PVS Theory for Continuity, Homeomorphisms, Connected and Compact Spaces, Borel sets/functions (2009). http://shemesh.larc.nasa.gov/fm/ftp/larc/PVS-library/pvslib.html. Accessed 22 Jan 2018
16. Lidl, R., Niederreiter, H.: Introduction to finite fields and their applications. Cambridge University Press, Cambridge (1994)
17. Schwarzweller, C.: The binomial theorem for algebraic structures. Formalized Math. **09**(3), 559–564 (2001)
18. Suárez, Y.G., Torres, E., Pereira, O., Pérez, C., Rodríguez, R.: Application of the ring theory in the segmentation of digital images. Int. J. Soft Comput. Math. Control **3**(4) (2014)

Software Tool Support for Modular Reasoning in Modal Logics of Actions

Samuel Balco[1]([✉]), Sabine Frittella[2], Giuseppe Greco[3], Alexander Kurz[1],
and Alessandra Palmigiano[4,5]

[1] Department of Informatics, University of Leicester, Leicester, England
`sb782@leicester.ac.uk`
[2] Laboratoire d'Informatique Fondamentale d'Orléans, Orléans, France
[3] Department of Languages, Literature and Communication,
Utrecht Institute of Linguistics, Utrecht, Netherlands
[4] Faculty of Technology, Policy and Management, Delft University of Technology,
Delft, Netherlands
[5] Department of Pure and Applied Mathematics, University of Johannesburg,
Johannesburg, South Africa

Abstract. We present a software tool for reasoning in and about propositional sequent calculi for modal logics of actions. As an example, we implement the display calculus D.EAK of dynamic epistemic logic. The tool generates embeddings of the calculus in the theorem prover Isabelle/HOL for formalising proofs about D.EAK. Integrating propositional reasoning in D.EAK with inductive reasoning in Isabelle/HOL, we verify the solution of the muddy children puzzle for any number of muddy children. There also is a set of meta-tools that allows us to adapt the software for a wide variety of user defined calculi.

1 Introduction

This paper is part of a long ranging project which aims at developing the proof theory of a wide variety modal logics of actions. The logical calculi to be developed should

- have 'good' proof theoretic properties,
- be built modularly from smaller calculi,
- have applications in a wide range of situations not confined to computer science.

In [18,19,22], we started from the observation that modal logics of actions such as propositional dynamic logic (PDL) or dynamic epistemic logic (DEL), despite having Hilbert style axiomatisations, typically did not have Gentzen systems with good proof theoretic properties. We found that a more expressive extension of sequent calculi, the *display calculi*, allow us to give a proof system D.EAK for a logic of action and knowledge, which enjoys the following properties that typically come with display calculi:

© Springer International Publishing AG, part of Springer Nature 2018
J. Avigad and A. Mahboubi (Eds.): ITP 2018, LNCS 10895, pp. 48–67, 2018.
https://doi.org/10.1007/978-3-319-94821-8_4

- a cut-elimination theorem that can be proved by instantiating a meta-theorem à la Belnap [7],
- modularity in the sense that—without endangering the cut-elimination theorem—connectives and rules can be added freely as long as one adheres to the 'display format'.

During this work we recognized that software tool support and interactive theorem provers will have to play an important role in our project in order to both

- perform proofs in which we reason in our calculi,
- formalise proofs about our calculi.

In this paper, we present a software tool providing the necessary infrastructure and an illustrating case study. The case study is of interest in its own right, even if Lescanne in [32] already provides a formalisation of the solution of the muddy-children puzzle. The propositional reasoning required is performed in the propositional calculus D.EAK and embedded into Isabelle in such a way that the necessary induction can be performed and verified in Isabelle.

Contributions. One aim of our software tool is to support research on the proof theory of modal logics of actions. The typical derivations may be relatively small, but they should be presented in a user interface in LATEX in a style familiar to the working proof theorist. Moreover, in order to facilitate experimenting with different rules and calculi, meta-tools are needed that construct a calculus toolbox from a calculus description file.

Second aim is to support investigations into the question whether a calculus is suited to reasoning in some application area. To perform relevant case studies, one must deal with much bigger derivations and additional features such as abbreviations and derived rules are necessary. Another challenge is that applications may require additional reasoning outside the given calculus (e.g. induction), for which we provide an interface with the theorem prover Isabelle.

More specifically, in the work presented in this paper, we focus on D.EAK and aim for applications to epistemic protocols. In detail, we provide the following.

- A calculus description language that allows the specification of the terms/rules and their typesetting in ASCII, Isabelle and LATEX in a calculus description file.
- A program creating from a calculus description file the calculus toolbox, which comprises the following.
  - A shallow embedding of the calculus in the theorem prover Isabelle. The shallow embedding encodes the terms and the rules of the calculus and allows us to verify in the theorem prover whether a sequent is derivable in D.EAK.
  - A deep embedding of the calculus in Isabelle. The deep embedding also has a datatype for derivations and allows us to prove theorems about derivations.
  - A user interface (UI) that supports

* interactive creation of proof trees,
* simple automatic proof search (currently only up to depth 5),
* export of proof trees to LaTeX and Isabelle,
* the use of derived rules, abbreviations, and tactics.

- A full formalisation of the proof system for dynamic epistemic logic of [22], which is the first display calculus of the logic of Baltag-Moss-Solecki [6] (without common knowledge).
- A fully formal proof of the solution of the muddy children puzzle for any number of dirty children. This is done by verifying in Isabelle that the solution of the muddy children puzzle can be derived in D.EAK for any number of muddy children.
- A set of meta-tools that enables a user to change the calculus.

Case Study: Muddy Children. The muddy children puzzle was chosen because it is a well-known example of an epistemic protocol and required us to extend the tool from one supporting short proofs of theoretical value to larger proofs in an application domain. On the UI side, this led us to add features including abbreviations, macros (derived rules), and two useful tactics. On the Isabelle side, we added a shallow embedding of D.EAK in which we do the inductive proof that the well-known solution of the muddy children puzzle holds for arbitrary number of children. Whereas most of the proof is done in D.EAK using the UI and then automatically translating to Isabelle, the induction itself is based on the higher order logic of Isabelle/HOL.

Related Work. The papers [13,14] pioneered the application of interactive theorem proving and Isabelle to the proof theory of display calculi of modal logic. A sequent calculus for dynamic epistemic logic was given by [15] and results on automatically proving correct the solution of the muddy children puzzle for small numbers of dirty children are reported in [39]. Tableau systems for automatic reasoning in epistemic logics with actions are studied in [1,2,35]. Even though tableaux are close relatives of sequent calculi, tableaux are designed towards efficient implementation whereas display calculi are designed to allow for a general meta-theory and uniform implementation. Their precise relationship needs further investigation in future work.

Comparison of Isabelle to Other Proof Assistants. The papers [31–33] implement epistemic logic in the proof assistant Coq. It would be interesting to conduct the work of this paper based on Coq to enable an in-depth comparison. Isabelle has some advantages for us, in particular, (1) the proof language Isar, (2) the sledgehammer method, and (3) export of theories into programming languages such as Scala. This allows us to build the user interface (UI) directly on the deep embedding of the calculus in Isabelle, thus reusing verified code. This will be important to us in Sect. 4, where we use (1) and (2) in order to write the mathematical parts of the proof of the solution of the muddy children puzzle in a mathematical style close to [34] and we use (3) and the UI to build the derivations in D.EAK.

Outline. Section 2 reviews what is needed about D.EAK. Section 3 presents the main components of the DEAK calculus toolbox. Section 4 discusses the implementation of the muddy children puzzle. Section 5 explains the efforts we have made to keep the tool parametric in the calculus. Section 6 concludes with lessons learned and directions of future research.

2   Display Calculi and D.EAK

In this section, we will introduce Belnap's display calculi [7], which are a refinement of Gentzen sequent calculi. Having received widespread application (e.g. [9,12,23,30,43]), we will argue that display calculi form a good framework for systematically studying modal logics. As a case study, we will also introduce a formalisation of the **D**isplay calculus for **E**pistemic **A**ctions and **K**nowledge (D.EAK) [22]; a display version of the the dynamic epistemic logic [42] of Baltag-Moss-Solecki [6] without common knowledge.

The current paper is part of a wider project that seeks to establish display calculi as a suitable framework for a wide variety of modal logics, both from the point of view of proof theory and from the point of view of tools supporting the reasoning in and about such logics. We use the D.EAK logic as a guiding example to show some of the features of the tools presented in the latter sections, but the tools and approaches presented in this paper do not depend on a detailed knowledge of display logics or D.EAK in particular. A complete description of D.EAK is available in [22] (where it is called D'.EAK).

Display Calculi. A tool or a framework for investigating modal logics in a systematic fashion should not only support as many of these logics as possible, it should also do this in a uniform way. That is, it should allow the user an easy way to define their own logic in a suitable calculus description language, as well as an easy way to combine already defined logics. After defining a new logic, either from scratch or by combining given logics, a reasoning tool for the new logic should be compiled automatically from the calculus description file.

Display calculi support such a modular and uniform approach in several ways. Firstly, the display format is restricted enough to allow us to define logics in a simple way: The language is given by a context free grammar and the inference rules by Horn clauses which only contain either logical connectives or their structural counterparts (and typically do not refer to external mechanisms such as side-conditions or labels encoding worlds in a Kripke model). Yet, display calculi are more *expressive* than sequent calculi (a wide array of logical properties in the language of display calculi is simply not expressible via Gentzen calculi [10]).

Secondly, the display format was invented because it supports a cut-elimination meta-theorem, that is, any display calculus will enjoy cut-elimination. Intuitively, this is due to the meaning of a connective changing in a controlled way when the logic is extended, which in turn makes *modularity* useful. Adding or removing so-called analytic structural rules to a display calculus captures different logics, whilst still preserving cut elimination [11,25].

Finally, display calculi aim to be *transparent*: the design of inference rules corresponds to a programming discipline and leads to a well-understood class of algebraic, and via correspondence theory, Kripke semantics [11, 25].

D.EAK is a proof system for (intuitionistic or classical) dynamic epistemic logic, the abstract syntax of formulas being defined by the grammar

$$\phi ::= p \mid \bot \mid \top \mid \phi \wedge \phi \mid \phi \vee \phi \mid \phi \rightarrow \phi \mid [\mathsf{a}]\phi \mid [\alpha]\phi \mid \langle \mathsf{a} \rangle \phi \mid \langle \alpha \rangle \phi \mid 1_\alpha \qquad (1)$$

where p ranges over atomic propositions, a ranges over agents, with $[\mathsf{a}]\phi$ standing for "agent a knows ϕ", and α ranges over actions with $[\alpha]\phi$ standing for "ϕ holds after α". 1_α represents the precondition of the action α in the sense of [6]. Negation is expressed by $\phi \rightarrow \bot$.

Operational Rules. Display calculi are sequent calculi in which the rules follow a particular format that guarantees good proof theoretic properties such as cut elimination. One of the major benefits is modularity: different calculi can be combined and rules can be added while the good properties are preserved.

The rules of the calculus are formulated in such a way that, in order to apply a rule to a formula, the formula needs to be 'in display'. For example, the following or-introduction on the left (where 'contexts' are denoted by W, X, Y, Z and formulas by A, B)

$$(\vee'_L) \quad \frac{W, A \vdash X \qquad Z, B \vdash Y}{W, Z, A \vee B \vdash X, Y} \qquad (2)$$

is not permitted in a display calculus, since the formula $A \vee B$ must be introduced in isolation as, for example, in our rule

$$(\vee_L) \quad \frac{A \vdash X \qquad B \vdash Y}{A \vee B \vdash X ; Y} \qquad (3)$$

where A and B are formulas and X, Y are arbitrary 'contexts'. Contexts are formalised below as 'structures', that is, as expressions formed from structural connectives. In sequent calculi, there is typically only one structural connective "," but display calculi generalise this. To emphasise that "," is now only one of many structural connectives we write it as ";".

Display Rules. In order to derive a rule such as $(\vee'_L)$ from the rule $(\vee_L)$, it becomes necessary to isolate formulas by moving contexts to the other side. This is achieved by pairing the structural connectives such as "," (written ';' in D.EAK) with so-called adjoint (aka residuated) operators such as ">" and adding bidirectional display rules

$$(;,>) \frac{X ; Y \vdash Z}{Y \vdash X > Z} \qquad \frac{Z \vdash X ; Y}{X > Z \vdash Y} (>,;)$$

which allow us to isolate, in this instance, Y on the left or right of the turnstile.

The name display calculus derives from the requirement that the so-called display property needs to hold: Each substructure can be isolated on the left-hand side, or, exclusively, on the right-hand side. This is the reason why we can confine ourselves, without loss of generality, to the special form of operational rules discussed above.

Structures. A systematic way of setting this up for the set of formulas (1) is to introduce *structural connectives* corresponding to the operational connectives as follows.

Structural	<	>	;	I	{α}	$\widehat{α}$	$Φ_α$	{a}	$\widehat{a}$
Operational	$\prec$	←	$\succ$ →	∧ ∨	⊤ ⊥	⟨α⟩ [α]	$\widehat{α}$ $\check{α}$	$1_α$	⟨a⟩ [a]	$\widehat{a}$ $\check{a}$

This leads to a two tiered calculus which has formulas and structures, with structures generalising contexts and being built from structural connectives. We briefly comment on the particular choice of structural connectives above. Keeping with the aim of modularity, D.EAK was designed in such a way that one can drop the exchange rule for ';' and treat non-commutative conjunction and disjunction, in which case we need two adjoints of ';' denoted by > and <.[1] Similarly, negation is formalised in terms of implication and bottom as in intuitionisitc and substructural logics. Following the symmetries inherent in this substructural analysis of logic [37] suggests to add the operational connectives $\prec$, ←, $\succ$ (but which are only needed if one does not have the rule of commutativity of ';'). Similarly, the modal operators [α] and [a] have structural counterparts {α} and {a} which in turn have adjoints $\widehat{α}$ and $\widehat{a}$. The formulas (1) do not have operational connectives corresponding to the structural connectives $\widehat{α}$ and $\widehat{a}$, but they can be added and are indeed useful (in terms of Kripke semantics, the adjoint of a box modality □ for a relation R is the diamond modality for the converse relation R^{-1} often denoted by ♦).

Structural Rules. The rules of D.EAK can be divided into operational rules and display rules, as discussed above, and structural rules, to which we turn now. The operational rules such as ($∨_L$) specify how to introduce a logical operation. Display rules such as (; >) are used to isolate formulas or structures to which we want to apply a specific rule. The logical axiomatisation sits in the structural rules. Apart from the structural rules like weakening, exchange, and contraction for ';' we have also other structural rules such as the display rules discussed above and rules that express properties such as 'actions are partial functions' axiomatised by the rule[2]

---

[1] For example, taking into account the correspondence between operational and structural connectives, the rule (; , >) above says precisely that the operation that maps C to $A → C$ is right-adjoint to the operation that maps B to $A ∧ B$. Similarly, (>, ;) expresses that $A \succ$ _ is left adjoint to $A ∨$ _.

[2] which implies that one can derive $⟨α⟩X ⊢ [α]X$.

$$\frac{X \vdash Y}{\{\alpha\}X \vdash \{\alpha\}Y} \tag{4}$$

and such as 'if a knows Y, then Y is true' axiomatised by the rule[3]

$$\frac{X \vdash \{a\}Y}{X \vdash Y} \tag{5}$$

The reason to axiomatise a logic via such structural rules instead of axioms is that then cut-elimination is preserved.

Modularity of D.EAK and Related Calculi. We have seen that D.EAK has a large number of connectives. But they arise according to clear principles: operational connectives have structural counterparts which in turn have adjoints. Similarly, the fact that D.EAK as we defined it, due to being built systematically from substructural connectives, has a large number of rules does not pose problems from a conceptual point as the rules fall into clearly delineated classes each serving their own purpose. It is exactly this feature which enables the modularity of the display logic approach to the proof theory of sequent calculi. But, from the practical point of view of creating proof trees or of composing a number of different calculi, this large number of connectives and rules makes working with these calculi difficult. Moreover, the encoding of terms and proof trees needed for automatic processing will not be readable to humans who would expect to manipulate proof trees displayed from LaTeX documents in an easy interactive way. How we propose to solve these problems will be discussed in the next section.

3   The D.EAK Calculus Toolbox

The aim of the D.EAK calculus toolbox[4] is to support research on the proof theory of dynamic epistemic logic as well as to conduct case studies exploring possible applications. It provides a shallow and a deep embedding of D.EAK into Isabelle and a user interface implemented in Scala.

The shallow embedding has an inductive datatype for the terms of the calculus and encodes the rules via a predicate describing which terms are derivable. It is used to prove correct the solution of the muddy children puzzle in Sect. 4.

The deep embedding also has datatypes for rules and derivations and provides functionality such as rule application (match and replace) as well as automatic proof search and tactics. The corresponding Isabelle code is exported to Scala and used in the user interface.

D.EAK proof trees can be constructed interactively in a graphical user interface by manipulating trees typeset in LaTeX. Proof trees can be exported to LaTeX/PDF and Isabelle. This was essential for creating the Isabelle proof in

---

[3] which implies that one can derive $[a]Y \vdash Y$.

[4] Compiled version available for download at: https://github.com/goodlyrottenapple/calculus-toolbox/raw/master/calculi/DEAK.jar.

Sect. 4. Examples of typeset LATEX proof trees can be found at [5]: The .cs files contain the proofs as done in the UI and the .tex-files the exported LATEX code. The tag cleaned_up was added after a small amount of manual post-processing of the .tex-files.

3.1 Shallow Embedding (SE) in Isabelle

The shallow embedding of the calculus D.EAK is available in the files DEAK_SE.thy and DEAK_SE_core.thy. The file DEAK_SE_core.thy contains the definitions of the terms via datatypes Atprop, Formula, Structure, Sequent. For example,

```
datatype Sequent = Sequent Structure Structure ("_ ⊢ _")
```

declares that an element of datatype Sequent consists of two structures. The annotation ("_ ⊢ _") allows us to use the familiar infix notation ⊢ in the Isabelle IDE.

The file DEAK_SE.thy encodes the rules of the calculus by defining a predicate derivable

```
inductive derivable :: "Locale list ⇒ Sequent ⇒ bool"  ("_ ⊢d _")
```

by induction over the rules of D.EAK. For example, the rule $(\vee_L)$ above is encoded as

```
Or_L:  "l ⊢d (B ⊢ Y) ⟹ l ⊢d (A ⊢  X) ⟹  l ⊢d (A ∨ B ⊢ X ; Y)"
```

which expresses in the higher-order logic of Isabelle/HOL that if $B \vdash Y$ and $A \vdash X$ are derivable, then $A \vee B \vdash X; Y$ is derivable. Note that A, B, X, Y are variables of Isabelle. The rule will be applied using the built-in reasoning mechanism of Isabelle/HOL which includes pattern matching.

The datatype Locale is used to carry around all the information needed in a proof that is not directly available, in a bottom up proof search, from the sequent on which we want to perform a rule.

For example, in order to perform a cut, we need to specify the cut formula. In the UI, when constructing a prooftree interactively, it will be given by the user. Internally, cut-formulas are of type Locale and the cut-rule is given by

```
"(CutFormula f)∈ set l ⟹ l⊢d(X ⊢  f) ⟹ l⊢d(f ⊢ Y) ⟹ l⊢d(X ⊢ Y)"
```

Similarly, the rules that describe the interaction of the knowledge of agents with epistemic actions depend on the so-called action structures, which define the actions, but are not part of the calculus itself. These action structures, therefore, are also encoded by data of type Locale.

Before coming to the deep embedding next, we would like to emphasise, that in order to prove in the shallow embedding, that a certain sequent is derivable in D.EAK, one shows in theorem prover Isabelle/HOL that the sequent is in the extension of the predicate derivable. The proof itself is not available as data that can be manipulated.

3.2   Deep Embedding (DE) in Isabelle

The deep embedding is available in the files DEAK.thy and DEAK_core.thy. The latter contains the encoding of the terms of D.EAK, which differs only slightly from the one of the shallow embedding. It also contains functions match and replace, plus some easy lemmas about their behaviour. The functions match and replace are used in DEAK.thy to define how rules are applied to sequents.

DEAK.thy starts out by defining the datatypes Rule and Prooftree. The function der implements how to reason backwards from a goal:

```
fun der :: "Locale ⇒ Rule ⇒ Sequent ⇒ (Rule * Sequent list)"
```

takes a locale, a rule r, and a sequent s and outputs the list of premises needed to prove the s via r.[5] This function is then used to define the predicate isProofTree and other functions that are used by the UI.

One reason to define the deep embedding in Isabelle (and not e.g. directly in the UI) is that we want to use the deep embedding in future work, to implement and prove correct the cut elimination for D.EAK and related calculi. Another is that the UI then uses reliable code compiled from its specification in Isabelle.

3.3   Functionality of the User Interface (UI)

The UI is an essential part of the toolbox and provides the following functionality:

- LATEX typesetting of the terms of the calculus, with user specified syntactic sugar.
- Graphical representation of proof trees in LATEX.
- Exporting proof trees to LATEX/PDF and to Isabelle SE.
- Automatic proof search (to a modest depth of 5).
- Interactive proof tree creation and modification, including merging proof trees, deleting portions of proof trees, and applying rules.
- Tactics for deriving the generalised identity/atom rules and applying display rules.
- User defined abbreviations and macros (derived rules).

The UI is implemented in Scala. There were several reasons for choosing Scala, one of which is Isabelle's code export functionality which translates functions written in Isabelle theory files to be exported into functional languages such as Scala or Haskell, amongst others. This meant that the underlying formalisation of terms, rules and proof trees of the deep embedding of the calculus and

---

[5] It is at this point where our implementation of the deep embedding is currently tailored towards substructural logics: For each rule r and each sequent s, there is only one list of premises to consider. Generalising the deep embedding to sequent calculi with rules such as (2) would require a modification: If we interpret the structure $W, X, A \lor B$ in (2) not as a structure (i.e. tree) but as a list, then matching the rule (2) against a sequent would typically not determine the sublists matching W and X in a unique way. More information is available at [3].

the functions necessary for building and verifying proof trees could be built in Isabelle and then exported for the UI into Scala.

Another advantage of using Scala is the fact that it is based on Java and runs on the JVM, which makes code execution fast enough, and, more importantly, it is cross platform and allows the use of Java libraries. This was especially useful when creating the graphical UI for manipulating proof trees, as the UI depends on two libraries, JLaTeXMath and abego TreeLayout, which allow for easy typesetting and pretty-printing of the proof trees as well as simple visual creation and modification of proof trees in the UI.

The UI was made more usable with the implementation of some simple tactics, such as the display tactic, which simplifies the often tedious application of display rules to isolate a particular structure in a sequent. This happens in a display calculus every time the user wants to apply an operational rule, and can be a bureaucratic pain point of little interest to the logician. Even though these tactics are unverified, once the generated proof tree is exported to Isabelle, it will be checked by the Isabelle core to ensure validity.

4   Case Study: The Muddy Children Puzzle

The muddy children puzzle is a classical example of reasoning in dynamic epistemic logic, since it highlights how epistemic actions such as public announcements modify the knowledge of agents. We will recall the puzzle in some detail below. The solution will state that, after k rounds of all agents announcing "I don't know", all agents do in fact know.

The correctness of the solution has been established, for all $k \in \mathbb{N}$ using induction, by informal mathematical proof [16] and by mathematical proofs about a formalisation in a Hilbert calculus [34]. It has also been automatically verified, for small values of k, using techniques from model checking [29] (see also [40] for related work) and automated theorem proving [15,39].

Here, we prove in Isabelle/HOL that for all k the solution is derivable in D.EAK.

4.1   The Muddy Children Puzzle

There are $n > 0$ children and $0 < k \leq n$ of them have mud on their foreheads. Each child sees (and hence knows) which of the others is dirty. But they cannot see (and therefore do not know at the beginning) whether they are dirty themselves (thus the number n is known to them but k is not). The first epistemic action is the father announcing (publicly and truthfully) that at least one of the children is dirty. From then on, the protocol proceeds in rounds. In each round, all children announce (simultaneously, publicly and truthfully) whether they know that they are dirty or not. How many rounds need to be played until the dirty children know that they are dirty?

In case $n = 1, k = 1$ the only child knows that it must be dirty, since the announcement by the father, as all announcements in this protocol, are assumed to be truthful. We write this as

$$[\mathsf{father}]\square_1 D_1,$$

where D_j is an atomic proposition encoding that child j is dirty, $\square_j p$ means child j knows p and [father]p means that p holds after father's announcement. We use $\square_j$ instead of $[j]$ for both stylistic purposes (the box $\square$, being the usual notation in the literature), as well as to distinguish it from announcements like [father].

The case $n > 1$ and $k = 1$ is similar. Let j be the dirty child. It sees, and therefore knows, that all the other children are clean. Since, after father's announcement, child j knows that there is at least one dirty child, it must be j, and j knows it.

In case $n > 1$ and $k = 2$ let $J = \{j, h\}$ be the set of dirty children. After father's announcement both j and h see one dirty child. But they do not know whether they are dirty themselves. So, according to the protocol, they announce that they do not know whether they are dirty. From the fact that h announced $\neg \square_h D_h$, child j can conclude D_j, that is, we have $\square_j D_j$. To see this, j reasons that if j was clean, then h would be in the situation of the previous paragraph, that is, we had $\square_h D_h$, in contradiction to the truthfulness of the announcement of h. Summarising, we have shown

$$[\mathsf{father}][\mathsf{no}]\square_j D_j,$$

where [no] is the modal operator corresponding to the children announcing that they don't know whether they are dirty.

The cases for $k > 2$ follow similarly, so that we obtain for all dirty children j

$$[\mathsf{father}][\mathsf{no}]^{k-1}\square_j D_j \tag{6}$$

For example, for $n = k = 100$, after 99 rounds of announcements "I don't know whether I am dirty" by the children, they all do know that they are dirty.

4.2 Muddy Children in Isabelle

Our proof in Isabelle follows [34, Prop.24], which gives a mathematical proof that for all $n, k > 0$ there is, in a Hilbert system equivalent to D.EAK, a derivation of (6) from the assumption

$$\mathsf{dirty}(n, J) \;\wedge\; \mathsf{E}(n)^k(\mathsf{vision}(n)) \tag{7}$$

which encodes the rules of the protocol. Specifically, $\mathsf{dirty}(n, J)$ encodes for each $J \subseteq \{1, \ldots n\}$ that precisely the children $j \in J$ are dirty, $\mathsf{vision}(n)$ expresses that each child knows whether any of the other children are dirty, $\mathsf{E}(n)(\phi)$ means that 'every one of the n children knows ϕ' and f^k indicates k-fold application of the function f so that $\mathsf{E}(n)^k(\mathsf{vision}(n))$ says that 'each child knowing whether the others are dirty' is common knowledge up to depth k.

This means that we need to prove by induction on n and k that for all n, k there is a derivation in the calculus D.EAK of the sequent

$$\mathsf{dirty}(n, J), \mathsf{E}(n)^k(\mathsf{vision}(n)) \vdash [\mathsf{father}][\mathsf{no}]^{k-1}\,\Box_j D_j \,. \tag{8}$$

where the actions father and no also depend on the parameter n.

For the cases $k = 1, 2$ the proofs can be done with a reasonable effort in the UI of the tool, filling in all the details of the proof of [34].

But as a propositional calculus, D.EAK does not allow us to do induction. Therefore we use the shallow embedding of D.EAK and do the induction in the logic of Isabelle. The expressions $\mathsf{dirty}(n, J)$ and $\mathsf{E}(n)^k(\mathsf{vision}(n))$ and $[\mathsf{father}][\mathsf{no}]^{k-1}\,\Box_j D_j$ then are Isabelle functions that map the parameters n, k to formulas (in the shallow embedding) of D.EAK, see the theory MuddyChildren [5].

The first part of the theory MuddyChildren contains the definitions of the formulas discussed above and establishes some of their basic properties. The actual proof is given as lemma dirtyChildren. We have taken care to follow [34] closely, so that the proof of its Proposition 24 can be read as a high-level specification of the proof in Isabelle of lemma dirtyChildren.

The proof in MuddyChildren differs from its specification in [34] only in a few minor ways. Instead of assuming the axiom of introspection $[\mathsf{a}]p \to p$, we added the corresponding structural rules to the calculus. This seems justified as it is a fundamental property of knowledge we are using and also illustrates a use of modularity. Instead of introducing separate atomic propositions for dirty and clean, we treat clean as an abbreviation for not dirty, which relieves us from axiomatising the relationship between dirty and clean explicitly. But if we want an intuitionistic proof, we need to add to our assumptions that 'not not dirty' implies dirty.

4.3   Conclusions from the Case Study

It took approximately 4 person-weeks to implement the proof of [34, Prop.24] in Isabelle. Part of this went into providing some 'infrastructure' contained in the theories NatToString and DEAKDerivedRules that could be reused for other case studies. On the other hand, we should say that it took maybe half a year to learn Isabelle and we couldn't have learned it from documentation and tutorials alone. At crucial points we had expert advice by Thomas Tuerk, Tom Ridge and Christian Urban.

For the construction of the proof in Isabelle, we made extensive use of the UI. Large parts of the Isabelle proof were constructed in the UI and exported to Isabelle.

One use one can make of the formal proof is to investigate which proof principles are actually needed. For example, examining the proof in MuddyChildren, it is easy to establish that the only point where a non-intuitionistic principle is used is to prove $\neg\neg D_j \to D_j$. Instead we could have added this formula (which only says that "not clean implies dirty") to the logical description of the puzzle (7).

It may be worth pointing out that this analysis is based on the substructural analysis of classical logic on which D.EAK is built. Accordingly, a proof in D.EAK is intuitionistic if and only if it does not use the so-called Grishin rules Grishin_L and Grishin_R (as defined in DEAK.json). Thus a simple text search for 'Grishin' in the theory MuddyChildren suffices to establish the claim that adding $\neg\neg D_j \rightarrow D_j$ to the assumptions, our proof of the muddy children puzzle follows the principles of intuitionistic logic.

5 *Meta-toolbox* - Building Your Own Calculus Toolbox

As discussed in Sect. 3, the D.EAK toolbox consists of a set of Isabelle theory files that formalize the terms and rules of this calculus, providing a base for reasoning about the properties of the calculus in the Isabelle theorem prover. The toolbox also includes a UI for building proof trees in the calculus.

As this paper is part of a wider program of the study of modal logics, the reader might naturally be interested in building her own calculus/logic, either building on top of D.EAK or starting from scratch. To do this, we provide a *meta-toolbox*, which consists of a set of scripts and utilities used for maintaining, modifying and building your own calculus toolbox (we will use italics when referring to the *meta-toolbox*, to avoid confusion with the toolbox generated by the *meta-toolbox* for a specific logic).

The *meta-toolbox* supports building tools for a wide range of propositional modal logics. Due to the display framework, constraints placed on the shape of the calculus terms and rules allow us to build large portions of the Isabelle theory files and the UI front-end in a generic, logic-agnostic way, from a calculus description file.

However, there are cases in which the language used to specify calculus description files may not be expressive enough to encode certain information about the logic. For example, in the case of rules with side-conditions, the user needs to implement the Isabelle code handling the side conditions manually. To avoid reimplementation on each change of the calculus description file, we provide the watcher utility, which builds a new calculus toolbox by weaving the user-specific manual changes into the generic automatically created code without breaking the user made modifications.

The main component of the *meta-toolbox* is the build script, which takes in a description file and expands it into multiple Isabelle theories and Scala code. Due to this centralised definition of the calculus, adding rules or logical connectives becomes easy as the user only needs to change the calculus description file and does not need to worry about how these changes affect multiple Isabelle and Scala files. The *meta-toolbox* thus allows for a more structured and uniform maintenance of the different encodings along with the UI. A detailed documentation and a tutorial is available [3].

5.1 Describing a Calculus

We highlight some elements of how to describe a calculus such as D.EAK in the format that can be read by the *meta-toolbox*.

The calculus is described in a *calculus description file* using the JavaScript Object Notation (JSON), in our example DEAK.json. This file specifies the types (Formula, Structure, Sequent, ...), the operational and structural connectives, and the rules. For example, linking up with the discussion in Sect. 3, in

```
"Sequent": {
  "type" : ["Structure", "Structure"],
  "isabelle" : "_ \\<turnstile> _",
  "ascii" : "_ |- _",
  "latex" : "_ {\\ {\\textcolor{magenta}\\boldsymbol{\\vdash}\\ } _",
  "precedence": [311,311,310]
}
```

"type" specifies that a sequent consists of two structures.[6] The next three lines specify how sequents will be typeset in Isabelle, ASCII and LATEX. To make proofs readable in the UI, the user can specify bespoke sugared notation using, for example, LATEX commands such as colours and fonts.

Next we explain how rules are encoded. The encoding is divided into two parts. In the first part, under the heading "calc_structure_rules" the rules are declared. For example, we find

```
"Or_L" : {
  "ascii" : "Or_L",
  "latex" : "\\vee_L"
}
```

telling us how the names of the rule are typeset in ASCII and LATEX. The rule (3) itself is described in the second part under the heading "rules" by

```
"Or_L" : ["F?A \\/ F?B |- ?X ; ?Y", "F?A |- ?X", "F?B |- ?Y"]
```

the first sequent of which is the conclusion, the following being the premises of the rule. The ? has been defined in DEAK.json to indicate the placeholders (a.k.a. free variables or meta-variables) that are instantiated when applying the rule. The F marks placeholders that can be instantiated by formulas only.

The description of Or_L above suffices to compile it to Isabelle. But some rules of D.EAK need to be implemented subject to restrictions expressed separately. For example, the so-called atom rule formalises that in D.EAK, actions do not change facts (but they may change knowledge). Thus, whereas the rule is encoded as

```
"Atom" : ["?X |- ?Y", ""]
```

---

[6] The presence of the \\ instead of just one \ is unfortunate but \ is a reserved character that needs to be escaped using \.

we need to enforce the condition that ?X |- ?Y is of the form $\Gamma p \vdash \Delta p$, where p is an atomic proposition and Γ, Δ are strings of action modalities. This is done by noting in the calculus description file the dependence on a condition called atom as follows.

```
"Atom" : {
  "condition" : "atom"
}
```

The condition atom itself is then implemented directly in Isabelle.

For bottom-up proof search, the deep embedding provides a function that, given a sequent and a rule, computes the list of premises (if the rule is applicable). For the cut rule, this is implemented by looking for a cut-formula in the corresponding Locale, see Sect. 3.1. (As stated earlier, whilst the calculus admits a cut free presentation, it is nonetheless useful to keep the cut rule in the calculus when manually constructing proofs.)

```
"RuleCut" : {
  "SingleCut" : {
    "locale" : "CutFormula f",
    "premise" : "(\\<lambda>x. Some [((?\\<^sub>S ''X'') ...",
    "se_rule" : "1 \\<turnstile>d (X \\<turnstile>\\<^sub>S f ..."
  }
}
```

After "premise" we find the Isabelle definition of the DE-version of the rule and after "se_rule" the SE-version of the rule.

The most complicated rules of D.EAK are those which describe the interaction of action and knowledge modalities and we are not going to describe them here. They need all of the additional components condition, locale, premise, se_rule, to deal with side conditions which depend on actions being agent-labeled relations on actions.

The ability to easily change the calculus description file will be useful in the future, but also appeared already in this work. Compared to the version of D.EAK from [22], we noticed during the work on the muddy children puzzle that we wanted to add rules Refl_ForwK expressing $[a]p \rightarrow p$ (i.e. that the knowledge-relation is reflexive) and rules Pre_L and Pre_R allowing us to replace in a proof the constant representing the precondition of an action by the actual formula expressing the precondition. Using the *meta-toolbox*, this change was a simple case of adding the rule to the JSON description file and recompiling the calculus.

5.2   The Build Script, the Template Files, and the Watcher Utility

To build the calculus toolbox from the calculus description file DEAK.json, one runs the Python script, passing the description file to the script via the --calculus flag. This produces the Isabelle code for the shallow and deep embedding and the Scala code for the UI. By default, this toolbox is output to a directory called gen_calc.

Template Files. The toolbox is generated from both the calculus description file and template files. Template files contain the code that cannot be directly compiled from the calculus description file, for example, the code of the UI. But whereas the code of the UI, in the folder gui, is independent of the particular calculus, the parser Parser.scala and the print class Print.scala consist of code written by the developer as well as code automatically generated from the calculus description file. Similarly, whereas parts of DEAK.thy are compiled from the calculus description file, other parts, such as the lemmas and their proofs are written by the developer.

The Isabelle and Scala Builder. In order to support the weaving of auto-matically generated code into the template files, there are two domain specific languages defined in the files isabuilder.py and scalabuilder.py. For example, in the template file Calc_core.thy, from which DEAK.thy is generated, the line

(*calc_structure*)

prompts the build script to call a method defined in isabuilder.py which inserts the Isabelle definition of the terms of the calculus into DEAK.thy.

The Watcher Utility. In order to make the maintenance of the template files easier there is a watcher utility which allows, instead of directly modifying the template files, to work on the generated code. For example, if we want to change how proof search works, we would make the changes to the Isabelle file DEAK.thy and not directly to the template file Calc_core.thy. The watcher utility, when launched, runs in the background and monitors the specified folder. Any changes made to a file inside this folder are registered and the utility decompiles this file back into its corresponding template, each time a modification occurs. The watcher utility decompiles a file by looking for any sections of the file that have been automatically generated, and replacing these definitions by the special comments that tell the build script where to put the auto-generated code. In order for the decompiling to work correctly, the auto-generated code must be enclosed by special delimiters. Looking back at the example of (*calc_structure*), when the template file is processed by the build script and expanded with the definitions from a specific calculus description file, the produced code is enclosed by the following delimiters:

(*calc_structure-BEGIN*)
auto-generated code ...
(*calc_structure-END*)

Hence, when the watcher utility decompiles a file into a template, it simply replaces anything of the form (*<identifier>-BEGIN*) ... (*<identifier>-END*) by the string (*<identifier>*).

6   Conclusion

We presented a software tool that makes interactive theorem proving available for the proof theoretic study of modal logics of actions. From a calculus description file, shallow and deep embeddings of deductions are generated. The deep embedding is used to automatically generate verified code for the user interface, which in turn allows us to make derivations in the calculus in a familiar proof theoretic environment and then export it to Isabelle. This has been used to develop a fully formalised proof of the correctness of the solution for the muddy children puzzle, making use of Isabelle's ability of inductive reasoning that goes beyond the expressiveness of (propositional) modal logic.

An interesting lesson learned from using interactive theorem proving in the proof theory of modal logics is that in our work the concerns of the software engineer and the proof theorist can be seen as two sides of the same coin as we will explain in the following.

From the point of view of proof theory, we are interested in developing 'good' calculi, which refers to, on the one hand, mathematical properties such as cut-elimination, and, on the other hand, to the design criteria developed in the area of proof theoretic semantics [36,38]. These design criteria include the following. (i) The meaning of a connective should be defined, in the spirit of introduction and elimination rules, in a way that renders their meaning independent of what other connectives and rules may be added to the calculus. (ii) The rules should be closed under uniform substitution and be free from extra-logical labels and side-conditions. (See (3) for an example.)

From a software engineering point of view, we want to (I) build software for a 'big' logical calculus comprising many connectives in a modular way, connective by connective and (II) provide the user with a domain specific language that allows for a user-friendly specification of a calculus and its rules. In particular, it should be possible to automatically build a set of tools that allows high-level reasoning and implementation independent use of the calculus from a single calculus description file.

A lesson learned is that (i) and (I) as well as (ii) and (II) are closely related. While we admit that our domain specific language is rudimentary and the calculus description files in Sect. 5.1 could be much more user-friendly, the main issues that need further research are extra-logical labels and side-conditions, see e.g. page 14 where we write "The condition atom itself is then implemented directly in Isabelle". Indeed, such side-conditions are not easily formulated without knowing the lower-level implementation of the logics (in our case their implementation in Isabelle) and therefore are in conflict with the software engineering principle of shielding the user from implementation details. While we solved this problem using a software engineering method (see Sect. 5.2), the next paragraph discusses a possible proof-theoretic solution, namely multi-type display calculi.

The move to multi-type display calculi is akin to the move from one-sorted algebras to many-sorted algebras. Multi-type display calculi, introduced in [19], allow us to absorb extra-logical labels and side conditions into the types. For example, by introducing a type for atoms, the condition on substitution becomes uniform substitution (of formulas of the correct type). Similarly, the extra-logical labels needed for actions can be eliminated. (In passing, we also note that the well-known side conditions for the rules of first-order quantifiers can be eliminated in this way.) This methodology has, by now, been successfully applied to a range of calculi [20,21,25–28].

Ongoing and future work arises from the discussion above. First, a new *meta-toolbox* for multi-type sequent and display calculi (an alpha version providing a more user-friendly interface to define calculi and manipulate derivations is already available [4]). Second, following the work of [13,14] on proving cut elimination of display calculi, a full formalisation of cut-elimination for D.EAK, or rather of the cut-elimination meta-theorems of [18,19,22]. Third, integrating interactive theorem proving with automatic proof search, much in the spirit of Isabelle's Sledgehammer. In particular, modal logics of actions can have tableau systems that do efficient automatic proof search [1] . One question here is, whether it will be possible to do this integration in a modular way: In the light of our discussion above, tableau systems are closely related to sequent calculi [17,24], but they typically do not do so well w.r.t. to property (II).

Acknowledgements. At several crucial points, we profited from expert advice on Isabelle by Tom Ridge, Thomas Tuerk and Christian Urban. We thank Roy Crole and Hans van Ditmarsch for valuable comments on an earlier draft.

References

1. Aucher, G., Schwarzentruber, F.: On the complexity of dynamic epistemic logic. In: Proceedings of the 14th Conference on Theoretical Aspects of Rationality and Knowledge (TARK 2013)
2. Balbiani, P., van Ditmarsch, H., Herzig, A., de Lima, T.: Tableaux for public announcement logic. J. Logic Comput. **20**(1), 55–76 (2010)
3. Balco, S.: The calculus toolbox. https://goodlyrottenapple.github.io/calculus-toolbox/
4. Balco, S.: The calculus toolbox 2. https://github.com/goodlyrottenapple/calculus-toolbox-2
5. Balco, S., Frittella, S.: Muddy children in Isabelle. https://goodlyrottenapple.github.io/muddy-children/
6. Baltag, A., Moss, L.S., Solecki, S.: The logic of public announcements, common knowledge and private suspicious. Technical Report SEN-R9922, CWI, Amsterdam (1999)
7. Belnap, N.: Display logic. J. Philos. Logic **11**, 375–417 (1982)
8. Blackburn, P., van Benthem, J., Wolter, F. (eds.): Handbook of Modal Logic. Elsevier, Amsterdam (2006)
9. Brotherston, J.: Bunched logics displayed. Stud. Logica. **100**(6), 1223–1254 (2012)
10. Ciabattoni, A., Galatos, N., Terui, K.: From axioms to analytic rules in nonclassical logics. In: Proceedings of the 23rd Annual IEEE Symposium on Logic in Computer Science (LICS 2008)
11. Ciabattoni, A., Ramanayake, R.: Power and limits of structural display rules. ACM Trans. Comput. Logic (TOCL) **17**(3), 17 (2016)
12. Ciabattoni, A., Ramanayake, R., Wansing, H.: Hypersequent and display calculi - a unified perspective. Stud. Logica. **102**(6), 1245–1294 (2014)
13. Dawson, J.E., Goré, R.: Embedding display calculi into logical frameworks: comparing twelf and Isabelle. Electr. Notes Theor. Comput. Sci. **42**, 89–103 (2001)

14. Dawson, J.E., Goré, R.: Formalised cut admissibility for display logic. In: Proceedings of 15th International Conference Theorem Proving in Higher Order Logics, TPHOLs (2002)
15. Dyckhoff, R., Sadrzadeh, M., Truffaut, J.: Algebra, proof theory and applications for an intuitionistic logic of propositions, actions and adjoint modal operators. ACM Trans. Comput. Logic **14**(4), 1–37 (2013)
16. Fagin, R., Halpern, J.Y., Moses, Y., Vardi, M.Y.: Reasoning About Knowledge. MIT Press, Cambridge (1995)
17. Fitting, M.: Proof Methods for Modal and Intuitionistic Logic. Springer, Netherlands (1983). https://doi.org/10.1007/978-94-017-2794-5
18. Frittella, S., Greco, G., Kurz, A., Palmigiano, A.: Multi-type display calculus for propositional dynamic logic. J. Log. Comput. **26**(6), 2067–2104 (2016)
19. Frittella, S., Greco, G., Kurz, A., Palmigiano, A., Sikimic, V.: Multi-type display calculus for dynamic epistemic logic. J. Log. Comput. **26**(6), 2017–2065 (2016)
20. Frittella, S., Greco, G., Kurz, A., Palmigiano, A., Sikimić, V.: Multi-type sequent calculi. In: Andrzej Indrzejczak, M.Z., Kaczmarek, J. (ed.) Trends in Logic XIII, pp. 81–93. Lodź University Press (2014). https://arxiv.org/abs/1609.05343
21. Frittella, S., Greco, G., Palmigiano, A., Yang, F.: A multi-type calculus for inquisitive logic. In: Väänänen, J., Hirvonen, Å., de Queiroz, R. (eds.) WoLLIC 2016. LNCS, vol. 9803, pp. 215–233. Springer, Heidelberg (2016). https://doi.org/10.1007/978-3-662-52921-8_14
22. Frittella, S., Greco, G., Kurz, A., Palmigiano, A., Sikimic, V.: A proof-theoretic semantic analysis of dynamic epistemic logic. J. Log. Comput. **26**(6), 1961–2015 (2016)
23. Goré, R.: Substructural logics on display. Logic J. IGPL **6**(3), 451–504 (1998)
24. Goré, R.: Tableau methods for modal and temporal logics. In: D'Agostino, M., Gabbay, D.M., Hähnle, R., Posegga, J. (eds.) Handbook of Tableau Methods. Springer, Dordrecht (1999). https://doi.org/10.1007/978-94-017-1754-0_6
25. Greco, G., Ma, M., Palmigiano, A., Tzimoulis, A., Zhao, Z.: Unified correspondence as a proof-theoretic tool. J. Log. Comput. (2016). https://doi.org/10.1093/logcom/exw022
26. Greco, G., Palmigiano, A.: Linear logic properly displayed. CoRR, abs/1611.04181 (2016)
27. Greco, G., Palmigiano, A.: Lattice logic properly displayed. In: Kennedy, J., de Queiroz, R.J.G.B. (eds.) WoLLIC 2017. LNCS, vol. 10388, pp. 153–169. Springer, Heidelberg (2017). https://doi.org/10.1007/978-3-662-55386-2_11
28. Greco, G., Liang, F., Moshier, M.A., Palmigiano, A.: Multi-type display calculus for semi de morgan logic. In: Kennedy, J., de Queiroz, R.J.G.B. (eds.) WoLLIC 2017. LNCS, vol. 10388, pp. 199–215. Springer, Heidelberg (2017). https://doi.org/10.1007/978-3-662-55386-2_14
29. Halpern, J.Y., Vardi, M.Y.: Model checking vs. theorem proving: a manifesto. In: Proceedings of the 2nd International Conference on Principles of Knowledge Representation and Reasoning (KR 1991), pp. 325–334 (1991)
30. Kracht, M.: Power and weakness of the modal display calculus. In: Proof Theory of Modal Logic, pp. 93–121. Kluwer, Netherlands (1996)
31. Lescanne, P.: Mechanizing common knowledge logic using COQ. Ann. Math. Artif. Intell. **48**(1–2), 15–43 (2006)
32. Lescanne, P.: Common knowledge logic in a higher order proof assistant. In: Programming Logics - Essays in Memory of Harald Ganzinger, pp. 271–284 (2013)
33. Lescanne, P., Puisségur, J.: Dynamic logic of common knowledge in a proof assistant. CoRR, abs/0712.3146 (2007)

34. Ma, M., Palmigiano, A., Sadrzadeh, M.: Algebraic semantics and model completeness for intuitionistic public announcement logic. Ann. Pure Appl. Logic **165**(4), 963–995 (2014)
35. Ma, M., Sano, K., Schwarzentruber, F., Velázquez-Quesada, F.R.: Tableaux for non-normal public announcement logic. In: Banerjee, M., Krishna, S.N. (eds.) ICLA 2015. LNCS, vol. 8923, pp. 132–145. Springer, Heidelberg (2015). https://doi.org/10.1007/978-3-662-45824-2_9
36. Piecha, T., Schroeder-Heister, P., (eds.): Advances in Proof-Theoretic Semantics. Springer, Heidelberg (2016). https://doi.org/10.1007/978-3-319-22686-6
37. Restall, G.: An Introduction to Substructural Logics. Routledge, London (2000)
38. Schroeder-Heister, P.: Proof-theoretic semantics. In: Zalta, E.N. (ed.) The Stanford Encyclopedia of Philosophy. Metaphysics Research Lab, Stanford University, winter 2016 edition (2016)
39. Truffaut, J.: Implementation and improvements of a cut-free sequent calculus for dynamic epistemic logic. M.Sc. thesis, University of Oxford (2011)
40. van Ditmarsch, H., van Eijck, J., Hernández-Antón, I., Sietsma, F., Simon, S., Soler-Toscano, F.: Modelling cryptographic keys in dynamic epistemic logic with DEMO. In: Proceedings of 10th International Conference on Practical Applications of Agents and Multi-Agent Systems, PAAMS (2012)
41. van Ditmarsch, H.P., Kooi, B.: One Hundred Prisoners and a Light Bulb. Springer, Switzerland (2015). https://doi.org/10.1007/978-3-319-16694-0
42. van Ditmarsch, H.P., van der Hoek, W., Kooi, B.: Dynamic Epistemic Logic, Springer, Netherlands (2007). https://doi.org/10.1007/978-1-4020-5839-4
43. Wansing, H.: Displaying Modal Logic. Kluwer, Netherlands (1998)

Backwards and Forwards
with Separation Logic

Callum Bannister[1,2], Peter Höfner[1,2(✉)], and Gerwin Klein[1,2]

[1] Data61, CSIRO, Sydney, Australia
`firstname.lastname@data61.csiro.au`
[2] Computer Science and Engineering, University of New South Wales,
Sydney, Australia

Abstract. The use of Hoare logic in combination with weakest pre-
conditions and strongest postconditions is a standard tool for program
verification, known as backward and forward reasoning. In this paper
we extend these techniques to allow backward and forward reasoning for
separation logic. While the former is derived directly from the standard
operators of separation logic, the latter uses a new one. We implement our
framework in the interactive proof assistant Isabelle/HOL, and enable
automation with several interactive proof tactics.

1 Introduction

The use of Hoare logic [19,21] in combination with weakest preconditions [16]
and strongest postconditions [19] is a standard tool for program verification,
known as backward and forward reasoning. These techniques are supported by
numerous tools, e.g. [1,6,33,36,37].

Although backward reasoning with weakest preconditions is more common
in practice, there are several applications where forward reasoning is more convenient, for example, for programs where the precondition is 'trivial', and the
postcondition either too complex or unknown. Moreover, "calculating strongest
postconditions by symbolic execution provides a smooth transition from testing to full verification: by weakening the initial precondition one can make the
verification cover more initial states" [20].

Hoare logic lacks expressiveness for mutable heap data structures. To overcome this deficiency, based on work of Burstall [8], Reynolds, O'Hearn and others
developed separation logic for reasoning about mutable data structures [38,40].
Separation logic allows for local reasoning by splitting memory into two halves:
the part the program interacts with, and the part which remains untouched,
called the *frame*.

The contribution of this paper is two-fold:

(i) Generic techniques for *backward and forward reasoning in separation logic*.
A kind of backward reasoning was already established by Reynolds [40].

The original version of this chapter was revised: On page 80 a typo was corrected. For
detailed information please see the erratum. The erratum to this chapter is available
at https://doi.org/10.1007/978-3-319-94821-8_38

Although he states that for each command rules for backward reasoning "can be given", he only lists rules for the assignment of variables (called mutation in [40]), and for the deallocation of memory. Reynolds does not present a general framework that can transform any given Hoare triple specification – enriched with a frame – into a rule that is ready to be used for backward reasoning. We present such a general framework. Since it is based on separation algebras [10] it not only applies to the standard heap model of separation logic, but to all instances of this algebra.

Using similar algebraic techniques we also derive a generic technique for forward reasoning in separation logic. To achieve this we introduce a new operator, separating coimplication, which algebraically completes the set of the standard operators of separating conjunction, separating implication, and septraction. To the best of our knowledge, we are the first who provide a technique for strongest postconditions in separation logic.

(ii) *Proof tactics* for the developed techniques in Isabelle/HOL [37].

To increase automation for both backward and forward reasoning we mechanise this framework in the interactive proof assistant Isabelle/HOL and provide automated proof tactics. In particular, we provide tactics that make it manageable to interactively reason about the separating implication, which is widely considered unwieldy [5,32].

To show feasibility of our techniques we not only present standard examples such as list reversal, but also look at a larger case study: a formally verified initialiser for component-based systems built on the formally verified seL4 microkernel [28]. A proof of this initialiser using 'standard' manual separation logic reasoning can be found in the literature [7]. Redoing parts of this proof illustrates the strength of our tactics, gives an indication of how much automation they achieve, and shows by how much they reduce manual proof effort.

2   Notation

In this section we present the notation of Isabelle/HOL [37] that deviate from standard mathematical notation.

We denote the space of total functions by $\Rightarrow$, and write type variables as 'a, 'b, etc. The option type

datatype 'a option = None | Some 'a

adjoins a new element None to type 'a. Hence 'a option models partial functions.

Separation logic assertions typically are functions from the state to bool, i.e., 's $\Rightarrow$ bool. We lift the standard logical connectives $\land$, $\lor$, $\neg$, and $\longrightarrow$ point-wise to the function space in the spirit of standard separation logic, e.g. (P $\implies$ Q) = ($\forall$s. P s $\longrightarrow$ Q s).

For the example programs in this paper, we use a *deterministic* state monad. Since we are interested in distinguishing failed executions we add a flag in the style of other monadic Hoare logic frameworks [14]. This means, a program execution has the type 's $\Rightarrow$ 'r $\times$'s $\times$ bool, i.e., a function that takes a state and returns as result a new state, and a flag indicating whether the execution was successful (`true`) or not (`false`). Sequential composition, denoted by $\gg=$, is defined as

```
f ≫= g ≡
  λs. let (r', s', c) = f s; (r'', s'', c') = g r' s'
      in (r'', s'', c ∧ c')
```

Since our theory is based on abstract separation algebra (see below), we can change the underlying monad without problems. In particular we use both a *nondeterministic* state monad and an error monad for our case study.

For larger programs we use do-notation for sequential composition, e.g.

$$\text{do } \{ \text{ x } \leftarrow \text{ f; g x; h x } \}$$

3 Hoare Logic and Separation Logic

Hoare logic or *Floyd-Hoare logic* [21], [19] is the standard logic for program analysis, based on the eponymous *Hoare triple*: $\{\!|P|\!\}$ m $\{\!|Q|\!\}$ (originally denoted by P {m} Q), where P and Q are assertions, called pre- and postcondition respectively, and m is a program or command.

Initially, Hoare logic considered *partial correctness* only [21], ignoring termination. In our monadic context, where we identify non-termination and failed execution, this translates to

$$\{\!|P|\!\} \text{ m } \{\!|Q|\!\} \equiv \forall h. \text{ P } h \longrightarrow (\text{let } (r', h', c) = m h \text{ in } c \longrightarrow Q r' h')$$

If the precondition P holds before the execution of m, *and* m terminates successfully (the flag c is true) then the postcondition Q holds afterwards. Successful termination needs to be proven separately. If m fails to terminate successfully under P, i.e., by non-termination or other program failure, then any postcondition forms a valid Hoare triple.

Total correctness combines termination with correctness.

$$\{\!|P|\!\} \text{ m } \{\!|Q|\!\}_t \equiv \forall h. \text{ P } h \longrightarrow (\text{let } (r', h', c) = m h \text{ in } Q r' h' \wedge c)$$

For total correctness, whenever P holds, m *will* terminate successfully, and the result satisfies Q.

Example 1. Assume the function `delete_ptr` p, which clears the allocated memory pointed to by p, and fails if p does not point to any location at all or to an address outside the current heap.

Let `emp` be the empty heap. Then the triple $\{\!|p \mapsto \_|\!\}$ `delete_ptr` p $\{\!|emp|\!\}$ describes the situation where the heap has a single location p, and is otherwise empty.[1] After successful termination the heap is empty.

However, this specification is limiting since it only allows one particular heap configuration as precondition. Consider two further scenarios, namely heap configurations where p does not point to any location in the heap (e.g. the empty heap), and heap configurations with additional memory.

---

[1] We will explain the heap model in detail later in this section.

In the first scenario, `delete_ptr` p fails. Hence {emp} `delete_ptr` p {Q} would hold under partial correctness for any Q, but not under total correctness. In the second scenario, with additional memory, that additional memory remains unchanged during the execution of `delete_ptr` p. This is the case separation logic deals with.   □

Separation logic (SL) (e.g. [40]) extends Hoare logic by assertions to express separation between memory regions, which allows reasoning about mutable data structures. It is built around *separating conjunction* _*_, which asserts that a heap can be split into two disjoint parts where its two argument predicates hold.

The usual convention in SL is to require that even in partial correctness the program is *abort*-free, in particular for pointer access. The semantics of our slightly more traditional setting does not distinguish between non-termination and failure. Hence partial correctness will not guarantee pointer safety, while total correctness will.

A standard ingredient of SL is the *frame rule*

$$\frac{\{P\}\ \text{m}\ \{Q\}}{\{P\ *\ R\}\ \text{m}\ \{Q\ *\ R\}}$$

The rule asserts that a program m that executes correctly in a state with a small heap satisfying its precondition P, with postcondition Q, can also execute in any state with a larger heap (satisfying P * R) and that the execution will not affect the additional part of the state. Traditionally, it requires as side condition that no variable occurring free in R is modified by m. In our shallow monadic setting, no such variables exist and hence no side condition is required. We differ from tradition by proving that the frame rule holds for particular specifications rather than over the program syntax as a whole. This allows us to talk about programs that are not *strictly* local, but may be local with regards to a particular precondition. When local specifications are given for the primitive operations of a program, it is easy to compose them to show the locality of larger programs.

SL can be built upon separation algebras, which are commutative partial monoids [10]. Such algebras offer a binary operation + and a neutral element 0, such that whenever x + y is defined, + is commutative and associative, and x + 0 = x. Our automation framework is built upon an existing Isabelle/HOL framework [29,30], which uses a total function + together with another commutative operation ## that weakly distributes over + [29], and expresses the aforementioned disjointness.

Using these operations, separating conjunction is defined as

$$P\ *\ Q\ \equiv\ \lambda h.\ \exists h_1\ h_2.\ h_1\ \#\#\ h_2\ \wedge\ h\ =\ h_1\ +\ h_2\ \wedge\ P\ h_1\ \wedge\ Q\ h_2 \qquad (1)$$

which implies associativity and commutativity of *.

The standard model of SL, and separation algebra, uses heaps. The term $(p \mapsto v)$ h indicates that the pointer p on heap h is allocated and points to value v. The term $p \mapsto$ _ indicates an arbitrary value at location p.

A heap is a partial function from addresses (pointers) to values. The operation $h_1\ \#\#\ h_2$ checks whether the domains of h_1 and h_2 are disjoint. When $h_1\ \#\#\ h_2$

evaluates to true, $h_1 + h_2$ 'merges' the heaps by forming their union. The formal definitions are straightforward and omitted here.

In separation algebras, the operations ## and + define a partial order, which formalises subheaps:

$$h_1 \preceq h \equiv \exists h_2.\ h_1\ \#\#\ h_2 \wedge h_1 + h_2 = h$$

SL usually leads to simple proofs of pointer manipulation for data structures. Classical examples of such data structures are singly- and doubly-linked lists, trees, as well as directed acyclic graphs (DAGs) [22,39].

Separating implication P $\longrightarrow\!\!*$ Q, also called *magic wand*, is another operator of SL. When applied to a heap h it asserts that extending h by a disjoint heap, satisfying P, guarantees that Q holds on the combined heap:

$$P \longrightarrow\!\!* Q \equiv \lambda h.\ \forall h_1.\ h\ \#\#\ h_1 \wedge P\ h_1 \longrightarrow Q\ (h + h_1) \tag{2}$$

Ishtiaq and O'Hearn use this operator for reasoning in the presence of sharing [25].

The operations $*$ and $\longrightarrow\!\!*$ are lower and upper adjoints of a Galois connection, e.g. [15]. This relationship implies useful rules, like currying (P $*$ Q $\Longrightarrow$ R) $\Longrightarrow$ (P $\Longrightarrow$ Q $\longrightarrow\!\!*$ R), decurrying (P $\Longrightarrow$ Q $\longrightarrow\!\!*$ R) $\Longrightarrow$ (P $*$ Q $\Longrightarrow$ R), and modus ponens Q $*$ (Q $\longrightarrow\!\!*$ P) $\Longrightarrow$ P. As we will see, separating implication is useful for backward reasoning.

The literature uses another 'basic' operator of SL, *septraction* [43]:

$$P \mathbin{-\!\circledast} Q \equiv \lambda h.\ \exists h_1.\ h_1\ \#\#\ h \wedge P\ h_1 \wedge Q\ (h_1 + h) \tag{3}$$

It is the dual of separating implication, i.e., $P \mathbin{-\!\circledast} Q = \neg(P \longrightarrow\!\!* \neg Q)$, and expresses that the heap can be extended with a state satisfying P, so that the extended state satisfies Q. Septraction plays a role in combining SL with rely-guarantee reasoning [43], and for shared data structures such as DAGs [22].

4   Separating Coimplication

While separating conjunction, separating implication, and septraction, as well as their relationships to each other are well studied and understood, one operation is missing in SL.

We define *separating coimplication*, denoted by $\leadsto\!\!*$, as

$$P \leadsto\!\!* Q \equiv \lambda h.\ \forall h_1\ h_2.\ h_1\ \#\#\ h_2 \wedge h = h_1 + h_2 \wedge P\ h_1 \longrightarrow Q\ h_2 \tag{4}$$

It states that whenever there is a subheap h_1 satisfying P then the remaining heap satisfies Q. To the best of our knowledge, we are the first to define this operator and explore its properties.

It is the dual of separating conjunction, i.e., $P \leadsto\!\!* Q = \neg(P * \neg Q)$, which is the same relationship as the one between separating implication and septraction.

Special instances of $\leadsto\!\!*$ (in the form of doubly negated conjunction) appear in the literature: the *dangling operator* of Vafeiadis and Parkinson [43] uses

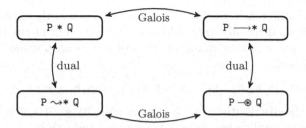

Fig. 1. Relationship between operators of separation logic

subterms of the form ¬(p ↦ _ * True), which equals p ↦ _ ⤳* False, and the *subtraction operator* by Calcagno et al. [9], used for comparing bi-abduction solutions, uses terms of the form P ⤳* emp. These occurrences indicate that separating coimplication is an important, yet unexplored operator for SL. As we will show, it is also the crucial ingredient to set up forward reasoning for SL.

Separating coimplication forms a Galois connection with septraction. Therefore, many useful theorems follow from abstract algebraic reasoning. For example, similar to the rules stated above for * and —*, we get rules for currying, decurrying and cancellation:

$$\frac{P -\circledast Q \implies R}{Q \implies P \leadsto* R} \text{ (CURRY)} \qquad \frac{Q \implies P \leadsto* R}{P -\circledast Q \implies R} \text{ (DECURRY)} \qquad \frac{Q -\circledast (Q \leadsto* P)}{P} \text{ (CANC)}$$

It follows that separating coimplication is isotone in one, and antitone in the other argument:

$$\frac{P' \implies P}{P \leadsto* Q \implies P' \leadsto* Q} \qquad \frac{Q \implies Q'}{P \leadsto* Q \implies P \leadsto* Q'}$$

Separating coimplication is not only interesting because it completes the set of 'nicely' connected operators for SL (see Fig. 1), it is also useful to characterise specific heap configurations. For example, (P ⤳* False) h states that no subheap of h satisfies P: P ⤳* False = λh. ∀h₁. h₁ ⪯ h ⟶ ¬ P h₁.

While properties concerning ⤳* and -⊛ mostly follow from the Galois connection, some need to be derived 'manually':

$$\frac{P \leadsto* Q \qquad P * R}{P * (Q \wedge R)} \qquad \frac{P \longrightarrow* (R \wedge (P \leadsto* \text{False}))}{\neg P * (P \longrightarrow* \neg R)}$$

The first rule states that whenever a heap satisfies P ⤳* Q, and can be split into two subheaps satisfying P and R, respectively, then the subheap satisfying R has to satisfy Q as well. The second rule connects separating implication and coimplication directly and states that if adding a heap satisfying P yields a heap with no subheap containing P, then the underlying heap cannot satisfy P.

SL considers different classes of assertions [40]; each of them plays an important role in SL, and usually gives additional information about the heap. For example, a *precise assertion* characterises a unique heap portion (when such a portion exists), i.e.,

$$\texttt{precise P} \equiv \forall h\ h_1\ h_2.\ h_1 \preceq h \wedge P\ h_1 \wedge h_2 \preceq h \wedge P\ h_2 \longrightarrow h_1 = h_2$$

P is precise iff the distributivity $\forall Q\ R.\ ((Q \wedge R) * P) = (Q * P \wedge R * P)$ holds [15]. Separating coimplication yields a nicer characterisation:

$$\texttt{precise P} = \forall R.\ P * R \implies (P \rightsquigarrow\!* R) \tag{5}$$

On the one hand this equivalence eliminates one of the $\forall$-quantifiers, which simplifies reasoning; on the other hand it directly relates separating conjunction with coimplication, stating that if P and R hold on a heap, and one pulls out an arbitrary subheap satisfying P, the remaining heap must satisfy R. Obviously, this relationship between $*$ and $\rightsquigarrow\!*$ does not hold in general since separating coimplication may pull out the 'wrong' subheap satisfying P.

As a consequence, using (CANC), we immediately get

$$\frac{\texttt{precise P}}{P -\circledast (P * R) \implies R} \tag{6}$$

Our Isabelle files [3] contain many more properties of separating coimplication. The most important use of separating coimplication, however, is its application in forward reasoning, as we will demonstrate in Sect. 6.

Example 2. Using separating coimplication we can fully specify `delete_ptr p` in a way that matches intuition: $\{\!|p \mapsto \_ \rightsquigarrow\!* R|\!\}$ `delete_ptr p` $\{\!|R|\!\}$. This rule states that the final state should satisfy R, when the pointer is deleted, and the pointer existed in the first place.   □

5   Walking Backwards

Backward reasoning [16] or *reasoning in weakest-precondition style* proceeds backwards from a given postcondition Q and a given program m by determining the *weakest* precondition `wp(m,Q)` such that $\{\!|\texttt{wp(m,Q)}|\!\}$ m $\{\!|Q|\!\}$ is a valid Hoare triple.

Backward reasoning is well established for formal programming languages, using classical logics. For example the weakest precondition `wp(m₁;m₂,Q)` for a sequential program equals `wp(m₁,wp(m₂,Q))`; the full set goes back to Dijkstra [16]. Using these equations, backward reasoning in Hoare logic is straightforward.

Avoiding Frame Calculations. In SL, however, it comes at a price, since reasoning has to work on Hoare triples of the form $\{\!|P * R|\!\}$ m $\{\!|Q * R|\!\}$ and has to consider the frame. Whenever an arbitrary postcondition X is given, one needs to split it up into the actual postcondition Q needed for reasoning about m, and the (untouched) frame R. That means for given X and Q one has to *calculate the frame* R such that $X = Q * R$. Frame calculations are often challenging in applications since X can be arbitrary complex. The same holds for a given precondition.

Example 3. Let `copy_ptr p p' = do { x ← get_ptr p; set_ptr p' x }` be the program
that copies the value at pointer p to the value at pointer p'. Its natural
specification is $\{p \mapsto x * p' \mapsto \_\}$ `copy_ptr p p'` $\{p \mapsto x * p' \mapsto x\}$. The specification
we use is

$$\forall R. \; \{p \mapsto x * p' \mapsto \_ * R\} \; \text{copy\_ptr p p'} \; \{p \mapsto x * p' \mapsto x * R\} \qquad (7)$$

In a larger program, the postcondition at the call site of `copy_ptr` will be more
complex than $Q = p \mapsto x * p' \mapsto x$. Say it is $\{p' \mapsto v * a \mapsto \_ * p \mapsto v * R'\}$, for
some heap R'. To determine the precondition, using Rule (7), the postcondition
needs to be in the form $Q * R$. One has to calculate the frame $R = a \mapsto \_ * R'$. $\square$

Phrasing specifications in the form $\{P * R\}$ m $\{Q * R\}$ (similar to Rule (7))
state that the frame rule holds for program m, i.e., that m only consists of local
actions with respect to P. In the monadic setting, where not all programs are
necessarily local, we find this form more convenient than a predicate on the
programs and a separate frame rule. That also means that our Isabelle/HOL
framework does not rely on the frame rule.

In the previous example the frame calculation uses only associativity and
commutativity of *, but in general such calculations can be arbitrarily complex. A solution to this problem follows directly from the Galois connection and
'rewrites' the pre- and postcondition.

$$(\forall R. \; \{P * R\} \; m \; \{Q * R\}) = (\forall X. \; \{P * (Q \longrightarrow\!\!* X)\} \; m \; \{X\}) \qquad (8)$$

The left-hand side coincides with the form we use to specify our programs. The
right-hand side has the advantage that it works for any postcondition X; no
explicit calculation of the frame is needed for the postcondition. Since $Q \longrightarrow\!\!* X$
is the *weakest* [9] choice of frame, the calculation happens implicitly and automatically in the precondition. This is a generalisation of what occurs in Reynolds'
work [40] for specific operations.

Since Rule (8) generates Hoare triples that can be applied to arbitrary postconditions, we can use these rules directly to perform backward reasoning in
the sense of Dijkstra [16]. That means that our calculations are similar to the
classical ones for reasoning with weakest preconditions, e.g. $\text{wp}(m_1, \text{wp}(m_2, Q))$. As
a consequence our framework can generate preconditions fully automatically. As
in the classical setting, applying the rules of Hoare logic is now separated from
reasoning about the content of the program and the proof engineer can focus
their effort on the part that requires creativity.

Example 4. Using Equivalence (8), the specification for `copy_ptr` (7) becomes

$$\{\exists x. \; p \mapsto x * p' \mapsto \_ * (p \mapsto x * p' \mapsto x \longrightarrow\!\!* X)\} \; \text{copy\_ptr p p'} \; \{X\} \qquad \square$$

Simplifying Preconditions. As mentioned above, Equivalence (8) allows us
to perform backward reasoning and to generate preconditions. However, the
generated formulas will often be large and hence automation for simplifying
generated preconditions is necessary. We provide such simplification tactics.

Both the right-hand side of (8) and the previous example show that generated preconditions contain interleavings of $*$ and $\longrightarrow\!*$. A simplifier suitable for our framework has to deal with such interleavings, in particular it should be able to handle formulas of the type P $*$ (Q $\longrightarrow\!*$ R), for any P, Q and R. Two rules that are indispensable here are cancellation and currying, as introduced in Sect. 3:

$$\frac{\text{R} \implies \text{R'}}{\text{P} * \text{R} \implies \text{P} * \text{R'}} \qquad \frac{\text{P} * \text{Q} \implies \text{R}}{\text{P} \implies \text{Q} \longrightarrow\!* \text{R}} \tag{9}$$

Currently, not many solvers support the separating implication operator [5, 32]. Some automatic solvers for separating implication exist for formulas over a restricted set of predicates [23]. Since we are aiming at a general framework for arbitrary specifications, we do not want to restrict the expressiveness of pre- and postconditions, and hence we cannot restrict our framework to such subsets. Moreover, we cannot hope to develop fully automatic solvers for the problem at hand at all, since it is undecidable for arbitrary pre- and postconditions [11].

We provide proof tactics for Isabelle/HOL that can simplify formulas of the form P $*$ (Q $\longrightarrow\!*$ R), for any P, Q and R, and hence can be used in the setting of backward reasoning. Although we cannot expect full automation, the simplification achieved by the tactics is significant, as we will show. Our tactics can make partial progress without fully solving the goal. As experience shows for standard proof methods in Isabelle, this is the most useful kind, e.g. the method simp, which rewrites the current goal and leaves a normal form, is much more frequently used than methods such as blast or force that either have to fully solve the goal or fail, but cannot make intermediate progress available to the user. What we provide is a simplifier, not an entailment solver or semi-solver.

Our framework [3] offers support for backward reasoning in SL, and builds on top of an existing library [30], which is based on separation algebras. This brings the advantage that abstract rules, such as Q $*$ (Q $\longrightarrow\!*$ P) $\implies$ P, which are indispensable for handling interleaving of $*$ and $\longrightarrow\!*$ are immediately available. Since the framework is independent of the concrete heap model, we can apply the tool to a wide range of problem domains. As usual, the tactics enable the user to give guidance to complete proofs where other methods fail, and to substantially reduce proof effort.

– The tactic sep_wp performs weakest-precondition reasoning on monads and automatically transforms specification Hoare triples provided as arguments into weakest-precondition format, using Equivalence (8). In addition to the transformations already described, it can also handle further combinations, e.g. with classical Hoare logic, or instances where the separation logic only operates on parts of the monad state. We integrate sep_wp into the existing tactic wp [14] of the seL4 proofs, which implements classical weakest-precondition reasoning with additional features such as structured decomposition of postconditions. The user sees a tactic that can handle both, SL and non-SL goals, gracefully.

– We develop the tactic `sep_mp` to support reasoning about separating implication, and `sep_lift` to support the currying rule of Sect. 3, eliminating separating implication. These are both integrated into the existing `sep_cancel` [30] method, for reducing formulas by means of cancellation rules.

Detailed Example. To illustrate backward reasoning in SL in more detail, we show the correctness of the program `swap_ptr p p'` that swaps the values p and p' point. Pointer programs are built from four basic operations that manipulate the heap: `new_ptr` allocates memory for a pointer, `delete_ptr` removes a pointer from the heap, `set_ptr` assigns a value, and `get_ptr` reads a value, respectively.

Their specifications are as follows:

$$\{R\} \; \texttt{new\_ptr} \; \{\lambda rv. \; rv \mapsto \_ \; * \; R\}$$
$$\{p \mapsto \_ \; * \; R\} \; \texttt{delete\_ptr} \; p \; \{R\}$$
$$\{p \mapsto \_ \; * \; R\} \; \texttt{set\_ptr} \; p \; v \; \{p \mapsto v \; * \; R\}$$
$$\{\exists x. \; p \mapsto x \; * \; R \; x\} \; \texttt{get\_ptr} \; p \; \{\lambda rv. \; p \mapsto rv \; * \; R \; rv\}$$

As before we use specifications with frames, avoiding the use of the frame rule.

Recall that in our monadic setting the postcondition R is a predicate over two parameters: the return value rv of the function, and the state s after termination. When there is no return value (e.g. for `set_ptr`) we omit the first parameter.

Using Rule (8), or the tactic `wp` (which includes `sep_wp`), we transform these specifications into a form to be used in backward reasoning (for partial and total correctness), except `delete_ptr`, which already has the appropriate form.

$$\{\forall x. \; x \mapsto \_ \; \longrightarrow\!\!* \; X \; x\} \; \texttt{new\_ptr} \; \{X\}$$
$$\{p \mapsto \_ \; * \; (p \mapsto v \; \longrightarrow\!\!* \; X)\} \; \texttt{set\_ptr} \; p \; v \; \{X\}$$
$$\{\exists x. \; p \mapsto x \; * \; (p \mapsto x \; \longrightarrow\!\!* \; X \; x)\} \; \texttt{get\_ptr} \; p \; \{X\}$$

$p \mapsto v \; * \; p' \mapsto v' \; * \; R \Longrightarrow$
$\forall x. \; x \mapsto \_ \; \longrightarrow\!\!*$
   $(\exists pv. \; p \mapsto pv \; *$
      $(p \mapsto pv \; \longrightarrow\!\!* \; x \mapsto \_ \; *$
      $(x \mapsto pv \; \longrightarrow\!\!*$
        $(\exists pv'. \; p' \mapsto pv' \; *$
          $(p' \mapsto pv' \; \longrightarrow\!\!* \; p \mapsto \_ \; *$
          $(p \mapsto pv' \; \longrightarrow\!\!*$
            $(\exists y. \; x \mapsto y \; *$
              $(x \mapsto y \; \longrightarrow\!\!* \; p' \mapsto \_ \; *$
              $(p' \mapsto y \; \longrightarrow\!\!* \; x \mapsto \_ \; * \; p \mapsto v' \; * \; p' \mapsto v \; * \; R))))))))$

Fig. 2. Backward reasoning: generated proof goal for `swap_ptr`

The program `swap_ptr`, which involves all heap operations, is given as

```
swap_ptr p p' = do {
  np ← new_ptr;
  copy_ptr p np;
  copy_ptr p' p;
  copy_ptr np p';
  delete_ptr np
}
```

where `copy_ptr p p' = do { x ← get_ptr p; set_ptr p' x }`, as before. We use the specifications of the basic operations to prove the specification

$$\{p \mapsto v * p' \mapsto v' * R\}\ \texttt{swap\_ptr}\ p\ p'\ \{p \mapsto v' * p' \mapsto v * R\}$$

Using equational reasoning of the form $\texttt{wp}(m_1;m_2,Q) = \texttt{wp}(m_1,\texttt{wp}(m_2,Q))$, and starting from the (given) postcondition our framework automatically derives a precondition `pre`. In case the given precondition $p \mapsto v * p' \mapsto v' * R$ implies `pre`, the specification of `swap_ptr` holds. The proof goal is depicted in Fig. 2.

Our tactics simplify this lengthy, unreadable formula, where major simplifications are based on the aforementioned rules (Eqs. (9)).

The tactic `sep_cancel` is able to simplify the generated goal automatically, but gets stuck at existential quantifiers. Although resolving existential quantifiers cannot be fully automated in general, our framework handles many common situations. The left-hand side of Fig. 3 shows an intermediate step illustrating the state before resolving the last existential quantifier. One of the assumptions is $x \mapsto v$ and hence the obvious choice for y is v. Here, Isabelle's simple existential introduction rule is sufficient to allow `sep_cancel` to perform the match without input. The tactic `sep_cancel` can then solve the proof goal fully automatically; for completeness we show a state where all occurrences of $\longrightarrow*$ have been eliminated.

Case Study: System Initialisation. Boyton et al. [7] present a formally verified initialiser for component-based systems built on the seL4 kernel. The safety and security of software systems depends heavily on their initialisation; if initialisation is broken all bets are off with regards to the system's behaviour. The previous proofs (about $15,000$ lines of proof script) were brittle, often manual, and involved frequent specification of the frame. Despite an early form of `sep_cancel` the authors note that "higher-level automation such as frame computation/matching would have improved productivity" [7].

$$p \mapsto v' * p' \mapsto v' * x \mapsto v * R \Longrightarrow$$
$$\exists y.\ x \mapsto y *$$
$$\quad (x \mapsto y \longrightarrow* p' \mapsto \_ *$$
$$\quad (p' \mapsto y \longrightarrow* x \mapsto \_ * p \mapsto v' *$$
$$\quad p' \mapsto v * R))$$

$$x \mapsto v * p \mapsto v' * p' \mapsto v * R \Longrightarrow$$
$$x \mapsto \_ * p \mapsto v' * p' \mapsto v * R$$

Fig. 3. Matching existential quantifier and eliminating $\longrightarrow*$ for `swap_ptr`

In contrast to our earlier examples, the initialiser proofs operate on a *non-deterministic* state monad, as well as the *non-deterministic error* state monad. Our tactics required only the addition of two trivial interface lemmas to adapt to this change of computation model, illustrating the genericity of our approach.

Substituting the previous mechanisation with our framework[2] we substantially reduce the proof effort: for commands specified in SL, the calculation of the weakest precondition is automatic, without any significant user interaction. Additionally, we find that calculating the frame indirectly via resolution of separating implications is significantly easier to automate, as the separating implication is the *weakest* choice of solution for in-place frame calculation. The general undecidability of separating implication did not pose a problem.

Figure 4 presents a sample of the entire proof script for an seL4 API function to give an indication of the improvements. For brevity Fig. 4 shortens some of the names in the proof. The separation algebra in this statement lets * be used inside larger heap objects, such as specifying the capabilities stored inside a

---

```
lemma restart_null_wp:
  {|(tcb, pop_slot) ↦c NullCap * (tcb, reply_slot) ↦c _ * R|}
    restart tcb
  {|(tcb, reply_slot) ↦c (MRCap tcb) * (tcb, pop_slot) ↦c RCap * R|}
```

```
apply (clarsimp simp:restart_def)        apply (clarsimp simp:restart_def)
apply (wp)                               apply (wp sep_wp:set_cap_wp ipc_cancel_ret)
apply (rule hoare_strengthen_post)       apply (sep_cancel | simp | safe)+
apply (rule set_cap_wp[where R=          done
      (tcb, reply_slot) ↦c MRCap tcb * R])
apply (sep_cancel)+
apply (rule hoare_strengthen_post)
apply (rule set_cap_wp[where
        R=(tcb, pop_slot) ↦c _ * R])
apply (sep_cancel)+
apply (rule hoare_strengthen_post)
apply (rule ipc_cancel_ret[where
        R=(tcb, reply_slot) ↦c _ * R])
apply (sep_cancel)+
apply (wp)
apply (clarsimp)
apply (intro conjI impI)
apply (drule opt_cap_sep_imp)
apply (clarsimp)
apply (drule opt_cap_sep_imp)
apply (clarsimp)
done
```

Fig. 4. Reducing user steps by a factor of 6 in system initialisation proofs

---

[2] Updated proofs at https://github.com/seL4/l4v/tree/seL4-7.0.0/sys-init.

Thread Control Block (TCB) object using $\mapsto$c. The lemma models which seL4 capabilities are available to the user after a `restart` operation.

The left-hand side of Fig. 4 shows the original proof. Each application of an SL specification rule required first a weakening of the postcondition to bring it into the expected form, and often a manual specification of the frame. Not only is this cumbersome and laborious for the proof engineer, it was highly brittle – any change of the functionality or specification requires a new proof.

The right-hand side shows the simplified proof. It shortens eighteen lines of proof script to three, without noticeable increase in prover time. By removing the manual term specification, the tactics also make the proof more robust to changes – we can rewrite parts of the code, while leaving the proof unchanged.

Another strength is that our tactics are incremental, i.e., we can use them alongside others. In our example we use `safe` and `simp`. This design allows us to attack arbitrary formulas of SL.

6 Walking Forwards

Forward reasoning uses strongest postconditions [19]. It proceeds forwards from a given precondition P and a given program m by calculating the strongest postcondition `sp(m,P)`. Although backward reasoning with weakest preconditions is more common in practice, there are several applications where forward reasoning is more convenient, for example, for programs where the precondition is 'trivial', and the postcondition either too complex or unknown.

Usually forward reasoning focuses on partial correctness. Recall that we admit memory failures in partial correctness. In larger proofs it is convenient to show absence of failure separately, e.g. during a refinement proof [14], and assume it in multiple partial-correctness proofs, thereby avoiding proof duplication.

To enable forward reasoning for SL it is desirable to transform a Hoare triple $\{P * R\}$ m $\{Q * R\}$ into the form $\{X\}$ m $\{post\}$, similar to Equivalence (8).

Avoiding Frame Calculations. In [22] Hobor and Villard present the rule FWRAMIFY:

$$\frac{\forall F.\ \{P * F\}\ m\ \{Q * F\} \qquad R \implies P * \text{True} \qquad Q * (P \twoheadrightarrow R) \implies R'}{\{R\}\ m\ \{R'\}}$$

At first glance this rule looks like the frame calculation could be avoided, since the conclusion talks about arbitrary preconditions R. It is a 'complification' of what we can more simply write as $(\forall F.\ \{P * F\}\ m\ \{Q * F\}) \wedge (R \implies P * \text{True}) \implies \{R\}\ m\ \{Q * (P \twoheadrightarrow R)\}$, which states that a terminating program m, specified by P * F and Q * F, will end up in a state satisfying $Q * (P \twoheadrightarrow R)$ if R contains a subheap satisfying P, which is characterised by $R \implies P * \text{True}$.

The reason FWRamify *cannot* be used to avoid the frame calculation is the subheap-test $R \implies P * \text{True}$, which includes a frame calculation itself, and is as hard to check as reasoning via weakening of the precondition.

In general it seems impossible to transform triples $\{P * R\}$ m $\{Q * R\}$ into strongest-postcondition form, without introducing additional proof burden, similar to FWRamify. As discussed in Sect. 4, the term $P \rightsquigarrow * R$ states that R holds, whenever P is removed from the heap – removal is only feasible if P exists.

Separating coimplication implies an equivalence similar to (8):

$$(\forall R. \ \{P \rightsquigarrow * R\} \text{ m } \{Q * R\}) = (\forall X. \ \{X\} \text{ m } \{Q * (P \multimap X)\}) \tag{10}$$

which we can use for forward reasoning. It is based on 'reverse modus ponens', $X \implies P \rightsquigarrow * (P \multimap X)$, which follows directly from the Galois connection. Intuitively, the postcondition is calculated from the heap satisfying X by subtracting the part satisfying the precondition P and replacing it with a heap satisfying Q.

In practice, specifications $\{P * R\}$ m $\{Q * R\}$ can almost always be rewritten into $\{P \rightsquigarrow * R\}$ m $\{Q * R\}$, especially if P is precise.

For example, the precondition of $\{p \mapsto \_ \rightsquigarrow * R\}$ set_ptr p v $\{p \mapsto v * R\}$ assumes the hypothetical case that if we had the required resource ($p \mapsto \_$), we would have a predicate R corresponding to the rest of the heap. In the postcondition, the resource does exist and is assigned to the correct value v.

Example 5. Using the specifications of the heap operations and Equivalence (10) yields the following Hoare triples (for partial correctness).

$$\{X\} \text{ new\_ptr } \{\lambda rv. \ rv \mapsto \_ * X\}$$
$$\{X\} \text{ delete\_ptr p } \{p \mapsto \_ \multimap X\}$$
$$\{X\} \text{ set\_ptr p v } \{p \mapsto v * (p \mapsto \_ \multimap X)\}$$
$$\{X\} \text{ get\_ptr p } \{\lambda rv. \ p \mapsto rv * (p \mapsto rv \multimap X)\}$$

□

Simplifying Postconditions. Since Equivalences (8) and (10) have the same shape, we can develop a framework for forward reasoning following the lines of backward reasoning. As for backward reasoning, forward reasoning generates lengthy postconditions that need simplification. This time we have to simplify interleavings of $*$ and $\multimap$.

$$\exists np. \ np \mapsto \_ \multimap$$
$$(\exists x. \ p' \mapsto x \ *$$
$$(p' \mapsto \_ \multimap np \mapsto x \ *$$
$$(np \mapsto x \multimap$$
$$(\exists x. \ p \mapsto x \ *$$
$$(p \mapsto \_ \multimap p' \mapsto x \ *$$
$$(p' \mapsto x \multimap$$
$$(\exists x. \ np \mapsto x \ *$$
$$(np \mapsto \_ \multimap p \mapsto x \ *$$
$$(p \mapsto x \multimap np \mapsto \_ \ * \ p \mapsto v \ * \ p' \mapsto v' \ * \ R)))))))))$$
$$\implies p \mapsto v' \ * \ p' \mapsto v \ * \ R$$

Fig. 5. Forward reasoning: generated proof goal for swap_ptr

Three laws are important for resolving interleavings of $*$ and $-\circledast$:

$$\frac{P \implies Q \longrightarrow * R}{P * Q \implies R} \qquad \frac{Q \implies P \rightsquigarrow * R}{P -\circledast Q \implies R} \qquad \frac{\text{precise } P}{P * R \implies P \rightsquigarrow * R}$$

The former two allow us to move subformulas from the antecedent to the consequent, while the latter one is a cancellation law. Depending on which term is precise different cancellation rules are needed.

We develop the following tactics for forward reasoning:

- `sep_invert` provides an 'inversion' simplification strategy, based on the aforementioned laws. It transforms interleavings of $*$ and $-\circledast$ into $\longrightarrow *$ and $\rightsquigarrow *$.
- `septract_cancel` simplifies $\rightsquigarrow *$ by means of cancellation rules.
- `sep_forward` integrates `septract_cancel` and `sep_cancel`, alongside a few other simple methods, to provide a simplification strategy for most formulas reasoning forwards.
- `sep_forward_solve` inverts, and then attempts to use `sep_forward` to fully solve the goal.

Figure 5 depicts the generated proof goal for `swap_ptr`. The first ten lines show the generated postcondition, whereas the last one is the given one. With the help of the developed tactics, our framework proves `swap_ptr`. As before instantiation of existential quantifiers is sometimes needed, and handled automatically for common cases.

Benchmark. One of the standard SL benchmarks is in-place list reversal:

```
list_rev p = do {
  (hd_ptr, rev) ← whileLoop (λ(hd_ptr, rev) s. hd_ptr ≠ NULL)
                    (λ(hd_ptr, rev). do {
                        next_ptr ← get_ptr hd_ptr;
                        set_ptr hd_ptr rev;
                        return (next_ptr, hd_ptr)
                      })
                    (p, NULL);
  return rev
}
```

For this example, the predicate `list` that relates pointers to abstract lists is defined in the standard, relational recursive way [35]. We used `septract_cancel` to verify the Hoare triple

$$\{\text{list } p \text{ ps} * R\} \text{ list\_rev } p \{\lambda rv. \text{ list } rv \text{ (rev ps)} * R\}$$

We only had to interact with our framework in a non-trivial way by adding the invariant that the list pointed to by the previous pointer is already reversed.

Case Study: System Initialisation. To investigate the robustness of our tactics in a real-world proof scenario, we again turn to the proof of system initialisation showcased earlier. We completed a portion of the proof, comprising of twenty function specifications, to demonstrate that a forward approach could achieve the same gains as our backward one, providing a degree of assurance that either approach could be taken without incurring costs.

As with weakest precondition, we are able to provide tactics enabling concise, highly automatic proofs. We give an example using the error monad, where the statement has a second postcondition {P} m {Q}, {E}. When the code throws an exception, E must hold. Leaving E free means no exception will occur. The following lemma models the result of invoking the seL4 API call move on two capabilities:

```
lemma invoke_cnode_move_cap:
  {dest ↦c _ * src ↦c cap * R}
    invoke_cnode (MoveCall cap' src dest)
  {dest ↦c cap' * src ↦c NullCap * R}, {E}
  apply (simp add:validE_def)
  apply (clarsimp simp:invoke_cnode_def liftE_bindE validE_def[symmetric])
  apply (sp sp:move_cap_sp)
  apply (sep_forward_solve)
  done
```

Most of the effort was in constructing the strongest-postcondition framework akin to wp. This is generic and can be used outside of our SL framework. The only work required to adapt our tactics to the proof was specialising the strongest postcondition Hoare triple transformation to the monads used, which was trivial.

7   Related Work

Separata [23, 24] is an Isabelle framework for separation logic based on a labelled sequent calculus. It is a general reasoning tool for separation logic, and supports separating conjunction, separating implication, and septraction. While it can prove a number of formulas that our tactics cannot, none of these formulas appear in our verification tasks using backward and forward reasoning, and they are unlikely to show up in these styles, owing to the highly regular shape of the generated formulas of our framework. Conversely, Separata was not able to solve the weakest-precondition formulas produced in our proof body. Our framework can integrate solvers such as Separata for the generated pre- and postconditions; hence we see these tools as additional support for our framework.

Other frameworks for reasoning in separation logic in an interactive setting such as the Coq-based VeriSmall [2] or CFML [12] stand in a similar relationship. One of the main strengths of our framework is its generality, which should allow it to be easily combined with other frameworks.

Many other tools such as Space Invader [18], Verifast [26], HOLfoot [42] offer a framework for forward reasoning within separation logic, based on variations of symbolic execution. Since they do not provide support for separating implication, they also do not perform backward reasoning in weakest-precondition style as presented. The few tools that do support separating implication are automatic full solvers [34] and do not provide a user-guided interactive framework.

The only approach using strongest postcondition we are aware of is the rule FWRamify by Hobor and Villard [22], which its authors find too difficult to use in practice. Using separating coimplication, we do not find septraction to be fundamentally more difficult than the existing well-known SL fragments.

Many existing interactive tools perform frame inference [1,4,13,41], which is the way SL was presented by O'Hearn et al. [38]. We take a different approach, and automatically divide logical reasoning from program text achieving the same separation of concerns standard Hoare logic enjoys. Our form of frame calculation is deferred to the purely logical part of the problem, where we can provide an interactive proof tactic for calculating the frame incrementally as needed.

Iris Proof Mode was developed in Coq by Krebbers et al. [31], on top of the Iris framework for Higher Order Concurrent Separation Logic [27]. They provide support for separating conjunction and implication, and use separating implication for performing weakest-precondition reasoning. As Iris is based on *affine* separation logic, where resource leaks are not reasoned about, it is unclear whether their tactics can be adopted in our *linear* setting.

Our framework operates entirely on the level of the abstract separation algebra, leaving it to the user to provide facts about model-dependent predicates, such as points-to predicates. This makes the tool highly adaptable. As we presented earlier, we have used it for ordinary heap models, for fine-grained 'partial' objects, as well as multiple different monad formalisations.

In the field of static analysis, bi-abduction is a promising technique in specification derivation, employed notably in the Infer tool [9], as well as in attempts to detect memory leaks automatically in Java programs [17]. Since the frame calculations happen in place, instead of separating logic from program as we do, it would be interesting to employ our framework to this space.

8   Summary

We have presented a methodology for backward and forward reasoning in separation logic. To support proof automation we have implemented our theoretical results in a framework for automation-assisted interactive proofs in Isabelle/HOL.

The more traditional backward reasoning works for both partial and total correctness. It makes use of the standard separating implication rule for weakest preconditions, which often counts as unwieldy. We, however, provide an interactive tactic that successfully resolves the separating implications we have encountered in sizeable practical applications.

The forward reasoning framework makes use of a new operator, the separating coimplication, which forms a nice algebraic completion of the existing operators of separating conjunction, separating implication, and septraction. The framework relies on the fact that specifications can be (re)written into the form $\{P \rightsquigarrow\!\!* R\}\; m\; \{Q * R\}$. This is always possible when P is precise. While we suspect that this weaker specification will usually be true for partial correctness, we leave the general case for future work.

We have demonstrated our new proof tactics in a case study for both forward and backward reasoning. For backward reasoning, we have achieved substantial improvements, reducing the number of user proof steps by a factor of up to six. For forward reasoning, we have taken a portion of the same proof and completed it with our strongest-postcondition framework, achieving similar gains. We believe this gives empirical grounds that users can decide which style of reasoning is suitable for their problem domain, without incurring costs in mechanisation.

References

1. Appel, A.W.: Verified software toolchain. In: Barthe, G. (ed.) ESOP 2011. LNCS, vol. 6602, pp. 1–17. Springer, Heidelberg (2011). https://doi.org/10.1007/978-3-642-19718-5_1
2. Appel, A.W.: VeriSmall: verified smallfoot shape analysis. In: Jouannaud, J.-P., Shao, Z. (eds.) CPP 2011. LNCS, vol. 7086, pp. 231–246. Springer, Heidelberg (2011). https://doi.org/10.1007/978-3-642-25379-9_18
3. Bannister, C., Höfner, P., Klein, G.: Forward and backward reasoning in separation logic. Isabelle theories (2018). https://github.com/sel4proj/Jormungand/tree/ITP18
4. Bengtson, J., Jensen, J.B., Birkedal, L.: Charge! A framework for higher-order. In: Beringer, L., Felty, A. (eds.) ITP 2012. LNCS, vol. 7406, pp. 315–331. Springer, Heidelberg (2012). https://doi.org/10.1007/978-3-642-32347-8_21
5. Berdine, J., Calcagno, C., O'Hearn, P.W.: Symbolic execution with separation logic. In: Yi, K. (ed.) APLAS 2005. LNCS, vol. 3780, pp. 52–68. Springer, Heidelberg (2005). https://doi.org/10.1007/11575467_5
6. Bertot, Y., Castéran, P.: Interactive Theorem Proving and Program Development. Coq'Art: The Calculus of Inductive Constructions. Texts in Theoretical Computer Science. An EATCS Series. Springer, Heidelberg (2004). https://doi.org/10.1007/978-3-662-07964-5
7. Boyton, A., Andronick, J., Bannister, C., Fernandez, M., Gao, X., Greenaway, D., Klein, G., Lewis, C., Sewell, T.: Formally verified system initialisation. In: Groves, L., Sun, J. (eds.) ICFEM 2013. LNCS, vol. 8144, pp. 70–85. Springer, Heidelberg (2013). https://doi.org/10.1007/978-3-642-41202-8_6
8. Burstal, R.: Some techniques for proving correctness of programs which alter data structures. In: Meltzer, B., Michie, D. (eds.) Machine Intelligence, vol. 7, pp. 23–50. Edinburgh University Press, Edinburgh (1972)
9. Calcagno, C., Distefano, D., O'Hearn, P.W., Yang, H.: Compositional shape analysis by means of bi-abduction. J. ACM 58(6), 26:1–26:66 (2011). https://doi.org/10.1145/2049697.2049700
10. Calcagno, C., O'Hearn, P.W., Yang, H.: Local action and abstract separation logic. In: Logic in Computer Science (LICS 2007), pp. 366–378. IEEE (2007). https://doi.org/10.1109/LICS.2007.30
11. Calcagno, C., Yang, H., O'Hearn, P.W.: Computability and complexity results for a spatial assertion language for data structures. In: Hariharan, R., Vinay, V., Mukund, M. (eds.) FSTTCS 2001. LNCS, vol. 2245, pp. 108–119. Springer, Heidelberg (2001). https://doi.org/10.1007/3-540-45294-X_10
12. Charguéraud, A.: Characteristic formulae for the verification of imperative programs. In: Chakravarty, M.M.T., Hu, Z., Danvy, O. (eds.) International Conference on Functional Programming (ICFP 2011), pp. 418–430. ACM (2011). https://doi.org/10.1145/2034773.2034828
13. Chlipala, A.: Mostly-automated verification of low-level programs in computational separation logic. In: Hall, M.W., Padua, D.A. (eds.) Programming Language Design and Implementation (PLDI 2011), pp. 234–245. ACM (2011). https://doi.org/10.1145/1993498.1993526

14. Cock, D., Klein, G., Sewell, T.: Secure microkernels, state monads and scalable refinement. In: Mohamed, O.A., Muñoz, C., Tahar, S. (eds.) TPHOLs 2008. LNCS, vol. 5170, pp. 167–182. Springer, Heidelberg (2008). https://doi.org/10.1007/978-3-540-71067-7_16

15. Dang, H.H., Höfner, P., Möller, B.: Algebraic separation logic. J. Logic Algebraic Programm. **80**(6), 221–247 (2011). https://doi.org/10.1016/j.jlap.2011.04.003

16. Dijkstra, E.W.: A Discipline of Programming. Prentice Hall, Upper Saddle River (1976)

17. Distefano, D., Filipović, I.: Memory leaks detection in java by bi-abductive inference. In: Rosenblum, D.S., Taentzer, G. (eds.) FASE 2010. LNCS, vol. 6013, pp. 278–292. Springer, Heidelberg (2010). https://doi.org/10.1007/978-3-642-12029-9_20

18. Distefano, D., O'Hearn, P.W., Yang, H.: A local shape analysis based on separation logic. In: Hermanns, H., Palsberg, J. (eds.) TACAS 2006. LNCS, vol. 3920, pp. 287–302. Springer, Heidelberg (2006). https://doi.org/10.1007/11691372_19

19. Floyd, R.W.: Assigning meanings to programs. Math. Aspects Comput. Sci. **19**, 19–32 (1967)

20. Gordon, M., Collavizza, H.: Forward with hoare. In: Roscoe, A.W., Jones, C.B., Wood, K.R. (eds.) Reflections on the Work of C.A.R. Hoare, pp. 101–121. Springer, London (2010). https://doi.org/10.1007/978-1-84882-912-1_5

21. Hoare, C.A.R.: An axiomatic basis for computer programming. Commun. ACM **12**(10), 576–580 (1969). https://doi.org/10.1145/363235.363259

22. Hobor, A., Villard, J.: The ramifications of sharing in data structures. In: Giacobazzi, R., Cousot, R. (eds.) Principles of Programming Languages (POPL 2013), pp. 523–536. ACM (2013). https://doi.org/10.1145/2429069.2429131

23. Hóu, Z., Goré, R., Tiu, A.: Automated theorem proving for assertions in separation logic with all connectives. In: Felty, A.P., Middeldorp, A. (eds.) CADE 2015. LNCS (LNAI), vol. 9195, pp. 501–516. Springer, Cham (2015). https://doi.org/10.1007/978-3-319-21401-6_34

24. Hóu, Z., Sanán, D., Tiu, A., Liu, Y.: Proof tactics for assertions in separation logic. In: Ayala-Rincón, M., Muñoz, C.A. (eds.) ITP 2017. LNCS, vol. 10499, pp. 285–303. Springer, Cham (2017). https://doi.org/10.1007/978-3-319-66107-0_19

25. Ishtiaq, S.S., O'Hearn, P.W.: BI as an assertion language for mutable data structures. In: Principles of Programming Languages (POPL 2001), vol. 36, pp. 14–26. ACM (2001). https://doi.org/10.1145/373243.375719

26. Jacobs, B., Smans, J., Piessens, F.: A quick tour of the verifast program verifier. In: Ueda, K. (ed.) APLAS 2010. LNCS, vol. 6461, pp. 304–311. Springer, Heidelberg (2010). https://doi.org/10.1007/978-3-642-17164-2_21

27. Jung, R., Swasey, D., Sieczkowski, F., Svendsen, K., Turon, A., Birkedal, L., Dreyer, D.: Iris: monoids and invariants as an orthogonal basis for concurrent reasoning. In: Principles of Programming Languages (POPL 2015), pp. 637–650. ACM (2015). https://doi.org/10.1145/2676726.2676980

28. Klein, G., Andronick, J., Elphinstone, K., Murray, T., Sewell, T., Kolanski, R., Heiser, G.: Comprehensive formal verification of an OS microkernel. Trans. Comput. Syst. **32**(1), 2:1–2:70 (2014). https://doi.org/10.1145/2560537

29. Klein, G., Kolanski, R., Boyton, A.: Mechanised separation algebra. In: Beringer, L., Felty, A. (eds.) ITP 2012. LNCS, vol. 7406, pp. 332–337. Springer, Heidelberg (2012). https://doi.org/10.1007/978-3-642-32347-8_22

30. Klein, G., Kolanski, R., Boyton, A.: Separation algebra. Archive of Formal Proofs, Formal proof development (2012). http://isa-afp.org/entries/Separation_Algebra.shtml

31. Krebbers, R., Timany, A., Birkedal, L.: Interactive proofs in higher-order concurrent separation logic. In: Castagna, G., Gordon, A.D. (eds.) Principles of Programming Languages (POPL 2017), pp. 205–217. ACM (2017). https://doi.org/10.1145/3009837.3009855

32. Lee, W., Park, S.: A proof system for separation logic with magic wand. In: Jagannathan, S., Sewell, P. (eds.) Principles of Programming Languages (POPL 2014), pp. 477–490. ACM (2014). https://doi.org/10.1145/2535838.2535871

33. Leino, K.R.M.: Dafny: an automatic program verifier for functional correctness. In: Clarke, E.M., Voronkov, A. (eds.) LPAR 2010. LNCS (LNAI), vol. 6355, pp. 348–370. Springer, Heidelberg (2010). https://doi.org/10.1007/978-3-642-17511-4_20

34. Maclean, E., Ireland, A., Grov, G.: Proof automation for functional correctness in separation logic. J. Logic Comput. **26**(2), 641–675 (2016). https://doi.org/10.1093/logcom/exu032

35. Mehta, F., Nipkow, T.: Proving pointer programs in higher-order logic. In: Baader, F. (ed.) CADE 2003. LNCS (LNAI), vol. 2741, pp. 121–135. Springer, Heidelberg (2003). https://doi.org/10.1007/978-3-540-45085-6_10

36. Nipkow, T.: Hoare logics in Isabelle/HOL. In: Schwichtenberg, H., Steinbrüggen, R. (eds.) Proof and System-Reliability, pp. 341–367. Springer, Heidelberg (2002). https://doi.org/10.1007/978-94-010-0413-8_11

37. Nipkow, T., Wenzel, M., Paulson, L.C. (eds.): Isabelle/HOL — A Proof Assistant for Higher-Order Logic. LNCS, vol. 2283. Springer, Heidelberg (2002). https://doi.org/10.1007/3-540-45949-9

38. O'Hearn, P., Reynolds, J., Yang, H.: Local reasoning about programs that alter data structures. In: Fribourg, L. (ed.) CSL 2001. LNCS, vol. 2142, pp. 1–19. Springer, Heidelberg (2001). https://doi.org/10.1007/3-540-44802-0_1

39. Reynolds, J.C.: Separation logic: a logic for shared mutable data structures. In: Logic in Computer Science (LICS 2002), pp. 55–74 (2002). https://doi.org/10.1109/LICS.2002.1029817

40. Reynolds, J.C.: An introduction to separation logic. In: Broy, M., Sitou, W., Hoare, T. (eds.) Engineering Methods and Tools for Software Safety and Security, NATO Science for Peace and Security Series - D: Information and Communication Security, vol. 22, pp. 285–310. IOS Press (2009). https://doi.org/10.3233/978-1-58603-976-9-285

41. Sergey, I., Nanevski, A., Banerjee, A.: Mechanized verification of fine-grained concurrent programs. In: Grove, D., Blackburn, S. (eds.) Programming Language Design and Implementation (PLDI 2015), pp. 77–87. ACM (2015). https://doi.org/10.1145/2737924.2737964

42. Tuerk, T.: A Separation Logic Framework for HOL. Ph.D. thesis, University of Cambridge, UK (2011)

43. Vafeiadis, V., Parkinson, M.: A marriage of rely/guarantee and separation logic. In: Caires, L., Vasconcelos, V.T. (eds.) CONCUR 2007. LNCS, vol. 4703, pp. 256–271. Springer, Heidelberg (2007). https://doi.org/10.1007/978-3-540-74407-8_18

A Coq Formalisation of SQL's Execution Engines

V. Benzaken[2], É. Contejean[1(✉)], Ch. Keller[2], and E. Martins[2]

[1] CNRS, Université Paris Sud, LRI, Orsay, France
evelyne.contejean@lri.fr
[2] Université Paris Sud, LRI, Orsay, France

Abstract. In this article, we use the Coq proof assistant to *specify* and *verify* the low level layer of SQL's execution engines. To reach our goals, we first design a high-level Coq specification for data-centric operators intended to capture their essence. We, then, provide two Coq implementations of our specification. The first one, the physical algebra, consists in the low level operators found in systems such as Postgresql or Oracle. The second, SQL algebra, is an extended relational algebra that provides a semantics for SQL. Last, we formally relate physical algebra and SQL algebra. By proving that the physical algebra implements SQL algebra, we give high level assurances that physical algebraic and SQL algebra expressions enjoy the same semantics. All this yields the first, to our best knowledge, formalisation and verification of the *low level layer of an RDBMS* as well as SQL's compilation's *physical optimisation*: fundamental steps towards mechanising SQL's compilation chain.

1 Introduction

Data-centric applications involve increasingly massive data volumes. An important part of such data is handled by relational database management systems (RDBMS's) through the SQL query language. Surprisingly, formal methods have not been broadly promoted for data-centric systems to ensure strong *safety* guarantees about their expected behaviours. Such guarantees can be obtained by using proof assistants like Coq [27] or Isabelle [28] for specifying, proving and testing (parts of) such systems. In this article, we use the Coq proof assistant to *specify* and *verify* the low level layer of an RDBMS as proposed in [26] and detailed in [18].

The theoretical foundations for RDBMS's go back to the 70's where relational algebra was originally defined by Codd [13]. Few years later, SQL, the standard domain specific language for manipulating relational data was designed [10]. SQL was dedicated to *efficiently* retrieve data stored on *secondary storage* in RDBMS's, as described in the seminal work [26] that addressed the low level layer as well as secondary memory access for such systems, known in the field as *physical algebra*, *access methods* and *iterator interface*. SQL and RDBMS's

Work funded by the DataCert ANR project: ANR-15-CE39-0009.

evolved over time but they still obey the principles described in those works and found in all textbooks on the topic (see [5,15,18,25] for instance). In particular, the *semantic analysis* of a SQL query could yield an expression, e_1, of an (extended) relational algebra. The *logical* optimisation step rewrites this expression into another algebraic expression e_2 (based on well-known rules that can be found in [18]). Then *physical* optimisation takes place and an evaluation strategy for expression e_2, called a *query execution plan (QEP)* in this setting, is produced. QEP's are composed by *physical algebra operators*. Yet there are no formal guarantees that the produced QEP and (optimised) algebraic expression do have the same semantics. One contribution of our work is to open the way to formally provide such evidences. To reach our goal, we adopt a very general approach that is not limited to our specific problem. It consists in providing a high-level pivotal specification that will be used to describe and relate several lower-level languages.

In our particular setting, we first design a high-level, very abstract, generic, thus extensible, Coq specification for data-centric operators intended to capture their essence (which will be useful to address other data models and languages than relational ones).

The first low-level language consists of physical operators as found in systems such as Postgresql and described in main textbooks on the topic [18,25]. One specificity and difficulty lied in the fact that, when evaluating a SQL query, all those operators are put together, and for efficiency purposes, database systems implement, as far as possible, on-line [22] versions of them through the iterator interface. At that point there is a discrepancy between the specifications that provide collection-at-a-time invariants and the implementations that account for value-at-a-time executions. To fill up the gap, we exhibit non trivial invariants to prove that our on-line algorithms do implement their high-level specification. Moreover, those operators are shown to be exhaustive and to terminate.

The second low-level language (actually mid-level specification) is SQL algebra (syntax and semantics), an algebra that hosts SQL. By hosting we mean that there is an embedding of SQL into this algebra which preserves SQL's semantics. Due to space limitations, such an embedding is out of the scope of this paper and is described in [7]. We relate each algebraic operator to our high level specification by proving adequacy lemmas providing strong guarantees that the operator at issue is a realization of the specification.

Last, we formally bridge both implementations. By proving that the physical algebra does implement SQL algebra, we give strong assurances that the QEP and the algebraic expression resulting from the semantics analysis and logical optimisation do have the same semantics. This last step has been eased thanks to the efforts devoted to the design of our high-level pivotal specification. All this yields the first, to our best knowledge, *executable* formalisation and verification of the *low level layer of an RDBMS* as well as SQL's compilation's *physical optimisation*: fundamental steps towards mechanising SQL's compilation chain.

Organisation. We briefly recall in Sect. 2 the key ingredients of SQL compilation and database engines: extended relational algebra, physical algebra operators and iterator interface. Section 3 presents our Coq high-level specification that captures the essence of data-centric operators. In Sect. 4, we formalise the iterator interface and physical algebra, detailing the necessary invariants. Section 5 presents the formal specification of SQL algebra. We formally establish, in Sect. 6, that any given physical operator does implement its corresponding logical operator. We draw lessons, compare our work, conclude and give perspectives in Sect. 7.

2   SQL's Compilation in a Nutshell

Following [18] SQL's compilation proceeds into three broad steps. First, the query is *parsed*, that is turned into a parse-tree representing its structure. Second, *semantics analysis* is performed transforming the parse tree into an expression tree of (extended) *relational algebra*. Third, the *optimisation* step is performed: using relational algebraic rewritings (logical optimisation) and based on a cost model[1], a physical *query execution plan (QEP)* is produced. It not only indicates the operations performed but also the order in which they will be evaluated, the algorithm chosen for each operation and the way stored data is obtained and passed from one operation to another. This last stage is *data dependent*.

We present the main concepts through the following example that models a movie database gathering information about movies (relation `movie`), movies' directors `director`, the movies they directed and relation `role` carrying information about who played (identified by his/her `pid`) which role in a given movie

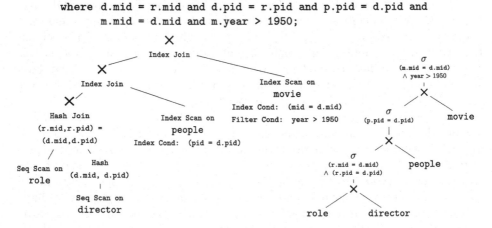

Fig. 1. A typical SQL query, its QEP and logical AST.

---

[1] The model exploits system collected statistics about the data stored in the database.

(identified by its mid). On Fig. 1 we give for a typical SQL query the corresponding (Postgresql)[2] QEP issued as well as the AST obtained after semantic analysis and logical optimisation.

The leaves (*i.e.*, relations) are treated by means of access methods such as Seq Scan or Index Scan (in case an index is available); a third access method usually provided by RDBMS's is the Sort Scan which orders the elements in the result according to a given criteria. In the example, relations role and director are accessed via Seq Scan, whereas people and movie are accessed thanks to Index Scan. The product of relations in the from part is reordered and the filtering condition is spread over the relevant (sub-product of) relations.

Intuitively, each physical operator corresponds to one or a combination of algebraic operators: σ (selection), $\times$ (product), completed with π (projection) and γ (grouping) (see Sect. 5.1 for their formal semantics).

Conversely, to each operator of the logical plan, $\sigma, \times, \ldots$, potentially corresponds one or more operators of the physical plan: the underlying database system provides several different algorithm's implementations. For the cross product, for instance, at least four such different algorithms are provided by mainstream systems: Nested Loop, Index Join, Sort Merge Join and Hash Join. For the selection operator the system may use the Filter physical operator.

The situation is made even more complex by the facts that a QEP contains some strategy (top-down, left-most evaluation) and that some physical operators are implemented via on-line algorithms. Hence a filtering condition which spans over a cross-product between two operands, in an algebraic expression, may be used in the corresponding QEP to filter the second one, by inlining the condition for each tuple of the first operand. This is the case for instance with the second

Table 1. Synthesis

Iterator interface operators				
Section 3 data centric operators	Section 4, ϕ algebra			Section 5 SQL algebra
	simple	index based	sort based	
map	Seq Scan	Index scan Bitmap index scan	Sort scan	r, π
join	Nested loop Block nested loop	Hash join Index join	Sort merge join	$\times$
filter	Filter			σ
group	Group			γ
bind	Subplan			env
accumulator	Aggregate, Hash, Hash aggregate			aggregate
	Intermediate results storage operators Materialize			

---

[2] The IJ nodes are expressed in Postqresql as Nested loop combined with an Index scan but corresponds to an index-based join.

join of Fig. 1 where the second operand is an Index-Scan. Therefore the pattern $x \times_{IJ}$ (Index Scan y Index Cond $:a = x.a'$) corresponds to $\sigma_{y.a=x.a'}(x \times y)$.

Unfortunately not all physical operators support the on-line approach and *materialising* partial results (*i.e.*, temporarily storing intermediate results) is needed: the Materialise physical operator allows to express this in Postgresql physical plans. Table 1 summarises our contributions where the colored cells indicate the Coq specified and implemented operators.

3   A High-Level Specification for Data-Centric Operators

In the data-centric setting, data are mainly collections of values. Such values can be combined and enjoy a decidable comparison. Operators allow for manipulating collections, that is to *extract* data from a collection according to a condition (filter), to *iterate* over a collection (map), to *combine* two collections (join) and, last, to *aggregate* results over a collection (group).

Since collections may be implemented by various means (lists with or without duplicates, AVL, etc.), in the following we shall call these implementations containerX 's. The content, that is the elements gathered in such a containerX, may be retrieved with the corresponding function contentX and we also make a last assumption, that there is a decidable equivalence equivX for elements. The function nb_occX is defined as the number of occurrences of an element in the contentX of a containerX modulo equivX[3].

We then characterise the essence of data centric operations performed on containers. Operators filter and map are a lifting of the usual operators on lists to containers.

```
Definition is_a_filter_op contentA contentA' (f: A → bool) (fltr: containerA → containerA')
  := ∀s, ∀t, nb_occA' t (fltr s) = (nb_occA t s) * (if f t then 1 else 0).
Definition is_a_map_op contentA contentB (f: A → B) (mp: containerA → containerB) :=
  ∀s, ∀t, nb_occB t (mp s) = nb_occ t (map f (contentA s)).
```

Unlike the first two operators which make no hypothesis on the nature of the elements of a containerX, joins manipulate *homogeneous* containers *i.e.*, their elements are equipped with a support supX which returns a set of attributes, and all elements in a containerX enjoy the same supX, which is called the sort of the container. Let us denote by A1 (resp. A2, resp. A) the type of the elements of the first operand (resp. the second operand, resp. the result) of a join operator j. Elements of type A are also equipped with two functions projA1 and projA2, which respectively project them over A1 and A2.

---

[3] X will be A, A', B, according to the various types of elements and various implementations for the collection. A particular case of nb_occX is nb_occ which denotes the number of occurrences in a list.

```
Definition is_a_join_op sa1 sa2 contentA1 contentA2 contentA
                        (j : containerA1 → containerA2 → containerA) :=
  ∀s1 s2, (∀t, 0 < nb_occA1 t s1 → supA1 t = sa1) →
            (∀t, 0 < nb_occA2 t s2 → supA2 t = sa2) →
  ((∀t, 0 < nb_occA t (j s1 s2) → supA t = (sa1 unionS sa2)) ∧
   (nb_occA t (j s1 s2) = nb_occA1 (projA1 t) s1 * nb_occA2 (projA2 t) s2))
                        * (if supA t = (sa1 unionS sa2) then 1 else 0).
```

Intuitively, joins allow for combining two homogeneous containers by taking the union of their **sort** and the product of their occurrence's functions.

The grouping operator, as presented in textbooks [18], partitions, using mk_g, a container into groups according to a grouping criteria g and then discards some groups that do not satisfy a filtering condition f. Last for the remaining groups it **builds** a new element.

```
Definition is_a_grouping_op (G : Type) (mk_g : G → containerA → list B) grp :=
  ∀(g : G) (f : B → bool) (build : B → A) (s : containerA) t,
    nb_occA t (grp g f build s) = nb_occ t (map build (filter f (mk_g g s))).
```

All the above definitions share a common pattern: they state that the number of occurences nb_occX t (o p s) of an element t in a container built from an operator o applied to some parameters p and some operands s, is equal to foopp(t, nb_occX (g t) s), where foopp is a function which depends only on the operator and the parameters. This implies that any two operators satisfying the same specification is_a_...._op are *interchangeable*. For grouping, the situation is slightly more subtle, however the same interchangeability property shall hold since nb_occA t (grp g f build s)) depends only on t and contentA s for the grouping criteria used in the following sections.

Tuning those definitions was really challenging: finding the relevant level of abstraction for containers and contents suitable to host both physical and logical operators was not intuitive. Even for the most simple one such as filter, we would have expected that the type of containers should be the same for input and output. It was not possible as we wanted a simple, concise and efficient implementation.

4 Physical Algebra

All physical operators that can be implemented by on-line algorithms rely on a common iterator interface that allows them to build the next tuple on demand.

4.1 Iterators

A key aspect in our formalisation of physical operators is a specification of such a common iterator interface together with the properties an iterator needs to satisfy. We validate this interface by implementing standard iterative physical operators, namely sequential scanning, filtering, and nested loop.

Abstract Iterator Interface. An iterator is a data structure that iterates over a collection of elements to provide them, on demand, one after the other. Following the iterator interface given in [18] and in the same spirit of the formalisation of cursors presented in [17], we define a `cursor` as an abstract object over some type `elt` of elements that must support three operations: `next`, that returns the next element of the iteration if it exists; `has_next`, that checks if such an element does exist; and `reset`, that restarts the cursor at its beginning. In Coq, this can be modelled as a record[4] named `Cursor` that contains (at least) an abstract type of `cursors` and these three operations:

```
Record Cursor (elt : Type) : Type :=        has_next : cursor → Prop;
  { cursor : Type;                          reset : cursor → cursor;
    next : cursor → result elt * cursor;    [...] (* Some properties, see below *) }.
```

Due to the immutable nature of Coq objects, the operations `next` and `reset` must return the modified cursor. Moreover, since `next` must be a total function, a monadic[5] construction is used to wrap the element of type `t` that it outputs:

```
Inductive result (A : Type) := Result : A → result A | No_Result | Empty_Cursor.
```

The constructor `Result` corresponds to the case where an element can be returned, and the two constructors `No_Result` and `Empty_Cursor` deal with the cases where an element cannot be returned, respectively because it does not match some selection condition (see Sect. 4.1) or because the cursor has been fully iterated over.

We designed a sufficient set of properties that a cursor should satisfy in order to be valid. These properties are expressed in terms of three high-level inspection functions (that are used for specification only, not for computation): `collection` returns all the elements of the cursor, `visited` returns the elements visited so far, and `coherent` states an invariant that the given cursor must preserve:

```
Record Cursor (elt : Type) : Type := { [...]
  collection : cursor → list elt;
  visited : cursor → list elt;
  coherent : cursor → Prop; [...] }.
```

collection {
Tuple C1 (A1,...,An) } visited
Tuple C2 (A1,...,An) } cursor
Tuple C3 (A1,...,An)
⋮
Tuple Cn (A1,...,An)

---

[4] We could also use a module type, but the syntax would be heavier and less general.

[5] This construction is similar to the exception monad. There is no interest to write the standard "return" and "bind" operators. The sequential scan and nested loop, respecitvely, can be seen as online versions of them.

Given these operations, the required properties are the following:

```
Record Cursor (elt : Type) : Type := { [...]
 (* next preserves the collection *)
 next_col : ∀c, coherent c → collection (snd (next c))) = collection c;
 (* next adds the returned element to visited *)
 next_visited_Result :
   ∀a c c', coherent c → next c = (Result a, c') → visited c' = a :: (visited c);
 next_visited_No_Result :
   ∀c c', coherent c → next c = (No_Result, c') → visited c' = visited c;
 next_visited_Empty_Cursor :
   ∀c c', coherent c → next c = (Empty_Cursor, c') → visited c' = visited c;
 (* next preserves coherence *) next_coherent : ∀c, coherent c → coherent (snd (next c));
 (* when a cursor has no element left, visited contains all the elements of the collection *)
 has_next_spec : ∀c, coherent c → ¬ has_next c → (collection c) = (rev (visited c));
 (* a cursor has new elements if and only if next may return something *)
 has_next_next_neg : ∀c, coherent c → (has_next c ↔ fst (next c) ≠ Empty_Cursor);
 (* reset preserves the collection *)
 reset_collection : ∀c, collection (reset c) = collection c;
 (* reset restarts the visited elements *) reset_visited : ∀c, visited (reset c) = nil;
 (* reset returns a coherent cursor *) reset_coherent : ∀c, coherent (reset c); [...]}.
```

The ..._coherent and ..._collection axioms ensure that coherent and the collection of elements are indeed invariants of the iterator. The ..._visited axioms explain how visited is populated. Finally, the has_next_spec axiom is the key property to express that all the elements have been visited at the end of the iteration.

Last, we require a progress property on cursors (otherwise next could return the No_Result value forever and still satisfy all the properties). Progress is stated in terms of an upper bound on the number of iterations of next before reaching an Empty_Cursor:

```
Record Cursor (elt : Type) : Type := { [...]
 (* an upper bound on the number of iterations before the cursor has been fully visited *)
 ubound : cursor → nat;
 (* this upper bound is indeed complete *)
 ubound_complete : ∀c acc, coherent c → ¬ has_next (fst (iter next (ubound c) c acc)); }.
```

where iter f n c acc iterates n times the function f on the cursor c, returning a pair of the resulting cursor and the accumulator acc augmented with the elements produced during the iteration. The upper bound is not only part of the specification (to state that cursors have a finite number of possible iterations) but can also be used in Coq to actually materialize them.

We will see that these properties are strong enough both to combine iterators and to derive their adequacy with respect to their algebraic counterparts.

First Instance: Sequential Scan. The base cursor implements sequential scan by returning, tuple by tuple, all the elements of a given relation, represented by a list in our high-level setting. It simply maintains a list of elements still to be visited named to_visit and its invariant expresses that the collection contains the elements visited so far and the elements that remain to be visited. A natural upper bound on the number of iterations is the number of elements to visit.

```
Definition coherent (c : cursor) := c.(collection) = rev c.(visited) ++ c.(to_visit).
Definition ubound (c : cursor) : nat := List.length c.(to_visit).
```

Second Instance: Filter. Filtering a cursor returns the same cursor, but with a different function **next** and accordingly different specification functions. Given a property on the elements **f : elt → bool**, the function **next** filters elements of the underlying cursor:

```
Definition next (c : cursor) : result elt * cursor :=
  match Cursor.next c with
  | (Result e, c') ⇒ if f e then (Result e, c') else (No_Result, c') | rc' ⇒ rc'
  end.
```

This is where **No_Result** is introduced when the condition is not met. Accordingly, the functions **collection** and **visited** are the filtered **collection** and **visited** of the underlying cursor and an upper bound on the number of iterations is the upper bound of the underlying cursor:

```
Definition collection (c : cursor) := List.filter f (Cursor.collection c).
Definition visited (c : cursor) := List.filter f (Cursor.visited c).
Definition ubound (q : cursor) : nat := Cursor.ubound q.
```

Third Instance: Nested Loop. The nested loop operator builds the cross-product between an outer cursor and an inner cursor: the **next** function returns either the combination of the current tuple of the outer cursor with the next tuple of the inner cursor (if this latter exists) or the combination of the next tuple of the outer cursor with the first tuple of the reset outer cursor (see Fig. 2).

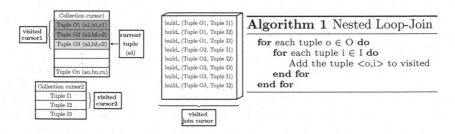

Fig. 2. Nested Loop-Join

Specifying such a cursor becomes slightly more involved. For correctness, one has to show the invariant stating that the elements visited so far contain (i) the last visited element of the outer cursor combined with all the visited elements of the inner cursor; and (ii) the other visited elements of the outer cursor combined with all the collection of the inner cursor.

```
Definition coherent (c: cursor) : Prop :=
  (* the two underlying cursors are coherent *)
  Cursor.coherent (outer c) ∧ Cursor.coherent (inner c) ∧
  match Cursor.visited (outer c) with
  (* if the outer cursor has not been visited yet, so as the inner cursor *)
  | nil ⇒ visited c = nil ∧ Cursor.visited (inner c) = nil
  (* otherwise, the visited elements are a partial cross-product *)
  | el :: li ⇒ visited c = (cross (el::nil) (Cursor.visited (inner c))) ++
                           (cross li (rev (Cursor.collection (inner c))))
  end.
```

where **cross** builds the cross product of two lists. For progress, an upper bound for the length of this partial cross-product is needed:

```
Definition ubound (c:cursor) : nat :=
  Cursor.ubound (inner c) +
  (Cursor.ubound (outer c) * (S (Cursor.ubound (Cursor.reset (inner c))))).
```

where a successor on the upper bound on the inner cursor has been added for simplicity reasons. The proof of completeness is elaborate and relies on key properties on bounds for cursors stating in particular that the bound decreases when **next** is applied to a non-empty cursor:

```
Lemma ubound_next_not_Empty:
  ∀c, coherent c → fst (next c) ≠ Empty_Cursor → ubound (snd (next c)) < ubound c;
```

Materialisation. Independently from any specific operator, materialising an iterator is achieved by resetting it, then iterating the upper bound number of times while accumulating the returned elements. We can show the key lemma for adequacy of operators: materialising an iterator produces all the elements of its collection.

```
Definition materialize c :=
  let c' := reset c in List.rev (snd (iter next (ubound c') c' nil)).
Lemma materialize_collection c : materialize c = collection c.
```

We used the same technique to implement the grouping operator by, instead of simply accumulating the elements, group them on the fly.

4.2 Index-Based Operators

Having an index on a given relation is modelled as a wrapper around cursors: such a relation must be able to provide a (possibly empty) cursor for each value of the index. The main components of an indexed relation are: (i) a type **containers** of the internal representation of data (which can be a hash table, a B-tree, a bitmap, ...), (ii) a function **proj**, representing the projection from tuples to their values on the attributes enjoying the index, (iii) a comparison function **P** on these attributes (which can be an equality for hash-indices, a comparison for tree-based indices, ...) and (iv) an indexing function **i** that, given a container and an index, returns the cursors of the elements of the container matched by the index (w.r.t. **P**). This is implemented as the following record:

```
Record Index (elt eltp : Type) : Type :=
  { containers : Type; (* representation of data *)
    proj : elt → eltp; (* projection on the index *)
    P : eltp → eltp → bool; (* comparison between two indices *)
    i : containers → eltp → Cursor.cursor; (* indexing function *)  [...] }.
```

As for sequential iterators, we state the main three properties that an index should satisfy. Again, these properties are expressed in terms of the collection of a container, used for specification purposes only.

```
Record Index (elt eltp : Type) : Type := { [...]
  ccollection : containers → list elt; (* the elements of a container *)
  (* the collection of an indexed cursor contains the filtered elements of the
     container w.r.t. P *)
  i_collection : ∀c x, Cursor.collection (i c x) =
                           List.filter (fun y ⇒ P x (proj y)) (ccollection c);
  (* a fresh indexed cursor has not been visited yet *)
  i_visited : ∀c x, Cursor.visited (i c x) = nil;
  (* a fresh indexed cursor is coherent *) i_coherent : ∀c x, Cursor.coherent (i c x) }.
```

First Instance: Sequential Scan. Let us start with a simple example: sequential scan can be seen as an index scan with a trivial comparison function that always returns true, and a trivial indexing function that returns a sequential cursor. It is thus sufficient to use the following definitions and the properties follow immediately:

```
Definition containers := list elt.
Definition P := fun _ _ ⇒ true.
Definition i := fun c _ ⇒ SeqScan.mk_cursor c.
```

Let us see how this setting models more interesting index-based algorithms.

Second Instance: Hash-Index Scan. In this case, the comparison function is an equality, and the underlying containers are hash tables whose keys are the attributes composing the index. To each key is associated the cursor whose collection contains elements whose projection on the index equals the key. In our development, we use the Coq `FMap` library to represent hash tables, but we are rather independent of the representation:

```
Record containers : Type := mk_containers
  { (* the hash table *) hash : FMapWeakList.Raw(Eltp) (cursor C);
    (* the elements are associated to the corresponding key *)
    keys : ∀x es, MapsTo x es hash → ∀e, List.In e (collection es) → P x (proj e) = true;
    noDup : NoDup hash (* the hash table has no key duplicate *) }.
```

where `MapsTo x es hash` means that `es` is the cursor associated to the key `x` in the hash table.

Given a particular index, the indexing function returns the cursor associated to the index in the hash table. Its properties follow from the properties of hash tables.

Third Instance: Bitmap-Index Scan In this case, the comparison function can be any predicate, and the containers are arrays of all the possible elements of the relation together with bitmaps (bit vectors) associated to each index, stating whether the nth element of the relation corresponds to the index. In our development, we use Coq vectors to represent this data structure:

```
Record containers : Type := mk_containers
{size : nat; (* the number of elements in the relation *)
 collection : Vector.t elt size; (* all the elements of the relation *)
 bitmap : eltp → Bvector size;(* a bitmap associated to every index *)
 (* each bitmap associates to true exactly the elements matching the corresponding index *)
 coherent : ∀n x0, nth (bitmap x0) n = P x0 (proj (nth collection n)) }.
```

Given a particular index, the indexing function returns the sequential cursor built from the elements for which the bitmap associated to the index returns true. Its properties follow by induction on the size of the relation.

Application: Index-Join Algorithm. The index-join algorithm is similar in principle to the nested loop algorithm but faster thanks to an exploitable index on the inner relation: for each tuple of the outer relation, only matching tuples of the inner relation are considered (see Fig. 3). Hence, our formal development is similar as the one for nested loop, but more involved: (i) in the function **next**, each time we get a new element from the outer relation, we need to generate the cursor corresponding to the index from the inner relation (instead of resetting the whole cursor) (ii) the **collection** is now a *dependent* cross-product between the outer relation and the matching inner tuples; the invariant predicate **coherent** has to be changed consequently (iii) the **ubound** is a *dependent* product of the bound of the outer relation with each bound of the matching cursors of the inner relation (obtained by materialising the outer relation).

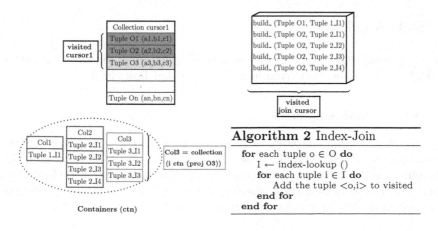

Fig. 3. Index-based nested loop

Derived Operators. Our high level of abstraction gives for free the specification of common variants of the physical operators. For instance, the Block Nested Loop algorithm is straightforwardly formalised by replacing, in the Nested Loop formalisation, the abstract type of elements by a type of "blocks" of elements (e.g., lists), and the function that combines two tuples by a function that combines two blocks of tuples.

4.3 Adequacy

All physical operators specified and implemented so far are shown to fulfil the high-level specification. For instance, if `C` is a cursor, `f` a filtering condition compatible with the equivalence of elements in `C`, then the corresponding filter iterator `F:= (Filter.build f f_eq C)` fulfils the specification of a filter:

```
Lemma mk_filter_is_a_filter_op :
  is_a_filter_op (Cursor.materialize C) (Cursor.materialize F) f (Filter.mk_filter F).
```

Sometimes, there are some additional side conditions: if `C1` and `C2` are two cursors, and `NL := (NestedLoop.build [...] C1 C2)` is the corresponding nested loop which combines elements thanks to the combination function `build_`, not only some hypotheses are needed to be able to build `NL`, but some extra ones are needed to prove `NL` is indeed a join operator:

```
Hypothesis [...]
Hypothesis build_split_eq_1 :
  ∀t1 u1 t2 u2, equivA (build_ t1 t2) (build_ u1 u2) → [...] → equivA1 t1 u1.
Hypothesis build_split_eq_2 :
  ∀t1 u1 t2 u2, equivA (build_ t1 t2) (build_ u1 u2) → [...] → equivA2 t2 u2.
Lemma NL_is_a_join_op :
  is_a_join_op [...] (Cursor.materialize C1) (Cursor.materialize C2) (Cursor.materialize NL)
           [...] (fun c1 c2 ⇒ NestedLoop.mk_cursor C1 C2 nil c1 c2).
```

5 SQL Algebra

We now present SQL algebra, our Coq formalisation of an algebra that satisfies the high-level specification given in Sect. 3 and that hosts SQL.

5.1 Syntax and Semantics

The extended relational algebra, as presented in textbooks, consists of the well-known operators π (projection), σ (selection) and $\times$ (join) completed with the γ (grouping) together with the set theoretic operators. We focus on the former four operators. In our formalisation, `formula` mimics the SQL's filtering conditions expressed in the `where` and `having` clauses of SQL.

```
Inductive query : Type :=
  | Q_Table : relname → query
  | Q_Set : set_op → query → query →
      query
  | Q_Join : query → query → query
  | Q_Pi : list select → query → query
  | Q_Sigma : formula → query → query
  | Q_Gamma :
      list term → formula → list select →
      query → query
with formula : Type :=
  | Q_Conj :
```

```
      and_or → formula → formula → formula
  | Q_Not : formula → formula
  | Q_Atom : atom → formula
with atom : Type :=
  | Q_True
  | Q_Pred : predicate → list term → atom
  | Q_Quant :
      quantifier → predicate → list term →
      query → atom
  | Q_In : list select → query → atom
  | Q_Exists : query → atom.
```

We assume that there is an **instance** which associates to each relation a multiset (**bagT**) of tuples, and that these multisets enjoy some list-like operators such as **empty**, **map**, **filter**, etc (see the additional material for more details and precise definitions). In order to support so-called SQL correlated queries, the notion of environment is necessary.

```
Fixpoint eval_query env q {struct q} : bagT :=
  match q with
    | Q_Table r ⇒ instance r
    | Q_Set o q1 q2 ⇒ if sort q1 = sort q2
                      then interp_set_op o (eval_query env q1) (eval_query env q2)
                      else empty
    | Q_Join q1 q2 ⇒ natural_join (eval_query env q1) (eval_query env q2)
    | Q_Pi s q ⇒  map (fun t ⇒ projection_ (env_t env t) s) (eval_query env q)
    | Q_Sigma f q ⇒ filter (fun t ⇒ eval_formula (env_t env t) f) (eval_query env q)
    | Q_Gamma lf f s q ⇒ let g := Group_By lf in
      mk_bag (map (fun l ⇒ projection_ (env_g env g l) s)
             (filter (fun l ⇒ eval_formula (env_g env g l) f)
             (make_groups_ env (eval_query env q) g)))
  end
with eval_formula env f := [ ... ]
with eval_atom env atm := [ ...]
end.
```

Let us detail the evaluation of **Q_Sigma f q** in environment **env**. It consists of the tuples **t** in the evaluation of **q** in **env** which satisfy the evaluation of formula **f** in **env**. In order to evaluate **f** one has to evaluate the expressions it contains. Such expressions are formed with attributes which are either bound in **env** or occur in tuple **t**'s support. This is why the evaluation of **f** takes place in environment **env_t env t** which corresponds to pushing **t** over **env** yielding

```
Q_Sigma f q ⇒ filter (fun t ⇒ eval_formula (env_t env t) f) (eval_query env q)
```

Similarly, we use **env_t env t** for the evaluation of expressions of **s** in the **Q_Pi s q** case. The grouping γ is expressed thanks to **Q_Gamma**. A group consists of elements which evaluate to the same values for a list of grouping expressions. Each group yields a tuple thanks to the **list select** part in which each (sub-)term either takes the same value for each tuple in the group, or consists in an aggregate expression. This usual definition (see for instance [18]) is not enough to handle SQL's **having** conditions, as **having** directly operates on the group

that carry more information than the corresponding tuple. This is why Q_Gamma
has also a formula operand. Thus the corresponding expression for query

```
    select avg(a1) as avg_a1, sum(b1) as sum_b1 from t1
      group by a1+b1, 3*b1 having a1 + b1 > 3 + avg(c1);
is Q_Gamma [a1 + b1; 3*b1] (Q_Atom (Q_Pred > [a1 + b1; 3 + avg(c1)]))
    [Select_As avg(a1) avg_a1; Select_As sum(b1) sum_b1] (Q_table t1)
```

5.2 Adequacy

The following lemmas assess that SQL algebra is a realisation of our high-level
specification. Note that, in the context of SQL algebra the notion of tuple corresponds to the high-level notion of elements' type X, finite bag corresponds to
the high-level notion of containerX and elements to contentX.

```
Lemma Q_Sigma_is_a_filter_op : ∀env f,
    is_a_filter_op [...]
      (* contentA := fun q ⇒ Febag.elements BTupleT (eval_query env q) *)
      (* contentA' := fun q ⇒ Febag.elements BTupleT (eval_query env q) *)
      (fun t ⇒ eval_formula (env_t env t) f)
      (fun q ⇒ Q_Sigma f q).
Lemma Q_Join_is_a_join_op : ∀env s1 s2,
    let Q_Join q1 q2 := Q_Join q1 q2 in
    is_a_join_op (* contentA1 := fun q ⇒ elements (eval_query env q) *)
                 (* contentA2 := fun q ⇒ elements (eval_query env q) *)
                 (* contentA := fun q ⇒ elements (eval_query env q) *) [...] s1 s2 Q_Join.
Lemma Q_Gamma_is_a_grouping_op : ∀env g f s ,
    let eval_s l := projection_ (env_g env (Group_By g) l) (Select_List s) in
    let eval_f l := eval_formula (env_g env (Group_By g) l) f in
    let mk_grp g q := partition_list_expr (elements (eval_query env q))
                            (map (fun f t ⇒ interp_funterm (env_t env t) f) g) in
    let Q_Gamma g f s q := eval_query env (Q_Gamma g f s q) in
    is_a_grouping_op [...] mk_grp g eval_f eval_s (Q_Gamma g f s).
```

6   Formally Bridging Logical and Physical Algebra

We now formally bridge physical algebra to SQL algebra. Figure 4 describes the
general picture. As pointed out in Sect. 3, any two operators which satisfy the
same high-level specification are interchangeable. This means in particular that
physical algebra's operators can be used to implement the evaluation of constructors of SQL algebra's inductive query.   The fundamental nature of the proof
of such facts is the transitivity of equality of number of occurences. However,
there are some additional hypotheses in both lemmas φ_..._op_is_a_..._op and
SQL_..._op_is_a_..._op. Some of them are trivially fulfilled when the elements
are tuples, while others cannot be discarded.

For instance, for proving that NestedLoop implements Q_Join, we have to
check that the hypotheses for NL_is_a_filter_op are fulfilled. Doing so, the condition that the queries to be joined must have disjoint sorts appeared mandatory
in order to prove the hypothesis build_split_eq_2 which assess that whenever

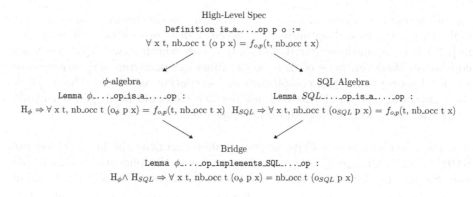

Fig. 4. Relating ϕ-algebra and SQL algebra.

two combined tuples are equivalent, their projections over the part corresponding to the inner relation also have to be equivalent.

```
Lemma NL_implements_Q_Join :
  (* Provided that the sorts are disjoined... *)
  ∀C1 C2 env q1 q2, (sort q1 interS sort q2) = emptysetS →
    (∀t, 0 < nb_occ t (eval_query env q1) → support t = sort q1) →
    (∀t, 0 < nb_occ t eval_query env q2 → support t = sort q2) →
    let NL := NestedLoop.build [...] C1 C2 in
    ∀c1 c2, (* ... if the two cursors implement the queries... *)
      (∀t, nb_occ t (eval_query env q1) = nb_occ t (Cursor.materialize C1 c1)) →
      (∀t, nb_occ t (eval_query env q2) = nb_occ t (Cursor.materialize C2 c2)) →
      (* ... then the nested loop implements the join *)
      ∀t, nb_occ t (eval_query env (Q_Join q1 q2)) =
        nb_occ t (Cursor.materialize NL (NestedLoop.mk_cursor C1 C2 nil c1 c2)).
```

This is an a posteriori justification that most systems implement combination of relations as cross-products whereas according to theory [1], combination should be the natural join.

7 Related Works, Lessons, Conclusions and Perspectives

Related work. Our work is rooted on the many efforts to use proof assistants to mechanise commercial languages' semantics and formally verify their compilation as done with the seminal work on Compcert [23]. The first attempt to formalise the relational data model using Agda is described in [19,20] and a first complete Coq formalisation of it is found in [8]. A SSreflect-based mechanisation of the Datalog language has been proposed in [9]. The very first Coq formalisation of RDBMSs' is detailed in [24] where the authors proposed a verified source to source compiler for (a small subset) of SQL. In [14], an approach which automatically compiles high-level SQL-like specifications down into performant, imperative, low-level code is presented. Our goal is different as we aim at verifying real-life RDBMS's execution strategies rather than producing imperative code. More recently, in [3,4] a Coq modelisation of the nested relational algebra is provided to assign a semantics to data-centric languages among which SQL. Regarding logical optimisation, the most in depth proposal is addressed in [12] where the authors describe a tool to decide whether two SQL queries are equivalent. However, none of these works consider specifying and verifying the low-level aspects of SQL's compilation and execution as we did. Our work is, thus, complementary to theirs and one perspective could be to join our efforts along the line of formalising data-centric systems.

Lessons, Conclusions and Perspectives. While formalising the *low level layer* of RDBMSs and SQL's *physical optimisation*, we learnt the following lessons: (i) not only finding the right invariants for physical operators was really involved but proving them (in particular termination for nested loop) was indeed subtle. This is due to the inherent difficulty to design on-line versions of even trivial off-line algorithms. (ii) we are even more convinced by the relevance of designing such a high-level specification that opens the way for accounting other data-centric languages. More precisely, we first formalised SQL algebra then the physical one, this implied revising the specification: in particular the introduction of `containersX` was made. Then, while bridging both formalisms we slightly modified the specification but without questionning our fundamental choices about abstracting over collections using `containersX`, only hypotheses were slightly tuned. (iii) The need for higher-order and polymorphism was mandatory both for the specification and physical algebra modelisation. This prevented us from using deductive verification tools such as Why3 [16] for instance: it was quite difficult to write down the algorithms and their invariants in this setting, even worse the automated provers were of no use to discharge the proof obligations. We tried tuning the invariants to help provers, without success. Hence our claim is that it is easier to directly use a proof assistant, where one has the control over the statements which have to be proven. (iv) The last point is that we experimented records versus modules: records are simpler to use than modules in our formalisation (no need of definitions' unfolding, no need of intermediate inductive types for technical reasons), the counterpart being that modules in the standard Coq library, such as `FSets` or `FMaps` were not directcy usable. The nice feature which allows to hide part of their contents through module subtyping was not needed here.

There are many points still to be addressed. In the very short term we plan to specify the missing operators of Table 1 and enrich the physical algebra with more fancy algorithms. Along this line two directions remain to be explored. In our development, the emphasis was put on specification rather than performance. Even if we carefully separated functions used in specification (such as `collection`, `coherent`, ...) from the concrete algorithms, these latter are defined in the functional language of Coq using higher-order data structures. We plan to refine these algorithms into more efficient versions, in particular that manipulate the memory. We plan to rely on CertiCoq [2] in order to produce fully certified C code. We are confident that our specification is modular enough to be plugged on other system components, such as buffer management, page allocation, disk access, already formalised in Coq as in [11,21]. Back to the general picture of designing a fully certified SQL compilation chain, in [6] we provided a Coq mechanised semantics pass that assigns any SQL query its executable SQL algebra expression. What remains to be done is to formally prove equivalence between SQL algebra expressions: those produced by the logical optimisation phase and the one corresponding to the query execution plan. Last, we are confident that our specification is general enough to host various data-centric languages and will provide a framework for data-centric languages interoperability which is our long term goal.

References

1. Abiteboul, S., Hull, R., Vianu, V.: Foundations of Databases. Addison-Wesley, Boston (1995)
2. Anand, A., Appel, A., Morrisett, G., Paraskevopoulou, Z., Pollack, R., Bélanger-Savary, O., Sozeau, M., Weaver, M.: Certicoq: a verified compiler for Coq. In: The Third International Workshop on Coq for Programming Languages (CoqPL) (2017)
3. Auerbach, J.S., Hirzel, M., Mandel, L., Shinnar, A., Siméon, J.: Handling environments in a nested relational algebra with combinators and an implementation in a verified query compiler. In: Salihoglu, S., Zhou, W., Chirkova, R., Yang, J., Suciu, D. (eds.) Proceedings of the 2017 ACM International Conference on Management of Data, SIGMOD Conference 2017, Chicago, IL, USA, 14–19 May 2017, pp. 1555–1569. ACM (2017). https://doi.org/10.1145/3035918.3035961, http://doi.acm.org/10.1145/3035918.3035961
4. Auerbach, J.S., Hirzel, M., Mandel, L., Shinnar, A., Siméon, J.: Q*cert: a platform for implementing and verifying query compilers. In: Proceedings of the 2017 ACM International Conference on Management of Data, SIGMOD Conference 2017, Chicago, IL, USA, 14–19 May 2017, pp. 1703–1706 (2017)
5. Bailis, P., Hellerstein, J.M., Stonebraker, M. (eds.): Readings in Database Systems, 5th edn. MIT-Press (2015). http://www.redbook.io/
6. Benzaken, V., Contejean, E.: SQLCert: Coq mechanisation of SQL's compilation: formally reconciling SQL and (relational) algebra, October 2016. Working paper available on demand
7. Benzaken, V., Contejean, E.: A Coq mechanised executable algebraic semantics for real life SQL queries (2018, Submitted for Publication)
8. Benzaken, V., Contejean, E., Dumbrava, S.: A Coq formalization of the relational data model. In: 23rd European Symposium on Programming (ESOP) (2014)
9. Benzaken, V., Contejean, É., Dumbrava, S.: Certifying standard and stratified datalog inference engines in SSReflect. In: Ayala-Rincón, M., Muñoz, C.A. (eds.) ITP 2017. LNCS, vol. 10499, pp. 171–188. Springer, Cham (2017). https://doi.org/10.1007/978-3-319-66107-0_12
10. Chamberlin, D.D., Boyce, R.F.: SEQUEL: a structured English query language. In: Rustin, R. (ed.) Proceedings of 1974 ACM-SIGMOD Workshop on Data Description, Access and Control, Ann Arbor, Michigan, 1–3 May 1974, 2 vols., pp. 249–264. ACM (1974). https://doi.org/10.1145/800296.811515, http://doi.acm.org/10.1145/800296.811515
11. Chen, H., Wu, X.N., Shao, Z., Lockerman, J., Gu, R.: Toward compositional verification of interruptible OS kernels and device drivers. In: Krintz, C., Berger, E.

(eds.) Proceedings of the 37th ACM SIGPLAN Conference on Programming Language Design and Implementation, PLDI 2016, Santa Barbara, CA, USA, 13–17 June 2016, pp. 431–447. ACM (2016).https://doi.org/10.1145/2908080.2908101, http://doi.acm.org/10.1145/2908080.2908101

12. Chu, S., Weitz, K., Cheung, A., Suciu, D.: HoTTSQL: proving query rewrites with univalent SQL semantics. In: Proceedings of the 38th ACM SIGPLAN Conference on Programming Language Design and Implementation, PLDI 2017, pp. 510–524. ACM, New York (2017)

13. Codd, E.F.: A relational model of data for large shared data banks. Commun. ACM **13**(6), 377–387 (1970). https://doi.org/10.1145/362384.362685, http://doi.acm.org/10.1145/362384.362685

14. Delaware, B., Pit-Claudel, C., Gross, J., Chlipala, A.: Fiat: Deductive synthesis of abstract data types in a proof assistant. In: Proceedings of the 42nd Annual ACM SIGPLAN-SIGACT Symposium on Principles of Programming Languages, POPL 2015, pp. 689–700 (2015)

15. Elmasri, R., Navathe, S.B.: Fundamentals of Database Systems, 2nd edn. Benjamin/Cummings, Redwood City (1994)

16. Filliâtre, J.-C., Paskevich, A.: Why3 — where programs meet provers. In: Felleisen, M., Gardner, P. (eds.) ESOP 2013. LNCS, vol. 7792, pp. 125–128. Springer, Heidelberg (2013). https://doi.org/10.1007/978-3-642-37036-6_8

17. Filliâtre, J.C., Pereira, M.: Itérer avec confiance. In: Journées Francophones des Langages Applicatifs. Saint-Malo, France, January 2016. https://hal.inria.fr/hal-01240891

18. Garcia-Molina, H., Ullman, J.D., Widom, J.: Database Systems - The Complete Book, 2nd edn. Pearson Education, Harlow (2009)

19. Gonzalía, C.: Towards a formalisation of relational database theory in constructive type theory. In: Berghammer, R., Möller, B., Struth, G. (eds.) RelMiCS 2003. LNCS, vol. 3051, pp. 137–148. Springer, Heidelberg (2004). https://doi.org/10.1007/978-3-540-24771-5_12

20. Gonzalia, C.: Relations in dependent type theory. Ph.D. thesis, Chalmers Göteborg University (2006)

21. Gu, R., Shao, Z., Chen, H., Wu, X.N., Kim, J., Sjöberg, V., Costanzo, D.: CertiKOS: an extensible architecture for building certified concurrent OS kernels. In: Keeton, K., Roscoe, T. (eds.) 12th USENIX Symposium on Operating Systems Design and Implementation, OSDI 2016, Savannah, GA, USA, 2–4 November 2016, pp. 653–669. USENIX Association (2016). https://www.usenix.org/conference/osdi16/technical-sessions/presentation/gu

22. Karp, R.M.: On-line algorithms versus off-line algorithms: how much is it worth to know the future? In: van Leeuwen, J. (ed.) Algorithms, Software, Architecture - Information Processing 1992, vol. 1, Proceedings of the IFIP 12th World Computer Congress, Madrid, Spain, 7–11 September 1992. IFIP Transactions, vol. A-12, pp. 416–429. North-Holland (1992)

23. Leroy, X.: A formally verified compiler back-end. J. Autom. Reason. **43**(4), 363–446 (2009)

24. Malecha, G., Morrisett, G., Shinnar, A., Wisnesky, R.: Toward a verified relational database management system. In: ACM International Conference on POPL (2010)

25. Ramakrishnan, R., Gehrke, J.: Database Management Systems, 3rd edn. McGraw-Hill, New York (2003)

26. Selinger, P.G., Astrahan, M.M., Chamberlin, D.D., Lorie, R.A., Price, T.G.: Access path selection in a relational database management system. In: Proceedings of the

1979 ACM SIGMOD International Conference on Management of Data, Boston, Massachusetts, 30 May–1 June 1979, pp. 23–34 (1979)

27. The Coq Development Team: The Coq Proof Assistant Reference Manual (2010). http://coq.inria.fr, http://coq.inria.fr

28. The Isabelle Development Team: The Isabelle Interactive Theorem Prover (2010). https://isabelle.in.tum.de/, https://isabelle.in.tum.de/

A Coq Tactic for Equality
Learning in Linear Arithmetic

Sylvain Boulmé[1]([⊠]) and Alexandre Maréchal[2]

[1] Univ. Grenoble-Alpes, CNRS, Grenoble INP, VERIMAG, 38000 Grenoble, France
sylvain.boulme@univ-grenoble-alpes.fr
[2] Sorbonne Université, CNRS, Laboratoire d'Informatique de Paris 6, LIP6,
75005 Paris, France
alexandre.marechal@lip6.fr

Abstract. Coq provides linear arithmetic tactics such as omega or lia. Currently, these tactics either fully prove the current goal in progress, or fail. We propose to improve this behavior: when the goal is not provable in linear arithmetic, we strengthen the hypotheses with new equalities discovered from the linear inequalities. These equalities may help other Coq tactics to discharge the goal. In other words, we apply – in interactive proofs – a seminal idea of SMT-solving: combining tactics by exchanging equalities. The paper describes how we have implemented equality learning in a new Coq tactic, dealing with linear arithmetic over rationals. It also illustrates how this tactic interacts with other Coq tactics.

1 Introduction

Several Coq tactics solve goals containing linear inequalities: omega and lia on integers; fourier or lra on reals and rationals [4,22]. This paper provides yet another tactic for proving such goals on *rationals*. This tactic – called vpl[1] – is built on the top of the *Verified Polyhedra Library* (VPL), a Coq-certified abstract domain of convex polyhedra [14,15]. Its main feature appears when it *cannot prove* the goal. In this case, whereas above tactics fail, our tactic "simplifies" the goal. In particular, it injects as hypotheses a *complete* set of linear equalities that are deduced from the linear inequalities in the context. Then, many Coq tactics – like congruence, field or even auto – can exploit these equalities, even if they cannot deduce them from the initial context by themselves. By simplifying the goal, our tactic both improves the user experience and proof automation.

Let us illustrate this feature on the following – almost trivial – Coq goal, where Qc is the type of rationals on which our tactic applies.

This work was partially supported by the European Research Council under the European Union's Seventh Framework Programme (FP/2007-2013)/ERC Grant Agreement nr. 306595 "STATOR".

[1] Available at http://github.com/VERIMAG-Polyhedra/VplTactic.

J. Avigad and A. Mahboubi (Eds.): ITP 2018, LNCS 10895, pp. 108–125, 2018.
https://doi.org/10.1007/978-3-319-94821-8_7

```
Lemma ex1 (x:Qc) (f:Qc → Qc):   x≤1 → (f x)<(f 1) → x<1.
```

This goal is valid on Qc and Z, but both omega and lia fail on the Z instance without providing any help to the user. Indeed, since this goal contains an uninterpreted function f, it does not fit into the pure linear arithmetic fragment. On the contrary, this goal is proved by two successive calls to the vpl tactic. As detailed below, equality learning plays a crucial role in this proof: the rewriting of a learned equality inside a non-linear term (because under symbol f) is interleaved between deduction steps in linear arithmetic. Of course, such a goal is also provable in Z by SMT-solving tactics: the verit tactic of SMTCoQ [2], the hammer tactic of CoQHAMMER [11], or the one of Besson et al. [5]. However, such SMT-tactics are also *"prove-or-fail"*: they do not simplify the goal when they cannot prove it. On the contrary, our tactic may help users in their interactive proofs, by simplifying goals that do not fully fit into the scope of existing SMT-solving procedures. Note that our tactic does not intend to compete in speed and power with SMT-based procedures. It mainly aims to ease *interactive proofs* which involve linear arithmetic.

In short, this paper provides three contributions. First, we provide a CoQ tactic with equality learning, which seems a new idea in the CoQ community. Second, we provide a simple and efficient algorithm which learns these equalities from conflicts between strict inequalities detected by a linear programming solver. On most cases, it is strictly more efficient than the naive equality learning algorithm previously implemented in the VPL [14]. In particular, our algorithm is cheap when there is no equality to learn. At last, we have implemented this algorithm in an OCAML oracle, able to produce *proof witnesses* for these equalities. The paper partially details this process, and in particular, how the *proof* of the learned equalities is computed in CoQ by reflection from these witnesses. Actually, we believe that our tactic could be easily adapted to other interactive provers, and, in particular, our oracle could be directly reused.

The paper follows a "top-down" presentation. Section 2 describes the specification of the vpl tactic. It also introduces a high-level specification of its underlying oracle. Section 3 illustrates our tactic on a non-trivial example and in particular how it collaborates with other tactics through equality learning. Section 4 details the certificate format produced by our oracle, and how it is applied in our CoQ tactic. At last, Sect. 5 details the algorithm we developed to produce such certificates.

2 Specification of the VPL Tactic

Let us now introduce the specification of the vpl tactic. As mentioned above, the core of the tactic is performed by an oracle programmed in OCAML, and called reduce. This oracle takes as input a *convex polyhedron* P and outputs a *reduced polyhedron* P' such that $P' \Leftrightarrow P$ and such that the *number of constraints* in P' is lower or equal to that of P.

Definition 1 (Polyhedron). *A (convex) polyhedron[2] on $\mathbb{Q}$ is a conjunction of linear (in)equalities of the form $\sum_i a_i x_i \bowtie b$ where a_i, b are constants in $\mathbb{Q}$, where x_i are variables ranging over $\mathbb{Q}$, and where $\bowtie$ represents a binary relation on $\mathbb{Q}$ among $\geq$, $>$ or $=$.*

A polyhedron may be suboptimally written. In particular, one of its constraints may be implied by the others: it is said *redundant* and can be discarded. Moreover, a set of inequalities can imply *implicit* equalities, such as $x = 0$ that can be deduced from $x \geq 0 \wedge -x \geq 0$. This notion of implicit equalities is standard and defined for instance in [19]. Definition 2 characterizes polyhedra without *implicit* equalities.

Definition 2 (Complete set of linear equalities). *Let E be a set of linear equalities and I be a set of linear inequalities. E is said* complete *w.r.t. I if any linear equality deduced from the conjunction $E \wedge I$ can also be deduced from E alone, meaning that I contains no equality, neither implicit nor explicit. Formally, E is complete iff for all linear terms $t_1 \, t_2$,*

$$(E \wedge I \Rightarrow t_1 = t_2) \text{ implies } (E \Rightarrow t_1 = t_2) \tag{1}$$

Definition 3 (Reduced Polyhedron). *A polyhedron P is* reduced *iff it satisfies the following conditions.*

- *If P is unsatisfiable, then P is a single constant constraint like $0 > 0$ or $0 \geq 1$. In other words, its unsatisfiability is checked by one comparison on $\mathbb{Q}$.*
- *Otherwise, P contains no redundant constraint and is syntactically given as a conjunction $E \wedge I$ where polyhedron I contains only inequalities and where polyhedron E is a complete set of equalities w.r.t. I.*

Having a reduced polyhedron ensures that any provable linear equality admits a pure equational proof which ignores the remaining inequalities.

2.1 The Three Steps of the Tactic

Roughly speaking, a COQ goal corresponds to a sequent $\Gamma \vdash T$ where context Γ represents a conjunction of hypotheses and T a conclusion. In other words, this goal is logically interpreted as the meta-implication $\Gamma \Rightarrow T$. The tactic transforms the current goal $\Gamma \vdash T$ through three successive steps.

1. First, constraints are retrieved from the goal: it is equivalently rewritten into $\Gamma', \llbracket P \rrbracket (m) \vdash T'$ where P is a polyhedron and m an assignment of P variables. For example, the `ex1` goal is rewritten as $\llbracket P_1 \rrbracket (m_1) \vdash$ `False`, where

$$P_1 := \; x_1 \leq 1 \wedge x_2 < x_3 \wedge x_1 \geq 1$$
$$m_1 := \{ \; x_1 \mapsto \texttt{x}; \; x_2 \mapsto (\texttt{f } \texttt{x}); \; x_3 \mapsto (\texttt{f } 1) \; \}$$

Here, constraint $x_1 \geq 1$ in P_1 comes from the negation of the initial `ex1` goal `x<1`. Hence, $\llbracket P \rrbracket (m)$ corresponds to a conjunction of inequalities on $\mathbb{Q}$ that

---

[2] Dealing only with convex polyhedra on $\mathbb{Q}$, we often omit the adjective "convex".

are not necessarily linear, because m may assign variables of P to arbitrary Coq terms on $\mathbb{Q}$. Actually, $[\![P]\!](m)$ contains at least all (in)equalities on $\mathbb{Q}$ that appear as hypotheses of Γ. Moreover, if T is an inequality on $\mathbb{Q}$, then an inequality equivalent to $\neg T$ appears in $[\![P]\!](m)$ and T' is proposition `False`.[3] This step is traditionally called *reification* in Coq tactics.

2. Second, polyhedron P is reduced. In other words, the goal is equivalently rewritten into $\Gamma', [\![P']\!](m) \vdash T'$ where P' is the *reduced polyhedron* computed from P by our `reduce` oracle. For instance, polyhedron P_1 found above is reduced into

$$P_1' := x_1 = 1 \wedge x_2 < x_3$$

3. At last, if P' is unsatisfiable, then so is $[\![P']\!](m)$, and the goal is finally discharged. Otherwise, given E the complete set of equalities in P', equalities of $[\![E]\!](m)$ are rewritten in the goal. For example, on the `ex1` goal, our tactic rewrites the learned equality "`x=1`" into the remaining hypothesis. In summary, a first call to the `vpl` tactic transforms the `ex1` goal into

$$\texttt{x=1, (f 1)<(f 1)} \vdash \texttt{False}$$

A second call to `vpl` detects that hypothesis `(f 1)<(f 1)` is unsatisfiable and finally proves the goal.

In the description above, we claim that our transformations on the goals are equivalences. This provides a guarantee to the user: the tactic can always be applied on the goal, without loss of information. However, in order to make the Coq proof checker accept our transformations, we only need to prove implications, as detailed in the next paragraph.

2.2 The Proof Built by the Tactic

The tactic mainly proves the following two implications which are verified by the Coq kernel:

$$\Gamma', [\![P]\!](m) \vdash T' \;\Rightarrow\; \Gamma \vdash T \tag{2}$$

$$\forall m, [\![P]\!](m) \;\Rightarrow\; [\![P']\!](m) \tag{3}$$

Semantics of polyhedron $[\![.]\!]$ is encoded as a Coq function, using binary integers to encode variables of polyhedra. After simple propositional rewritings in the initial goal $\Gamma \vdash T$, an Ocaml oracle provides m and P to the Coq kernel, which simply computes $[\![P]\!](m)$ and checks that it is syntactically equal to the expected part of the context. Hence, verifying implication (2) is mainly syntactical.

For implication (3), our `reduce` oracle actually produces a Coq AST, that represents a *proof witness* allowing to build each constraint of P' as a nonnegative linear combination of P constraints. Indeed, such a combination is necessarily a logical consequence of P. In practice, this proof witness is a value of a Coq inductive type. A Coq function called `reduceRun` takes as input a polyhedron

---

[3] Here, $T \Leftrightarrow (\neg T \Rightarrow \texttt{False})$ because comparisons on $\mathbb{Q}$ are decidable.

P and its associated witness, and computes P'. A COQ theorem ensures that any result of `reduceRun` satisfies implication (3). Thus, this implication is ensured by construction, while – for the last step of the tactic described above – the COQ kernel computes P' by applying `reduceRun`.

3   Using the vpl Tactic

Combining solvers by exchanging equalities is one of the basis of modern SMT-solving, as pioneered by approaches of Nelson-Oppen [17,18] and Shostak [20]. This section illustrates how equality learning in an interactive prover mimics such equality exchange, in order to combine independent tactics. While much less automatic than standard SMT-solving, our approach provides opportunities for the user to compensate by "hand" for the weaknesses of a given tactic.

The main aspects of the `vpl` tactic are illustrated on the following single goal. This goal contains two uninterpreted functions f and g such that f domain and g codomain are the same uninterpreted type A. As we will see below, in order to prove this goal, we need to use its last hypothesis – of the form "$g(..)\ <>\ g(13)$" – by combining equational reasoning on g and on Qc field. Of course, we also need linear arithmetic on Qc order.

```
Lemma ex2 (A:Type) (f:A → Qc) (g:Qc → A) (v1 v2 v3 v4:Qc) :
   6*v1 - v2 - 10*v3 + 7*(f(g v1)+1) ≤ -1
   → 3*(f(g v1)-2*v3)+4 ≥ v2-4*v1
   → 8*v1 - 3*v2 - 4*v3 - f(g v1) ≤ 2
   → 11*v1 - 4*v2 > 3
   → v3 > -1
   → v4 ≥ 0
   → g((11-v2+13*v4)/(v3+v4)) <> g(13)
   → 3 + 4*v2 + 5*v3 + f(g v1) > 11*v1.
```

The `vpl` tactic reduces this goal to the equivalent one given below (where typing of variables is omitted).

```
H5: g((11-(11-13*v3)+13*v4)/(v3+v4))=(g 13)  → False
vpl:  v1 = 4-4*v3
vpl0: v2 = 11-13*v3
vpl1: f(g(4-4*v3)) = -3+3*v3
-------------------------------------- (1/1)
0 ≤ v4  →  (3#8) < v3  → False
```

Here, three equations `vpl`, `vpl0` and `vpl1` have been learned from the goal. Two irredundant inequalities remain in the hypotheses of the conclusion – where $(3\#8)$ is the COQ notation for $\frac{3}{8}$. The bound $v3 > -1$ has disappeared because it is implied by $(3\#8) < v3$. By taking $v3 = 1$, we can build a model satisfying all the hypotheses of the goal – including $(3\#8) < v3$ – except `H5`. Thus, using `H5` is necessary to prove `False`.

Actually, we provide another tactic which automatically proves the remaining goal. This tactic (called `vpl_post`) combines equational reasoning on Qc field with a bit of congruence.[4] Let us detail how it works on this example. First, in backward reasoning, hypothesis H5 is applied to eliminate `False` from the conclusion. We get the following conclusion (where previous hypotheses have been omitted).

```
----------------------------------------(1/1)
g((11-(11-13*v3)+13*v4)/(v3+v4))=(g 13)
```

Here, backward congruence reasoning reduces this conclusion to

```
----------------------------------------(1/1)
(11-(11-13*v3)+13*v4)/(v3+v4)=13
```

Now, the `field` tactic reduces the conclusion to

```
----------------------------------------(1/1)
v3+v4 <> 0
```

Indeed, the `field` tactic mainly applies ring rewritings on Qc while generating subgoals for checking that denominators are not zero. Here, because we have a linear denominator, we discharge the remaining goal using the `vpl` tactic again. Indeed, it gets the following polyhedron in hypotheses – which is unsatisfiable.

$$v4 \geq 0 \quad \wedge \quad v3 > \frac{3}{8} \quad \wedge \quad v3 + v4 = 0$$

Let us remark that lemma `ex2` is also valid when the codomain of `f` and types of variables `v1 ... v4` are restricted to $\mathbb{Z}$ and operator "/" means the Euclidean division. However, both `omega` and `lia` fail on this goal without providing any help to the user. This is also the case of the `verit` tactic of SMTCoQ because it deals with "/" as a non-interpreted symbol and can only deal with uninterpreted types A providing a decidable equality. By assuming a decidable equality on type A and by turning the hypothesis involving "/" into "g((11-v2+13*v4)) <> g(13*(v3+v4))", we get a slightly weaker version of `ex2` goal which is proved by `verit`. CoQHAMMER is currently not designed to solve such a complex arithmetic goal [11].

This illustrates that our approach is complementary to SMT-solving: it provides less automation than SMT-solving, but it may still help to progress in an interactive proof when SMT-solvers fail.

4   The Witness Format in the Tactic

Section 4.3 below presents our proof witness format in CoQ to build a reduced polyhedron P' as a logical consequence of P. It also details the implementation of `reduceRun` and its correctness property, formalizing property (3) given in Sect. 2.2. In preliminaries, Sect. 4.1 recalls the Farkas operations of the VPL, at the basis of our proof witness format, itself illustrated in Sect. 4.2.

---

[4] It is currently implemented on the top of `auto` with a dedicated basis of lemmas.

4.1   Certified Farkas Operations

The tactic uses the linear constraints defined in the VPL [13], that we recall here. Type `var` is the type of variables in polyhedra. Actually, it is simply defined as type `positive`, the positive integers of COQ. Module `Cstr` provides an efficient representation for linear constraints on `Qc`, the COQ type for $\mathbb{Q}$. Type `Cstr.t` handles constraints of the form "$t \bowtie 0$" where t is a linear term and $\bowtie \in \{=, \geq$, $>\}$. Hence, each input constraint "$t_1 \bowtie t_2$" will be encoded as "$t_1 - t_2 \bowtie 0$". Linear terms are themselves encoded as radix trees over `positive` with values in `Qc`.

The semantics of `Cstr.t` constraints is given by predicate (`Cstr.sat c m`), expressing that model `m` of type `var → Qc` satisfies constraint `c`. Module `Cstr` provides also the following operations

Add: $(t_1 \bowtie_1 0) + (t_2 \bowtie_2 0) \triangleq (t_1 + t_2) \bowtie 0$ where $\bowtie \triangleq \max(\bowtie_1, \bowtie_2)$ for the total increasing order induced by the sequence $=, \geq, >$;

Mul: $n \cdot (t \bowtie 0) \triangleq (n \cdot t) \bowtie 0$ assuming $n \in \mathbb{Q}$ and, if $\bowtie \in \{\geq, >\}$ then $n \geq 0$;

Merge: $(t \geq 0) \ \& \ (-t \geq 0) \triangleq t = 0$.

It is easy to prove that each of these operations returns a constraint that is satisfied by the models of its inputs. For example, given constraints `c1` and `c2` and a model `m` such that (`sat c1 m`) and (`sat c2 m`), then (`sat (c1+c2) m`) holds. When invoked on a wrong precondition, these operations actually return "$0 = 0$" which is satisfied by any model. Still, this precondition violation only appears if there is a bug in the `reduce` oracle. These operations are called *Farkas operations*, in reference to Farkas' lemma recalled on page 11.

In the following, we actually handle each constraint with a proof that it satisfies a given set `s` of models (encoded here by its characteristic function). The type of such a constraint is (`wcstr s`), as defined below.

```
Record wcstr (s: (var → Qc) → Prop) := {
  rep: Cstr.t;
  rep_sat: ∀ m, s m → Cstr.sat rep m
}.
```

Hence, all the Farkas operations are actually lifted to type (`wcstr s`), for all `s`.

4.2   Example of Proof Witness

We introduce our syntax for proof witnesses on Fig. 1. Our oracle detects that P is satisfiable, and thus returns the "proof script" of Fig. 1. This script instructs `reduceRun` to produce P' from P. By construction, we have $P \Rightarrow P'$.

This script has three parts. In the first part – from line 1 to 5 – the script considers each constraint of P and binds it to a name, or skips it. For instance, $x_1 \geq -10$ is skipped because it is redundant: it is implied by P' and thus not necessary to build P' from P. In the second part – from line 6 to 9 – the script builds intermediate constraints: their value is detailed on the right hand-side of

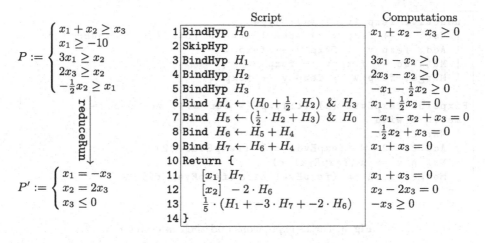

Fig. 1. Example of a proof script and its interpretation by reduceRun

```
Definition pedra := list Cstr.t.
Definition [[l]] m := List.Forall (fun c ⇒ Cstr.sat c m) l.
Definition answ (o: option pedra) m :=
  match o with
  | Some l ⇒ [[l]] m
  | None ⇒ False
  end.

Definition reduceRun(l:pedra)(p:∀ v,script v): option pedra
  := scriptEval (s:=[[l]]) (p _) l (*...*).
Lemma reduceRun_correct l m p:[[l]] m → answ (reduceRun l p) m.
```

Fig. 2. Definition of reduceRun and its Correctness

the figure. Each of these constraints is bound to a name. Hence, when a constraint
– like H_4 – is used several times, we avoid a duplication of its computation.

In the last part – from line 10 to 14 – the script returns the constraints of
P'. As further detailed in Sect. 5, each equation defines one variable in terms of
the others. For each equation, this variable is explicitly given between brackets
"[.]" in the script of Fig. 1, such as x_1 at line 11 and x_2 at line 12. This instructs
reduceRun to rewrite equations in the form "$x = t$".

4.3 The HOAS of Proof Witnesses

Our reduceRun function and its correctness are defined in Coq, as shown on
Fig. 2. The input polyhedron is given as a list of constraints l of type pedra.
The output is given as type (option pedra) where a None value corresponds
to the case where l is unsatisfiable.

```
Inductive fexp (v: Type): Type :=
 | Var: v → fexp v (* name bound to [Bind] or [BindHyp] *)
 | Add: fexp v → fexp v → fexp v
 | Mul: Qc → fexp v → fexp v
 | Merge: fexp v → fexp v → fexp v.

Fixpoint fexpEval {s} (c:fexp (wcstr s)): wcstr s :=
 match c with
 | Var c ⇒ c
 | Add c1 c2 ⇒ (fexpEval c1)+(fexpEval c2)
 | Mul n c ⇒ n·(fexpEval c)
 | Merge c1 c2 ⇒ (fexpEval c1)&(fexpEval c2)
 end.
```

Fig. 3. Farkas expressions and their interpreter

Given a value l: pedra, its semantics – noted $[\![l]\!]$ – is a predicate of type $(var \to Qc) \to Prop$ which is defined from Cstr.sat. This semantics is extended to type (option pedra) by the predicate answ. Property (3) of page 4 is hence formalized by lemma reduceRun_correct with a minor improvement: when the input polyhedron is unsatisfiable, a proof of False is directly generated.

The proof witness in input of reduceRun is a value of type $\forall$ v, script v. Here, script – defined at Fig. 5 page 10 – is the type of a Higher-Order Abstract Syntax (HOAS) parameterized by the type v of variables [9]. A HOAS avoids the need to handle explicit variable substitutions when interpreting binders: those are encoded as functions, and variable substitution is delegated to the COQ engine.[5] The universal quantification over v avoids exposing the representation of v – used by reduceRun – in the proof witness p.

The bottom level of our HOAS syntax is given by type fexp defined at Fig. 3 and representing "Farkas expressions". Each constructor in this type corresponds to a Farkas operation, except constructor Var that represents a constraint name which is bound to a Bind or a BindHyp binder (see Fig. 1). The function fexpEval computes any such Farkas expression c into a constraint of type (wcstr s) – for some given s – where type v is itself identified with type (wcstr s).

Farkas expressions are combined in order to compute polyhedra. This is expressed through "polyhedral expressions" of type pexp on Fig. 4 which are

---

[5] For a prototype like our tactic, such a HOAS has mainly the advantage of simplicity: it avoids formalizing in COQ the use of a substitution mechanism. The impact on the efficiency at runtime remains unclear. On one side, typechecking a higher-order term is more expensive than typechecking a first-order term. On the other side, implementing an efficient substitution mechanism in COQ is currently not straightforward: purely functional data-structures induce a non-negligible logarithmic factor over imperative ones. Imperative arrays with a purely functional API have precisely been introduced by [3] in an experimental version of COQ with this motivation. But this extension is not yet integrated into the stable release of COQ.

```
Inductive pexp (v: Type): Type :=
 | Bind: fexp v → (v → pexp v) → pexp v
 | Contrad: (fexp v) → pexp v
 | Return: list ((option var)*(fexp v)) → pexp v.

Fixpoint pexpEval {s} (p:pexp (wcstr s)): option pedra :=
 match p with
 | Bind c bp ⇒ pexpEval (bp (fexpEval c))
 | Contrad c ⇒ contrad c
 | Return l ⇒ Some (ret l nil)
 end.

Lemma pexpEval_correct s (p:pexp (wcstr s)) m:
 s m → answ (pexpEval p) m.
```

Fig. 4. Polyhedral computations and their interpreter

```
Inductive script (v: Type): Type :=
 | SkipHyp: script v → script v
 | BindHyp: (v → script v) → script v
 | Run: (pexp v) → script v.

Fixpoint scriptEval {s} (p: script(wcstr s)) (l: pedra):
 (∀ m, s m → ⟦l⟧ m) →option pedra := (*...*)

Lemma scriptEval_correct s (p:script(wcstr s)) l m:
 (∀ m, s m → ⟦l⟧ m) → s m →  answ (scriptEval p l) m.
```

Fig. 5. Script expressions and their interpreter

computed by `pexpEval` into (`option pedra`) values. Type `pexp` has 3 constructors. First, constructor (`Bind c (fun H ⇒ p)`) is a higher-order binder of our HOAS: it computes an intermediate Farkas expression `c` and stores the result in a variable `H` bound in the polyhedral expression `p`. Second, constructor (`Contrad c`) returns an *a priori* unsatisfiable constant constraint, which is verified by function `contrad` in `pexpEval`. At last, constructor (`Return l`) returns an *a priori* satisfiable reduced polyhedron, which is encoded as a list of Farkas expressions associated to an optional variable of type `var` (indicating a variable defined by an equation, see example of Fig. 1).

Finally, a witness of type `script` first starts by naming useful constraints of the input (given as a value `l: pedra`) and then runs a polyhedral expression in this naming context. This semantics is given by function `scriptEval` specified at Fig. 5. On a script (`SkipHyp p'`), interpreter `scriptEval` simply skips the first constraint by running recursively (`scriptEval p' (List.tl l)`). Similarly, on a script (`BindHyp (fun H ⇒ p')`), it pops the first constraint of `l` in variable `H` and then runs itself on `p'`. Technically, function `scriptEval` assumes the following precondition on polyhedron `l`: it must satisfy all models `m` characterized by `s`. As shown on Fig. 2, (`reduceRun l p`) is a simple instance of (`scriptEval (p (wcstr s)) l`) where `s:=[[l]]`. Hence, this precondition is trivially satisfied.

5   The Reduction Algorithm

The specification of the `reduce` oracle is given in introduction of the paper: it transforms a polyhedron P into a reduced polyhedron P' with a smaller number of constraints and such that $P' \leftrightarrow P$. Sections 5.3 and 5.4 describe our implementation. In preliminaries, Sect. 5.1 gives a sufficient condition, through Lemma 2, for a polyhedron to be reduced. This condition lets us learn equalities from conflicts between strict inequalities as detailed in Sect. 5.2. In our proofs and algorithms, we only handle linear constraints in the restricted form "$t \bowtie 0$". But, for readability, our examples use the arbitrary form "$t_1 \bowtie t_2$".

5.1   A Refined Specification of the Reduction

Definition 4 (Echelon Polyhedron). *An echelon polyhedron is written as a conjunction $E \wedge I$ where polyhedron I contains only inequalities and where polyhedron E is written "$\bigwedge_{i \in \{1,...,k\}} x_i - t_i = 0$" such that each x_i is a variable and each t_i is a linear term, and such that the following two conditions are satisfied. First, no variable x_i appears in polyhedron I. Second, for all integers $i, j \in \{1, \ldots, k\}$ with $i \leq j$ then x_i does not appear in t_j.*

Intuitively, in such a polyhedron, each equation "$x_i - t_i = 0$" actually *defines* variable x_i as t_i. As a consequence, $E \wedge I$ is satisfiable iff I is satisfiable.

   We recall below the Farkas' lemma [10,12] which reduces the unsatisfiability of a polyhedron to the one of a constant constraint, like $0 > 0$. The unsatisfiability of such a constraint is checked by a simple comparison on $\mathbb{Q}$.

Lemma 1 (Farkas). *Let I be a polyhedron containing only inequalities. I is unsatisfiable if and only if there is an unsatisfiable constraint $-\lambda \bowtie 0$, computable from a nonnegative linear combination of constraints of I (i.e. using operators "$+$" and "$\cdot$" defined in Sect. 4.1), and such that $\bowtie \in \{\geq, >\}$ and $\lambda \in \mathbb{Q}^+$.*

   From Farkas' lemma, we derive the following standard corollary which reduces the verification of an implication $I \Rightarrow t \geq 0$ to the verification of a syntactic equality between linear terms.

Corollary 1 (Implication Witness). *Let t be a linear term and let I be a satisfiable polyhedron written $\bigwedge_{j \in \{1,...,k\}} t_j \bowtie_j 0$ with $\bowtie_j \in \{\geq, >\}$.*
   If $I \Rightarrow t \geq 0$ then there are $k + 1$ nonnegative rationals $(\lambda_j)_{j \in \{0,...,k\}}$ such that $t = \lambda_0 + \Sigma_{j \in \{1,...,k\}} \lambda_j t_j$.

   In the following, we say that the nonnegative coefficients $(\lambda_j)_{j \in \{0,...,k\}}$ define a "*Farkas combination of t in terms of I*".

Definition 5 (Strict Version of Inequalities). *Let I be a polyhedron with only inequalities. We note $I^>$ the polyhedron obtained from I by replacing each non-strict inequality "$t \geq 0$" by its strict version "$t > 0$". Strict inequalities of I remain unchanged in $I^>$.*

Geometrically, polyhedron $I^>$ is the interior of polyhedron I. Hence if $I^>$ is satisfiable (i.e. the interior of I is non empty), then polyhedron I does not fit inside a hyperplane. Lemma 2 formalizes this geometrical intuition as a consequence of Farkas' lemma. Its proof invokes the following corollary of Farkas' lemma, which is really at the basis of our equality learning algorithm.

Corollary 2 (Witness of Empty Interior). *Let us consider a* satisfiable *polyhedron I written $\bigwedge_{j \in \{1,\ldots,k\}} t_j \bowtie_j 0$ with $\bowtie_j \in \{\geq,>\}$. Then, $I^>$ is unsatisfiable if and only if there exists k nonnegative rationals $(\lambda_j)_{j \in \{1,\ldots,k\}} \in \mathbb{Q}^+$ such that $\Sigma_{j \in \{1,\ldots,k\}} \lambda_j t_j = 0$ and $\exists j \in \{1,\ldots,k\}, \lambda_j > 0$.*

Proof.
$\Leftarrow$: Suppose k nonnegative rationals $(\lambda_j)_{j \in \{1,\ldots,k\}}$ such that $\Sigma_{j \in \{1,\ldots,k\}} \lambda_j t_j = 0$ and some index j such that $\lambda_j > 0$. It means that there is a Farkas combination of $0 > 0$ in terms of $I^>$. Thus by Farkas' lemma, $I^>$ is unsatisfiable.
$\Rightarrow$: Let us assume that $I^>$ is unsatisfiable. By Farkas' lemma, there exists an unsatisfiable constant constraint $-\lambda \bowtie 0$, where $-\lambda = \Sigma_{j \in \{1,\ldots,k\}} \lambda_j t_j$, with all $\lambda_j \in \mathbb{Q}^+$, and such that there exists some j with $\lambda_j > 0$. Let m be an assignment of I variables such that $[\![I]\!] m$. By definition, we have $[\![\Sigma_{j \in \{1,\ldots,k\}} \lambda_j t_j]\!] m = \lambda'$ with $\lambda' \in \mathbb{Q}^+$. Thus, $-\lambda = \lambda' = 0$.

Lemma 2 (Completeness from Strict Satisfiability). *Let us assume an echelon polyhedron $E \wedge I$ without redundant constraints, and such that $I^>$ is satisfiable. Then, $E \wedge I$ is a reduced polyhedron.*

Proof. Let us prove property (1) of Definition 2, i.e. that E is complete w.r.t. I. Because $t_1 = t_2 \Leftrightarrow t_1 - t_2 = 0$, without loss of generality, we only prove property (1) in the case where $t_2 = 0$ and t_1 is an arbitrary linear term t. Let t be a linear term such that $E \wedge I \Rightarrow t = 0$.

In particular, $E \wedge I \Rightarrow t \geq 0$. By Corollary 1, there are $k' + 1$ nonnegative rationals $(\lambda_j)_{j \in \{0,\ldots,k'\}}$ such that $t = \lambda_0 + \Sigma_{j \in \{1,\ldots,k'\}} \lambda_j t_j$ where E is written $\bigwedge_{j \in \{1,\ldots,k\}} t_j = 0$ and I is written $\bigwedge_{j \in \{k+1,\ldots,k'\}} t_j \bowtie_j 0$. Suppose that there exists $j \in \{k+1,\ldots,k'\}$, such that $\lambda_j \neq 0$. Since $I^>$ is satisfiable, by Corollary 2, we deduce that $\Sigma_{j \in \{k+1,\ldots,k'\}} \lambda_j t_j \neq 0$. Thus, we have $E \wedge I^> \Rightarrow t > 0$ with $E \wedge I^>$ satisfiable. This contradicts the initial hypothesis $E \wedge I \Rightarrow t = 0$. Thus, $t = \lambda_0 + \Sigma_{j \in \{1,\ldots,k\}} \lambda_j t_j$ which proves $E \Rightarrow t \geq 0$.
A similar reasoning from $E \wedge I \Rightarrow -t \geq 0$ finishes the proof that $E \Rightarrow t = 0$.

Lemma 2 gives a strategy to implement the **reduce** oracle. If the input polyhedron P is satisfiable, then try to rewrite P as an echelon polyhedron $E \wedge I$ where $I^>$ is satisfiable. The next step is to see that from an echelon polyhedron $E \wedge I$ where $I^>$ is unsatisfiable, we can *learn* new equalities from a subset of $I^>$

inequalities that is unsatisfiable. The inequalities in such a subset are said "*in conflict*". The Farkas witness proving the conflict is used to deduce new equalities from I. This principle can be viewed as an instance of "*conflict driven clause learning*" – at the heart of modern DPLL procedures [21].

5.2 Building Equality Witnesses from Conflicts

Consider a satisfiable set of inequalities I, from which we wish to extract implicit equalities. First, let us build $I^>$ the strict version of I as described in Definition 5. Then, an oracle runs the simplex algorithm to decide whether $I^>$ is satisfiable. If so, then we are done: there is no implicit equality to find in I. Otherwise, by Corollary 2, the oracle finds that the unsatisfiable constraint $0 > 0$ can be written $\Sigma_{j \in J} \lambda_j t_j > 0$ where for all $j \in J$, $\lambda_j > 0$ and $(t_j > 0) \in I^>$. Since $\bigwedge_{j \in J} t_j > 0$ is unsatisfiable, we can learn that $\bigwedge_{j \in J} t_j = 0$. Indeed, since $\Sigma_{j \in J} \lambda_j t_j = 0$ (by Corollary 2) and $\forall j \in J$, $\lambda_j > 0$, then each term t_j of this sum must be 0. Thus, $\forall j \in J$, $t_j = 0$.

Let us now detail our algorithm to compute equality witnesses. Let I be a satisfiable inequality set such that $I^>$ is unsatisfiable. The oracle returns a witness combining $n+1$ constraints of $I^>$ (for $n \geq 1$) that implies a contradiction:

$$\sum_{i=1}^{n+1} \lambda_i \cdot I_i^> \text{ where } \lambda_i > 0$$

By Corollary 2, this witness represents a contradictory constraint $0 > 0$. Moreover, each inequality I_i is non-strict (otherwise, I would be unsatisfiable). We can thus turn each inequality I_i into an equality written $I_i^=$ – proved by

$$I_i \ \& \ \frac{1}{\lambda_i} \cdot \sum_{\substack{j \in \{1...n+1\} \\ j \neq i}} \lambda_j \cdot I_j$$

Hence, each equality $I_i^=$ is proved by combining $n + 1$ constraints. Proving $(I_i^=)_{i \in \{1,...,n+1\}}$ in this naive approach combines $\Theta(n^2)$ constraints.

We rather propose a more symmetric way to build equality witnesses which leads to a simple linear algorithm. Actually, we build a system of n equalities noted $(E_i)_{i \in \{1,...,n\}}$, where – for $i \in \{1, ..., n\}$ – each E_i corresponds to the unsatisfiability witness where the i-th "+" has been replaced by a "&":

$$\left(\sum_{j=1}^{i} \lambda_j \cdot I_j \right) \ \& \ \left(\sum_{j=i+1}^{n+1} \lambda_j \cdot I_j \right)$$

This system of equations is proved equivalent to system $(I_i^=)_{i \in \{1,...,n+1\}}$ thanks to the following correspondence.

$$\begin{cases} I_1^= = \frac{1}{\lambda_1} \cdot E_1 \\ I_{n+1}^= = -\frac{1}{\lambda_n} \cdot E_n \\ \text{for } i \in \{2, ..., n\}, I_i^= = \frac{1}{\lambda_i} \cdot (E_i - E_{i-1}) \end{cases}$$

This also shows that one equality $I_i^=$ is redundant, because $(I_i^=)_{i \in \{1,...,n+1\}}$ contains one more equality than $(E_i)_{i \in \{1,...,n\}}$.

In order to use a linear number of combinations, we build $(E_i)_{i \in \{1,...,n\}}$ thanks to two lists of intermediate constraints $(A_i)_{i \in \{1,...,n\}}$ and $(B_i)_{i \in \{2,...,n+1\}}$ defined by

$$\begin{cases} A_1 := \lambda_1 \cdot I_1 & \text{for } i \text{ from 2 up to } n, \ A_i := A_{i-1} + \lambda_i \cdot I_i \\ B_{n+1} := \lambda_{n+1} \cdot I_{n+1} & \text{for } i \text{ from } n \text{ down to 2, } B_i := B_{i+1} + \lambda_i \cdot I_i \end{cases}$$

Then, we build $E_i := A_i$ & B_{i+1} for $i \in \{1, \ldots, n\}$.

5.3 Illustration on the Running Example

Let us detail how to compute the reduced form of polyhedron P from Fig. 1.

$$P := \left\{ I_1 : x_1 + x_2 \geq x_3, \ I_2 : x_1 \geq -10, \ I_3 : 3x_1 \geq x_2, \ I_4 : 2x_3 \geq x_2, \ I_5 : -\frac{1}{2}x_2 \geq x_1 \right\}$$

P is a satisfiable set of inequalities. Thus, we first extract a complete set of equalities E from constraints of P by applying the previous ideas. We ask a Linear Programming (LP) solver for a point satisfying $P^>$, the strict version of P. Because there is no such point, the solver returns the unsatisfiability witness $I_1^> + \frac{1}{2} \cdot I_4^> + I_5^>$ (which reduces to $0 > 0$). By building the two sequences (A_i) and (B_i) defined previously, we obtain the two equalities

$$E_1 : x_1 + x_2 = x_3 \textbf{ proved by } \underbrace{(x_1 + x_2 \geq x_3)}_{A_1:\ I_1} \ \& \ \underbrace{(x_3 \geq x_1 + x_2)}_{B_2:\ \frac{1}{2}\cdot I_4 + I_5}$$

$$E_2 : x_1 = -\frac{1}{2}x_2 \textbf{ proved by } \underbrace{(x_1 \geq -\frac{1}{2}x_2)}_{A_2:\ I_1 + \frac{1}{2}\cdot I_4} \ \& \ \underbrace{(-\frac{1}{2}x_2 \geq x_1)}_{B_3:\ I_5}$$

Thus, P is rewritten into $E \wedge I$ with

$$E := \left\{ E_1 : x_1 + x_2 = x_3, \ E_2 : x_1 = -\frac{1}{2}x_2 \right\}, I := \left\{ I_2 : x_1 \geq 10, \ I_3 : 3x_1 \geq x_2 \right\}$$

To be reduced, the polyhedron must be in echelon form, as explained in Definition 4. This implies that each equality of E must have the form $x_i - t_i = 0$, and each such x_i must not appear in I. Here, let us consider that E_1 defines x_2: we rewrite E_1 into $x_2 - (x_3 - x_1) = 0$. Then, x_2 is eliminated from E_2, leading to $E_2' : x_1 + x_3 = 0$. In practice, we go one step further by rewriting x_1 (using its definition in E_2') into E_1 to get a *reduced* echelon system E' of equalities:

$$E' := \{ E_1' : x_2 - 2 \cdot x_3 = 0, \ E_2' : x_1 + x_3 = 0 \}$$

Moreover, the variables defined in E' (i.e. x_1 and x_2) are eliminated from I, which is rewritten into

$$I' := \{ I_2' : -x_3 \geq -10, \ I_3' : -x_3 \geq 0 \}$$

The last step is to detect that I_2' is redundant w.r.t. I_3' with a process which is indicated in the next section.

```
function reduce(E∧I) =
    (E,I) ← echelon(E,I)
    match is_sat(I) with
    | Unsat(λ) -> return Contrad(λ^T·I)
    | Sat(_) ->
        loop
            match is_sat(I^>) with
            | Unsat(λ) ->
                (E',I') ← learn(I,λ)
                (E,I) ← echelon(E∧E',I')
            | Sat(m) ->
                I ← rm_redundancies(I,m)
                return Reduced(E∧I)
```

Fig. 6. Pseudo-code of the reduce oracle

5.4   Description of the Algorithm

The pseudo-code of Fig. 6 describes the reduce algorithm. For simplicity, the construction of proof witnesses is omitted from the pseudo-code. To summarize, the result of reduce is either "Contrad(c)" where c is a contradictory constraint or "Reduced(P')" where P' is a satisfiable reduced polyhedron. The input polyhedron is assumed to be given in the form $E \wedge I$, where E contains only equalities and I contains only inequalities. First, polyhedron $E \wedge I$ is echeloned: function echelon returns a new system $E \wedge I$ where E is an echelon system of equalities without redundancies (they have been detected as $0 = 0$ during echeloning and removed) and without contradiction (they have been detected as $1 = 0$ during echeloning). Second, the satisfiability of I is tested by function is_sat. If is_sat returns "Unsat (λ)", then λ is a Farkas witness allowing to return a contradictory constant constraint written $\lambda^T \cdot I$. Otherwise, I is satisfiable and reduce enters into a loop to learn all implicit equalities.

At each step of the loop, the satisfiability of $I^>$ is tested. If is_sat returns "Unsat (λ)", then a new set E' of equalities is learned from λ and I' contains the inequalities of I that do not appear in the conflict. After echeloning the new system, the loop continues.

Otherwise, is_sat returns "Sat(m)" where m is a model of $I^>$. Geometrically, m is a point in the interior of polyhedron I. Point m helps rm_redundancies to detect and remove redundant constraints of I, by a ray-tracing method described in [16]. At last, reduce returns $E \wedge I$, which is a satisfiable reduced polyhedron because of Lemma 2.

Variant. In a variant of this algorithm, we avoid to test the satisfiability of I before entering the loop (i.e. the first step of the algorithm). Indeed, the satisfiability of I can be directly deduced from the witness returned by is_sat($I^>$). If the combination of the linear terms induced by the witness gives a negative number instead of 0, it means that I is unsatisfiable. However, we could make several loop executions before finding that I is unsatisfiable: polyhedron I may contain several implicit equalities which do not imply the unsatisfiability of I and which may be discovered first. We do not know which version is the most efficient one. It probably differs according to applications.

6   Conclusion and Related Works

This paper describes a Coq tactic that learns equalities from a set of linear rational inequalities. It is less powerful than Coq SMT tactics [2,5,11] and than the famous sledgehammer of ISABELLE [6,7]. But, it may help users to progress on goals that do not exactly fit into the scope of existing SMT-solving procedures.

This tactic uses a simple algorithm – implemented in the new VPL [15] – that follows a kind of conflict driven clause learning. This equality learning algorithm only relies on an efficient SAT-solver on inequalities able to generate nonnegativity witnesses. Hence, we may hope to generalize it to polyhedra on $\mathbb{Z}$.

The initial implementation of the VPL [14] also reduces polyhedra as defined in Definition 3. Its equality learning is more naive: for each inequality $t \geq 0$ of the current (satisfiable) inequalities I, the algorithm checks whether $I \wedge t > 0$ is satisfiable. If not, equality $t = 0$ is learned. In other words, each learned equality derives from one satisfiability test. Our new algorithm is more efficient, since it may learn several equalities from a single satisfiability test. Moreover, when there is no equality to learn, this algorithm performs only one satisfiability test.

We have implemented this algorithm in an OCAML oracle, able to produce *proof witnesses* for these equalities. The format of these witnesses is very similar to the one of micromega [4], except that it provides a bind operator which avoids duplication of computations (induced by rewriting of learned equalities). In the core of our oracle, the production of these witnesses follows a lightweight, safe and evolutive design, called *polymorphic* LCF *style* [8]. This style makes the implementation of VPL oracles much simpler than in the previous VPL implementation. Our implementation thus illustrates how to instantiate "polymorphic witnesses" of polymorphic LCF style in order to generate Coq abstract syntax trees, and thus to prove the equalities in Coq by computational reflection.

The previous Coq frontend of the VPL [13] would also allow to perform such proofs by reflection. Here, we believe that the HOAS approach followed in Sect. 4.3 is much simpler and more efficient than this previous implementation (where substitutions were very inefficiently encoded with lists of constraints).

Our tactic is still a prototype. Additional works are required to make it robust in interactive proofs. For example, the user may need to stop the tactic before that the rewritings of the learned equalities are performed, for instance when some rewriting interferes with dependent types. Currently, the user can invoke instead a subtactic vpl_reduce, and apply these rewritings by "hand". The maintainability of such user scripts thus depends on the stability of the generated equalities and their order w.r.t. small changes in the input goal. However, we have not yet investigated these stability issues. A first step toward stability would be to make our tactic idempotent by keeping the goal unchanged on a already reduced polyhedron.

Another library, called COQ-POLYHEDRA [1], now formalizes a large part of the convex polyhedra theory without depending on external oracles. Our work is based on the VPL, because it wraps efficient external solvers [14]. In particular, computations in VPL oracles mix floating-points and GMP numbers, which are far more efficient than COQ numbers. However, the usability of the VPL would probably increase by being linked to such a general library.

Acknowledgements. We thank anonymous referees for their useful feedback on a preliminary version of this paper.

References

1. Allamigeon, X., Katz, R.D.: A formalization of convex polyhedra based on the simplex method. In: Ayala-Rincón, M., Muñoz, C.A. (eds.) ITP 2017. LNCS, vol. 10499, pp. 28–45. Springer, Cham (2017). https://doi.org/10.1007/978-3-319-66107-0_3
2. Armand, M., Faure, G., Grégoire, B., Keller, C., Théry, L., Werner, B.: A modular integration of SAT/SMT solvers to Coq through proof witnesses. In: Jouannaud, J.-P., Shao, Z. (eds.) CPP 2011. LNCS, vol. 7086, pp. 135–150. Springer, Heidelberg (2011). https://doi.org/10.1007/978-3-642-25379-9_12
3. Armand, M., Grégoire, B., Spiwack, A., Théry, L.: Extending COQ with imperative features and its application to SAT verification. In: Kaufmann, M., Paulson, L.C. (eds.) ITP 2010. LNCS, vol. 6172, pp. 83–98. Springer, Heidelberg (2010). https://doi.org/10.1007/978-3-642-14052-5_8
4. Besson, F.: Fast reflexive arithmetic tactics the linear case and beyond. In: Altenkirch, T., McBride, C. (eds.) TYPES 2006. LNCS, vol. 4502, pp. 48–62. Springer, Heidelberg (2007). https://doi.org/10.1007/978-3-540-74464-1_4
5. Besson, F., Cornilleau, P.-E., Pichardie, D.: Modular SMT proofs for fast reflexive checking inside Coq. In: Jouannaud, J.-P., Shao, Z. (eds.) CPP 2011. LNCS, vol. 7086, pp. 151–166. Springer, Heidelberg (2011). https://doi.org/10.1007/978-3-642-25379-9_13
6. Blanchette, J.C., Böhme, S., Paulson, L.C.: Extending sledgehammer with SMT solvers. J. Autom. Reason. **51**(1), 109–128 (2013). https://doi.org/10.1007/s10817-013-9278-5
7. Böhme, S., Nipkow, T.: Sledgehammer: judgement day. In: Giesl, J., Hähnle, R. (eds.) IJCAR 2010. LNCS (LNAI), vol. 6173, pp. 107–121. Springer, Heidelberg (2010). https://doi.org/10.1007/978-3-642-14203-1_9
8. Boulmé, S., Maréchal, A.: Toward Certification for Free! July 2017. https://hal.archives-ouvertes.fr/hal-01558252, preprint
9. Chlipala, A.: Parametric higher-order abstract syntax for mechanized semantics. In: International Conference on Functional Programming (ICFP). ACM Press (2008)
10. Cook, W.J., Cunningham, W.H., Pulleyblank, W.R., Schrijver, A.: Combinatorial Optimization. Wiley, New York (1998)
11. Czajka, L., Kaliszyk, C.: Goal translation for a hammer for Coq. In: Proceedings First International Workshop on Hammers for Type Theories, HaTT@IJCAR 2016. EPTCS, vol. 210, pp. 13–20 (2016)
12. Farkas, J.: Theorie der einfachen Ungleichungen. Journal für die Reine und Angewandte Mathematik, 124 (1902)

13. Fouilhe, A., Boulmé, S.: A certifying frontend for (sub)polyhedral abstract domains. In: Giannakopoulou, D., Kroening, D. (eds.) VSTTE 2014. LNCS, vol. 8471, pp. 200–215. Springer, Cham (2014). https://doi.org/10.1007/978-3-319-12154-3_13

14. Fouilhe, A., Monniaux, D., Périn, M.: Efficient generation of correctness certificates for the abstract domain of polyhedra. In: Logozzo, F., Fähndrich, M. (eds.) SAS 2013. LNCS, vol. 7935, pp. 345–365. Springer, Heidelberg (2013). https://doi.org/10.1007/978-3-642-38856-9_19

15. Maréchal, A.: New Algorithmics for Polyhedral Calculus via Parametric Linear Programming. Theses, UGA - Université Grenoble Alpes, December 2017. https://hal.archives-ouvertes.fr/tel-01695086

16. Maréchal, A., Périn, M.: Efficient elimination of redundancies in polyhedra by ray-tracing. In: Bouajjani, A., Monniaux, D. (eds.) VMCAI 2017. LNCS, vol. 10145, pp. 367–385. Springer, Cham (2017). https://doi.org/10.1007/978-3-319-52234-0_20

17. Nelson, G., Oppen, D.C.: Simplification by cooperating decision procedures. ACM Trans. Program. Lang. Syst. 1(2), 245–257 (1979)

18. Oppen, D.C.: Complexity, convexity and combinations of theories. Theor. Comput. Sci. 12, 291–302 (1980)

19. Schrijver, A.: Theory of Linear and Integer Programming. Wiley, Chichester, New York (1986)

20. Shostak, R.E.: Deciding combinations of theories. J. ACM 31(1), 1–12 (1984)

21. Silva, J.P.M., Lynce, I., Malik, S.: Conflict-driven clause learning SAT solvers. In: Handbook of Satisfiability, Frontiers in Artificial Intelligence and Applications, vol. 185, pp. 131–153. IOS Press (2009)

22. The Coq Development Team: The Coq proof assistant reference manual - version 8.7. INRIA (2017)

The Coinductive Formulation
of Common Knowledge

Colm Baston and Venanzio Capretta[(✉)]

Functional Programming Lab, School of Computer Science,
University of Nottingham, Nottingham, UK
{colm.baston,venanzio.capretta}@nottingham.ac.uk

Abstract. We study the coinductive formulation of common knowledge
in type theory. We formalise both the traditional relational semantics and
an operator semantics, similar in form to the epistemic system S5, but at
the level of events on possible worlds rather than as a logical derivation
system. We have two major new results. Firstly, the operator semantics is
equivalent to the relational semantics: we discovered that this requires a
new hypothesis of *semantic entailment* on operators, not known in previous literature. Secondly, the coinductive version of common knowledge is
equivalent to the traditional transitive closure on the relational interpretation. All results are formalised in the proof assistants Agda and Coq.

1 Introduction

Common knowledge is a modality in epistemic logic: a group of agents has common knowledge of an event if everyone knows it, everyone knows that everyone
knows it, everyone knows that everyone knows that everyone knows it, and so on
ad infinitum. Some famous logical puzzles (the *muddy children* or the *cheating
husbands* problem [8,9]) involve clever uses of this notion: the solution is based
on some information shared by the agents and on their ability to deduce other
people's reasoning in a potentially unlimited reflection.

Type-theoretic logical systems allow the direct definition of coinductive types
which may contain infinite objects, constructed by guarded corecursion [2,7,11].
By the propositions-as-types interpretation that is standard in type theory, coinductive types are propositions whose proofs may be infinite. Common knowledge
can be naturally expressed as a coinductive operator: common knowledge of an
event is recursively defined as universal knowledge of it in conjunction with common knowledge of the universal knowledge of it. Although it is well-known that
the common knowledge modality is a greatest fixed point [1,8], and some coinductive methods have been tried with it before [4], our work [6] is the first direct
formalisation of this approach.

The traditional *frame semantics* [12,14] account of knowledge modalities
interprets them as equivalence relations on the set of possible states of the world:
the knowledge of an agent is represented as a relation identifying states that

© Springer International Publishing AG, part of Springer Nature 2018
J. Avigad and A. Mahboubi (Eds.): ITP 2018, LNCS 10895, pp. 126–141, 2018.
https://doi.org/10.1007/978-3-319-94821-8_8

cannot be distinguished by the agent. Common knowledge is interpreted as the transitive closure of the union of the knowledge relations for all agents.

The authors of [4] develop an infinitary deductive system with the aim that the system's derivations serve as justification terms for common knowledge. The derivations are finitely branching trees but, along coinductive lines, branches may be infinitely deep. The authors establish the soundness and completeness of this system with respect to a relational semantics where common knowledge is treated as a transitive closure, but do not employ coinduction as a proof technique which can generate such an infinite proof term from a finite specification.

We instead take an entirely semantic approach, which is *shallowly* embedded in the logic of a type theory with coinductive types. This is in contrast to a previous type-theoretic formalisation of common knowledge [15] which is *deeply* embedded in the logic of the Coq proof assistant. This allows Coq to be used as a metatheory to experiment with the target logic, but does not allow for all of the features of Coq's own logic, such as coinduction, to be used from within the target logic.

Our formulation of epistemic logic is based on an underlying set of states of the world; epistemic propositions are interpreted as events, that is, predicates on states (Sect. 2). Knowledge modalities are then functions over events. Our treatment of these knowledge modalities is similar to the syntactic encoding of epistemic logic in the modal logic system S5 [16]. Its study on the semantics side, as the algebraic structure of knowledge operators, is new (Sect. 3).

We give two main original contributions. Firstly, we prove that the operator semantics is equivalent to the relational semantics (Sect. 4). In formalising the equivalence, we discovered that it is necessary to assume a previously unknown property for operators: *preservation of semantic entailment*, which states that the knowledge operator preserves the consequences of a possibly infinite set of events (S5 gives this only for finite sets). Secondly, we prove that the coinductive formulation of common knowledge is equivalent to the relational representation as transitive closure, and that the coinductive operator itself satisfies the properties of a knowledge operator (Sect. 5). All these results are formalised in two type-theoretic proof assistants: Agda[1] and Coq[2].

2 Possible Worlds and Events

In this section, we present the semantic framework in which we will be working throughout the paper. Our formalisation of epistemic logic is not axiomatic, but definitional. Instead of postulating a set of axioms for a knowledge modality, we define it using the logic of a proof assistant, like Agda or Coq, along the semantic lines of possible-worlds models. In other words, we work with a shallow

---

[1] https://colmbaston.co.uk/files/Common-Knowledge.agda.
[2] http://www.duplavis.com/venanzio/publications/common_knowledge.v.

embedding of epistemic logic in type theory. (The *deep* and *shallow embedding* approaches to the formalisation of logics and domain-specific languages are well-known and widespread. The first exposition of the concepts, but not the terminology, that we can find is by Reynolds [17]. See, for example, [13] for a clear explanation.)

We postulate a set of possible worlds, which we call states. A state encodes all relevant information about the world we are modelling. In the semantics of epistemic logic, a proposition may be true in some states, but false in others, so we interpret a proposition as a predicate over states, called an *event.*

To avoid confusion with the propositions that are native to the type theory, we shall refer to epistemic propositions only as events from this point on. By the standard propositions-as-types interpretation, we identify type-theoretic propositions with the type of their proofs, adopting an Agda-like notation. That is, we use the type universe Set for both data types and propositions.

$$\text{State : Set} \qquad \text{Event} = \text{State} \to \text{Set}$$

An event can be seen extensionally as the set of states in which that event occurs, or is true. It is convenient to define set-like operators to combine events and make logical statements about them. In the following, the variable e ranges over events and the variable w ranges over states. We obtain the truth assignment of an event e in a state w by simply applying the event to the state, written $e\,w$.

$\_\sqcap\_ : \text{Event} \to \text{Event} \to \text{Event}$
$e_1 \sqcap e_2 = \lambda w.e_1\,w \wedge e_2\,w$

$\_\sqcup\_ : \text{Event} \to \text{Event} \to \text{Event}$
$e_1 \sqcup e_2 = \lambda w.e_1\,w \vee e_2\,w$

$\_\sqsubset\_ : \text{Event} \to \text{Event} \to \text{Event}$
$e_1 \sqsubset e_2 = \lambda w.e_1\,w \to e_2\,w$

$\sim\_ : \text{Event} \to \text{Event}$
$\sim e = \lambda w.\neg(e\,w)$

$\_\subset\_ : \text{Event} \to \text{Event} \to \text{Set}$
$e_1 \subset e_2 = \forall w.e_1\,w \to e_2\,w$

$\_\equiv\_ : \text{Event} \to \text{Event} \to \text{Set}$
$e_1 \equiv e_2 = (e_1 \subset e_2) \wedge (e_2 \subset e_1)$

$\mathbb{W} : \text{Event} \to \text{Set}$
$\mathbb{W}\,e = \forall w.e\,w$

The first three operators, $\sqcap$, $\sqcup$, and $\sqsubset$, are binary operations on events: they map two events to the event that is their conjunction, disjunction, and implication, respectively; we can see them set-theoretically as intersection, union, and exponent of sets of states. The fourth, $\sim$, is a unary operator expressing event negation, set theoretically it is the complement.

The next two operators are relations between two events. The first of the two, $\subset$, states that the first event logically implies the second, set-theoretically the first is a subset of the second. The next, $\equiv$, states that two events are logically equivalent, their set extensions being equal. Finally, the operator $\mathbb{W}$ expresses the fact that an event is true in all states: that it is semantically forced to be true. On the logical system side, it corresponds to a tautology. The operator $\subset$ can also be expressed in terms of the equivalence: $e_1 \subset e_2 \leftrightarrow \mathbb{W}(e_1 \sqsubset e_2)$.

As a simple example, imagine that we are modelling a scenario in which a coin has been tossed and a six-sided die has been rolled. We have these primitive events:

$$D_1 = \text{"the die rolled a 1"}$$

$$C_H = \text{"the coin landed heads side up"}$$
$$C_T = \text{"the coin landed tails side up"}$$

$$\vdots$$

$$D_6 = \text{"the die rolled a 6"}$$

Then, for example, $D_3 \sqcup D_4$ is the event which is true in those states where the die rolled a 3 or a 4, we might assume $\mathbb{W}(\sim(C_H \sqcap C_T))$ so that there cannot be a state in which the coin landed both heads side up and tails side up, and so on.

Now we come to introducing modal operators for knowledge, so let us introduce two agents in our example, Alice and Bob. A modal operator in this setting is a function Event $\to$ Event, so we give each agent an operator of this type, K_A and K_B respectively. This allows us to also express events such as the following:

$$\mathsf{K}_A\, D_1 \qquad = \text{"Alice knows that the die rolled a 1"}$$
$$\mathsf{K}_B\,(\mathsf{K}_A\, D_2) \qquad = \text{"Bob knows that Alice knows that the die rolled a 2"}$$
$$\sim(\mathsf{K}_B\, C_H \sqcup \mathsf{K}_B\, C_T) = \text{"Bob does not know on which side the coin landed"}$$

But not all operators on events are suitable to represent the knowledge of an agent. In the next section, we will define a class of operators on events that can be considered possible descriptions of an agent's knowledge. Then, assuming there is a set of agents, each with their own knowledge operator, we give a coinductive definition of another operator expressing their common knowledge.

3 Knowledge Operator Semantics

For the moment we do not consider a set of agents, but just fix a single operator K : Event $\to$ Event and specify a set of properties that it must satisfy to be a possible interpretation of the knowledge of an agent. Traditionally, the modal logic system S5 [16] is employed to provide an idealised model for knowledge (modern presentations and historical overviews can be found in [3,8]). In short, its properties state that agents are perfect reasoners, can only know events which are actually true, and are aware of what they do and do not know. We posit a version of this logic as the properties that the knowledge operator K must satisfy.

We discovered that an extra infinitary deduction rule is required to obtain a perfect correspondence with the traditional relational interpretation which we describe in Sect. 4. This cannot be expressed at the level of the syntactic logical system, but it becomes essential at the semantic level of operators on events. It states that the knowledge operator must preserve semantic entailments, even if the conclusion follows from an infinite set of premises. Like other epistemic postulates in the standard literature, this is a strong principle which may be unrealistic to assume for real-world agents, but our discovery shows that it is already an implicit feature of the classical frame semantics.

Definition 1. *A family of events indexed on a type X is a function $E : X \to$* Event. *Given a family of events E, we can generate the event $\bigsqcap E$ that is true in those states where all members of the family are true:*

$$\textstyle\bigsqcap E = \lambda w. \forall (x : X). E\, x\, w$$

We can map K *onto the whole family by applying it to every member: We write* K E *for the family $\lambda x.$K $(E\, x)$. We say that* K *preserves semantic entailment if, for every family $E : X \to$* Event *and every event e, we have:*

$$\textstyle\bigsqcap E \subset e \to \bigsqcap (\mathsf{K}\, E) \subset \mathsf{K}\, e.$$

We require that K has this property and also satisfies the properties of S5. Some of these are derivable from semantic entailment, but we formulate them all in the definition to clearly reflect the relation with traditional epistemic logic.

Definition 2. *An operator on events,* K : Event $\to$ Event, *is called a* knowledge operator *if it preserves semantic entailment and satisfies the event-based version of the properties of S5:*

1. $\mathbb{W}\, e \to \mathbb{W}\, \mathsf{K}\, e$
   This principle is known as knowledge generalisation *(or* necessitation *in modal logics with an operator $\square$ that is interpreted as "it is necessary that"). It states that all derivable theorems are known, that is, the agent is capable of applying pure logic to derive tautologies. Here we work on the semantics side: instead of logical formulas, the objects of the knowledge operators are events, that is, predicates on states. We understand an event to be a tautology if it is true in every state. The unfolding of the principle is: $(\forall w. e\, w) \to \forall v. \mathsf{K}\, e\, v$.*
2. $\mathsf{K}\, (e_1 \sqsubset e_2) \subset (\mathsf{K}\, e_1 \sqsubset \mathsf{K}\, e_2)$
   Corresponding to Axiom K, this states that the knowledge operator distributes over implication: The agent is capable of applying modus ponens *to what they know. Notice the use of the two operators $\sqsubset$, mapping two events to the event expressing the implication between the two, and $\subset$, stating that the second event is true whenever the first one is. If we unfold the definitions, this states that: $\forall w. \mathsf{K}\, (e_1 \sqsubset e_2)\, w \to \mathsf{K}\, e_1\, w \to \mathsf{K}\, e_2\, w$. That is, if in a state w the agent knows that e_1 implies e_2 and also knows e_1, then they know e_2.*
3. $\mathsf{K}\, e \subset e$
   Corresponding to Axiom T, this states that knowledge is true: what distinguishes knowledge from belief or opinion is that when an agent knows an event, that event must actually hold in the present state.
4. $\mathsf{K}\, e \subset \mathsf{K}\, (\mathsf{K}\, e)$
   Corresponding to Axiom 4, this is a principle of self-awareness of knowledge: agents know when they know something.
5. $\sim\! \mathsf{K}\, e \subset \mathsf{K}\, (\sim\! \mathsf{K}\, e)$
   Corresponding to Axiom 5, this negative version of the principle of self-awareness could be called the Socratic Principle: *When an agent does not know something, they at least know that they do not know it.*

Lemma 1. *The first two properties in the definition of knowledge operator (knowledge generalisation and Axiom K) are consequences of preservation of semantic entailment.*

Proof. Assume that K preserves semantic entailment.

- Knowledge generalisation is immediate once we see that $\mathbb{W}\, e$ is equivalent to the semantic entailment from the empty family: $\bigsqcap \varnothing \sqsubset e$.
- Axiom K follows from applying preservation of the semantic entailment version of modus ponens (using a family with just two elements): $\bigsqcap\{e_1 \sqsubset e_2, e_1\} \sqsubset e_2$.

(We have used set notation for the families: they indicate the trivial family indexed on the empty type and a family indexed on the Booleans.) $\qquad\square$

Let us now return to a setting with a non-empty set of agents, ranged over by the variable a. Each agent has an individual knowledge operator K_a satisfying Definition 2. Recall that common knowledge of an event intuitively means that everyone knows it, everyone knows that everyone knows it, everyone knows that everyone knows that everyone knows it, and so on ad infinitum.

We define EK to be the "everyone knows" operator expressing universal knowledge of an event:

$$\mathsf{EK} : \mathsf{Event} \to \mathsf{Event}$$
$$\mathsf{EK}\, e = \lambda w.\forall a.\mathsf{K}_a\, e\, w$$

EK is not itself a knowledge operator. It is possible to show that it satisfies knowledge generalisation, Axiom K, and, with at least one agent, Axiom T. It also preserves semantic entailment. The two introspective properties of Axioms 4 and 5, however, are not satisfied in general: if they were, there would be no distinction between universal knowledge and common knowledge.

Common knowledge of an event e intuitively means the infinite conjunction:

$$\mathsf{EK}\, e \sqcap \mathsf{EK}\,(\mathsf{EK}\, e) \sqcap \mathsf{EK}\,(\mathsf{EK}\,(\mathsf{EK}\, e)) \sqcap \ldots$$

This infinite conjunction can be expressed by a *coinductive definition* saying that common knowledge of e means the conjunction of EK e and, corecursively, common knowledge of EK e. In Agda or Coq, this can be defined directly by a coinductive operator:

$$\mathsf{CoInductive\ cCK} : \mathsf{Event} \to \mathsf{Event}$$
$$\mathsf{cCK-intro} : \forall e.\mathsf{EK}\, e \sqcap \mathsf{cCK}\,(\mathsf{EK}\, e) \sqsubset \mathsf{cCK}\, e$$

This defines common knowledge at a high level without mentioning states, naturally corresponding to the informal recursive notion. If we unfold the definitions so that we can see the constructor's type in full, it becomes evident that the definition satisfies the positivity condition of (co)inductive types:

$$\mathsf{cCK-intro} : \forall e.\forall w.(\mathsf{EK}\, e\, w) \wedge (\mathsf{cCK}\,(\mathsf{EK}\, e)\, w) \to \mathsf{cCK}\, e\, w$$

The meaning of the definition is that a proof of cCK e must have the form of the constructor cCK−intro applied to proofs of EK e and cCK (EK e). The latter must in turn be obtained by another application of cCK−intro. This process proceeds infinitely, without end. To obtain such a proof, we can give a finite *corecursive* definition that, when unfolded, generates the infinite structure.

The idea is that when proving that an event e is common knowledge, we must prove EK e without any extra assumption, but we can recursively use the statement that we are proving in the derivation of cCK (EK e). This apparently circular process must satisfy a *guardedness condition*, ensuring that the unfolding is productive. See, for an introduction, Chap. 13 of the Coq book by Bertot and Casteran [2] or the application to general recursion by one of us [5]. We will soon give an example in the proof of Lemma 4.

Since a proof of cCK e must be constructed by a proof of EK $e \sqcap$ cCK (EK e), we can derive either conjunct if we have that e is common knowledge. That is, we obtain the following trivial lemmas:

Lemma 2. *For every event e we have:* cCK $e \subset$ EK e.

Lemma 3. *For every event e we have:* cCK $e \subset$ cCK (EK e).

We now illustrate a proof by coinduction as a first simple example, showing that common knowledge is equivalent to the family of events expressing finite iterations of EK:

$$\text{recEK} : \text{Event} \to \mathbb{N} \to \text{Event}$$
$$\text{recEK}\, e\, 0 = \text{EK}\, e$$
$$\text{recEK}\, e\, (n+1) = \text{EK}\, (\text{recEK}\, e\, n)$$

Lemma 4. *For every event e, the family* recEK e *semantically entails* cCK e:

$$\sqcap(\text{recEK}\, e) \subset \text{cCK}\, e$$

Proof. In a coinductive proof, we are allowed to assume the statement we are proving and use it in a restricted way:

CoInductive Hypothesis CH: $\forall e. \sqcap(\text{recEK}\, e) \subset \text{cCK}\, e$.

Of course, we cannot just use the assumption CH directly to prove the theorem. We must make at least one step in the proof without circularity.

Unfolding the statement, we need to prove that for every state w we have:

$$(\forall(n : \mathbb{N}).\text{recEK}\, e\, n\, w) \to \text{cCK}\, e\, w$$

So let us assume that for every natural number n, recEK $e\, n\, w$ holds.

We must now prove cCK $e\, w$, which can be derived using the constructor cCK−intro from EK $e\, w$ and cCK (EK e) w.

− EK $e\, w$ is just recEK $e\, 0\, w$, which is true by assumption;
− To prove cCK (EK e) w, we now invoke CH, instantiated for the event EK e:

$$\sqcap(\text{recEK}\, (\text{EK}\, e)) \subset \text{cCK}\, (\text{EK}\, e)$$

That is:

$$\forall w.(\forall n.\mathsf{recEK}\,(\mathsf{EK}\,e)\,n\,w) \rightarrow \mathsf{cCK}\,(\mathsf{EK}\,e)\,w$$

So we need to prove that for every n, $\mathsf{recEK}\,(\mathsf{EK}\,e)\,n\,w$. This is trivially equivalent to $\mathsf{recEK}\,e\,(n+1)\,w$, which is true by assumption. Therefore, Assumption CH allows us to conclude $\mathsf{cCK}\,(\mathsf{EK}\,e)\,w$, as desired.   □

Let us observe the structure of this proof. We allowed ourselves to assume the statement of the theorem as a hypothesis. But it can only be used in a limited way. We used it immediately after applying the constructor $\mathsf{cCK}-\mathsf{intro}$, to prove the recursive branch of it. This is the typical way in which *guarded corecursion* works: we can make a circular call to the object we are defining immediately under the application of the constructor.

The proof of the implication in the other direction, omitted here, is simply by induction over natural numbers, repeatedly unfolding the definition of common knowledge.

Lemma 5. *For every event e and $n : \mathbb{N}$: $\mathsf{cCK}\,e \subset \mathsf{recEK}\,e\,n$.*

The equivalence of common knowledge with the family recEK gives an immediate proof of the following useful property corresponding to Axiom 4 of S5.

Lemma 6. *For every event e, we have: $\mathsf{cCK}\,e \subset \mathsf{cCK}\,(\mathsf{cCK}\,e)$.*

Finally, the coinductive definition of common knowledge satisfies the properties of knowledge operators. We must prove all the S5 properties and preservation of semantic entailment for cCK.

Theorem 1. *Common knowledge, cCK, is itself a knowledge operator.*

Proof. We can give a direct proof of the statement by deriving all the properties of knowledge operators for cCK. Lemma 6 shows that Axiom 4 holds. Proofs of all other S5 properties and of preservation of semantic entailment are interesting applications of coinductive methods. These proofs are omitted here, but are used in the Coq formalisation.

This theorem is also a consequence of Theorem 4 (equivalence of cCK with the relational characterisation) and Theorem 3 (equivalence relations define knowledge operators). This proof is used in the Agda formalisation.   □

4   Relational Semantics

In this section, we present the traditional frame semantics of epistemic logic, the knowledge operators being introduced through equivalence relations on states. We prove that a knowledge operator semantics can be generated from an equivalence relation and vice versa, additionally showing that these transformations form an isomorphism.

Two states may differ by a number of events: some events may be true in one of the states, but false in the other. If an agent has no knowledge of any of these discriminating events, only knowing events which are common to both states, then those states are indistinguishable as far as the agent is aware. We say that these states are *epistemically accessible* from one another: if the world were in one of those states, the agent would consider either state to be plausible, not having sufficient knowledge to inform them precisely in which state the world is actually in.

To say that an agent has knowledge of an event in a particular state is then to say that the event holds in all states that the agent finds epistemically accessible from that state. We formalise this notion by defining a transformation from relations on states to unary operators on events:

$$K_{[]} : (\text{State} \to \text{State} \to \text{Set}) \to (\text{Event} \to \text{Event})$$
$$K_{[R]} = \lambda e.\lambda w.\forall v.w\, R\, v \to e\, v$$

Care must be taken to distinguish this notation from the notation of earlier sections where each agent a had a knowledge operator K_a directly postulated. When talking in terms of the relational semantics, we do not take these operators as primitive. Here, $K_{[R]}$ refers to the operator generated when transforming some relation R : State $\to$ State $\to$ Set.

It is a well known result in modal logic that applying this transformation to an equivalence relation yields a knowledge operator satisfying the properties of S5. We establish this fact here, assuming only the needed properties of the relation (see [10] for a more extensive listing of which relational properties imply which modal axioms). The proofs are omitted, but can be adapted from standard expositions. They are also present in the Agda and Coq formalisations.

Lemma 7. *If R is a relation on states, then the operator $K_{[R]}$ has the following properties.*

- *$K_{[R]}$ satisfies knowledge generalisation: $\mathbb{W}\, e \to \mathbb{W}\, K_{[R]}\, e$*
- *$K_{[R]}$ satisfies Axiom K: $K_{[R]}\, (e_1 \sqsubseteq e_2) \subset K_{[R]}\, e_1 \sqsubseteq K_{[R]}\, e_2$*
- *If R is reflexive, then $K_{[R]}$ satisfies Axiom T: $K_{[R]}\, e \subset e$*
- *If R is transitive, then $K_{[R]}$ satisfies Axiom 4: $K_{[R]}\, e \subset K_{[R]}\, (K_{[R]}\, e)$*
- *If R is symmetric and transitive, then $K_{[R]}$ satisfies Axiom 5: $\sim K_{[R]}\, e \subset K_{[R]}\, (\sim K_{[R]}\, e)$*

To complete a proof that $K_{[R]}$ is a knowledge operator, we have to show in addition that it preserves semantic entailment. As is the case with knowledge generalisation and Axiom K, this does not require any hypothesis on the properties of R.

Lemma 8. *For every family $E : X \to$ Event and every event e, we have:*

$$\bigsqcap E \subset e \to \bigsqcap (K_{[R]}\, E) \subset K_{[R]}\, e$$

Proof. Let us assume that $\bigsqcap E \subset e$ (ASSUMPTION 1).

We must prove $\bigsqcap (K_{[R]}\, E) \subset K_{[R]}\, e$, that is, unfolding the definitions of $\bigsqcap$ and $\subset$, for every state w, $\forall x.K_{[R]}\, (E\, x)\, w \to K_{[R]}\, e\, w$, where x ranges over the index of the family E. So let us assume that $\forall x.\mathsf{K}_{[R]}\,(E\,x)\,w$ (ASSUMPTION 2). We must then prove that $\mathsf{K}_{[R]}\,e\,w$.

Unfolding the definition of $\mathsf{K}_{[R]}$, our goal becomes $\forall v.w\,R\,v \rightarrow e\,v$. So let v be any state such that $w\,R\,v$ (ASSUMPTION 3). We must prove that $e\,v$.

To prove this goal we apply directly Assumption 1, which states (when unfolded) that $\forall v.(\forall x.(E\,x)\,v) \rightarrow e\,v$. Therefore, to prove the goal, we just have to show that $\forall x.(E\,x)\,v$.

For any index x, Assumption 2 tells us that $\mathsf{K}_{[R]}\,(E\,x)\,w$, that is, by definition of $\mathsf{K}_{[R]}$, $\forall v.w\,R\,v \rightarrow (E\,x)\,v$. But since our choice of v satisfies $w\,R\,v$ by Assumption 3, we have that $(E\,x)\,v$, as desired. $\qquad\square$

We can then put the two lemmas together to satisfy Definition 2.

Theorem 2. *If R is an equivalence relation on states, then $\mathsf{K}_{[R]}$ is a knowledge operator.*

The inverse transformation, taking a knowledge operator and returning a relation on states is:

$$R_{[]} : (\mathsf{Event} \rightarrow \mathsf{Event}) \rightarrow (\mathsf{State} \rightarrow \mathsf{State} \rightarrow \mathsf{Set})$$
$$R_{[\mathsf{K}]} = \lambda w.\lambda v.\forall e.\mathsf{K}\,e\,w \leftrightarrow \mathsf{K}\,e\,v$$

This transformation always results in an equivalence relation, as $\leftrightarrow$ is itself an equivalence relation. In fact, if we admit classical reasoning, one direction of the implication is sufficient.

Lemma 9. *If K is a knowledge operator, then $\lambda w.\lambda v.\forall e.\mathsf{K}\,e\,w \rightarrow \mathsf{K}\,e\,v$ is an equivalence relation.*

As an immediate corollary, it is equivalent to $\lambda w.\lambda v.\forall e.\mathsf{K}\,e\,w \leftrightarrow \mathsf{K}\,e\,v$.

Proof. Reflexivity and transitivity are trivial. To show symmetry, we first assume, for some states w and v, that $\forall e.\mathsf{K}\,e\,w \rightarrow \mathsf{K}\,e\,v$ and, for some event e, $\mathsf{K}\,e\,v$. We want to prove that $\mathsf{K}\,e\,w$.

Suppose, towards a contradiction, that $\neg(\mathsf{K}\,e\,w)$, which can also be written as $(\sim\mathsf{K}\,e)\,w$. By Axiom 5, we have $\mathsf{K}\,(\sim\mathsf{K}\,e)\,w$. By instantiating the first assumption with event $\sim\mathsf{K}\,e$, we deduce that $\mathsf{K}\,(\sim\mathsf{K}\,e)\,v$. By Axiom T, this implies $(\sim\mathsf{K}\,e)\,v$, which can be written as $\neg(\mathsf{K}\,e\,v)$, contradicting the second assumption: our supposition $\neg(\mathsf{K}\,e\,w)$ must be false. We conclude, by excluded middle, that $\mathsf{K}\,e\,w$ is true, as desired. $\qquad\square$

We now prove that the mappings of knowledge operators to equivalence relations and vice versa are actually inverse: we can equivalently work with either representation of knowledge. The proofs are mostly straightforward applications of the properties of S5 and equivalence relations, except one direction, for which we added the assumption of preservation of semantic entailment. We give the proof of this.

In order to do this we first characterise the transformations of K using event families generated by K on a fixed state w. Choose as index set the set of events that are known in w: $X = \{e \mid \mathsf{K}\,e\,w\}$ (in Coq or Agda, we use the dependent sum type $\Sigma e.\mathsf{K}\,e\,w$ whose elements are pairs $\langle e, h \rangle$ of an event e and a proof h of $\mathsf{K}\,e\,w$); the family itself is just the application of K. Formally:

$$\mathsf{KFam}^w : (\Sigma e.\mathsf{K}\,e\,w) \to \mathsf{Event}$$
$$\mathsf{KFam}^w \langle e, h \rangle = \mathsf{K}\,e$$

Intuitively, KFam^w is the total amount of knowledge in state w. Set-theoretically it is $\{\mathsf{K}\,e \mid \mathsf{K}\,e\,w\}$. One observation, whose proof we omit here, is that $R_{[\mathsf{K}]}\,w\,v$ is equivalent to $\bigsqcap(\mathsf{KFam}^w) \equiv \bigsqcap(\mathsf{KFam}^v)$. Another observation will allow us to replace $\mathsf{K}_{[R_{[\mathsf{K}]}]}\,e\,w$ with an expression involving KFam^w.

Lemma 10. *For every event e and state w, the proposition $\mathsf{K}_{[R_{[\mathsf{K}]}]}\,e\,w$ is equivalent to $\bigsqcap(\mathsf{KFam}^w) \subset e$.*

Proof. We just unfold the definitions and use the previous lemma:

$$
\begin{aligned}
\mathsf{K}_{[R_{[\mathsf{K}]}]}\,e\,w &\Leftrightarrow \forall v.w\,R_{[\mathsf{K}]}\,v \to e\,v & \text{by definition of } \mathsf{K}_{[]} \\
&\Leftrightarrow \forall v.(\forall e'.\mathsf{K}\,e'\,w \leftrightarrow \mathsf{K}\,e'\,v) \to e\,v & \text{by definition of } R_{[]} \\
&\Leftrightarrow \forall v.(\forall e'.\mathsf{K}\,e'\,w \to \mathsf{K}\,e'\,v) \to e\,v & \text{by Lemma 9} \\
&\Leftrightarrow \bigsqcap(\mathsf{KFam}^w) \subset e & \text{by definition of } \mathsf{KFam}.
\end{aligned}
$$

$\square$

Lemma 11. *For every knowledge operator K and every event e, we have:*

$$\mathsf{K}_{[R_{[\mathsf{K}]}]}\,e \subset \mathsf{K}\,e$$

Proof. Assume, for some state w, that $\mathsf{K}_{[R_{[\mathsf{K}]}]}\,e\,w$. We must prove $\mathsf{K}\,e\,w$.

By Lemma 10, the assumption is equivalent to $\bigsqcap(\mathsf{KFam}^w) \subset e$. Since K preserves semantic entailment, we also have $\bigsqcap(\mathsf{K}\,\mathsf{KFam}^w) \subset \mathsf{K}\,e$.

We just need to prove that all elements of the family $\mathsf{K}\,\mathsf{KFam}^w$ are true in state w to deduce that $\mathsf{K}\,e\,w$ holds, as desired. But in fact, given an index $\langle e', h \rangle$ for the family KFam^w, with h being a proof of $\mathsf{K}\,e'\,w$, this goal becomes $\mathsf{K}\,(\mathsf{KFam}^w \langle e', h \rangle)\,w = \mathsf{K}\,(\mathsf{K}\,e')\,w$ which can be dispatched by applying Axiom 4 to h. $\square$

The other three directions of the isomorphisms are straightforward applications of the properties of knowledge operators and equivalence relations.

Theorem 3. *For every knowledge operator K, $\mathsf{K}_{[R_{[\mathsf{K}]}]}$ is equivalent to K: for every event e and every state w, $\mathsf{K}_{[R_{[\mathsf{K}]}]}\,e\,w \leftrightarrow \mathsf{K}\,e\,w$.*

For every equivalence relation on states R, $R_{[\mathsf{K}_{[R]}]}$ is equivalent to R: for every pair of states w and v, $R_{[\mathsf{K}_{[R]}]}\,w\,v \leftrightarrow R\,w\,v$.

In this section we proved that the traditional frame semantics of epistemic logic is equivalent with our notion of knowledge operator. This isomorphism validates our discovery of the property of preservation of semantic entailments and shows that it was already implicitly present in the relational view.

5 Equivalence with Relational Common Knowledge

This section shows that the coinductive definition of common knowledge is equivalent to the traditional characterisation as transitive closure of the union of all the agents' accessibility relations. We use the isomorphism of Theorem 3 to treat equivalence relations on states and their corresponding knowledge operators interchangeably.

We first equip our agents with individual knowledge operators by postulating an equivalence relation $\simeq_a$: State $\to$ State $\to$ Set for each agent a as their epistemic accessibility relation. The knowledge operator for an agent a is then $K_{[\simeq_a]}$, which we shall write in shorthand as K_a.

Our formulation of the "everyone knows" operator, EK, and the coinductive common knowledge operator, cCK, are as they appear in Sect. 3. The only difference is in the underlying definition of K_a, which had previously been taken as primitive and assumed to satisfy the knowledge operator properties outlined in Definition 2. The relations $\simeq_a$ are equivalence relations, so we can conclude that this new formulation of K_a also satisfies these properties by Theorem 2.

The relational definition of the common knowledge operator is given by its own relation: the transitive closure of the union of all accessibility relations $\simeq_a$. We write this relation as $\propto$. It is defined inductively as follows:

$$\text{Inductive} \_ \propto \_ : \text{State} \to \text{State} \to \text{Set}$$
$$\propto\text{-union} : \forall a. \forall w. \forall v. w \simeq_a v \to w \propto v$$
$$\propto\text{-trans} : \forall w. \forall v. \forall u. w \propto v \to v \propto u \to w \propto u$$

Lemma 12. $\propto$ *is an equivalence relation.*

Proof. Transitivity is immediate by definition of constructor $\propto$–trans. Reflexivity follows from the reflexivity of the agents' underlying accessibility relations included in $\propto$ by constructor $\propto$–union (it is essential that there is at least one agent). Symmetry is proved by induction on the proof of $\propto$: the base case follows from the symmetry of the single agents' accessibility relation, while the recursive case is straightforward from the proof of transitivity and the inductive hypotheses. $\square$

We can intuitively grasp how it gets us to common knowledge in the following way. Observe that in an agent a's accessibility relation, if each state were alone in its own equivalence class, then a would be omniscient, able to perfectly distinguish each state from all others. If a were to forget an event, however, then all of those states which differ only by that event would collapse into an equivalence class together. In general, the fewer the number of equivalence classes in $\simeq_a$, the fewer the number of events a knows.

Taking the union of all agents' accessibility relations is essentially taking the union of their ignorance. This gets us as far as a relational interpretation of EK, which is not necessarily transitive. We take the transitive closure to reobtain an equivalence relation, further expanding the ignorance represented by the relation, but ensuring that we have the introspective properties of Axioms 4 and 5 that are essential to common knowledge.

It is as if there were a virtual, maximally-ignorant agent whose accessibility relation is $\propto$, knowing only those events which are common knowledge among all agents and nothing more. With this in mind, we can define the relational common knowledge operator, rCK, in the same way that we defined each agent's knowledge operator:

$$\mathsf{rCK} : \mathsf{Event} \rightarrow \mathsf{Event}$$
$$\mathsf{rCK} = \mathsf{K}_{[\propto]}$$

By Theorem 3 and Lemma 12, we can conclude that rCK satisfies all of the knowledge operator properties: We can also verify that it has properties corresponding to the two trivial properties of cCK, Lemmas 2 and 3.

Lemma 13. *For every event e we have:* $\mathsf{rCK}\, e \subset \mathsf{EK}\, e$.

Proof. Unfolding the statement, we need to prove that for every state w we have:

$$(\forall v.w \propto v \rightarrow e\, v) \rightarrow \forall a.\forall u.w \simeq_a u \rightarrow e\, u$$

So we assume the first statement, $\forall v.w \propto v \rightarrow e\, v$, and we also assume we have an agent a and state u such that $w \simeq_a u$.

We are left to show that e holds in u. By the definition of constructor $\propto$−union, given $w \simeq_a u$, we can derive that $w \propto u$, and then, instantiating our first assumption with state u, we obtain $e\, u$ as desired.   □

Lemma 14. *For every event e we have:* $\mathsf{rCK}\, e \subset \mathsf{rCK}\, (\mathsf{EK}\, e)$.

Proof. Unfolding the statement, we need to prove that for every state w we have:

$$(\forall v.w \propto v \rightarrow e\, v) \rightarrow \forall u.w \propto u \rightarrow \forall a.\forall t.u \simeq_a t \rightarrow e\, t$$

As in the previous proof, we have the assumption that for any state v such that $w \propto v$, e holds in v, so to reach our conclusion $e\, t$ we can prove that $w \propto t$. We have the additional assumptions $w \propto u$ and $u \simeq_a t$.

From the latter, by $\propto$−union we derive $u \propto t$. Then by the transitive property of $\propto$, constructor $\propto$−trans, we conclude $w \propto t$.   □

With these results, we are now able to prove the first direction of the equivalence of rCK and cCK.

Lemma 15. *For every event e we have:* $\mathsf{rCK}\, e \subset \mathsf{cCK}\, e$.

Proof. The conclusion of this theorem is an application of the coinductive predicate cCK, so we may proceed by coinduction, assuming the statement as our COINDUCTIVE HYPOTHESIS CH : $\forall e.\mathsf{rCK}\, e \subset \mathsf{cCK}\, e$

Unfolding the application of $\subset$, we need to prove that, for every state w, $\mathsf{rCK}\, e\, w \rightarrow \mathsf{cCK}\, e\, w$. So we assume $\mathsf{rCK}\, e\, w$, and use constructor cCK−intro to derive the conclusion $\mathsf{cCK}\, e\, w$, generating the proof obligations $\mathsf{EK}\, e\, w$ and $\mathsf{cCK}\, (\mathsf{EK}\, e)\, w$.

– EK $e\,w$ comes from Lemma 13 applied to assumption rCK $e\,w$.
– To prove cCK $(EK\,e)\,w$ we invoke assumption CH, instantiating it with event EK e, leaving us to prove rCK $(EK\,e)\,w$. This is the conclusion of Lemma 14, which we can apply to assumption rCK $e\,w$ to complete the proof. $\square$

We need an additional property of cCK before we are able to complete the other direction of the equivalence proof. The property is related to Axiom 4 of knowledge operators, for example, for an agent a's knowledge operator K_a:

$$\forall e.K_a\,e \subset K_a\,(K_a\,e)$$

Unfolding $\subset$ and the outermost application of K_a in the conclusion yields the principle:

$$\forall e.\forall w.K_a\,e\,w \to \forall v.w \simeq_a v \to K_a\,e\,v$$

That is, if we have that $K_a\,e$ holds at some state w, and we also have that $w \simeq_a v$ for some state v, then we can conclude that $K_a\,e$ holds at state v too. We call this *transporting* the agent's knowledge across the relation $\simeq_a$.

Since rCK is also a knowledge operator, defined in the same way as K_a but for a different equivalence relation, this transportation principle must hold for it too: relational common knowledge of an event can be transported from one state to another provided that those states are bridged by $\propto$. The additional property of cCK that we are to prove is that it too can be transported across $\propto$.

Lemma 16. *For every two states w and v and event e we have:*

$$cCK\,e\,w \to w \propto v \to cCK\,e\,v$$

Proof. We assume cCK $e\,w$ and proceed by induction on $w \propto v$:

– If $w \propto v$ is constructed by $\propto$–union, then there is some agent a for whom $w \simeq_a v$ holds. We can apply Lemma 6, the Axiom 4 property for cCK, to obtain cCK $(cCK\,e)\,w$, and then, by Lemma 2, it follows that EK $(cCK\,e)\,w$. Since all agents know this, we can instantiate this fact with agent a to conclude that a must know it: $K_a\,(cCK\,e)\,w$. We use the transportation principle of K_a to transport $K_a\,(cCK\,e)$ from state w to state v as these states are bridged by $w \simeq_a v$. Then, as a knows cCK e at state v, by Axiom T, it must actually hold in state v.
– If $w \propto v$ is constructed by $\propto$–trans, then there is some state u for which $w \propto u$ and $u \propto v$ hold. By induction hypothesis, we also have cCK $e\,w \to$ cCK $e\,u$ and cCK $e\,u \to$ cCK $e\,v$. We can simply use the transitivity of implication, induction hypotheses, and our assumption cCK $e\,w$ to reach our goal. $\square$

Lemma 17. *For every event e we have:* cCK $e \subset$ rCK e.

Proof. Unfolding the statement we are to prove, for every event e and state w:
cCK $e\,w \to \forall v.w \propto v \to e\,v$.

We assume cCK $e\,w$ and $w \propto v$. By Lemma 16 and these assumptions, we can then transport cCK e from state w to v: cCK $e\,v$. From this, we can derive

EK $e\,v$ by Lemma 2. Since everyone knows e at state v, and our set of agents is non-empty, there must be some agent who knows e at v. By Axiom T, e must actually hold at v.   □

Combining Lemmas 15 and 17 gives the full equivalence.

Theorem 4. *For all events e, $rCK\,e \equiv cCK\,e$, that is, $K_{[\alpha]}\,e \equiv cCK\,e$.*

6   Conclusion

We presented a type-theoretic formalisation of epistemic logic and a coinductive implementation of the common knowledge operator. This was done through a shallow embedding: we formulated knowledge operators as functions on events, which are predicates on a set of possible worlds or states.

The coinductive version of common knowledge has some advantages with respect to the traditional relational version.

- It is a straightforward formulation of the intuitive definition: common knowledge of an event means that everyone knows it and the fact that everyone knows it is itself common knowledge.
- It can be formulated at a higher level, using only the knowledge operators of each agent and the connectives of epistemic logic: the coinductive definition of cCK does not mention states.
- It gives us a new reasoning tool in the form of guarded corecursion. We demonstrated its power in several proofs in this paper and in the previous work on Aumann's Theorem.

We proved that our coinductive formulation is equivalent to two other versions:

- The traditional one as transitive closure of the union of the accessibility relations of all agents;
- The recursive family of iterations of the "everyone knows" operator.

In the process of investigating this subject we discovered that knowledge operators obtained from equivalence relations satisfy a previously unknown property of *preservation of semantic entailment* in addition to the properties of S5. We proved that this fully characterises knowledge operators and gives an isomorphism between them and equivalence relations.

References

1. Barwise, J.: Three views of common knowledge. In: Vardi, M.Y. (ed.) Proceedings of the 2nd Conference on Theoretical Aspects of Reasoning about Knowledge, Pacific Grove, CA, March 1988, pp. 365–379. Morgan Kaufmann (1988)
2. Bertot, Y., Castéran, P.: Interactive Theorem Proving and Program Development. Coq'Art: The Calculus of Inductive Constructions. Springer, Heidelberg (2004). https://doi.org/10.1007/978-3-662-07964-5

3. Blackburn, P., de Rijke, M., Venema, Y.: Modal Logic. Cambridge University Press, New York (2001)
4. Bucheli, S., Kuznets, R., Struder, T.: Two ways to common knowledge. Electron. Notes Theor. Comput. Sci. **262**, 83–98 (2010)
5. Capretta, V.: General recursion via coinductive types. Log. Methods Comput. Sci. **1**(2), 1–18 (2005). https://doi.org/10.2168/LMCS-1(2:1)2005
6. Capretta, V.: Common knowledge as a coinductive modality. In: Barendsen, E., Geuvers, H., Capretta, V., Niqui, M. (eds.) Reflections on Type Theory, Lambda Calculus, and the Mind, pp. 51–61. ICIS, Faculty of Science, Radbout University Nijmegen (2007). Essays Dedicated to Henk Barendregt on the Occasion of his 60th Birthday
7. Coquand, T.: Infinite objects in type theory. In: Barendregt, H., Nipkow, T. (eds.) TYPES 1993. LNCS, vol. 806, pp. 62–78. Springer, Heidelberg (1994). https://doi.org/10.1007/3-540-58085-9_72
8. Fagin, R., Halpern, J.Y., Vardi, M.Y., Moses, Y.: Reasoning About Knowledge. MIT Press, Cambridge (1995)
9. Gamow, G., Stern, M.: Puzzle Math. Viking Press, New York (1958)
10. Garson, J.: Modal logic. In: Zalta, E.N. (ed.) The Stanford Encyclopedia of Philosophy. Metaphysics Research Lab, Stanford University (2016)
11. Giménez, E.: Codifying guarded definitions with recursive schemes. In: Dybjer, P., Nordström, B., Smith, J. (eds.) TYPES 1994. LNCS, vol. 996, pp. 39–59. Springer, Heidelberg (1995). https://doi.org/10.1007/3-540-60579-7_3
12. Hintikka, J.: Knowledge and Belief. Cornell University Press, Ithaca (1962)
13. Keller, C., Werner, B.: Importing HOL light into Coq. In: Kaufmann, M., Paulson, L.C. (eds.) ITP 2010. LNCS, vol. 6172, pp. 307–322. Springer, Heidelberg (2010). https://doi.org/10.1007/978-3-642-14052-5_22
14. Kripke, S.A.: A completeness theorem in modal logic. J. Symb. Logic **24**(1), 1–14 (1959)
15. Lescanne, P.: Common knowledge logic in a higher order proof assistant. In: Voronkov, A., Weidenbach, C. (eds.) Programming Logics. LNCS, vol. 7797, pp. 271–284. Springer, Heidelberg (2013). https://doi.org/10.1007/978-3-642-37651-1_11
16. Lewis, C.I., Langford, C.H.: Symbolic Logic. The Century Co., New York (1932)
17. Reynolds, J.C.: User-defined types and procedural data structures as complementary approaches to data abstraction. In: Gries, D. (ed.) Programming Methodology. MCS, pp. 309–317. Springer, New York (1978). https://doi.org/10.1007/978-1-4612-6315-9_22

Tactics and Certificates in Meta Dedukti

Raphaël Cauderlier[✉]

University Paris Diderot, Irif, Paris, France
raphael.cauderlier@irif.fr

Abstract. Tactics are often featured in proof assistants to simplify the interactive development of proofs by allowing domain-specific automation. Moreover, tactics are also helpful to check the output of automatic theorem provers because they can rebuild details that the provers omit.

We use meta-programming to define a tactic language for the Dedukti logical framework which can be used both for checking certificates produced by automatic provers and for developing proofs interactively.

More precisely, we propose a dependently-typed tactic language for first-order logic in Meta Dedukti and an untyped tactic language built on top of the typed one. We show the expressivity of these languages on two applications: a transfer tactic and a resolution certificate checker.

1 Introduction

Dedukti [23] is a logical framework implementing the $\lambda\Pi$-calculus modulo theories. It has been proposed as a universal proof checker [7]. In the tradition of the Edinburgh Logical Framework, Dedukti is based on the Curry-Howard isomorphism: it reduces the problem of checking proofs in an embedded logic to the problem of type-checking terms in a given signature. In order to express complex logical systems such as the Calculus of Inductive Constructions, Dedukti features rewriting: the user can declare rewrite rules handling the computational part of the system.

Proof translators from the proof assistants HOL Light, Coq, Matita, FoCaLiZe, and PVS to Dedukti have been developed and used to recheck proofs of these systems [1,2,10,16]. Moreover, Zenon Modulo [12] and iProver Modulo [9], two automatic theorem provers for an extension of classical first-order logic with rewriting known as Deduction modulo, are able to output proofs in Dedukti.

These proof-producing provers are helpful in the context of proof interoperability between proof assistants. Independently developed formal libraries often use equivalent but non identical definitions and these equivalences can often be proved by automatic theorem provers [11]. Hence the stronger proof automation in Dedukti is, the easiest it is to exchange a proof between proof assistants.

Dedukti is a mere type checker and it is intended to check machine-generated proofs, not to assist human users in the formalisation of mathematics. It lacks

This work has been supported in part by the VECOLIB project of the French national research organization ANR (grant ANR-14-CE28-0018).

J. Avigad and A. Mahboubi (Eds.): ITP 2018, LNCS 10895, pp. 142–159, 2018.
https://doi.org/10.1007/978-3-319-94821-8_9

many features found in proof assistants to help the human user such as meta variables, implicit arguments, and a tactic language. However these features, especially tactics implementing decision procedures for some fragments of the considered logic, can be very helpful to check less detailed proof *certificates* produced by automatic theorem provers and SMT solvers.

Fortunately, Dedukti already has all the features required to turn it into a powerful meta-programming language in which tactics and certificates can be transformed into proof objects. In this article, we propose a dependently typed monadic tactic language similar to Mtac [25]. This tactic language can be used for interactive proof development and certificate checking but because of the lack of implicit arguments in Dedukti, it is still very verbose. For this reason, we also introduce an untyped tactic language on top of the typed one to ease the writing of tactics.

Since our goal is to check certificates from automatic theorem provers and to construct proof object out of them, we focus in this article on the Dedukti encoding of classical first-order logic. In Sect. 2, we present Dedukti and the encoding of classical first-order logic. The typed and untyped tactic languages are respectively presented in Sects. 3 and 4. Their applications to interactive proof development, theorem transfer, and certificate checking are shown in Sects. 5, 6, and 7.

2   First-Order Logic in Dedukti

In this section, we present Dedukti by taking as example the encoding of first-order logic. We consider a multisorted first-order logic similar to the logics of the TPTP-TFF1 [5] and SMTLIB [3] problem formats; its syntax of terms, and formulae is given in Fig. 1. The logic is parameterized by a possibly infinite set of sorts S. Each function symbol f has to be declared with a domain – a list of sorts $[A_1, \ldots, A_n]$ – and with a codomain $A \in S$. A term of sort A is either a variable of sort A or a function symbol f of domain $[A_1, \ldots, A_n]$ and codomain A applied to terms $t_1, \ldots, t_n$ such that each t_i has sort A_i. Similarly, each predicate symbol P has to be declared with a domain $[A_1, \ldots, A_n]$. A formula is either an atom, that is a predicate symbol P of domain $[A_1, \ldots, A_n]$ applied to terms $t_1, \ldots, t_n$ such that each t_i has sort A_i or is obtained from the first-order logical connectives $\bot$ (falsehood), $\wedge$ (conjunction), $\vee$ (disjunction), $\Rightarrow$ (implication) and the quantifiers $\forall$ (universal) and $\exists$ (existential). As usual, we define negation by $\neg\varphi := \varphi \Rightarrow \bot$, truth by $\top := \neg\bot$, and equivalence by $\varphi_1 \Leftrightarrow \varphi_2 := (\varphi_1 \Rightarrow \varphi_2) \wedge (\varphi_2 \Rightarrow \varphi_1)$.

Terms $t := x \mid f(t_1, \ldots, t_n)$
Formulae $\varphi := P(t_1, \ldots, t_n)$
$\qquad\quad \mid \bot \mid \varphi_1 \wedge \varphi_2 \mid \varphi_1 \vee \varphi_2 \mid \varphi_1 \Rightarrow \varphi_2 \mid \forall x : A.\ \varphi \mid \exists x : A.\ \varphi$

Fig. 1. Syntax of multisorted first-order logic

In Dedukti, we declare symbols for each syntactic class to represent: sorts, lists of sorts, terms, lists of terms, function symbols, predicate symbols, and formulae.

```
sort : Type.
sorts : Type.
term : sort -> Type.
terms : sorts -> Type.
function : Type.
predicate : Type.
prop : Type.
```

Type is Dedukti's builtin kind of types so the declaration sort : Type. means that sort is a Dedukti type and the declaration term : sort -> Type. means that term is a type family indexed by a sort.

Then we require domain and codomains for the symbols.

```
def fun_domain : function -> sorts.
def fun_codomain : function -> sort.
def pred_domain : predicate -> sorts.
```

The def keyword is used in Dedukti to indicate that the declared symbol is *definable*: this means that it is allowed to appear as head symbol in rewrite rules. In the case of the fun_domain, fun_codomain, and pred_domain functions, we do not give any rewrite rule now but each theory declaring new symbols is in charge of extending the definitions of these functions for the new symbols by adding the appropriate rewrite rules.

We then provide all the syntactic constructs, binding is represented using higher-order abstract syntax:

```
nil_sort : sorts.
cons_sort : sort -> sorts -> sorts.

nil_term : terms nil_sort.
cons_term : A : sort -> term A -> As : sorts -> terms As ->
  terms (cons_sort A As).

fun_apply : f : function -> terms (fun_domain f) -> term (fun_codomain f).

pred_apply : p : predicate -> terms (pred_domain p) -> prop.
false : prop.
and : prop -> prop -> prop.
or : prop -> prop -> prop.
imp : prop -> prop -> prop.
all : A : sort -> (term A -> prop) -> prop.
ex : A : sort -> (term A -> prop) -> prop.

def not (a : prop) := imp a false.
def eqv (a : prop) (b : prop) := and (imp a b) (imp b a).
```

The types of cons_term, fun_apply, pred_apply, all, and ex use the dependent product $\Pi x : A.B$ where x might occur in B; it is written x : A -> B in Dedukti.

Finally, we define what it means to be a proof of some proposition. For this we could declare symbols corresponding to the derivation rules of some proof system such as natural deduction or sequent calculus. However, the standard way to do this for first-order logic in Dedukti is to use the second-order definition of connectives and then derive the rules of natural deduction.

```
def proof : prop -> Type.
[] proof false --> a : prop -> proof a
[a,b] proof (and a b) -->
  c : prop -> (proof a -> proof b -> proof c) -> proof c
[a,b] proof (or a b) -->
  c : prop -> (proof a -> proof c) -> (proof b -> proof c) -> proof c
[a,b] proof (imp a b) --> proof a -> proof b
[A,p] proof (all A p) --> x : term A -> proof (p x)
[A,p] proof (ex A p) -->
  c: prop -> (x : term A -> proof (p x) -> proof c) -> proof c.
```

Each rewrite rule in this definition of proof has the form [context] lhs --> rhs The context lists the free variables appearing in the left-hand side, the left-hand side is a pattern (a first-order pattern in this case but higher-order patterns in the sense of Miller [21] are also supported by Dedukti) and the right-hand side is a term whose free variables are contained in the context.

All the rules of natural deduction can now be proved, here is for example, the introduction rule for conjunction:

```
def and_intro (a : prop) (b : prop) (Ha : proof a) (Hb : proof b)
  : proof (and a b) :=
  c : prop => f : (proof a -> proof b -> proof c) => f Ha Hb.
```

The syntax x : A => b is used in Dedukti for the λ-abstraction $\lambda x : A.b$.

To check the certificates found by automatic theorem provers for classical logic, we need two axiom schemes: the law of excluded middle and the assumption that all sorts are inhabited.

```
excluded_middle : a : prop -> proof (or a (not a)).
default_value : A : sort -> term A.
```

The Dedukti signature that we have described in this section is a faithful encoding of classical first-order logic [14]: a first-order formula φ is provable in classical natural deduction if and only if the Dedukti type proof φ is inhabited.

3 A Typed Tactic Language for Meta Dedukti

Unfortunately, writing Dedukti terms in the signature of the previous section is tedious not only for human users but also for automated tools which typically reason at a higher level than natural deduction proofs. In this section, we propose a first tactic language to ease the creation of terms in this signature.

Since Dedukti does not check for termination, it is very easy to encode a Turing-complete language in Dedukti. For example, the untyped λ-calculus can be encoded with only one declaration `def A : Type.` and one rewrite rule `[ ] A --> A -> A.`

Thanks to Turing-completeness, we can use Dedukti as a dependently-typed programming language based on rewriting. The results of these programs are Dedukti terms that need to be checked in a trusted Dedukti signature such as the one of Sect. 2 if we want to interpret them as proofs. We distinguish two different Dedukti signatures: the trusted signature of Sect. 2 and an untrusted signature extending the one of Sect. 2 and used to elaborate terms to be checked in the trusted one. Unless otherwise precised, all the Dedukti excerpts from now on are part of this second, untrusted signature.

When using Dedukti as a meta-programming language, we are not so much interested in the type-checking problem than in the normal forms (with respect to the untrusted system) of some terms. For this reason, we use a fork of Dedukti called Meta Dedukti [13] that we developed with Thiré. This tool outputs a copy of its input Dedukti file in which each term is replaced by its normal form. The produced file can then be sent to Dedukti to be checked in the trusted signature.

```
exc : Type.
mtactic : prop -> Type.

mret : A : prop -> proof A -> mtactic A.
mraise : A : prop -> exc -> mtactic A.
def mrun : A : prop -> mtactic A -> proof A.
def mbind : A : prop -> B : prop ->
  mtactic A -> (proof A -> mtactic B) -> mtactic B.
def mtry : A : prop -> mtactic A -> (exc -> mtactic A) -> mtactic A.
def mintro_term : A : sort -> B : (term A -> prop) ->
                    (x : term A -> mtactic (B x)) -> mtactic (all A B).
def mintro_proof : A : prop -> B : prop ->
                    (proof A -> mtactic B) -> mtactic (imp A B).
```

Fig. 2. The typed tactic language: declarations

In Figs. 2 and 3 we define our typed tactic language for Meta Dedukti inspired by the MTac tactic language for Coq [25]. The main type of this development is the type `mtactic a` (for monadic tactic) where a is a proposition. We call *tactical* any function returning a term of type `mtactic a` for some a. A term t of type `mtactic a` contains instructions to attempt a proof of the proposition a. Each tactic can either fail, in which case its normal form is `mraise a e` where e is of type `exc`, an extensible type of exceptions or succeed in which case its normal form is `mret a p` where p is a proof of a. The tacticals `mret` and `mraise` can be seen as the two constructors of the inductive type family `mtactic`. When evaluating a tactic is successful, we can extract the produced proof using the `mrun` partial function which is undefined in the case of the `mraise` constructor. Tactics can be chained using the `mbind` tactical and backtracking points can be set using the `mtry` tactical.

```
[a] mrun _ (mret _ a) --> a.

[f,t] mbind _ _ (mret _ t) f --> f t
[B,t] mbind _ B (mraise _ t) _ --> mraise B t.

[A,t] mtry A (mret _ t) _ --> mret A t
[t,f] mtry _ (mraise _ t) f --> f t.

[A,B,b] mintro_term A B (x => mret (B x) (b x)) -->
    mret (all A B) (all_intro A B (x => b x))
[A,B,e] mintro_term A B (x => mraise (B x) e) --> mraise (all A B) e.

[A,B,b] mintro_proof A B (x => mret B (b x)) -->
    mret (imp A B) (imp_intro A B (x => b x))
[A,B,e] mintro_proof A B (x => mraise _ e) --> mraise (imp A B) e.
```

Fig. 3. The typed tactic language: rewrite rules

The `mbind` tactical is enough to define tactics corresponding to all the rules of natural deduction that do not change the proof context. As a simple example, we can define a `msplit` tactical attempting to prove goals of the form `and a b` from tactics `t1` and `t2` attempting to prove `a` and `b` respectively.

```
def msplit (a : prop) (b : prop) (t1 : mtactic a) (t2 : mtactic b)
    : mtactic (and a b) :=
    mbind a (and a b) t1 (Ha =>
    mbind b (and a b) t2 (Hb =>
    mret (and a b) (and_intro a b Ha Hb))).
```

To handle the natural deduction rules that do modify the rule context such as the introduction rules for implication and universal quantification, we add two new tacticals `mintro_term` and `mintro_proof`. These tacticals are partial functions only defined if their argument is a tactical that uniformly succeed on all arguments or uniformly fail on all arguments.

4 An Untyped Tactic Language for Meta Dedukti

The main limitation of the typed tactic language presented in Sect. 3 is its verbosity. Since Dedukti does not feature implicit arguments, each time the user applies the `msplit` tactical, she has to provide propositions a and b such that the goal to be proved is convertible with `and a b`. Another issue is that this tactic language does not permit to automate search among assumptions; new assumptions can be introduced by the `mintro_proof` tactical but the user of the typed tactic language then has to refer explicitly to the introduced assumption.

The untyped[1] tactic language that we now consider solves both issues. Tactics are interpreted in a proof context, a list of terms and proofs, by the `eval` function returning a typed tactic. For the common case of evaluating a tactic in the empty context, we define the `prove` function.

```
context : Type.
nil_ctx : context.
cons_ctx_var : A : sort -> term A -> context -> context.
cons_ctx_proof : A : prop -> proof A -> context -> context.

tactic : Type.
def eval : context -> goal : prop -> tactic -> mtactic goal.

def prove (a : prop) (t : tactic) : proof a :=
  mrun a (eval nil_ctx a t).
```

Some of the most fundamental tacticals of the untyped language are defined in Fig. 4 by the way `eval` behaves on them. The `with_goal` tactical is used to get access to the current goal, it takes another tactical as argument and evaluates it on the goal. The `with_assumption` tactical tries a tactical on each assumption of the context until one succeeds. The `exact`, `raise`, `try`, `bind` and `intro` tacticals are wrapper around the constructs of the typed language. The full definitions of these tacticals and many other are available in the file https://gitlab.math. univ-paris-diderot.fr/cauderlier/dktactics/blob/master/meta/tactic.dk.

On top of these basic tacticals, we have implemented tacticals corresponding to the rules of intuitionistic sequent calculus. For example, Fig. 5 presents the definitions of the tacticals about conjunction: `match_and` deconstructs formulae of the form `and a b`, `split` performs the right rule of conjunction in sequent calculus and is defined very similarly to `msplit`, its typed variant of Sect. 3. The tactical `destruct_and` implements the following generalisation of the left rule for conjunction:

$$\frac{\Gamma \vdash A \wedge B \quad \Gamma, A, B \vdash C}{\Gamma \vdash C}$$

The axiom rule of sequent calculus is implemented by the `assumption` tactic defined as `with_assumption exact`.

5   Example of Interactive Proof Development

Before considering sophisticated applications of our tactic languages in Sects. 6 and 7, we illustrate the interactive use of our untyped tactic language on a simple example: commutativity of conjunction.

---

[1] By "untyped" we do not mean that no type is assigned to the Dedukti terms of the language but that typing is trivial: all the tactics have the same type (`tactic`).

```
with_goal : (prop -> tactic) -> tactic.
[ctx,goal,F] eval ctx goal (with_goal F) --> eval ctx goal (F goal).

with_assumption : (A : prop -> proof A -> tactic) -> tactic.
[ctx,goal,F] eval ctx goal (with_assumption F) --> ...

exact_mismatch : exc.
exact : a : prop -> proof a -> tactic.
[a,H] eval _ a (exact a H) --> mret a H
[a] eval _ a (exact _ _) --> mraise a exact_mismatch.

raise : exc -> tactic.
[a,e] eval _ a (raise e) --> mraise a e.

try : tactic -> (exc -> tactic) -> tactic.
[ctx,goal,t,f] eval ctx goal (try t f) --> mtry ...

bind : A : prop -> tactic -> (proof A -> tactic) -> tactic.
[ctx,goal,A,t,f] eval ctx goal (bind A t f) --> mbind ...

intro_failure : exc.
intro : tactic -> tactic.
[ctx,a,b,t] eval ctx (imp a b) (intro t) --> mintro_proof ...
[ctx,A,p,t] eval ctx (all A p) (intro t) --> mintro_term ...
[goal] eval _ goal (intro _) --> mraise goal intro_failure.
```

Fig. 4. Low-level untyped tacticals

```
matching_failure : exc.

def match_and : prop -> (prop -> prop -> tactic) -> tactic.
[a,b,t] match_and (and a b) t --> t a b
[] match_and _ _ --> raise matching_failure.

def split (t1 : tactic) (t2 : tactic) :=
  with_goal (goal => match_and goal
    (a => b => bind a t1 (Ha =>
             bind b t2 (Hb =>
             exact (and a b) (and_intro a b Ha Hb)))))).

def destruct_and (a : prop) (b : prop) (tab : tactic) (t : tactic) :=
  with_goal (goal => bind (and a b) tab (Hab =>
    bind (imp a (imp b goal)) (intro (intro t)) (Hf =>
    exact goal (Hf (and_elim_1 a b Hab) (and_elim_2 a b Hab)))))).
```

Fig. 5. Conjunction tacticals

We start with the following Dedukti file:

```
def t0 : tactic.

def and_commutes (a : prop) (b : prop) : proof (imp (and a b) (and b a))
:= prove (imp (and a b) (and b a)) t0.
```

The undefined constant t0 is a placeholder for an unsolved goal. The interactive process consists in looking into the normal form of this file for blocked applications of the eval function, adding some lines after the declaration of t0, and repeating until the definition of and_commutes is a term of the encoding of Sect. 2.

At the first iteration, Meta Dedukti answers

```
def t0 : tactic.

def and_commutes : a:prop -> b:prop ->
  (c:prop -> ((proof a) -> (proof b) -> proof c) -> proof c) ->
  c:prop -> ((proof b) -> (proof a) -> proof c) -> proof c
  := a:prop => b:prop =>
  mrun (imp (and a b) (and b a))
    (eval nil_ctx (imp (and a b) (and b a)) t0).
```

We have one blocked call to eval on the last line:

`eval nil_ctx (imp (and a b) (and b a)) t0`; this means we have to prove $\vdash (a \land b) \Rightarrow (b \land a)$. To apply the intro tactical, we introduce a new undefined subgoal t1 and define t0 as intro t1 by adding the following line in the middle of our file.

<code>def t1 : tactic. [ ] t0 --> intro t1.</code>

Normalising again produces a file containing the term

`eval (cons_ctx_proof (and a b) a0 nil_ctx) (and b a) t1` which means we now have to prove $a \land b \vdash b \land a$. To do this we add the following lines right after the previously added line:

```
def t2 : tactic.
def t3 : tactic.
[t1] t1 --> with_assumption (c => H => match_and c
    (a => b => destruct_and a b t2 t3)).
```

In other words we try to apply the destruct_and tactical successively to all assumptions of the proof context. Since we have only one assumption and it is indeed a conjunction, the call reduces and Meta Dedukti tells us that we are left with `eval (cons_ctx_proof (and a b) a0 nil_ctx) (and a b) t2` and `eval (cons_ctx_proof b a3 (cons_ctx_proof a a2 (cons_ctx_proof (and a b) a0 nil_ctx))) (and b a) t3` corresponding respectively to $a \land b \vdash a \land b$ and $b, a, a \land b \vdash b \land a$. The first subgoal is trivial and can be solved with the assumption tactic that succeeds when the goal matches one of the assumptions. For the second subgoal, we introduce the conjunction.

```
[] t2 --> assumption.
def t4 : tactic.
def t5 : tactic.
[] t3 --> split t4 t5.
```

We again have two subgoals, **eval (cons_ctx_proof b a2 (cons_ctx_proof a a1 (cons_ctx_proof (and a b) a0 nil_ctx)))** b t4 and **eval (cons_ctx_proof b a2 (cons_ctx_proof a a1 (cons_ctx_proof (and a b) a0 nil_ctx)))** a t5 corresponding to $b, a, a \wedge b \vdash b$ and $b, a, a \wedge b \vdash a$. In both cases, the goal corresponds to one of the assumptions so the **assumption** tactic does the job.

```
[] t4 --> assumption.
[] t5 --> assumption.
```

Our theorem is now proved; the following definition of **and_commutes** given by Meta Dedukti is accepted by Dedukti:

```
def and_commutes : a:prop -> b:prop ->
  (c:prop -> ((proof a) -> (proof b) -> proof c) -> proof c) ->
  c:prop -> ((proof b) -> (proof a) -> proof c) -> proof c
  :=
  a:prop => b:prop => x => c:prop =>
  f:((proof b) -> (proof a) -> proof c) =>
  f (x b (x0 => y => y)) (x a (x0 => y => x0)).
```

6   Theorem Transfer

When translating independently developed formal libraries in Dedukti, we end up with two isomorphic copies A and B of the same notions. Contrary to the mathematical habit of identifying isomorphic structures, in formal proof systems a theorem φ_A on the structure A cannot be used without justification as a theorem φ_B on the structure B. However this justification, a proof of $\varphi_A \Rightarrow \varphi_B$, can be automated in tactic based proof assistants. The automation of such goals of the form $\varphi_A \Rightarrow \varphi_B$ is called theorem transfer [17,26] and the tactic implementing it is called a transfer tactic.

In Fig. 6, we adapt the higher-order transfer calculi of [17,26] to first-order logic. The notations $P(\mathcal{R}_1 \times \ldots \times \mathcal{R}_n)Q$ abbreviates the formula $\forall x_1, \ldots x_n, y_1, \ldots y_n . x_1 \mathcal{R}_1 y_1 \Rightarrow \ldots \Rightarrow x_n \mathcal{R}_n y_n \Rightarrow P(x_1, \ldots, x_n) \Rightarrow Q(y_1, \ldots, y_n)$ and the notation $f(\mathcal{R}_1 \times \ldots \times \mathcal{R}_n \to \mathcal{R})g$ abbreviates the formula $\forall x_1, \ldots x_n, y_1, \ldots y_n . x_1 \mathcal{R}_1 y_1 \Rightarrow \ldots \Rightarrow x_n \mathcal{R}_n y_n \Rightarrow f(x_1, \ldots, x_n)\mathcal{R}g(y_1, \ldots, y_n)$.

Implementing a proof search algorithm for this calculus in our untyped tactic language is straightforward once we have proved the formula schemes $\bot \Rightarrow \bot$, $(\varphi_1 \Rightarrow \psi_1) \Rightarrow (\varphi_2 \Rightarrow \psi_2) \Rightarrow (\varphi_1 \wedge \varphi_2) \Rightarrow (\psi_1 \wedge \psi_2)$, $(\varphi_1 \Rightarrow \psi_1) \Rightarrow (\varphi_2 \Rightarrow \psi_2) \Rightarrow (\varphi_1 \vee \varphi_2) \Rightarrow (\psi_1 \vee \psi_2)$, $\ldots$ corresponding to the rules of the calculus.

Instead of deriving the proofs of these formula schemes in natural deduction directly, we take benefit of our tactic language to define an **auto** tactic following a rather naive strategy for sequent calculus: it applies right rules for all connectives but the existential quantifier as long as possible and then applies left rules for all connectives but universal quantification until the goal matches one of the assumptions. The auto tactic is able to prove the four first rules of our transfer calculus. The four remaining rules require to instantiate universal assumptions and are hence out of its scope but they are easy to prove directly.

Our implementation is available at the following URL: https://gitlab.math.univ-paris-diderot.fr/cauderlier/dktransfer.

$$\frac{}{\Gamma \vdash \bot \Rightarrow \bot} \qquad \frac{\Gamma \vdash \varphi_1 \Rightarrow \psi_1 \quad \Gamma \vdash \varphi_2 \Rightarrow \psi_2}{\Gamma \vdash (\varphi_1 \wedge \varphi_2) \Rightarrow (\psi_1 \wedge \psi_2)}$$

$$\frac{\Gamma \vdash \varphi_1 \Rightarrow \psi_1 \quad \Gamma \vdash \varphi_2 \Rightarrow \psi_2}{\Gamma \vdash (\varphi_1 \vee \varphi_2) \Rightarrow (\psi_1 \vee \psi_2)} \qquad \frac{\Gamma \vdash \psi_1 \Rightarrow \varphi_1 \quad \Gamma \vdash \varphi_2 \Rightarrow \psi_2}{\Gamma \vdash (\varphi_1 \Rightarrow \varphi_2) \Rightarrow (\psi_1 \Rightarrow \psi_2)}$$

$$\frac{\Gamma, a : A, c : C, a \, \mathcal{R} \, c \vdash \varphi_a \Rightarrow \psi_c \qquad \vdash \forall c : C. \; \exists a : A. \; a \, \mathcal{R} \, c}{\Gamma \vdash (\forall a : A. \; \varphi_a) \Rightarrow (\forall c : C. \; \psi_c)}$$

$$\frac{\Gamma, a : A, c : C, a \, \mathcal{R} \, c \vdash \varphi_a \Rightarrow \psi_c \qquad \vdash \forall a : A. \; \exists c : C. \; a \, \mathcal{R} \, c}{\Gamma \vdash (\exists a : A. \; \varphi_a) \Rightarrow (\exists c : C. \; \psi_c)}$$

$$\frac{\Gamma \vdash t_1 \, \mathcal{R}_1 \, u_1 \quad \ldots \quad \Gamma \vdash t_n \, \mathcal{R}_n \, u_n \qquad \vdash P \, (\mathcal{R}_1 \times \ldots \times \mathcal{R}_n) \, Q}{\Gamma \vdash P(t_1, \ldots, t_n) \Rightarrow Q(u_1, \ldots, u_n)}$$

$$\frac{\Gamma \vdash t_1 \, \mathcal{R}_1 \, u_1 \quad \ldots \quad \Gamma \vdash t_n \, \mathcal{R}_n \, u_n \qquad \vdash f \, (\mathcal{R}_1 \times \ldots \times \mathcal{R}_n \to \mathcal{R}) \, g}{\Gamma \vdash f(t_1, \ldots, t_n) \, \mathcal{R} \, g(u_1, \ldots, u_n)}$$

Fig. 6. A first-order transfer calculus

7   Resolution Certificates

Robinson's resolution calculus [22] is a popular proof calculus for first-order automatic theorem provers. It is a clausal calculus; this means that it does not handle the full syntax of first-order formulae but only the CNF (clausal normal form) fragment.

A *literal* is either an atom (a positive literal) or the negation of an atom (a negative literal). We denote by $\bar{l}$ the opposite literal of l defined by $\bar{a} := \neg a$ and $\overline{\neg a} := a$ where a is any atom. A *clause* is a possibly empty disjunction of literals. The empty clause corresponds to falsehood. Literals and clauses may contain free variables which are to be interpreted as universally quantified. We make this explicit by considering *quantified clauses* (qclauses for short) which are formulae of the form $\forall x_1, \ldots, x_k.l_1 \vee \ldots \vee l_n$.

A resolution proof is a derivation of the empty clause from a set of clauses assumed as axioms. The rules of the resolution calculus are given in Fig. 7. The (*Factorisation*) and (*Resolution*) rules are standard, the (*Unquantification*) rule is required to remove useless quantifications in the clauses produced by the two other rules. Note that the correctness of this (*Unquantification*) rule requires the `default_value` axiom that we introduced in Sect. 2.

We consider resolution certificates in which the assumed and derived clauses are numbered and each line of the certificate indicates:

1. the name of the derivation rule (either "Factorisation" or "Resolution"),
2. the numbers identifying one or two (depending on the chosen derivation rule) previously assumed or derived clauses,
3. the indexes i and j of the literals to unify, and
4. the number of the newly derived clause.

$$\frac{\forall \vec{x}.\ l_1 \vee \ldots \vee l_n \qquad \sigma = mgu(l_i, l_j)}{\forall \vec{x}.\ \sigma(l_1 \vee \ldots \vee l_{j-1} \vee l_{j+1} \vee \ldots \vee l_n)}(Factorisation)$$

$$\frac{\forall \vec{x}.\ l_1 \vee \ldots \vee l_n \qquad \forall \vec{y}.\ l'_1 \vee \ldots \vee l'_m \qquad \sigma = mgu(l_i, \overline{l_j})}{\forall \vec{x}, \vec{y}.\ \sigma(l_1 \ldots l_{i-1} \vee l_{i+1} \ldots l_n \vee l'_1 \ldots l'_{j-1} \vee l'_{j+1} \ldots l'_m)}(Resolution)$$

$$\frac{\forall \vec{x}.\ C \qquad FV(C) = \vec{y}}{\forall \vec{y}.\ C}(Unquantification)$$

Fig. 7. The resolution calculus with quantified clauses

```
atom : Type.
mk_atom : p : predicate -> terms (pred_domain p) -> atom.

literal : Type.
pos_lit : atom -> literal.
neg_lit : atom -> literal.

clause : Type.
empty_clause : clause.
cons_clause : literal -> clause -> clause.

qclause : Type.
qc_base : clause -> qclause.
qc_all : A : sort -> (term A -> qclause) -> qclause.
```

Fig. 8. Syntactic definitions for the CNF fragment of first-order logic

This level of detail is not unreasonable to ask from a resolution prover; Prover9 [20] for example is able to produce such certificates. To express these certificates in Meta Dedukti, we have extended the trusted signature of first-order logic with the definitions of the syntactic notions of atoms, literals, clauses, and qclauses (see Fig. 8) and we have defined functions `factor`, `resolve`, and unquantify returning the qclause resulting respectively from the (*Factorisation*), (*Resolution*), and (*Unquantification*) rules and tacticals factor_correct, resolve_correct, and unquantify_correct attempting to prove the resulting clauses from proofs of the initial clauses. Moreover, we defined a partial function qclause_of_prop mapping propositions in the clausal fragment to quantified clauses and we proved it correct on this fragment. The signature of these functions is given in Fig. 9.

```
def cprop : qclause -> prop.

def qclause_of_prop : prop -> qclause.
def qclause_of_prop_correct : a : prop -> proof a ->
  mtactic (cprop (qclause_of_prop a)).

def factor : nat -> nat -> qclause -> qclause.
def resolve : nat -> nat -> qclause -> qclause -> qclause.

def factor_correct : i : nat -> j : nat -> C : qclause ->
  proof (cprop C) -> mtactic (cprop (factor i j C)).
def resolve_correct : i : nat -> j : nat -> C : qclause -> D : qclause ->
  proof (cprop C) -> proof (cprop D) -> mtactic (cprop (resolve i j C D)).

def unquantify : qclause -> qclause.
def unquantify_correct : C : qclause -> proof (cprop C) ->
  mtactic (cprop (unquantify C)).
```

Fig. 9. Signature of the resolution tacticals

As a small example illustrating the use of these tacticals, we consider the problem NUM343+1 from the TPTP benchmark [24]. Among the clauses resulting from the clausification of the problem, two of them are used in the proof found by Prover9: $x \leq y \vee y \leq x$ and $\neg(x \leq n)$. The translation of this problem in Dedukti is given in Fig. 10. Here is a resolution certificate of the empty clause from these axioms:

1.	$x \leq y \vee y \leq x$	Axiom
2.	$\neg(x \leq n)$	Axiom
3.	$x \leq x$	Factorisation at positions 0 and 1 in clause 1
4.	$\bot$	Resolution at positions 0 and 0 in clauses 2 and 3

This certificate can be translated in our formalism by adding an (*Unquantification*) step after each other step. The Meta Dedukti version of this certificate is given in Fig. 11, once normalized by Meta Dedukti, we obtain a Dedukti file of 518 lines that is successfully checked by Dedukti in the trusted signature.

During the definition of the tacticals of Fig. 9, we were happily surprised to discover that the tacticals qclause_of_prop_correct, factor_correct, resolve_correct, and unquantify_correct were not much harder to define than the corresponding clause computing functions because we did not prove the soundness of the resolution calculus. In particular, we did not prove the correctness of our unification algorithm but we check *a posteriori* that the returned substitution is indeed an unifier of the given literals. The main difficulty comes from the application of substitution to qclauses which can be isolated in a rule called specialisation:

$$\frac{\forall x_1 \ldots \forall x_n.C}{\forall x_1 \ldots \forall x_n.\sigma(C)}\ (Specialisation)$$

If c is a proof of $\forall \overrightarrow{x}.C$, then a proof of $\forall \overrightarrow{x}.\sigma(C)$ can be obtained by first introducing all the quantifiers then applying c to $\sigma(x_1), \ldots, \sigma(x_n)$.

```
(; Signature ;)
A : sort.
LEQ : predicate.
[] pred_domain LEQ --> cons_sort A (cons_sort A nil_sort).
N : function.
[] fun_domain N --> nil_sort.
[] fun_codomain N --> A.
def leq (a : term A) (b : term A) : prop :=
  pred_apply LEQ
     (cons_term A a (cons_sort A nil_sort)
        (cons_term A b nil_sort nil_term)).
def n : term A := fun_apply N nil_term.

(; Axioms ;)
def A0 := all A (x => all A (y => or (leq x y) (leq y x))).
a0 : proof A0.
def A1 := all A (x => not (leq x n)).
a1 : proof A1.
```

Fig. 10. The TPTP problem NUM343+1 in Meta Dedukti

```
(; Clauses ;)
def C0 := qclause_of_prop A0.
def c0 := mrun (cprop C0) (qclause_of_prop_correct A0 a0).
def C1 := qclause_of_prop A1.
def c1 := mrun (cprop C1) (qclause_of_prop_correct A1 a1).
def C2' := factor 0 1 C0.
def c2' := mrun (cprop C2') (factor_correct 0 1 C0 c0).
def C2 := unquantify C2'.
def c2 := mrun (cprop C2) (unquantify_correct C2' c2').
def C3' := resolve 0 0 C1 C2.
def c3' := mrun (cprop C3') (resolve_correct 0 0 C1 C2 c1 c2).
def C3 := unquantify C3'.
def c3 : proof false := mrun (cprop C3) (unquantify_correct C3' c3').
```

Fig. 11. A resolution certificate for TPTP problem NUM343+1 in Meta Dedukti

From our tactic languages, it is not easy to do this because the number n of introduction and elimination rules to apply is unknown. To solve this problem, we defined an alternative form of quantified clauses were instead of quantifying over terms one by one, we quantify over lists of terms: `def lqclause (As : sorts) : Type := terms As -> clause`. We proved the specialisation rule on this type `lqclause` and the equivalence between `lqclause` and `qclause`.

The tactic languages of Sects. 3 and 4 and the resolution certificate checker of this section are available at the following URL: https://gitlab.math.univ-paris-diderot.fr/cauderlier/dktactics.

8   Related Works

The main source of inspiration for the typed tactic language that we have proposed in Sect. 3 is MTac [25], a typed monadic language for the Coq proof assistant. Our language is a fragment of MTac; the missing MTac primitives provide non-termination (the **mfix** construct) and give access to operations of Coq refiner such as syntactic deconstruction of terms (**mmatch**), higher-order unification, and handling of meta variables. To provide these features, the operational semantics of MTac is implemented inside Coq refiner. In this work in contrast, we did not modify Dedukti at all. The **mfix** and **mmatch** operations are not needed in our tactic languages because the user already has access to Dedukti higher-order rewrite engine. Since Dedukti is not a full proof assistant but only a proof checker, it does not currently feature a refiner from which we could leverage higher-order unification or meta variables. However, as we have seen in Sect. 5, we can simulate meta variables by definable symbols of type `tactic` and as we have seen in Sect. 7 in the first-order case we can also define the unification algorithm.

A second version of MTac is in preparation [19]. In MTac2, an untyped tactic language is built on top of the MTac monad but contrary to our untyped language in which tactics promise proofs of the current goal, MTac2 tactics promise lists of subgoals and the actual proof is built by instanciation of meta variables. This gives MTac2 the flexibility to define tactics generating a number of subgoals that is not known statically.

Exceptions and backtracking are also implemented by a monad in the meta language of Lean which is used to implement Lean tactics [15]. However, Lean meta language is poorly typed making this tactic language closer to our untyped tactic language: the way tactics manipulate the proof state in Lean is not made explicit in their types and terms are all reified in the same type `expr`.

The tactics of the Idris [8] system, which are used to elaborate terms from the full Idris syntax to the syntax of Idris' kernel, are also implemented by a monad in Haskell. However, this tactic monad is not reflected in Idris so Idris users do not have access to an extensible tactic language.

To bridge the gap between automatic and interactive theorem proving, a lot of efforts has been put to check the certificates of automatic theorem provers.

iProver Modulo [9], Zenon Modulo [12], and Metis [18] are first-order theorem provers able to produce independently checkable proofs. Metis in particular can be used as a tactic in Isabelle/HOL. The Sledgehammer tool [4] checks certificates from first-order provers and SMT solvers using Isabelle tactics implementing decision procedures and the Metis tactic. These works have in common an access to a deep representation of terms, typically using De Bruijn indices or named variables, at proof producing time whereas our tactics for the resolution calculus only use higher-order abstract syntax. Recently, the Foundational Proof Certificate framework has been used to add enough details to Prover9 resolution certificates that they can be checked by a simple tool that does not need to compute the unifiers [6]. In our context, we have found that is was actually easier to perform the unification in the certificate checker than to extend the format of certificates to include the substitutions because the naming of free variables in clauses (or the order in which variables are implicitly quantified) is hard to predict.

9 Conclusion

We have shown that Dedukti could be used as an expressive meta language for writing tactics and checking proof certificates. We have proposed two tactic languages for Dedukti, a typed one and an untyped one and shown applications of these languages to interactive proof development, automated theorem transfer, and checking of resolution certificates.

For interactive proof development and tactic debugging, our languages would greatly benefit from pretty-printing functions. We believe such functions can be defined in a second meta signature used to transform blocked `eval` calls to something more readable.

Our tactic and certificate languages are defined specifically for first-order logic. Since it was inspired by tactic languages for the Calculus of Inductive Constructions, we believe that most of the work presented in this article can be adapted straightforwardly to richer logics with the notable exception of the unification algorithm used to check resolution certificates.

Most clausal first-order theorem provers use an extra rule called paramodulation to handle equality. We would like to extend our resolution certificate language to take this rule into account. This would allow us to benchmark our certificate checker on large problem sets such as TPTP.

References

1. Assaf, A.: A framework for defining computational higher-order logics. Ph.D. thesis, École Polytechnique (2015). https://tel.archives-ouvertes.fr/tel-01235303
2. Assaf, A., Burel, G.: Translating HOL to Dedukti. In: Kaliszyk, C., Paskevich, A. (eds.) Proceedings Fourth Workshop on Proof eXchange for Theorem Proving, Berlin, Germany, 2–3 August 2015. Electronic Proceedings in Theoretical Computer Science, vol. 186, pp. 74–88. Open Publishing Association, Berlin, August 2015. https://doi.org/10.4204/EPTCS.186.8

3. Barrett, C., Fontaine, P., Tinelli, C.: The Satisfiability Modulo Theories Library (SMT-LIB) (2016). http://smtlib.cs.uiowa.edu
4. Blanchette, J.C., Bulwahn, L., Nipkow, T.: Automatic proof and disproof in Isabelle/HOL. In: Tinelli, C., Sofronie-Stokkermans, V. (eds.) FroCoS 2011. LNCS (LNAI), vol. 6989, pp. 12–27. Springer, Heidelberg (2011). https://doi.org/10.1007/978-3-642-24364-6_2
5. Blanchette, J.C., Paskevich, A.: TFF1: the TPTP typed first-order form with rank-1 polymorphism. In: Bonacina, M.P. (ed.) CADE 2013. LNCS (LNAI), vol. 7898, pp. 414–420. Springer, Heidelberg (2013). https://doi.org/10.1007/978-3-642-38574-2_29
6. Blanco, R., Chihani, Z., Miller, D.: Translating between implicit and explicit versions of proof. In: de Moura, L. (ed.) CADE 2017. LNCS (LNAI), vol. 10395, pp. 255–273. Springer, Cham (2017). https://doi.org/10.1007/978-3-319-63046-5_16
7. Boespflug, M., Carbonneaux, Q., Hermant, O.: The $\lambda\Pi$-calculus modulo as a universal proof language. In: David Pichardie, T.W. (ed.) The Second International Workshop on Proof Exchange for Theorem Proving (PxTP 2012), Manchester, vol. 878, pp. 28–43, June 2012. https://hal-mines-paristech.archives-ouvertes.fr/hal-00917845
8. Brady, E.: Idris, a general-purpose dependently typed programming language: design and implementation. J. Funct. Program. **23**(5), 552–593 (2013). https://doi.org/10.1017/S095679681300018X
9. Burel, G.: A shallow embedding of resolution and superposition proofs into the $\lambda\Pi$-calculus modulo. In: Blanchette, J.C., Urban, J. (eds.) Third International Workshop on Proof Exchange for Theorem Proving, PxTP 2013, Lake Placid, NY, USA, 9–10 June 2013. EPiC Series in Computing, vol. 14, pp. 43–57. EasyChair, Lake Placid, June 2013. http://www.easychair.org/publications/paper/141241
10. Cauderlier, R., Dubois, C.: ML pattern-matching, recursion, and rewriting: from FoCaLiZe to Dedukti. In: Sampaio, A., Wang, F. (eds.) ICTAC 2016. LNCS, vol. 9965, pp. 459–468. Springer, Cham (2016). https://doi.org/10.1007/978-3-319-46750-4_26
11. Cauderlier, R., Dubois, C.: FoCaLiZe and Dedukti to the rescue for proof interoperability. In: Ayala-Rincón, M., Muñoz, C.A. (eds.) ITP 2017. LNCS, vol. 10499, pp. 131–147. Springer, Cham (2017). https://doi.org/10.1007/978-3-319-66107-0_9
12. Cauderlier, R., Halmagrand, P.: Checking Zenon Modulo proofs in Dedukti. In: Kaliszyk, C., Paskevich, A. (eds.) Proceedings 4th Workshop on Proof eXchange for Theorem Proving, Berlin, Germany, 2–3 August 2015. Electronic Proceedings in Theoretical Computer Science, vol. 186, pp. 57–73. Open Publishing Association, Berlin, August 2015. https://doi.org/10.4204/EPTCS.186.7
13. Cauderlier, R., Thiré, F.: Meta Dedukti. http://deducteam.gforge.inria.fr/metadedukti/
14. Dorra, A.: Équivalence Curry-Howard entre le lambda-Pi-calcul et la logique intuitionniste (2010). Undergrad research intership report
15. Ebner, G., Ullrich, S., Roesch, J., Avigad, J., de Moura, L.: A metaprogramming framework for formal verification. PACMPL **1**(ICFP), 34:1–34:29 (2017). https://doi.org/10.1145/3110278
16. Gilbert, F.: Proof certificates in PVS. In: Ayala-Rincón, M., Muñoz, C.A. (eds.) ITP 2017. LNCS, vol. 10499, pp. 262–268. Springer, Cham (2017). https://doi.org/10.1007/978-3-319-66107-0_17
17. Huffman, B., Kunčar, O.: Lifting and transfer: a modular design for quotients in Isabelle/HOL. In: Gonthier, G., Norrish, M. (eds.) CPP 2013. LNCS, vol. 8307, pp. 131–146. Springer, Cham (2013). https://doi.org/10.1007/978-3-319-03545-1_9

18. Hurd, J.: First-order proof tactics in higher-order logic theorem provers. In: Archer, M., Vito, B.D., Muñoz, C. (eds.) Design and Application of Strategies/Tactics in Higher Order Logics (STRATA 2003), pp. 56–68. No. NASA/CP-2003-212448 in NASA Technical Reports, September 2003. http://www.gilith.com/papers
19. Kaiser, J.O., Ziliani, B., Krebbers, R., Régis-Gianas, Y., Dreyer, D.: Mtac2: Typed tactics for backward reasoning in Coq (2018, submitted for publication)
20. McCune, W.: Prover9 and Mace4 (2005–2010). http://www.cs.unm.edu/~mccune/prover9/
21. Miller, D.: A logic programming language with lambda-abstraction, function variables, and simple unification. J. Log. Comput. 1(4), 497–536 (1991). https://doi.org/10.1093/logcom/1.4.497
22. Robinson, J.A.: A machine-oriented logic based on the resolution principle. J. ACM 12(1), 23–41 (1965). https://doi.org/10.1145/321250.321253
23. Saillard, R.: Type checking in the Lambda-Pi-Calculus Modulo: theory and practice. Ph.D. thesis, MINES Paritech (2015). https://pastel.archives-ouvertes.fr/tel-01299180
24. Sutcliffe, G.: The TPTP problem library and associated infrastructure. J. Autom. Reasoning 43(4), 337–362 (2009). https://doi.org/10.1007/s10817-009-9143-8
25. Ziliani, B., Dreyer, D., Krishnaswami, N.R., Nanevski, A., Vafeiadis, V.: Mtac: a monad for typed tactic programming in Coq. J. Funct. Program. 25 (2015). https://doi.org/10.1017/S0956796815000118
26. Zimmermann, T., Herbelin, H.: Automatic and Transparent Transfer of Theorems along Isomorphisms in the Coq Proof Assistant. CoRR abs/1505.05028 (2015). http://arxiv.org/abs/1505.05028

A Formalization of the LLL Basis Reduction Algorithm

Jose Divasón[1], Sebastiaan Joosten[2], René Thiemann[3(⊠)],
and Akihisa Yamada[4]

[1] University of La Rioja, Logroño, Spain
[2] University of Twente, Enschede, The Netherlands
[3] University of Innsbruck, Innsbruck, Austria
rene.thiemann@uibk.ac.at
[4] National Institute of Informatics, Tokyo, Japan

Abstract. The LLL basis reduction algorithm was the first polynomial-time algorithm to compute a reduced basis of a given lattice, and hence also a short vector in the lattice. It thereby approximates an NP-hard problem where the approximation quality solely depends on the dimension of the lattice, but not the lattice itself. The algorithm has several applications in number theory, computer algebra and cryptography.

In this paper, we develop the first mechanized soundness proof of the LLL algorithm using Isabelle/HOL. We additionally integrate one application of LLL, namely a verified factorization algorithm for univariate integer polynomials which runs in polynomial time.

1 Introduction

The LLL basis reduction algorithm by Lenstra, Lenstra and Lovász [8] is a remarkable algorithm with numerous applications. There even exists a 500-page book solely about the LLL algorithm [10]. It lists applications in number theory and cryptology, and also contains the best known polynomial factorization algorithm that is used in today's computer algebra systems.

The LLL algorithm plays an important role in finding short vectors in lattices: Given some list of linearly independent integer vectors $f_0, \ldots, f_{m-1} \in \mathbb{Z}^n$, the corresponding lattice L is the set of integer linear combinations of the f_i; and the shortest vector problem is to find some non-zero element in L which has the minimum norm.

Example 1. Consider $f_1 = (1, 1\,894\,885\,908, 0)$, $f_2 = (0, 1, 1\,894\,885\,908)$, and $f_3 = (0, 0, 2\,147\,483\,648)$. The lattice of f_1, f_2, f_3 has a shortest vector $(-3, 17, 4)$. It is the linear combination $(-3, 17, 4) = -3f_1 + 5\,684\,657\,741f_2 + 5\,015\,999\,938f_3$.

Whereas finding a shortest vector is NP-hard [9], the LLL algorithm is a polynomial time algorithm for approximating a shortest vector: The algorithm is parametric by some $\alpha > \frac{4}{3}$ and computes a *short vector*, i.e., a vector whose norm is at most $\alpha^{\frac{m-1}{2}}$ times as large than the norm of any shortest vector.

© The Author(s) 2018
J. Avigad and A. Mahboubi (Eds.): ITP 2018, LNCS 10895, pp. 160–177, 2018.
https://doi.org/10.1007/978-3-319-94821-8_10

In this paper, we provide the first mechanized soundness proof of the LLL algorithm: the functional correctness is formulated as a theorem in the proof assistant Isabelle/HOL [11]. Regarding the complexity of the LLL algorithm, we did not include a formal statement which would have required an instrumentation of the algorithm by some instruction counter. However, from the termination proof of our Isabelle implementation of the LLL algorithm, one can easily infer a polynomial bound on the number of arithmetic operations.

In addition to the LLL algorithm, we also verify one application, namely a polynomial-time[1] algorithm for the factorization of univariate integer polynomials, that is: factorization into the content and a product of irreducible integer polynomials. It reuses most parts of the formalization of the Berlekamp–Zassenhaus factorization algorithm, where the main difference is the replacement of the exponential-time reconstruction phase [1, Sect. 8] by a polynomial-time one based on the LLL algorithm.

The whole formalization is based on definitions and proofs from a textbook on computer algebra [16, Chap. 16]. Thanks to the formalization work, we figured out that the factorization algorithm in the textbook has a serious flaw.

The paper is structured as follows. We present preliminaries in Sect. 2. In Sect. 3 we describe an extended formalization about the Gram–Schmidt orthogonalization procedure. This procedure is a crucial sub-routine of the LLL algorithm whose correctness is verified in Sect. 4. Since our formalization of the LLL algorithm is also executable, in Sect. 5 we compare it against a commercial implementation. We present our verified polynomial-time algorithm to factor integer polynomials in Sect. 6, and describe the flaw in the textbook. Finally, we give a summary in Sect. 7.

Our formalization is available in the archive of formal proofs (AFP) [2,3].

2 Preliminaries

We assume some basic knowledge of linear algebra, but recall some notions and notations. The inner product of two real vectors $v = (c_0, \ldots, c_n)$ and $w = (d_0, \ldots, d_n)$ is $v \cdot w = \sum_{i=0}^{n} c_i d_i$. Two real vectors are orthogonal if their inner product is zero. The Euclidean norm of a real vector v is $\|v\| = \sqrt{v \cdot v}$. A linear combination of vectors $v_0, \ldots, v_m$ is $\sum_{i=0}^{m} c_i v_i$ with $c_0, \ldots, c_m \in \mathbb{R}$, and we say it is an integer linear combination if $c_0, \ldots, c_m \in \mathbb{Z}$. Vectors are linearly independent if none of them is a linear combination of the others. The span – the set of all linear combinations – of linearly independent vectors $v_0, \ldots, v_{m-1}$ forms a vector space of dimension m, and $v_0, \ldots, v_{m-1}$ are its basis. The *lattice* generated by linearly independent vectors $v_0, \ldots, v_{m-1} \in \mathbb{Z}^n$ is the set of integer linear combinations of $v_0, \ldots, v_{m-1}$.

The degree of a polynomial $f(x) = \sum_{i=0}^{n} c_i x^i$ is *degree* $f = n$, the leading coefficient is *lc* $f = c_n$, the content is the GCD of coefficients $\{c_0, \ldots, c_n\}$, and the norm $\|f\|$ is the norm of its corresponding coefficient vector, i.e., $\|(c_0, \ldots, c_n)\|$.

---

[1] Again, we only mechanized the correctness proof and not the proof of polynomial complexity.

If $f = f_0 \cdot \ldots \cdot f_m$, then each f_i is a factor of f, and is a proper factor if f is not a factor of f_i. Units are the factors of 1, i.e., ± 1 in integer polynomials and non-zero constants in field polynomials. By a factorization of polynomial f we mean a decomposition $f = c \cdot f_0 \cdot \ldots \cdot f_m$ into the content c and irreducible factors $f_0, \ldots, f_m$; here irreducibility means that each f_i is not a unit and admits only units as proper factors.

Our formalization has been carried out using Isabelle/HOL, and we follow its syntax to state theorems, functions and definitions. Isabelle's keywords are written in **bold**. Throughout Sects. 3 and 4, we present Isabelle sources in a way where we are inside some context which fixes a parameter n, the dimension of the vector space.

3   Gram–Schmidt Orthogonalization

The Gram–Schmidt orthogonalization (GSO) procedure takes a list of linearly independent vectors $f_0, \ldots, f_{m-1}$ from $\mathbb{R}^n$ or $\mathbb{Q}^n$ as input, and returns an orthogonal basis $g_0, \ldots, g_{m-1}$ for the space that is spanned by the input vectors. In this case, we also write that $g_0, \ldots, g_{m-1}$ is the *GSO* of $f_0, \ldots, f_{m-1}$.

We already formalized this procedure in Isabelle as a function *gram_schmidt* when proving the existence of Jordan normal forms [15]. That formalization uses an explicit carrier set to enforce that all vectors are of the same dimension. For the current formalization task, the usage of a carrier-based vector and matrix library is important: Harrison's encoding of dimensions via types [5] is not expressive enough for our application; for instance for a given square matrix of dimension n we need to multiply the determinants of all submatrices that only consider the first i rows and columns for all $1 \le i \le n$.

Below, we summarize the main result that is formally proven about *gram_schmidt* [15]. To this end, we open a context assuming common conditions for invoking the Gram–Schmidt procedure, namely that *fs* is a list of linearly independent vectors, and that *gs* is the GSO of *fs*. Here, we also introduce our notion of linear independence for lists of vectors, based on an existing definition of linear independence for sets coming from a formalization of vector spaces [7].

definition *lin_indpt_list fs =*
  (*set fs ⊆ carrier_vec n ∧ distinct fs ∧ lin_indpt (set fs)*)

context
  fixes *fs gs m*
  assumes *lin_indpt_list fs* **and** *length fs = m* **and** *gram_schmidt n fs = gs*
begin
lemma *gram_schmidt:*
  shows *span (set fs) = span (set gs)* **and** *orthogonal gs*
    and *set gs ⊆ carrier_vec n* **and** *length gs = m*

Unfortunately, lemma *gram_schmidt* does not suffice for verifying the LLL algorithm: We need to know how *gs* is computed, that the connection between *fs* and *gs* can be expressed via a matrix multiplication, and we need recursive equations to compute *gs* and the matrix. In the textbook the Gram–Schmidt orthogonalization on input $f_0, \ldots, f_{m-1}$ is defined via mutual recursion.

$$g_i := f_i - \sum_{j<i} \mu_{i,j} g_j \tag{1}$$

where

$$\mu_{i,j} := \begin{cases} 1 & \text{if } i = j \\ 0 & \text{if } j > i \\ \frac{f_i \cdot g_j}{\|g_j\|^2} & \text{if } j < i \end{cases} \tag{2}$$

and the connection between these values is expressed as the equality

$$\begin{bmatrix} f_0 \\ \vdots \\ f_{m-1} \end{bmatrix} = \begin{bmatrix} \mu_{0,0} & \cdots & \mu_{0,m-1} \\ \vdots & \ddots & \vdots \\ \mu_{m-1,0} & \cdots & \mu_{m-1,m-1} \end{bmatrix} \cdot \begin{bmatrix} g_0 \\ \vdots \\ g_{m-1} \end{bmatrix} \tag{3}$$

by interpreting the f_i's and g_i's as row vectors.

Whereas there is no conceptual problem in expressing these definitions and proving the equalities in Isabelle/HOL, there still is some overhead because of the conversion of types. For instance in lemma *gram_schmidt*, *gs* is a list of vectors; in (1), *g* is a recursively defined function from natural numbers to vectors; and in (3), the list of g_i's is seen as a matrix.

Consequently, the formalized statement of (3) contains these conversions like *mat* and *mat_of_rows* which convert a function and a list of vectors into matrices, respectively. Note that in the formalization an implicit parameter to μ – the input vectors $f_0, \ldots, f_{m-1}$ – is made explicit as *fs*.

lemma *mat_of_rows n fs = mat m m ($\lambda(i, j)$. μ fs i j) · mat_of_rows n gs*

A key ingredient in reasoning about the LLL algorithm are projections. We say $w \in \mathbb{R}^n$ is a projection of $v \in \mathbb{R}^n$ into the orthocomplement of $S \subseteq \mathbb{R}^n$, or just w is an *oc-projection* of v and S, if $v - w$ is in the span of S and w is orthogonal to every element of S:

definition *is_oc_projection w S v =*
   ($w \in$ carrier_vec n $\land$ v $-$ w $\in$ span S $\land$ ($\forall u \in S$. w • u = 0))

A nice property of oc-projections is that they are unique up to v and the span of S. Back to GSO, since g_i is the oc-projection of f_i and $\{f_0, \ldots, f_{i-1}\}$, we conclude that g_i is uniquely determined in terms of f_i and the span of $\{f_0, \ldots, f_{i-1}\}$. Put differently, we obtain the same g_i even if we modify some of the first i input vectors of the GSO: only the span of these vectors must be preserved.

The connection between the Gram–Schmidt procedure and short vectors becomes visible in the following lemma: some vector in the orthogonal basis *gs* is shorter than any non-zero vector *h* in the lattice generated by *fs*. Here, 0_v *n* denotes the zero-vector of dimension *n*.

lemma *gram_schmidt_short_vector:*
  assumes $h \in$ *lattice_of fs* $- \{0_v\ n\}$
  shows $\exists\ g_i \in$ *set gs.* $\|g_i\|^2 \leq \|h\|^2$

Whereas this short-vector lemma requires only a half of a page in the textbook, in the formalization we had to expand the condensed paper-proof into 170 lines of more detailed Isabelle source, plus several auxiliary lemmas.

For instance, on papers one easily multiplies two sums $(\sum \ldots \cdot \sum \ldots = \sum \ldots)$ and directly omits quadratically many neutral elements by referring to orthogonality, whereas we first had to prove this auxiliary fact in 34 lines.

The short-vector lemma is the key to obtaining a short vector in the lattice. It tells us that the minimum value of $\|g_i\|^2$ is a lower bound for the norms of the non-zero vectors in the lattice. If $\|g_0\|^2 \leq \alpha \|g_1\|^2 \leq \ldots \leq \alpha^{m-1} \|g_{m-1}\|^2$ for some $\alpha \in \mathbb{R}$, then the basis is *weakly reduced w.r.t.* α. If moreover $\alpha \geq 1$, then $f_0 = g_0$ is a short vector in the lattice generated by $f_0, \ldots, f_{m-1}$: $\|f_0\|^2 = \|g_0\|^2 \leq \alpha^{m-1} \|g_i\|^2 \leq \alpha^{m-1} \|h\|^2$ for any non-zero vector *h* in the lattice.

In the formalization, we generalize the property of being weakly reduced by adding an argument *k*, and only demand that the first *k* vectors satisfy the required inequality. This is important for stating the invariant of the LLL algorithm. Moreover, we also define a partially *reduced* basis which additionally demands that the first *k* elements of the basis are nearly orthogonal, i.e., the μ-values are small.

definition *weakly_reduced* α *k gs* $= (\forall i.\ Suc\ i < k \longrightarrow \|gs\ !\ i\|^2 \leq \alpha \cdot \|gs\ !\ Suc\ i\|^2)$
definition *reduced* α *k gs* $\mu = ($*weakly_reduced* α *k gs* $\wedge \forall j < i < k.\ |\mu\ i\ j| \leq \frac{1}{2})$

lemma *weakly_reduced_imp_short_vector:*
  assumes *weakly_reduced* α *m gs*
    and $h \in$ *lattice_of fs* $- \{0_v\ n\}$
    and $1 \leq \alpha$
  shows $\|fs\ !\ 0\|^2 \leq \alpha^{m-1} \cdot \|h\|^2$
end (* close context about fs, gs, and m *)

The GSO of some basis $f_0, \ldots, f_{m-1}$ will not generally be (weakly) reduced, but this problem can be solved with the LLL algorithm.

4   The LLL Basis Reduction Algorithm

The LLL algorithm modifies the input $f_0, \ldots, f_{m-1} \in \mathbb{Z}^n$ until the corresponding GSO is reduced, while preserving the generated lattice. It is parametrized by some *approximation factor*[2] $\alpha > \frac{4}{3}$.

Algorithm 1. The LLL basis reduction algorithm, readable version

---

Input: A list of linearly independent vectors $f_0, \ldots, f_{m-1} \in \mathbb{Z}^n$ and $\alpha > \frac{4}{3}$
Output: A basis that generates the same lattice as $f_0, \ldots, f_{m-1}$ and has
reduced GSO w.r.t. α

1 $i := 0$; $g_0, \ldots, g_{m-1} := $ *gram_schmidt* $f_0, \ldots, f_{m-1}$
2 **while** $i < m$ **do**
3   **for** $j = i - 1, \ldots, 0$ **do**
4          $f_i := f_i - \lfloor \mu_{i,j} \rceil \cdot f_j$; $g_0, \ldots, g_{m-1} := $ *gram_schmidt* $f_0, \ldots, f_{m-1}$
5   **if** $i > 0 \wedge \|g_{i-1}\|^2 > \alpha \cdot \|g_i\|^2$ **then**
6          $(i, f_{i-1}, f_i) := (i - 1, f_i, f_{i-1})$; $g_0, \ldots, g_{m-1} := $ *gram_schmidt* $f_0, \ldots, f_{m-1}$
       else
7          $i := i + 1$
8 **return** $f_0, \ldots, f_{m-1}$

A readable, but inefficient, implementation of the LLL algorithm is given by Algorithm 1, which mainly corresponds to the algorithm in the textbook [16, Chap. 16.2–16.3]: the textbook fixes $\alpha = 2$ and $m = n$. Here, it is important to know that whenever the algorithm mentions $\mu_{i,j}$, it is referring to μ as defined in (2) for the *current* values of $f_0, \ldots, f_{m-1}$ and $g_0, \ldots, g_{m-1}$. In the algorithm, $\lfloor x \rceil$ denotes the integer closest to x, i.e., $\lfloor x \rceil := \lfloor x + \frac{1}{2} \rfloor$.

Let us have a short informal view on the properties of the LLL algorithm. The first required property is maintaining the lattice of the original input $f_0, \ldots, f_{m-1}$. This is obvious, since the basis is only changed in lines (4) and (6), and since swapping two basis elements, and adding a multiple of one basis element to a different basis element will not change the lattice. Still, the formalization of these facts required 190 lines of Isabelle code.

The second property is that the resulting GSO should be weakly reduced. This requires a little more argumentation, but is also not too hard: the algorithm maintains the invariant of the while-loop that the first i elements of the GSO are weakly reduced w.r.t. approximation factor α. This invariant is trivially maintained in line (7), since the condition in line (5) precisely checks whether the weakly reduced property holds between elements g_{i-1} and g_i. Moreover, being weakly reduced up to the first i vectors is not affected by changing the value of f_i, since the first i elements of the GSO only depend on $f_0, \ldots, f_{i-1}$. So, the modification of f_i in line (4) can be ignored w.r.t. being weakly reduced.

---

[2] The formalization also shows soundness for $\alpha = \frac{4}{3}$, but then the polynomial runtime is not guaranteed.

Hence, formalizing the partial correctness of Algorithm 1 w.r.t. being weakly reduced is not too difficult. What makes it interesting, are the remaining properties we did not yet discuss. The most difficult part is the termination of the algorithm, and it is also nontrivial that the final basis is reduced. Both of these properties require equations which precisely determine how the GSO will change by the modification of $f_0, \ldots, f_{m-1}$ in lines 4 and 6.

Once having these equations, we can drop the recomputation of the full GSO within the while-loop and replace it by local updates. Algorithm 2 is a more efficient algorithm to perform basis reduction, incorporating this improvement. Note that the GSO does not change in line 4, which can be shown with the help of oc-projections.

Algorithm 2. The LLL basis reduction algorithm, verified version

Input: A list of linearly independent vectors $f_0, \ldots, f_{m-1} \in \mathbb{Z}^n$ and $\alpha > \frac{4}{3}$
Output: A basis that generates the same lattice as $f_0, \ldots, f_{m-1}$ and has reduced GSO w.r.t. α

1   $i := 0$; $g_0, \ldots, g_{m-1} := gram\_schmidt\ f_0, \ldots, f_{m-1}$
2   **while** $i < m$ **do**
3   **for** $j = i - 1, \ldots, 0$ **do**
4         $f_i := f_i - \lceil \mu_{i,j} \rfloor \cdot f_j$
5   **if** $i > 0 \land \|g_{i-1}\|^2 > \alpha \cdot \|g_i\|^2$ **then**
6         $g'_{i-1} := g_i + \mu_{i,i-1} \cdot g_{i-1}$
7         $g'_i := g_{i-1} - \frac{f_{i-1} \cdot g'_{i-1}}{\|g'_{i-1}\|^2} \cdot g'_{i-1}$
8         $(i, f_{i-1}, f_i, g_{i-1}, g_i) := (i - 1, f_i, f_{i-1}, g'_{i-1}, g'_i)$
     else
9         $i := i + 1$
10 **return** $f_0, \ldots, f_{m-1}$

Concerning the complexity, each $\mu_{i,j}$ can be computed with $\mathcal{O}(n)$ arithmetic operations using its defining Eq. (2). Also, the updates of the GSO after swapping basis elements require $\mathcal{O}(n)$ arithmetic operations. Since there are at most m iterations in the for-loop, each iteration of the while-loop in Algorithm 2 requires $\mathcal{O}(n \cdot m)$ arithmetic operations. As a consequence, if $n = m$ and I is the number of iterations of the while-loop, then Algorithm 2 requires $\mathcal{O}(n^3 + n^2 \cdot I)$ many arithmetic operations, where the cubic number stems from the initial computation of the GSO in line 1.

To verify that Algorithm 2 computes a weakly reduced basis, it "only" remains to verify its termination, and the invariant that the updates of g_i's indeed correspond to the recomputation of the GSOs. These parts are closely related, because the termination argument depends on the explicit formula for the new value of g_{i-1} as defined in line 6 as well as on the fact that the GSO remains unchanged in lines 3–4.

Since the termination of the algorithm is not at all obvious, and since it depends on valid inputs (e.g., it does not terminate if $\alpha \leq 0$) we actually modeled the while-loop as a partial function in Isabelle [6]. Then in the soundness proof we only consider valid inputs and use induction via some measure which in turn gives us an upper bound on the number of loop iterations.

The soundness proof is performed via the following invariant. It is a simplified version of the actual invariant in the formalization, which also includes a property w.r.t. data refinement. It is defined in a context which fixes the lattice L, the number of basis elements m, and an approximation factor $\alpha \geq \frac{4}{3}$. Here, the function RAT converts a list of integer vectors into a list of rational vectors.

context ... **begin**
definition *LLL_invariant* (*i*, *fs*, *gs*) = (
  gram_schmidt *n* *fs* = *gs* $\wedge$
  lin_indpt_list (*RAT fs*) $\wedge$
  lattice_of fs = *L* $\wedge$
  length fs = *m* $\wedge$
  reduced α *i gs* (μ (*RAT fs*)) $\wedge$
  i $\leq$ *m*)

Using this invariant, the soundness of lines 3–4 is expressed as follows.

lemma *basis_reduction_add_rows:*
  assumes *LLL_invariant* (*i*, *fs*, *gs*)
    and *i* < *m*
    and *basis_reduction_add_rows* (*i*, *fs*, *gs*) = (*i'*, *fs'*, *gs'*)
  shows *LLL_invariant* (*i'*, *fs'*, *gs'*) **and** *i'* = *i* **and** *gs'* = *gs*
    and $\forall$ *j* < *i*. $|\mu$ *fs'* *i* *j*$| \leq \frac{1}{2}$

In the lemma, *basis_reduction_add_rows* is a function which implements lines 3–4 of the algorithm. The lemma shows that the invariant is maintained and the GSO is unchanged, and moreover expresses the sole purpose of lines 3–4: they make the values $\mu_{i,j}$ small.

For the total correctness of the algorithm, we must prove not only that the invariant is preserved in every iteration, but also that some measure decreases for proving termination. This measure is defined below using *Gramian determinants*, a generalization of determinants which also works for non-square matrices. This is also the point where the condition $\alpha > \frac{4}{3}$ becomes important: it ensures that the base $\frac{4\alpha}{4+\alpha}$ of the logarithm is strictly larger than 1.[3]

definition *Gramian_determinant fs k* =
  (**let** *M* = *mat_of_rows* *n* (*take k fs*) **in** *det* (*M* $\cdot$ *M*$^\mathsf{T}$))
definition *D fs* = ($\prod$ *k* < *m*. *Gramian_determinant fs k*)
definition *LLL_measure* (*i*, *fs*, *gs*) = *max* 0 (2 $\cdot$ $\lfloor$ *log* ($\frac{4\cdot\alpha}{4+\alpha}$) (*D fs*)$\rfloor$ + *m* − *i*)

---

[3] $\frac{4\alpha}{4+\alpha} = 1$ for $\alpha = \frac{4}{3}$ and in that case one has to drop the logarithm from the measure.

In the definition, the matrix M is the $k \times n$ submatrix of fs corresponding to the first k elements of fs. Note that the measure solely depends on the index i and the basis fs. However, for lines 3–4 we only proved that i and gs remain unchanged. Hence the following lemma is important: it implies that the measure can also be defined purely from i and gs, and that D fs will be positive.

lemma *Gramian_determinant:*

  assumes *LLL_invariant* (i, fs, gs) **and** $k \leq m$

  shows *Gramian_determinant* fs $k = (\prod j{<}k.\ \|gs\ !\ j\|^2)$

    and *Gramian_determinant* fs $k > 0$

With these preparations we are able to prove the most important property of the LLL algorithm, namely that each loop iteration – implemented as function *basis_reduction_step* – preserves the invariant and decreases the measure.

lemma *basis_reduction_step:*

  assumes *LLL_invariant* (i, fs, gs) **and** $i < m$

    and *basis_reduction_step* α $(i, fs, gs) = (i', fs', gs')$

  shows *LLL_invariant* (i', fs', gs')

    and *LLL_measure* $(i', fs', gs') < $ *LLL_measure* (i, fs, gs)

Our correctness proofs for *basis_reduction_add_rows* and *basis_reduction_step* closely follows the description in the textbook, and we mainly refer to the formalization and the textbook for more details: the presented lemmas are based on a sequence of technical lemmas that we do not expose at this point.

Here, we only sketch the termination proof: The value of the Gramian determinant for parameter $k \neq i$ stays identical when swapping f_i and f_{i-1}, since it just corresponds to an exchange of two rows which will not modify the absolute value of the determinant. The Gramian determinant for parameter $k = i$ will decrease by using the first statement of lemma *Gramian_determinant*, the explicit formula for the updated g_{i-1} in line 6, the condition $\|g_{i-1}\|^2 > \alpha \cdot \|g_i\|^2$, and the fact that $|\mu_{i,i-1}| \leq \frac{1}{2}$.

The termination proof together with the measure function shows that the implemented algorithm requires a polynomial number of arithmetic operations: we prove an upper bound on the number of iterations which in total shows that executing Algorithm 2 for $n = m$ requires $\mathcal{O}(n^4 \cdot \log A)$ many arithmetic operations for $A = \max\{\|f_i\|^2 \mid i < m\}$.

lemma *LLL_measure_approx:*

  assumes *LLL_invariant* (i, fs, gs) **and** $\alpha > 4/3$

    and $\forall\ i < m.\ \|fs\ !\ i\|^2 \leq A$

  shows *LLL_measure* $(i, fs, gs) \leq m + 2 \cdot m \cdot m \cdot\ log\ (\frac{4 \cdot \alpha}{4+\alpha})\ A$

end (* context of L, m, and α *)

We did not formally prove the polynomial-time complexity in Isabelle. This task would at least require two further steps: since Isabelle/HOL functions are mathematical functions, we would need to instrument them by an instruction counter and hence make its usage more cumbersome; and we would need to formally prove that each arithmetic operation can be performed in polynomial time by giving bounds for the numerators and denominators in f_i, g_i, and $\mu_{i,j}$.

Note that the reasoning on the number bounds should not be under-estimated. To illustrate this, consider the following modification to the algorithm, which we described in the submitted version of this paper: Since the termination proof only requires that $|\mu_{i,i-1}|$ must be small, for obtaining a weakly reduced basis one may replace the for-loop in lines 3–4 by a single update $f_i := f_i - \lfloor \mu_{i,i-1} \rceil \cdot f_{i-1}$. Then the total number of arithmetic operations will reduce to $\mathcal{O}(n^3 \cdot \log A)$. However, we figured out experimentally that this change is a bad idea, since then the bit-lengths of the norms of f_i are no longer polynomially bounded: some input lattice of dimension 20 with 20 digit numbers led to the consumption of more than 64 GB of memory so that we had to abort the computation.

This indicates that formally verified bounds would be valuable. And indeed, the textbook contains informal proofs for bounds, provided that each $\mu_{i,j}$ is small after executing lines 3–4. Here, a weakly reduced basis does not suffice.

With the help of *basis_reduction_step* it is now trivial to formally verify the correctness of the LLL algorithm, which is defined as *reduce_basis* in the sources.

Finally, recall that the first element of a (weakly) reduced basis *fs* is a short vector in the lattice. Hence, it is easy to define a wrapper function *short_vector* around *reduce_basis*, which is a verified algorithm to compute short vectors.

lemma *short_vector:*
  assumes $\alpha \geq 4/3$ **and** *lin_indpt_list* (*RAT fs*)
    and *short_vector* α *fs* $= v$ **and** *length fs* $= m$ **and** $m \neq 0$
  shows $v \in$ *lattice_of fs* $- \{0_v\ n\}$
    and $h \in$ *lattice_of fs* $- \{0_v\ n\} \longrightarrow \|v\|^2 \leq \alpha^{m-1} \cdot \|h\|^2$

5   Experimental Evaluation of the Verified LLL Algorithm

We formalized *short_vector* in a way that permits code generation [4]. Hence, we can evaluate the efficiency of our verified LLL algorithm. Here, we use a fixed approximation factor $\alpha = 2$, we map Isabelle's integer operations onto the unbounded integer operations of the target language Haskell, and we compile the code with ghc version 8.2.1 using the -O2 parameter.

We consider the LatticeReduce procedure of Mathematica version 11.2 as an alternative way to compute short vectors. The documentation does not specify the value of α, but mentions that Storjohann's variant [12] of the LLL basis reduction algorithm is implemented.

For the input, we use lattices of dimension n where each of the n basis vectors has n-digit random numbers as coefficients. So, the size of the input basis is cubic in n; for instance the bases for $n = 1$, 10, and 100 are stored in files measured in bytes, kilo-bytes, and mega-bytes, respectively.

Figure 1 displays the execution times in seconds for increasing n. The experiments have all been run on an iMacPro with a 3.2 GHz Intel Xeon W running macOS 10.13.4. The execution times of both algorithms can both be approximated by a polynomial $c \cdot n^6$ – the gray lines behind the dots – where the ratio between the constant factors c is 306.5, which is also reflected in the diagrams by using different scales for the time-axis.

Note that for $n \geq 50$, our implementation spends between 28–67 % of its time to compute the initial GSO on rational numbers, and 83–85 % of the total time of each run is used in integer GCD-computations. The GCDs are required for the verified implementation of rational numbers in Isabelle/HOL, which always normalizes rational numbers. Hence, optimizing our verified implementation for random lattices is not at all trivial, since a significant amount of time is spent in the cubic number of rational-number operations required in line 1 of Algorithm 2. Integrating and verifying known optimizations of GSO computations and LLL [10, Chap. 4] also looks challenging, since they depend on floating-point arithmetic. A corresponding implementation outperforms our verified algorithm by far: the tool *fplll* version 5.2.0 [13] can reduce each of our examples in less than 0.01 s.

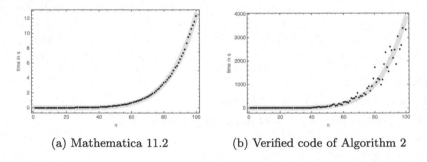

(a) Mathematica 11.2   (b) Verified code of Algorithm 2

Fig. 1. Execution time of short vector computation on random lattices

Besides efficiency, it is worth mentioning that we did not find bugs in fplll's or Mathematica's implementation: the short vectors that are generated by both tools have always been as short as our verified ones, but not much shorter: the average ratio between the norms is 0.74 for fplll and 0.93 for Mathemathica.

Under http://cl-informatik.uibk.ac.at/isafor/LLL_src.tgz one can access the sources and the input lattices for the experiments.

6   Factorization of Polynomials in Polynomial Time

In this section we first describe how the LLL algorithm helps to factor integer polynomials, by following the textbook [16, Chap. 16.4–16.5].

We only summarize how we tried to verify the corresponding factorization Algorithm 16.22 of the textbook. Indeed, we almost finished it: after 1 500 lines of code we had only one remaining goal to prove. However, we were unable to figure out how to discharge this goal and then also started to search for inputs where the algorithm delivers wrong results. After a while we realized that Algorithm 16.22 indeed has a serious flaw, with details provided in Sect. 6.2.

Therefore, we derive another algorithm based on ideas from the textbook, which also runs in polynomial-time, and prove its soundness in Isabelle/HOL.

6.1   Short Vectors for Polynomial Factorization

In order to factor an integer polynomial f, we may assume a *modular* factorization of f into several monic factors u_i: $f \equiv lc \; f \cdot \prod_i u_i$ modulo m where $m = p^l$ is some prime power for user-specified l. In Isabelle, we just reuse our verified modular factorization algorithm [1] to obtain the modular factorization of f.

We briefly explain how to compute non-trivial integer polynomial factors h of f based on Lemma 1, as also informally described in the textbook.

Lemma 1 (*[16, Lemma 16.20]*).   *Let f, g, u be non-constant integer polynomials. Let u be monic. If u divides f modulo m, u divides g modulo m, and $\|f\|^{degree\ g} \cdot \|g\|^{degree\ f} < m$, then $h = gcd\ f\ g$ is non-constant.*

Let f be a polynomial of degree n. Let u be any degree-d factor of f modulo m. Now assume that f is reducible, so $f = f_1 \cdot f_2$ where w.l.o.g., we assume that u divides f_1 modulo m and that $0 < degree\ f_1 < n$. Let the lattice $L_{u,k}$ be the set of all polynomials of degree below $d + k$ which are divisible by u modulo m. As $degree\ f_1 < n$, clearly $f_1 \in L_{u,n-d}$.

In order to instantiate Lemma 1, it now suffices to take g as the polynomial corresponding to any short vector in $L_{u,k}$: u will divide g modulo m by definition of $L_{u,k}$ and moreover $degree\ g < n$. The short vector requirement will provide an upper bound to satisfy the assumption $\|f_1\|^{degree\ g} \cdot \|g\|^{degree\ f_1} < m$.

$$\|g\| \leq 2^{(n-1)/2} \cdot \|f_1\| \leq 2^{(n-1)/2} \cdot 2^{n-1}\|f\| = 2^{3(n-1)/2}\|f\| \quad (4)$$

$$\|f_1\|^{degree\ g} \cdot \|g\|^{degree\ f_1} \leq (2^{n-1}\|f\|)^{n-1} \cdot (2^{3(n-1)/2}\|f\|)^{n-1} \quad (5)$$

$$= \|f\|^{2(n-1)} \cdot 2^{5(n-1)^2/2}$$

Here, the first inequality in (4) is the short vector approximation ($f_1 \in L_{u,k}$). The second inequality in (4) is Mignotte's factor bound (f_1 is a factor of f). Finally, Mignotte's factor bound and (4) are used as approximations of $\|f_1\|$ and $\|g\|$ in (5), respectively.

Hence, if l is chosen large enough so that $m = p^l > \|f\|^{2(n-1)} \cdot 2^{5(n-1)^2/2}$ then all preconditions of Lemma 1 are satisfied, and $h_1 := gcd\ f_1\ g$ will be a non-constant factor of f. Since f_1 divides f, also $h := gcd\ f\ g$ will be a non-constant factor of f. Moreover, the degree of h will be strictly less than n, and so h is a proper factor of f.

6.2   Bug in Modern Computer Algebra

In the previous section we have chosen the lattice $L_{u,k}$ for $k = n - d$ to find a polynomial h that is a proper factor of f. This has the disadvantage that h is not necessarily irreducible. In contrast, Algorithm 16.22 tries to directly find *irreducible* factors by iteratively searching for factors w.r.t. the lattices $L_{u,k}$ for increasing k from 1 up to $n - d$.

We do not have the space to present Algorithm 16.22 in detail, but just state that the arguments in the textbook and the provided invariants all look reasonable. Luckily, Isabelle was not so convinced: We got stuck with the goal that the content of the polynomial g corresponding to the short vector is not divisible by the chosen prime p, and this is not necessarily true.

The first problem occurs if the content of g is divisible by p. Consider $f_1 = x^{12} + x^{10} + x^8 + x^5 + x^4 + 1$ and $f_2 = x$. When trying to factor $f = f_1 \cdot f_2$, then $p = 2$ is chosen, and at a certain point the short vector computation is invoked for a modular factor u of degree 9 where $L_{u,4}$ contains f_1. Since f_1 itself is a shortest vector, $g = p \cdot f_1$ is a short vector: the approximation quality permits any vector of $L_{u,4}$ of norm at most $\alpha^{degree\ f_1/2} \cdot \|f_1\| = 64 \cdot \|f_1\|$. For this valid choice of g, the result of Algorithm 16.22 will be the non-factorization $f = f_1 \cdot 1$.

We informed the authors of the textbook about this first problem. They admitted the flaw and it is easy to fix.

There is however a second potential problem. If g is even divisible by p^l, then Algorithm 16.22 will again return wrong results. In the formalization we therefore integrate the check $|lc\ g| < p^l$ into the factorization algorithm[4], and then this modified version of Algorithm 16.22 is correct.

We could not conclude the question whether the additional check is really required, i.e., whether $|lc\ g| \geq p^l$ can ever happen, and just give some indication that the question is non-trivial. For instance, when factoring f_1 above, then p^l is a number with 124 digits, $\|u\| > p^l$, so in particular all basis elements of $L_{u,1}$ will have a norm of at least p^l. Note that $L_{u,1}$ also does not have quite *short* vectors: any vector in $L_{u,1}$ will have norm of at least 111 digits. However, since the approximation factor in this example is only two digits long, the short vector computation must result in a vector whose norm has at most 113 digits, which is not enough for permitting p^l with its 124 digits as leading coefficient of g.

6.3   A Verified Factorization Algorithm

To verify the factorization algorithm of Sect. 6.1, we formalize the two key facts to relate lattices and factors of polynomials: Lemma 1 and the lattice $L_{u,k}$.

To prove Lemma 1, we partially follow the textbook, although we do the final reasoning by means of some properties of resultants which were already proved in the previous development of algebraic numbers [14]. We also formalize Hadamard's inequality, which states that for any square matrix A having rows

---

[4] When discussing the second problem with the authors, they proposed an even more restrictive check.

v_i, then $|det\ A| \leq \prod \|v_i\|$. Essentially, the proof of Lemma 1 consists of showing that the resultant of f and g is 0, and then deduce $degree\ (gcd\ f\ g) > 0$. We omit the full-detailed proof, the interested reader can see it in the sources.

To define the lattice $L_{u,k}$ for a degree d polynomial u and integer k, we define the basis $v_0, \ldots, v_{k+d-1}$ of the lattice $L_{u,k}$ such that each v_i is the $(k+d)$-dimensional vector corresponding to polynomial $u(x) \cdot x^i$ if $i < k$, and to monomial $m \cdot x^{k+d-i}$ if $k \leq i < k+d$.

We define the basis in Isabelle/HOL as *factorization_lattice u k m* as follows:

definition *factorization_lattice u k m* = (**let** *d = degree u* **in**
   map ($\lambda i.\ vec\_of\_poly\_n\ (u \cdot monom\ 1\ i)\ (d + k))\ [k >..0]$ @
   map ($\lambda i.\ vec\_of\_poly\_n\ (monom\ m\ i)\ (d + k))\ [d >..0]$)

Here *vec_of_poly_n p n* is a function that transforms a polynomial p into a vector of dimension n with coefficients in the reverse order and completing with zeroes if necessary. We use it to identify an integer polynomial f of degree $< n$ with its coefficient vector in $\mathbb{Z}^n$. We also define its inverse operation, which transforms a vector into a polynomial, as *poly_of_vec*. *[a>..b]* denotes the reversed sorted list of natural elements from b to $a-1$ (with $b < a$) and *monom a b* denotes the monomial ax^b. To visualize the definition, for $u(x) = \sum_{i=0}^{d} u_i x^i$ we have

$$
\begin{bmatrix}
v_0^{\mathsf{T}} \\
\vdots \\
v_{k-1}^{\mathsf{T}} \\
v_k^{\mathsf{T}} \\
\vdots \\
v_{k+d-1}^{\mathsf{T}}
\end{bmatrix}
=
\begin{bmatrix}
u_d & u_{d-1} & \cdots & u_0 & & & \\
& \ddots & \ddots & & \ddots & & \\
& & u_d & u_{d-1} & \cdots & u_0 \\
& & & m & & & \\
& & & & \ddots & & \\
& & & & & & m
\end{bmatrix}
=: S
\tag{6}
$$

and *factorization_lattice* $(x + 1\,894\,885\,908)\ 2\ 2^{31}$ is precisely the basis (f_1, f_2, f_3) of Example 1.

There are some important facts that we must prove about *factorization_lattice*.

- *factorization_lattice u k m* is a list of linearly independent vectors as required for applying the LLL algorithm to find a short vector in $L_{u,k}$.
- $L_{u,k}$ characterizes the polynomials which have u as a factor modulo m:

$$g \in poly\_of\_vec\ (L_{u,k}) \iff degree\ g < k+d\ \text{and}\ u\ \text{divides}\ g\ \text{modulo}\ m$$

That is, any polynomial that satisfies the right hand side can be transformed into a vector that can be expressed as integer linear combination of the vectors of *factorization_lattice*. Similarly, any vector in the lattice $L_{u,k}$ can be expressed as integer linear combination of *factorization_lattice* and corresponds to a polynomial of degree $< k+d$ which is divisible by u modulo m.

The first property is a consequence of obvious facts that the matrix S is upper triangular, and its diagonal entries are non-zero if both u and m are non-zero. Thus, the vectors in *factorization_lattice u k m* are linearly independent.

Now we look at the second property. For one direction, we see the matrix S in (6) as (a generalization of) the *Sylvester matrix* of polynomial u and constant polynomial m. Then we generalize an existing formalization about Sylvester matrices as follows:

lemma *sylvester_sub_poly:*
  assumes *degree $u \le d$* **and** *degree $q \le k$* **and** *$c \in$ carrier_vec $(k+d)$*
  shows *poly_of_vec $((sylvester\_mat\_sub\ d\ k\ u\ q)^T *_v c) =$*
    poly_of_vec (vec_first c k) · u + poly_of_vec (vec_last c d) · q

We instantiate q by the constant polynomial m. So for every $c \in \mathbb{Z}^{k+d}$ we get

$$poly\_of\_vec\ (S^\mathsf{T} c) = r \cdot u + m \cdot s \equiv ru \text{ modulo } m$$

for some polynomials r and s. As every $g \in L_{u,k}$ is represented as $S^\mathsf{T} c$ for some integer coefficient vector $c \in \mathbb{Z}^{k+d}$, we conclude that every $g \in L_{u,k}$ is divisible by u modulo m. The other direction requires the use of the division with remainder by the monic polynomial u. Although we closely follow the textbook, the actual formalization of these reasonings requires some more tedious work, namely the connection between the matrix-to-vector multiplication ($*_v$) of `Matrix.thy` and linear combinations (*lincomb*) of HOL-Algebra. The former is naturally described as a summation over lists (*sumlist* which we define via *foldr*), while the latter is set-based *finsum*. We follow the existing connection between *sum_list* and *sum* of the class-based world to the locale-based world, which demands some generalizations of the standard library.

Once those properties are proved, we implement an algorithm for the reconstruction of factors within a context that fixes p and l:[5]

function *LLL_reconstruction f us =*
(*let u = choose_u us;*   (* pick any element of us *)
      g = LLL_short_polynomial (degree f) u;
      f2 = gcd f g   (* candidate factor *)
  in if degree f2 = 0 then [f]   (* f is irreducible *)
    else let f1 = f div f2;   (* f = f1 * f2 *)
        (us1, us2) = partition (λ ui. poly_mod.dvdm p ui f1) us
      in LLL_reconstruction f1 us1 @ LLL_reconstruction f2 us2)

LLL_reconstruction is a recursive function which receives two parameters: the polynomial *f* that has to be factored and *us*, which is the list of modular factors of the polynomial *f*. *LLL_short_polynomial* computes a short vector (and transforms it into a polynomial) in the lattice generated by a basis for $L_{u,k}$ and suitable k, that is, *factorization_lattice u (degree f - degree u)*. *us1* is the list of elements of *us* that divide *f1* modulo *p*, and *us2* contains the rest of elements

---

[5] The corresponding Isabelle/HOL implementation contains some sanity checks which are solely used to ensure termination. We present here a simplified version.

of *us*. *LLL_reconstruction* returns the list of irreducible factors of *f*. Termination follows from the fact that the degree decreases, that is, in each step the degree of both *f1* and *f2* is strictly less than the degree of *f*.

In order to formally verify the correctness of the reconstruction algorithm for a polynomial *F* we use the following invariants. They consider invocations *LLL_reconstruction f us* for every intermediate factor *f* of *F*.

1. *f* divides *F*
2. *degree f* > 0
3. *lc f* $\cdot \prod us$ is the unique modular factorization of *f* modulo p^l
4. *lc F* and *p* are coprime, and *F* is square-free in $\mathbb{Z}_p[x]$
5. p^l is sufficently large: $\|F\|^{2(N-1)} 2^{5(N-1)^2/2} < p^l$ where $N = degree\ F$

Concerning complexity, it is easy to see that if *f* splits into *i* factors, then *LLL_reconstruction* invokes the short vector computation for exactly $i + (i - 1)$ times: $i - 1$ invocations are used to split *f* into the *i* irreducible factors, and for each of these factors one invocation is required to finally detect irreducibility.

Finally, we combine the new reconstruction algorithm with existing results (the algorithms for computing an appropriate prime *p*, the corresponding exponent *l*, the factorization in $\mathbb{Z}_p[x]$ and its Hensel-lifting to $\mathbb{Z}_{p^l}[x]$) presented in the Berlekamp–Zassenhaus development to get a polynomial-time factorization algorithm for square-free and content-free polynomials.

lemma *LLL_factorization:*
   assumes *LLL_factorization f = gs*
     and *square_free f* **and** *content_free f* **and** *degree f* $\neq 0$
   shows *f = prod_list gs* **and** $\forall g \in set\ gs.\ irreducible\ g$

We further combine this algorithm with a pre-processing algorithm of earlier work [1]. This pre-processing splits a polynomial *f* into $c \cdot f_1^1 \cdot \ldots \cdot f_k^k$ where *c* is the content of *f* which is not further factored. Each f_i is square-free and content-free, and will then be passed to *LLL_factorization*. The combined algorithm factors arbitrary univariate integer polynomials into its content and a list of irreducible polynomials.

When experimentally comparing our verified LLL-based factorization algorithm with the verified Berlekamp–Zassenhaus factorization algorithm [1] we see no surprises. On the random polynomials from the experiments in [1], Berlekamp–Zassenhaus's algorithm performs much better: it can factor each polynomial within a minute, whereas the LLL-based algorithm already fails on the smallest example. It is an irreducible polynomial with 100 coefficients where the LLL algorithm was aborted after a day when trying to compute a reduced basis for a lattice of dimension 99 with coefficients having up to 7 763 digits.

7   Summary

We formalized the LLL algorithm for finding a basis with short, nearly orthogonal vectors of an integer lattice, as well as its most famous application to get a verified factorization algorithm for integer polynomials which runs in polynomial time. The work is based on our previous formalization of the Berlekamp–Zassenhaus factorization algorithm, where the exponential reconstruction phase is replaced by the polynomial-time lattice-reduction algorithm. The whole formalization consists of about 10 000 lines of code, including a 2 200-line theory which contains generalizations and theorems that are not exclusive to our development. This theory can extend the Isabelle standard library and up to six different AFP entries. As far as we know, this is the first formalization of the LLL algorithm and its application to factor polynomials in any theorem prover. This formalization led us to find a major flaw in a textbook.

There are some possibilities to extend the current formalization, e.g., by verifying faster variants of the LLL algorithm or by integrating other applications like the more efficient factorization algorithm of van Hoeij [10, Chap. 8]: it uses simpler lattices to factor polynomials, but its verification is much more intricate.

Acknowledgments. This research was supported by the Austrian Science Fund (FWF) project Y757. Jose Divasón is partially funded by the Spanish projects MTM2014-54151-P and MTM2017-88804-P. Akihisa Yamada is supported by ERATO HASUO Metamathematics for Systems Design Project (No. JPMJER1603), JST. Some of the research was conducted while Sebastiaan Joosten and Akihisa Yamada were working in the University of Innsbruck. We thank Jürgen Gerhard and Joachim von zur Gathen for discussions on the problems described in Sect. 6.2, and Bertram Felgenhauer for discussions on gaps in the paper proofs. The authors are listed in alphabetical order regardless of individual contributions or seniority.

References

1. Divasón, J., Joosten, S.J.C., Thiemann, R., Yamada, A.: A formalization of the Berlekamp–Zassenhaus factorization algorithm. In: CPP 2017, pp. 17–29. ACM (2017)
2. Divasón, J., Joosten, S., Thiemann, R., Yamada, A.: A verified factorization algorithm for integer polynomials with polynomial complexity. Archive of Formal Proofs, Formal proof development, February 2018. http://isa-afp.org/entries/LLL_Factorization.html
3. Divasón, J., Joosten, S., Thiemann, R., Yamada, A.: A verified LLL algorithm. Archive of Formal Proofs, Formal proof development, February 2018. http://isa-afp.org/entries/LLL_Basis_Reduction.html
4. Haftmann, F., Nipkow, T.: Code generation via higher-order rewrite systems. In: Blume, M., Kobayashi, N., Vidal, G. (eds.) FLOPS 2010. LNCS, vol. 6009, pp. 103–117. Springer, Heidelberg (2010). https://doi.org/10.1007/978-3-642-12251-4_9
5. Harrison, J.: The HOL light theory of Euclidean space. J. Autom. Reason. **50**(2), 173–190 (2013)
6. Krauss, A.: Recursive definitions of monadic functions. In: PAR 2010. EPTCS, vol. 43, pp. 1–13 (2010)
7. Lee, H.: Vector spaces. Archive of Formal Proofs, Formal proof development, August 2014. http://isa-afp.org/entries/VectorSpace.html

8. Lenstra, A.K., Lenstra, H.W., Lovász, L.: Factoring polynomials with rational coefficients. Math. Ann. **261**, 515–534 (1982)
9. Micciancio, D.: The shortest vector in a lattice is hard to approximate to within some constant. SIAM J. Comput. **30**(6), 2008–2035 (2000)
10. Nguyen, P.Q., Vallée, B. (eds.): The LLL Algorithm - Survey and Applications. Information Security and Cryptography. Springer, Heidelberg (2010). https://doi.org/10.1007/978-3-642-02295-1
11. Nipkow, T., Wenzel, M., Paulson, L.C. (eds.): Isabelle/HOL. LNCS, vol. 2283. Springer, Heidelberg (2002). https://doi.org/10.1007/3-540-45949-9
12. Storjohann, A.: Faster algorithms for integer lattice basis reduction. Technical report 249, Department of Computer Science, ETH Zurich (1996)
13. The FPLLL development team. fplll, a lattice reduction library (2016). https://github.com/fplll/fplll
14. Thiemann, R., Yamada, A.: Algebraic numbers in Isabelle/HOL. In: Blanchette, J.C., Merz, S. (eds.) ITP 2016. LNCS, vol. 9807, pp. 391–408. Springer, Cham (2016). https://doi.org/10.1007/978-3-319-43144-4_24
15. Thiemann, R., Yamada, A.: Formalizing Jordan normal forms in Isabelle/HOL. In: CPP 2016, pp. 88–99. ACM (2016)
16. von zur Gathen, J., Gerhard, J.: Modern Computer Algebra, 3rd edn. Cambridge University Press, New York (2013)

A Formal Proof of the Minor-Exclusion Property for Treewidth-Two Graphs

Christian Doczkal$^{(\boxtimes)}$, Guillaume Combette, and Damien Pous

Univ Lyon, CNRS, ENS de Lyon, UCB Lyon 1, LIP, Lyon, France
{christian.doczkal,guillaume.combette,damien.pous}@ens-lyon.fr

Abstract. We give a formal and constructive proof in Coq/Ssreflect of the known result that the graphs of treewidth two are exactly those that do not admit K4 as a minor. This result is a milestone towards a formal proof of the recent result that isomorphism of treewidth-two graphs can be finitely axiomatized. The proof is based on a function extracting terms from K4-free graphs in such a way that the interpretation of an extracted term yields a treewidth-two graph isomorphic to the original graph.

Keywords: Graph theory · Graph minor theorem · Coq · Ssreflect

1 Introduction

The notion of *treewidth* [6] measures how close a graph is to a forest. Graph homomorphism (and thus k-coloring) becomes polynomial-time for classes of graphs of bounded treewidth [1,10,13], so does model-checking of Monadic Second Order (MSO) formulae, and satisfiability of MSO formulae becomes decidable, even linear [4,5].

Robertson and Seymour's graph minor theorem [18], a cornerstone of algorithmic graph theory, states that graphs are well-quasi-ordered by the *minor* relation. As a consequence, the classes of graphs of bounded treewidth, which are closed under taking minors, can be characterized by finite sets of excluded minors. Two standard instances are the following ones: the graphs of treewidth at most one (the forests) are precisely those excluding the cycle with three vertices ($\mathsf{C_3}$); those of treewidth at most two are those excluding the complete graph with four vertices ($\mathsf{K_4}$) [8].

$(\mathsf{C_3})$ $\qquad\qquad\qquad$ $(\mathsf{K_4})$

This work has been funded by the European Research Council (ERC) under the European Union's Horizon 2020 programme (CoVeCe, grant agreement No. 678157), and was supported by the LABEX MILYON (ANR-10-LABX-0070) of Université de Lyon, within the program "Investissements d'Avenir" (ANR-11-IDEX-0007) operated by the French National Research Agency (ANR).

J. Avigad and A. Mahboubi (Eds.): ITP 2018, LNCS 10895, pp. 178–195, 2018.
https://doi.org/10.1007/978-3-319-94821-8_11

We present a constructive and formal proof of the latter result in Coq/Ssreflect.

Amongst the open problems related to treewidth, there is the question of finding finite axiomatisations of isomorphism for graphs of a given treewidth [5, p. 118]. This question was recently answered positively for treewidth two [14]:

$$K_4\text{-free graphs form the free } 2p\text{-algebra,} \qquad (\dagger)$$

where $2p$-algebras are algebraic structures characterized by twelve equational axioms. The proof is rather technical; it builds on a precise analysis of the structure of K_4-free graphs and contains the specific form of the graph minor theorem for treewidth two which we present here. Further, invalid proofs of related claims have already been published in the literature (see [14]). Our long term goal is to formalize ($\dagger$): not only will this give us assurance about the validity of the proof in [14], it will also allow for the development of automation tactics for certain algebraic theories (e.g., 2p-algebra, allegories [11]). The Coq development accompanying the present paper [7] is a milestone for this project.

Independently from the aforementioned specific objective, formalizing the graph minor theorem for treewidth two requires us to develop a general Coq library for graph theory which should also be useful in other contexts. This library currently includes basic notions like paths, trees, subgraphs, and isomorphisms and also a few more advanced ones: minors, tree decompositions, and checkpoints (a variant of cut vertices).

We had to design this library from scratch. Indeed, there are very few formalizations of graph theory results in Coq, and none of them were applicable. Gonthier's formal proof of the Four-Color Theorem [12] is certainly the most advanced, but it restricts (by design) to planar graphs so that it cannot be used as a starting point for graph theory in the large. Similarly, Durfourd and Bertot's study of Delaunay triangulation [9] employs a notion of graphs based on hypermaps embedded in a plane. There are more formalizations in other interactive theorem provers. For instance, planar graphs were formalized in Isabelle/HOL for the Flyspeck project [16]. Noschinski recently developed a library for both simple and multi-graphs in Isabelle/HOL [17]. Chou developed a large part of undirected graph theory in HOL [2]. Euler's theorem was formalized in Mizar [15]. To the best of our knowledge, the theory of minors and tree decompositions was never formalized.

Overview of the proof. We focus on connected graphs: the general case follows by decomposing any given graph into connected components. The overall strategy of our proof of the minor exclusion theorem for treewidth two is depicted in Fig. 1.

We first prove that treewidth two graphs exclude K_4 as a minor (i). This proof is standard and relatively easy. For proving the converse implication, we introduce a notion of *term* that allow us to denote graphs. We prove that graphs of terms have treewidth at most two (ii) using properties of tree decompositions and a simple induction on terms. The main difficulty then consists in proving that every K_4-free graph can be represented by a term (iii).

Term

(ii)   (iii)

$TW_2 \xrightarrow{\text{(i)}} K_4$-free

Fig. 1. Structure of the proof.

(a)   (b)   (c)

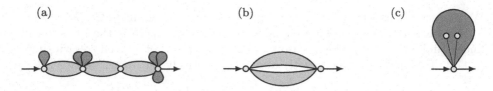

Fig. 2. The three main cases for extracting a term from a K_4-free graph.

Due to our long-term objective (†), the syntax we use for those terms is that of 2p-algebras [14];

$$u, v, w ::= u{\cdot}v \mid u \,\|\, v \mid u^\circ \mid \mathrm{dom}(u) \mid 1 \mid \top \mid a \qquad (a \in \Sigma)$$

This syntax makes it possible to denote directed multi-graphs, with edges labeled by letters from an alphabet Σ and with two designated vertices (the *input* and the *output*). The binary operations in the syntax correspond to series and parallel composition. The first unary operation, *converse*, exchanges input and output; the second one, *domain*, relocates the output to the input. The constant 1 represents the graph with just a single vertex; $\top$ is the disconnected graph with just two vertices. Letters represent single edges. For instance the graphs of the terms $a{\cdot}(b \,\|\, c^\circ) \,\|\, d$ and $1 \,\|\, a{\cdot}b$ are the following ones:

The second graph is also represented by the term $\mathrm{dom}(a \,\|\, b^\circ)$.

We use the concept of *checkpoint* to extract terms from graphs; those are the vertices which every path between input and output must visit. Using those, we get that every connected graph with distinct input and output has the shape depicted in Fig. 2(a), where the checkpoints are the only depicted vertices. One can parse such a graph as a sequential composition and proceed recursively once we have proved that the green and yellow components are K_4-free whenever the starting graph is so.

If there are no proper checkpoints between input and output, we exploit a key property of K_4-free graphs: in such a case, either the graph is just an edge, or it consists of at least two parallel components, which make it possible to proceed recursively. This is case (b) in Fig. 2. Establishing this property requires a deep analysis of the structure of K_4-free graphs.

The last case to consider (c) is when the input and the output of the graph coincide. One can recursively extract a term for the graph obtained by relocating the output to one of the neighbors of the input, and use the domain operation to recover the starting graph.

Outline. We first discuss our representation of simple graphs and the associated library about paths; there we make use of the support for finite types from the

Ssreflect library [20], and we rely on dependent types to provide a user-friendly interface (Sect. 2). Then we proceed with our formalization of tree decompositions, minors, and associated results. This leads to implication (i) in Fig. 1, as a special instance of the fact that treewidth at most i graphs are K_{i+2}-free (Sect. 3)

Once this basic infrastructure has been set up, we move to the formalization of the concepts and results that are specific to our objective. This includes terms as well as directed labeled and possibly pointed multigraphs. We prove the implication (ii) there: terms denote graphs of treewidth at most two (Sect. 4).

As explained above, the remaining implication (iii) is the most delicate. We first establish preliminary lemmas about checkpoints and the structure of K_4-free graphs (Sect. 5), which are then used to define an extraction function from graphs to terms (Sect. 6). Proving that this function is appropriate amounts to exhibiting a number of isomorphisms (Sect. 7).

We conclude with general remarks and statistics about the development (Sect. 8).

2 Simple Graphs

In this section we briefly describe how we represent finite simple graphs in Coq. The representation is based on finite types as defined in the Mathematical Component Libraries [20]. We start by briefly introducing finite types and the notations we are going to use in the mathematical development.

If X and Y are types, we write $X + Y$ for the *sum type* (with elements $\mathsf{inl}\,x$ and $\mathsf{inr}\,y$) and $X_\perp$ for the *option type* (with elements $\mathsf{Some}\,x$ and None). As usual, we write $g \circ f$ for the composition of f and g. If $f : X \to Y_\perp$ and $g : Y \to Z_\perp$, we also write $g \circ f$ for the result of the monadic bind operation (with type $X \to Z_\perp$). For functions f and g, we write $f \equiv g$ to mean that f and g agree on all arguments.

A *finite type* is a type X together with a list enumerating its elements. Finite types are closed under many type constructors (e.g., sum types and option types). If X is a finite type, we write 2^X for the (finite) type of sets (with decidable membership) over X. If $A : 2^X$ is a set, we write $\overline{A}$ for complement of A (in X). We slightly abuse notation and also write X for the full set over some type X. Finite sets come with an operation $\mathsf{pick} : 2^X \to X_\perp$ yielding elements of nonempty sets and None for empty sets. Moreover, if X is a finite type and $\approx\, : X \to X \to \mathbb{B}$ is a boolean equivalence relation, the *quotient* [3] of X with respect to $\approx$, written $X_{/\approx}$, is a finite type as well. The type $X_{/\approx}$ comes with functions $\pi : X \to X_{/\approx}$ and $\overline{\pi} : X_{/\approx} \to X$ such that $\pi(\overline{\pi}\,x) = x$ for all $x : X_{/\approx}$ and $\overline{\pi}(\pi\,x) \approx x$ for all $x : X$.

We use finite types as the basic building block for defining finite simple graphs.

Definition 1. *A (finite) simple graph is a structure $\langle V, R \rangle$ where V is a finite type of vertices and $R : V \to V \to \mathbb{B}$ is a symmetric and irreflexive edge relation.*

In Coq, we represent finite graphs using dependently typed records where the last two fields are propositions:

Record sgraph := SGraph { svertex : fin Type;
                             sedge   : rel svertex;
                             sg_sym : symmetric sedge;
                             sg_irrefl : irreflexive sedge}.

We introduce a coercion from graphs to the underlying type of vertices allowing us to write $x : G$ to denote that x is a vertex of G. For vertices $x, y : G$ we write $x{-}y$ if there is an edge between x and y. We write $G + xy$ for the graph G with an additional xy-edge.

For sets $U : 2^G$ of vertices of G, we write $G|_U$ for the subgraph of G induced by U. This is formalized by taking the type $\Sigma x : G.\ x \in U$ of (dependent) pairs of vertices $x : G$ and proofs of $x \in U$ and lifting the edge relation accordingly. Note that while, technically, the vertices of G and $G|_U$ have different types, we will ignore this in the mathematical presentation. In Coq, we have a generic projection from $G|_U$ to G. For the converse direction we, of course, need to construct dependent pairs of vertices $x : G$ and proofs of $x \in U$.

Definition 2. *Let G be a simple graph. An xy-path is a nonempty sequence of vertices p beginning with x and ending with y such that $z{-}z'$ for all adjacent elements z and z' of p (if any). A path is* irredundant *if all vertices on the path are distinct (i.e., the path contains no cycles). A set of vertices U is* connected *if there exists a path in U between any two vertices of U.*

The Mathematical Component Libraries include a predicate and a function

path : $(\forall\ T : \mathsf{Type},\ \mathsf{rel}\ T \rightarrow T \rightarrow \mathsf{seq}\ T \rightarrow \mathsf{bool})$ last : $\forall\ T,\ T \rightarrow \mathsf{list}\ T \rightarrow T$

such that path $e\,x\,q$ holds if the list $x :: q$ represents a path in the relation e, and last $x\,q$ returns the last element of $x :: q$. Note that path and last account for the nonemptiness of paths though the use of two arguments: the first vertex x and the (possibly empty) list of remaining vertices q. This asymmetric treatment makes symmetry reasoning (using path reversal) rather cumbersome. We therefore package the path predicate and a check for the last vertex into an indexed family of types Path $x\,y$ whose elements represent xy-paths. Doing so abstracts from the asymmetry in the definition of path, makes it possible to write more compact (and thus readable) statements, helps us keeping the local context of proofs shorter, and facilitates without loss of generality reasoning.

On these packaged paths we provide (dependently typed) concatenation and reversal operations as well as an indexing operation yielding the position of the first occurrence of a vertex on the path. We define a number of splitting lemmas for packaged paths as exemplified by the lemma below.

Lemma 3. *Let p be an irredundant xy-path such that z_1 occurs before z_2 on p. Then there exists a z_2-avoiding xz_1-path, a z_1z_2-path and a z_1-avoiding z_2y-path) such that $p = p_1p_2p_3$.*

While the lemma above may seem overly specific, it is used in five different proofs (usually following some without loss of generality reasoning to order z_1 and z_2).

3 Treewidth and Minors

We now define the notions of treewidth and minors in order to state our main result. Both notions appear in the literature with slight (but equivalent) variations. We choose variants that yield reasonable proof principles.

Definition 4. *A* forest *is a simple graph where there is at most one irredundant path between any two nodes.*

Definition 5. *A* tree decomposition *of a simple graph G is a forest T together with a function $B : T \to 2^G$ such that:*

T1. for every vertex $x : G$, there exists some $t : T$, such that $x \in B(t)$.
T2. for every x, the set of nodes $t : T$ such that $x \in B(t)$ is connected in T.
T3. if $x{-}y$, then there exists a node t, such that $\{x, y\} \subseteq B(t)$;

The width *of a tree decomposition is the size of the largest set $B(t)$ minus one; the* treewidth *of a graph is the minimal width of a tree decomposition.*

Note that we define the notion of tree decomposition using forests rather than trees. The two notions are equivalent since every forest can be turned into a tree by connecting arbitrary nodes of disconnected trees. Using forests rather than trees has the advantage that tree decompositions for the disjoint union of two graphs G and G' can be obtained as the disjoint union of tree decompositions for G and G'.

The minors of a graph G are customarily defined to be those graphs that can be obtained by a series of the following operations: remove a vertex, remove an edge, or contract an edge. We use instead a monolithic definition in terms of partial functions inspired by [6].

Definition 6. *Let G and G' be simple graphs. A function $\phi : G \to G'_\perp$ is called a* minor map *if:*

M1. For every $y : G'$, there exists some $x : G$ such that $\phi x = \mathsf{Some}\, y$.
M2. For every $y : G'$, $\phi^{-1}(\mathsf{Some}\, y)$ is connected in G.
M3. If $x{-}y$ for $x, y : G'$, there exist $x_0 \in \phi^{-1}(\mathsf{Some}\, x)$ and $y_0 \in \phi^{-1}(\mathsf{Some}\, y)$ such that $x_0{-}y_0$.

G' is a minor *of G, written $G' \prec G$ if there exists a minor map $\phi : G \to G'_\perp$.*

Intuitively, the (nonempty) preimage $\phi^{-1}(\mathsf{Some}\, x)$ of a given vertex x is the (connected) set of vertices being contracted to x and the vertices mapped to None are the vertices that are removed. We sometimes use (total) minor maps $\phi : G \to G'$ corresponding to minor maps that do not delete nodes, allowing us to avoid option types in certain cases.

Making the notion of minor map explicit is convenient in that it allows us to easily construct minor maps for a given graph, starting from minor maps (with extra properties) for some of its subgraphs (cf. Lemma 29 and Proposition 30).

Definition 7. *We write* K_4 *for the complete graph with 4 vertices. A simple graph* G *is* K_4-*free if* K_4 *is not a minor of* G.

Our main result is a formal proof that a simple graph is K_4-free iff if it has treewidth at most two. We first sketch the proof that graphs of treewidth at most two are always K_4-free.

Lemma 8. *If* $\phi : G \to H_\perp$ *and* $\psi : H \to I_\perp$ *are minor maps, then* $\psi \circ \phi$ *is a minor map.*

As a consequence of the lemma above, we obtain that $\prec$ is transitive.

Lemma 9. *If* $H \prec G$, *then the treewidth of* H *is at most the treewidth of* G.

Lemma 10. *Let* T *be a forest and let* $B : T \to G$ *be a tree decomposition of* G. *Then every clique of* G *is contained in* $B(t)$ *for some* $t : T$.

The proof of Lemma 10 proceeds by induction on the size of (nonempty) cliques. For cliques of size larger than two, the proof boils down to an analysis of the set of nodes in the tree decomposition containing all vertices of the clique but one (which is nonempty by induction hypothesis) and then arguing that (due to condition T2) the removed vertex must also be present. As a consequence of Lemma 10, we have:

Proposition 11. *If* G *has treewidth at most two, then* G *is* K_4-*free.*

This corresponds to the arrow (i) in the overall proof structure (Fig. 1).

4 Graphs

In this section we define labeled directed graphs following [6]. Then we show how to interpret terms as such graphs and prove that the graphs of terms have treewidth at most two. We fix some countably infinite type of *symbols* Σ.

Definition 12. *A graph is a structure* $G = \langle V, E, s, t, l \rangle$, *where* V *is a finite type of vertices,* E *is a finite type of edges,* $s, t : E \to V$ *are functions indicating the* source *and* target *of each edge, and* $l : E \to \Sigma$ *is function indicating the* label *of each edge. If* G *is a graph, we write* $x : G$ *to denote that* x *is a vertex of* G. *A two-pointed graph (or 2p-graph for short) is a structure* $\langle G, \iota, o \rangle$ *where* $\iota : G$ *and* $o : G$ *are two vertices called* input *and* output *respectively.*

Note that self-loops are allowed, as well parallel edges with the same label.
    Recall the syntax of *terms* from the introduction:

$$u, v, w ::= u \cdot v \mid u \parallel v \mid u^\circ \mid \mathrm{dom}(u) \mid 1 \mid \top \mid a \qquad (a \in \Sigma)$$

For each term constructor we define an operation on 2p-graphs. Those operations are depicted informally on the right of Fig. 3. For instance, $G \parallel H$, the parallel composition of G and H, consists of (disjoint) copies of G and H with the respective inputs and outputs identified. Formally, we express these graph operations in terms disjoint unions and quotients of graphs.

$$\underline{a} \triangleq \langle\langle\{0,1\},\{*\},\lambda_{-}.0,\lambda_{-}.1,\lambda_{-}.a\rangle,0,1\rangle$$

$$\top \triangleq \langle\langle\{0,1\},\emptyset,\emptyset,\emptyset,\emptyset\rangle,0,1\rangle$$

$$1 \triangleq \langle\langle\{*\},\emptyset,\emptyset,\emptyset,\emptyset\rangle,*,*\rangle$$

$$\langle G,\iota,o\rangle^{\circ} \triangleq \langle G,o,\iota\rangle$$

$$\mathrm{dom}(\langle G,\iota,o\rangle) \triangleq \langle G,\iota,\iota\rangle$$

$$\langle G,\iota,o\rangle \parallel \langle G',\iota',o'\rangle \triangleq \langle(G+G')_{/\approx},\pi(\mathsf{inl}\,\iota),\pi(\mathsf{inl}\,o)\rangle$$

$$\text{where } \approx \triangleq \big\{(\mathsf{inl}\,\iota,\mathsf{inr}\,\iota'),(\mathsf{inl}\,o,\mathsf{inr}\,o')\big\}^{\mathsf{eqv}}$$

$$\langle G,\iota,o\rangle\cdot\langle G',\iota',o'\rangle \triangleq \langle(G+G')_{/\approx},\pi(\mathsf{inl}\,\iota),\pi(\mathsf{inr}\,o')\rangle$$

$$\text{where } \approx \triangleq \big\{(\mathsf{inl}\,o,\mathsf{inr}\,\iota')\big\}^{\mathsf{eqv}}$$

Fig. 3. The algebra of 2p-graphs.

Definition 13. *Let* $G = \langle V,E,s,t,l\rangle$ *and* $G' = \langle V',E',s',t',l'\rangle$. *The* disjoint union *of* G *and* G', *written* $G + G'$, *is defined to be the graph*

$$\langle V+V',E+E',s+s',t+t',l+l'\rangle$$

Here, $s+s'$ *is the pointwise lifting of* s *and* s' *to the sum type* $E+E'$.

Definition 14. *Let* $G = \langle V,E,s,t,l\rangle$ *and let* $\approx : G \to G \to \mathbb{B}$ *be an equivalence relation. The* quotient *of* G *modulo* $\approx$, *written* $G_{/\approx}$, *is defined to be the graph*

$$\langle V_{/\approx},E,\pi \circ s,\pi \circ t,l\rangle$$

The precise definitions of the graph operations are given on the left side of Fig. 3 (A^{eqv} denotes the equivalence relation generated by the pairs in A). This allows us to interpret every term t as a 2p-graph $\mathbf{g}(t)$, recursively. We now have to prove that every 2p-graph of a term has treewidth at most two. In order to use the definition of treewidth, we first need to abstract 2p-graphs into simple graphs. This is achieved through the notion of a skeleton.

Definition 15. *Let* $G = \langle V,E,s,t,l\rangle$. *The* (weak) skeleton *of* G *is the simple graph* $\langle V,R\rangle$ *where* xRy *iff* $x \neq y$ *and there exists an edge* $e : E$ *such that* $s(e) = x$ *and* $t(e) = y$ *or vice versa. The weak skeleton of the 2p-graph* $\langle G,\iota,o\rangle$ *is the skeleton of* G. *The* strong skeleton *of a 2p-graph* $\langle G,\iota,o\rangle$ *is the skeleton of* G *with an additional* ιo-*edge.*

We remark that the operation of taking the weak or strong skeleton does not change the type of vertices. This greatly simplifies lifting properties of the skeleton to the graph and vice versa. In practice, we turn the construction of taking the weak skeleton into a coercion from graphs to simple graphs (leaving extractions of strong skeletons explicit).

   The following lemma makes it possible to show that both series and parallel composition preserve treewidth two.

Lemma 16. *Let $G_1 = \langle G_1', \iota, o \rangle$ and $G_2 = \langle G_2', \iota', o' \rangle$ be 2p-graphs and let $\langle T_i, B_i \rangle$ ($i \in \{1, 2\}$) be tree decompositions of the strong skeletons of G_1 and G_2 respectively. Further let $\approx$ be an equivalence relation on $G_1 + G_2$ identifying at least two vertices from the set $P \triangleq \{\text{inl } \iota, \text{inr } \iota', \text{inl } o, \text{inr } o'\}$ and no other vertices. Then there exists a tree decomposition of the skeleton of $(G1 + G2)_{/\approx}$ of width at most two having a node t such that $P_{/\approx} \subseteq B(t)$.*

Proof. We use the three following facts. (1) A tree decomposition for a disjoint union of simple graphs can be obtained by taking the disjoint union of tree decompositions for those graphs. (2) Two trees of a tree decomposition can be joined through a new node containing the vertices of its neighbors. (3) A tree decomposition can be quotiented (to give a tree decomposition of a quotiented graph) as soon as it has nodes for all equivalence classes.   □

Proposition 17. *For all terms u, the strong skeleton of $\mathbf{g}(u)$ has a tree decomposition of width at most two.*

Proof. By induction on u. The cases for $\parallel$ and $\cdot$ follow with Lemma 16. All other cases are trivial.   □

This finishes arrow (ii) of the overall proof structure (Fig. 1). The rest of the paper is concerned with arrow (iii), i.e., extracting for every 2p-graph G whose skeleton is K_4-free a term whose graph is isomorphic to G.

5   Checkpoints

Before we can define the function extracting terms from graphs, we need a number of results on simple graphs. These will allow us to analyze the structure of graphs (via their skeletons), facilitating the termination and correctness arguments for the extraction function.

For the remainder of this section, G refers to some *connected* simple graph.

Definition 18. *The* checkpoints *between two vertices x, y are the vertices which any xy-path must visit:*

$$\mathsf{cp}\, x\, y \triangleq \{z \mid \text{every } xy\text{-path crosses } z\}$$

Two vertices x, y are linked, *written $x \Diamond y$, when $x \neq y$ and $\mathsf{cp}\, x\, y = \{x, y\}$, i.e., when there are no proper checkpoints between x and y. The* link graph *of G is the graph of linked vertices.*

Consider the graph on the left in Fig. 4; its link graph is obtained by adding the three dotted edges to the existing ones.

Note that every proper checkpoint z between vertices x and y (i.e., a vertex $z \in \mathsf{cp}\, x\, y \setminus \{x, y\}$) is a cut vertex (i.e., removing z disconnects G) and vice versa. Also note that membership in cp is decidable (i.e., $\mathsf{cp}\, x\, y$ can be defined as a finite set in the Ssreflect sense) since it suffices to check whether the finitely many irredundant paths cross z.

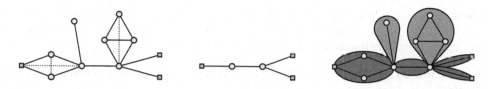

Fig. 4. Link graph, checkpoint graph, and decomposition into intervals and bags.

Lemma 19. *1.* $\mathsf{cp}\,x\,x = \{x\}$
2. $\{x,y\} \subseteq \mathsf{cp}\,x\,y = \mathsf{cp}\,y\,x$

Lemma 20. *Every irredundant cycle in the link graph is a clique.*

For a set of vertices $U \subseteq G$, we take $G+U$ to be the graph G with one additional vertex, denoted $\bullet$, whose neighbors are exactly the elements of U.

Lemma 21. *If $\{x,y,z\}$ is a triangle in the link graph, then $\mathsf{K}_4 \prec G + \{x,y,z\}$.*

Lemma 21 is first in a series of nontrivial lemmas required to justify the splitting of graphs into parallel components. Its proof boils down to an elaborate construction on paths between x, y, and z that yields a minor map from G to C_3 (the cycle with three vertices), which is subsequently extended to a minor map from $G + \{x,y,z\}$ to K_4. This is one instance where our definition of minors using minor maps pays off.

Definition 22. *Let U be a set of vertices of G. The* checkpoints *of U, written* $\mathsf{CP}\,U$, *are the vertices which are checkpoints of some pair in U.*

$$\mathsf{CP}\,U \triangleq \bigcup_{x,y \in U} \mathsf{cp}\,x\,y$$

The checkpoint graph *of U is the subgraph of the link graph induced by this set. We also denote this graph by* $\mathsf{CP}\,U$.

The graph in the middle of Fig. 4 is the checkpoint graph of the one of the left, when U consists of the blue square vertices.

Lemma 23. *Let $x,y \in \mathsf{CP}\,U$. Then $\mathsf{cp}\,x\,y \subseteq \mathsf{CP}\,U$.*

We give the proof of this lemma below. It is relatively simple, but indicative of the type of reasoning required to prove checkpoint properties. Those proofs usually contain a combination of the following: splitting paths at vertices, concatenating paths, and without loss of generality reasoning. For the latter, Ssreflects `wlog`-tactic proved extremely helpful.

Proof. We have $x \in \mathsf{cp}\,x_1\,x_2$ and $y \in \mathsf{cp}\,y_1\,y_2$ for some vertices $\{x_1,x_2,y_1,y_2\} \subseteq U$ by the definition of CP. Fix some $z \in \mathsf{cp}\,x\,y$. If $z \in \{x,y\}$, the claim is trivial, so assume $z \notin \{x,y\}$. Hence, we obtain either an xx_1-path or an xx_2-path not containing z by splitting some irredundant x_1x_2-path at x. Without loss of generality, the xx_1-path avoids z. Similarly, we obtain, again w.l.o.g., a yy_1-path avoiding z. Thus $z \in \mathsf{cp}\, x_1\, y_1$ since the existence of an $x_1 y_1$-path avoiding z would contradict $z \in \mathsf{cp}\, x\, y$ (by concatenation with the paths obtained above).   □

Definition 24. *Let $x, y : G$. The* strict interval *$]x; y[$ is the following set of vertices.*

$$]x; y[\,\triangleq\, \{p \mid \text{there is an } xp\text{-path avoiding } y$$
$$\text{and a } py\text{-path avoiding } x\}$$

The interval $[x; y]$ is obtained by adding x and y to that set. We abuse notation and also write $[x; y]$ for the subgraph of G induced by the set $[x; y]$.

Definition 25. *The* bag *of a checkpoint $x \in \mathsf{CP}\, U$ is the set of vertices that need to cross x in order to reach the other checkpoints.*

$$[x]_U \,\triangleq\, \{p \mid \forall y \in \mathsf{CP}\, U.\ \text{every } py\text{-path crosses } x\}.$$

As before, we also write $[x]_U$ for the induced subgraph of G.

Note that $[x]_U$ depends on U and differs from $[x; x]$ (which is always the singleton $\{x\}$). The main purpose of bags and intervals is to aid in decomposing graphs for the term extraction function, as depicted on the right in Fig. 4. We first show that distinct bags and adjacent bags and strict intervals are disjoint.

Lemma 26. *1. If $y \in \mathsf{CP}\, U$, then $[x]_U \cap\,]x; y[\,= \emptyset$.*
2. If $x, y \in \mathsf{CP}\, U$ and $x \neq y$, then $[x]_U \cap [y]_U = \emptyset$.
3. If $z \in \mathsf{cp}\, x\, y$, then $[x; y] = [x; z] \cup [z]_{\{x,y\}} \cup [z; y]$.
4. If $z \in \mathsf{cp}\, x\, y$, then $]x; z[$, $[z]_{\{x,y\}}$ and $]z; y[$ are pairwise disjoint.

Lemma 27. *Let $x, y \in \mathsf{CP}\, U$. Then there exist $x_0 \in U$ and $y_0 \in U$ such that $\{x, y\} \subseteq \mathsf{cp}\, x_0\, y_0$.*

Lemma 28. *Let $\{x, y, z\}$ be a triangle in $\mathsf{CP}\, U$. Then there exist $x_0, y_0, z_0 \in U$ such that $x_0 \in [x]_{\{x,y,z\}}$, $y_0 \in [y]_{\{x,y,z\}}$, and $z_0 \in [z]_{\{x,y,z\}}$.*

Proof. Follows with Lemma 27.   □

Lemma 29. *Let U be nonempty and let $T \triangleq U \cup (G \setminus \bigcup_{x \in U}[x]_U)$. Then there exists a minor map $\phi : G \to G|_T$ such that ϕ maps the elements of each bag $[x]_U$ to x and every other vertex to itself.*

The above series of lemmas leads us to the following proposition, that corresponds to [14, Proposition 20(i)]; the proof given here is significantly simpler than the proof given in [14].

Proposition 30. *Let $U \subseteq G$ such that $G + U$ is K_4-free. Then $\mathsf{CP}\, U$ is a tree.*

Proof. Assume that $\mathsf{CP}\, U$ is not a tree. Then $\mathsf{CP}\, U$ contains a triangle $\{x, y, z\}$ (Lemma 20). Let x_0, y_0, z_0 as given by Lemma 28. We obtain a minor map collapsing the bags for x, y, and z (Lemma 29 with $U = \{x, y, z\}$). This identifies x and x_0 and likewise for y and z. Since x, y, z is still a triangle in the link graph of the collapsed graph and since $\bullet$ is adjacent to x, y, z in the collapsed graph, Lemma 21 yields $K_4 \prec G + U$, a contradiction. □

The following proposition establishes the key property of K_4-free graphs we alluded to in the introduction. Its proof is particularly tricky to formalize due to the number of different graphs with shared vertices (we have G, $G' \triangleq G|_{\overline{\{i\}}}$ and $G' + U$ (the graph Proposition 30 is instantiated with). Consequently, we often need to cast vertices from one graph to another.

Proposition 31. *Let $\iota, o : G$ such that $G + \iota o$ is K_4-free, $\llbracket \iota \rrbracket_{\{\iota, o\}} = \{\iota\}$, and $\iota \Diamond o$, but not $\iota - o$. Then $\rrbracket \iota; o \llbracket$ has at least two connected components.*

Proof. Let G' be the graph G with ι removed and let $U \subseteq G'$ be the set of neighbors of ι (in G) plus o. By Proposition 30 (on G' and U), $\mathsf{CP}\, U$ is a tree in G'. The vertex o cannot be a leaf in $\mathsf{CP}\, U$ since if it were, its unique neighbor would be a proper checkpoint between ι and o. Moreover, o is a checkpoint between any distinct neighbors of o. Removing o yields that $\rrbracket \iota; o \llbracket$ has at least two components. □

The above proposition is used for splitting paths into parallel components (case (b) in Fig. 2); the one below allows us to proceed recursively in case (a).

Proposition 32. *Let $\iota, o : G$ such that $G + \iota o$ is K_4-free and let $x, y \in \mathsf{cp}\, \iota o$ such $x \neq y$. Then $\llbracket x; y \rrbracket + xy$ is K_4-free.*

Proof. Without loss of generality x appears before y on every ιo-path. We obtain that $\llbracket x; y \rrbracket + xy$ is a minor of $G + \iota o$ by collapsing $\llbracket x \rrbracket_{\{x,y\}}$ (which contains ι) to x and $\llbracket y \rrbracket_{\{x,y\}}$ (which contains o) to y (Lemma 29). □

6 Extracting Terms from K_4-free Graphs

We say that a 2p-graph G is *CK4F* if its skeleton is connected and its strong skeleton is K_4-free. We now define a function extracting terms from CK4F graphs. Defining this function in Coq is challenging for a number of reasons. First, its definition involves ten cases, most with multiple recursive calls. Second, we need to argue that all the recursive calls are made on smaller graphs which are CK4F.

To facilitate the definition, we construct our own operator for bounded recursion. The reason for this is that none of the facilities for defining functions in Coq (e.g., Fixpoint, Function and Program) are suited to deal with the kind of complex function definition we require. We define a bounded recursion operator with the following type:

Fix : ∀ aT rT : Type, rT → (aT → ℕ) → ((aT → rT) → aT → rT) → aT→ rT

Here the argument of type aT → ℕ is a measure on the input to bound the number of recursive calls, and the argument of type rT is the default value to be returned when no more recursive calls are allowed.

We only need one lemma about the recursion operator, namely that the operator satisfies the usual fixpoint equation provided that the functional it is applied to calls its argument only on smaller arguments in the desired domain of the function (here, CK4F).[1] That is, we have the following lemma:

Fix_eq : ∀ (aT rT : Type) (P : aT → Prop) (x0 : rT) (m : aT → ℕ)
  (F : (aT → rT) → aT → rT),
  (∀ (f g : aT → rT) (x : aT),
    P x → (∀ y : aT, P y → m y < m x → f y = g y) → F f x = F g x) →
  ∀ x : aT, P x → Fix x0 m F x = F (Fix x0 m F) x

While its proof is straightforward, this lemma is useful in that it allows us to abstract from the fact that we are using bounded recursion (i.e., neither the default result nor the recursion on ℕ are visible in the proofs).

We now define the extraction function using the recursion operator. The various cases of the definition roughly correspond to the cases outlined in Fig. 2. The main difference is that in case (a), rather than partitioning the graph as shown in the picture, we only identify a single nontrivial bag or a single proper checkpoint between input and output. This is sufficient to make recursive calls on smaller graphs. In the case where input and output coincide (case (c)), we relocate the output and proceed recursively. This requires a measure that treats graphs with shared input and output as larger than those with distinct input and output. We use the measure below to justify termination.

Definition 33. *Let* $G = \langle\langle V, E, s, t, l\rangle, \iota, o\rangle$ *be a 2p-graph. The measure of G is* $2|E|$ *if* $\iota \neq o$ *and* $2|E| + 1$ *if* $\iota = o$.

The term extraction is then defined as follows:

$$t \triangleq \text{Fix 1 measure F}$$

where the definition of F is given in Fig. 5. This definition makes use of a number of auxiliary constructions which we define below. For a set of vertices U and a set of edges E (of some graph G) such that $\{s(e), t(e)\} \subseteq U$ for all e, the *subgraph of* G *with vertices* U *and edges* E is written $G[U, E]$. We write $\mathcal{E}(U)$ for the set of edges with source and target in U and the *induced subgraph for* U, written $G[U]$, is defined as $G[U, \mathcal{E}(U)]$. For 2p-graphs G, $G[U]$ and $G[U, E]$ are only defined if $\{\iota, o\} \subseteq U$. In this case, $G[U]$ and $G[U, E]$ have the same input and output as G.

When instantiating the definitions above, U will sometimes be an interval or a bag. In this case, the intervals and bags are computed on the weak skeleton

---

[1] To be precise, F may call its argument on anything. However, the result of F may only depend on calls to smaller arguments in the domain.

```
1: Definition F(t : 2p-graph → term)(G : 2p-graph) ≜
2:     let ⟨⟨V, E', s, t, l⟩, ι, o⟩ := G in
3:     if ι = o then
4:         let E := ℰ({ι}) in
5:         if E = ∅ then
6:             if pick(components(V ∖ {i})) is Some C then
7:                 dom(t(redirect C)) ∥ t(G[C̄])
8:             else 1
9:         else (* E ≠ ∅ *)
10:             (∥_{e∈E} tm(e))  ∥  G[V, Ē]
11:     else (* i ≠ o *)
12:         if ℰ(⟦ι⟧_{ι,o}) = ∅ ∧ ℰ(⟦o⟧_{ι,o}) = ∅ ∧ cp ι o = {ι, o} then
13:             let P := components(⟧ι; o⟦) in
14:             let E := ℰ({ι, o}) in
15:             if E = ∅ then
16:                 if pick P is Some C then
17:                     t(component(C)) ∥ t(G[C̄])
18:                 else 1 (* never reached *)
19:             else (* E ≠ ∅ *)
20:                 if P = ∅ then
21:                     ∥_{e∈E} tm(e)
22:                 else
23:                     (∥_{e∈E} tm(e)) ∥ t(G[V, Ē])
24:         else (* nontrivial ι or o-bag or proper checkpoint between ι and o *)
25:             if ℰ(⟦ι⟧_{ι,o}) ≠ ∅ ∨ ℰ(⟦o⟧_{ι,o}) ≠ ∅ then
26:                 t(G[ι])·t(G[ι, o])·t(G[o])
27:             else
28:                 if pick (cp ι o ∖ {ι, o}) is Some z then
29:                     t(G[ι, z])·t(G[z])·t(G[z, o])
30:                 else 1 (* never reached *)
```

Fig. 5. The term extraction function

of G (not the strong skeleton). For a given 2p-graph $G = \langle G', \iota, o\rangle$, we also define:

$$\mathsf{components}(U) \triangleq \{C \mid C \text{ connected component of } U \text{ in the skeleton of } G\}$$

$$\mathsf{component}(C) \triangleq G[C \cup \{\iota, o\}]$$

$$\mathsf{redirect}(C) \triangleq \langle G'[C \cup \{\iota\}], i, x\rangle \text{ where } x \text{ is some neighbor of } \iota \text{ in } C$$

$$G[x, y] \triangleq \langle G'[\llbracket x; y\rrbracket], \mathcal{E}(\llbracket x; y\rrbracket) \setminus (\mathcal{E}(\{x\}) \cup \mathcal{E}(\{y\}))\rangle, x, y\rangle$$

$$G[x] \triangleq \langle G'[\llbracket x\rrbracket_{\{\iota,o\}}], x, x\rangle$$

$$\mathsf{tm}(e) \triangleq \begin{cases} l(e) & s(e) = \iota \wedge t(e) = o \\ l(e)^{\circ} & \text{otherwise} \end{cases}$$

Note that component(C) is obtained as induced subgraph of G whereas the other constructions are obtained as subgraphs of G' (with new inputs and outputs).

Before we can establish properties of t, we need to establish that all (relevant) calls to t in F are made on CK4F graphs with smaller measure.

Lemma 34. *Let t, t' be functions from graphs to terms. If t and t' agree on all CK4F graphs with measure smaller than a CK4F graph G, then $\mathsf{F}\,t\,G = \mathsf{F}\,t'\,G$.*

The proof of this lemma boils down to a number of lemmas for the various branches of F. For each recursive call, we need to establish both that the measure decreases and that the graph is indeed CK4F. When splitting of a parallel component (line 17), Proposition 31 ensures that there are at least two nonempty components, thus ensuring that the remainder of the graph is both smaller and connected. Note that the case distinction in line 20 is required since if $P = \emptyset$, removing the ιo-edges disconnects the graph (the remaining graph would be isomorphic to $\top$). In the case where there is a proper checkpoint z between input and output (line 29), Proposition 32 ensures that the strong skeletons of $G[\iota, z]$ and $G[z, o]$ are $\mathsf{K_4}$-free.

As a consequence of Lemma 34, we obtain:

Proposition 35. *Let G be CK4F. Then $\mathsf{t}\,G = \mathsf{F}\,\mathsf{t}\,G$.*

7 Isomorphism Properties

In this section we establish that interpreting the terms extracted from a 2p-graph G yields a graph that is isomorphic to G. This is the part of the proof where the difference of what one would find in a detailed paper proof and what is required in order to obtain a formal proof is greatest.

Definition 36. *A homomorphism from the graph $G = \langle V, E, s, t, l \rangle$ to the graph $G' = \langle V', E', s', t', l' \rangle$ is a pair $\langle f, g \rangle$ of functions $f : V \to V'$ and $g : E \to E'$ that respect the various components: $s' \circ g \equiv f \circ s$, $t' \circ g \equiv f \circ t$, and $l \equiv l' \circ g$. A homomorphism from $\langle G, \iota, o \rangle$ to $\langle G', \iota', o' \rangle$ is a graph homomorphism $\langle f, g \rangle$ from G to G' respecting inputs and outputs: $f(\iota) = \iota'$ and $f(o) = o'$.*

An isomorphism is a homomorphism whose two components are bijective functions. We write $G \simeq G'$ when there exists an isomorphism between graphs G and G'.

The extraction function decomposes the graph into smaller graphs in order to extract a term. The interpretation of this term then joins the graphs extracted by the recursive calls back together using the graph operations $\|$ and $\cdot$. We need to establish that the decomposition performed during extraction is indeed correct (i.e., that no vertices or edges are lost or misplaced). This requires establishing a number of isomorphism properties.

We first establish that all graph operations respect isomorphism classes.

Lemma 37. *Let $G_1 \simeq G_1'$ and $G_2 \simeq G_2'$. Then we have $G_1 \parallel G_2 \simeq G_1' \parallel G_2'$, $G_1 \cdot G_2 \simeq G_1' \cdot G_2'$, and $\mathrm{dom}(G_1) \simeq \mathrm{dom}(G_1')$.*

Lemma 37 allows rewriting with isomorphisms underneath the graph operations using Coq's generalized (setoid) rewriting tactic [19].

The proofs for establishing that two graphs (of some known shape) are isomorphic generally follow the same pattern: define the pair of functions $\langle f, g \rangle$ (cf. Definition 36) as well as their respective inverses and then show all the required equalities (including that the proposed inverses are indeed inverses). This amounts to 9 equalities per isomorphism that all need to be verified. Additional complexity is introduced by the fact that we are almost exclusively interested in isomorphism properties involving $\parallel$ and $\cdot$ which are defined using quotient constructions. Among others, we establish the following isomorphism lemmas:

Lemma 38. *Let $G = \langle G', \iota, o \rangle$ such that $\iota \neq o$ and the skeleton of G is connected. Then $G \simeq G[\iota] \cdot G[\iota, o] \cdot G[o]$.*

Lemma 39. *Let $G = \langle G', \iota, o \rangle$ such that $\mathcal{E}(\llbracket \iota \rrbracket_{\{\iota, o\}}) = \emptyset$, $\mathcal{E}(\llbracket o \rrbracket_{\{\iota, o\}}) = \emptyset$, and $\iota \neq o$, and let $z \in \mathrm{cp}\, \iota\, o \setminus \{\iota, o\}$. Then $G \simeq G[\iota, z] \cdot G[z] \cdot G[z, o]$.*

Lemma 40. *Let $G = \langle G', \iota, o \rangle$ with $\mathcal{E}(\{\iota, o\}) = \emptyset$ and let $C \in$ components$(\overline{\{\iota, o\}})$. Then $G \simeq$ component$(C) \parallel G[\overline{C}]$.*

For the following, let $E_{x,y} \triangleq \{e \mid s(e) = x, t(e) = y\}$.

Lemma 41. *Let $G = \langle V, E, s, t, l \rangle$, let $x, y : G$ and let $E' \triangleq E_{x,y} \cup E_{y,x}$ Then $G \simeq G[\{x, y\}, E'] \parallel G[V, \overline{E'}]$.*

Theorem 42. *Let G be a 2p-graph. Then $\mathsf{g}(\mathsf{t}\, G) \simeq G$.*

Proof. By induction on the measure of G. We use Proposition 35 to unfold the definition of t. Each of the cases follows with the induction hypothesis (using the lemmas underlying the proof of Lemma 34 to justify that the induction hypothesis applies) and some isomorphism lemmas (e.g., Lemmas 37 to 41).   □

Note that Lemma 40 justifies both the split in line 7 and the split in line 17 (in the latter case $\rrbracket \iota; o \llbracket = \overline{\{\iota, o\}}$).

Putting everything together, we obtain our main result.

Theorem 43. *A simple graph is K_4-free iff if it has treewidth at most two.*

Proof. Fix some simple graph G. The direction from right to left follows with Proposition 11. For the converse direction we proceed by induction on $|G|$. If G is connected (and nonempty; otherwise the claim is trivial), we construct a 2p-graph (with $\iota = o$) whose (strong) skeleton is isomorphic to G. By Theorem 42, the skeleton of $\mathsf{g}(\mathsf{t}\, G)$ is isomorphic to G and, hence, K_4-free by Proposition 17. If G contains disconnected vertices x and y, then G is isomorphic to the disjoint union of the connected component containing x and the rest of the graph (which must contain y). The claim then follows by induction hypothesis using the fact that treewidth is preserved under disjoint union.   □

Note that Theorem 42 is significantly stronger than what is needed to establish Theorem 43. To prove the latter, it would be sufficient to extract terms that can be interpreted as simple graphs, thus avoiding the complexity introduced by labels, edge multiplicities and loops. The fine-grained analysis we formalize here is however crucial for our long-term objective (†).

8   Conclusion

We have developed a library for graph theory based on finite types as provided by the Mathematical Components Libraries. As a major step towards proving that K_4-free 2p-graphs form the free 2p-algebra (†), we gave a proof of the graph-minor theorem for treewidth two, using a function extracting terms from K_4-free graphs.

The Coq development accompanying this paper [7] consists of about 6700 lines of code, with a ratio of roughly 1:2 between specifications and proofs. It contains about 200 definitions and about 550 lemmas. Many of these have short proofs, but some proofs (e.g., the proof of Proposition 31) are long intricate constructions without any obvious lemmas to factor out. As mentioned before, the isomorphism proofs for Sect. 7 mostly follows the same pattern. Hence, we hope that they can be automated to some degree.

As it comes to proving (†), there are two main challenges to be solved. First we should prove that the choices made by the extraction function are irrelevant modulo the axioms of 2p-algebras (e.g., which neighbor is chosen in redirect(C)). This is why we were careful to define this function as deterministically as possible. Second, we should prove that it is a homomorphism (again, modulo the axioms of 2p-algebras). Those two steps seem challenging: their paper proofs require a lot of reasoning modulo graph isomorphism [14].

References

1. Chekuri, C., Rajaraman, A.: Conjunctive query containment revisited. Theoret. Comput. Sci. **239**(2), 211–229 (2000). https://doi.org/10.1016/S0304-3975(99)00220-0
2. Chou, C.-T.: A formal theory of undirected graphs in higher-order logc. In: Melham, T.F., Camilleri, J. (eds.) HUG 1994. LNCS, vol. 859, pp. 144–157. Springer, Heidelberg (1994). https://doi.org/10.1007/3-540-58450-1_40
3. Cohen, C.: Pragmatic quotient types in Coq. In: Blazy, S., Paulin-Mohring, C., Pichardie, D. (eds.) ITP 2013. LNCS, vol. 7998, pp. 213–228. Springer, Heidelberg (2013). https://doi.org/10.1007/978-3-642-39634-2_17
4. Courcelle, B.: The monadic second-order logic of graphs. I: recognizable sets of finite graphs. Inf. Comput. **85**(1), 12–75 (1990). https://doi.org/10.1016/0890-5401(90)90043-H
5. Courcelle, B., Engelfriet, J.: Graph Structure and Monadic Second-Order Logic - A Language-Theoretic Approach. Encyclopedia of Mathematics and Its Applications, vol. 138. Cambridge University Press, Cambridge (2012)

6. Diestel, R.: Graph Theory, Graduate Texts in Mathematics. Springer, New York (2005)
7. Doczkal, C., Combette, G., Pous, D.: Coq formalization accompanying this paper. https://perso.ens-lyon.fr/damien.pous/covece/k4tw2
8. Duffin, R.: Topology of series-parallel networks. J. Math. Anal. Appl. **10**(2), 303–318 (1965). https://doi.org/10.1016/0022-247X(65)90125-3
9. Dufourd, J.-F., Bertot, Y.: Formal study of plane delaunay triangulation. In: Kaufmann, M., Paulson, L.C. (eds.) ITP 2010. LNCS, vol. 6172, pp. 211–226. Springer, Heidelberg (2010). https://doi.org/10.1007/978-3-642-14052-5_16
10. Freuder, E.C.: Complexity of k-tree structured constraint satisfaction problems. In: NCAI, pp. 4–9. AAAI Press/The MIT Press (1990)
11. Freyd, P., Scedrov, A.: Categories, Allegories. North Holland, Elsevier, Amsterdam (1990)
12. Gonthier, G.: Formal proof – the four-color theorem. Notices Amer. Math. Soc. **55**(11), 1382–1393 (2008)
13. Grohe, M.: The complexity of homomorphism and constraint satisfaction problems seen from the other side. J. ACM **54**(1), 1:1–1:24 (2007). https://doi.org/10.1145/1206035.1206036
14. Llópez, E.C., Pous, D.: K4-free graphs as a free algebra. In: MFCS. LIPIcs, vol. 83, pp. 76:1–76:14. Schloss Dagstuhl - Leibniz-Zentrum für Informatik (2017). https://doi.org/10.4230/LIPIcs.MFCS.2017.76
15. Nakamura, Y., Rudnicki, P.: Euler circuits and paths. Formalized Math. **6**(3), 417–425 (1997)
16. Nipkow, T., Bauer, G., Schultz, P.: Flyspeck I: tame graphs. In: Furbach, U., Shankar, N. (eds.) IJCAR 2006. LNCS (LNAI), vol. 4130, pp. 21–35. Springer, Heidelberg (2006). https://doi.org/10.1007/11814771_4
17. Noschinski, L.: A graph library for Isabelle. Math. Comput. Sci. **9**(1), 23–39 (2015). https://doi.org/10.1007/s11786-014-0183-z
18. Robertson, N., Seymour, P.: Graph minors. XX. Wagner's conjecture. J. Comb. Theor. Ser. B **92**(2), 325–357 (2004). https://doi.org/10.1016/j.jctb.2004.08.001
19. Sozeau, M.: A new look at generalized rewriting in type theory. J. Form. Reason. **2**(1), 41–62 (2009). https://doi.org/10.6092/issn.1972-5787/1574
20. The Mathematical Components Team: Mathematical Components (2017). http://math-comp.github.io/math-comp/

Verified Analysis of Random Binary Tree Structures

Manuel Eberl$^{(\boxtimes)}$ ⓘ, Max W. Haslbeck ⓘ, and Tobias Nipkow ⓘ

Technische Universität München, 85748 Garching bei München, Germany
eberlm@in.tum.de

Abstract. This work is a case study of the formal verification and complexity analysis of some famous probabilistic algorithms and data structures in the proof assistant Isabelle/HOL. In particular, we consider the expected number of comparisons in randomised quicksort, the relationship between randomised quicksort and average-case deterministic quicksort, the expected shape of an unbalanced random Binary Search Tree, and the expected shape of a Treap. The last two have, to our knowledge, not been analysed using a theorem prover before and the last one is of particular interest because it involves continuous distributions.

1 Introduction

This paper conducts verified analyses of a number of classic probabilistic algorithms and data structures related to binary search trees. It is part of a continuing research programme to formalise classic data structure analyses [1–3], especially for binary search trees, by adding randomisation. The key novel contributions of the paper are readable (with one caveat, discussed in the conclusion) formalised analyses of

- the *precise* expected number of comparisons in randomised quicksort
- the relationship between the average-case behaviour of deterministic quicksort and the distribution of randomised quicksort
- the expected path length and height of a random binary search tree
- the expected shape of a treap, which involves *continuous* distributions.

The above algorithms are shallowly embedded and expressed using the Giry monad, which allows for a natural and high-level presentation. All verifications were carried out in Isabelle/HOL [4,5].

After an introduction to the representation of probability theory in Isabelle/HOL, the core content of the paper consists of three sections that analyse quicksort, random binary search trees, and treaps, respectively. The corresponding formalisations can be found in the *Archive of Formal Proofs* [6–8].

© Springer International Publishing AG, part of Springer Nature 2018
J. Avigad and A. Mahboubi (Eds.): ITP 2018, LNCS 10895, pp. 196–214, 2018.
https://doi.org/10.1007/978-3-319-94821-8_12

2 Probability Theory in Isabelle/HOL

2.1 Measures and Probability Mass Functions

The foundation for measure theory (and thereby probability theory) in Isabelle/ HOL was laid by Hölzl [9]. This approach is highly general and flexible, allowing also measures with uncountably infinite support (e. g. normal distributions on the reals) and has been used for a number of large formalisation projects related to randomisation, e. g. Ergodic theory [10], compiling functional programs to densities [11], Markov chains and decision processes [12], and cryptographic algorithms [13].

Initially we shall only consider probability distributions over *countable* sets. In Isabelle, these are captured as *probability mass functions* (PMFs). A PMF is simply a function that assigns a probability to each element, with the property that the probabilities are non-negative and sum up to 1. For any HOL type α, the type α *pmf* denotes the type of all probability distributions over values of type α with countable support.

Working with PMFs is quite pleasant, since we do not have to worry about measurability of sets or functions. Since everything is countable, we can always choose the power set as the measurable space, which means that everything is always trivially measurable.

Later, however, we will also need continuous distributions. For these, there exists a type α *measure*, which describes a measure-theoretic measure over elements of type α. Such a measure is formally defined as a triple consisting of a carrier set Ω, a σ-algebra on Ω (which we call the set of measurable sets), and a measure function $\mu : \alpha \to$ *ennreal*, where *ennreal* is the type of extended non-negative real numbers $\mathbb{R}_{\geq 0} \cup \{\infty\}$. Of course, since we only consider probability measures here, our measures will always return values between 0 and 1.

One problem with these general measures (which are only relevant for Sect. 5), is that we often need to annotate the corresponding σ-algebras and prove that everything we do with a distribution is in fact measurable. These details are unavoidable on a formal level, but typically very uninteresting to a human: There is usually a 'natural' choice for these σ-algebras and any set or operation that can be written down is typically measurable in some adequate sense. For the sake of readability, we will therefore omit everything related to measurability in this presentation.

Table 1 gives an overview of the notation that we use for PMFs and general measures. Although we allow ourselves some more notational freedoms in this paper, these are purely syntactical changes designed to make the presentation easier without introducing additional notation.

2.2 The Giry Monad

Specifying probabilistic algorithms compositionally requires a way to express sequential composition of randomised choice. The standard way to do this is the *Giry monad* [14]. A detailed explanation of this (especially in the context

Table 1. Basic operations on PMFs and general measures. The variables $p :: \alpha$ *pmf* and $M :: \alpha$ *measure* denote an arbitrary PMF (resp. measure)

PMFs	Measures	Meaning
pmf p x		probability of x in distribution p
set_pmf p		support of p, i. e. $\{x \mid p(x) > 0\}$
measure_pmf.prob p X	emeasure M X	probability of set X
measure_pmf.expectation p f	expectation M f	expectation of $f :: \alpha \to \mathbb{R}$
map_pmf g p	distr g M	image measure under $g :: \alpha \to \beta$
pmf_of_set A	uniform_measure A	uniform distribution over A
pair_pmf p q	$M \otimes N$	binary product measure
	$\bigotimes_{x \in A} M(x)$	indexed product measure

of Isabelle/HOL) can be found in an earlier paper by Eberl *et al.* [11]. For the purpose of this paper, we only need to know that the Giry monad provides functions

$$\text{return} :: \alpha \to \alpha \text{ pmf} \qquad \text{bind} :: \alpha \text{ pmf} \to (\alpha \to \beta \text{ pmf}) \to \beta \text{ pmf}$$

(and analogously for α *measure*) where *return* x gives us the singleton distribution where x is chosen with probability 1 and *bind* p f composes two distributions in the intuitive sense of randomly choosing a value x according to the distribution p and then returning a value randomly chosen according to the distribution $f(x)$.

For better readability, Isabelle supports a Haskell-like do-notation as syntactic sugar for *bind* operations where e. g.

$$\text{bind } A \ (\lambda x. \ \text{bind } B \ (\lambda y. \ \text{return } (x + y)))$$

can be written succinctly as

$$\textbf{do } \{x \leftarrow A; \ y \leftarrow B; \ \text{return } (x + y)\}.$$

3   Quicksort

We now show how to define and analyse quicksort [15,16] (in its functional representation) within this framework. Since all of the randomisation is discrete in this case, we can restrict ourselves to PMFs for the moment.

For the sake of simplicity (and because it relates to binary search trees, which we will treat later), we shall only treat the case of sorting lists without repeated elements. (See the end of this section for further discussion of this point.)

As is well known, quicksort has quadratic worst-case performance if the pivot is chosen poorly. Using the true median as the pivot would solve this, but is impractical. Instead, a simple alternative is to choose the pivot randomly, which is the variant that we shall analyse first.

3.1   Randomised Quicksort

Intuitively, the good performance of randomised quicksort can be explained by the fact that a random pivot will usually not be among the most extreme values of the list, but somewhere in the middle, which means that, on average, the size of the lists is reduced significantly in every recursion step.

To make this more rigorous, let us look at the definition of the algorithm in Isabelle:

Definition 1 (Randomised quicksort)

$$\text{rquicksort } R \text{ } xs =$$
$$\quad \textbf{if } xs = [\,] \textbf{ then}$$
$$\quad\quad \text{return } [\,]$$
$$\quad \textbf{else do } \{$$
$$\quad\quad i \leftarrow \text{pmf\_of\_set } \{0 \ldots |xs| - 1\}$$
$$\quad\quad \textbf{let } x = xs\,!\,i$$
$$\quad\quad \textbf{let } xs' = \text{delete\_index } i \text{ } xs$$
$$\quad\quad ls \leftarrow \text{rquicksort } R \text{ } [y \mid y \leftarrow xs', (y, x) \in R]$$
$$\quad\quad rs \leftarrow \text{rquicksort } R \text{ } [y \mid y \leftarrow xs', (y, x) \notin R]$$
$$\quad\quad \text{return } (ls \text{ @ } [x] \text{ @ } rs)$$
$$\quad \}$$

Here, @ denotes list concatenation and $xs\,!\,i$ denotes the i-th element of the list xs, where $0 \leq i < |xs|$. The *delete_index* function removes the i-th element of a list, and the parameter R is a linear ordering represented as a set of pairs.

It is easy to prove that all of the lists that can be returned by the algorithm are sorted w. r. t. R. To analyse its running time, its actual Isabelle definition was extended to also count the number of element comparisons made, i. e. to return an (α *list* $\times$ *nat*) *pmf*. The base case makes 0 comparisons and the recursive case makes $|xs| - 1 + n_1 + n_2$ comparisons, where n_1 and n_2 are the numbers of comparisons made by the recursive calls. This could easily be encapsulated in a resource monad (as we have done elsewhere [3] for more complex code), but it is not worth the effort in this case.

For an element x and some list xs, we call the number of elements of xs that are smaller than x the *rank* of x w. r. t. xs. In lists with distinct elements, each element can clearly be uniquely identified by either its index in the list or its rank w. r. t. that list, so choosing an element uniformly at random, choosing its index uniformly at random, or choosing a rank uniformly at random are all interchangeable.

In the above algorithm, the length of ls is simply the rank r of the pivot, and the length of rs is simply $|xs| - 1 - r$, so choosing the pivot uniformly at random means that the length of ls is also distributed uniformly between 0 and $|xs| - 1$. From this, we can see that the distribution of the number of comparisons does not actually depend on the content of the list or the ordering R at all, but only on the *length* of the list, and we can find the following recurrence for it:

Definition 2 (Cost of randomised quicksort)

$$rqs\_cost\ 0\ =\ return\ 0$$
$$rqs\_cost\ (n+1)\ =$$
$$\textbf{do}\ \{$$
$$r \leftarrow pmf\_of\_set\ \{0 \ldots n\}$$
$$a \leftarrow rqs\_cost\ r$$
$$b \leftarrow rqs\_cost\ (n-r)$$
$$return\ (n+a+b)$$
$$\}$$

For any list xs with no repeated elements and a linear ordering R, we can easily show the equation

$$map\_pmf\ snd\ (rquicksort\ R\ xs)\ =\ rqs\_cost\ |xs|\ ,$$

i. e. projecting out the number of comparisons from our cost-aware randomised quicksort yields the distribution given by rqs_cost.

Due to the recursive definition of rqs_cost, we can easily show that its expected value, which we denote by $Q(n)$, satisfies the characteristic recurrence

$$Q(n+1) = n + \frac{1}{n+1}\left(\sum_{i=0}^{n} Q(i) + Q(n-i)\right),$$

or, equivalently,

$$Q(n+1) = n + \frac{2}{n+1}\left(\sum_{i=0}^{n} Q(i)\right).$$

This is often called the *quicksort recurrence*. Cichoń [17] gave a simple way of solving this by turning it into a linear recurrence

$$\frac{Q(n+1)}{n+2} = \frac{2n}{(n+1)(n+2)} + \frac{Q(n)}{n+1},$$

which gives us (by telescoping)

$$\frac{Q(n)}{n+1} = 2\sum_{k=1}^{n} \frac{k-1}{k(k+1)} = 4H_{n+1} - 2H_n - 4$$

and thereby the closed-form solution

$$Q(n) = 2(n+1)H_n - 4n,$$

where H_n is the n-th harmonic number. We can use the well-known asymptotics $H_n \sim \ln n + \gamma$ (where $\gamma \approx 0.5772$ is the Euler–Mascheroni constant) from the Isabelle library and obtain $Q(n) \sim 2n \ln n$, which shows that the expected number of comparisons is logarithmic.

Remember, however, that we only considered lists with no repeated elements. If there are any repeated elements, the performance of the above algorithm can deteriorate to quadratic time. This can be fixed easily by using a three-way partitioning function instead, although this makes things slightly more complicated since the number of comparisons made now depends on the *content* of the list and not just its *length*. The only real difference in the cost analysis is that the lists in the recursive call no longer simply have lengths r and $n - r - 1$, but can also be shorter if the pivot is contained in the list more than once. We can still show that the expected number of comparisons is at most $Q(n)$ in much the same way as before (and our entry [6] in the *Archive of Formal Proofs* does contain that proof), but we shall not go into more detail here.

Comparing our proof to those in the literature, note that both Cormen *et al.* [18] and Knuth [19] also restrict their analysis to distinct elements. Cormen *et al.* use a non-compositional approach with indicator variables and only derive the logarithmic upper bound, whereas Knuth's analysis counts the detailed number of different operations made by a particular implementation of the algorithm in MIX. His general approach is very similar to the one presented here.

3.2   Average-Case of Non-randomised Quicksort

The above results carry over directly to the average-case analysis of non-randomised quicksort (again, we will only consider lists with distinct elements). Here, the pivot is chosen deterministically; we always choose the first element for simplicity. This gives us the following definitions of quicksort and its cost:

Definition 3 (Deterministic quicksort and its cost)

quicksort $R\ [] = []$

quicksort $R\ (x \mathbin{\#} xs) = $ quicksort $R\ [y \mid y \leftarrow xs, (y, x) \in R]\ @$
$\quad [x]\ @$ quicksort $R\ [y \mid y \leftarrow xs, (y, x) \notin R]$

qs_cost $R\ [] = 0$

qs_cost $R\ (x \mathbin{\#} xs) = |xs| +$
$\quad$ qs_cost $R\ [y \mid y \leftarrow xs, (y, x) \in R] +$ qs_cost $R\ [y \mid y \leftarrow xs, (y, x) \notin R]$

Interestingly, the number of comparisons made on a randomly-permuted input list has exactly the same distribution as the number of comparisons in randomised quicksort from before. The underlying idea is that when randomly permuting the input, the randomness can be 'deferred' to the first point where an element is actually inspected, which means that choosing the first element of a randomly-permuted list still makes the pivot essentially random.

The formal proof of this starts by noting that choosing a random permutation of a non-empty finite set A is the same as first choosing the first list element $x \in A$ uniformly at random and then choosing a random permutation of $A \setminus \{x\}$ as the remainder of the list, allowing us to pull out the pivot selection. Then, we note that taking a random permutation of $A \setminus \{x\}$ and partitioning it into elements that are smaller and bigger than x is the same as first partitioning the set $A \setminus \{x\}$ into $\{y \in A \setminus \{x\} \mid (y, x) \in R\}$ and $\{y \in A \setminus \{x\} \mid (y, x) \notin R\}$ and choosing random permutations of these sets independently.

This last step, which interchanges partitioning and drawing a random permutation, is probably the most crucial one and one that we will need again later, so we present the corresponding lemma in full here. Let *partition P xs* be the function that splits the list *xs* into the pair of sub-sequences that satisfy (resp. do not satisfy) the predicate P. Then, we have:

Lemma 1 (Partitioning a randomly permuted list)

> **assumes** finite A
>
> **shows** map_pmf (partition P) (pmf_of_set (permutations_of_set A)) =
>
> pair_pmf (pmf_of_set (permutations_of_set $\{x \in A.\ P\ x\}$))
>
> (pmf_of_set (permutations_of_set $\{x \in A.\ \neg P\ x\}$))

This lemma is easily proven directly by extensionality, i.e. fixing permutations *xs* of $\{x \in A.\ P\ x\}$ and *ys* of $\{x \in A.\ \neg P\ x\}$ and computing their probabilities in the two distributions and noting that they are the same.

With this, the proof of the following theorem is just a straightforward induction on the recursive definition of *rqs_cost*:

Theorem 1 (Cost distribution of randomised quicksort). *For every linear order R on a finite set A, we have:*

$$\text{map\_pmf (qs\_cost } R\text{) (pmf\_of\_set (permutations\_of\_set } A\text{))} = \text{rqs\_cost } |A|$$

Thus, the cost distribution of deterministic quicksort applied to a randomly-permuted list is the same as that of randomised quicksort. In particular, the results about the logarithmic expectation of *rqs_cost* carry over directly.

4   Random Binary Search Trees

4.1   Preliminaries

We now turn to another average-case complexity problem that is somewhat related to quicksort, though not in an obvious way. We consider node-labelled binary trees, defined by the algebraic datatype

datatype α tree = Leaf | Node (α tree) α (α tree) .

We denote *Leaf* by $\langle \rangle$ and *Node l x r* by $\langle l, x, r \rangle$. When the values of the tree have some linear ordering, we say that the tree is a *binary search tree* (BST) if, for every node with some element x, all of the values in the left sub-tree are smaller than x and all of the values in the right sub-tree are larger than x.

Inserting elements can be done by performing a search and, if the element is not already in the tree, adding a node at the leaf at which the search ends. We denote this operation by *bst_insert*. Note that these are simple, unbalanced BSTs and our analysis will focus on what happens when elements are inserted into them in *random order*. We call the tree that results from adding elements of a set A to an initially empty BST in random order a *random BST*. This can also be seen as a kind of 'average-case' analysis of BSTs.

To analyse random BSTs, let us first examine what happens when we insert a list of elements into an empty BST from left to right; formally:

Definition 4 (Inserting a list of elements into a BST)

$$\text{bst\_of\_list } xs \;=\; \text{fold bst\_insert } xs \; \langle\rangle$$

Let x be the first element of the list. This element will become the root of the tree and will never move again. Similarly, the next element will become either the left or right child of x and will then also never move again and so on. It is also clear that no elements greater than x will end up in the left sub-tree of x at any point in the process, and no elements smaller in the right sub-tree. This leads us to the following recurrence for *bst_of_list*:

Lemma 2 (Recurrence for *bst_of_list*)

$$\text{bst\_of\_list } [] \;=\; \langle\rangle$$
$$\text{bst\_of\_list } (x \,\#\, xs) \;=$$
$$\langle\text{bst\_of\_list } [y \mid y \leftarrow xs, y < x], \; x, \; \text{bst\_of\_list } [y \mid y \leftarrow xs, y > x]\rangle$$

We can now formally define our notion of 'random BST':

Definition 5 (Random BSTs)

$$\text{random\_bst } A \;=$$
$$\text{map\_pmf bst\_of\_list } (\text{pmf\_of\_set } (\text{permutations\_of\_set } A))$$

By re-using Lemma 1, we easily get the following recurrence:

Lemma 3 (A recurrence for random BSTs)

$$\text{random\_bst } A \;=$$
$$\textbf{if } A = \{\} \textbf{ then } \text{return } \langle\rangle \textbf{ else do } \{$$
$$x \leftarrow \text{pmf\_of\_set } A$$
$$l \leftarrow \text{random\_bst } \{y \in A \mid y < x\}$$
$$r \leftarrow \text{random\_bst } \{y \in A \mid y > x\}$$
$$\text{return } \langle l, x, r\rangle$$
$$\}$$

We can now analyse some of the properties of such a random BST. In particular, we will look at the expected height and the expected internal path length, and we will start with the latter since it is easier.

4.2   Internal Path Length

The internal path length (IPL) is essentially the sum of the lengths of all the paths from the root of the tree to each node. Alternatively, one can think of it the sum of all the *level numbers* of the nodes in the tree, where the root is on the 0-th level, its immediate children are on the first level etc.

One reason why this number is important is that it is related to the time it takes to access a random element in the tree: the number of steps required to access some particular element x is equal to the number of that element's level, so if one chooses a random element in the tree, the average number of steps needed to access it is exactly the IPL divided by the size of the tree.

The IPL can be defined recursively by noting that $ipl \langle \rangle = 0$ and $ipl \langle l, x, r \rangle = ipl\ l + ipl\ r + |l| + |r|$. With this, we can show the following theorem by a simple induction over the recurrence for *random_bst*:

Theorem 2 (Internal path length of a random BST)

$$\text{map\_pmf ipl (random\_bst } A) = \text{rqs\_cost } |A|$$

Thus, the IPL of a random BST has the exact same distribution as the number of comparisons in randomised quicksort, which we already analysed before. This analysis was also carried out by Ottman and Widmayer [20], who also noted its similarity to the analysis of quicksort.

4.3   Height

The height of a random BST is more difficult to analyse. By our definition, an empty tree (i. e. a leaf) has height 0, and the height of a non-empty tree is the maximum of the heights of its left and right sub-trees, plus one. It is easy to show that the height distribution only depends on the *number* of elements and not their actual content, so let $H(n)$ denote the height of a random BST with n nodes.

The asymptotics of its expectation and variance were found by Reed [21], who showed that $E[H(n)] = \alpha \ln n - \beta \ln \ln n + O(1)$ and $\text{Var}[H(n)] \in O(1)$ where $\alpha \approx 4.311$ is the unique real solution of $\alpha \ln(2e/\alpha) = 1$ with $\alpha \geq 2$ and $\beta = 3\alpha/(2\alpha - 2) \approx 1.953$. The proof of this is quite intricate, so we will restrict ourselves to showing that $E[H(n)] \leq \frac{3}{\ln 2} \ln n \approx 4.328 \ln n$, which is enough to see that the expected height is logarithmic.

Before going into a precise discussion of the proof, let us first undertake a preliminary exploration of how we can analyse the expectation of $H(n)$. The base cases $H(0) = 0$ and $H(1) = 1$ are obvious. For any $n > 1$, the recursion formula for *random_bst* suggests:

$$E[H(n)] = 1 + \frac{1}{n} \sum_{k=0}^{n-1} E[\max(H(k), H(n - k - 1))]$$

The *max* term is somewhat problematic, since the expectation of the maximum of two random variables is, in general, difficult to analyse. A relatively obvious bound is $E[\max(A, B)] \leq E[A] + E[B]$, but that will only give us

$$E[H(n)] \leq 1 + \frac{1}{n} \sum_{k=0}^{n-1} (E[H(k)] + E[H(n-k-1)])$$

and if we were to use this to derive an explicit upper bound on $E[H(n)]$ by induction, we would only get the trivial upper bound $E[H(n)] \leq n$.

A trick suggested e. g. by Cormen *et al.* [18] (which they attribute to Aslam [22]) is to instead use the *exponential height* (which we shall denote by *eheight*) of the tree, which, in terms of our height, is defined as 0 for a leaf and $2^{height(t)-1}$ for a non-empty tree. The advantage of this is that it decreases the relative error that we make when we bound $E[\max(A, B)]$ by $E[A] + E[B]$: this error is precisely $E[\min(A, B)]$, and if A and B are heights, these heights only differ by a small amount in many cases. However, even a height difference of 1 will lead to a relative error in the exponential height of at most $\frac{1}{2}$, and considerably less than that in many cases. This turns out to be enough to obtain a relatively precise upper bound.

Let $H'(n)$ be the exponential height of a random BST. Since $x \mapsto 2^x$ is convex, any upper bound on $H'(n)$ can be used to derive an upper bound on $H(n)$ by Jensen's inequality:

$$2^{E[H(n)]} = 2 \cdot 2^{E[H(n)-1]} \leq 2E[2^{H(n)-1}] = 2E[H'(n)]$$

Therefore, we have

$$E[H(n)] \leq \log_2 E[H'(n)] + 1 .$$

In particular, a polynomial upper bound on $E[H'(n)]$ directly implies a logarithmic upper bound on $E[H(n)]$.

It remains to analyse $H'(n)$ and find a polynomial upper bound for it. As a first step, note that if l and r are not both empty, the exponential height satisfies the recurrence *eheight* $\langle l, x, r \rangle = 2 \cdot \max (eheight\ l)\ (eheight\ r)$. When we combine this with the recurrence for *random_bst*, the following recurrence for $H'(n)$ suggests itself:

Definition 6 (The exponential height of a random BST)

$$\text{eheight\_rbst } 0 = \text{return } 0$$
$$\text{eheight\_rbst } 1 = \text{return } 1$$
$$n > 1 \Longrightarrow \text{eheight\_rbst } n =$$
$$\textbf{do } \{$$
$$\quad k \leftarrow \text{pmf\_of\_set } \{0 \ldots n - 1\}$$
$$\quad h_1 \leftarrow \text{eheight\_rbst } k$$
$$\quad h_2 \leftarrow \text{eheight\_rbst } (n - k - 1)$$
$$\quad \text{return } (2 \cdot \max h_1 \ h_2)$$
$$\}$$

Showing that this definition is indeed the correct one can be done by a straight-forward induction following the recursive definition of *random_bst*:

Lemma 4 (Correctness of *eheight_rbst*)

$$\text{finite } A \Longrightarrow \text{eheight\_rbst } |A| = \text{map\_pmf eheight (random\_bst } A)$$

Using this, we note that for any $n > 1$:

$$E[H'(n)] = \frac{2}{n} \sum_{k=0}^{n-1} E[\max(H'(k), H'(n - k - 1))]$$

$$\leq \frac{2}{n} \sum_{k=0}^{n-1} E[H'(k) + H'(n - k - 1)]$$

$$= \frac{2}{n} \left(\sum_{k=0}^{n-1} E[H'(k)] + \sum_{k=0}^{n-1} E[H'(n - k - 1)] \right) = \frac{4}{n} \sum_{k=0}^{n-1} E[H'(k)]$$

However, we still have to find a suitable polynomial upper bound to complete the induction argument. If we had some polynomial $P(n)$ that fulfils $0 \leq P(0)$, $1 \leq P(1)$, and the recurrence $P(n) = \frac{4}{n} \sum_{k=0}^{n-1} P(k)$, the above recursive estimate for $E[H'(n)]$ would directly imply $E[H'(n)] \leq P(n)$ by induction. Cormen *et al.* give the following polynomial, which satisfies all these conditions and makes everything work out nicely:

$$P(n) = \frac{1}{4} \binom{n + 3}{3} = \frac{1}{24}(n + 1)(n + 2)(n + 3)$$

Putting all of these together gives us the following theorem:

Theorem 3 (Asymptotic expectation of $H(n)$)

$$E[H(n)] \ \leq \ \log_2 E[H'(n)] + 1 \ \leq \ \log_2 P(n) + 1 \ \sim \ \frac{3}{\ln 2} \ln n$$

5 Treaps

As we have seen, BSTs have the nice property that even without any explicit balancing, they tend to be fairly balanced if elements are inserted into them in random order. However, if, for example, the elements are instead inserted in ascending order, the tree degenerates into a list and no longer has logarithmic height. One interesting way to prevent this is to use a *treap* instead, which we shall introduce and analyse now.

5.1 Definition

A treap is a binary tree in which every node contains both an element and an associated priority and which is a BST w. r. t. the elements and a heap w. r. t. the priorities (i. e. the root is always the node with the lowest priority). This kind of structure was first described by Vuillemin [23], who called it a *cartesian tree*, and independently studied further by Seidel and Aragon [24], who noticed its relationship to random BSTs. Due to space constraints, we shall not go into how insertion of elements (denoted by *ins*) works, but it is fairly easy to implement.

An interesting consequence of these treap conditions is that, as long as all of the priorities are distinct, the shape of a treap is uniquely determined by the set of its elements and their priorities. Since the sub-trees of a treap must also be treaps, this uniqueness property follows by induction and we can construct this unique treap for a given set using the following simple algorithm:

Lemma 5 (Constructing the unique treap for a set). *Let A be a set of pairs of type $\alpha \times \mathbb{R}$ where the second components are all distinct. Then there exists a unique treap* treap_of *A whose elements are precisely A, and it satisfies the recurrence*

treap_of A =
  if $A = \{\}$ **then** $\langle\rangle$ **else**
    let x = arg_min_on snd A
    in $\langle$treap_of $\{y \in A \mid$ fst $y <$ fst $x\}, x,$ treap_of $\{y \in A \mid$ fst $y >$ fst $x\}\rangle$

where arg_min_on f A *is some $a \in A$ such that $f(a)$ is minimal on A. In our case the choice of a is unique by assumption.*

This is very similar to the recurrence for *bst_of_list* that we saw earlier. In fact, it is easy to prove that if we forget about the priorities in the treap and consider it as a simple BST, the resulting tree is exactly the same as if we had first sorted the keys by increasing priority and then inserted them into an empty BST in that order. Formally, we have the following lemma:

Lemma 6 (Connection between treaps and BSTs). *Let p be an injective function that associates a priority to each element of a list xs. Then*

$$\text{map\_tree fst (treap\_of } \{(x, p(x)) \mid x \in \text{set } xs\}) = \text{bst\_of\_list (sort\_key } p \; xs),$$

where sort_key sorts a list in ascending order w. r. t. the given priority function.

Proof. By induction over $xs' := sort\_key \; p \; xs$, using the fact that sorting w. r. t. distinct priorities can be seen as a selection sort: The list xs' consists of the unique minimum-priority element x, followed by $sort\_key \; p \; (remove1 \; x \; xs)$, where $remove1$ deletes the first occurrence of an element from a list.

With this and Lemma 2, the recursion structure of the right-hand side is exactly the same as that of the *treap_of* from Lemma 5. □

This essentially allows us to build a BST that behaves as if we inserted the elements by ascending priority regardless of the order in which they were actually inserted. In particular, we can assign each element a *random* priority upon its insertion, which turns our treap (a deterministic data structure for values of type $(\alpha \times \mathbb{R})$ *set*) into a randomised treap, which is a *randomised* data structure for values of type α that has the same distribution as a random BST with the same content.

One caveat is that for all the results so far, we assumed that no two distinct elements have the same priority, and, of course, without that assumption, we lose all these nice properties. If the priorities in our randomised treap are chosen from some discrete probability distribution, there will always be some non-zero probability that they are not distinct. For this reason, treaps are usually described in the literature as using a continuous distribution (e. g. uniformly-distributed real numbers between 0 and 1), even though this cannot be implemented faithfully on an actual computer. We shall do the same here, since it makes the analysis much easier.[1]

The argument goes as follows:

1. Choosing the priority of each element randomly when we insert it is the same as choosing all the priorities beforehand (i. i. d. at random) and then inserting the elements into the treap deterministically.
2. By the theorems above, this is the same as choosing the priorities i. i. d. at random, sorting the elements by increasing priority, and then inserting them into a BST in that order.
3. By symmetry considerations, choosing priorities i. i. d. for all elements and then looking at the linear ordering defined by these priorities is the same as choosing one of the $n!$ possible linear orderings uniformly at random.
4. Thus, inserting a list of elements into a randomised treap is the same as inserting them into a BST in random order.

---

[1] In fact, any non-discrete probability distribution works, where by 'non-discrete' we mean that all singleton sets have probability 0. In the formalisation, however, we restricted ourselves to the case of a uniform distribution over a real interval.

5.2   The Measurable Space of Trees

One complication when formalising treaps that is typically not addressed in pen-and-paper accounts is that since we will randomise over priorities, we need to talk about continuous distributions of trees, i.e. distributions of type $(\alpha \times \mathbb{R})$ *tree measure*. For example, if we insert the element x into an empty treap with a priority that is uniformly distributed between 0 and 1, we get a distribution of trees with the shape $\langle\langle\rangle, (x, p), \langle\rangle\rangle$ where p is uniformly distributed between 0 and 1.

In order to be able to express this formally, we need a way to lift some measurable space M to a measurable space $\mathcal{T}(M)$ of trees with elements from M attached to their nodes. Of course, we cannot just pick *any* measurable space: for our treap operations to be well-defined, all the basic tree operations need to be measurable w.r.t. $\mathcal{T}(M)$; in particular:

- the constructors *Leaf* and *Node*, i.e. we need $\{Leaf\} \in \mathcal{T}(M)$ and *Node* must be $\mathcal{T}(M) \otimes M \otimes \mathcal{T}(M)$–$\mathcal{T}(M)$-measurable
- the projection functions, i.e. selecting the value/left sub-tree/right sub-tree of a node; e.g. selecting a node's value must be $(\mathcal{T}(M) \setminus \{\langle\rangle\})$–$M$-measurable
- primitively recursive functions involving only measurable operations must also be measurable; we will need this to define e.g. the insertion operation

We can construct such a measurable space by taking the σ-algebra that is generated by certain *cylinder sets*: consider a tree whose nodes each have an M-measurable set attached to them. Then this tree can be 'flattened' into the set of trees of the same shape where each node has a single value from the corresponding set in t attached to it. Then we define $\mathcal{T}(M)$ to be the measurable space generated by all these cylinder sets, and prove that all the above-mentioned operations are indeed measurable.

5.3   Randomisation

In order to achieve a good height distribution on average, the priorities of a treap need to be chosen randomly. Since we do not know how many elements will be inserted into the tree in advance, we need to draw the priority to assign to an element when we insert it, i.e. insertion is now a randomised operation.

Definition 7 (Randomised insertion into a treap)

> rins :: $\alpha \rightarrow \alpha$ treap $\rightarrow \alpha$ treap measure
>
> rins $x\ t = \textbf{do}\ \{p \leftarrow \text{uniform\_measure}\ \{0 \ldots 1\};\ \text{return (ins } x\ p\ t)\}$

Since we would like to analyse what happens when we insert a large number of elements into an initially empty treap, we also define the following 'bulk insert' operation that inserts a list of elements into the treap from left to right:

> rinss :: α list $\rightarrow \alpha$ treap $\rightarrow \alpha$ treap measure
>
> rinss $[]\ t = \text{return } t$
>
> rinss $(x \mathbin{\#} xs)\ t = \textbf{do}\ \{t' \leftarrow \text{rins } x\ t;\ \text{rinss } xs\ t'\}$

Note that, from now on, we will again assume that all of the elements that we insert are *distinct*. This is not really a restriction, since inserting duplicate elements does not change the tree, so we can just drop any duplicates from the list without changing the result. Similarly, the uniqueness property of treaps means that after deleting an element, the resulting treap is exactly the same as if the element had never been inserted in the first place, so even though we only analyse the case of insertions without duplicates, this extends to any sequence of insertion and deletion operations (although we do not show this explicitly).

The main result, as sketched above, shall be that after inserting a certain number of distinct elements into the treap and then forgetting about their priorities, we get a BST that is distributed identically to a random BST with the same elements, i.e. the treap behaves as if we had inserted the elements in random order. Formally, this can be expressed like this:

Theorem 4 (Connecting randomised treaps to random BSTs)

$$\text{distr (rinss } xs \; \langle \rangle) \; (\text{map\_tree fst}) = \text{random\_bst (set } xs)$$

Proof. Let $\mathcal{U}$ denote the uniform distribution of real numbers between 0 and 1 and $\mathcal{U}^A$ denote a vector of i.i.d. distributions $\mathcal{U}$, indexed by A:

$$\mathcal{U} := \text{uniform\_measure } \{0 \ldots 1\} \qquad \mathcal{U}^A := \bigotimes_A \mathcal{U}$$

The first step is to show that our bulk-insertion operation *rinss* is equivalent to *first* choosing random priorities for *all* the elements at once and then inserting them all (with their respective priorities) deterministically:

$$\text{rinss } xs \; t = \text{distr } \mathcal{U}^{\text{set } xs} \; (\lambda p. \; \text{foldl } (\lambda t \; x. \; \text{ins } x \; (p(x)) \; t) \; t \; xs)$$
$$= \text{distr } \mathcal{U}^{\text{set } xs} \; (\lambda p. \; \text{treap\_of } [(x, p(x)) \mid x \leftarrow xs])$$

The first equality is proved by induction over xs, pulling out one insertion in the induction step and moving the choice of the priority to the front. This is intuitively obvious, but the formal proof is nonetheless rather tedious, mostly because of the issue of having to prove measurability in every single step. The second equality follows from the uniqueness of treaps.

Next, we note that the priority function returned by $\mathcal{U}^{\text{set } xs}$ is almost surely injective, so we can apply Lemma 6 and get:

$$\text{distr (rinss } xs \; \langle \rangle) \; (\text{map\_tree fst}) =$$
$$\text{distr } \mathcal{U}^{\text{set } xs} \; (\lambda p. \; \text{bst\_of\_list (sort\_key } p \; xs))$$

The next key lemma is the following, which holds for any finite set A:

$$\text{distr } \mathcal{U}^A \; (\text{linorder\_from\_keys } A) = \text{uniform\_measure (linorders\_on } A)$$

This essentially says that choosing priorities for all elements of A and then looking at the ordering on A that these priorities induce will give us the uniform distribution on all the $|A|!$ possible linear ordering relations on A. In particular, this means that that relation will be linear with probability 1, i.e. the priorities will almost surely be injective. The proof of this is a simple symmetry argument: given any two linear orderings R and R' of A, we can find some permutation π of A that maps R' to R. However, $\mathcal{U}^A$ is stable under permutation. Therefore, R and R' have the same probability, and since this holds for all R, R', the distribution must be the uniform distribution.

This brings us to the last step: Proving that sorting our list of elements by random priorities and then inserting them to a BST is the same as inserting them in random order (in the sense of inserting them in the order given by a randomly-permuted list):

$$\text{distr } \mathcal{U}^{\text{set } xs} \ (\lambda p. \ \text{bst\_of\_list } (\text{sort\_key } p \ xs) =$$
$$\text{distr } (\text{uniform\_measure } (\text{permutations\_of\_set } (\text{set } xs))) \ \text{bst\_of\_list}$$

Here we use the fact that priorities chosen uniformly at random induce a uniformly random linear ordering, and that sorting a list with such an ordering produces permutations of that list uniformly at random. The proof of this involves little more than rearranging and using some obvious lemmas on *sort_key* etc. Now the right-hand side is exactly the definition of a random BST (up to a conversion between *pmf* and *measure*), which concludes the proof.   □

6   Related Work

The earliest analysis of randomised algorithms in a theorem prover was probably by Hurd [25] in the HOL system, who modelled them by assuming the existence of an infinite sequence of random bits which programs can consume. He used this approach to formalise the Miller–Rabin primality test.

Audebaud and Paulin-Mohring [26] created a shallowly-embedded formalisation of (discrete) randomised algorithms in Coq and demonstrate its usage on two examples. Barthe *et al.* [27] used this framework to implement the *CertiCrypt* system to write machine-checked cryptographic proofs for a deeply embedded imperative language. Petcher and Morrisett [28] developed a similar framework but based on a monadic embedding. Another similar framework was developed for Isabelle/HOL by Lochbihler [29].

The expected running time of randomised quicksort (possibly including repeated elements) was first analysed in a theorem prover by van der Weegen and McKinna [30] using Coq. They proved the upper bound $2n\lceil\log_2 n\rceil$, whereas we actually proved the closed-form result $2(n+1)H_n - 4n$ and its precise asymptotics. Although their paper's title mentions "average-case complexity', they, in fact, only treat the expected running time of the randomised algorithm in their paper. They did, however, later add a separate proof of an upper bound for the average-case of deterministic quicksort to their GitHub repository. Unlike us, they allow lists to have repeated elements even in the average case, but they proved the expectation bounds separately and independently, while we assumed that there are no repeated elements, but showed something stronger, namely that the distributions are exactly the same, allowing us to reuse the results from the randomised case.

Kaminski *et al.* [31] presented a Hoare-style calculus for analysing the expected running time of imperative programs and used it to analyse a one-dimensional random walk and the Coupon Collector's problem. Hölzl [32] formalised this approach in Isabelle and found a mistake in their proof of the random walk in the process.

At the same time as our work and independently, Tassarotti and Harper [33] gave a Coq formalisation of a cookbook-like theorem based on work by Karp [34] that is able to provide tail bounds for a certain class of randomised recurrences such as the number of comparisons in quicksort and the height of a random BST. In contrast to the expectation results we proved, such bounds are very difficult to obtain on a case-by-case basis, which makes such a cookbook-like result particularly useful.

Outside the world of theorem provers, other approaches exist for automating the analysis of such algorithms: Probabilistic model checkers like PRISM [35] can check safety properties and compute expectation bounds. The $\Lambda\Upsilon\Omega$ system by Flajolet *et al.* [36] conducts fully automatic analysis of average-case running time for a restricted variety of (deterministic) programs. Chatterjee *et al.* [37] developed a method for deriving bounds of the shape $O(\ln n)$, $O(n)$, or $O(n \ln n)$ for certain recurrences that are relevant to average-case analysis automatically and applied it to a number of interesting examples, including quicksort.

7   Future Work

We have closed a number of important gaps in the formalisation of classic probabilistic algorithms related to binary search trees, including the thorny case of treaps, which requires measure theory. Up to that point we claim that these formalisations are readable (the definitions thanks to the Giry monad and the proofs thanks to Isar [38]), but for treaps this becomes debatable: the issue of measurability makes proofs and definitions significantly more cumbersome and less readable. Although existing automation for measurability is already very helpful, there is still room for improvement. Also, the construction of the measurable space of trees generalises to other data types and could be automated.

All of our work so far has been at the functional level, but it would be desirable to refine it to the imperative level in a modular way. The development of the necessary theory and infrastructure is future work.

Acknowledgement. This work was funded by DFG grant NI 491/16-1. We thank Johannes Hölzl and Andreas Lochbihler for helpful discussions, Johannes Hölzl for his help with the construction of the tree space, and Bohua Zhan and Maximilian P. L. Haslbeck for comments on a draft. We also thank the reviewers for their suggestions.

References

1. Nipkow, T.: Amortized complexity verified. In: Urban, C., Zhang, X. (eds.) ITP 2015. LNCS, vol. 9236, pp. 310–324. Springer, Cham (2015). https://doi.org/10.1007/978-3-319-22102-1_21
2. Nipkow, T.: Automatic functional correctness proofs for functional search trees. In: Blanchette, J.C., Merz, S. (eds.) ITP 2016. LNCS, vol. 9807, pp. 307–322. Springer, Cham (2016). https://doi.org/10.1007/978-3-319-43144-4_19
3. Nipkow, T.: Verified root-balanced trees. In: Chang, B.-Y.E. (ed.) APLAS 2017. LNCS, vol. 10695, pp. 255–272. Springer, Cham (2017). https://doi.org/10.1007/978-3-319-71237-6_13
4. Nipkow, T., Wenzel, M., Paulson, L.C. (eds.): Isabelle/HOL. LNCS, vol. 2283. Springer, Heidelberg (2002). https://doi.org/10.1007/3-540-45949-9
5. Nipkow, T., Klein, G.: Concrete Semantics. With Isabelle/HOL. Springer, Cham (2014). https://doi.org/10.1007/978-3-319-10542-0
6. Eberl, M.: The number of comparisons in QuickSort. Archive of Formal Proofs, Formal proof development, March 2017. http://isa-afp.org/entries/Quick_Sort_Cost.html
7. Eberl, M.: Expected shape of random binary search trees. Archive of Formal Proofs, Formal proof development, April 2017. http://isa-afp.org/entries/Random_BSTs.html
8. Haslbeck, M., Eberl, M., Nipkow, T.: Treaps. Archive of Formal Proofs, Formal proof development, March 2018. http://isa-afp.org/entries/Treaps.html
9. Hölzl, J., Heller, A.: Three chapters of measure theory in Isabelle/HOL. In: van Eekelen, M., Geuvers, H., Schmaltz, J., Wiedijk, F. (eds.) ITP 2011. LNCS, vol. 6898, pp. 135–151. Springer, Heidelberg (2011). https://doi.org/10.1007/978-3-642-22863-6_12
10. Gouëzel, S.: Ergodic theory. Archive of Formal Proofs, Formal proof development, December 2015. http://isa-afp.org/entries/Ergodic_Theory.html
11. Eberl, M., Hölzl, J., Nipkow, T.: A verified compiler for probability density functions. In: Vitek, J. (ed.) ESOP 2015. LNCS, vol. 9032, pp. 80–104. Springer, Heidelberg (2015). https://doi.org/10.1007/978-3-662-46669-8_4
12. Hölzl, J.: Markov chains and Markov decision processes in Isabelle/HOL. J. Autom. Reason. **59**, 345–387 (2017)
13. Basin, D.A., Lochbihler, A., Sefidgar, S.R.: Crypthol: game-based proofs in higher-order logic. Cryptology ePrint Archive, report 2017/753 (2017). https://eprint.iacr.org/2017/753
14. Giry, M.: A categorical approach to probability theory. In: Banaschewski, B. (ed.) Categorical Aspects of Topology and Analysis. LNM, vol. 915, pp. 68–85. Springer, Heidelberg (1982). https://doi.org/10.1007/BFb0092872
15. Hoare, C.A.R.: Quicksort. Comput. J. **5**(1), 10 (1962)
16. Sedgewick, R.: The analysis of Quicksort programs. Acta Inf. **7**(4), 327–355 (1977)
17. Cichoń, J.: Quick Sort - average complexity. http://cs.pwr.edu.pl/cichon/Math/QSortAvg.pdf
18. Cormen, T.H., Stein, C., Rivest, R.L., Leiserson, C.E.: Introduction to Algorithms, 2nd edn. McGraw-Hill Higher Education, Boston (2001)
19. Knuth, D.E.: The Art of Computer Programming. Sorting and Searching, vol. 3. Addison Wesley Longman Publishing Co., Redwood City (1998)
20. Ottmann, T., Widmayer, P.: Algorithmen und Datenstrukturen, 5th edn. Spektrum Akademischer Verlag, Auflage (2012)

21. Reed, B.: The height of a random binary search tree. J. ACM **50**(3), 306–332 (2003)
22. Aslam, J.A.: A simple bound on the expected height of a randomly built binary search tree. Technical report TR2001-387, Dartmouth College, Hanover, NH (2001). Abstract and paper lost
23. Vuillemin, J.: A unifying look at data structures. Commun. ACM **23**(4), 229–239 (1980)
24. Seidel, R., Aragon, C.R.: Randomized search trees. Algorithmica **16**(4), 464–497 (1996)
25. Hurd, J.: Formal verification of probabilistic algorithms. Ph.D. thesis, University of Cambridge (2002)
26. Audebaud, P., Paulin-Mohring, C.: Proofs of randomized algorithms in COQ. Sci. Comput. Program. **74**(8), 568–589 (2009)
27. Barthe, G., Grégoire, B., Béguelin, S.Z.: Formal certification of code-based cryptographic proofs. In: Proceedings of the 36th ACM SIGPLAN-SIGACT Symposium on Principles of Programming Languages, POPL 2009, pp. 90–101 (2009)
28. Petcher, A., Morrisett, G.: The foundational cryptography framework. In: Focardi, R., Myers, A. (eds.) POST 2015. LNCS, vol. 9036, pp. 53–72. Springer, Heidelberg (2015). https://doi.org/10.1007/978-3-662-46666-7_4
29. Lochbihler, A.: Probabilistic functions and cryptographic oracles in higher order logic. In: Thiemann, P. (ed.) ESOP 2016. LNCS, vol. 9632, pp. 503–531. Springer, Heidelberg (2016). https://doi.org/10.1007/978-3-662-49498-1_20
30. van der Weegen, E., McKinna, J.: A machine-checked proof of the average-case complexity of quicksort in Coq. In: Berardi, S., Damiani, F., de'Liguoro, U. (eds.) TYPES 2008. LNCS, vol. 5497, pp. 256–271. Springer, Heidelberg (2009). https://doi.org/10.1007/978-3-642-02444-3_16
31. Kaminski, B.L., Katoen, J.-P., Matheja, C., Olmedo, F.: Weakest Precondition reasoning for expected run–times of probabilistic programs. In: Thiemann, P. (ed.) ESOP 2016. LNCS, vol. 9632, pp. 364–389. Springer, Heidelberg (2016). https://doi.org/10.1007/978-3-662-49498-1_15
32. Hölzl, J.: Formalising semantics for expected running time of probabilistic programs. In: Blanchette, J.C., Merz, S. (eds.) ITP 2016. LNCS, vol. 9807, pp. 475–482. Springer, Cham (2016). https://doi.org/10.1007/978-3-319-43144-4_30
33. Tassarotti, J., Harper, R.: Verified tail bounds for randomized programs. In: Avigad, J., Mahboubi, A. (eds.) Interactive Theorem Proving. Springer International Publishing, Cham (2018)
34. Karp, R.M.: Probabilistic recurrence relations. J. ACM **41**(6), 1136–1150 (1994)
35. Kwiatkowska, M.Z., Norman, G., Parker, D.: Quantitative analysis with the probabilistic model checker PRISM. Electr. Notes Theor. Comput. Sci. **153**(2), 5–31 (2006)
36. Flajolet, P., Salvy, B., Zimmermann, P.: Lambda-Upsilon-Omega: an assistant algorithms analyzer. In: Mora, T. (ed.) AAECC 1988. LNCS, vol. 357, pp. 201–212. Springer, Heidelberg (1989). https://doi.org/10.1007/3-540-51083-4_60
37. Chatterjee, K., Fu, H., Murhekar, A.: Automated recurrence analysis for almost-linear expected-runtime bounds. In: Majumdar, R., Kunčak, V. (eds.) CAV 2017. LNCS, vol. 10426, pp. 118–139. Springer, Cham (2017). https://doi.org/10.1007/978-3-319-63387-9_6
38. Wenzel, M.: Isabelle/Isar – a versatile environment for human-readable formal proof documents. Ph.D. thesis, Institut für Informatik, Technische Universität München (2002). https://mediatum.ub.tum.de/node?id=601724

HOL Light QE

Jacques Carette, William M. Farmer$^{(\boxtimes)}$, and Patrick Laskowski

Computing and Software, McMaster University, Hamilton, Canada
wmfarmer@mcmaster.ca
http://www.cas.mcmaster.ca/~carette
http://imps.mcmaster.ca/wmfarmer

Abstract. We are interested in algorithms that manipulate mathematical expressions in mathematically meaningful ways. Expressions are syntactic, but most logics do not allow one to discuss syntax. $\mathrm{CTT_{qe}}$ is a version of Church's type theory that includes quotation and evaluation operators, akin to quote and eval in the Lisp programming language. Since the HOL logic is also a version of Church's type theory, we decided to add quotation and evaluation to HOL Light to demonstrate the implementability of $\mathrm{CTT_{qe}}$ and the benefits of having quotation and evaluation in a proof assistant. The resulting system is called HOL Light QE. Here we document the design of HOL Light QE and the challenges that needed to be overcome. The resulting implementation is freely available.

1 Introduction

A *syntax-based mathematical algorithm (SBMA)* manipulates mathematical expressions in a meaningful way. SBMAs are commonplace in mathematics. Examples include algorithms that compute arithmetic operations by manipulating numerals, linear transformations by manipulating matrices, and derivatives by manipulating functional expressions. Reasoning about the mathematical meaning of an SBMA requires reasoning about the relationship between how the expressions are manipulated by the SBMA and what the manipulations mean.

We argue in [25] that the combination of quotation and evaluation, along with appropriate inference rules, provides the means to reason about the interplay between syntax and semantics, which is what is needed for reasoning about SBMAs. *Quotation* is an operation that maps an expression e to a special value called a *syntactic value* that represents the syntax tree of e. Quotation enables expressions to be manipulated as syntactic entities. *Evaluation* is an operation that maps a syntactic value s to the value of the expression that is represented by s. Evaluation enables meta-level reasoning via syntactic values to be reflected into object-level reasoning. Quotation and evaluation thus form an infrastructure for integrating meta-level and object-level reasoning. Quotation gives a form of *reification* of object-level values which allows introspection. Along with inference rules, this gives a certain amount of *logical reflection*; evaluation adds to this some aspects of *computational reflection* [23,35].

This research was supported by NSERC.

Incorporating quotation and evaluation operators — like quote and eval in the Lisp programming language — into a traditional logic like first-order logic or simple type theory is not a straightforward task. Several challenging design problems stand in the way. The three design problems that most concern us are the following. We will write the quotation and evaluation operators applied to an expression e as $\ulcorner e \urcorner$ and $[\![e]\!]$, respectively.

1. *Evaluation Problem.* An evaluation operator is applicable to syntactic values that represent formulas and thus is effectively a truth predicate. Hence, by the proof of Tarski's theorem on the undefinability of truth [53], if the evaluation operator is total in the context of a sufficiently strong theory (like first-order Peano arithmetic), then it is possible to express the liar paradox. Therefore, the evaluation operator must be partial and the law of disquotation cannot hold universally (i.e., for some expressions e, $[\![\ulcorner e \urcorner]\!] \neq e$). As a result, reasoning with evaluation can be cumbersome and leads to undefined expressions.

2. *Variable Problem.* The variable x is not free in the expression $\ulcorner x + 3 \urcorner$ (or in any quotation). However, x is free in $[\![\ulcorner x + 3 \urcorner]\!]$ because $[\![\ulcorner x + 3 \urcorner]\!] = x + 3$. If the value of a constant c is $\ulcorner x + 3 \urcorner$, then x is free in $[\![c]\!]$ because $[\![c]\!] = [\![\ulcorner x + 3 \urcorner]\!] = x + 3$. Hence, in the presence of an evaluation operator, whether or not a variable is free in an expression may depend on the values of the expression's components. As a consequence, the substitution of an expression for the free occurrences of a variable in another expression depends on the semantics (as well as the syntax) of the expressions involved and must be integrated with the proof system for the logic. That is, a logic with quotation and evaluation requires a semantics-dependent form of substitution in which side conditions, like whether a variable is free in an expression, are proved within the proof system. This is a major departure from traditional logic.

3. *Double Substitution Problem.* By the semantics of evaluation, the value of $[\![e]\!]$ is the *value* of the expression whose syntax tree is represented by the *value* of e. Hence the semantics of evaluation involves a double valuation. This is most apparent when the value of a variable involves a syntax tree that refers to the name of that same variable. For example, if the value of a variable x is $\ulcorner x \urcorner$, then $[\![x]\!] = [\![\ulcorner x \urcorner]\!] = x = \ulcorner x \urcorner$. Hence the substitution of $\ulcorner x \urcorner$ for x in $[\![x]\!]$ requires one substitution inside the argument of the evaluation operator and another substitution after the evaluation operator is eliminated. This double substitution is another major departure from traditional logic.

CTT$_{qe}$ [26,27] is version of Church's type theory [18] with quotation and evaluation that solves these three design problems. It is based on $\mathcal{Q}_0$ [3], Peter Andrews' version of Church's type theory. We believe CTT$_{qe}$ is the first readily implementable version of simple type theory that includes *global* quotation and evaluation operators. We show in [27] that it is suitable for defining, applying, and reasoning about SBMAs.

To demonstrate that CTT$_{qe}$ is indeed implementable, we have done so by modifying HOL Light [36], a compact implementation of the HOL proof assistant [32]. The resulting version of HOL Light is called HOL Light QE. Here we present its design, implementation, and the challenges encountered. (HOL2P [54] is another example of a logical system built by modifying HOL Light.)

The rest of the paper is organized as follows. Section 2 presents the key ideas underlying CTT_{qe} and explains how CTT_{qe} solves the three design problems. Section 3 offers a brief overview of HOL Light. The HOL Light QE implementation is described in Sect. 4, and examples of how quotation and evaluation are used in it are discussed in Sect. 5. Section 6 is devoted to related work. And the paper ends with some final remarks including a brief discussion on future work.

The major contributions of the work presented here are:

1. We show that the logical machinery for quotation and evaluation embodied in CTT_{qe} can be straightforwardly implemented by modifying HOL Light.
2. We produce an HOL-style proof assistant with a built-in global reflection infrastructure for defining, applying, and proving properties about SBMAs.
3. We demonstrate how this reflection infrastructure can be used to express formula schemas, such as the induction schema for first-order Peano arithmetic, as single formulas.

2 $\mathbf{CTT}_{qe}$

The syntax, semantics, and proof system of CTT_{qe} are defined in [27]. Here we will only introduce the definitions and results of that are key to understanding how HOL Light QE implements CTT_{qe}. The reader is encouraged to consult [27] when additional details are required.

2.1   Syntax

CTT_{qe} has the same machinery as $\mathcal{Q}_0$ plus an inductive type ϵ of syntactic values, a partial quotation operator, and a typed evaluation operator.

A *type* of CTT_{qe} is defined inductively by the following formation rules:

1. *Type of individuals*: ι is a type.
2. *Type of truth values*: o is a type.
3. *Type of constructions*: ϵ is a type.
4. *Function type*: If α and β are types, then $(\alpha \to \beta)$ is a type.

Let $\mathcal{T}$ denote the set of types of CTT_{qe}. A *typed symbol* is a symbol with a subscript from $\mathcal{T}$. Let $\mathcal{V}$ be a set of typed symbols such that, for each $\alpha \in \mathcal{T}$, $\mathcal{V}$ contains denumerably many typed symbols with subscript α. A *variable of type* α of CTT_{qe} is a member of $\mathcal{V}$ with subscript α. $\mathbf{x}_\alpha, \mathbf{y}_\alpha, \mathbf{z}_\alpha, \ldots$ are syntactic variables ranging over variables of type α. Let $\mathcal{C}$ be a set of typed symbols disjoint from $\mathcal{V}$. A *constant of type* α of CTT_{qe} is a member of $\mathcal{C}$ with subscript α. $\mathbf{c}_\alpha, \mathbf{d}_\alpha, \ldots$ are syntactic variables ranging over constants of type α. $\mathcal{C}$ contains a set of *logical constants* that include $\mathsf{app}_{\epsilon \to \epsilon \to \epsilon}$, $\mathsf{abs}_{\epsilon \to \epsilon \to \epsilon}$, and $\mathsf{quo}_{\epsilon \to \epsilon}$.

An *expression of type* α of CTT_{qe} is defined inductively by the formation rules below. $\mathbf{A}_\alpha, \mathbf{B}_\alpha, \mathbf{C}_\alpha, \ldots$ are syntactic variables ranging over expressions of type α. An expression is *eval-free* if it is constructed using just the first five rules.

1. *Variable*: $\mathbf{x}_\alpha$ is an expression of type α.
2. *Constant*: $\mathbf{c}_\alpha$ is an expression of type α.
3. *Function application*: $(\mathbf{F}_{\alpha \to \beta} \, \mathbf{A}_\alpha)$ is an expression of type β.
4. *Function abstraction*: $(\lambda \, \mathbf{x}_\alpha \, . \, \mathbf{B}_\beta)$ is an expression of type $\alpha \to \beta$.
5. *Quotation*: $\ulcorner \mathbf{A}_\alpha \urcorner$ is an expression of type ϵ if $\mathbf{A}_\alpha$ is eval-free.
6. *Evaluation*: $\llbracket \mathbf{A}_\epsilon \rrbracket_{\mathbf{B}_\beta}$ is an expression of type β.

The sole purpose of the second component $\mathbf{B}_\beta$ in an evaluation $\llbracket \mathbf{A}_\epsilon \rrbracket_{\mathbf{B}_\beta}$ is to establish the type of the evaluation; we will thus write $\llbracket \mathbf{A}_\epsilon \rrbracket_{\mathbf{B}_\beta}$ as $\llbracket \mathbf{A}_\epsilon \rrbracket_\beta$.

A *construction* of $\mathrm{CTT}_{\mathrm{qe}}$ is an expression of type ϵ defined inductively by:

1. $\ulcorner \mathbf{x}_\alpha \urcorner$ is a construction.
2. $\ulcorner \mathbf{c}_\alpha \urcorner$ is a construction.
3. If $\mathbf{A}_\epsilon$ and $\mathbf{B}_\epsilon$ are constructions, then $\mathsf{app}_{\epsilon \to \epsilon \to \epsilon} \, \mathbf{A}_\epsilon \, \mathbf{B}_\epsilon$, $\mathsf{abs}_{\epsilon \to \epsilon \to \epsilon} \, \mathbf{A}_\epsilon \, \mathbf{B}_\epsilon$, and $\mathsf{quo}_{\epsilon \to \epsilon} \, \mathbf{A}_\epsilon$ are constructions.

The set of constructions is thus an inductive type whose base elements are quotations of variables and constants, and whose constructors are $\mathsf{app}_{\epsilon \to \epsilon \to \epsilon}$, $\mathsf{abs}_{\epsilon \to \epsilon \to \epsilon}$, and $\mathsf{quo}_{\epsilon \to \epsilon}$. As we will see shortly, constructions serve as syntactic values.

Let $\mathcal{E}$ be the function mapping eval-free expressions to constructions that is defined inductively as follows:

1. $\mathcal{E}(\mathbf{x}_\alpha) = \ulcorner \mathbf{x}_\alpha \urcorner$.
2. $\mathcal{E}(\mathbf{c}_\alpha) = \ulcorner \mathbf{c}_\alpha \urcorner$.
3. $\mathcal{E}(\mathbf{F}_{\alpha \to \beta} \, \mathbf{A}_\alpha) = \mathsf{app}_{\epsilon \to \epsilon \to \epsilon} \, \mathcal{E}(\mathbf{F}_{\alpha \to \beta}) \, \mathcal{E}(\mathbf{A}_\alpha)$.
4. $\mathcal{E}(\lambda \, \mathbf{x}_\alpha \, . \, \mathbf{B}_\beta) = \mathsf{abs}_{\epsilon \to \epsilon \to \epsilon} \, \mathcal{E}(\mathbf{x}_\alpha) \, \mathcal{E}(\mathbf{B}_\beta)$.
5. $\mathcal{E}(\ulcorner \mathbf{A}_\alpha \urcorner) = \mathsf{quo}_{\epsilon \to \epsilon} \, \mathcal{E}(\mathbf{A}_\alpha)$.

When $\mathbf{A}_\alpha$ is eval-free, $\mathcal{E}(\mathbf{A}_\alpha)$ is the unique construction that represents the syntax tree of $\mathbf{A}_\alpha$. That is, $\mathcal{E}(\mathbf{A}_\alpha)$ is a syntactic value that represents how $\mathbf{A}_\alpha$ is syntactically constructed. For every eval-free expression, there is a construction that represents its syntax tree, but not every construction represents the syntax tree of an eval-free expression. For example, $\mathsf{app}_{\epsilon \to \epsilon \to \epsilon} \, \ulcorner \mathbf{x}_\alpha \urcorner \ulcorner \mathbf{x}_\alpha \urcorner$ represents the syntax tree of $(\mathbf{x}_\alpha \, \mathbf{x}_\alpha)$ which is not an expression of $\mathrm{CTT}_{\mathrm{qe}}$ since the types are mismatched. A construction is *proper* if it is in the range of $\mathcal{E}$, i.e., it represents the syntax tree of an eval-free expression.

The purpose of $\mathcal{E}$ is to define the semantics of quotation: the meaning of $\ulcorner \mathbf{A}_\alpha \urcorner$ is $\mathcal{E}(\mathbf{A}_\alpha)$.

2.2   Semantics

The semantics of $\mathrm{CTT}_{\mathrm{qe}}$ is based on Henkin-style general models [38]. An expression $\mathbf{A}_\epsilon$ of type ϵ denotes a construction, and when $\mathbf{A}_\epsilon$ is a construction, it denotes itself. The semantics of the quotation and evaluation operators are defined so that the following two theorems hold:

Theorem 2.21 (Law of Quotation). $\ulcorner \mathbf{A}_\alpha \urcorner = \mathcal{E}(\mathbf{A}_\alpha)$ *is valid in* $\mathrm{CTT}_{\mathrm{qe}}$.

Corollary 2.22. $\ulcorner A_\alpha \urcorner = \ulcorner B_\alpha \urcorner$ *iff* A_α *and* B_α *are identical expressions.*

Theorem 2.23 (Law of Disquotation). $\llbracket \ulcorner A_\alpha \urcorner \rrbracket_\alpha = A_\alpha$ *is valid in* CTT$_{qe}$.

Remark 2.24. Notice that this is not the full Law of Disquotation, since only eval-free expressions can be quoted. As a result of this restriction, the liar paradox is not expressible in CTT$_{qe}$ and the Evaluation Problem mentioned above is effectively solved.

2.3   Quasiquotation

Quasiquotation is a parameterized form of quotation in which the parameters serve as holes in a quotation that are filled with expressions that denote syntactic values. It is a very powerful syntactic device for specifying expressions and defining macros. Quasiquotation was introduced by Willard Van Orman Quine in 1940 in the first version of his book *Mathematical Logic* [51]. It has been extensively employed in the Lisp family of programming languages [5][1], and from there to other families of programming languages, most notably the ML family.

In CTT$_{qe}$, constructing a large quotation from smaller quotations can be tedious because it requires many applications of the syntax constructors $\mathsf{app}_{\epsilon \to \epsilon \to \epsilon}$, $\mathsf{abs}_{\epsilon \to \epsilon \to \epsilon}$, and $\mathsf{quo}_{\epsilon \to \epsilon}$. Quasiquotation alleviates this problem. It can be defined straightforwardly in CTT$_{qe}$. However, quasiquotation is not part of the official syntax of CTT$_{qe}$; it is just a notational device used to write CTT$_{qe}$ expressions in a compact form.

As an example, consider $\ulcorner \neg(A_o \wedge \lfloor B_\epsilon \rfloor) \urcorner$. Here $\lfloor B_\epsilon \rfloor$ is a *hole* or *antiquotation*. Assume that A_o contains no holes. $\ulcorner \neg(A_o \wedge \lfloor B_\epsilon \rfloor) \urcorner$ is then an abbreviation for the verbose expression

$$\mathsf{app}_{\epsilon \to \epsilon \to \epsilon} \ulcorner \neg_{o \to o} \urcorner (\mathsf{app}_{\epsilon \to \epsilon \to \epsilon} (\mathsf{app}_{\epsilon \to \epsilon \to \epsilon} \ulcorner \wedge_{o \to o \to o} \urcorner \ulcorner A_o \urcorner) B_\epsilon).$$

$\ulcorner \neg(A_o \wedge \lfloor B_\epsilon \rfloor) \urcorner$ represents the syntax tree of a negated conjunction in which the part of the tree corresponding to the second conjunct is replaced by the syntax tree represented by B_ϵ. If B_ϵ is a quotation $\ulcorner C_o \urcorner$, then the quasiquotation $\ulcorner \neg(A_o \wedge \lfloor \ulcorner C_o \urcorner \rfloor) \urcorner$ is *equivalent* to the quotation $\ulcorner \neg(A_o \wedge C_o) \urcorner$.

2.4   Proof System

The proof system for CTT$_{qe}$ consists of the axioms for $\mathcal{Q}_0$, the single rule of inference for $\mathcal{Q}_0$, and additional axioms [27, B1–B13] that define the logical constants of CTT$_{qe}$ (B1–B4, B5, B7), specify ϵ as an inductive type (B4, B6), state the properties of quotation and evaluation (B8, B10), and extend the rules for beta-reduction (B9, B11–13). We prove in [27] that this proof system is sound for all formulas and complete for eval-free formulas.

---

[1] In Lisp, the standard symbol for quasiquotation is the backquote (') symbol, and thus in Lisp, quasiquotation is usually called *backquote*.

The axioms that express the properties of quotation and evaluation are:

B8 (Properties of Quotation)

1. $\ulcorner \mathbf{F}_{\alpha \to \beta} \mathbf{A}_\alpha \urcorner = \mathsf{app}_{\epsilon \to \epsilon \to \epsilon} \ulcorner \mathbf{F}_{\alpha \to \beta} \urcorner \ulcorner \mathbf{A}_\alpha \urcorner$.
2. $\ulcorner \lambda \mathbf{x}_\alpha \cdot \mathbf{B}_\beta \urcorner = \mathsf{abs}_{\epsilon \to \epsilon \to \epsilon} \ulcorner \mathbf{x}_\alpha \urcorner \ulcorner \mathbf{B}_\beta \urcorner$.
3. $\ulcorner \ulcorner \mathbf{A}_\alpha \urcorner \urcorner = \mathsf{quo}_{\epsilon \to \epsilon} \ulcorner \mathbf{A}_\alpha \urcorner$.

B10 (Properties of Evaluation)

1. $\llbracket \ulcorner \mathbf{x}_\alpha \urcorner \rrbracket_\alpha = \mathbf{x}_\alpha$.
2. $\llbracket \ulcorner \mathbf{c}_\alpha \urcorner \rrbracket_\alpha = \mathbf{c}_\alpha$.
3. $(\mathsf{is\text{-}expr}_{\epsilon \to o}^{\alpha \to \beta} \mathbf{A}_\epsilon \wedge \mathsf{is\text{-}expr}_{\epsilon \to o}^\alpha \mathbf{B}_\epsilon) \supset \llbracket \mathsf{app}_{\epsilon \to \epsilon \to \epsilon} \mathbf{A}_\epsilon \mathbf{B}_\epsilon \rrbracket_\beta = \llbracket \mathbf{A}_\epsilon \rrbracket_{\alpha \to \beta} \llbracket \mathbf{B}_\epsilon \rrbracket_\alpha$.
4. $(\mathsf{is\text{-}expr}_{\epsilon \to o}^\beta \mathbf{A}_\epsilon \wedge \neg(\mathsf{is\text{-}free\text{-}in}_{\epsilon \to \epsilon \to o} \ulcorner \mathbf{x}_\alpha \urcorner \ulcorner \mathbf{A}_\epsilon \urcorner)) \supset$
   $\llbracket \mathsf{abs}_{\epsilon \to \epsilon \to \epsilon} \ulcorner \mathbf{x}_\alpha \urcorner \mathbf{A}_\epsilon \rrbracket_{\alpha \to \beta} = \lambda \mathbf{x}_\alpha \cdot \llbracket \mathbf{A}_\epsilon \rrbracket_\beta$.
5. $\mathsf{is\text{-}expr}_{\epsilon \to o}^\epsilon \mathbf{A}_\epsilon \supset \llbracket \mathsf{quo}_{\epsilon \to \epsilon} \mathbf{A}_\epsilon \rrbracket_\epsilon = \mathbf{A}_\epsilon$.

The axioms for extending the rules for beta-reduction are:

B9 (Beta-Reduction for Quotations)

$(\lambda \mathbf{x}_\alpha \cdot \ulcorner \mathbf{B}_\beta \urcorner) \mathbf{A}_\alpha = \ulcorner \mathbf{B}_\beta \urcorner$.

B11 (Beta-Reduction for Evaluations)

1. $(\lambda \mathbf{x}_\alpha \cdot \llbracket \mathbf{B}_\epsilon \rrbracket_\beta) \mathbf{x}_\alpha = \llbracket \mathbf{B}_\epsilon \rrbracket_\beta$.
2. $(\mathsf{is\text{-}expr}_{\epsilon \to o}^\beta ((\lambda \mathbf{x}_\alpha \cdot \mathbf{B}_\epsilon) \mathbf{A}_\alpha) \wedge \neg(\mathsf{is\text{-}free\text{-}in}_{\epsilon \to \epsilon \to o} \ulcorner \mathbf{x}_\alpha \urcorner ((\lambda \mathbf{x}_\alpha \cdot \mathbf{B}_\epsilon) \mathbf{A}_\alpha))) \supset$
   $(\lambda \mathbf{x}_\alpha \cdot \llbracket \mathbf{B}_\epsilon \rrbracket_\beta) \mathbf{A}_\alpha = \llbracket (\lambda \mathbf{x}_\alpha \cdot \mathbf{B}_\epsilon) \mathbf{A}_\alpha \rrbracket_\beta$.

B12 ("Not Free In" means "Not Effective In")

$\neg \mathsf{IS\text{-}EFFECTIVE\text{-}IN}(\mathbf{x}_\alpha, \mathbf{B}_\beta)$
where $\mathbf{B}_\beta$ is eval-free and $\mathbf{x}_\alpha$ is not free in $\mathbf{B}_\beta$.

B13 (Beta-Reduction for Function Abstractions)

$(\neg \mathsf{IS\text{-}EFFECTIVE\text{-}IN}(\mathbf{y}_\beta, \mathbf{A}_\alpha) \vee \neg \mathsf{IS\text{-}EFFECTIVE\text{-}IN}(\mathbf{x}_\alpha, \mathbf{B}_\gamma)) \supset$
   $(\lambda \mathbf{x}_\alpha \cdot \lambda \mathbf{y}_\beta \cdot \mathbf{B}_\gamma) \mathbf{A}_\alpha = \lambda \mathbf{y}_\beta \cdot ((\lambda \mathbf{x}_\alpha \cdot \mathbf{B}_\gamma) \mathbf{A}_\alpha)$
where $\mathbf{x}_\alpha$ and $\mathbf{y}_\beta$ are distinct.

Substitution is performed using the properties of beta-reduction as Andrews does in the proof system for $\mathcal{Q}_0$ [3, p. 213]. The following three beta-reduction cases require discussion:

1. $(\lambda \mathbf{x}_\alpha \cdot \lambda \mathbf{y}_\beta \cdot \mathbf{B}_\gamma) \mathbf{A}_\alpha$ where $\mathbf{x}_\alpha$ and $\mathbf{y}_\beta$ are distinct.
2. $(\lambda \mathbf{x}_\alpha \cdot \ulcorner \mathbf{B}_\beta \urcorner) \mathbf{A}_\alpha$.
3. $(\lambda \mathbf{x}_\alpha \cdot \llbracket \mathbf{B}_\epsilon \rrbracket_\beta) \mathbf{A}_\alpha$.

The first case can normally be reduced when either (1) $\mathbf{y}_\beta$ is not free in $\mathbf{A}_\alpha$ or (2) $\mathbf{x}_\alpha$ is not free in $\mathbf{B}_\gamma$. However, due to the Variable Problem mentioned before, it is only possible to syntactically check whether a "variable is not free in an expression" when the expression is eval-free. Our solution is to replace the syntactic notion of "a variable is free in an expression" by the semantic notion of "a variable is effective in an expression" when the expression is not necessarily eval-free, and use Axiom B13 to perform the beta-reduction.

"$\mathbf{x}_\alpha$ is effective in $\mathbf{B}_\beta$" means the value of $\mathbf{B}_\beta$ depends on the value of $\mathbf{x}_\alpha$. Clearly, if $\mathbf{B}_\beta$ is eval-free, "$\mathbf{x}_\alpha$ is effective in $\mathbf{B}_\beta$" implies "$\mathbf{x}_\alpha$ is free in $\mathbf{B}_\beta$". However, "$\mathbf{x}_\alpha$ is effective in B_β" is a refinement of "$\mathbf{x}_\alpha$ is free in $\mathbf{B}_\beta$" on eval-free expressions since $\mathbf{x}_\alpha$ is free in $\mathbf{x}_\alpha = \mathbf{x}_\alpha$, but $\mathbf{x}_\alpha$ is not effective in $\mathbf{x}_\alpha = \mathbf{x}_\alpha$. "$\mathbf{x}_\alpha$ is effective in $\mathbf{B}_\beta$" is expressed in $\mathrm{CTT}_{\mathrm{qe}}$ as $\mathsf{IS\text{-}EFFECTIVE\text{-}IN}(\mathbf{x}_\alpha, \mathbf{B}_\beta)$, an abbreviation for

$$\exists \mathbf{y}_\alpha \cdot ((\lambda \mathbf{x}_\alpha \cdot \mathbf{B}_\beta) \mathbf{y}_\alpha \neq \mathbf{B}_\beta)$$

where $\mathbf{y}_\alpha$ is any variable of type α that differs from $\mathbf{x}_\alpha$.

The second case is simple since a quotation cannot be modified by substitution — it is effectively the same as a constant. Thus beta-reduction is performed without changing $\ulcorner \mathbf{B}_\beta \urcorner$ as shown in Axiom B9 above.

The third case is handled by Axioms B11.1 and B11.2. B11.1 deals with the trivial case when $\mathbf{A}_\alpha$ is the bound variable $\mathbf{x}_\alpha$ itself. B11.2 deals with the other much more complicated situation. The condition

$$\neg(\mathsf{is\text{-}free\text{-}in}_{\epsilon \to \epsilon \to o} \ulcorner \mathbf{x}_\alpha \urcorner ((\lambda \mathbf{x}_\alpha \cdot \mathbf{B}_\epsilon) \mathbf{A}_\alpha))$$

guarantees that there is no *double substitution*. $\mathsf{is\text{-}free\text{-}in}_{\epsilon \to \epsilon \to o}$ is a logical constant of $\mathrm{CTT}_{\mathrm{qe}}$ such that $\mathsf{is\text{-}free\text{-}in}_{\epsilon \to \epsilon \to o} \ulcorner \mathbf{x}_\alpha \urcorner \ulcorner \mathbf{B}_\beta \urcorner$ says that the variable $\mathbf{x}_\alpha$ is free in the (eval-free) expression $\mathbf{B}_\beta$.

Thus we see that substitution in $\mathrm{CTT}_{\mathrm{qe}}$ in the presence of evaluations may require proving semantic side conditions of the following two forms:

1. $\neg\mathsf{IS\text{-}EFFECTIVE\text{-}IN}(\mathbf{x}_\alpha, \mathbf{B}_\beta)$.
2. $\mathsf{is\text{-}free\text{-}in}_{\epsilon \to \epsilon \to o} \ulcorner \mathbf{x}_\alpha \urcorner \ulcorner \mathbf{B}_\beta \urcorner$.

2.5   The Three Design Problems

To recap, $\mathrm{CTT}_{\mathrm{qe}}$ solves the three design problems given in Sect. 1. The Evaluation Problem is avoided by restricting the quotation operator to eval-free expressions and thus making it impossible to express the liar paradox. The Variable Problem is overcome by modifying Andrews' beta-reduction axioms. The Double Substitution Problem is eluded by using a beta-reduction axiom for evaluations that excludes beta-reductions that embody a double substitution.

3 HOL Light

HOL Light [36] is an open-source proof assistant developed by John Harrison. It implements a logic (HOL) which is a version of Church's type theory. It is a simple implementation of the HOL proof assistant [32] written in OCaml and hosted on GitHub at https://github.com/jrh13/hol-light/. Although it is a relatively small system, it has been used to formalize many kinds of mathematics and to check many proofs including the lion's share of Tom Hales' proof of the Kepler conjecture [1].

HOL Light is very well suited to serve as a foundation on which to build an implementation of $\mathrm{CTT_{qe}}$: First, it is an open-source system that can be freely modified as long as certain very minimal conditions are satisfied. Second, it is an implementation of a version of simple type theory that is essentially $\mathcal{Q}_0$, the version of Church's type theory underlying $\mathrm{CTT_{qe}}$, plus (1) polymorphic type variables, (2) an axiom of choice expressed by asserting that the Hilbert ϵ operator is a choice (indefinite description) operator, and (3) an axiom of infinity that asserts that **ind**, the type of individuals, is infinite [36]. The type variables in the implemented logic are not a hindrance; they actually facilitate the implementation of $\mathrm{CTT_{qe}}$. The presence of the axioms of choice and infinity in HOL Light alter the semantics of $\mathrm{CTT_{qe}}$ without compromising in any way the semantics of quotation and evaluation. And third, HOL Light supports the definition of inductive types so that ϵ can be straightforwardly defined.

4 Implementation

4.1 Overview

HOL Light QE was implemented in four stages:

1. The set of terms was extended so that $\mathrm{CTT_{qe}}$ expressions could be mapped to HOL Light terms. This required the introduction of **epsilon**, the type of constructions, and term constructors for quotations and evaluations. See Subsect. 4.2.
2. The proof system was modified to include the machinery in $\mathrm{CTT_{qe}}$ for reasoning about quotations and evaluations. This required adding new rules of inference and modifying the **INST** rule of inference that simultaneously substitutes terms $t_1, \ldots, t_n$ for the free variables $x_1, \ldots, x_n$ in a sequent. See Subsect. 4.3.
3. Machinery — consisting of HOL function definitions, tactics, and theorems — was created for supporting reasoning about quotations and evaluations in the new system. See Subsect. 4.4.
4. Examples were developed in the new system to test the implementation and to demonstrate the benefits of having quotation and evaluation in higher-order logic. See Sect. 5.

The first and second stages have been completed; both stages involved modifying the kernel of HOL Light. The third stage is sufficiently complete to enable our examples in Sect. 5 to work well, and did not involve any further changes to the HOL Light kernel. We do expect that adding further examples, which is ongoing, will require additional machinery but no changes to the kernel.

The HOL Light QE system was developed by the third author under the supervision of the first two authors on an undergraduate NSERC USRA research project at McMaster University and is available at

https://github.com/JacquesCarette/hol-light.

It should be further remarked that our fork, from late April 2017, is not fully up-to-date with respect to HOL Light. In particular, this means that it is best to compile it with OCAML 4.03.0 and CAMLP5 6.16, both available from OPAM.

To run HOL Light QE, execute the following commands in HOL Light QE top-level directory named hol_light:

```
1) install opam
2) opam init --comp 4.03.0
3) opam install "camlp5=6.16"
5) opam 'eval config env'
5) cd hol_light
6) make
7) run ocaml via
      ocaml -I 'camlp5 -where' camlp5o.cma
8) #use "hol.ml";;
   #use "Constructions/epsilon.ml";;
   #use "Constructions/pseudoquotation.ml";;
   #use "Constructions/QuotationTactics.ml";;
```

Each test can be run by an appropriate further #use statement.

4.2   Mapping of CTT$_{qe}$ Expressions to HOL Terms

Tables 1 and 2 illustrate how the CTT$_{qe}$ types and expressions are mapped to the HOL types and terms, respectively. The HOL types and terms are written in the internal representation employed in HOL Light QE. The type epsilon and the term constructors Quote and Eval are additions to HOL Light explained below. Since CTT$_{qe}$ does not have type variables, it has a logical constant $=_{\alpha \to \alpha \to o}$

Table 1. Mapping of CTT$_{qe}$ Types to HOL Types

CTT$_{qe}$ Type α	HOL Type $\mu(\alpha)$	Abbreviation for $\mu(\alpha)$
o	Tyapp("bool",[])	bool
ι	Tyapp("ind",[])	ind
ϵ	Tyapp("epsilon",[])	epsilon
$\beta \to \gamma$	Tyapp("fun",[$\mu(\beta),\mu(\gamma)$])	$\mu(\beta)$->$\mu(\gamma)$

Table 2. Mapping of CTT$_{qe}$ Expressions to HOL Terms

CTT$_{qe}$ Expression e	HOL Term $\nu(e)$
x_α	Var("x",$\mu(\alpha)$)
c_α	Const("c",$\mu(\alpha)$)
$=_{\alpha\to\alpha\to o}$	Const("=",a_ty_var->a_ty_var->bool)
$(F_{\alpha\to\beta} \, A_\alpha)$	Comb($\nu(F_{\alpha\to\beta})$,$\nu(A_\alpha)$)
$(\lambda x_\alpha \, . \, B_\beta)$	Abs(Var("x",$\mu(\alpha)$),$\nu(B_\beta)$)
$\ulcorner A_\alpha \urcorner$	Quote($\nu(A_\alpha)$,$\mu(\alpha)$)
$\llbracket A_\epsilon \rrbracket_{B_\beta}$	Eval($\nu(A_\epsilon)$,$\mu(\beta)$)

representing equality for each $\alpha \in \mathcal{T}$. The members of this family of constants are all mapped to a single HOL constant with the polymorphic type a_ty_var->a_ty_var->bool where a_ty_var is any chosen HOL type variable.

The other logical constants of CTT$_{qe}$ [27, Table 1] are not mapped to primitive HOL constants. app$_{\epsilon\to\epsilon\to\epsilon}$, abs$_{\epsilon\to\epsilon\to\epsilon}$, and quo$_{\epsilon\to\epsilon}$ are implemented by App, Abs, and Quo, constructors for the inductive type epsilon given below. The remaining logical constants are predicates on constructions that are implemented by HOL functions. The CTT$_{qe}$ type ϵ is the type of constructions, the syntactic values that represent the syntax trees of eval-free expressions. ϵ is formalized as an inductive type epsilon. Since types are components of terms in HOL Light, an inductive type type of syntax values for HOL Light QE types (which are the same as HOL types) is also needed. Specifically:

```
define_type "type = TyVar string
              | TyBase string
              | TyMonoCons string type
              | TyBiCons string type type"
```

```
define_type "epsilon = QuoVar string type
               | QuoConst string type
               | App epsilon epsilon
               | Abs epsilon epsilon
               | Quo epsilon"
```

Terms of type type denote the syntax trees of HOL Light QE types, while the terms of type epsilon denote the syntax trees of those terms that are eval-free.

The OCaml type of HOL types in HOL Light QE

```
type hol_type = Tyvar of string
              | Tyapp of string * hol_type list
```

is the same as in HOL Light, but the OCaml type of HOL terms in HOL Light QE

```
type term = Var of string * hol_type
          | Const of string * hol_type
          | Comb of term * term
          | Abs of term * term
          | Quote of term * hol_type
          | Hole of term * hol_type
          | Eval of term * hol_type
```

has three new constructors: Quote, Hole, and Eval.

Quote constructs a quotation of type epsilon with components t and α from a term t of type α that is eval-free. Eval constructs an evaluation of type α with components t and α from a term t of type epsilon and a type α. Hole is used to construct "holes" of type epsilon in a quasiquotation as described in [27]. A quotation that contains holes is a quasiquotation, while a quotation without any holes is a normal quotation. The construction of terms has been modified to allow a hole (of type epsilon) to be used where a term of some other type is expected.

The external representation of a quotation Quote(t,ty) is Q_ t _Q. Similarly, the external representation of a hole Hole(t,ty) is H_ t _H. The external representation of an evaluation Eval(t,ty) is eval t to ty.

4.3   Modification of the HOL Light Proof System

The proof system for CTT$_{qe}$ is obtained by extending $\mathcal{Q}_0$'s with additional axioms B1–B13 (see Sect. 2.4). Since $\mathcal{Q}_0$ and HOL Light are both complete (with respect to the semantics of Henkin-style general models), HOL Light includes the reasoning capabilities of the proof system for $\mathcal{Q}_0$ but not the reasoning capabilities embodied in the B1–B13 axioms, which must be implemented in HOL Light QE as follows. First, the logical constants defined by Axioms B1–B4, B5, and B7 are defined in HOL Light QE as HOL functions. Second, the no junk (B6) and

Table 3. New Inference Rules in HOL Light QE

CTT$_{qe}$ Axioms	NewRules of Inference
B8 (Properties of Quotation)	LAW_OF_QUO
B10 (Properties of Evaluation)	
B10.1	VAR_DISQUO
B10.2	CONST_DISQUO
B10.3	APP_SPLIT
B10.4	ABS_SPLIT
B10.5	QUOTABLE
B11 (Beta-Reduction for Evaluations)	
B11.1	BETA_EVAL
B11.2	BETA_REVAL
B12 ("Not Free In" means "Not Effective In")	NOT_FREE_OR_EFFECTIVE_IN
B13 (Beta-Reduction for Function Abstractions)	NEITHER_EFFECTIVE

no confusion (B4) requirements for ϵ are automatic consequences of defining epsilon as an inductive type. Third, Axiom B9 is implemented directly in the HOL Light code for substitution. Fourth, the remaining axioms, B8 and B10–B13 are implemented by new rules of inference in as shown in Table 3.

The INST rule of inference is also modified. This rule simultaneously substitutes a list of terms for a list of variables in a sequent. The substitution function vsubst defined in the HOL Light kernel is modified so that it works like substitution (via beta-reduction rules) does in CTT_{qe}. The main changes are:

1. A substitution of a term t for a variable x in a function abstraction Abs(y,s) is performed as usual if (1) t is eval-free and x is not free in t, (2) there is a theorem that says x is not effective in t, (3) s is eval-free and x is not free in s, or (4) there is a theorem that says x is not effective in s. Otherwise, if s or t is not eval-free, the substitution fails and if s and t are eval-free, the variable x is renamed and the substitution is continued.
2. A substitution of a term t for a variable x in a quotation Quote(e,ty) where e does not contain any holes (i.e., terms of the form Hole(e',ty')) returns Quote(e,ty) unchanged (as stated in Axiom B9). If e does contain holes, then t is substituted for the variable x in the holes in Quote(e,ty).
3. A substitution of a term t for a variable x in an evaluation Eval(e,ty) returns (1) Eval(e,ty) when t is x and (2) the function abstraction application Comb(Abs(x,Eval(e,ty)),t) otherwise. (1) is valid by Axiom B11.1. When (2) happens, this part of the substitution is finished and the user can possibly continue it by applying BETA_REVAL, the rule of inference corresponding to Axiom B11.2.

4.4   Creation of Support Machinery

The HOL Light QE system contains a number of HOL functions, tactics, and theorems that are useful for reasoning about constructions, quotations, and evaluations. An important example is the HOL function isExprType that implements the CTT_{qe} family of logical constants is-expr$^{\alpha}_{\epsilon \to o}$ where α ranges over members of $\mathcal{T}$. This function takes terms s_1 and s_1 of type epsilon and type, respectively, and returns true iff s_1 represents the syntax tree of a term t, s_2 represents the syntax tree of a type α, and t is of type α.

4.5   Metatheorems

We state three important metatheorems about HOL Light QE. The proofs of these metatheorems are straightforward but also tedious. We label the metatheorems as conjectures since their proofs have not yet been fully written down.

Conjecture 1. Every formula provable in HOL Light's proof system is also provable in HOL Light QE's proof system.

Proof sketch. HOL Light QE's proof system extends HOL Light's proof system with new machinery for reasoning about quotations and evaluations. Thus every HOL Light proof remains valid in HOL Light QE. □

Note: All the proofs loaded with the HOL Light system continue to be valid when loaded in HOL Light QE. A further test for the future would be to load a variety of large HOL Light proofs in HOL Light QE to check that their validity is preserved.

Conjecture 2. The proof system for HOL Light QE is sound for all formulas and complete for all eval-free formulas.

Proof sketch. The analog of this statement for CTT_{qe} is proved in [27]. It should be possible to prove this conjecture by just imitating the proof for CTT_{qe}.   □

Conjecture 3. HOL Light QE is a model-theoretic conservative extension of HOL Light.

Proof sketch. A model of HOL Light QE is a model of HOL Light with definitions of the type ϵ and several constants and interpretations for the (quasi)quotation and evaluation operators. These additions do not impinge upon the semantics of HOL Light; hence every model of HOL Light can be expanded to a model of the HOL Light QE, which is the meaning of the conjecture.   □

5 Examples

We present two examples that illustrate its capabilities by expressing, instantiating, and proving formula schemas in HOL Light QE.

5.1 Law of Excluded Middle

The *law of excluded middle (LEM)* is expressed as the formula schema $A \vee \neg A$ where A is a syntactic variable ranging over all formulas. Each instance of LEM is a theorem of HOL, but LEM cannot be expressed in HOL as a single formula. However, LEM can be formalized in CTT_{qe} as the universal statement

$$\forall x_\epsilon \, . \, \text{is-expr}^o_{\epsilon \to o} \, x_\epsilon \supset [\![x_\epsilon]\!]_o \vee \neg [\![x_\epsilon]\!]_o.$$

An instance of LEM may be written in HOL Light QE as

```
'!x:epsilon. isExprType (x:epsilon) (TyBase "bool")
   ==> ((eval x to bool) \/ ~(eval x to bool))'
```

that is readily proved. Instances of this are obtained by applying INST followed by BETA_REVAL, the second beta-reduction rule for evaluations.

5.2 Induction Schema

The (first-order) *induction schema for Peano arithmetic* is usually expressed as the formula schema

$$(P(0) \wedge \forall x \, . \, (P(x) \supset P(S(x)))) \supset \forall x \, . \, P(x)$$

where $P(x)$ is a parameterized syntactic variable that ranges over all formulas of first-order Peano arithmetic. If we assume that the domain of the type ι is the natural numbers and $\mathcal{C}$ includes the usual constants of natural number arithmetic (including a constant $S_{\iota \to \iota}$ representing the successor function), then this schema can be formalized in $\mathrm{CTT_{qe}}$ as

$$\forall f_\epsilon \, . \, ((\text{is-expr}^{\iota \to o}_{\epsilon \to o} \, f_\epsilon \wedge \text{is-peano}_{\epsilon \to o} \, f_\epsilon) \supset$$
$$((\llbracket f_\epsilon \rrbracket_{\iota \to o} 0 \wedge (\forall x_\iota \, . \, \llbracket f_\epsilon \rrbracket_{\iota \to o} x_\iota \supset \llbracket f_\epsilon \rrbracket_{\iota \to o} (S_{\iota \to \iota} x_\iota))) \supset \forall x_\iota \, . \, \llbracket f_\epsilon \rrbracket_{\iota \to o} x_\iota))$$

where $\text{is-peano}_{\epsilon \to o} f_\epsilon$ holds iff f_ϵ represents the syntax tree of a predicate of first-order Peano arithmetic. The *induction schema for Presburger arithmetic* is exactly the same as the induction schema for Peano arithmetic except that the predicate $\text{is-peano}_{\epsilon \to o}$ is replaced by an appropriate predicate $\text{is-presburger}_{\epsilon \to o}$.

It should be noted that the induction schemas for Peano and Presburger arithmetic are weaker that the full induction principle for the natural numbers:

$$\forall p_{\iota \to o} \, . \, ((p_{\iota \to o} 0 \wedge (\forall x_\iota \, . \, p_{\iota \to o} x_\iota \supset p_{\iota \to o} (S_{\iota \to \iota} x_\iota))) \supset \forall x_\iota \, . \, p_{\iota \to o} x_\iota)$$

The full induction principle states that induction holds for all properties of the natural numbers (which is an uncountable set), while the induction schemas for Peano and Presburger arithmetic hold only for properties that are definable in Peano and Presburger arithmetic (which are countable sets).

The full induction principle is expressed in HOL Light as the theorem

```
'!P. P(_0) / (!n. P(n) ==> P(SUC n)) ==> !n. P n'
```

named `num_INDUCTION`. However, it is not possible to directly express the Peano and Presburger induction schemas in HOL Light without adding new rules of inference to its kernel.

The induction schema for Peano arithmetic can be written in HOL Light QE just as easily as in $\mathrm{CTT_{qe}}$:

```
'!f:epsilon.
  (isExprType (f:epsilon) (TyBiCons "fun" (TyVar "num")
    (TyBase "bool")))
  /\ (isPeano f)
  ==>
  (eval (f:epsilon) to (num->bool)) 0
  /\ (!n:num. (eval (f:epsilon) to (num->bool)) n
    ==> (eval (f:epsilon) to (num->bool)) (SUC n))
  ==> (!n:num. (eval (f:epsilon) to (num->bool)) n)'
```

`peanoInduction` is proved from `num_INDUCTION` in HOL Light QE by:

1. Instantiate `num_INDUCTION` with 'P:num->bool' to obtain `indinst`.
2. Prove and install the theorem `nei_peano` that says the variable (n:num) is not effective in (eval (f:epsilon) to (num->bool)).
3. Logically reduce `peanoInduction`, then prove the result by instantiating 'P:num->bool' in `indinst` with 'eval (f:epsilon) to (num->bool)' using the INST rule, which requires the previously proved theorem `nei_peano`.

The induction schema for Presburger arithmetic is stated and proved in the same way. By being able to express the Peano and Presburger induction schemas, we can properly define the first-order theories of Peano arithmetic and Presburger arithmetic in HOL Light QE.

6 Related Work

Quotation, evaluation, reflection, reification, issues of intensionality versus extensionality, metaprogramming and metareasoning each have extensive literature — sometimes in more than one field. For example, one can find a vast literature on reflection in logic, programming languages, and theorem proving. Due to space restrictions, we cannot do justice to the full breadth of issues. For a full discussion, please see the related work section in [27]. The surveys of Costantini [23], Harrison [35] are excellent. From a programming perspective, the discussion and extensive bibliography of Kavvos' D.Phil. thesis [44] are well worth reading.

Focusing just on interactive proof assistants, we find that Boyer and Moore developed a global infrastructure [7] for incorporating symbolic algorithms into Nqthm [8]. This approach is also used in ACL2 [43], the successor to Nqthm; see [41]. Over the last 30 years, the Nuprl group has produced a large body of work on metareasoning and reflection for theorem proving [2,4,20,40,45,47,57] that has been implemented in the Nuprl [21] and MetaPRL [39] systems. Proof by reflection has become a mainstream technique in the Coq [22] proof assistant with the development of tactics based on symbolic computations like the Coq ring tactic [6,33] and the formalizations of the *four color theorem* [29] and the *Feit-Thompson odd-order theorem* [30]. See [6,9,14,31,33,42,49] for a selection of the work done on using reflection in Coq. Many other systems also support metareasoning and reflection: Agda [48,55,56], Idris [15–17] Isabelle/HOL [13], Lean [24], Maude [19], PVS [37], reFLect [34,46], and Theorema [10,28].

The semantics of the quotation operator $\ulcorner \cdot \urcorner$ is based on the *disquotational theory of quotation* [11]. According to this theory, a quotation of an expression e is an expression that denotes e itself. In $\mathrm{CTT_{qe}}$, $\ulcorner \mathbf{A}_\alpha \urcorner$ denotes a value that represents the syntactic structure of $\mathbf{A}_\alpha$. Polonsky [50] presents a set of axioms for quotation operators of this kind. Other theories of quotation have been proposed — see [11] for an overview. For instance, quotation can be viewed as an operation that constructs literals for syntactic values [52].

It is worth quoting Boyer and Moore [7] here:

> The basic premise of all work on extensible theorem-provers is that it should be possible to add new proof techniques to a system without endangering the soundness of the system. It seems possible to divide current work into two broad camps. In the first camp are those systems that allow the introduction of arbitrary new procedures, coded in the implementation language, but require that each application of such a procedure produce a formal proof of the correctness of the transformation performed. In the

second camp are those systems that contain a formal notion of what it means for a proof technique to be sound and require a machine-checked proof of the soundness of each new proof technique. Once proved, the new proof technique can be used without further justification.

This remains true to this day. The systems in the LCF tradition (Isabelle/HOL, Coq, HOL Light) are in the "first camp", while Nqthm, ACL2, Nuprl, MetaPRL, Agda, Idris, Lean, Maude and Theorema, as well as our approach broadly fall in the "second camp". However, all systems in the first camp have started to offer some reflection capabilities on top of their tactic facilities. Below we give some additional details for each system, leveraging information from the papers already cited above as well as the documentation of each system[2].

SSReflect [31] (*small scale reflection*) is a Coq extension that works by locally reflecting the syntax of particular kinds of objects — such as decidable predicates and finite structures. It is the pervasive use of decidability and computability which gives SSReflect its power, and at the same time, its limitations. An extension to PVS allows reasoning much in the style of SSReflect. Isabelle/HOL offers a nonlogical `reify` function (aka quotation), while its `interpret` function is in the logic; it uses global datatypes to represent HOL terms.

The approach for the second list of systems also varies quite a bit. Nqthm, ACL2, Theorema (as well as now HOL Light QE) have global quotation and evaluation operators in the logic, as well as careful restrictions on their use to avoid paradoxes. Idris also has global quotation and evaluation, and the *totality checker* is used to avoid paradoxes. MetaPRL has evaluation but no global quotation. Agda has global quotation and evaluation, but their use are mediated by a built-in TC (TypeChecking) monad which ensures soundness. Lean works similarly: all reflection must happen in the `tactic` monad, from which one cannot escape. Maude appears to offer a global quotation operator, but it is unclear if there is a global evaluation operator; quotations are offered by a built-in module, and those are extra-logical.

7   Conclusion

CTT$_{qe}$ [26,27] is a version of Church's type theory with global quotation and evaluation operators that is intended for defining, applying, proving properties about syntax-based mathematical algorithms (SBMAs), algorithms that manipulate expressions in a mathematically meaningful ways. HOL Light QE is an implementation of CTT$_{qe}$ obtained by modifying HOL Light [36], a compact implementation of the HOL proof assistant [32]. In this paper, we have presented the design and implementation of HOL Light QE. We have discussed the challenges that needed to be overcome. And we have given some examples that test the implementation and show the benefits of having quotation and evaluation in higher-order logic.

---

[2] And some personal communication with some of system authors.

The implementation of HOL Light QE was very straightforward since the logical issues were worked out in CTT_{qe} and HOL Light provides good support for inductive types. Pleasingly, and surprisingly, no new issues arose during the implementation. HOL Light QE works in exactly the same way as HOL Light except that, in the presence of evaluations, the instantiation of free variables may require proving side conditions that say (1) a variable is not effective in a term or (2) that a variable represented by a construction is not free in a term represented by a construction (see Subsects. 2.4 and 4.3). This is the only significant cost we see for using HOL Light QE in place of HOL Light.

HOL Light QE provides a built-in global reflection infrastructure [27]. This infrastructure can be used to reason about the syntactic structure of terms and, as we have shown, to express formula schemas as single formulas. More importantly, the infrastructure provides the means to define, apply, and prove properties about SBMAs. An SBMA can be defined as a function that manipulates constructions. The *meaning formula* that specifies its mathematical meaning can be stated using the evaluation of constructions. And the SBMA's meaning formula can be proved from the SBMA's definition. In other words, the infrastructure provides a unified framework for formalizing SBMAs in a proof assistant.

We plan to continue the development of HOL Light QE and to show that it can be effectively used to develop SBMAs as we have just described. In particular, we intend to formalize in HOL Light QE the example on the symbolic differentiation we formalized in CTT_{qe} [27]. This will require defining the algorithm for symbolic differentiation, writing its meaning formula, and finally proving the meaning formula from the algorithm's definition and properties about derivatives. We also intend, down the road, to formalize in HOL Light QE the graph of biform theories encoding natural number of arithmetic described in [12].

References

1. Hales, T., et al.: A formal proof of the Kepler conjecture. Forum Math. Pi **5** (2017)
2. Allen, S.F., Constable, R.L., Howe, D.J., Aitken, W.E.: The semantics of reflected proof. In: Proceedings of the Fifth Annual Symposium on Logic in Computer Science (LICS 1990), pp. 95–105. IEEE Computer Society (1990)
3. Andrews, P.B.: An Introduction to Mathematical Logic and Type Theory: To Truth Through Proof, 2nd edn. Kluwer, Dordrecht (2002)
4. Barzilay, E.: Implementing Reflection in Nuprl. Ph.D. thesis. Cornell University (2005)
5. Bawden, A.: Quasiquotation in Lisp. In: Danvy, O. (ed.) Proceedings of the 1999 ACM SIGPLAN Symposium on Partial Evaluation and Semantics-Based Program Manipulation, pp. 4–12 (1999), Technical report BRICS-NS-99-1, University of Aarhus (1999)
6. Boutin, S.: Using reflection to build efficient and certified decision procedures. In: Abadi, M., Ito, T. (eds.) TACS 1997. LNCS, vol. 1281, pp. 515–529. Springer, Heidelberg (1997). https://doi.org/10.1007/BFb0014565
7. Boyer, R., Moore, J.: Metafunctions: proving them correct and using them efficiently as new proof procedures. In: Boyer, R., Moore, J. (eds.) The Correctness Problem in Computer Science, pp. 103–185. Academic Press, New York (1981)

8. Boyer, R., Moore, J.: A Computational Logic Handbook. Academic Press, San Diego (1988)
9. Braibant, T., Pous, D.: Tactics for reasoning modulo AC in Coq. In: Jouannaud, J.-P., Shao, Z. (eds.) CPP 2011. LNCS, vol. 7086, pp. 167–182. Springer, Heidelberg (2011). https://doi.org/10.1007/978-3-642-25379-9_14
10. Buchberger, B., Craciun, A., Jebelean, T., Kovacs, L., Kutsia, T., Nakagawa, K., Piroi, F., Popov, N., Robu, J., Rosenkranz, M., Windsteiger, W.: Theorema: towards computer-aided mathematical theory exploration. J. Appl. Logic **4**, 470–504 (2006)
11. Cappelen, H., LePore, E.: Quotation. In: Zalta, E.N. (ed.) The Stanford Encyclopedia of Philosophy, Spring 2012
12. Carette, J., Farmer, W.M.: Formalizing mathematical knowledge as a biform theory graph: a case study. In: Geuvers, H., England, M., Hasan, O., Rabe, F., Teschke, O. (eds.) CICM 2017. LNCS (LNAI), vol. 10383, pp. 9–24. Springer, Cham (2017). https://doi.org/10.1007/978-3-319-62075-6_2
13. Chaieb, A., Nipkow, T.: Proof synthesis and reflection for linear arithmetic. J. Autom. Reason. **41**, 33–59 (2008)
14. Chlipala, A.: Certified Programming with Dependent Types: A Pragmatic Introduction to the Coq Proof Assistant. MIT Press, Cambridge (2013)
15. Christiansen, D., Brady, E.: Elaborator reflection: Extending Idris in Idris. SIGPLAN Not. **51**, 284–297 (2016). https://doi.org/10.1145/3022670.2951932, http://doi.acm.org/10.1145/3022670.2951932
16. Christiansen, D.R.: Type-directed elaboration of quasiquotations: a high-level syntax for low-level reflection. In: Proceedings of the 26nd 2014 International Symposium on Implementation and Application of Functional Languages, IFL 2014, pp. 1:1–1:9. ACM, New York (2014). https://doi.org/10.1145/2746325.2746326, http://doi.acm.org/10.1145/2746325.2746326
17. Christiansen, D.R.: Practical Reflection and Metaprogramming for Dependent Types. Ph.D. thesis. IT University of Copenhagen (2016)
18. Church, A.: A formulation of the simple theory of types. J. Symbolic Logic **5**, 56–68 (1940)
19. Clavel, M., Meseguer, J.: Reflection in conditional rewriting logic. Theoret. Comput. Sci. **285**, 245–288 (2002)
20. Constable, R.L.: Using reflection to explain and enhance type theory. In: Schwichtenberg, H. (ed.) Proof and Computation. NATO ASI Series, vol. 139, pp. 109–144. Springer, Heidelberg (1995). https://doi.org/10.1007/978-3-642-79361-5_3
21. Constable, R.L., Allen, S.F., Bromley, H.M., Cleaveland, W.R., Cremer, J.F., Harper, R.W., Howe, D.J., Knoblock, T.B., Mendler, N.P., Panangaden, P., Sasaki, J.T., Smith, S.F.: Implementing Mathematics with the Nuprl Proof Development System. Prentice-Hall, Englewood Cliffs (1986)
22. Coq Development Team: The Coq Proof Assistant Reference Manual, Version 8.5 (2016). https://coq.inria.fr/distrib/current/refman/
23. Costantini, S.: Meta-reasoning: a survey. In: Kakas, A.C., Sadri, F. (eds.) Computational Logic: Logic Programming and Beyond. LNCS (LNAI), vol. 2408, pp. 253–288. Springer, Heidelberg (2002). https://doi.org/10.1007/3-540-45632-5_11
24. Ebner, G., Ullrich, S., Roesch, J., Avigad, J., de Moura, L.: A metaprogramming framework for formal verification. Proc. ACM Program. Lang. **1**, 34 (2017)
25. Farmer, W.M.: The formalization of syntax-based mathematical algorithms using quotation and evaluation. In: Carette, J., Aspinall, D., Lange, C., Sojka, P., Windsteiger, W. (eds.) CICM 2013. LNCS (LNAI), vol. 7961, pp. 35–50. Springer, Heidelberg (2013). https://doi.org/10.1007/978-3-642-39320-4_3

26. Farmer, W.M.: Incorporating quotation and evaluation into church's type theory: syntax and semantics. In: Kohlhase, M., Johansson, M., Miller, B., de de Moura, L., Tompa, F. (eds.) CICM 2016. LNCS (LNAI), vol. 9791, pp. 83–98. Springer, Cham (2016). https://doi.org/10.1007/978-3-319-42547-4_7

27. Farmer, W.M.: Incorporating quotation and evaluation into Church's type theory. Inf. Comput. **260C**, 9–50 (forthcoming, 2018)

28. Giese, M., Buchberger, B.: Towards practical reflection for formal mathematics. RISC Report Series 07–05, Research Institute for Symbolic Computation (RISC). Johannes Kepler University (2007)

29. Gonthier, G.: The four colour theorem: engineering of a formal proof. In: Kapur, D. (ed.) ASCM 2007. LNCS (LNAI), vol. 5081, pp. 333–333. Springer, Heidelberg (2008). https://doi.org/10.1007/978-3-540-87827-8_28

30. Gonthier, G., et al.: A machine-checked proof of the odd order theorem. In: Blazy, S., Paulin-Mohring, C., Pichardie, D. (eds.) ITP 2013. LNCS, vol. 7998, pp. 163–179. Springer, Heidelberg (2013). https://doi.org/10.1007/978-3-642-39634-2_14

31. Gonthier, G., Mahboubi, A.: An introduction to small scale reflection in Coq. J. Formalized Reason. **3**, 95–152 (2010)

32. Gordon, M.J.C., Melham, T.F.: Introduction to HOL: A Theorem Proving Environment for Higher Order Logic. Cambridge University Press, New York (1993)

33. Grégoire, B., Mahboubi, A.: Proving equalities in a commutative ring done right in Coq. In: Hurd, J., Melham, T. (eds.) TPHOLs 2005. LNCS, vol. 3603, pp. 98–113. Springer, Heidelberg (2005). https://doi.org/10.1007/11541868_7

34. Grundy, J., Melham, T., O'Leary, J.: A reflective functional language for hardware design and theorem proving. Funct. Program. **16** (2006)

35. Harrison, J.: Metatheory and reflection in theorem proving: a survey and critique. Technical Report CRC-053. SRI Cambridge (1995). http://www.cl.cam.ac.uk/~jrh13/papers/reflect.ps.gz

36. Harrison, J.: HOL light: an overview. In: Berghofer, S., Nipkow, T., Urban, C., Wenzel, M. (eds.) TPHOLs 2009. LNCS, vol. 5674, pp. 60–66. Springer, Heidelberg (2009). https://doi.org/10.1007/978-3-642-03359-9_4

37. von Henke, F.W., Pfab, S., Pfeifer, H., Rueß, H.: Case studies in meta-level theorem proving. In: Grundy, J., Newey, M. (eds.) TPHOLs 1998. LNCS, vol. 1479, pp. 461–478. Springer, Heidelberg (1998). https://doi.org/10.1007/BFb0055152

38. Henkin, L.: Completeness in the theory of types. J. Symbolic Logic **15**, 81–91 (1950)

39. Hickey, J., Nogin, A., Constable, R.L., Aydemir, B.E., Barzilay, E., Bryukhov, Y., Eaton, R., Granicz, A., Kopylov, A., Kreitz, C., Krupski, V.N., Lorigo, L., Schmitt, S., Witty, C., Yu, X.: MetaPRL – a modular logical environment. In: Basin, D., Wolff, B. (eds.) TPHOLs 2003. LNCS, vol. 2758, pp. 287–303. Springer, Heidelberg (2003). https://doi.org/10.1007/10930755_19

40. Howe, D.: Reflecting the semantics of reflected proof. In: Aczel, P., Simmons, H., Wainer, S. (eds.) Proof Theory, pp. 229–250. Cambridge University Press, Cambridge (1992)

41. Hunt Jr., W.A., Kaufmann, M., Krug, R.B., Moore, J.S., Smith, E.W.: Meta reasoning in ACL2. In: Hurd, J., Melham, T. (eds.) TPHOLs 2005. LNCS, vol. 3603, pp. 163–178. Springer, Heidelberg (2005). https://doi.org/10.1007/11541868_11

42. James, D.W.H., Hinze, R.: A reflection-based proof tactic for lattices in Coq. In: Horváth, Z., Zsók, V., Achten, P., Koopman, P.W.M. (eds.) Proceedings of the Tenth Symposium on Trends in Functional Programming (TFP 2009). Trends in Functional Programming, vol. 10, pp. 97–112. Intellect (2009)

43. Kaufmann, M., Moore, J.S.: An industrial strength theorem prover for a logic based on Common Lisp. IEEE Trans. Softw. Eng. **23**, 203–213 (1997)
44. Kavvos, G.A.: On the Semantics of Intensionality and Intensional Recursion, December 2017. http://arxiv.org/abs/1712.09302
45. Knoblock, T.B., Constable, R.L.: Formalized metareasoning in type theory. In: Proceedings of the Symposium on Logic in Computer Science (LICS 1986), pp. 237–248. IEEE Computer Society (1986)
46. Melham, T., Cohn, R., Childs, I.: On the semantics of ReFLect as a basis for a reflective theorem prover. Computing Research Repository (CoRR) abs/1309.5742 (2013). http://arxiv.org/abs/1309.5742
47. Nogin, A., Kopylov, A., Yu, X., Hickey, J.: A computational approach to reflective meta-reasoning about languages with bindings. In: Pollack, R. (ed.) ACM SIGPLAN International Conference on Functional Programming, Workshop on Mechanized Reasoning about Languages with Variable Binding (MERLIN 2005), pp. 2–12. ACM (2005)
48. Norell, U.: Dependently typed programming in Agda. In: Kennedy, A., Ahmed, A. (eds.) Proceedings of TLDI 2009, pp. 1–2. ACM (2009)
49. Oostdijk, M., Geuvers, H.: Proof by computation in the Coq system. Theor. Comput. Sci. **272** (2002)
50. Polonsky, A.: Axiomatizing the quote. In: Bezem, M. (ed.) Computer Science Logic (CSL 2011) – 25th International Workshop/20th Annual Conference of the EACSL. Leibniz International Proceedings in Informatics (LIPIcs), vol. 12, pp. 458–469. Schloss Dagstuhl – Leibniz-Zentrum für Informatik (2011)
51. Quine, W.V.O.: Mathematical Logic, Revised edn. Harvard University Press, Cambridge (2003)
52. Rabe, F.: Generic literals. In: Kerber, M., Carette, J., Kaliszyk, C., Rabe, F., Sorge, V. (eds.) CICM 2015. LNCS (LNAI), vol. 9150, pp. 102–117. Springer, Cham (2015). https://doi.org/10.1007/978-3-319-20615-8_7
53. Tarski, A.: The concept of truth in formalized languages. In: Corcoran, J. (ed.) Logic, Semantics, Meta-Mathematics, 2nd edn., pp. 152–278. Hackett (1983)
54. Völker, N.: HOL2P - a system of classical higher order logic with second order polymorphism. In: Schneider, K., Brandt, J. (eds.) TPHOLs 2007. LNCS, vol. 4732, pp. 334–351. Springer, Heidelberg (2007). https://doi.org/10.1007/978-3-540-74591-4_25
55. van der Walt, P.: Reflection in Agda. Master's thesis, Universiteit Utrecht (2012)
56. van der Walt, P., Swierstra, W.: Engineering proof by reflection in agda. In: Hinze, R. (ed.) IFL 2012. LNCS, vol. 8241, pp. 157–173. Springer, Heidelberg (2013). https://doi.org/10.1007/978-3-642-41582-1_10
57. Yu, X.: Reflection and Its Application to Mechanized Metareasoning about Programming Languages. Ph.D. thesis. California Institute of Technology (2007)

Efficient Mendler-Style
Lambda-Encodings in Cedille

Denis Firsov[✉], Richard Blair, and Aaron Stump

Department of Computer Science, The University of Iowa, Iowa City, IA, USA
{denis-firsov,richard-blair,aaron-stump}@uiowa.edu

Abstract. It is common to model inductive datatypes as least fixed points of functors. We show that within the Cedille type theory we can relax functoriality constraints and generically derive an induction principle for Mendler-style lambda-encoded inductive datatypes, which arise as least fixed points of covariant schemes where the morphism lifting is defined only on identities. Additionally, we implement a destructor for these lambda-encodings that runs in constant-time. As a result, we can define lambda-encoded natural numbers with an induction principle and a constant-time predecessor function so that the normal form of a numeral requires only linear space. The paper also includes several more advanced examples.

Keywords: Type theory · Lambda-encodings · Cedille
Induction principle · Predecessor function · Inductive datatypes

1 Introduction

It is widely known that inductive datatypes may be defined in pure impredicative type theory. For example, Church encodings identify each natural number n with its iterator $\lambda\ s.\ \lambda\ z.\ s^n\ z$. The Church natural numbers can be typed in System F by means of impredicative polymorphism:

$$\texttt{cNat} \blacktriangleleft \star = \forall\ X\ :\ \star.\ (X \to X) \to X \to X.$$

The first objection to lambda-encodings is that it is provably impossible to derive an induction principle in second-order dependent type theory [1]. As a consequence, most languages come with a built-in infrastructure for defining inductive datatypes and their induction principles. Here are the definitions of natural numbers in Agda and Coq:

```
data Nat : Set          Inductive nat : Type :=
  zero : Nat              | 0 : nat
  suc  : Nat → Nat        | S : nat → nat.
```

Coq will automatically generate the induction principle for `nat`, and in Agda it can be derived by pattern matching and explicit structural recursion.

© Springer International Publishing AG, part of Springer Nature 2018
J. Avigad and A. Mahboubi (Eds.): ITP 2018, LNCS 10895, pp. 235–252, 2018.
https://doi.org/10.1007/978-3-319-94821-8_14

Therefore, we can ask if it is possible to extend the Calculus of Constructions with **typing constructs** that make induction derivable for lambda-encoded datatypes. Stump gave a positive answer to this question by introducing the Calculus of Dependent Lambda Eliminations (CDLE) [2]. CDLE is a Curry-style Calculus of Constructions extended with implicit products, intersection types, and primitive heterogeneous equality. Stump proved that natural number induction is derivable in this system for lambda-encoded natural numbers. Later, we generalized this work by deriving induction for lambda-encodings of inductive datatypes which arise as least fixed points of functors [3]. Moreover, we observed that the proof of induction for Mendler-style lambda-encoding relied only on the identity law of functors. In this paper, we exploit this observation to define a new class of covariant schemes, which includes functors, and induces a larger class of inductive datatypes supporting derivable induction.

Another objection to lambda-encodings is their computational inefficiency. For example, computing the predecessor of a Church encoded Peano natural provably requires linear time [4]. The situation was improved by Parigot who proposed a new lambda-encoding of numerals with a constant-time predecessor, but the size of the number n is exponential $O(2^n)$ [5]. Later, the situation was improved further by the Stump-Fu encoding, which supports a constant-time predecessor and reduces the size of the natural number n to $O(n^2)$ [6]. In this paper, we show how to develop a constant-time predecessor within CDLE for a Mendler-style lambda-encoded naturals that are linear in space.

This paper makes the following technical contributions:

1. We introduce a new kind of parameterized scheme using **identity mappings** (function lifting defined only on identities). We show that every functor has an associated identity mapping, but not vice versa.
2. We use a Mendler-style lambda-encoding to prove that every scheme with an identity mapping induces an inductive datatype. Additionally, we generically derive an induction principle for these datatypes.
3. We implement a generic constant-time destructor of Mendler-style lambda-encoded inductive datatypes. To the best of our knowledge, we offer a first example of typed lambda-encoding of inductive datatypes with derivable induction and a constant-time destructor where normal forms of data require linear space.
4. We give several examples of concrete datatypes defined using our development. We start by giving a detailed description of lambda-encoded naturals with an induction principle and a constant-time predecessor function that only requires linear space to encode a numeral. We also give examples of infinitary datatypes. Finally, we present an inductive datatype that arises as a least fixed point of a scheme that is not a functor, but has an identity mapping.

2   Background

In this section, we briefly summarize the main features of Cedille's type theory. For full details on CDLE, including semantics and soundness results, please see the previous papers [2, 7]. The main metatheoretic property proved in the previous work is logical consistency: there are types which are not inhabited. CDLE is an extrinsic (i.e., Curry-style) type theory, whose terms are exactly those of the pure untyped lambda calculus (with no additional constants or constructs). The type-assignment system for CDLE is not subject-directed, and thus cannot be used directly as a typing algorithm. Indeed, since CDLE includes Curry-style System F as a subsystem, type assignment is undecidable [8]. To obtain a usable type theory, Cedille thus has a system of annotations for terms, where the annotations contain sufficient information to type terms algorithmically. But true to the extrinsic nature of the theory, these annotations play no computational role. Indeed, they are erased both during compilation and before formal reasoning about terms within the type theory, in particular by definitional equality (see Fig. 1).

$$\frac{\Gamma, x : T' \vdash t : T \quad x \notin FV(|t|)}{\Gamma \vdash \Lambda x{:}T'.t : \forall x{:}T'.T} \qquad \frac{\Gamma \vdash t : \forall x{:}T'.T \quad \Gamma \vdash t' : T'}{\Gamma \vdash t - t' : [t'/x]T}$$

$$\frac{\Gamma \vdash t : T}{\Gamma \vdash \beta : t \simeq t} \qquad \frac{\Gamma \vdash t' : t_1 \simeq t_2 \quad \Gamma \vdash t : [t_1/x]T}{\Gamma \vdash \rho\, t' - t : [t_2/x]T}$$

$$\frac{\Gamma \vdash t_1 : T \quad \Gamma \vdash t_2 : [t_1/x]T' \quad \Gamma \vdash p : |t_1| \simeq |t_2|}{\Gamma \vdash [t_1, t_2\{p\}] : \iota x{:}T.\,T'}$$

$$\frac{\Gamma \vdash t : \iota x{:}T.T'}{\Gamma \vdash t.1 : T} \qquad \frac{\Gamma \vdash t : \iota x{:}T.T'}{\Gamma \vdash t.2 : [t.1/x]T'}$$

$$
\begin{aligned}
|\Lambda x{:}T.\,t| &= |t| \\
|t - t'| &= |t| \\
|\beta| &= \lambda x.\,x \\
|\rho\, t' - t| &= |t| \\
|[t_1, t_2\{p\}]| &= |t_1| \\
|t.1| &= |t| \\
|t.2| &= |t|
\end{aligned}
$$

Fig. 1. Introduction, elimination, and erasure rules for additional type constructs

CDLE extends the (Curry-style) Calculus of Constructions (CC) with primitive heterogeneous equality, intersection types, and implicit products:

- $t_1 \simeq t_2$, a heterogeneous equality type. The terms t_1 and t_2 are required to be typed, but need not have the same type. We introduce this with a constant β which erases to λ x. x (so our type-assignment system has no additional constants, as promised); β proves t $\simeq$ t for any typeable term t . Combined with definitional equality, β proves $t_1 \simeq t_2$ for any $\beta\eta$-equal t_1 and t_2 whose free variables are all declared in the typing context. We eliminate the equality type by rewriting, with a construct ρ t' $-$ t. Suppose t' proves $t_1 \simeq t_2$ and we synthesize a type T for t, where T has several occurrences of terms definitionally equal to t_1. Then the type synthesized for ρ t' $-$ t is T except with those occurrences replaced by t_2. Note that the types of the terms are not part of the equality type itself, nor does the elimination rule require that the types of the left-hand and right-hand sides are the same to do an elimination.

- ι x : T. T', the dependent intersection type of Kopylov [9]. This is the type for terms t which can be assigned both the type T and the type [t/x]T', the substitution instance of T' by t. There are constructs t.1 and t.2 to select either the T or [t.1/x]T' view of a term t of type ι x : T. T'. We introduce a value of ι x : T. T' by construct $[t_1, t_2 \{p\}]$, where t_1 has type T (algorithmically), t_2 has type [t/x]T', and p proves $t_1 \simeq t_2$.
- $\forall$ x : T. T', the implicit product type of Miquel [10]. This can be thought of as the type for functions which accept an erased input of type x : T, and produce a result of type T'. There are term constructs Λ x. t for introducing an implicit input x, and t -t' for instantiating such an input with t'. The implicit arguments exist just for purposes of typing so that they play no computational role and equational reasoning happens on terms from which the implicit arguments have been erased.

It is important to understand that the described constructs are erased before the formal reasoning, according to the erasure rules in Fig. 1.

We have implemented CDLE in a tool called Cedille, which we have used to typecheck the developments of this paper. The pre-release version of Cedille and the Cedille code accompanying this paper could be found here:

http://firsov.ee/efficient-lambda/

3 Preliminaries

We skip the details of the lambda-encoded implementation of basic datatypes like Unit, Empty, sums (X + Y), and dependent sums (Σ x : X. Y x), for which the usual introduction and elimination rules are derivable in Cedille.

In this paper, we use syntactical simplifications to improve readability. In particular, we hide the type arguments in the cases when they are unambiguous. For example, if x : X and y : Y then we write pair x y instead of fully type-annotated pair X Y x y. The current version of Cedille requires fully annotated terms.

3.1 Multiple Types of Terms

CDLE's dependent intersection types allow propositionally equal values to be intersected. Given x : X, y : Y x, and a proof p of x $\simeq$ y, we can introduce an intersection value v := [x, y {p}] of type ι x : X. Y x. Every intersection has two "views": the first view v.1 has type X and the second view v.2 has type Y v.1. The term [x, y {p}] erases to x according to the erasure rules in Fig. 1. This allows us to see x as having two distinct types, namely X and Y x:

phi ◄ $\forall$ X: $\star$. $\forall$ Y: X → $\star$. Π x: X. $\forall$ y: Y x. $\forall$ p: x $\simeq$ y. Y x
= Λ X. Λ Y. λ x. Λ y. Λ p. [x, y {p}].2.

Indeed, the definition of phi erases to the term λ x. x. Hence, phi x -y -p beta-reduces to x and has type Y x (dash denotes the application of implicitly quantified arguments). Note that Π x. X. T is the usual "explicit" dependent function space.

In our setting, we implemented the phi combinator by utilizing propositional intersection types that allow us to intersect provably equal terms. The converse is also possible: phi can be taken as a language primitive and then used to derive propositional intersection types from definitional intersection types, where only definitionally equal values are allowed to be intersected (this variation of CDLE is described and used in [11]).

3.2 Identity Functions

In our setting, it is possible to implement a function of type X → Y so that it erases to term λ x. x where X is different from Y. The simplest example is the first (or second) "view" from an intersection value:

$$\text{view1} \blacktriangleleft \forall \text{ X} : \star. \ \forall \text{ Y} : \text{X} \to \star. \ (\iota \text{ x} : \text{X. Y x}) \to \text{X}$$
$$= \Lambda \text{ X. } \Lambda \text{ Y. } \lambda \text{ x. x.1.}$$

Indeed, according to the erasure rules view1 erases to the term λ x. x. We introduce a type Id X Y, which is the set of all functions from X to Y that erase to the identity function (λ x. x):

$$\text{id} \blacktriangleleft \forall \text{ X} : \star. \ \text{X} \to \text{X} = \Lambda \text{ X. } \lambda \text{ x. x.}$$

$$\text{Id} \blacktriangleleft \star \to \star \to \star = \lambda \text{ X. } \lambda \text{ Y. } \Sigma \text{ f} : \text{X} \to \text{Y. f} \simeq \text{id.}$$

Introduction. The importance of the previously implemented combinator phi is that it allows to introduce an identity function Id X Y from any extensional identity f : X → Y (i.e., x ≃ f x for any x):

$$\text{intrId} \blacktriangleleft \forall \text{ X Y} : \star. \ \Pi \text{ f} : \text{X} \to \text{Y. } (\Pi \text{ x} : \text{X. x} \simeq \text{f x}) \to \text{Id X Y}$$
$$= \Lambda \text{ X. } \Lambda \text{ Y. } \lambda \text{ f. } \lambda \text{ prf. pair } (\lambda \text{ x. phi x -(f x) -(prf x)) } \beta.$$

Elimination. The identity function Id X Y allows us to see values of type X as also having type Y. The implementation is in terms of phi combinator:

$$\text{elimId} \blacktriangleleft \forall \text{ X Y} : \star. \ \forall \text{ c} : \text{Id X Y. X} \to \text{Y} =$$
$$= \Lambda \text{ X. } \Lambda \text{ Y. } \Lambda \text{ c. } \lambda \text{ x. phi x -}(\pi_1 \text{ c x}) \text{ -}(\rho \ (\pi_2 \text{ c}) \text{ - } \beta).$$

The subterm $\rho \ (\pi_2 \text{ c}) - \beta$ proves x ≃ π_1 c x, where π_i is the i-th projections from a dependent sum. Observe that elimId itself erases to λ x. x, hence elimId -c x ≃ x by beta-reduction. In other words, an identity function Id X Y allows x : X to be seen as having types X and Y at the same time.

3.3 Identity Mapping

A scheme F : ⋆ → ⋆ is a functor if it comes equipped with a function fmap that satisfies the identity and composition laws:

```
Functor ◀ (⋆ → ⋆) → ⋆ = λ F.
  Σ fmap : ∀ X Y : ⋆. (X → Y) → F X → F Y.
  IdentityLaw fmap × CompositionLaw fmap.
```

However, it is simple to define a covariant scheme for which the function `fmap` cannot be implemented (below, $x_1 \neq x_2$ is shorthand for $x_1 \simeq x_2 \to$ Empty):

```
UneqPair ◀ ⋆ → ⋆ = λ X. Σ x₁ : X. Σ x₂ : X. x₁ ≠ x₂.
```

We introduce schemes with *identity mappings* as a new class of parameterized covariant schemes. An identity mapping is a lifting of identity functions:

```
IdMapping ◀ (⋆ → ⋆) → ⋆ = λ F.
   ∀ X Y : ⋆. Id X Y → Id (F X) (F Y).
```

Intuitively, `IdMapping F` is similar to `fmap` of functors, but it needs to be defined only on identity functions. The identity law is expressed as a requirement that an identity function of type `Id X Y` is mapped to an identity function of type `Id (F X) (F Y)`.

Clearly, every functor induces an identity mapping (by the application of `intrId` to `fmap` and its identity law):

```
fm2im ◀ ∀ F : ⋆ → ⋆. Functor F → IdMapping F = <..>
```

However, `UneqPair` is an example of scheme which is not a functor, but has an identity mapping (see example in Sect. 6.3).

In the rest of the paper we show that every identity mapping `IdMapping F` induces an inductive datatype which is a least fixed point of F. Additionally, we generically derive an induction principle and implement a constant-time destructor for these datatypes.

4   Inductive Datatypes from Identity Mappings

In our previous paper, we used Cedille to show how to generically derive an induction principle for Mendler-style lambda-encoded datatypes that arise as least fixed points of functors [3]. In this section, we revisit this derivation to show that it is possible to relax functoriality constraints and only assume that the underlying signature scheme is accompanied by an identity mapping.

4.1   Basics of Mendler-Style Encoding

In this section, we investigate the standard definitions of Mendler-style F-algebras that are well-defined for any unrestricted scheme $F : \star \to \star$. To reduce the notational clutter, we assume that $F : \star \to \star$ is a global (module) parameter:

```
module _ (F : ⋆ → ⋆).
```

In the abstract setting of category theory, a Mendler-style F-algebra is a pair (X, Φ) where X is an object (i.e., the *carrier*) in $\mathcal{C}$ and $\Phi : \mathcal{C}(-, X) \to \mathcal{C}(F-, X)$ is a natural transformation. In the concrete setting of Cedille, objects are types, arrows are functions, and natural transformations are polymorphic functions. Therefore, Mendler-style F-algebras are defined as follows:

AlgM ◀ ⋆ → ⋆ = λ X. ∀ R : ⋆. (R → X) → F R → X.

Uustalu and Vene showed that initial Mendler-style F-algebras offer an alternative categorical model of inductive datatypes [12]. The carrier of an initial F-algebra is an inductive datatype that is a least fixed point of F. It is known that if F is a positive scheme then the least fixed point of it may be implemented in terms of universal quantification [13]:

FixM ◀ ⋆ = ∀ X : ⋆. AlgM X → X.

foldM ◀ ∀ X : ⋆. AlgM X → FixM → X = Λ X. λ alg. λ x. x alg.

In essence, this definition identifies inductive datatypes with iterators and every function on FixM is to be computed by iteration.

The natural transformation of the initial Mendler-style F-algebra denotes the collection of constructors of its carrier [12]. In our setting, the initial Mendler-style F-algebra AlgM FixM is not definable because F is not a functor [3]. Instead, we express the collection of constructors of datatype FixM as a conventional F-algebra F FixM → FixM:

inFixM ◀ F FixM → FixM = λ x. Λ X. λ alg. alg (foldM alg) x.

The function inFixM is of crucial importance because it expresses constructors of FixM without requirements of functoriality on F : ⋆ → ⋆.

It is provably impossible to define the mutual inverse of inFixM (destructor of FixM) without introducing additional constraints on F. Assume the existence of function outFixM (which need not be an inverse of inFixM), typed as follows:

outFixM ◀ ∀ F : ⋆ → ⋆. FixM F → F (FixM F) = <..>

Next, recall that in the impredicative setting the empty type is encoded as ∀ X : ⋆. X (its inhabitant implies any equation). Then, we instantiate F with the negative polymorphic scheme NegF X := ∀ Y : ⋆. X → Y, and exploit the function outFixM to construct a witness of the empty type:

T ◀ ⋆ = FixM NegF.

ty ◀ ∀ Y : ⋆. T → Y = Λ Y. λ t. outFixM -NegF t -Y t.

t ◀ T = ty -T (inFixM -NegF ty).

unsound ◀ ∀ X : ⋆. X = Λ X. ty -X t.

Hence, the existence of function outFixM contradicts the consistency of Cedille. Hence, the inverse of inFixM can exist only for some restricted class of schemes F : ⋆ → ⋆.

4.2 Inductive Subset

From this point forward we assume that the scheme F is also accompanied by an identity mapping imap:

$$\text{module \_ (F : } \star \to \star) \text{ (imap : IdMapping F).}$$

In our previous work we assumed that F is a functor and showed how to specify the "inductive" subset of the type FixM F. Then, we generically derived induction for this subset. In this section, we update the steps of our previous work to account for F : $\star \to \star$ not being a functor.

The dependent intersection type ι x : X. Y x can be understood as a subset of X defined by a predicate Y. However, to construct the value of this type we must provide x : X and a proof p : Y x so that x and p are provably equal (x $\simeq$ p). Hence, to align with this constraint we use implicit products to express inductivity of FixM as its "dependently-typed" version. Recall that FixM is defined in terms of Mendler-style F-algebras:

$$\text{AlgM} \blacktriangleleft \star \to \star = \lambda \text{ X. } \forall \text{ R : } \star. \text{ (R} \to \text{X)} \to \text{F R} \to \text{X.}$$

In our previous work, we introduced the *Q-proof F-algebras* as a "dependently-typed" counterpart of AlgM. The value of type PrfAlgM X Q alg should be understood as an inductive proof that predicate Q holds for every X where X is a least fixed point of F and alg : F X $\to$ X is a collection of constructors of X.

$$\text{PrfAlgM} \blacktriangleleft \Pi \text{ X : } \star. \text{ (X} \to \star) \to \text{(F X} \to \text{X)} \to \star$$
$$= \lambda \text{ X. } \lambda \text{ Q. } \lambda \text{ alg.}$$
$$\forall \text{ R : } \star. \forall \text{ c : Id R X. } (\Pi \text{ r : R. Q (elimId -c r))} \to$$
$$\Pi \text{ fr : F R. Q (alg (elimId -(imap c) fr)).}$$

Mendler-style F-algebras (AlgM) allow recursive calls to be explicitly stated by providing arguments R $\to$ X and F R, where the polymorphically quantified type R ensures termination. Similarly, Q-proof F-algebras allow the inductive hypotheses to be explicitly stated for every R by providing an implicit identity function c : Id R X, and a dependent function of type Π r : R. Q (elimId $-$ c r) (recall that elimId -c r reduces to r and has type X). Given the inductive hypothesis for every R, the proof algebra must conclude that the predicate Q holds for every X, which is produced by constructors alg from any given F R that has been "casted" to F X.

Next, recall that FixM is defined as a function from AlgM X to X for every X.

$$\text{FixM} \blacktriangleleft \star = \forall \text{ X : } \star. \text{ AlgM X} \to \text{X.}$$

To retain the analogy of definitions, we express the inductivity of value x : FixM as a dependent function from a *Q-proof F-algebra* to Q x.

$$\text{IsIndFixM} \blacktriangleleft \text{FixM} \to \star = \lambda \text{ x : FixM.}$$
$$\forall \text{ Q : FixM} \to \star. \text{ PrfAlgM FixM Q inFixM} \to \text{Q x.}$$

Now, we employ intersection types to define a type FixIndM as a subset of FixM carved out by the "inductivity" predicate IsIndFixM:

$$\text{FixIndM} \blacktriangleleft \star = \iota \text{ x : FixM. IsIndFixM x.}$$

Finally, we must explain how to construct the values of this type. As in the case of FixM, the set of constructors of FixIndM is expressed by a conventional F-algebra F FixIndM → FixIndM. The implementation is divided into three steps:

First, we define a function from F FixIndM to FixM:

```
tc1 ◄ F FixIndM → FixM = λ x.
let c ◄ Id (F FixIndM) (F FixM)
   = imap (intrId (λ x. x.1) (λ x. β)) in
inFixM (elimId -c x).
```

The implementation simply "casts" its argument to F FixM and then applies the previously implemented constructor of FixM (inFixM). Because elimId -c x reduces to x, the erasure of tc1 is the same as the erasure of inFixM which is a term λ x. λ q. q (λ r. r q) x.

Second, we show that the same lambda term could also be typed as a proof that every tc1 x is inductive:

```
tc2 ◄ Π x : F FixIndM. IsIndFixM (tc1 x) = λ x.
(Λ Q. λ q. (q -(intrId  (λ x. x.1) (λ x. β)) (λ r. r.2 q) x)).
```

Indeed, functions tc1 and tc2 are represented by the same pure lambda term.

Finally, given any value x : F FixInd we can intersect tc1 x and the proof of its inductivity tc2 x to construct an element of an inductive subset FixIndM:

```
inFixIndM ◄ F FixIndM → FixIndM = λ x. [ tc1 x, tc2 x { β } ].
```

Recall that erasure of intersection [x, y {p}] equals the erasure of x. Therefore, functions inFixM, tc1, tc2, and inFixIndM all erase to the same pure lambda term. In other words, in Cedille the term λ x. λ q. q (λ r. r q) x can be extrinsically typed as any of these functions.

4.3   Induction Principle

We start by explaining why we need to derive induction for FixIndM, even though it is definitionally an inductive subset of FixM. Indeed, every value x : FixIndM can be "viewed" as a proof of its own inductivity. More precisely, the term x.2 is a proof of the inductivity of x.1. Moreover, the equational theory of CDLE gives us the following equalities $x.1 \simeq x \simeq x.2$ (due to the rules of erasure). But recall that the inductivity proof provided by the second view x.2 is typed as follows:

$$\forall \text{ Q : Fix} \rightarrow \star. \text{ PrfAlgM FixM Q inFixM} \rightarrow \text{Q x.1}$$

Note that Q is a predicate on FixM and not FixIndM! This form of inductivity does not allow properties of FixIndM to be proven.

Therefore, our goal is to prove that every x : FixIndM is inductive in its own right. We phrase this in terms of proof-algebras parameterized by FixIndM, a predicate on FixIndM, and its constructors (inFixIndM):

$$\forall \text{ Q : FixIndM} \rightarrow \star. \text{ PrfAlgM FixIndM Q inFixIndM} \rightarrow \text{Q x}.$$

In our previous work, we already made an observation that the derivation of induction for Mendler-style encodings relies only on the identity law of functors [3]. Therefore, the current setting only requires minor adjustments of our previous proof. For the sake of completeness, we present a main idea of this derivation.

The key insight is that we can convert predicates on FixIndM to logically equivalent predicates on FixM by using heterogeneous equality:

```
Lift ◀ (FixInd → ⋆) → Fix → ⋆
    = λ Q. λ y. Σ x : FixInd. x ≃ y × Q x.
```

```
eqv1 ◀ Π x: FixIndM. ∀ Q: FixIndM → ⋆. Q x → Lift Q x.1 = <..>
```

```
eqv2 ◀ Π x: FixIndM. ∀ Q: FixIndM → ⋆. Lift Q x.1 → Q x = <..>
```

These properties allow us to convert a Q-proof algebra to a proof algebra for a lifted predicate Lift Q, and then derive the generic induction principle:

```
convIH ◀ ∀ Q : FixIndM → ⋆. PrfAlgM FixIndM Q inFixIndM
    → PrfAlgM FixM (Lift Q) inFixM = <..>
```

```
induction ◀ ∀ Q: FixIndM → ⋆. PrfAlgM FixIndM Q inFixIndM →
    Π e: FixIndM. Q e = Λ Q. λ p. λ e. eqv2 e (e.2 (convIH p)).
```

Let Q be a predicate on FixIndM and p be a Q-proof algebra: we show that Q holds for any e : FixIndM. Recall that every e : FixIndM can be viewed as a proof of inductivity of e.1 via e.2 : IsIndFixM e.1. We use this to get a proof of the lifted predicate Lift Q e.1 from the proof algebra delivered by convIH p. Finally, we get Q e by using eqv2.

5   Constant-Time Destructors

An induction principle is needed to prove properties about programs, but practical functional programming also requires constant-time destructors (also called accessors) of inductive datatypes. Let us illustrate the problem using the datatype of natural numbers. In Agda it is easy to implement the predecessor function by pattern matching:

```
pred : Nat → Nat
pred zero = zero
pred (suc n) = n
```

The correctness of pred trivially follows by beta-reduction:

```
predProp : (n : Nat) → pred (suc n) ≡ n
predProp n = refl
```

Let us switch to Cedille and observe that it is much less trivial to implement the predecessor for the impredicative encoding of Peano numerals. Here is the definition of Church encoded Peano naturals and their constructors:

```
cNat ◄ ⋆ = ∀ X : ⋆. (X → X) → X → X.
```

```
zero ◄ cNat = Λ X. λ s. λ z. z.
suc  ◄ cNat → cNat = λ n. Λ X. λ s. λ z. s (n s z).
```

Next, we implement the predecessor for cNat which is due to Kleene:

```
zCase ◄ cNat × cNat = pair zero zero
sCase ◄ cNat × cNat → cNat × cNat
    = λ n. pair (π₂ n) (suc (π₂ n)).
```

```
predK ◄ Nat → Nat = λ n. π₁ (n sCase zCase).
```

The key to the Kleene predecessor is the function sCase, which ignores the first item of the input pair, moves the second natural to the first position, and then applies the successor of the second element within the second position. Hence, folding a natural number n with sCase and zCase produces a pair (n-1, n). In the end, predK n projects the first element of a pair.

Kleene predecessor runs in linear time. Also, predK (suc n) gets stuck after reducing to π_1 (pair (π_2 (n sCase zCase)) (suc (π_2 (n sCase zCase)))). Hence, we must use induction to prove that predK (suc n) computes to n.

Furthermore, Parigot proved that any definition of predecessor for the Church-style lambda-encoded numerals requires linear time [4].

5.1   Constant-Time Destructor for Mendler-Style Encoding

In previous sections we defined a datatype FixIndM for every scheme F that has an identity mapping. Then, we implemented the constructors of the datatype as the function inFixIndM, and defined an induction principle phrased in terms of this function. In this section, we develop a mutual inverse of inFixIndM that runs in constant time. As a simple consequence, we prove that FixIndM is a least fixed point of F.

Let us start by exploring the computational behaviour of the function foldM. The following property is a variation of the *cancellation law* for Mendler-style encoded data [12], and its proof is simply by beta-reduction.

```
foldHom ◄ ∀ X : ⋆. Π x : F FixM. Π alg : AlgM X.
    foldM alg (inFixM x) ≃ alg (foldM alg) x = Λ X. λ x. λ a. β.
```

In other words, folding the inductive value inFixM x replaces its outermost "constructor" inFixM with the partially applied F-algebra alg (foldM alg).

It is well-known that (computationally) induction can be reduced to iteration (folding). Therefore, we can state the cancellation law for the induction rule in terms of proof algebras.

```
indHom ◄ ∀ Q : FixIndM → ⋆. Π alg : PrfAlgM FixIndM Q inFixIndM.
Π x : F FixIndM. ∀ c : Id FixIndM FixIndM.
 induction alg (inFixInd x) ≃ alg -c (induction alg) x
= Λ Q. λ p. λ x. β.
```

Crucially, the proof of indHom is by reflexivity (β), which ensures that the left-hand side of equality beta-reduces to the right-hand side in a constant number of beta-reductions.

Next, we implement a proof algebra for the predicate λ _. F FixIndM.

```
outAlgM ◄ PrfAlgM FixIndM (λ _. F FixIndM) inFixIndM
= Λ R. Λ c. λ f. λ y.  elimId -(imap c) y.
```

The identity mapping of F lifts the identity function c : Id R X to an identity function Id (F R) (F FixIndM), which is then applied to the argument y : F R to get the desired value of F FixIndM.

The proof algebra outAlgM induces the constant-time inverse of inFixIndM:

```
outFixIndM ◄ FixInd → F FixInd = induction outAlgM.
```

We defined outFixIndM (inFixIndM x) as induction outAlgM (inFixInd x), which reduces to outAlgM -c (induction outAlgM) x in a constant number of steps (indHom). Because outAlgM -c erases to λ f. λ y. y, it follows that outFixIndM computes an inverse of inFixIndM in a constant number of beta-reductions:

```
lambek1 ◄ Π x: F FixInd. outFixIndM (inFixIndM x) ≃ x = λ x. β.
```

Furthermore, we show that outFixIndM is a post-inverse:

```
lambek2 ◄ Π x: FixIndM. inFixIndM (outFixIndM x) ≃ x
= λ x. induction (Λ R. Λ c. λ ih. λ fr. β) x.
```

This direction requires us to invoke induction to "pattern match" on the argument value to get x := inFixIndM y for some value y of type F FixIndM. Then, inFixIndM (outFixIndM (inFixIndM y)) ≃ inFixIndM y because the inner term outFixIndM (inFixIndM y) is just y by beta reduction (lambek1).

6   Examples

In this section, we demonstrate the utility of our derivations on three examples. First, we present a detailed implementation of natural numbers with a constant-time predecessor function. Second, we show examples of infinitary datatypes. Finally, we give an example of a datatype arising as a least fixed point of a scheme that is not a functor, but has an identity mapping.

6.1   Natural Numbers with Constant-Time Predecessor

Natural numbers arise as a least fixed point of the functor NF:

$$\text{NF} \blacktriangleleft \star \to \star = \lambda \text{ X. Unit + X.}$$

$$\text{nfmap} \blacktriangleleft \text{Functor NF} = \texttt{<..>}$$

Because every functor induces an identity mapping, we can use our framework to define the natural numbers as follows:

$$\text{nfimap} \blacktriangleleft \text{IdMapping NF} = \text{fm2im nfmap.}$$

$$\text{Nat} \blacktriangleleft \star = \text{FixInd NF nfimap.}$$

$$\text{zero} \blacktriangleleft \text{Nat} = \text{inFixIndM (in1 unit).}$$
$$\text{suc} \blacktriangleleft \text{Nat} \to \text{Nat} = \lambda \text{ n. inFixIndM (in2 n).}$$

If we assume that injections in1 and in2 erase to λ a. λ i. λ j. i a and λ a. λ i. λ j. j a, then the natural number constructors have the following erasures:

```
zero ≃ λ alg. (alg (λ f. (f alg)) (λ i. λ j. (i (λ x. x))))
suc n ≃ λ alg. (alg (λ f. (f alg)) (λ i. λ j. (j n)))
```

Intuitively, Mendler-style numerals have a constant-time predecessor because every natural number suc n contains the previous natural n as its direct subpart (which is not true for the Church encoding).

We implement the predecessor for Nat in terms of the generic constant-time destructor outFixIndM:

```
pred ◄ Nat → Nat = λ n. case (outFixIndM n) (λ _. zero) (λ m. m).
```

Because elimination of disjoint sums (case) and outFixIndM are both constant-time operations, pred is also a constant-time function and its correctness is immediate (i.e., by beta-reduction):

$$\text{predSuc} \blacktriangleleft \Pi \text{ n : Nat. pred (suc n)} \simeq \text{n} = \lambda \text{ n. } \beta.$$

We also show that the usual "flat" induction principle can be derived from our generic induction principle (induction) by dependent elimination of NF:

```
indNat ◄ ∀ P : Nat → ⋆. (Π n : Nat. P n → P (suc n)) → P zero
 → Π n : Nat. P n = Λ P. λ s. λ z. λ n. induction
(Λ R. Λ c. λ ih. λ v. case v (λ u. ρ (etaUnit u) - z)
                          (λ r. s (elimId -c r) (ih r))) n.
```

6.2   Infinitary Trees

In Agda, we can give the following inductive definition of infinitary trees:

```
data ITree : Set where
    node : (Nat → Unit + ITree) → ITree
```

ITree is a least fixed point of functor IF X := Nat → Unit + X. In Cedille, we can implement a functorial function lifting for IF:

```
itfmap ◄ ∀ X Y : *. (X → Y) → IF X → IF Y
      = λ f. λ t. λ n. case (t n) (λ u. in1 u) (λ x. in2 (f x)).
```

To our best knowledge, it is impossible to prove that itfmap satisfies the functorial laws without functional extensionality (which is unavailable in Cedille). However, it is possible to implement an *identity mapping* for the scheme IF:

```
itimap ◄ IdMapping IF
      = λ c. pair (λ x. λ n. elimId -(nfimap -c) (x n)) β.
```

The first element of a pair erases to λ x. λ n. λ n. x n, which is λ x. x by the eta law. Because we showed that IF has an identity mapping, our generic development induces the datatype ITree with its constructor, destructor, and induction principle.

```
ITree ◄ * = FixIndM IF itimap.
```

```
inode ◄ (Nat → Unit + ITree) → ITree = λ f. inFixIndM f.
```

Below, the specialized induction is phrased in terms of the "empty tree" iempty, which acts as a base case (projR "projects" a tree from a disjoint sum or returns iempty):

```
iempty ◄ ITree = inode (λ _. in1 unit).
```

```
indITree ◄ ∀ P : ITree → *. P iempty →
       (Π f: Nat → Unit + ITree. (Π n : Nat. P (projR (f n))
       → P (inode f)) → Π t: ITree. P t = <..>
```

Next, let us look at another variant of infinitary datatypes in Agda:

```
data PTree : Set where
     pnode : ((PTree → Bool) → Unit + PTree) → PTree
```

This definition will be rejected by Agda (and Coq) since it arises as a least fixed point of the scheme PF X := Unit + ((X → Bool) → X) → X, which is positive but not strictly positive. The definition is rejected because it is currently unclear if non-strict definitions are sound in Agda. For the Coq setting, there is a proof by Coquand and Paulin that non-strict positivity combined with an impredicative universe and a predicative universe hierarchy leads to inconsistency [14]. In Cedille, we can implement an identity mapping for the scheme PF in a similar fashion as the previously discussed UF. Hence, the datatype induced by PF exists in the type theory of Cedille.

6.3 Unbalanced Trees

Consider the following definition of "unbalanced" binary trees in Agda:

```
data UTree : Set where
  leaf : Bool → UTree
  node : (b₁ : UTree) → (b₂ : UTree) → b₁ ≠ b₂ → UTree
```

The datatype UTree arises as a least fixed point of the following scheme:

UF ◀ ⋆ → ⋆ = λ X. Bool + (Σ x_1 : X. Σ x_2 : X. x_1 ≠ x_2).

Because the elements x_1 and x_2 must be different, lifting an arbitrary function X → Y to UF X → UF Y is impossible. Hence, the scheme UF is not a functor.

However, we can show that UF has an identity mapping. We start by producing a function UF X → UF Y from an identity Id X Y:

```
uimap' ◀ ∀ X Y : ⋆. Π i : Id X Y. UF X → UF Y = λ i. λ uf.
  case uf (λ u. in1 u)
          (λ u. in2 (pair (elimId -i (π₁ u))
                    (pair (elimId -i (π₁ (π₂ u))) (π₂ (π₂ u)))))
```

We prove that uimap' -i is extensionally an identity function:

uimP ◀ ∀ X Y: ⋆. Π i: Id X Y. Π u: UF X. u ≃ uimap' i u = <..>

This is enough to derive an identity mapping for UF by using the previously implemented combinator intrId:

```
uimap ◀ IdMapping UF = λ i. intrId (uimap' i) (uimP i).
```

Therefore, we conclude that the datatype of unbalanced trees exists in Cedille and can be defined as a least fixed point of the scheme UF:

UTree ◀ ⋆ = FixIndM UF uimap.

The specialized constructors, induction principle, and a destructor function for UTree are easily derived from their generic counterparts (inFixIndM, induction, outFixIndM).

7 Related Work

Pfenning and Paulin-Mohring show how to model inductive datatypes using impredicative encodings in the Calculus of Constructions (CC) [15]. Because induction is not provable in the CC, the induction principles are generated and added as axioms. This approach was adopted by initial versions of the Coq proof assistant, but later Coq switched to the Calculus of Inductive Constructions (CIC), which has built-in inductive datatypes.

Delaware et al. derived induction for impredicative lambda-encodings in Coq as a part of their framework for modular definitions and proofs (using the à la carte technique [16]). They showed that a value v : Fix F is inductive if it is accompanied by a proof of the universal property of folds [17].

Similarly, Torrini introduced the *predicatisation* technique, reducing dependent induction to proofs that only rely on non-dependent Mendler induction (by requiring the inductive argument to satisfy an extra predicatisation hypothesis) [18].

Traytel et al. present a framework for constructing (co)datatypes in HOL [19,20]. The main ingredient is a notion of a bounded natural functor (BNF), or a binary functor with additional structure. BNFs are closed under composition and fixed points, which enables support for both mutual and nested (co)recursion with mixed combinations of datatypes and codatatypes. The authors developed a package that can generate (co)datatypes with their associated proof-principles from user specifications (including custom bounded natural functors). In contrast, our approach provides a single generic derivation of induction within the theory of Cedille, but does not address codatatypes. It would be interesting to further investigate the exact relationship between schemes with identity mappings and BNFs.

Church encodings are typeable in System F and represent datatypes as their own iterators. Parigot proved that the lower bound of the predecessor function for Church numerals has linear time complexity [4].

Parigot designed an impredicative lambda-encoding that is typeable in System F_ω with positive-recursive type definitions. The encoding identifies datatypes with their own recursors, allowing constant time destructors to be defined, but the drawback is that the representation of a natural number n is exponential in the call-by-value setting [5].

The Stump-Fu encoding is also typeable in System F_ω with positive-recursive type definitions. It improves upon the Parigot representation by requiring only quadratic space, and it also supports constant-time destructors [6].

8   Conclusions and Future Work

In this work, we showed that the Calculus of Dependent Lambda Eliminations is a compact pure type theory that allows a general class of Mendler-style lambda-encoded inductive datatypes to be defined as least fixed points of schemes with identity mappings. We also gave a generic derivation of induction and implemented a constant-time destructor for these datatypes. We used our development to give the first example (to the best of our knowledge) of lambda-encoded natural numbers with: provable induction, a constant-time predecessor function, and a linear size (in the numeral n) term representation. Our formal development is around 700 lines of Cedille code.

For future work, we plan to explore coinductive definitions and to use the categorical model of Mendler-style datatypes to investigate histomorphisms and inductive-recursive datatypes in Cedille [12].

Additionally, we are investigating approaches to generic "proof reuse" in Cedille. The key idea is that implicit products allow indexed and non-indexed datatypes to be represented by the same pure lambda term. For example, this allows the proof of associativity of append for lists to be reused as the proof of associativity of append for vectors [11].

Acknowledgments. The first author is thankful to Anna, Albert, and Eldar for all the joy and support. Authors are thankful to Larry Diehl for proof reading and numerous grammatical adjustments. We gratefully acknowledge NSF support under award 1524519, and DoD support under award FA9550-16-1-0082 (MURI program).

References

1. Geuvers, H.: Induction is not derivable in second order dependent type theory. In: Abramsky, S. (ed.) TLCA 2001. LNCS, vol. 2044, pp. 166–181. Springer, Heidelberg (2001). https://doi.org/10.1007/3-540-45413-6_16
2. Stump, A.: The calculus of dependent lambda eliminations. J. Funct. Program. **27**, e14 (2017)
3. Firsov, D., Stump, A.: Generic derivation of induction for impredicative encodings in cedille. In: Proceedings of the 7th ACM SIGPLAN International Conference on Certified Programs and Proofs, CPP 2018, New York, NY, USA, pp. 215–227. ACM (2018)
4. Parigot, M.: On the representation of data in lambda-calculus. In: Börger, E., Büning, H.K., Richter, M.M. (eds.) CSL 1989. LNCS, vol. 440, pp. 309–321. Springer, Heidelberg (1990). https://doi.org/10.1007/3-540-52753-2_47
5. Parigot, M.: Programming with proofs: a second order type theory. In: Ganzinger, H. (ed.) ESOP 1988. LNCS, vol. 300, pp. 145–159. Springer, Heidelberg (1988). https://doi.org/10.1007/3-540-19027-9_10
6. Stump, A., Fu, P.: Efficiency of lambda-encodings in total type theory. J. Funct. Program. **26**, e3 (2016)
7. Stump, A.: From realizability to induction via dependent intersection. Ann. Pure Appl. Logic **169**, 637–655 (2018)
8. Wells, J.B.: Typability and type checking in system F are equivalent and undecidable. Ann. Pure Appl. Logic **98**(1–3), 111–156 (1999)
9. Kopylov, A.: Dependent intersection: a new way of defining records in type theory. In: 18th IEEE Symposium on Logic in Computer Science (LICS), pp. 86–95 (2003)
10. Miquel, A.: The implicit calculus of constructions extending pure type systems with an intersection type binder and subtyping. In: Abramsky, S. (ed.) TLCA 2001. LNCS, vol. 2044, pp. 344–359. Springer, Heidelberg (2001). https://doi.org/10.1007/3-540-45413-6_27
11. Diehl, L., Firsov, D., Stump, A.: Generic zero-cost reuse for dependent types. CoRR abs/1803.08150 (2018)
12. Uustalu, T., Vene, V.: Mendler-style inductive types, categorically. Nordic J. Comput. **6**(3), 343–361 (1999)
13. Wadler, P.: Recursive types for free! (1990)
14. Coquand, T., Paulin, C.: Inductively defined types. In: Martin-Löf, P., Mints, G. (eds.) COLOG 1988. LNCS, vol. 417, pp. 50–66. Springer, Heidelberg (1990). https://doi.org/10.1007/3-540-52335-9_47
15. Pfenning, F., Paulin-Mohring, C.: Inductively defined types in the Calculus of Constructions. In: Main, M., Melton, A., Mislove, M., Schmidt, D. (eds.) MFPS 1989. LNCS, vol. 442, pp. 209–228. Springer, New York (1990)
16. Swierstra, W.: Data types à la carte. J. Funct. Program. **18**(4), 423–436 (2008)
17. Delaware, B., d. S. Oliveira, B.C., Schrijvers, T.: Meta-theory à la carte. In: Proceedings of the 40th Annual ACM SIGPLAN-SIGACT Symposium on Principles of Programming Languages, POPL 2013, New York, NY, USA, pp. 207–218. ACM (2013)

18. Torrini, P.: Modular dependent induction in Coq, Mendler-style. In: Blanchette, J.C., Merz, S. (eds.) ITP 2016. LNCS, vol. 9807, pp. 409–424. Springer, Cham (2016). https://doi.org/10.1007/978-3-319-43144-4_25
19. Traytel, D., Popescu, A., Blanchette, J.C.: Foundational, compositional (Co)datatypes for higher-order logic: category theory applied to theorem proving. In: Proceedings of the 27th Annual IEEE Symposium on Logic in Computer Science, LICS 2012, Dubrovnik, Croatia, 25–28 June 2012, pp. 596–605. IEEE Computer Society (2012)
20. Biendarra, J., et al.: Foundational (Co)datatypes and (Co)recursion for higher-order logic. In: Dixon, C., Finger, M. (eds.) FroCoS 2017. LNCS (LNAI), vol. 10483, pp. 3–21. Springer, Cham (2017)

Verification of PCP-Related Computational Reductions in Coq

Yannick Forster$^{(\boxtimes)}$, Edith Heiter, and Gert Smolka

Saarland University, Saarbrücken, Germany
{forster,heiter,smolka}@ps.uni-saarland.de

Abstract. We formally verify several computational reductions concerning the Post correspondence problem (PCP) using the proof assistant Coq. Our verification includes a reduction of the halting problem for Turing machines to string rewriting, a reduction of string rewriting to PCP, and reductions of PCP to the intersection problem and the palindrome problem for context-free grammars.

Keywords: Post correspondence problem · String rewriting
Context-free grammars · Computational reductions · Undecidability
Coq

1   Introduction

A problem P can be shown undecidable by giving an undecidable problem Q and a computable function reducing Q to P. There are well known reductions of the halting problem for Turing machines (TM) to the Post correspondence problem (PCP), and of PCP to the intersection problem for context-free grammars (CFI). We study these reductions in the formal setting of Coq's type theory [16] with the goal of providing elegant correctness proofs.

Given that the reduction of TM to PCP appears in textbooks [3,9,15] and in the standard curriculum for theoretical computer science, one would expect that rigorous correctness proofs can be found in the literature. To our surprise, this is not the case. Missing is the formulation of the inductive invariants enabling the necessary inductive proofs to go through. Speaking with the analogue of imperative programs, the correctness arguments in the literature argue about the correctness of programs with loops without stating and verifying loop invariants.

By inductive invariants we mean statements that are shown inductively and that generalise the obvious correctness statements one starts with. Every substantial formal correctness proof will involve the construction of suitable inductive invariants. Often it takes ingenuity to generalise a given correctness claim to one or several inductive invariants that can be shown inductively.

It took some effort to come up with the missing inductive invariants for the reductions leading from TM to PCP. Once we had the inductive invariants, we had rigorous and transparent proofs explaining the correctness of the reductions in a more satisfactory way than the correctness arguments we found in the literature.

© Springer International Publishing AG, part of Springer Nature 2018
J. Avigad and A. Mahboubi (Eds.): ITP 2018, LNCS 10895, pp. 253–269, 2018.
https://doi.org/10.1007/978-3-319-94821-8_15

Reduction of problems is transitive. Given a reduction $P \preceq Q$ and a reduction $Q \preceq R$, we have a reduction $P \preceq R$. This way, complex reductions can be factorised into simpler reductions. Following ideas in the literature, we will establish the reduction chain

$$\text{TM} \preceq \text{SRH} \preceq \text{SR} \preceq \text{MPCP} \preceq \text{PCP}$$

where TM is the halting problem of single-tape Turing machines, SRH is a generalisation of the halting problem for Turing machines, SR is the string rewriting problem, and MPCP is a modified version of PCP fixing a first card. The most interesting steps are $\text{SR} \preceq \text{MPCP}$ and $\text{MPCP} \preceq \text{PCP}$.

We also consider the intersection problem (CFI) and the palindrome problem (CFP) for a class of linear context-free grammars we call Post grammars. CFP asks whether a Post grammar generates a palindrome, and CFI asks whether for two Post grammars there exists a string generated by both grammars. We will verify reductions $\text{PCP} \preceq \text{CFI}$ and $\text{PCP} \preceq \text{CFP}$, thus showing that CFP and CFI are both undecidable.

Coq's type theory provides an ideal setting for the formalisation and verification of the reductions mentioned. The fact that all functions in Coq are total and computable makes the notion of computable reductions straightforward.

The correctness arguments coming with our approach are inherently constructive, which is verified by the underlying constructive type theory. The main inductive data types we use are numbers and lists, which conveniently provide for the representation of strings, rewriting systems, Post correspondence problems, and Post grammars.

The paper is accompanied by a Coq development covering all results of this paper. The definitions and statements in the paper are hyperlinked with their formalisations in the HTML presentation of the Coq development at http://www.ps.uni-saarland.de/extras/PCP.

Organisation

We start with the necessary formal definitions covering all reductions we consider in Sect. 2. We then present each of the six reductions and conclude with a discussion of the design choices underlying our formalisations. Sections 3, 4, 5, 6, 7 and 8 on the reductions are independent and can be read in any order.

We only give definitions for the problems and do not discuss the underlying intuitions, because all problems are covered in a typical introduction to theoretical computer science and the interested reader can refer to various textbooks providing good intuitions, e.g. [3,9,15].

Contribution

Our reduction functions follow the ideas in the literature. The main contributions of the paper are the formal correctness proofs for the reduction functions. Here some ingenuity and considerable elaboration of the informal arguments in the literature were needed. As one would expect, the formal proofs heavily rely on inductive techniques. In contrast, the informal proof sketches in the literature do not introduce the necessary inductions (in fact, they don't even mention inductive proofs). To the best of our knowledge, the present paper is the first paper providing formal correctness proofs for basic reductions to and from PCP.

2 Definitions

Formalising problems and computable reductions in constructive type theory is straightforward. A *problem* consists of a type X and a unary predicate p on X, and a *reduction* of (X,p) to (Y,q) is a function $f : X \to Y$ such that $\forall x.\ px \leftrightarrow q(fx)$. Note that the usual requirement that f is total and computable can be dropped since it is satisfied by every function in a constructive type theory. We write $p \preceq q$ and say that p *reduces to* q if a reduction of (X,p) to (Y,q) exists.

Fact 1. *If $p \preceq q$ and $q \preceq r$, then $p \preceq r$.*

The basic inductive data structures we use are *numbers* ($n ::= 0 \mid Sn$) and *lists* ($L ::= [\,] \mid s :: L$). We write $L_1 \mathbin{+\!\!\!+} L_2$ for the *concatenation* of two lists, $\overline{L}$ for the reversal of a list, $[\,fs \mid s \in A\,]$ for a map over a list, and $[\,fs \mid s \in A \wedge ps\,]$ for a map and filter over a list. Moreover, we write $s \in L$ if s is a member of L, and $L_1 \subseteq L_2$ if every member of L_1 is a member of L_2.

A *string* is a list of symbols, and a *symbol* is a number. The letters x, y, z, u, and v range over strings, and the letters a, b, c range over symbols. We write xy for $x \mathbin{+\!\!\!+} y$ and ax for $a :: x$. We use ϵ to denote the empty string. A *palindrome* is a string x such that $x = \overline{x}$.

Fact 2. $\overline{xy} = \overline{y}\,\overline{x}$ *and* $\overline{\overline{x}} = x$.

Fact 3. *If $xay = uav$, $a \notin x$, and $a \notin u$, then $x = u$ and $y = v$.*

Proof. By induction on x.   □

A *card* x/y or a *rule* x/y a is a pair (x,y) of two strings. When we call x/y a card we see x as the upper and y as the lower string of the card. When we call x/y a rule we see x as the left and y as the right side of the rule.

The letters A, B, C, P, R range over list of cards or rules.

2.1 Post Correspondence Problem

A *stack* is a list of cards. The *upper trace* A^1 and the *lower trace* A^2 of a stack A are strings defined as follows:

$$[\,]^1 := \epsilon \qquad\qquad\qquad\qquad [\,]^2 := \epsilon$$
$$(x/y :: A)^1 := x(A^1) \qquad\qquad (x/y :: A)^2 := y(A^2)$$

Note that A^1 is the concatenation of the upper strings of the cards in A, and that A^2 is the concatenation of the lower strings of the cards in A. We say that a stack A *matches* if $A^1 = A^2$ and a *match* is a matching stack. An example for a match is the list $A = [\epsilon/ab,\ a/c,\ bc/\epsilon]$, which satisfies $A^1 = A^2 = abc$.

We can now define the predicate for the *Post correspondence problem*:

$$\mathsf{PCP}\,(P) \;:=\; \exists A \subseteq P.\ A \neq [\,]\wedge A^1 = A^2$$

Note that $\mathsf{PCP}\,(P)$ holds iff there exists a nonempty match $A \subseteq P$. We then say that A is a *solution* of P. For instance,

$$P = [a/\epsilon,\ b/a,\ \epsilon/bb]$$

is solved by the match

$$A = [\epsilon/bb,\ b/a,\ b/a,\ a/\epsilon,\ a/\epsilon].$$

While it is essential that A is a list providing for order and duplicates, P may be thought of as a finite set of cards.

We now define the predicate for the *modified Post correspondence problem*:

$$\mathsf{MPCP}\,(x/y, P) \;:=\; \exists A \subseteq x/y :: P.\ xA^1 = yA^2$$

Informally, $\mathsf{MPCP}\,(x/y, P)$ is like $\mathsf{PCP}\,(x/y :: P)$ with the additional constraint that the solution for $x/y :: P$ starts with the first card x/y.

Note that in contrary to most text books we leave open whether x/y is an element of P and instead choose A as subset of $x/y :: P$. While this might first seem more complicated, it actually eases formalisation. Including x/y into P would require MPCP to be a predicate on arguments of the form $(P, x/y, H : x/y \in P)$, i.e. dependent pairs containing a proof.

2.2   String Rewriting

Given a list R of rules, we define *string rewriting* with two inductive predicates $x \succ_R y$ and $x \succ_R^* y$:

$$\frac{x/y \in R}{uxv \succ_R uyv} \qquad \frac{}{z \succ_R^* z} \qquad \frac{x \succ_R y \quad y \succ_R^* z}{x \succ_R^* z}$$

Note that $\succ_R^*$ is the reflexive transitive closure of $\succ_R$, and that $x \succ_R y$ says that y can be obtained from x with a single rewriting step using a rule in R.

Fact 4. *The following hold:*

1. *If $x \succ_R^* y$ and $y \succ_R^* z$, then $x \succ_R^* z$.*
2. *If $x \succ_R^* y$, then $ux \succ_R^* uy$.*
3. *If $x \succ_R^* y$ and $R \subseteq P$, then $x \succ_P^* y$.*

Proof. By induction on $x \succ_R^* y$.   □

Note that the induction lemma for string rewriting can be stated as

$$\forall z.\ Pz \rightarrow (\forall xy.\ x \succ_R y \rightarrow Py \rightarrow Px) \rightarrow \forall x.\ x \succ_R^* z \rightarrow Px.$$

This is stronger than the lemma Coq infers, because of the quantification over z on the outside. The quantification is crucial for many proofs that do induction on derivations $x \succ_R z$, and we use the lemma throughout the paper without explicitly mentioning it.

We define the predicates for the *string rewriting problem* and the *generalised halting problem* as follows:

$$\mathsf{SR}\,(R,x,y) \ := \ x \succ_R^* y$$
$$\mathsf{SRH}\,(R,x,a) \ := \ \exists y.\ x \succ_R^* y \land a \in y$$

We call the second problem *generalised halting problem*, because it covers the halting problem for deterministic single-tape Turing machines, but also the halting problems for nondeterministic machines or for more exotic machines that e.g. have a one-way infinite tape or can read multiple symbols at a time.

We postpone the definition of Turing machines and of the halting problem TM to Sect. 8.

2.3   Post Grammars

A *Post grammar* is a pair (R, a) of a list R of rules and a symbol a. Informally, a Post grammar (R, a) is a special case of a context-free grammar with a single nonterminal S and two rules $S \rightarrow xSy$ and $S \rightarrow xay$ for every rule $x/y \in R$, where $S \neq a$ and S does not occur in R. We define the *projection* $\sigma_a A$ of a list of rules A with a symbol a as follows:

$$\sigma_a[\,] := a$$
$$\sigma_a(x/y::A) := x(\sigma_a A)y$$

We say that a Post grammar (R, a) *generates* a string u if there exists a nonempty list $A \subseteq R$ such that $\sigma_a A = u$. We then say that A is a *derivation of* u in (R, a).

We can now define the predicates for the problems CFP and CFI:

$$\mathsf{CFP}\,(R,a) \ := \ \exists A \subseteq R.\ A \neq [\,] \land \sigma_a A = \overline{\sigma_a A}$$
$$\mathsf{CFI}\,(R_1,R_2,a) \ := \ \exists A_1 \subseteq R_1\ \exists A_2 \subseteq R_2.$$
$$A_1 \neq [\,] \land A_2 \neq [\,] \land \sigma_a A_1 = \sigma_a A_2$$

Informally, $\mathsf{CFP}\,(R, a)$ holds iff the grammar (R, a) generates a palindrome, and $\mathsf{CFI}\,(R_1, R_2, a)$ holds iff there exists a string that is generated by both grammars (R_1, a) and (R_2, a). Note that as Post grammars are special cases of context-free grammars, the reduction of PCP to CFG and CFI can be trivially extended to reductions to the respective problems for context-free grammars. We prove this formally in the accompanying Coq development.

2.4 Alphabets

For some proofs it will be convenient to fix a finite set of symbols. We represent such sets as lists and speak of *alphabets*. The letter Σ ranges over alphabets. We say that an alphabet Σ *covers* a string, card, or stack if Σ contains every symbol occurring in the string, card, or stack. We may write $x \subseteq \Sigma$ to say that Σ covers x since both x and Σ are lists of symbols.

2.5 Freshness

At several points we will need to pick fresh symbols from an alphabet. Because we model symbols as natural numbers, a very simple definition of freshness suffices. We define a function fresh such that fresh $\Sigma \notin \Sigma$ for an alphabet Σ as follows:

$$\mathsf{fresh}\ [\,] = 0$$
$$\mathsf{fresh}\ (a :: \Sigma) = 1 + a + \mathsf{fresh}\ \Sigma$$

fresh has the following characteristic property:

Lemma 5. *For all $a \in \Sigma$, fresh $\Sigma > a$.*

Proof. By induction on Σ, with a generalised. $\square$

The property is most useful when exploited in the following way:

Corollary 6. *For all $a \in \Sigma$, fresh $\Sigma \neq a$.*

An alternative approach to this is to formalise alphabets explicitly as types Σ. This has the advantage that arbitrarily many fresh symbols can be introduced simultaneously using definitions like $\Gamma := \Sigma + X$, and symbols in Γ stemming from Σ can easily be shown different from fresh symbols stemming from X by inversion. However, this means that strings $x : \Sigma^*$ have to be explicitly embedded pointwise when used as strings of type Γ^*, which complicates proofs.

In general, both approaches have benefits and tradeoffs. Whenever proofs rely heavily on inversion (as e.g. our proofs in Sect. 8), the alternative approach is favorable. If proofs need the construction of many strings, as most of our proofs do, modelling symbols as natural numbers shortens proofs.

3 SRH to SR

We show that SRH (the generalised halting problem) reduces to SR (string rewriting). We start with the definition of the reduction function. Let R, x_0, and a_0 be given.

We fix an alphabet Σ covering R, x_0, and a_0. We now add rules to R that allow $x \succ_R^* a_0$ if $a_0 \in x$.

$$P := R + [\,aa_0/a_0 \mid a \in \Sigma\,] + [\,a_0a/a_0 \mid a \in \Sigma\,]$$

Lemma 7. *If $a_0 \in x \subseteq \Sigma$, then $x \succ_P^* a_0$.*

Proof. For all $y \subseteq \Sigma$, $a_0 y \succ_P^* a_0$ and $y a_0 \succ_P^* a_0$ follow by induction on y. The claim now follows with Fact 4 (1, 2). □

Lemma 8. $\mathsf{SRH}\,(R, x_0, a_0) \leftrightarrow \mathsf{SR}\,(P, x_0, a_0)$.

Proof. Let $x_0 \succ_R^* y$ and $a_0 \in y$. Then $y \succ_P^* a_0$ by Lemma 7. Moreover, $x_0 \succ_P^* y$ by Fact 4 (3). Thus $x_0 \succ_P^* a_0$ by Fact 4 (1).

Let $x_0 \succ_P^* a_0$. By induction on $x_0 \succ_P^* a_0$ it follows that there exists y such that $x_0 \succ_R^* y$ and $a_0 \in y$. □

Theorem 9. *SRH reduces to SR.*

Proof. Follows with Lemma 8. □

4   SR to MPCP

We show that SR (string rewriting) reduces to MPCP (the modified Post correspondence problem). We start with the definition of the reduction function.

Let R, x_0 and y_0 be given. We fix an alphabet Σ covering R, x_0, and y_0. We also fix two symbols $\$, \# \notin \Sigma$ and define:

$$
\begin{aligned}
d &:= \$ \,/\, \$x_0\# \\
e &:= y_0\#\$ \,/\, \$ \\
P &:= [d, e] + \!\!\!+ R + \!\!\!+ [\#/\#] + \!\!\!+ [a/a \mid a \in \Sigma]
\end{aligned}
$$

The idea of the reduction is as follows: Assume $\Sigma = [a, b, c]$ and rules bc/a and aa/b in R. Then $abc \succ_R aa \succ_R b$ and we have $d = \$/\$abc\#$, $e = b\#\$/\$$, and $P = [d, e, bc/a, aa/b, \ldots, a/a, b/b, c/c]$, omitting possible further rules in R. Written suggestively, the following stack matches:

$\$$	a	bc	$\#$	aa	$\#$	$b\#\$$
$\$abc\#$	a	a	$\#$	b	$\#$	$\$$

And, vice versa, every matching stack starting with d will yield a derivation of $abc \succ_R^* b$.

We now go back to the general case and state the correctness lemma for the reduction function.

Lemma 10. $x_0 \succ_R^* y_0$ *if and only if there exists a stack $A \subseteq P$ such that $d :: A$ matches.*

From this lemma we immediately obtain the reduction theorem (Theorem 13). The proof of the lemma consists of two *translation lemmas*: Lemmas 11 and 12. The translation lemmas generalise the two directions of Lemma 10 such that they can be shown with canonical inductions.

Lemma 11. *Let* $x \subseteq \Sigma$ *and* $x \succ_R^* y_0$. *Then there exists* $A \subseteq P$ *such that* $A^1 = x \# A^2$.

Proof. By induction on $x \succ_R^* y_0$. In the first case, $x = y_0$ and $[e]^1 = x\#[e]^2$. In the second case, $x \succ y$ and $y \succ^* y_0$. By induction hypothesis there is $A \subseteq P$ such that $A^1 = y\#A^2$. Let $x = (a_1 \ldots a_n)u(b_1 \ldots b_n)$ and $y = (a_1 \ldots a_n)v(b_1 \ldots b_n)$ for $u/v \in R$. We define $B := (a_1/a_1) \ldots (a_n/a_n) :: (u/v) :: (b_1/b_1) \ldots (b_n/b_n) :: (\#/\#) :: A$. Now $B^1 = x\#A^1 = x\#y\#A^2 = x\#B^2$. □

Lemma 12. *Let* $A \subseteq P$, $A^1 = x\#yA^2$, *and* $x, y \subseteq \Sigma$. *Then* $yx \succ_R^* y_0$.

Proof. By induction on A with x and y generalised. We do all cases in detail:

- The cases where $A = [\,]$ or $A = d :: B$ are contradictory.
- Let $A = e :: B$. By assumption, $y_0\#\$B^1 = x\#y\$B^2$. Then $x = y_0$, $y = \epsilon$ and $yx = y_0 \succ_R^* y_0$.
- Let $A = u/v :: B$ for $u/v \in R$. Because $\#$ is not in u and by assumption $uB^1 = x\#yvB^2$, $x = u \mathbin{+\mkern-10mu+} x'$. And $yx = yux' \succ yvx' \succ^* y_0$ by induction hypothesis.
- Let $A = \#/\# :: B$. By assumption, $\#B^1 = x\#y\#B^2$. Then $x = \epsilon$ and we have $B^1 = y\#\epsilon B^2$. By induction hypothesis, this yields $yx = \epsilon y \succ_R^* y_0$ as needed.
- Let $A = a/a :: B$ for $a \in \Sigma$ and assume $aB^1 = x\#yaB^2$. Then $x = ax'$ and $B^1 = x'\#yaB^2$. By induction hypothesis, this yields $yx = yax' \succ_R^* y_0$ as needed. □

Theorem 13. *SR reduces to MPCP.*

Proof. Follows with Lemma 10. □

The translation lemmas formulate what we call the *inductive invariants* of the reduction function. The challenge of proving the correctness of the reduction function is finding strong enough inductive invariants that can be verified with canonical inductions.

5   MPCP to PCP

We show that MPCP (modified PCP) reduces to PCP.

The idea of the reduction is that for a stack $A = [x_1/y_1, \ldots, x_n, y_n]$ and a first card x_0/y_0 where $x_i = a_i^0 \ldots a_i^{m_i}$ we have

$$(a_0^0 \ldots a_0^{m_0})(a_1^0 \ldots a_1^{m_1}) \ldots (a_n^0 \ldots a_n^{m_n})$$
$$= (b_0^0 \ldots b_0^{m_0})(b_1^0 \ldots b_1^{m_1}) \ldots (b_n^0 \ldots b_n^{m_n})$$

if and only if we have

$$\$(\#a_0^0 \ldots \#a_0^{m_0})(\#a_1^0 \ldots \#a_1^{m_1}) \ldots (\#a_n^0 \ldots \#a_n^{m_n})\#\$$$
$$= \$\#(b_0^0\# \ldots b_0^{m_0}\#)(b_1^0\# \ldots b_1^{m_1}\#) \ldots (b_n^0\# \ldots b_n^{m_n}\#)\$.$$

The reduction function implements this idea by constructing a dedicated first and a dedicated last card and by inserting #-symbols into the MPCP cards:

Let x_0/y_0 and R be given. We fix an alphabet Σ covering x_0/y_0 and R. We also fix two symbols $\$, \# \notin \Sigma$. We define two functions ${}^\#x$ and $x^\#$ inserting the symbol # before and after every symbol of a string x:

$$^\#\epsilon := \epsilon \qquad\qquad\qquad \epsilon^\# := \epsilon$$

$$^\#(ax) := \#a(^\#x) \qquad\qquad (ax)^\# := a\#(x^\#)$$

We define:

$$d := \$(^\#x_0)\,/\,\$\#(y_0^\#)$$
$$e := \#\$\,/\,\$$$
$$P := [d,e] + \!+\, [\,^\#x\,/\,y^\#\mid x/y \in x_0/y_0 :: R \wedge (x/y) \neq (\epsilon/\epsilon)\,]$$

We now state the correctness lemma for the reduction function.

Lemma 14. *There exists a stack $A \subseteq x_0/y_0 :: R$ such that $x_0 A^1 = y_0 A^2$ if and only if there exists a nonempty stack $B \subseteq P$ such that $B^1 = B^2$.*

From this lemma we immediately obtain the desired reduction theorem (Theorem 19). The proof of the lemma consists of two translation lemmas (Lemmas 17 and 18) and a further auxiliary lemma (Lemma 15).

Lemma 15. *Every nonempty match $B \subseteq P$ starts with d.*

Proof. Let B be a nonempty match $B \subseteq P$. Then e cannot be the first card of B since the upper string and lower string of e start with different symbols. For the same reason $^\#x\,/\,y^\#$ cannot be the first card of B if $x/y \in R$ and both x and y are nonempty.

Consider $\epsilon/ay \in R$. Then $\epsilon/(ay)^\#$ cannot be the first card of B since no card of P has an upper string starting with a.

Consider $ax/\epsilon \in R$. Then $^\#(ay)/\epsilon$ cannot be the first card of B since no card of P has a lower string starting with #. $\square$

For the proofs of the translation lemmas we need a few facts about $^\#x$ and $x^\#$.

Lemma 16. *The following hold:*

1. $(^\#x)\# = \#(x^\#)$.
2. $^\#(xy) = (^\#x)(^\#y)$.
3. $(xy)^\# = (x^\#)(y^\#)$.
4. $^\#x \neq \#(y^\#)$.
5. $x^\# = y^\# \to x = y$.

Proof. By induction on x. $\square$

Lemma 17. *Let $A \subseteq x_0/y_0 :: R$ and $xA^1 = yA^2$. Then there exists a stack $B \subseteq P$ such that $(^\#x)B^1 = \#(y^\#)B^2$.*

Proof. By induction on A with x and y generalised. The case for $A = [\,]$ follows from Lemma 16 (1) by choosing $[e]$.

For the other case, let $A = x'/y' :: A'$. Then by assumption $xx'A'^1 = yy'A'^2$. And thus by induction hypothesis there exists $B \subseteq P$ such that $^\#(xx')B^1 = \#(yy')^\# B^2$. By Lemma 16 (2) and (3), $(^\#x)(^\#x')B^1 = \#(y^\#)(y'^\#)B^2$.

If $(x'/y') \neq (\epsilon/\epsilon)$, then choosing $^\#x'/y'^\# :: B \subseteq P$ works. Otherwise, $B \subseteq P$ works. □

Lemma 18. *Let $B \subseteq P$ such that $(^\#x)B^1 = \#(y^\#)B^2$ and $x, y \subseteq \Sigma$. Then there exists a stack $A \subseteq x_0/y_0 :: R$ such that $xA^1 = yA^2$.*

Proof. By induction on B. The cases $B = [\,]$ and $B = d :: B'$ yield contradictions using Lemma 16 (4). For $B = e :: B'$, choosing $A = [\,]$ works by Lemma 16 (5).

The interesting case is $B =^\# x'/y'^\# :: B'$ for $x'/y' \in x_0/y_0 :: R$ with $(x'/y') \neq (\epsilon/\epsilon)$. By assumption and Lemma 16 (2) and (3) we know that $^\#(xx')B'^1 = \#(yy')^\# B'^2$. Now by induction hypothesis, where all premises follow easily, there is $A \subseteq x_0/y_0 :: R$ with $xx'A^1 = yy'A^2$ and thus $x'/y' :: A$ works. □

Theorem 19. *MPCP reduces to PCP.*

Proof. Follows with Lemma 14. □

6 PCP to CFP

We show that PCP reduces to CFP (the palindrome problem for Post grammars).
Let $\#$ be a symbol.

Fact 20. *Let $\# \notin x, y$. Then $x \# y$ is a palindrome iff $y = \overline{x}$.*

Proof. Follows with Facts 2 and 3. □

There is an obvious connection between matching stacks and palindromes: A stack

$$[x_1/y_1, \ldots, x_n/y_n]$$

matches if and only if the string

$$x_1 \cdots x_n \# \overline{y_n} \cdots \overline{y_1}$$

is a palindrome, provided the symbol $\#$ does not appear in the stack (follows with Facts 2 and 20 using $\overline{y_n} \cdots \overline{y_1} = \overline{y_1 \cdots y_n}$). Moreover, strings of the form $x_1 \cdots x_n \# \overline{y_n} \cdots \overline{y_1}$ with $n \geq 1$ may be generated by a Post grammar having a rule $x/\overline{y}$ for every card x/y in the stack. The observations yield a reduction of PCP to CFP.

We formalise the observations with a function

$$\gamma A := [\, x/\overline{y} \mid x/y \in A \,].$$

Lemma 21. $\sigma_\#(\gamma A) = A^1 \# \overline{A^2}$.

Proof. By induction on A using Fact 2. □

Lemma 22. *Let A be a stack and $\#$ be a symbol not occurring in A. Then A is a match if and only if $\sigma_\#(\gamma A)$ is a palindrome.*

Proof. Follows with Lemma 21 and Facts 20 and 2. □

Lemma 23. $\gamma(\gamma A) = A$ *and* $A \subseteq \gamma B \to \gamma A \subseteq B$.

Proof. By induction on A using Fact 2. □

Theorem 24. *PCP reduces to CFP.*

Proof. Let P be a list of cards. We fix a symbol $\#$ that is not in P and show $\mathsf{PCP}\,(P) \leftrightarrow \mathsf{CFP}\,(\gamma P, \#)$.

Let $A \subseteq P$ be a nonempty match. It suffices to show that $\gamma A \subseteq \gamma P$ and $\sigma_\#(\gamma A)$ is a palindrome. The first claim follows with Lemma 23, and the second claim follows with Lemma 22.

Let $B \subseteq \gamma P$ be a nonempty stack such that $\sigma_\# B$ is a palindrome. By Lemma 23 we have $\gamma B \subseteq P$ and $B = \gamma(\gamma B)$. Since γB matches by Lemma 22, we have $\mathsf{PCP}\,(P)$. □

7   PCP to CFI

We show that PCP reduces to CFI (the intersection problem for Post grammars). The basic idea is that a stack $A = [x_1/y_1, \ldots, x_n/y_n]$ with $n \geq 1$ matches if and only if the string

$$x_1 \cdots x_n \# \, x_n \# y_n \# \cdots \# x_1 \# y_1 \#$$

equals the string

$$y_1 \cdots y_n \# \, x_n \# y_n \# \cdots \# x_1 \# y_1 \#$$

provided the symbol $\#$ does not occur in A. Moreover, strings of these forms can be generated by the Post grammars $([\,x/x\#y\# \mid x/y \in A\,], \#)$ and $([\,y/x\#y\# \mid x/y \in A\,], \#)$, respectively.

We fix a symbol $\#$ and formalise the observations with two functions

$$\gamma_1 A := [\,x/x\#y\# \mid x/y \in A\,] \qquad \gamma_2 A := [\,y/x\#y\# \mid x/y \in A\,]$$

and a function γA defined as follows:

$$\gamma[\,] := [\,]$$
$$\gamma(x/y :: A) := (\gamma A)x\#y\#$$

Lemma 25. $\sigma_\#(\gamma_1 A) = A^1 \# (\gamma A)$ *and* $\sigma_\#(\gamma_2 A) = A^2 \# (\gamma A)$.

Proof. By induction on A. □

Lemma 26. *Let $B \subseteq \gamma_i C$. Then there exists $A \subseteq C$ such that $\gamma_i A = B$.*

Proof. By induction on B using Fact 3.   □

Lemma 27. *Let $\#$ not occur in A_1 and A_2. Then $\gamma A_1 = \gamma A_2$ implies $A_1 = A_2$.*

Proof. By induction on A_1 using Fact 3.   □

Theorem 28. *PCP reduces to CFI.*

Proof. Let P be a list of cards. We fix a symbol $\#$ not occurring in P and define $R_1 := \gamma_1 P$ and $R_2 := \gamma_2 P$. We show $\mathsf{PCP}\,(P) \leftrightarrow \mathsf{CFI}\,(R_1, R_2, \#)$.

Let $A \subseteq P$ be a nonempty match. Then $\gamma_1 A \subseteq R_1$, $\gamma_2 A \subseteq R_2$, and $\sigma_\#(\gamma_1 A) = \sigma_\#(\gamma_2 A)$ by Lemma 25.

Let $B_1 \subseteq R_1$ and $B_2 \subseteq R_2$ be nonempty lists such that $\sigma_\# B_1 = \sigma_\# B_2$. By Lemma 26 there exist nonempty stacks $A_1, A_2 \subseteq P$ such that $\gamma_i(A_i) = B_i$. By Lemma 25 we have $A_1^1 \#(\gamma A_1) = A_2^2 \#(\gamma A_2)$. By Fact 3 we have $A_1^1 = A_1^2$ and $\gamma A_1 = \gamma A_2$. Thus $A_1 = A_2$ by Lemma 27. Hence $A_1 \subseteq P$ is a nonempty match. □

Hopcroft et al. [9] give a reduction of PCP to CFI by using grammars equivalent to the following Post grammars:

$$\gamma_1 A := [\,x/i \mid x/y \in A \text{ at position } i\,] \qquad \gamma_2 A := [\,y/i \mid x/y \in A \text{ at position } i\,]$$

While being in line with the presentation of PCP with indices, it complicates both the formal definition and the verification.

Hesselink [8] directly reduces CFP to CFI for general context-free grammars, making the reduction PCP to CFI redundant. The idea is that a context-free grammar over Σ contains a palindrome if and only if its intersection with the context-free grammar of all palindromes over Σ is non-empty. We give a formal proof of this statement using a definition of context-free rewriting with explicit alphabets.

For Post grammars, CFP is not reducible to CFI, because the language of all palindromes is not expressible by a Post grammar.

8   TM to SRH

A Turing machine, independent from its concrete type-theoretic definition, always consists of an alphabet Σ, a finite collection of states Q, an initial state q_0, a collection of halting states $H \subseteq Q$, and a step function which controls the behaviour of the head on the tape. The halting problem for Turing machines TM then asks whether a Turing machine M reaches a final state when executed on a tape containing a string x.

In this section, we briefly report on our formalisation of a reduction from TM to SRH following ideas from Hopcroft et al. [9]. In contrast to the other sections, we omit the technical details of the proof, because there are abundantly many, and none of them is interesting from a mathematical standpoint. We refer the interested reader to [7] for all details.

In the development, we use a formal definition of Turing machines from Asperti and Ricciotti [1].

To reduce TM to SRH, a representation of configurations c of Turing machines as strings $\langle c \rangle$ is needed. Although the content of a tape can get arbitrarily big over the run of a machine, it is finite in every single configuration. It thus suffices to represent only the part of the tape that the machine has previously written to.

We write the current state to the left of the currently read symbol and, following [1], distinguish four non-overlapping situations: The tape is empty $(q(|))$, the tape contains symbols and the head reads one of them $((|xqay|))$, the tape contains symbols and the head reads none of them, because it is in a left-overflow position where no symbol has been written before $(q(|ax|))$ or the right-overflow counterpart of the latter situation $((|xaq|))$. Note the usage of left and right markers to indicate the end of the previously written part.

The reduction from TM to SRH now works in three steps. Given a Turing machine M, one can define whether a configuration c' is reachable from a configuration c using its transition function [1,7]. First, we translate the transition function of the Turing machine into a string rewriting system using the translation scheme depicted in Table 1.

Table 1. Rewriting rules x/y in R if the machine according to its transition function in state q_1 continues in q_2 and reads, writes and moves as indicated. For example, if the transition function of the machine indicates that in state q_1 if symbol a is read, the machine proceeds to state q_2, writes nothing and moves to the left, we add the rule $(|q_1 a \,/\, q_2 (|a$ and rules $cq_1 a \,/\, q_2 ca$ for every c in the alphabet.

Read	Write	Move	x	y	x	y	x	y
$\perp$	$\perp$	L	$q_1 ($	$q_2 ($	$a\,q_1)$	$q_2\,a)$		
$\perp$	$\perp$	N	$q_1 ($	$q_2 ($	$q_1)$	$q_2)$		
$\perp$	$\perp$	R	$q_1 (()$	$q_2 (()$	$q_1)$	$q_2)$	$q_1 (a$	$(q_1\,a$
$\perp$	$\lfloor b \rfloor$	L	$q_1 ($	$q_2 (b$	$a\,q_1)$	$q_2\,a\,b)$		
$\perp$	$\lfloor b \rfloor$	N	$q_1 ($	$(q_2\,b$	$q_1)$	$q_2\,b)$		
$\perp$	$\lfloor b \rfloor$	R	$q_1 ($	$(b\,q_2$	$q_1)$	$b\,q_2)$		
$\lfloor a \rfloor$	$\perp$	L	$(q_1\,a$	$q_2 (a$	$c\,q_1\,a$	$q_2\,c\,a$		
$\lfloor a \rfloor$	$\perp$	N	$q_1\,a$	$q_2\,a$				
$\lfloor a \rfloor$	$\perp$	R	$q_1\,a$	$a\,q_2$				
$\lfloor a \rfloor$	$\lfloor b \rfloor$	L	$(q_1\,a$	$q_2 (b$	$c\,q_1\,a$	$q_2\,c\,b$		
$\lfloor a \rfloor$	$\lfloor b \rfloor$	N	$q_1\,a$	$q_2\,b$				
$\lfloor a \rfloor$	$\lfloor b \rfloor$	R	$q_1\,a$	$b\,q_2$				

Lemma 29. *For all Turing machines M and configurations c and c' there is a SRS R such that $\langle c \rangle \succ_R^* \langle c' \rangle$ if and only if the configuration c' is reachable from the configuration c by the machine M.*

In the development, we first reduce to a version of string rewriting with explicit alphabets, and then reduce this version to string rewriting as defined before.

This proof is by far the longest in our development. In its essence, it is only a shift of representation, making explicit that transition functions encode a rewriting relation on configurations. The proof is mainly a big case distinction over all possible shapes of configurations of a machine, which leads to a combinatorial explosion and a vast amount of subcases. The proof does, however, not contain any surprises or insights.

Note that, although we work with deterministic machines in the Coq development, the translation scheme described in Table 1 also works for nondeterministic Turing machines.

The second step of the reduction is to incorporate the set of halting states H. We define an intermediate problem SRH', generalising the definition of SRH to strings:

$$\mathsf{SRH}'(R, x, z) := \exists y.\ x \succ_R^* y \wedge \exists a \in z.\ a \in y$$

Note that $\mathsf{SRH}(R, x, a) \leftrightarrow \mathsf{SRH}'(R, x, [a])$. TM can then easily be reduced to SRH':

Lemma 30. TM *reduces to* SRH'.

Proof. Given a Turing machine M and a string x, M accepts x if and only if $\mathsf{SRH}(R, q_0 (\!| x |\!), z)$, where R is the system from the last lemma, q_0 is the starting state of M and z is a string containing exactly all halting states of M.   □

Third, we can reduce SRH' to SRH:

Lemma 31. SRH' *reduces to* SRH.

Proof. Given a SRS R, a string x and a string z, we first fix an alphabet Σ covering R and x, and a fresh symbol $\#$. We then have $\mathsf{SRH}'(R, x, z)$ if and only if $\mathsf{SRH}(R +\!\!+ [a/\# \mid a \in z], x, \#)$.   □

All three steps combined yield:

Theorem 32. TM *reduces to* SRH.

9   Discussion

We have formalised and verified a number of computational reductions to and from the Post correspondence problem based on Coq's type theory. Our goal was to come up with a development as elegant as possible. Realising the design presented in this paper in Coq yields an interesting exercise practising the verification of list-processing functions. If the intermediate lemmas are hidden and just the reductions and accompanying correctness statements are given, the exercise gains difficulty since the correctness proofs for the reductions $\mathsf{SR} \preceq \mathsf{MPCP} \preceq \mathsf{PCP}$ require the invention of general enough inductive invariants (Lemmas 11, 12,

17, 18). To our surprise, we could not find rigorous correctness proofs for the reductions TM $\preceq$ SR $\preceq$ MPCP $\preceq$ PCP in the literature (e.g., [3,9,15]). Teaching these reductions without rigorous correctness proofs in theoretical computer science classes seems bad practice. As the paper shows, elegant and rigorous correctness proofs using techniques generally applicable in program verification are available.

The ideas for the reductions TM $\preceq$ SRH $\preceq$ SR $\preceq$ MPCP $\preceq$ PCP are taken from Hopcroft et al. [9]. They give a monolithic reduction of the halting problem for Turing machines to MPCP. The decomposition TM $\preceq$ SRH $\preceq$ SR $\preceq$ MPCP is novel. Davis et al. [3] give a monolithic reduction SR $\preceq$ PCP based on different ideas. The idea for the reduction PCP $\preceq$ CFP is from Hesselink [8], and the idea for the reduction PCP $\preceq$ CFI appears in Hopcroft et al. [9].

There are several design choices we faced when formalising the material presented in this paper.

1. We decided to formalise PCP without making use of the positions of the cards in the list P. Most presentations in the literature (e.g., [9,15]) follow Post's original paper [13] in using positions (i.e., indices) rather than cards in matches. An exception is Davis et al. [3]. We think formulating PCP with positions is an unnecessary complication.
2. We decided to represent symbols as numbers rather than elements of finite types serving as alphabets. Working with implicit alphabets represented as lists rather than explicit alphabets represented as finite types saves bureaucracy.
3. We decided to work with Post grammars (inspired by Hesselink [8]) rather than general context-free grammars since Post grammars sharpen the result and enjoy a particularly simple formalisation. In the Coq development, we show that Post grammars are an instance of context-free grammars.

Furthermore, we decided to put the focus of this paper on the elegant reductions and not to cover Turing machines in detail. While being a wide-spread model of computation, even the concrete formal definition of Turing machines contains dozens of details, all of them not interesting from a mathematical perspective.

The Coq development verifying the results of Sects. 3, 4, 5, 6 and 7 consists of about 850 lines of which about one third realises specifications. The reduction SR $\preceq$ SRH takes 70 lines, SR $\preceq$ MPCP takes 105 lines, MPCP $\preceq$ PCP takes 206 lines, PCP $\preceq$ CFP takes 60 lines, and PCP $\preceq$ CFI takes 107 lines. The reduction TM $\preceq$ SRH takes 610 lines, 230 of them specification, plus a definition of Turing machines taking 291 lines.

Future Work

Undecidability proofs for logics are often done by reductions from PCP or related tiling problems. We thus want to use our work as a stepping stone to build a library of reductions which can be used to verify more undecidability proofs. We want to reduce PCP to the halting problem of Minsky machines to prove the undecidability of intuitionistic linear logic [11]. Another possible step would be to reduce PCP to validity for first-order logic [2], following the reduction from e.g. [12]. Many other undecidability proofs are also done by direct reductions from PCP, like the intersection problem for two-way-automata [14], unification in third-order logic [10], typability in the $\lambda \Pi$-calculus [4], satisfiability for more applied logics like HyperLTL [5], or decision problems of first order theories [17].

In this paper, we gave reductions directly as functions in Coq instead of appealing to a concrete model of computation. Writing down concrete Turing machines computing the reductions is possible in principle, but would be very tedious and distract from the elegant arguments our proofs are based on.

In previous work [6] we studied an explicit model of computation based on a weak call-by-value calculus L in Coq. L would allow an implementation of all reduction functions without much overhead, which would also formally establish the computability of all reductions.

Moreover, it should be straightforward to reduce PCP to the termination problem for L. Reducing the termination problem of L to TM would take considerable effort. Together, the two reductions would close the loop and verify the computational equivalence of TM, SRH, SR, PCP, and the termination problem for L. Both reducing PCP to L and implementing all reductions in L is an exercise in the verification of deeply embedded functional programs, and orthogonal in the necessary methods to the work presented in this paper.

References

1. Asperti, A., Ricciotti, W.: A formalization of multi-tape turing machines. Theor. Comput. Sc. **603**, 23–42 (2015)
2. Church, A.: A note on the Entscheidungsproblem. J. Symb. Log. **1**(1), 40–41 (1936)
3. Davis, M.D., Sigal, R., Weyuker, E.J.: Computability, Complexity, and Languages: Fundamentals of Theoretical Computer Science, 2nd edn. Academic Press, San Diego (1994)
4. Dowek, G.: The undecidability of typability in the Lambda-Pi-calculus. In: Bezem, M., Groote, J.F. (eds.) TLCA 1993. LNCS, vol. 664, pp. 139–145. Springer, Heidelberg (1993). https://doi.org/10.1007/BFb0037103
5. Finkbeiner, B., Hahn, C.: Deciding hyperproperties. In: CONCUR 2016, pp. 13:1–13:14 (2016)
6. Forster, Y., Smolka, G.: Weak call-by-value lambda calculus as a model of computation in Coq. In: Ayala-Rincón, M., Muñoz, C.A. (eds.) ITP 2017. LNCS, vol. 10499, pp. 189–206. Springer, Cham (2017). https://doi.org/10.1007/978-3-319-66107-0_13
7. Heiter, E.: Undecidability of the Post correspondence problem in Coq. Bachelor's Thesis, Saarland University (2017). https://www.ps.uni-saarland.de/~heiter/bachelor.php
8. Hesselink, W.H.: Post's correspondence problem and the undecidability of context-free intersection. Manuscript, University of Groningen (2015). http://wimhesselink.nl/pub/whh513.pdf
9. Hopcroft, J.E., Motwani, R., Ullman, J.D.: Introduction to Automata Theory, Languages, and Computation, 3rd edn. Addison-Wesley, Boston (2006)

10. Huet, G.P.: The undecidability of unification in third order logic. Inf. Control **22**(3), 257–267 (1973)
11. Larchey-Wendling, D., Galmiche, D.: The undecidability of Boolean BI through phase semantics. In: LICS 2010, pp. 140–149. IEEE (2010)
12. Manna, Z.: Mathematical theory of computation. Dover Publications Incorporated, Mineola (2003)
13. Post, E.L.: A variant of a recursively unsolvable problem. Bull. Am. Math. Soc. **52**(4), 264–268 (1946)
14. Rabin, M.O., Scott, D.: Finite automata and their decision problems. IBM J. Res. Dev. **3**(2), 114–125 (1959)
15. Sipser, M.: Introduction to the Theory of Computation. Cengage Learning, Boston (2012). International edition
16. The Coq Proof Assistant (2017). http://coq.inria.fr
17. Treinen, R.: A new method for undecidability proofs of first order theories. J. Symbolic Comput. **14**(5), 437–457 (1992)

ProofWatch: Watchlist Guidance
for Large Theories in E

Zarathustra Goertzel[1]($\boxtimes$), Jan Jakubův[1], Stephan Schulz[2], and Josef Urban[1]

[1] Czech Technical University in Prague, Prague, Czech Republic
goertzar@fel.cvut.cz
[2] DHBW Stuttgart, Stuttgart, Germany

Abstract. Watchlist (also hint list) is a mechanism that allows related proofs to guide a proof search for a new conjecture. This mechanism has been used with the Otter and Prover9 theorem provers, both for interactive formalizations and for human-assisted proving of open conjectures in small theories. In this work we explore the use of watchlists in large theories coming from first-order translations of large ITP libraries, aiming at improving hammer-style automation by smarter internal guidance of the ATP systems. In particular, we (i) design watchlist-based clause evaluation heuristics inside the E ATP system, and (ii) develop new proof guiding algorithms that load many previous proofs inside the ATP and focus the proof search using a dynamically updated notion of proof matching. The methods are evaluated on a large set of problems coming from the Mizar library, showing significant improvement of E's standard portfolio of strategies, and also of the previous best set of strategies invented for Mizar by evolutionary methods.

1 Introduction: Hammers, Learning and Watchlists

Hammer-style automation tools connecting interactive theorem provers (ITPs) with automated theorem provers (ATPs) have recently led to a significant speedup for formalization tasks [5]. An important component of such tools is *premise selection* [1]: choosing a small number of the most relevant facts that are given to the ATPs. Premise selection methods based on machine learning from many proofs available in the ITP libraries typically outperform manually specified heuristics [1,2,4,7,17,19]. Given the performance of such *ATP-external guidance* methods, learning-based *internal proof search guidance* methods have started to be explored, both for ATPs [8,15,18,23,36] and also in the context of tactical ITPs [10,12].

In this work we develop learning-based internal proof guidance methods for the E [30] ATP system and evaluate them on the large Mizar Mathematical Library [11]. The methods are based on the *watchlist* (also *hint list*) technique

---

Z. Goertzel, J. Jakubův and J. Urban—Supported by the *AI4REASON* ERC Consolidator grant number 649043, and by the Czech project AI&Reasoning CZ.02.1.01/0.0/0.0/15_003/0000466 and the European Regional Development Fund.

J. Avigad and A. Mahboubi (Eds.): ITP 2018, LNCS 10895, pp. 270–288, 2018.
https://doi.org/10.1007/978-3-319-94821-8_16

developed by Veroff [37], focusing proof search towards lemmas (*hints*) that were useful in related proofs. Watchlists have proved essential in the AIM project [21] done with Prover9 [25] for obtaining very long and advanced proofs of open conjectures. Problems in large ITP libraries however differ from one another much more than the AIM problems, making it more likely for unrelated watchlist lemmas to mislead the proof search. Also, Prover9 lacks a number of large-theory mechanisms and strategies developed recently for E [13,15,16].

Therefore, we first design watchlist-based clause evaluation heuristics for E that can be combined with other E strategies. Second, we complement the internal watchlist guidance by using external statistical machine learning to preselect smaller numbers of watchlist clauses relevant for the current problem. Finally, we use the watchlist mechanism to develop new proof guiding algorithms that load many previous proofs inside the ATP and focus the search using a *dynamically* updated heuristic representation of *proof search state* based on matching the previous proofs.

The rest of the paper is structured as follows. Section 2 briefly summarizes the work of saturation-style ATPs such as E. Section 3 discusses heuristic representation of search state and its importance for learning-based proof guidance. We propose an abstract vectorial representation expressing similarity to other proofs as a suitable evolving characterization of saturation proof searches. We also propose a concrete implementation based on *proof completion ratios* tracked by the watchlist mechanism. Section 4 describes the standard (*static*) watchlist mechanism implemented in E and Sect. 5 introduces the new *dynamic* watchlist mechanisms and its use for guiding the proof search. Section 6 evaluates the static and dynamic watchlist guidance combined with learning-based pre-selection on the Mizar library. Section 7 shows several examples of nontrivial proofs obtained by the new methods, and Sect. 8 discusses related work and possible extensions.

2   Proof Search in Saturating First-Order Provers

The state of the art in first-order theorem proving is a saturating prover based on a combination of resolution/paramodulation and rewriting, usually implementing a variant of the superposition calculus [3]. In this model, the *proof state* is represented as a set of first-order clauses (created from the axioms and the negated conjecture), and the system systematically adds logical consequences to the state, trying to derive the empty clause and hence an explicit contradiction.

All current saturating first-order provers are based on variants of the *given-clause algorithm*. In this algorithm, the proof state is split into two subsets of clauses, the processed clauses P (initially empty) and the unprocessed clauses U. On each iteration of the algorithm, the prover picks one unprocessed clause g (the so-called *given clause*), performs all inferences which are possible with g and all clauses in P as premises, and then moves g into P. The newly generated consequences are added to U. This maintains the core invariant that all inferences between clauses in P have been performed. Provers differ in how they integrate simplification and redundancy into the system, but all enforce the variant that

P is maximally simplified (by first simplifying g with clauses in P, then back-simplifying P with g) and that P contains neither tautologies nor subsumed clauses.

The core choice point of the given-clause algorithm is the selection of the next clause to process. If theoretical completeness is desired, this has to be *fair*, in the sense that no clause is delayed forever. In practice, clauses are ranked using one or more heuristic evaluation functions, and are picked in order of increasing evaluation (i.e. small values are good). The most frequent heuristics are based on symbol counting, i.e., the evaluation is the number of symbol occurrences in the clause, possibly weighted for different symbols or symbols types. Most provers also support interleaving a symbol-counting heuristic with a first-in-first-out (FIFO) heuristic. E supports the dynamic specification of an arbitrary number of differently parameterized priority queues that are processed in weighted round-robbin fashion via a small *domain-specific language* for heuristics.

Previous work [28,31] has both shown that the choice of given clauses is critical for the success rate of a prover, but also that existing heuristics are still quite bad - i.e. they select a large majority of clauses not useful for a given proof. Positively formulated, there still is a huge potential for improvement.

3   Proof Search State in Learning Based Guidance

A good representation of the current *state* is crucial for learning-based guidance. This is quite clear in theorem proving and famously so in Go and Chess [32,33]. For example, in the TacticToe system [10] proofs are composed from pre-programmed HOL4 [34] tactics that are chosen by statistical learning based on similarity of the evolving *goal state* to the goal states from related proofs. Similarly, in the learning versions of leanCoP [26] – (FE)MaLeCoP [18,36] – the tableau extension steps are guided by a trained learner using similarity of the evolving tableau (the ATP *proof search state*) to many other tableaux from related proofs.

Such intuitive and compact notion of proof search state is however hard to get when working with today's high-performance saturation-style ATPs such as E [30] and Vampire [22]. The above definition of saturation-style proof state (Sect. 2) as either one or two (processed/unprocessed) large sets of clauses is very unfocused. Existing learning-based guiding methods for E [15,23] practically ignore this. Instead, they use only the original conjecture and its features for selecting the relevant given clauses throughout the whole proof search.

This is obviously unsatisfactory, both when compared to the evolving search state in the case of tableau and tactical proving, and also when compared to the way humans select the next steps when they search for proofs. The proof search state in our mind is certainly an evolving concept based on the search done so far, not a fixed set of features extracted just from the conjecture.

3.1 Proof Search State Representation for Guiding Saturation

One of the motivations for the work presented here is to produce an intuitive, compact and evolving heuristic representation of proof search state in the context of learning-guided saturation proving. As usual, it should be a vector of (real-valued) features that are either manually designed or learned. In a high-level way, our proposed representation is a *vector expressing an abstract similarity of the search state to (possibly many) previous related proofs*. This can be implemented in different ways, using both statistical and symbolic methods and their combinations. An example and motivation comes again from the work of Veroff, where a search is considered promising when the given clauses frequently match hints. The gaps between the hint matchings may correspond to the more brute-force bridges between the different proof ideas expressed by the hints.

Our first practical implementation introduced in Sect. 5 is to load upon the search initialization N related proofs P_i, and for each P_i keep track of the ratio of the clauses from P_i that have already been subsumed during the search. The subsumption checking is using E's watchlist mechanism (Sect. 4). The N-long vector p of such *proof completion ratios* is our heuristic representation of the proof search state, which is both compact and typically evolving, making it suitable for both hard-coded and learned clause selection heuristics.

In this work we start with fast hard-coded watchlist-style heuristics for focusing inferences on clauses that progress the more finished proofs (Sect. 5). However training e.g. a statistical ENIGMA-style [15] clause evaluation model by adding p to the currently used ENIGMA features is a straightforward extension.

4 Static Watchlist Guidance and Its Implementation in E

E originally implemented a watchlist mechanism as a means to force direct, constructive proofs in first order logic. For this application, the watchlist contains a number of goal clauses (corresponding to the hypotheses to be proven), and all newly generated and processed clauses are checked against the watchlist. If one of the watchlist clauses is subsumed by a new clause, the former is removed from the watchlist. The proof search is complete, once all clauses from the watchlist have been removed. In contrast to the normal proof by contradiction, this mechanism is not complete. However, it is surprisingly effective in practice, and it produces a proof by forward reasoning.

It was quickly noted that the basic mechanism of the watchlist can also be used to implement a mechanism similar to the *hints* successfully used to guide Otter [24] (and its successor Prover9 [25]) in a semi-interactive manner [37]. Hints in this sense are intermediate results or lemmas expected to be useful in a proof. However, they are not provided as part of the logical premises, but have to be derived during the proof search. While the hints are specified when the prover is started, they are only used to guide the proof search - if a clause matches a hint, it is prioritized for processing. If all clauses needed for a proof are provided as hints, in theory the prover can be guided to prove a theorem without any search, i.e. it can *replay* a previous proof. A more general idea, explored in this paper, is to fill the watchlist with a large number of clauses useful in proofs of similar problems.

In E, the watchlist is loaded on start-up, and is stored in a feature vector index [29] that allows for efficient retrieval of subsumed (and subsuming) clauses. By default, watchlist clauses are simplified in the same way as processed clauses, i.e. they are kept in normal form with respect to clauses in P. This increases the chance that a new clause (which is always simplified) can match a similar watchlist clause. If used to control the proof search, subsumed clauses can optionally remain on the watchlist.

We have extended E's domain-specific language for search heuristics with two priority functions to access information about the relationship of clauses to the watchlist - the function `PreferWatchlist` gives higher rank to clauses that subsume at least one watchlist clause, and the dual function `DeferWatchlist` ranks them lower. Using the first, we have also defined four built-in heuristics that preferably process watchlist clauses. These include a pure watchlist heuristic, a simple interleaved watch list function (picking 10 out of every eleven clauses from the watchlist, the last using FIFO), and a modification of a strong heuristic obtained from a genetic algorithm [27] that interleaves several different evaluation schemes and was modified to prefer watchlist clauses in two of its four sub-evaluation functions.

5   Dynamic Watchlist Guidance

In addition to the above mentioned *static watchlist guidance*, we propose and experiment with an alternative: *dynamic watchlist guidance*. With dynamic watchlist guidance, several watchlists, as opposed to a single watchlist, are loaded on start-up. Separate watchlists are supposed to group clauses which are more likely to appear together in a single proof. The easiest way to produce watchlists with this property is to collect previously proved problems and use their proofs as watchlists. This is our current implementation, i.e., each watchlist corresponds to a previous proof. During a proof search, we maintain for each watchlist its *completion status*, i.e. the number of clauses that were already encountered. The main idea behind our dynamic watchlist guidance is to prefer clauses which appear on watchlists that are closer to completion. Since watchlists now exactly correspond to previous refutational proofs, completion of any watchlist implies that the current proof search is finished.

5.1   Watchlist Proof Progress

Let watchlists $W_1, \ldots, W_n$ be given for a proof search. For each watchlist W_i we keep a *watchlist progress counter*, denoted $progress(W_i)$, which is initially set to 0. Whenever a clause C is generated during the proof search, we have to check whether C subsumes some clause from some watchlist W_i. When C subsumes a clause from W_i we increase $progress(W_i)$ by 1. The subsumed clause from

W_i is then marked as encountered, and it is not considered in future watchlist subsumption checks.[1] Note that a single generated clause C can subsume several clauses from one or more watchlists, hence several progress counters might be increased multiple times as a result of generating C.

5.2 Standard Dynamic Watchlist Relevance

The easiest way to use progress counters to guide given clause selection is to assign the *(standard) dynamic watchlist relevance* to each generated clause C, denoted $relevance_0(C)$, as follows. Whenever C is generated, we check it against all the watchlists for subsumption and we update watchlist progress counters. Any clause C which does not subsume any watchlist clause is given $relevance_0(C) = 0$. When C subsumes some watchlist clause, its relevance is the maximum watchlist completion ratio over all the matched watchlists. Formally, let us write $C \sqsubseteq W_i$ when clause C subsumes some clause from watchlist W_i. For a clause C matching at least one watchlist, its relevance is computed as follows.

$$relevance_0(C) = \max_{W \in \{W_i : C \sqsubseteq W_i\}} \left(\frac{progress(W)}{|W|} \right)$$

The assumption is that a watchlist W that is matched more is more relevant to the current proof search. In our current implementation, the relevance is computed at the time of generation of C and it is not updated afterwards. As future work, we propose to also update the relevance of all generated but not yet processed clauses from time to time in order to reflect updates of the watchlist progress counters. Note that this is expensive, as the number of generated clauses is typically high. Suitable indexing could be used to lower this cost or even to do the update immediately just for the affected clauses.

To use the watchlist relevance in E, we extend E's domain-specific language for search heuristics with two priority functions `PreferWatchlistRelevant` and `DeferWatchlistRelevant`. The first priority function ranks higher the clauses with higher watchlist relevance[2], and the other function does the opposite. These priority functions can be used to build E's heuristics just like in the case of the static watchlist guidance. As a results, we can instruct E to process watchlist-relevant clauses in advance.

5.3 Inherited Dynamic Watchlist Relevance

The previous standard watchlist relevance prioritizes only clauses subsuming watchlist clauses but it behaves indifferently with respect to other clauses. In

---

[1] Alternatively, the subsumed watchlist clause $D \in W_i$ can be considered for future subsumption checks but the watchlist progress counter $progress(W_i)$ should not be increased when D is subsumed again. This is because we want the progress counter to represent the number of *different* clauses from W_i encountered so far.

[2] Technically, E's priority function returns an integer priority, and clauses with smaller values are preferred. Hence we compute the priority as $1000 * (1 - relevance_0(C))$.

order to provide some guidance even for clauses which do not subsume any watchlist clause, we can examine the watchlist relevance of the parents of each generated clause, and prioritize clauses with watchlist-relevant parents. Let $parents(C)$ denote the set of previously processed clauses from which C have been derived. *Inherited dynamic watchlist relevance*, denoted $relevance_1$, is a combination of the standard dynamic relevance with the average of parents relevances multiplied by a *decay* factor $\delta < 1$.

$$relevance_1(C) = relevance_0(C) + \delta * \operatorname*{avg}_{D \in parents(C)} \left(relevance_1(D) \right)$$

Clearly, the inherited relevance equals to the standard relevance for the initial clauses with no parents. The decay factor (δ) determines the importance of parents watchlist relevances.[3] Note that the inherited relevances of $parents(C)$ are already precomputed at the time of generating C, hence no recursive computation is necessary.

With the above $relevance_1$ we compute the average of parents *inherited* relevances, hence the inherited watchlist relevance accumulates relevance of all the ancestors. As a result, $relevance_1(C)$ is greater than 0 if and only if C has some ancestor which subsumed a watchlist clause at some point. This might have an undesirable effect that clauses unrelated to the watchlist are completely ignored during the proof search. In practice, however, it seems important to consider also watchlist-unrelated clauses with some degree in order to prove new conjectures which do not appear on the input watchlist. Hence we introduce two *threshold* parameters α and β which resets the relevance to 0 as follows. Let $length(C)$ denote the length of clause C, counting occurrences of symbols in C.

$$relevance_2(C) = \begin{cases} 0 & \text{iff } relevance_1(C) < \alpha \text{ and } \frac{relevance_1(C)}{length(C)} < \beta \\ relevance_1(C) & \text{otherwise} \end{cases}$$

Parameter α is a threshold on the watchlist inherited relevance while β combines the relevance with the clause length.[4] As a result, shorter watchlist-unrelated clauses are preferred to longer (distantly) watchlist-related clauses.

6 Experiments with Watchlist Guidance

For our experiments we construct watchlists from the proofs found by E on a benchmark of 57897 Mizar40 [19] problems in the MPTP dataset [35][5,6]. These

---

[3] In our experiments, we use $\delta = 0.1$.

[4] In our experiments, we use $\alpha = 0.03$ and $\beta = 0.009$. These values have been found useful by a small grid search over a random sample of 500 problems.

[5] Precisely, we have used the small (*bushy*, re-proving) versions, but without ATP minimization. They can be found at http://grid01.ciirc.cvut.cz/~mptp/7.13.01_4. 181.1147/MPTP2/problems_small_consist.tar.gz.

[6] Experimental results and code can be found at https://github.com/ai4reason/ eprover-data/tree/master/ITP-18.

initial proofs were found by an evolutionarily optimized [14] ensemble of 32 E strategies each run for 5 s. These are our *baseline* strategies. Due to limited computational resources, we do most of the experiments with the top 5 strategies that (greedily) cover most solutions (*top 5 greedy cover*). These are strategies number 2, 8, 9, 26 and 28, henceforth called *A, B, C, D, E*. In 5 s (in parallel) they together solve 21122 problems. We also evaluate these five strategies in 10 s, jointly solving 21670 problems. The 21122 proofs yield over 100000 unique proof clauses that can be used for watchlist-based guidance in our experiments. We also use smaller datasets randomly sampled from the full set of 57897 problems to be able to explore more methods. All problems are run on the same hardware[7] and with the same memory limits.

Each E strategy is specified as a frequency-weighted combination of parameterized *clause evaluation functions* (CEF) combined with a selection of inference rules. Below we show a simplified example strategy specifying the term ordering *KBO*, and combining (with weights 2 and 4) two CEFs made up of weight functions *Clauseweight* and *FIFOWeight* and priority functions *DeferSOS* and *PreferWatchlist*.

```
-tKBO -H(2*Clauseweight(DeferSoS,20,9999,4),4*FIFOWeight(PreferWatchlist))
```

6.1 Watchlist Selection Methods

We have experimented with several methods for creation of static and dynamic watchlists. Typically we use only the proofs found by a particular baseline strategy to construct the watchlists used for testing the guided version of that strategy. Using all 100000+ proof clauses as a watchlist slows E down to 6 given clauses per second. This is comparable to the speed of Prover9 with similarly large watchlists, but there are indexing methods that could speed this up. We have run several smaller tests, but do not include this method in the evaluation due to limited computational resources. Instead, we select a smaller set of clauses. The methods are as follows:

(art) Use all proof clauses from theorems in the problem's Mizar article[8]. Such watchlist sizes range from 0 to 4000, which does not cause any significant slowdown of E.

(freq) Use high-frequency proof clauses for static watchlists, i.e., clauses that appear in many proofs.

(kNN-st) Use k-nearest neighbor (k-NN) learning to suggest useful static watchlists for each problem, based on symbol and term-based features [20] of the conjecture. This is very similar to the standard use of k-NN and other learners for premise selection. In more detail, we use symbols, walks of length 2 on formula trees and common subterms (with variables and skolem symbols unified). Each proof is turned into a multi-label training example, where the labels are

---

[7] Intel(R) Xeon(R) CPU E5-2698 v3 @ 2.30 GHz with 256G RAM.
[8] Excluding the current theorem.

the (serially numbered) clauses used in the proof, and the features
are extracted from the conjecture.

(**kNN-dyn**) Use k-NN in a similar way to suggest the most related proofs for
dynamic watchlists. This is done in two iterations.

  (i) In the first iteration, only the conjecture-based similarity is used
  to select related problems and their proofs.

  (ii) The second iteration then uses data mined from the proofs
  obtained with dynamic guidance in the first iteration. From
  each such proof P we create a training example associating P's
  conjecture features with the names of the proofs that matched
  (i.e., guided the inference of) the clauses needed in P. On this
  dataset we again train a k-NN learner, which recommends the
  most useful related proofs for guiding a particular conjecture.

6.2 Using Watchlists in E Strategies

As described in Sect. 4, watchlist subsumption defines the `PreferWatchlist`
priority function that prioritizes clauses that subsume at least one watchlist
clause. Below we describe several ways to use this priority function and the newly
defined dynamic `PreferWatchlistRelevant` priority function and its relevance-
inheriting modifications. Each of them can additionally take the "no-remove"
option, to keep subsumed watchlist clauses in the watchlist, allowing repeated
matching by different clauses. Preliminary testing has shown that just adding a
single watchlist-based clause evaluation function (*CEF*) to the baseline CEFs[9]
is not as good as the methods defined below. In the rest of the paper we provide
short names for the methods, such as *prefA* (baseline strategy A modified by the
pref method described below).

1. *evo*: the default heuristic strategy (Sect. 4) evolved (genetically [27]) for static
   watchlist use.
2. *pref*: replace all priority functions in a baseline strategy with the `Prefer-`
   `Watchlist` priority function. The resulting strategies look as follows:
   ```
   -H(2*Clauseweight(PreferWatchlist,20,9999,4),
      4*FIFOWeight(PreferWatchlist))
   ```
3. *const*: replace all priority functions in a baseline strategy with `ConstPrio`,
   which assigns the same priority to all clauses, so all ranking is done by weight
   functions alone.
4. *uwl*: always prefer clauses that match the watchlist, but use the baseline
   strategy's priority function otherwise[10].
5. *ska*: modify watchlist subsumption in E to treat all skolem symbols of the
   same arity as equal, thus widening the watchlist guidance. This can be used
   with any strategy. In this paper it is used with *pref*.

---

[9] Specifically we tried adding Defaultweight(PreferWatchlist) and ConjectureRela-
tiveSymbolWeight(PreferWatchlist) with frequencies $1, 2, 5, 10, 20$ times that of the
rest of the CEFs in the strategy.

[10] *uwl* is implemented in E's source code as an option.

6. *dyn*: replace all priority functions in a baseline strategy with `PreferWatch-listRelevant`, which dynamically weights watchlist clauses (Sect. 5.2).
7. *dyndec*: add the relevance inheritance mechanisms to *dyn* (Sect. 5.3).

6.3   Evaluation

First we measure the slowdown caused by larger static watchlists on the best baseline strategy and a random sample of 10000 problems. The results are shown in Table 1. We see that the speed significantly degrades with watchlists of size 10000, while 500-big watchlists incur only a small performance penalty.

Table 1. Tests of the watchlist size influence (ordered by frequency) on a random sample of 10000 problems using the "no-remove" option and one static watchlist with strategy *prefA*. PPS is average processed clauses per second, a measure of E's speed.

Size	10	100	256	512	1000	10000
Proved	3275	3275	3287	3283	3248	2912
PPS	8935	9528	8661	7288	4807	575

Table 2 shows the 10 s evaluation of several static and dynamic methods on a random sample of 5000 problems using article-based watchlists (method **art** in Sect. 6.1). For comparison, E's *auto* strategy proves 1350 of the problems in 10 s and its *auto-schedule* proves 1629. Given 50 s the *auto-schedule* proves 1744 problems compared to our top 5 cover's 1964.

The first surprising result is that *const* significantly outperforms the *baseline*. This indicates that the old-style simple E priority functions may do more harm than good if they are allowed to override the more recent and sophisticated weight functions. The *ska* strategy performs best here and a variety of strategies provide better coverage. It's interesting to note that *ska* and *pref* overlap only on 1893 problems. The original *evo* strategy performs well, but lacks diversity.

Table 2. Article-based watchlist benchmark. A top 5 greedy cover proves 1964 problems (in bold).

Strategy	baseline	const	pref	ska	dyn	evo	uwl
A	1238	1493	1503	**1510**	1500	1303	1247
B	1255	1296	1315	1330	1316	1300	1277
C	1075	1166	**1205**	1183	1201	1068	1097
D	1102	1133	1176	**1190**	1175	**1330**	1132
E	1138	**1141**	1141	1153	1139	1070	1139
Total	1853	1910	1931	1933	1922	1659	1868

Table 3 briefly evaluates k-NN selection of watchlist clauses (method **kNN-st** in Sect. 6.1) on a single strategy *prefA*. Next we use k-NN to suggest watchlist proofs[11] (method **kNN-dyn.i**) for *pref* and *dyn*. Table 4 evaluates the influence of the number of related proofs loaded for the dynamic strategies. Interestingly, *pref* outperforms *dyn* almost everywhere but *dyn*'s ensemble of strategies A-E generally performs best and the top 5 cover is better. We conclude that *dyn*'s dynamic relevance weighting allows the strategies to diversify more.

Table 3. Evaluation of **kNN-st** on prefA

Watchlist size	16	64	256	1024	2048
Proved	1518	1531	1528	1532	1520

Table 5 evaluates the top 5 greedy cover from Table 4 on the full Mizar dataset, already showing significant improvement over the 21670 proofs produced by the 5 baseline strategies. Based on proof data from a full-run of the top-5 greedy cover in Table 5, new k-NN proof suggestions were made (method **kNN-dyn.ii**) and *dyn*'s grid search re-run, see Table 6 and Table 7 for k-NN round 2 results.

We also test the relevance inheriting dynamic watchlist feature (*dyndec*), primarily to determine if different proofs can be found. The results are shown in Table 8. This version adds 8 problems to the top 5 greedy cover of all the strategies run on the 5000 problem dataset, making it useful in a schedule despite lower performance alone. Table 9 shows this greedy cover, and then its evaluation on the full dataset. The 23192 problems proved by our new greedy cover is a 7% improvement over the top 5 baseline strategies.

7   Examples

The Mizar theorem `YELLOW_5:36`[12] states De Morgan's laws for Boolean lattices:

```
theorem Th36: :: YELLOW_5:36
for L being non empty Boolean RelStr for a, b being Element of L
holds ( 'not' (a "\/" b) = ('not' a) "/\" ('not' b)
        & 'not' (a "/\" b) = ('not' a) "\/" ('not' b) )
```

Using 32 related proofs results in 2220 clauses placed on the watchlists. The dynamically guided proof search takes 5218 (nontrivial) given clause loops done in 2 s and the resulting ATP proof is 436 inferences long. There are 194 given clauses that match the watchlist during the proof search and 120 (61.8%) of them end up being part of the proof. I.e., 27.5% of the proof consists of steps guided by the watchlist mechanism. The proof search using the same settings,

---

[11] All clauses in suggested proofs are used.
[12] http://grid01.ciirc.cvut.cz/~mptp/7.13.01_4.181.1147/html/yellow_5#T36.

Table 4. k-NN proof recommendation watchlists (**kNN-dyn.i**) for *dyn pref*. Size is number of proofs, averaging 40 clauses per proof. A top 5 greedy cover of *dyn* proves 1972 and *pref* proves 1959 (in bold).

Size	dynA	dynB	dynC	dynD	dynE	Total
4	1531	1352	1235	1194	**1165**	1957
8	1543	1366	**1253**	1188	1170	1956
16	1529	1357	1224	**1218**	1185	1951
32	**1546**	1373	1240	1218	1188	1962
64	1535	**1376**	1216	1215	1166	1935
128	1506	1351	1195	1214	1147	1907
1024	1108	963	710	943	765	1404
Size	prefA	prefB	prefC	prefD	prefE	Total
4	1539	1369	1210	1220	**1159**	1944
8	1554	1385	**1219**	1240	1168	1941
16	**1572**	1405	1225	1254	1180	1952
32	1568	1412	1231	**1271**	1190	1958
64	1567	1402	1228	1262	1172	1952
128	**1552**	1388	1210	1248	1160	1934
1024	1195	1061	791	991	806	1501

Table 5. K-NN round 1 greedy cover on full dataset and proofs added by each successive strategy for a total of 22579. dynA_32 means strategy *dynA* using 32 proof watchlists.

	dynA_32	dynC_8	dynD_16	dynE_4	dynB_64
Added	17964	2531	1024	760	282
Total	17964	14014	14294	13449	16175

Table 6. Problems proved by round 2 k-NN proof suggestions (**kNN-dyn.ii**). The top 5 greedy cover proves 1981 problems (in bold). *dyn2A* means *dynA* run on the 2nd iteration of k-NN suggestions.

Size	dyn2A	dyn2B	dyn2C	dyn2D	dyn2E	Total	Round 1 total
4	1539	**1368**	1235	1209	**1179**	1961	1957
8	1554	1376	1253	1217	1183	1971	1956
16	**1565**	1382	**1256**	1221	1181	1972	1951
32	1557	1383	1252	**1227**	1182	1968	1962
64	1545	1385	1244	1222	1171	1963	1935
128	1531	1374	1221	1227	1171	1941	1907

Table 7. K-NN round 2 greedy cover on full dataset and proofs added by each successive strategy for a total of 22996

	dyn2A_16	dyn2C_16	dyn2D_32	dyn2E_4	dyn2B_4
Total	18583	14486	14720	13532	16244
Added	18583	2553	1007	599	254

Table 8. Problems proved by round 2 k-NN proof suggestions with *dyndec*. The top 5 greedy cover proves 1898 problems (in bold).

Size	dyndec2A	dyndec2B	dyndec2C	dyndec2D	dyndec2E	Total
4	**1432**	1354	**1184**	**1203**	**1152**	1885
16	1384	1316	1176	1221	1140	1846
32	1381	1309	1157	1209	1133	1820
128	1326	**1295**	1127	1172	1082	1769

Table 9. Top: Cumulative sum of the 5000 test set greedy cover. The k-NN based dynamic watchlist methods dominate, improving by 2.1% over the baseline and article-based watchlist strategy greedy cover of 1964 (Table 2). Bottom: Greedy cover run on the full dataset, cumulative and total proved.

Total	dyn2A_16	dyn2C_16	dyndec2D_16	dyn2E_4	dyndec2A_128
2007	1565	230	97	68	47
23192	18583	2553	1050	584	422
23192	18583	14486	14514	13532	15916

but without the watchlist takes 6550 nontrivial given clause loops (25.5% more). The proof of the theorem WAYBEL_1:85[13] is considerably used for this guidance:

```
theorem :: WAYBEL_1:85
for H being non empty lower-bounded RelStr st H is Heyting holds
for a, b being Element of H holds 'not' (a "/\" b) >= ('not' a) "\/" ('not' b)
```

Note that this proof is done under the weaker assumptions of H being lower bounded and Heyting, rather than being Boolean. Yet, 62 (80.5%) of the 77 clauses from the proof of WAYBEL_1:85 are eventually matched during the proof search. 38 (49.4%) of these 77 clauses are used in the proof of YELLOW_5:36. In Table 10 we show the final state of proof progress for the 32 loaded proofs after the last non empty clause matched the watchlist. For each we show both the computed ratio and the number of matched and all clauses.

An example of a theorem that can be proved in 1.2 s with guidance but cannot be proved in 10 s with any unguided method is the following theorem BOOLEALG:62[14] about the symmetric difference in Boolean lattices:

---

[13] http://grid01.ciirc.cvut.cz/~mptp/7.13.01_4.181.1147/html/waybel_1#T85.
[14] http://grid01.ciirc.cvut.cz/~mptp/7.13.01_4.181.1147/html/boolealg#T62.

Table 10. Final state of the proof progress for the (serially numbered) 32 proofs loaded to guide the proof of `YELLOW_5:36`. We show the computed ratio and the number of matched and all clauses.

0	0.438	42/96	1	0.727	56/77	2	0.865	45/52	3	0.360	9/25
4	0.750	51/68	5	0.259	7/27	6	0.805	62/77	7	0.302	73/242
8	0.652	15/23	9	0.286	8/28	10	0.259	7/27	11	0.338	24/71
12	0.680	17/25	13	0.509	27/53	14	0.357	10/28	15	0.568	25/44
16	0.703	52/74	17	0.029	8/272	18	0.379	33/87	19	0.424	14/33
20	0.471	16/34	21	0.323	20/62	22	0.333	7/21	23	0.520	26/50
24	0.524	22/42	25	0.523	45/86	26	0.462	6/13	27	0.370	20/54
28	0.411	30/73	29	0.364	20/55	30	0.571	16/28	31	0.357	10/28

```
for L being B_Lattice
for X, Y being Element of L holds (X \+\ Y) \+\ (X "^" Y) = X "v" Y
```

Using 32 related proofs results in 2768 clauses placed on the watchlists. The proof search then takes 4748 (nontrivial) given clause loops and the watchlist-guided ATP proof is 633 inferences long. There are 613 given clauses that match the watchlist during the proof search and 266 (43.4%) of them end up being part of the proof. I.e., 42% of the proof consists of steps guided by the watchlist mechanism. Among the theorems whose proofs are most useful for the guidance are the following theorems `LATTICES:23`[15], `BOOLEALG:33`[16] and `BOOLEALG:54`[17] on Boolean lattices:

```
theorem Th23: :: LATTICES:23
for L being B_Lattice
for a, b being Element of L holds (a "^" b)‘ = a‘ "v" b‘

theorem Th33: :: BOOLEALG:33
for L being B_Lattice for X, Y being Element of L holds X \ (X "^" Y) = X \ Y

theorem :: BOOLEALG:54
for L being B_Lattice for X, Y being Element of L
st X‘ "v" Y‘ = X "v" Y & X misses X‘ & Y misses Y‘
holds  X = Y‘ & Y = X‘
```

Finally, we show several theorems[18,19,20,21] with nontrivial Mizar proofs and relatively long ATP proofs obtained with significant guidance. These theorems cannot be proved by any other method used in this work.

---

[15] http://grid01.ciirc.cvut.cz/~mptp/7.13.01_4.181.1147/html/lattices#T23.

[16] http://grid01.ciirc.cvut.cz/~mptp/7.13.01_4.181.1147/html/boolealg#T33.

[17] http://grid01.ciirc.cvut.cz/~mptp/7.13.01_4.181.1147/html/boolealg#T54.

[18] http://grid01.ciirc.cvut.cz/~mptp/7.13.01_4.181.1147/html/boolealg#T68.

[19] http://grid01.ciirc.cvut.cz/~mptp/7.13.01_4.181.1147/html/closure1#T21.

[20] http://grid01.ciirc.cvut.cz/~mptp/7.13.01_4.181.1147/html/bcialg_4#T44.

[21] http://grid01.ciirc.cvut.cz/~mptp/7.13.01_4.181.1147/html/xxreal_3#T67.

```
theorem :: BOOLEALG:68
for L being B_Lattice for X, Y being Element of L
holds (X \+\ Y)' = (X "∧" Y) "∨" ((X') "∧" (Y'))

theorem :: CLOSURE1:21
for I being set for M being ManySortedSet of I
for P, R being MSSetOp of M st P is monotonic & R is monotonic
holds P ** R is monotonic

theorem :: BCIALG_4:44
for X being commutative BCK-Algebra_with_Condition(S)
for a, b, c being Element of X st Condition_S (a,b) c= Initial_section c holds
for x being Element of Condition_S (a,b) holds x <= c \ ((c \ a) \ b)

theorem :: XXREAL_3:67
for f, g being ext-real number holds (f * g)"=(f") * (g")
```

8   Related Work and Possible Extensions

The closest related work is the hintguidance in Otter and Prover9. Our focus is however on large ITP-style theories with large signatures and heterogeneous facts and proofs spanning various areas of mathematics. This motivates using machine learning for reducing the size of the static watchlists and the implementation of the dynamic watchlist mechanisms. Several implementations of internal proof search guidance using statistical learning have been mentioned in Sects. 1 and 3. In both the tableau-based systems and the tactical ITP systems the statistical learning guidance benefits from a compact and directly usable notion of proof state, which is not immediately available in saturation-style ATP.

By delegating the notion of similarity to subsumption we are relying on fast, crisp and well-known symbolic ATP mechanisms. This has advantages as well as disadvantages. Compared to the ENIGMA [15] and neural [23] statistical guiding methods, the subsumption-based notion of clause similarity is not feature-based or learned. This similarity relation is crisp and sparser compared to the similarity relations induced by the statistical methods. The proof guidance is limited when no derived clauses subsume any of the loaded proof clauses. This can be countered by loading a high number of proofs and widening (or softening) the similarity relation in various approximate ways. On the other hand, subsumption is fast compared to the deep neural methods (see [23]) and enjoys clear guarantees of the underlying symbolic calculus. For example, when all the (non empty) clauses from a loaded related proof have been subsumed in the current proof search, it is clear that the current proof search is successfully finished.

A clear novelty is the focusing of the proof search towards the (possibly implausible) inferences needed for completing the loaded proofs. Existing statistical guiding methods will fail to notice such opportunities, and the static watchlist guidance has no way of distinguishing the watchlist matchers that lead faster to proof completion. In a way this mechanism resembles the feedback obtained by Monte Carlo exploration, where a seemingly statistically unlikely decision can be made, based on many rollouts and averaging of their results. Instead, we rely here on a database of previous proofs, similar to previously played and finished games. The newly introduced heuristic proof search (proof progress) representation may however enable further experiments with Monte Carlo guidance.

8.1 Possible Extensions

Several extensions have been already discussed above. We list the most obvious.

More Sophisticated Progress Metrics: The current proof-progress criterion may be too crude. Subsuming all the *initial* clauses of a related proof is unlikely until the empty clause is derived. In general, a large part of a related proof may not be needed once the right clauses in the "middle of the proof" are subsumed by the current proof search. A better proof-progress metric would compute the smallest number of proof clauses that are still needed to entail the contradiction. This is achievable, however more technically involved, also due to issues such as rewriting of the watchlist clauses during the current proof search.

Clause Re-evaluation Based on the Evolving Proof Relevance: As more and more watchlist clauses are matched, the proof relevance of the clauses generated earlier should be updated to mirror the current state. This is in general expensive, so it could be done after each N given clause loops or after a significant number of watchlist matchings. An alternative is to add corresponding indexing mechanisms to the set of generated clauses, which will immediately reorder them in the evaluation queues based on the proof relevance updates.

More Abstract/Approximate Matching: Instead of the strict notion of subsumption, more abstract or heuristic matching methods could be used. An interesting symbolic method to consider is matching modulo symbol alignments [9]. A number of approximate methods are already used by the above mentioned statistical guiding methods.

Adding Statistical Methods for Clause Guidance: Instead of using only hard-coded watchlist-style heuristics for focusing inferences, a statistical (e.g. ENIGMA-style) clause evaluation model could be trained by adding the vector of proof completion ratios to the currently used ENIGMA features.

9 Conclusion

The portfolio of new proof guiding methods developed here significantly improves E's standard portfolio of strategies, and also the previous best set of strategies invented for Mizar by evolutionary methods. The best combination of five new strategies run in parallel for 10 s (a reasonable hammering time) will prove over 7% more Mizar problems than the previous best combination of five non-watchlist strategies. Improvement over E's standard portfolio is much higher. Even though we focus on developing the strongest portfolio rather than a single best method, it is clear that the best guided versions also significantly improve over their non-guided counterparts. This improvement for the best new strategy (`dyn2A` used with 16 most relevant proofs) is 26.5% (=18583/14693). These are relatively high improvements in automated theorem proving.

We have shown that the new dynamic methods based on the idea of proof completion ratios improve over the static watchlist guidance. We have also shown that as usual with learning-based guidance, iterating the methods to produce more proofs leads to stronger methods in the next iteration. The first experiments with widening the watchlist-based guidance by relatively simple inheritance mechanisms seem quite promising, contributing many new proofs. A number of extensions and experiments with guiding saturation-style proving have been opened for future research. We believe that various extensions of the compact and evolving heuristic representation of saturation-style proof search as introduced here will turn out to be of great importance for further development of learning-based saturation provers.

Acknowledgments. We thank Bob Veroff for many enlightening explanations and discussions of the watchlist mechanisms in Otter and Prover9. His "industry-grade" projects that prove open and interesting mathematical conjectures with hints and proof sketches have been a great sort of inspiration for this work.

References

1. Alama, J., Heskes, T., Kühlwein, D., Tsivtsivadze, E., Urban, J.: Premise selection for mathematics by corpus analysis and kernel methods. J. Autom. Reasoning **52**(2), 191–213 (2014)
2. Alemi, A.A., Chollet, F., Eén, N., Irving, G., Szegedy, C., Urban, J.: DeepMath - deep sequence models for premise selection. In: Lee, D.D., Sugiyama, M., Luxburg, U.V., Guyon, I., Garnett, R. (eds.) Advances in Neural Information Processing Systems 29: Annual Conference on Neural Information Processing Systems 2016, 5–10 December 2016, Barcelona, Spain, pp. 2235–2243 (2016)
3. Bachmair, L., Ganzinger, H.: Rewrite-based equational theorem proving with selection and simplification. J. Logic Comput. **3**(4), 217–247 (1994)
4. Blanchette, J.C., Greenaway, D., Kaliszyk, C., Kühlwein, D., Urban, J.: A learning-based fact selector for Isabelle/HOL. J. Autom. Reasoning **57**(3), 219–244 (2016)
5. Blanchette, J.C., Kaliszyk, C., Paulson, L.C., Urban, J.: Hammering towards QED. J. Formalized Reasoning **9**(1), 101–148 (2016)
6. Eiter, T., Sands, D. (eds.): LPAR-21, 21st International Conference on Logic for Programming, Artificial Intelligence and Reasoning, Maun, Botswana, 7–12 May 2017, EPiC Series in Computing, vol. 46. EasyChair (2017)
7. Färber, M., Kaliszyk, C.: Random forests for premise selection. In: Lutz, C., Ranise, S. (eds.) FroCoS 2015. LNCS (LNAI), vol. 9322, pp. 325–340. Springer, Cham (2015). https://doi.org/10.1007/978-3-319-24246-0_20
8. Färber, M., Kaliszyk, C., Urban, J.: Monte carlo tableau proof search. In: de Moura, L. (ed.) CADE 2017. LNCS (LNAI), vol. 10395, pp. 563–579. Springer, Cham (2017). https://doi.org/10.1007/978-3-319-63046-5_34
9. Gauthier, T., Kaliszyk, C.: Matching concepts across HOL libraries. In: Watt, S.M., Davenport, J.H., Sexton, A.P., Sojka, P., Urban, J. (eds.) CICM 2014. LNCS (LNAI), vol. 8543, pp. 267–281. Springer, Cham (2014). https://doi.org/10.1007/978-3-319-08434-3_20

10. Gauthier, T., Kaliszyk, C., Urban, J.: TacticToe: learning to reason with HOL4 tactics. In: Eiter and Sands [6], pp. 125–143
11. Grabowski, A., Korniłowicz, A., Naumowicz, A.: Mizar in a nutshell. J. Formalized Reasoning 3(2), 153–245 (2010)
12. Gransden, T., Walkinshaw, N., Raman, R.: SEPIA: search for proofs using inferred automata. In: Felty, A.P., Middeldorp, A. (eds.) CADE 2015. LNCS (LNAI), vol. 9195, pp. 246–255. Springer, Cham (2015). https://doi.org/10.1007/978-3-319-21401-6_16
13. Jakubův, J., Urban, J.: Extending E prover with similarity based clause selection strategies. In: Kohlhase, M., Johansson, M., Miller, B., de de Moura, L., Tompa, F. (eds.) CICM 2016. LNCS (LNAI), vol. 9791, pp. 151–156. Springer, Cham (2016). https://doi.org/10.1007/978-3-319-42547-4_11
14. Jakubův, J., Urban, J.: BliStrTune: hierarchical invention of theorem proving strategies. In: Bertot, Y., Vafeiadis, V. (eds.) Proceedings of the 6th ACM SIGPLAN Conference on Certified Programs and Proofs, CPP 2017, Paris, France, 16–17 January 2017, pp. 43–52. ACM (2017)
15. Jakubův, J., Urban, J.: ENIGMA: efficient learning-based inference guiding machine. In: Geuvers, H., England, M., Hasan, O., Rabe, F., Teschke, O. (eds.) CICM 2017. LNCS (LNAI), vol. 10383, pp. 292–302. Springer, Cham (2017). https://doi.org/10.1007/978-3-319-62075-6_20
16. Kaliszyk, C., Schulz, S., Urban, J., Vyskočil, J.: System description: E.T. 0.1. In: Felty, A.P., Middeldorp, A. (eds.) CADE 2015. LNCS (LNAI), vol. 9195, pp. 389–398. Springer, Cham (2015). https://doi.org/10.1007/978-3-319-21401-6_27
17. Kaliszyk, C., Urban, J.: Learning-assisted automated reasoning with Flyspeck. J. Autom. Reasoning 53(2), 173–213 (2014)
18. Kaliszyk, C., Urban, J.: FEMaLeCoP: fairly efficient machine learning connection prover. In: Davis, M., Fehnker, A., McIver, A., Voronkov, A. (eds.) LPAR 2015. LNCS, vol. 9450, pp. 88–96. Springer, Heidelberg (2015). https://doi.org/10.1007/978-3-662-48899-7_7
19. Kaliszyk, C., Urban, J.: MizAR 40 for Mizar 40. J. Autom. Reasoning 55(3), 245–256 (2015)
20. Kaliszyk, C., Urban, J., Vyskočil, J.: Efficient semantic features for automated reasoning over large theories. In: Yang, Q., Wooldridge, M. (eds.) IJCAI 2015, pp. 3084–3090. AAAI Press (2015)
21. Kinyon, M., Veroff, R., Vojtěchovský, P.: Loops with abelian inner mapping groups: an application of automated deduction. In: Bonacina, M.P., Stickel, M.E. (eds.) Automated Reasoning and Mathematics. LNCS (LNAI), vol. 7788, pp. 151–164. Springer, Heidelberg (2013). https://doi.org/10.1007/978-3-642-36675-8_8
22. Kovács, L., Voronkov, A.: First-order theorem proving and VAMPIRE. In: Sharygina, N., Veith, H. (eds.) CAV 2013. LNCS, vol. 8044, pp. 1–35. Springer, Heidelberg (2013). https://doi.org/10.1007/978-3-642-39799-8_1
23. Loos, S.M., Irving, G., Szegedy, C., Kaliszyk, C.: Deep network guided proof search. In: Eiter and Sands [6], pp. 85–105
24. McCune, W., Wos, L.: Otter: the CADE-13 competition incarnations. J. Autom. Reasoning 18(2), 211–220 (1997). Special Issue on the CADE 13 ATP System Competition
25. McCune, W.W.: Prover9 and Mace4 (2005–2010). http://www.cs.unm.edu/~mccune/prover9/. Accessed 29 Mar 2016
26. Otten, J., Bibel, W.: leanCoP: lean connection-based theorem proving. J. Symb. Comput. 36(1–2), 139–161 (2003)

27. Schäfer, S., Schulz, S.: Breeding theorem proving heuristics with genetic algorithms. In: Gottlob, G., Sutcliffe, G., Voronkov, A. (eds.) Global Conference on Artificial Intelligence, GCAI 2015, Tbilisi, Georgia, 16–19 October 2015, EPiC Series in Computing, vol. 36, pp. 263–274. EasyChair (2015)

28. Schulz, S.: Learning search control knowledge for equational theorem proving. In: Baader, F., Brewka, G., Eiter, T. (eds.) KI 2001. LNCS (LNAI), vol. 2174, pp. 320–334. Springer, Heidelberg (2001). https://doi.org/10.1007/3-540-45422-5_23

29. Schulz, S.: Simple and efficient clause subsumption with feature vector indexing. In: Bonacina, M.P., Stickel, M.E. (eds.) Automated Reasoning and Mathematics. LNCS (LNAI), vol. 7788, pp. 45–67. Springer, Heidelberg (2013). https://doi.org/10.1007/978-3-642-36675-8_3

30. Schulz, S.: System description: E 1.8. In: McMillan, K., Middeldorp, A., Voronkov, A. (eds.) LPAR 2013. LNCS, vol. 8312, pp. 735–743. Springer, Heidelberg (2013). https://doi.org/10.1007/978-3-642-45221-5_49

31. Schulz, S., Möhrmann, M.: Performance of clause selection heuristics for saturation-based theorem proving. In: Olivetti, N., Tiwari, A. (eds.) IJCAR 2016. LNCS (LNAI), vol. 9706, pp. 330–345. Springer, Cham (2016). https://doi.org/10.1007/978-3-319-40229-1_23

32. Silver, D., Huang, A., Maddison, C.J., Guez, A., Sifre, L., van den Driessche, G., Schrittwieser, J., Antonoglou, I., Panneershelvam, V., Lanctot, M., Dieleman, S., Grewe, D., Nham, J., Kalchbrenner, N., Sutskever, I., Lillicrap, T.P., Leach, M., Kavukcuoglu, K., Graepel, T., Hassabis, D.: Mastering the game of go with deep neural networks and tree search. Nature **529**(7587), 484–489 (2016)

33. Silver, D., Hubert, T., Schrittwieser, J., Antonoglou, I., Lai, M., Guez, A., Lanctot, M., Sifre, L., Kumaran, D., Graepel, T., Lillicrap, T.P., Simonyan, K., Hassabis, D.: Mastering Chess and Shogi by self-play with a general reinforcement learning algorithm. CoRR, abs/1712.01815 (2017)

34. Slind, K., Norrish, M.: A brief overview of HOL4. In: Mohamed, O.A., Muñoz, C., Tahar, S. (eds.) TPHOLs 2008. LNCS, vol. 5170, pp. 28–32. Springer, Heidelberg (2008). https://doi.org/10.1007/978-3-540-71067-7_6

35. Urban, J.: MPTP 0.2: design, implementation, and initial experiments. J. Autom. Reasoning **37**(1–2), 21–43 (2006)

36. Urban, J., Vyskočil, J., Štěpánek, P.: MaLeCoP machine learning connection prover. In: Brünnler, K., Metcalfe, G. (eds.) TABLEAUX 2011. LNCS (LNAI), vol. 6793, pp. 263–277. Springer, Heidelberg (2011). https://doi.org/10.1007/978-3-642-22119-4_21

37. Veroff, R.: Using hints to increase the effectiveness of an automated reasoning program: case studies. J. Autom. Reasoning **16**(3), 223–239 (1996)

Reification by Parametricity
Fast Setup for Proof by Reflection, in Two Lines of Ltac

Jason Gross[✉], Andres Erbsen, and Adam Chlipala

MIT CSAIL, Cambridge, MA, USA
{jgross,andreser}@mit.edu, adamc@csail.mit.edu

Abstract. We present a new strategy for performing reification in Coq. That is, we show how to generate first-class abstract syntax trees from "native" terms of Coq's logic, suitable as inputs to verified compilers or procedures in the *proof-by-reflection* style. Our new strategy, based on simple generalization of subterms as variables, is straightforward, short, and fast. In its pure form, it is only complete for constants and function applications, but "let" binders, eliminators, lambdas, and quantifiers can be accommodated through lightweight coding conventions or preprocessing.

We survey the existing methods of reification across multiple Coq metaprogramming facilities, describing various design choices and tricks that can be used to speed them up, as well as various limitations. We report benchmarking results for 18 variants, in addition to our own, finding that our own reification outperforms 16 of these methods in all cases, and one additional method in some cases; writing an OCaml plugin is the only method tested to be faster. Our method is the most concise of the strategies we considered, reifying terms using only two to four lines of LTAC—beyond lists of the identifiers to reify and their reified variants. Additionally, our strategy automatically provides error messages that are no less helpful than Coq's own error messages.

1 Introduction

Proof by reflection [2] is an established method for employing verified proof procedures, within larger proofs. There are a number of benefits to using verified functional programs written in the proof assistant's logic, instead of tactic scripts. We can often prove that procedures always terminate without attempting fallacious proof steps, and perhaps we can even prove that a procedure gives logically complete answers, for instance telling us definitively whether a proposition is true or false. In contrast, tactic-based procedures may encounter runtime errors or loop forever. As a consequence, those procedures must output proof terms, justifying their decisions, and these terms can grow large, making for

Electronic supplementary material The online version of this chapter (https://doi.org/10.1007/978-3-319-94821-8_17) contains supplementary material, which is available to authorized users.

© Springer International Publishing AG, part of Springer Nature 2018
J. Avigad and A. Mahboubi (Eds.): ITP 2018, LNCS 10895, pp. 289–305, 2018.
https://doi.org/10.1007/978-3-319-94821-8_17

slower proving and requiring transmission of large proof terms to be checked slowly by others. A verified procedure need not generate a certificate for each invocation.

The starting point for proof by reflection is *reification*: translating a "native" term of the logic into an explicit abstract syntax tree. We may then feed that tree to verified procedures or any other functional programs in the logic. The benefits listed above are particularly appealing in domains where goals are very large. For instance, consider verification of large software systems, where we might want to reify thousands of lines of source code. Popular methods turn out to be surprisingly slow, often to the point where, counter-intuitively, the majority of proof-execution time is spent in reification – unless the proof engineer invests in writing a plugin directly in the proof assistant's metalanguage (e.g., OCaml for Coq).

In this paper, we show that reification can be both simpler and faster than with standard methods. Perhaps surprisingly, we demonstrate how to reify terms almost entirely through reduction in the logic, with a small amount of tactic code for setup and no ML programming. Though our techniques should be broadly applicable, especially in proof assistants based on type theory, our experience is with Coq, and we review the requisite background in the remainder of this introduction. In Sect. 2, we summarize our survey into prior approaches to reification and provide high-quality implementations and documentation for them, serving a tutorial function independent of our new contributions. Experts on the subject might want to skip directly to Sect. 3, which explains our alternative technique. We benchmark our approach against 18 competitors in Sect. 4.

1.1 Proof-Script Primer

Basic Coq proofs are often written as lists of steps such as `induction` on some structure, `rewrite` using a known equivalence, or `unfold` of a definition. Very quickly, proofs can become long and tedious, both to write and to read, and hence Coq provides LTAC, a scripting language for proofs. As theorems and proofs grow in complexity, users frequently run into performance and maintainability issues with LTAC. Consider the case where we want to prove that a large algebraic expression, involving many `let ... in ...` expressions, is even:

```
Inductive is_even : nat -> Prop :=
| even_O : is_even O
| even_SS : forall x, is_even x -> is_even (S (S x)).
Goal is_even (let x := 100 * 100 * 100 * 100 in
              let y := x * x * x * x in
              y * y * y * y).
```

Coq stack-overflows if we try to reduce this goal. As a workaround, we might write a lemma that talks about evenness of `let ... in ...`, plus one about evenness of multiplication, and we might then write a tactic that composes such lemmas.

Even on smaller terms, though, proof size can quickly become an issue. If we give a naive proof that 7000 is even, the proof term will contain all of the even numbers between 0 and 7000, giving a proof-term-size blow-up at least quadratic in size (recalling that natural numbers are represented in unary; the challenges remain for more efficient base encodings). Clever readers will notice that Coq could share subterms in the proof tree, recovering a term that is linear in the size of the goal. However, such sharing would have to be preserved very carefully, to prevent size blow-up from unexpected loss of sharing, and today's Coq version does not do that sharing. Even if it did, tactics that rely on assumptions about Coq's sharing strategy become harder to debug, rather than easier.

1.2 Reflective-Automation Primer

Enter reflective automation, which simultaneously solves both the problem of performance and the problem of debuggability. Proof terms, in a sense, are traces of a proof script. They provide Coq's kernel with a term that it can check to verify that no illegal steps were taken. Listing every step results in large traces.

The idea of reflective automation is that, if we can get a formal encoding of our goal, plus an algorithm to *check* the property we care about, then we can do much better than storing the entire trace of the program. We can prove that our checker is correct once and for all, removing the need to trace its steps.

A simple evenness checker can just operate on the unary encoding of natural numbers (Fig. 1). We can use its correctness theorem to prove goals much more quickly:

```
Fixpoint check_is_even
  (n : nat) : bool
:= match n with
   | 0 => true
   | 1 => false
   | S (S n)
       => check_is_even n
   end.
```

Fig. 1. Evenness checking

```
Theorem soundness : forall n, check_is_even n = true -> is_even n.
Goal is_even 2000.
  Time repeat (apply even_SS || apply even_0). (* 1.8 s *)
  Undo.
  Time apply soundness; vm_compute; reflexivity. (* 0.004 s *)
```

The tactic vm_compute tells Coq to use its virtual machine for reduction, to compute the value of check_is_even 2000, after which reflexivity proves that true = true. Note how much faster this method is. In fact, even the asymptotic complexity is better; this new algorithm is linear rather than quadratic in n.

However, even this procedure takes a bit over three minutes to prove is_even (10 * 10 * 10 * 10 * 10 * 10 * 10 * 10 * 10). To do better, we need a formal representation of terms or expressions.

1.3 Reflective-Syntax Primer

Sometimes, to achieve faster proofs, we must be able to tell, for example, whether we got a term by multiplication or by addition, and not merely whether its normal form is 0 or a successor.

A reflective automation procedure generally has two steps. The first step is to *reify* the goal into some abstract syntactic representation, which we call the *term language* or an *expression language*. The second step is to run the algorithm on the reified syntax.

```
Inductive expr :=
| Nat0 : expr
| NatS (x : expr) : expr
| NatMul (x y : expr) : expr.
```

Fig. 2. Simple expressions

What should our expression language include? At a bare minimum, we must have multiplication nodes, and we must have `nat` literals. If we encode S and O separately, a decision that will become important later in Sect. 3, we get the inductive type of Fig. 2.

Before diving into methods of reification, let us write the evenness checker.

```
Fixpoint check_is_even_expr (t : expr) : bool
  := match t with
     | Nat0 => true
     | NatS x => negb (check_is_even_expr x)
     | NatMul x y => orb (check_is_even_expr x) (check_is_even_expr y)
     end.
```

Before we can state the soundness theorem (whenever this checker returns `true`, the represented number is even), we must write the function that tells us what number our expression represents, called *denotation* or *interpretation*:

```
Fixpoint denote (t : expr) : nat
  := match t with
     | Nat0 => 0
     | NatS x => S (denote x)
     | NatMul x y => denote x * denote y
     end.

Theorem check_is_even_expr_sound (e : expr)
  : check_is_even_expr e = true -> is_even (denote e).
```

Given a tactic `Reify` to produce a reified term from a `nat`, we can time `check_is_even_expr`. It is instant on the last example.

Before we proceed to reification, we will introduce one more complexity. If we want to support our initial example with `let ... in ...` efficiently, we must also have `let`-expressions. Our current procedure that inlines `let`-expressions takes 19 s, for example, on `let x0 := 10 * 10 in let x1 := x0 * x0 in ... let x24 := x23 * x23 in x24`. The choices of representation include higher-order abstract syntax (HOAS) [11], parametric higher-order abstract syntax (PHOAS) [4], and de Bruijn indices [3]. The PHOAS representation is particularly convenient. In PHOAS, expression binders are represented by binders in Gallina, the functional language of Coq, and the expression language is parameterized over the type of the binder. Let us define a constant and notation for `let` expressions as definitions (a common choice in real Coq developments, to block Coq's default behavior of inlining `let` binders silently;

the same choice will also turn out to be useful for reification later). We thus have:

```
Inductive expr {var : Type} :=
| Nat0 : expr
| NatS : expr -> expr
| NatMul : expr -> expr -> expr
| Var : var -> expr
| LetIn : expr -> (var -> expr) -> expr.
Definition Let_In {A B} (v : A) (f : A -> B) := let x := v in f x.
Notation "'dlet' x := v 'in' f" := (Let_In v (fun x => f)).
Notation "'elet' x := v 'in' f" := (LetIn v (fun x => f)).
Fixpoint denote (t : @expr nat) : nat
  := match t with
     | Nat0 => 0
     | NatS x => S (denote x)
     | NatMul x y => denote x * denote y
     | Var v => v
     | LetIn v f => dlet x := denote v in denote (f x)
     end.
```

A full treatment of evenness checking for PHOAS would require proving well-formedness of syntactic expressions; for a more complete discussion of PHOAS, we refer the reader elsewhere [4]. Using Wf to denote the well-formedness predicate, we could prove a theorem

```
Theorem check_is_even_expr_sound (e : ∀ var, @expr var) (H : Wf e)
: check_is_even_expr (e bool) = true -> is_even (denote (e nat)).
```

To complete the picture, we would need a tactic Reify which took in a term of type nat and gave back a term of type forall var, @expr var, plus a tactic prove_wf which solved a goal of the form Wf e by repeated application of constructors. Given these, we could solve an evenness goal by writing[1]

```
match goal with
| [ |- is_even ?v ]
  => let e := Reify v in
     refine (check_is_even_expr_sound e _ _);
     [ prove_wf | vm_compute; reflexivity ]
end.
```

2 Methods of Reification

We implemented reification in 18 different ways, using 6 different metaprogramming facilities in the Coq ecosystem: Ltac, Ltac2, Mtac [8], type classes [12],

---

[1] Note that for the refine to be fast, we must issue something like Strategy -10 [denote] to tell Coq to unfold denote before Let_In.

canonical structures [7], and reification-specific OCaml plugins (quote [5], template-coq [1], ours). Figure 3 displays the simplest case: an Ltac script to reify a tree of function applications and constants. Unfortunately, all methods we surveyed become drastically more complicated or slower (and usually both) when adapted to reify terms with variable bindings such as let-in or λ nodes.

We have made detailed walk-throughs and source code of these implementations available[2] in hope that they will be useful for others considering implementing reification using one of these metaprogramming mechanisms, instructive as nontrivial examples of multiple metaprogramming facilities, or helpful as a case study in Coq performance engineering. However, we do *not* recommend reading these

```
Ltac f v x := (* reify var term *)
  lazymatch x with
  | O => constr:(@NatO v)
  | S ?x => let X := f v x in
            constr:(@NatS v X)
  | ?x*?y => let X := f v x in
             let Y := f v y in
             constr:(@NatMul v X Y)
  end.
```

Fig. 3. Reification without binders in Ltac

out of general interest: most of the complexity in the described implementations strikes us as needless, with significant aspects of the design being driven by surprising behaviors, misfeatures, bugs, and performance bottlenecks of the underlying machinery as opposed to the task of reification.

3   Reification by Parametricity

We propose factoring reification into two passes, both of which essentially have robust, built-in implementations in Coq: *abstraction* or *generalization*, and *substitution* or *specialization*.

The key insight to this factoring is that the shape of a reified term is essentially the same as the shape of the term that we start with. We can make precise the way these shapes are the same by abstracting over the parts that are different, obtaining a function that can be specialized to give either the original term or the reified term.

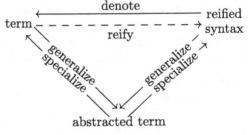

Fig. 4. Abstraction and reification

That is, we have the commutative triangle in Fig. 4.

3.1   Case-By-Case Walkthrough

Function Applications and Constants. Consider the example of reifying 2×2. In this case, the *term* is 2×2 or (mul (S (S O)) (S (S O))).

---

[2] https://github.com/mit-plv/reification-by-parametricity.

To reify, we first *generalize* or *abstract* the term 2×2 over the successor function S, the zero constructor O, the multiplication function mul, and the type $\mathbb{N}$ of natural numbers. We get a function taking one type argument and three value arguments:

$$\Lambda N.\ \lambda(\text{MUL} : N \to N \to N)\ (O : N)\ (S : N \to N).\ \text{MUL}\ (S\ (S\ O))\ (S\ (S\ O))$$

We can now specialize this term in one of two ways: we may substitute $\mathbb{N}$, mul, O, and S, to get back the term we started with; or we may substitute `expr`, `NatMul`, `NatO`, and `NatS` to get the reified syntax tree

`NatMul (NatS (NatS NatO)) (NatS (NatS NatO))`

This simple two-step process is the core of our algorithm for reification: abstract over all identifiers (and key parts of their types) and specialize to syntax-tree constructors for these identifiers.

Wrapped Primitives: "Let" Binders, Eliminators, Quantifiers. The above procedure can be applied to a term that contains "let" binders to get a PHOAS syntax tree that represents the original term, but doing so would not capture sharing. The result would contain native "let" bindings of subexpressions, not PHOAS let expressions. Call-by-value evaluation of any procedure applied to the reification result would first substitute the let-bound subexpressions – leading to potentially exponential blowup and, in practice, memory exhaustion.

The abstraction mechanisms in all proof assistants (that we know about) only allow abstracting over terms, not language primitives. However, primitives can often be wrapped in explicit definitions, which we *can* abstract over. For example, we already used a wrapper for "let" binders, and terms that use it can be reified by abstracting over that definition. If we start with the expression

$$\text{dlet } a := 1 \text{ in } a \times a$$

and abstract over (@Let_In $\mathbb{N}$ $\mathbb{N}$), S, O, mul, and $\mathbb{N}$, we get a function of one type argument and four value arguments:

$$\Lambda N.\ \lambda\ (\text{MUL} : N \to N \to N).\ \lambda(O : N).\ \lambda(S : N \to N).$$
$$\lambda(\text{LETIN} : N \to (N \to N) \to N).\ \text{LETIN}\ (S\ O)\ (\lambda a.\ \text{MUL}\ a\ a)$$

We may once again specialize this term to obtain either our original term or the reified syntax. Note that to obtain reified PHOAS syntax, we must include a `Var` node in the `LetIn` expression; we substitute $(\lambda x\ f.\ \text{LetIn}\ x\ (\lambda v.\ f\ (\text{Var}\ v)))$ for LETIN to obtain the PHOAS syntax tree

`LetIn (NatS NatO) (`λv`. NatMul (Var`v`) (Var`v`))`

Wrapping a metalanguage primitive in a definition in the code to be reified is in general sufficient for reification by parametricity. Pattern matching and recursion cannot be abstracted over directly, but if the same code is expressed using eliminators, these can be handled like other functions. Similarly, even though $\forall/\Pi$ cannot be abstracted over, proof automation that itself introduces universal quantifiers before reification can easily wrap them in a marker definition (_forall T P := forall (x:T), P x) that can be. Existential quantifiers are not primitive in Coq and can be reified directly.

Lambdas. While it would be sufficient to require that, in code to be reified, we write all lambdas with a named wrapper function, that would significantly clutter the code. We can do better by making use of the fact that a PHOAS object-language lambda (Abs node) consists of a metalanguage lambda that binds a value of type var, which can be used in expressions through constructor Var : var → expr. Naive reification by parametricity would turn a lambda of type $N \to N$ into a lambda of type expr → expr. A reification procedure that explicitly recurses over the metalanguage syntax could just precompose this recursive-call result with Var to get the desired object-language encoding of the lambda, but handling lambdas specially does not fit in the framework of abstraction and specialization.

First, let us handle the common case of lambdas that appear as arguments to higher-order functions. One easy approach: while the parametricity-based framework does not allow for special-casing lambdas, it is up to us to choose how to handle functions that we expect will take lambdas as arguments. We may replace each higher-order function with a metalanguage lambda that wraps the higher-order arguments in object-language lambdas, inserting Var nodes as appropriate. Code calling the function sum_upto n $f := f(0) + f(1) + \cdots + f(n)$ can be reified by abstracting over relevant definitions and substituting $(\lambda n\, f.\, \mathsf{SumUpTo}\, n\, (\mathsf{Abs}\, (\lambda v.\, f\, (\mathsf{Var}\, v))))$ for sum_upto. Note that the expression plugged in for sum_upto differs from the one plugged in for Let_In only in the use of a deeply embedded abstraction node. If we wanted to reify LetIn as just another higher-order function (as opposed to a distinguished wrapper for a primitive), the code would look identical to that for sum_upto.

It would be convenient if abstracting and substituting for functions that take higher-order arguments were enough to reify lambdas, but here is a counterexample.

$$\lambda\, x\, y.\, x \times ((\lambda\, z.\, z \times z)\, y)$$

$$\Lambda N.\, \lambda(\mathrm{MUL} : N \to N \to N).\, \lambda\, (x\, y : N).\, \mathsf{Mul}\, x\, ((\lambda\, (z : N).\, \mathsf{Mul}\, z\, z)\, y)$$

$$\lambda\, (x\, y : \mathsf{expr}).\, \mathsf{NatMul}\, x\, (\mathsf{NatMul}\, y\, y)$$

The result is not even a PHOAS expression. We claim a desirable reified form is

$$\mathsf{Abs}(\lambda\, x.\, \mathsf{Abs}(\lambda\, y.\, \mathsf{NatMul}\, (\mathsf{Var}\, x)\, (\mathsf{NatMul}\, (\mathsf{Var}\, y)\, (\mathsf{Var}\, y))))$$

Admittedly, even our improved form is not quite precise: $\lambda\, z.\, z \times z$ has been lost. However, as almost all standard Coq tactics silently reduce applications of lambdas, working under the assumption that functions not wrapped in definitions will be arbitrarily evaluated during scripting is already the norm. Accepting that limitation, it remains to consider possible occurrences of metalanguage lambdas in normal forms of outputs of reification as described so far. As lambdas in `expr` nodes that take metalanguage functions as arguments (`LetIn`, `Abs`) are handled by the rules for these nodes, the remaining lambdas must be exactly at the head of the expression. Manipulating these is outside of the power of abstraction and specialization; we recommend postprocessing using a simple recursive tactic script.

3.2   Commuting Abstraction and Reduction

Sometimes, the term we want to reify is the result of reducing another term. For example, we might have a function that reduces to a term with a variable number of `let` binders.[3] We might have an inductive type that counts the number of `let ... in ...` nodes we want in our output.

```
Inductive count := none | one_more (how_many : count).
```

It is important that this type be syntactically distinct from $\mathbb{N}$ for reasons we will see shortly.

We can then define a recursive function that constructs some number of nested `let` binders:

```
Fixpoint big (x:nat) (n:count)
   : nat
   := match n with
     | none => x
     | one_more n'
       => dlet x' := x * x in
           big x' n'
     end.
```

Fig. 5. Abstraction, reification, reduction

Our commutative diagram in Fig. 4 now has an additional node, becoming Fig. 5. Since generalization and specialization are proportional in speed to the size of the term begin handled, we can gain a significant performance boost by performing generalization before reduction. To explain why, we split apart the commutative diagram a bit more; in reduction, there is a δ or unfolding step, followed by a $\beta\iota$ step that reduces applications of λs and evaluates recursive calls. In specialization, there is an application step, where the λ is applied to arguments, and a β-reduction step, where the arguments are substituted. To obtain reified syntax, we may perform generalization after δ-reduction (before $\beta\iota$-reduction), and we are not required

---

[3] More realistically, we might have a function that represents big numbers using multiple words of a user-specified width. In this case, we may want to specialize the procedure to a couple of different bitwidths, then reifying the resulting partially reduced term.

to perform the final β-reduction step of specialization to get a well-typed term. It is important that unfolding big results in exposing the body for generalization, which we accomplish in Coq by exposing the anonymous recursive function; in other languages, the result may be a primitive eliminator applied to the body of the fixpoint. Either way, our commutative diagram thus becomes

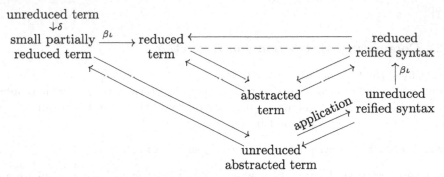

Let us step through this alternative path of reduction using the example of the unreduced term big 1 100, where we take 100 to mean the term represented by $\underbrace{(\text{one\_more} \cdots (\text{one\_more}}_{100} \underbrace{\text{none}) \cdots)}_{100}$.

Our first step is to unfold big, rendered as the arrow labeled δ in the diagram. In Coq, the result is an anonymous fixpoint; here we will write it using the recursor count_rec of type $\forall T.\ T \to (\text{count} \to T \to T) \to \text{count} \to T$. Performing δ-reduction, that is, unfolding big, gives us the small partially reduced term

$(\lambda(x : \mathbb{N}).\ \lambda(n : \text{count}).$
count_rec $(\mathbb{N} \to \mathbb{N})\ (\lambda x.\ x)\ (\lambda n'.\ \lambda\text{big}_{n'}.\ \lambda x.\ \text{dlet}\ x' := x \times x\ \text{in}\ \text{big}_{n'}\ x'))\ 1\ 100$

We call this term small, because performing $\beta\iota$ reduction gives us a much larger reduced term:

$$\text{dlet}\ x_1 := 1 \times 1\ \text{in}\ \cdots\ \text{dlet}\ x_{100} := x_{99} \times x_{99}\ \text{in}\ x_{100}$$

Abstracting the small partially reduced term over (@Let_In $\mathbb{N}$ $\mathbb{N}$), S, O, mul, and $\mathbb{N}$ gives us the abstracted unreduced term

$\Lambda N.\ \lambda(\text{MUL} : N \to N \to N)(\text{O} : N)(\text{S} : N \to N)(\text{LETIN} : N \to (N \to N) \to N).$
$\quad (\lambda(x : N).\ \lambda(n : \text{count}).\ \text{count\_rec}\ (N \to N)\ (\lambda x.\ x)$
$\quad\quad\quad (\lambda n'.\ \lambda\text{big}_{n'}.\ \lambda x.\ \text{LETIN}\ (\text{MUL}\ x\ x)\ (\lambda x'.\ \text{big}_{n'}\ x')))$
$\quad (\text{S O})\ 100$

Note that it is essential here that count is not syntactically the same as $\mathbb{N}$; if they were the same, the abstraction would be ill-typed, as we have not abstracted over count_rec. More generally, it is essential that there is a clear separation between types that we reify and types that we do not, and we must reify *all* operations on the types that we reify.

We can now apply this term to `expr`, `NatMul`, `NatS`, `NatO`, and, finally, $(\lambda v\ f.\ \text{LetIn } v\ (\lambda x.\ f\ (\text{Var } x)))$. We get an unreduced reified syntax tree of type `expr`. If we now perform $\beta\iota$ reduction, we get our fully reduced reified term.

We take a moment to emphasize that this technique is not possible with any other method of reification. We could just as well have not specialized the function to the `count` of 100, yielding a function of type `count` → `expr`, despite the fact that our reflective language knows nothing about `count`!

This technique is especially useful for terms that will not reduce without concrete parameters, but which should be reified for many different parameters. Running reduction once is slightly faster than running OCaml reification once, and it is more than twice as fast as running reduction followed by OCaml reification. For sufficiently large terms and sufficiently many parameter values, this performance beats even OCaml reification.[4]

3.3 Implementation in Ltac

`ExampleMoreParametricity.v` in the code supplement mirrors the development of reification by parametricity in Subsect. 3.1.

Unfortunately, Coq does not have a tactic that performs abstraction.[5] However, the `pattern` tactic suffices; it performs abstraction followed by application, making it a sort of one-sided inverse to β-reduction. By chaining `pattern` with an LTAC-match statement to peel off the application, we can get the abstracted function.

```
Ltac Reify x :=
match(eval pattern nat, Nat.mul, S, O, (@Let_In nat nat) in x)with
| ?rx _ _ _ _ _ =>
 constr:( fun var => rx (@expr var) NatMul NatS NatO
                    (fun v f => LetIn v (fun x => f (Var x))) )
end.
```

Note that if `@expr var` lives in `Type` rather than `Set`, an additional step involving retyping the term is needed; we refer the reader to `Parametricity.v` in the code supplement.

The error messages returned by the `pattern` tactic can be rather opaque at times; in `ExampleParametricityErrorMessages.v`, we provide a procedure for decoding the error messages.

---

[4] We discovered this method in the process of needing to reify implementations of cryptographic primitives [6] for a couple hundred different choices of numeric parameters (e.g., prime modulus of arithmetic). A couple hundred is enough to beat the overhead.

[5] The `generalize` tactic returns ∀ rather than λ, and it only works on types.

Open Terms. At some level it is natural to ask about generalizing our method to reify open terms (i.e., with free variables), but we think such phrasing is a red herring. Any lemma statement about a procedure that acts on a representation of open terms would need to talk about how these terms would be closed. For example, solvers for algebraic goals without quantifiers treat free variables as implicitly universally quantified. The encodings are invariably ad-hoc: the free variables might be assigned unique numbers during reification, and the lemma statement would be quantified over a sufficiently long list that these numbers will be used to index into. Instead, we recommend directly reifying the natural encoding of the goal as interpreted by the solver, e.g. by adding new explicit quantifiers. Here is a hypothetical goal and a tactic script for this strategy:

```
(a b : nat) (H : 0 < b) |- ∃ q r, a = q × b + r ∧ r < b
```

```
repeat match goal with
       | n : nat |- ?P =>
         match eval pattern n in P with
         | ?P' _ => revert n; change (_forall nat P')
         end
       | H : ?A  |- ?B => revert H; change (impl A B)
       | |- ?G => (* ∀ a b, 0 < b -> ∃ q r, a = q × b + r ∧ r < b *)
         let rG := Reify G in
         refine (nonlinear_integer_solver_sound rG _ _);
         [ prove_wf | vm_compute; reflexivity ]
       end.
```

Briefly, this script replaced the context variables a and b with universal quantifiers in the conclusion, and it replaced the premise H with an implication in the conclusion. The syntax-tree datatype used in this example can be found in ExampleMoreParametricity.v.

3.4   Advantages and Disadvantages

This method is faster than all but LTAC2 and OCaml reification, and commuting reduction and abstraction makes this method faster even than the low-level LTAC2 reification in many cases. Additionally, this method is much more concise than nearly every other method we have examined, and it is very simple to implement.

We will emphasize here that this strategy shines when the initial term is small, the partially computed terms are big (and there are many of them), and the operations to evaluate are mostly well-separated by types (e.g., evaluate all of the count operations and none of the nat ones).

This strategy is not directly applicable for reification of match (rather than eliminators) or let ... in ... (rather than a definition that unfolds to let ... in ...), forall (rather than a definition that unfolds to forall), or when reification should not be modulo $\beta\iota\zeta$-reduction.

4   Performance Comparison

We have done a performance comparison of the various methods of reification to the PHOAS language @expr var from Fig. 1.3 in Coq 8.7.1. A typical reification routine will obtain the term to be reified from the goal, reify it, run `transitivity (denote reified_term)` (possibly after normalizing the reified term), and solve the side condition with something like `lazy [denote];` `reflexivity`. Our testing on a few samples indicated that using `change` rather than `transitivity; lazy [denote]; reflexivity` can be around 3X slower; note that we do not test the time of `Defined`.

There are two interesting metrics to consider: (1) how long does it take to reify the term? and (2) how long does it take to get a normalized reified term, i.e., how long does it take both to reify the term and normalize the reified term? We have chosen to consider (1), because it provides the most fine-grained analysis of the actual reification method.

4.1   Without Binders

We look at terms of the form $1 * 1 * 1 * \ldots$ where multiplication is associated to create a balanced binary tree. We say that the *size of the term* is the number of 1s. We refer the reader to the attached code for the exact test cases and the code of each reification method being tested.

We found that the performance of all methods is linear in term size.

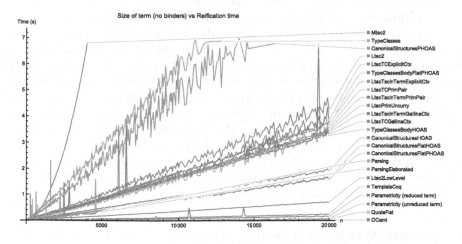

Fig. 6. Performance of reification without binders

Sorted from slowest to fastest, most of the labels in Fig. 6 should be self-explanatory and are found in similarly named .v files in the associated code; we call out a few potentially confusing ones:

- The "Parsing" benchmark is "reification by copy-paste": a script generates a .v file with notation for an already-reified term; we benchmark the amount of time it takes to parse and typecheck that term. The "ParsingElaborated" benchmark is similar, but instead of giving notation for an already-reified term, we give the complete syntax tree, including arguments normally left implicit. Note that these benchmarks cut off at around 5000 rather than at around 20 000, because on large terms, Coq crashes with a stack overflow in parsing.
- We have four variants starting with "CanonicalStructures" here. The Flat variants reify to @expr nat rather than to forall var, @expr var and benefit from fewer function binders and application nodes. The HOAS variants do not include a case for let ... in ... nodes, while the PHOAS variants do. Unlike most other reification methods, there is a significant cost associated with handling more sorts of identifiers in canonical structures.

We note that on this benchmark our method is slightly faster than template-coq, which reifies to de Bruijn indices, and slightly slower than the quote plugin in the standard library and the OCaml plugin we wrote by hand.

4.2 With Binders

We look at terms of the form dlet a_1 := 1 * 1 in dlet a_2 := a_1 * a_1 in ... dlet a_n := a_{n-1} * a_{n-1} in a_n, where n is the size of the term. The first graph shown here includes all of the reification variants at linear scale, while the next step zooms in on the highest-performance variants at log-log scale.

In addition to reification benchmarks, the graph in Fig. 7 includes as a reference (1) the time it takes to run lazy reduction on a reified term already in normal form ("identity lazy") and (2) the time it takes to check that the reified term matches the original native term ("lazy Denote"). The former is just barely faster than OCaml reification; the latter often takes longer than reification itself. The line for the template-coq plugin cuts off at around 10 000 rather than around 20 000 because at that point template-coq starts crashing with stack overflows.

A nontrivial portion of the cost of "Parametricity (reduced term)" seems to be due to the fact that looking up the type of a binder is linear in the number of binders in the context, thus resulting in quadratic behavior of the retyping step that comes after abstraction in the pattern tactic. In Coq 8.8, this lookup will be $\log n$, and so reification will become even faster [10].

5 Future Work, Concluding Remarks

We identify one remaining open question with this method that has the potential of removing the next largest bottleneck in reification: using reduction to show that the reified term is correct.

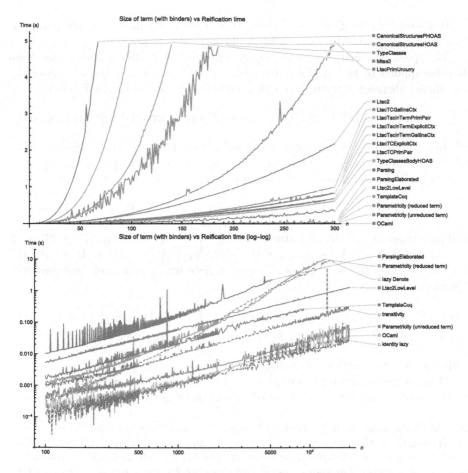

Fig. 7. Performance of reification with binders

Recall our reification procedure and the associated diagram, from Fig. 3.2. We perform δ on an unreduced term to obtain a small, partially reduced term; we then perform abstraction to get an abstracted, unreduced term, followed by application to get unreduced reified syntax.

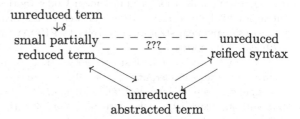

Fig. 8. Completing the commutative triangle

These steps are all fast. Finally, we perform $\beta\iota$-reduction to get reduced, reified syntax and perform $\beta\iota\delta$ reduction to get back a reduced form of our original term. These steps are slow, but we must do them if we are to have verified reflective automation.

It would be nice if we could prove this equality without ever reducing our term. That is, it would be nice if we could have the diagram in Fig. 8.

The question, then, is how to connect the small partially reduced term with `denote` applied to the unreduced reified syntax. That is, letting F denote the unreduced abstracted term, how can we prove, without reducing F, that

$$F \, \mathbb{N} \, \text{Mul} \, O \, S \, (@\text{Let\_In} \, \mathbb{N} \, \mathbb{N}) = \text{denote} \, (F \, \text{expr} \, \text{NatMul} \, \text{Nat0} \, \text{NatS} \, \text{LetIn})$$

We hypothesize that a form of internalized parametricity would suffice for proving this lemma. In particular, we could specialize F's type argument with $\mathbb{N} \times \text{expr}$. Then we would need a proof that for any function F of type

$$\forall (T : \text{Type}), (T \to T \to T) \to T \to (T \to T) \to (T \to (T \to T) \to T) \to T$$

and any types A and B, and any terms $f_A : A \to A \to A$, $f_B : B \to B \to B$, $a : A$, $b : B$, $g_A : A \to A$, $g_B : B \to B$, $h_A : A \to (A \to A) \to A$, and $h_B : B \to (B \to B) \to B$, using $f \times g$ to denote lifting a pair of functions to a function over pairs:

$$\text{fst} \, (F \, (A \times B) \, (f_A \times f_B) \, (a, b) \, (g_A \times g_B) \, (h_A \times h_B)) = F \, A \, f_A \, a \, g_A \, h_A \, \wedge$$
$$\text{snd} \, (F \, (A \times B) \, (f_A \times f_B) \, (a, b) \, (g_A \times g_B) \, (h_A \times h_B)) = F \, B \, f_B \, b \, g_B \, h_B$$

This theorem is a sort of parametricity theorem.

Despite this remaining open question, we hope that our performance results make a strong case for our method of reification; it is fast, concise, and robust.

Acknowledgments and Historical Notes. We would like to thank Hugo Herbelin for sharing the trick with `type of` to propagate universe constraints (https://github.com/coq/coq/issues/5996#issuecomment-338405694) as well as useful conversations on Coq's bug tracker that allowed us to track down performance issues (https://github.com/coq/coq/issues/6252). We would like to thank Pierre-Marie Pédrot for conversations on Coq's Gitter and his help in tracking down performance bottlenecks in earlier versions of our reification scripts and in Coq's tactics. We would like to thank Beta Ziliani for his help in using Mtac2, as well as his invaluable guidance in figuring out how to use canonical structures to reify to PHOAS. We also thank John Wiegley for feedback on the paper.

For those interested in history, our method of reification by parametricity was inspired by the `evm_compute` tactic [9]. We first made use of `pattern` to allow `vm_compute` to replace cbv-with-an-explicit-blacklist when we discovered cbv was too slow and the blacklist too hard to maintain. We then noticed that in the sequence of doing abstraction; `vm_compute`; application; β-reduction; reification, we could move β-reduction to the end of the sequence if we fused reification with application, and thus reification by parametricity was born.

This work was supported in part by a Google Research Award and National Science Foundation grants CCF-1253229, CCF-1512611, and CCF-1521584.

References

1. Anand, A., Boulier, S., Tabareau, N., Sozeau, M.: Typed Template Coq. CoqPL 2018, January 2018. https://popl18.sigplan.org/event/coqpl-2018-typed-template-coq

2. Boutin, S.: Using reflection to build efficient and certified decision procedures. In: Abadi, M., Ito, T. (eds.) TACS 1997. LNCS, vol. 1281, pp. 515–529. Springer, Heidelberg (1997). https://doi.org/10.1007/BFb0014565

3. de Bruijn, N.G.: Lambda-calculus notation with nameless dummies: a tool for automatic formal manipulation with application to the Church-Rosser theorem. Indagationes Mathematicae (Proceedings) **34**(5), 381–392 (1972). https://doi.org/10.1016/1385-7258(72)90034-0. http://www.sciencedirect.com/science/article/pii/1385725872900340

4. Chlipala, A.: Parametric higher-order abstract syntax for mechanized semantics. In: Proceedings of the 13th ACM SIGPLAN International Conference on Functional Programming, ICFP 2008, September 2008. http://adam.chlipala.net/papers/PhoasICFP08/

5. Coq Development Team: The Coq Proof Assistant Reference Manual, chap. 10.3 Detailed examples of tactics (quote). INRIA, 8.7.1 edn. (2017). https://coq.inria.fr/distrib/V8.7.1/refman/tactic-examples.html#quote-examples

6. Erbsen, A., Philipoom, J., Gross, J., Sloan, R., Chlipala, A.: Simple high-level code for cryptographic arithmetic - with proofs, without compromises. In: Proceedings of IEEE Symposium on Security & Privacy, May 2019

7. Gonthier, G., Mahboubi, A., Tassi, E.: A small scale reflection extension for the Coq system. Technical report, Inria Saclay Ile de France, November 2016. https://hal.inria.fr/inria-00258384/

8. Gonthier, G., Ziliani, B., Nanevski, A., Dreyer, D.: How to make ad hoc proof automation less ad hoc. J. Funct. Programm. **23**(4), 357–401 (2013). https://doi.org/10.1017/S0956796813000051. https://people.mpi-sws.org/~beta/lessadhoc/lessadhoc-extended.pdf

9. Malecha, G., Chlipala, A., Braibant, T.: Compositional computational reflection. In: Proceedings of the 5th International Conference on Interactive Theorem Proving, ITP 2014 (2014). http://adam.chlipala.net/papers/MirrorShardITP14/

10. Pédrot, P.M.: Fast REL lookup #6506, December 2017. https://github.com/coq/coq/pull/6506

11. Pfenning, F., Elliot, C.: Higher-order abstract syntax. In: Proceedings of PLDI, pp. 199–208 (1988). https://www.cs.cmu.edu/~fp/papers/pldi88.pdf

12. Sozeau, M., Oury, N.: First-class type classes. In: Mohamed, O.A., Muñoz, C., Tahar, S. (eds.) TPHOLs 2008. LNCS, vol. 5170, pp. 278–293. Springer, Heidelberg (2008). https://doi.org/10.1007/978-3-540-71067-7_23. https://www.irif.fr/~sozeau/research/publications/

Verifying the LTL to Büchi Automata Translation via Very Weak Alternating Automata

Simon Jantsch[1](✉) and Michael Norrish[2]

[1] TU Dresden, Dresden, Germany
simon.jantsch@gmail.com
[2] Data61, CSIRO and Australian National University, Canberra, Australia

Abstract. We present a formalization of a translation from LTL formulae to generalized Büchi automata in the HOL4 theorem prover. Translations from temporal logics to automata are at the core of model checking algorithms based on automata-theoretic techniques. The translation we verify proceeds in two steps: it produces very weak alternating automata at an intermediate stage, and then ultimately produces a generalized Büchi automaton. After verifying both transformations, we also encode both of these automata models using a generic, functional graph type, and use the CakeML compiler to generate fully verified machine code implementing the translation.

1 Introduction

As the goal of verification techniques is to give the user of a system guarantees about its behaviour, bugs in verification tools can potentially have severe consequences and considerably reduce the trust of users in the techniques. While new verification algorithms are usually proven correct on paper, the gap between the abstract proof and any actual implementation can be large. Many times different representations are used and optimizations are added that are not considered in the proofs.

Our aim is to bridge this gap for one standard algorithm used for automata-based LTL model checking. The algorithm, by Gastin and Oddoux [7] (G&O henceforth), improves on the efficiency of the translation of LTL formulae into automata. Rather than moving directly from such formulae into generalized Büchi automata (GBA), it introduces an intermediate step, the rather complicated *alternating automata*. Whereas the efficient translation from LTL to alternating automata was known before, G&O showed that a property, namely very weakness, of the resulting automata can be exploited for the translation to GBA.

This new step represents an advantage on earlier techniques in part because automata-optimizations can be applied in both phases. Optimizing the alternating automaton is especially interesting as it is linear in the size of the formula,

The author was supported by the European Master's Program in Computational Logic (EMCL).

even though the final GBA may still be of exponential size. As noted in Schimpf *et al.* [15], the original tool implementing this algorithm contained a bug that went unnoticed for several years despite widespread use.

Translations of LTL formulae to automata play a core role in LTL model checking. In the usual approach, an LTL formula φ is given together with a labeled transition system S, and the questions is whether all executions of S satisfy the formula φ. To check this, an automaton is constructed for $\neg\varphi$, which is then combined with an automaton describing all executions of S. If the combined automaton is empty, S indeed satisfies the property specified by φ, otherwise a counterexample to this claim can be given.

To obtain a formally verified implementation of the algorithm in G&O, we proceed as follows: first we formalize the procedure in an abstract way, using set notation and mathematical functions. We prove correctness of this function, which is a mechanization of the proof given in G&O. Then we implement another version of the algorithm, now defined on concrete data structures that represent the automata in a compact way. In contrast to the first function, this second version describes an algorithm: a step-by-step expansion of a graph.

The relation between our two versions is established by defining abstraction functions from our concrete automata to their abstract counterparts. Using these functions, we show that the automata we obtain in our concrete algorithm coincide with the abstract automata, for which we have proved the desired property. One strength of this approach is that it lets us separate the correctness proofs of the main function and the restriction to reachable states on the abstract level, while still combining the two functions on the concrete level in a single expansion algorithm. We believe that this idea can be extended to add optimizations to the translation in a manageable way by defining them as seperate transformation steps on the abstract level, and efficiently embedding them into the expansion algorithm on the concrete level.

Finally, we compile our function into machine code using the CakeML compiler. This adds another guarantee to our implementation, as we do not have to trust the translation of the algorithm as expressed in HOL4 into SML, nor the correctness of an SML compiler. The proof scripts and definitions for our translation are available as part of the HOL4 system, and the scripts to compile the algorithm with CakeML are available on `Gitlab`.[1]

The paper is structured as follows: Sect. 2 introduces LTL and the automata models we consider. Section 3 recalls the algorithm in G&O, and Sect. 4 discusses our formalization in HOL4. Section 5 gives an overview of related work, and we conclude in Sect. 6.

---

[1] For the abstract and concrete algorithms, see the `examples/logic/ltl` directory in HOL4 after commit `b4576ed`, and see https://gitlab.com/simon-jantsch/ltl2baHol-paper/tree/master/cmlltl for our CakeML translations, which in turn depend on CakeML commit `891cbf4a`.

2   Preliminaries

2.1   Linear Temporal Logic

Linear Temporal Logic (LTL) is a logic that extends propositional logic with temporal operators. We define it using unary **X** ("next") and binary **U** ("until").

Definition 1 (Syntax of LTL). *Given a set of atomic propositions AP, the set of LTL formulae over AP is defined with the following grammar:*

$$\varphi ::= p \mid \neg\varphi \mid \varphi \wedge \varphi \mid \mathbf{X}\varphi \mid \varphi\mathbf{U}\varphi$$

where $p \in AP$

An interpretation of an LTL formula is a sequence of propositional valuations over AP, one for each point in time. This sequence is viewed as an infinite word over $\mathscr{P}(AP)$ (we write $\mathscr{P}(S)$ to mean the powerset of S). The symbol of w at position i is denoted by $w[i]$ and the suffix of w starting at position i by $w[i..]$. Given $w \in (\mathscr{P}(AP))^{\omega}$, we define

Definition 2 (Semantics of LTL)

$$
\begin{array}{ll}
w \models p & \textit{iff } p \in w[0], \textit{ for all } p \in AP \\
w \models \neg\varphi & \textit{iff } w \not\models \varphi \\
w \models \varphi_1 \wedge \varphi_2 & \textit{iff } w \models \varphi_1 \textit{ and } w \models \varphi_2 \\
w \models \mathbf{X}\varphi & \textit{iff } w[1..] \models \varphi \\
w \models \varphi_1\mathbf{U}\varphi_2 & \textit{iff } \exists i.\ w[i..] \models \varphi_2 \textit{ and } \forall j < i.\ w[j..] \models \varphi_1
\end{array}
$$

As we want to use a negation normal form we introduce the dual operators $\vee$ and $\varphi_1\mathbf{R}\varphi_2 = \neg(\neg\varphi_1\mathbf{U}\neg\varphi_2)$. An LTL formula φ is in negation normal form if all occurrences of $\neg$ are directly in front of an atomic proposition. We call a formula a *temporal formula* if it is a (possibly negated) atomic proposition or if its outermost operator is **X**, **U** or **R**. We use $\mathcal{L}(\varphi) = \{w \in (\mathscr{P}(AP))^{\omega} \mid w \models \varphi\}$ to denote the *language* of an LTL formula.

As the semantics of LTL is defined using infinite words, questions about LTL formulae can often be formulated as word problems. This is where automata, in our case recognizing languages of infinite words, come into play. In the following sections we introduce the two automata types used in G&O, beginning with alternating automata.

2.2   Co-Büchi Alternating Automata

In an *alternating automaton*, each state nondeterministically chooses between *sets* of successor states. Intuitively, a word $w = a_0a_1\ldots$ is accepted from a state q if there *exists* a successor set S reachable via the symbol a_0 such that $a_1a_2\ldots$ is accepted from *all* states in S.

Definition 3. *A co-Büchi alternating automaton is a tuple* $\mathcal{A} = (Q, \Sigma, \delta, I, F)$, *where Q is a finite set of states, Σ is a finite alphabet, $\delta : Q \rightarrow \mathscr{P}(\mathscr{P}(\Sigma) \times \mathscr{P}(Q))$, $I \subseteq \mathscr{P}(Q)$ is the set of initial sets and $F \subseteq Q$ is the set of final states.*

Alternating automata can be defined with different acceptance conditions but we will always mean co-Büchi alternating automata in what follows. In HOL4 we use the following datatype for abstract alternating automata:

$$(\alpha, \ \sigma)\ ALTER\_A \ = \ <|$$

$$\text{states} \ : \ \sigma \ set;$$

$$\text{alphabet} \ : \ \alpha \ set;$$

$$\text{trans} \ : \ \sigma \ \rightarrow \ (\alpha \ set \ \times \ \sigma \ set) \ set;$$

$$\text{initial} \ : \ \sigma \ set \ set;$$

$$\text{final} \ : \ \sigma \ set$$

$$|>$$

The transition function δ assigns to each state in the automaton a set of pairs (A, S), where $A \subseteq \Sigma$ and $S \subseteq Q$. Such a pair stands for a transition that is active for every symbol in A and has successor set S. This definition of alternating automata was introduced in G&O and differs from the more usual definition, where the transition function is defined using positive boolean formulae over the states (e.g. Löding [11] or Vardi [17]). As noted in G&O, the two can easily be transformed into each other: the presented definition corresponds closely to the disjunctive normal form of the positive boolean formula.

Following Löding [11] we define a run of an alternating automaton $\mathcal{A}$ on a word $w \in \Sigma^\omega$ as a directed acyclic graph $\rho = (V, E)$, where $V \subseteq Q \times \mathbb{N}$, $E \subseteq \bigcup_{i \geq 0}(Q \times \{i\}) \times (Q \times \{i+1\})$ and

- $\{q \mid (q, 0) \in V\} \in I$;
- for all $(q, i) \in V$ there exists $(A, S) \in \delta(q)$ such that $w[i] \in A$ and $\{q' \mid ((q, i), (q', i+1)) \in E\} = S$; and
- for all $(q, i) \in V$ where $i > 0$, there exists some $(q_p, i - 1) \in V$ such that $((q_p, i - 1), (q, i)) \in E$.

For co-Büchi automata, acceptance is defined as follows: a run ρ is accepting if there is no path through ρ that visits a state in F infinitely often. The language of a co-Büchi alternating automaton is defined as $\mathcal{L}(\mathcal{A}) = \{w \in \Sigma^\omega \mid$ there exists an accepting run of $\mathcal{A}$ on $w\}$.

Note that the transition function allows empty successor sets. Such a transition corresponds to the empty conjunction (i.e. *true*) and leads to direct acceptance of any suffix word for which it is active.

An alternating automaton is *very weak* if there is a partial order R on Q, such that whenever $(A, S) \in \delta(q)$ and $q' \in S$ then $R\,q'\,q$. As Q is finite, this implies that all loops in the automaton are self-loops and every path in a run ρ ultimately stabilizes on some state.

2.3 Generalized Büchi Automata

The algorithm we consider produces generalized Büchi automata (GBA), where the acceptance condition is defined using the edges, rather than the states, of the automaton.

Definition 4. *A generalized Büchi automaton is a tuple* $\mathcal{G} = (Q, \Sigma, \delta, I, T)$, *where* Q *is a finite set of states,* Σ *is a finite alphabet,* $\delta : Q \to \mathcal{P}(\mathcal{P}(\Sigma) \times Q)$ *is the transition function,* $I \subseteq Q$ *is the set of initial states and* $T = \{T_1, T_2, \ldots\}$, *with* $T_i \subseteq Q \times \mathcal{P}(\Sigma) \times Q$, *is a set of sets of accepting edges.*

A run $r = q_0 q_1 \ldots \in Q^\omega$ of a GBA $\mathcal{G}$ on a word $w \in \Sigma^\omega$ is a sequence of states such that $q_0 \in I$ and for all i there exists a pair $(A, q_{i+1}) \in \delta(q_i)$ such that $w[i] \in A$. It is accepting if for all $T \in \mathcal{T}$ there exist infinitely many positions i such that for some A: $(A, q_{i+1}) \in \delta(q_i)$, $w[i] \in A$ and $(q_i, A, q_{i+1}) \in T$. The language of a GBA is defined accordingly: $\mathcal{L}(\mathcal{G}) = \{w \in \Sigma^\omega \mid \text{there exists an accepting run of } \mathcal{G} \text{ on } w\}$.

GBA can be transformed into ordinary Büchi automata via a standard linear transformation called degeneralization. The emptiness check, which is required for LTL model checking, can be done on Büchi automata efficiently [2]. However, approaches have been developed to use the GBA directly to check emptiness, thereby omitting degeneralization [3].

3 Translating LTL to GBA

We now recall the translation presented in G&O. The algorithm proceeds in two steps: it first translates an LTL formula into an equivalent very weak alternating automaton (VWAA), and then translates that VWAA into a GBA. By "equivalent", we mean that the words accepted by the VWAA are exactly the words that satisfy the formula, as per Definition 2.

We introduce two functions that we need for the definition, $\overline{\varphi}$ gives an approximation of the DNF of φ without simplifying temporal subformulae. $\otimes$ is an operation on the transitions of the VWAA that corresponds to conjunction on the formula level. From now on we consider all formulae to be in negation normal form.

Definition 5. *Let* φ *be an LTL formula.* $\overline{\varphi} = \{\{\varphi\}\}$ *if* φ *is a temporal formula,* $\overline{\varphi \wedge \psi} = \{S_1 \cup S_2 \mid S_1 \in \overline{\varphi} \text{ and } S_2 \in \overline{\psi}\}$ *and* $\overline{\varphi \vee \psi} = \overline{\varphi} \cup \overline{\psi}$.
Let $D_1, D_2 \in \mathcal{P}(\mathcal{P}(\Sigma) \times \mathcal{P}(Q))$. $D_1 \otimes D_2 = \{(A_1 \cap A_2, S_1 \cup S_2) \mid (A_1, S_1) \in D_1 \text{ and } (A_2, S_2) \in D_2\}$.

Now we can define the first step of the translation. It models the boolean structure of the formulae with the transitions of the VWAA and makes use of the equalities $\varphi \mathbf{U} \psi = \psi \vee (\varphi \wedge \mathbf{X}(\varphi \mathbf{U} \psi))$ and $\varphi \mathbf{R} \psi = \psi \wedge (\varphi \vee \mathbf{X}(\varphi \mathbf{R} \psi))$.

Definition 6. *Let φ be an LTL formula over AP. We define $\mathcal{A}_\varphi = (Q, \Sigma, \delta, I, F)$, where Q is the set of temporal subformulae of φ, $\Sigma = \mathscr{P}(AP)$, $I = \overline{\varphi}$, F is the set of subformulae of φ of the type $\psi_1 \mathbf{U} \psi_2$ and δ is defined by:*

$$\delta(p) = \{(\Sigma_p, \emptyset)\}, \text{ where } \Sigma_p = \{A \in \Sigma \mid p \in A\}$$
$$\delta(\neg p) = \{(\Sigma_{\neg p}, \emptyset)\}, \text{ where } \Sigma_{\neg p} = \Sigma \setminus \Sigma_p$$
$$\delta(\mathbf{X}\psi) = \{(\Sigma, S) \mid S \in \overline{\psi}\}$$
$$\delta(\psi_1 \mathbf{U} \psi_2) = \Delta(\psi_2) \cup (\Delta(\psi_1) \otimes \{(\Sigma, \{\psi_1 \mathbf{U} \psi_2\})\})$$
$$\delta(\psi_1 \mathbf{R} \psi_2) = \Delta(\psi_2) \otimes (\Delta(\psi_1) \cup \{(\Sigma, \{\psi_1 \mathbf{R} \psi_2\})\})$$
$$\Delta(\psi) = \delta(\psi), \text{ if } \psi \text{ is a temporal formula}$$
$$\Delta(\psi_1 \wedge \psi_2) = \Delta(\psi_1) \otimes \Delta(\psi_2)$$
$$\Delta(\psi_1 \vee \psi_2) = \Delta(\psi_1) \cup \Delta(\psi_2)$$

As every transition contains only subformulae of the considered formula, we see that $\mathcal{A}_\varphi$ is very weak. In G&O the following theorem is stated without a complete proof. We discuss our proof and its mechanization in Sect. 4. A proof for the standard setting, which simplifies the proof, can be found in Vardi [17].

Theorem 1. $\mathcal{L}(\mathcal{A}_\varphi) = \mathcal{L}(\varphi)$

The second step of the algorithm is a translation of a VWAA into a GBA. We first define a relation $\preceq$ on transitions that we use in the later definition. Let $t_1 = (S, A_1, S_1')$ and $t_2 = (S, A_2, S_2')$ be transitions of the GBA. Then $t_1 \preceq t_2$ if $A_2 \subseteq A_1$, $S_1' \subseteq S_2'$ and for all $T \in \mathcal{T}$: $t_2 \in T \Rightarrow t_1 \in T$.

Definition 7. *Let $\mathcal{A} = (Q, \Sigma, \delta, I, F)$ be a VWAA. We define $\mathcal{G}_\mathcal{A} = (\mathscr{P}(Q), \Sigma, \delta', I, \mathcal{T})$, where*

- *$\delta'(\{q_0, q_1, \ldots, q_n\})$ is the set of $\preceq$-minimal transitions in $\bigotimes_{i=0}^{n} \delta(q_i)$*
- *$\mathcal{T} = \{T_f \mid f \in F\}$, where*
  $T_f = \{(S, A, S') \mid f \notin S' \text{ or there is } (B, X) \in \delta(f) \text{ such that } A \subseteq B \text{ and } f \notin X \subseteq S'\}$

An example of the translations to VWAA and GBA is given in Fig. 1.

Theorem 2. $\mathcal{L}(\mathcal{G}_\mathcal{A}) = \mathcal{L}(\mathcal{A})$

Proof. See G&O for a proof.

4   Verifying the Algorithm

Note that the way the translation is presented is far from an actual implementation. In particular the worst case complexity is always exhibited as nonreachable states are not excluded. Also the way the transitions are defined, where the first

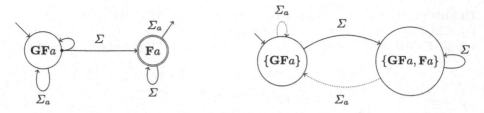

Fig. 1. Translation of the formula **GF**a into a VWAA (left) and a GBA (right). Here **F** φ (eventually) and **G** φ (always) abbreviate *true***U**φ and $\neg$**F**$\neg\varphi$ respectively. Transitions conjoined with • are conjunctive transitions to multiple successors. Recall that Σ_a is the set of all elements in Σ that contain a. Arrows with no successor node indicate transitions to the empty set. Final states in the VWAA are indicated by doubled circles and accepting transitions (of the single acceptance set) in the GBA are indicated by a dotted line.

component is a set of subsets of AP, is prohibitively inefficient. These representations are convenient for the proofs, but the question is how exactly any concrete algorithm relates to this abstract description. We introduce a more compact representation and define its relation to the abstract one.

Figure 2 visualizes our approach. As in G&O, we do not worry about reachable states in our main correctness proof; rather we implement the restriction on the reachable states as a separate function (restr_states in Fig. 2).

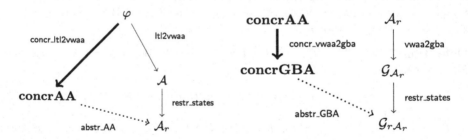

Fig. 2. Dividing the formalization into abstract and concrete parts. Thick arrows represent concrete functions, thin arrows represent abstract functions, and dotted arrows are abstractions from concrete to abstract automata. An r in the subscript stands for a restriction to reachable states.

4.1   Mechanizing the Abstract Proofs

Our abstract formalizations in HOL4 are basically identical to the mathematical definitions given in Sect. 2. This allows us to closely follow the proof of Theorem 2 from G&O. First, we discuss the proof of Theorem 1, which is not presented in G&O:

$$\vdash \mathcal{L}\, \phi \;=\; \mathcal{L}_{AA}\,(\mathsf{ltl2vwaa}\,\phi)$$

In the proof we fix a formula ϕ and with it the alphabet we are considering, namely $\mathscr{P}(\text{props } \phi)$, where props is the function that collects all atomic propositions that occur in a formula. Then we show the claim for all subformulae of ϕ by structural induction on LTL formulae.

The base case is the translation of an atomic proposition $p \in$ props ϕ. The corresponding automaton $\mathcal{A}_p$ has one state with transitions to the empty set for all elements in $\mathscr{P}(\text{props } \phi)$ that contain p. Thus the automaton accepts exactly the words w for which such a transition is active, which is the case exactly if $p \in w[0]$.

In the other cases, we show how accepting runs of the sub-automata can be used to build accepting runs of the automata of the current case. Consider the case $\mathbf{X}\psi$. For any word w such that $w \models \mathbf{X}\psi$ we get $w[1..] \models \psi$ and by induction hypothesis an accepting run of $\mathcal{A}_\psi$ on $w[1..]$. By shifting this run by one and adding the vertex $(\mathbf{X}\psi, 0)$, we get an accepting run of $\mathcal{A}_{\mathbf{X}\psi}$. For the other direction we start with an accepting run of $\mathcal{A}_{\mathbf{X}\psi}$ on w. By the structure of $\mathcal{A}_{\mathbf{X}\psi}$ we can extract a run of $\mathcal{A}_\psi$ on the word $w[1..]$. This is done by again shifting the run by one, but now in the other direction. Applying the induction hypothesis yields $w[1..] \models \psi$, from which we can conclude $w \models \mathbf{X}\psi$.

The existence of these two runs is shown in the proofs for the following lemmata:[2]

$$\vdash \text{runOfAA } (\text{ltl2vwaa}_\phi \ \psi) \ r \ w[1..] \ \wedge \ \text{word\_range } w \subseteq \mathscr{P}(\text{props } \phi) \Rightarrow$$
$$\exists \, r'. \ \text{runOfAA } (\text{ltl2vwaa}_\phi \ (\mathbf{X} \ \psi)) \ r' \ w$$
$$\vdash \text{runOfAA } (\text{ltl2vwaa}_\phi \ (\mathbf{X} \ \psi)) \ r \ w \Rightarrow \exists \, r'. \ \text{runOfAA } (\text{ltl2vwaa}_\phi \ \psi) \ r' \ w[1..]$$

The expression $\text{ltl2vwaa}_\phi \ \psi$ denotes the automaton for ψ, as defined by Definition 6, with respect to the alphabet $\mathscr{P}(\text{props } \phi)$. (In particular, $\text{ltl2vwaa } \phi = \text{ltl2vwaa}_\phi \ \phi$.) The condition runOfAA *aut r w* states that r is a run of *aut* on w.

To show acceptance of the runs we construct in this case we use the fact that the final states of the automata $\mathcal{A}_\psi$ and $\mathcal{A}_{\mathbf{X}\psi}$ are the same, as no "until"-formula is added to the automaton in the $\mathbf{X}$ case. So it is enough if we can map every path in the run we construct to some path in the old run that visits the same set of nodes infinitely often. This is clearly possible as the only way we transformed the runs was to shift them by one.

The most interesting cases are the temporal operators $\mathbf{U}$ and $\mathbf{R}$, where the acceptance conditions become important. In $\varphi\mathbf{U}\psi$, for example, we first show that the automaton cannot stay in the state $\varphi\mathbf{U}\psi$ forever, as this would lead to a rejecting path in the corresponding run. This is because all "until"-formulae are final states in our automata, and the co-Büchi condition requires an accepting run to have no paths visiting infinitely many final states. At the position where $\varphi\mathbf{U}\psi$ no longer loops, its next transition needs to be a transition of ψ, by Definition 6. Thus we can extract an accepting run of $\mathcal{A}_\psi$ for the suffix word

---

[2] We need the precondition word_range $w \subseteq \mathscr{P}(\text{props } \phi)$ to make sure that w is a word over the alphabet $\mathscr{P}(\text{props } \phi)$ as we have no restriction on $w[0]$ otherwise.

starting at that position. For all positions until that point we can extract runs of $\mathcal{A}_\varphi$ and thus, via induction hypothesis, show that the word satisfies $\varphi \mathbf{U} \psi$.

The correctness of the second part, from VWAA to GBA, is captured in the following theorem.

$$\vdash \text{isVeryWeakAA } a_{AA} \wedge \text{ FINITE } a_{AA}.\text{alphabet} \wedge \text{ FINITE } a_{AA}.\text{states} \wedge$$
$$\text{isValidAA } a_{AA} \Rightarrow$$
$$\mathcal{L}_{GB} (\text{vwaa2gba } a_{AA}) = \mathcal{L}_{AA} \, a_{AA}$$

We have to show that for every accepting run of the VWAA on a word w, there exists an accepting run of the GBA on w, and vice versa. By the way the GBA transitions are defined it can be seen that the sequence of layers in a run of the VWAA corresponds to a run of the GBA. The two main difficulties are to cope with the reduction of transitions by $\preceq$ in Definition 7 and to show acceptance of the runs. As our formalization follows the proof in G&O closely, we omit the details here.

Finally we show that we can restrict our automata to reachable states, by proving that no state that is not reachable can appear in any run of the corresponding automaton. We define a function for each automata model, with the overloaded name restr_states, that implements this restriction.

$$\vdash \mathcal{L}_{AA} \, a_{AA} = \mathcal{L}_{AA} (\text{restr\_states } a_{AA})$$
$$\vdash \text{isValidGBA } a_{GB} \Rightarrow \mathcal{L}_{GB} \, a_{GB} = \mathcal{L}_{GB} (\text{restr\_states } a_{GB})$$

4.2 Concrete Data Structures

We use the following generic finite graph type to implement concrete representations of our automata in HOL4:

$$
\begin{aligned}
(\alpha, \epsilon) \, gfg = <&| \\
&\text{node\_info} \, : \, \alpha \, spt; \\
&\text{followers} \, : \, (\epsilon \times num) \, list \, spt; \\
&\text{preds} \, : \, (\epsilon \times num) \, list \, spt; \\
&\text{next} \, : \, num \\
|&>
\end{aligned}
$$

The $\alpha \, spt$ type implements a dictionary with keys that are natural numbers and values of type α. Thus, a graph contains a set of nodes uniquely labeled with natural numbers. Each node is associated with "node information" (the α type parameter). In addition, dictionaries map each node label to outgoing and incoming edges, where each edge connects to another node (identified by the num), and "edge information" (the ϵ parameter). Finally, the next field tracks the next node label, to be used when a node is inserted. This representation is inspired by Erwig [4], and is readily translated into CakeML.

The types used to capture node and edge information are given in Fig. 3. As the transition structure of alternating automata allows conjunctive transitions to several successors we cannot directly map it into the transition structure of the graph. To solve this we extend the edge labels by a field called edge_grp. Multiple edges with the same value of edge_grp are meant to belong to the same conjunctive edge of the alternating automaton. The set of symbols of Σ for which the transition is active is represented using two lists of atomic propositions, one for positive and one for negative occurrences. This is possible because the first component of any transition is always the result of intersecting sets Σ, Σ_p and $\Sigma_{\neg p}$, by Definition 6, which was observed in G&O. This explains the type of our edge labels α edge_labelAA as defined in Fig. 3.

> α edge_labelAA = <| edge_grp : num; pos_lab : α list; neg_lab : α list |>
> α node_labelAA =
>   <| frml : α ltl_frml; is_final : bool; true_labels : α edge_labelAA list |>
> α concrAA = <|
>   graph : (α node_labelAA, α edge_labelAA) gfg;
>   init : num list list;
>   atomic_prop : α list
> |>

Fig. 3. Encoding the concrete representation of alternating automata.

Another aspect of alternating automata that cannot be captured immediately in the graph are transitions to an empty set of successors. One way to handle them is to add a state representing *true* from which any suffix word is accepted. As we do not have this state in our abstract automata in general, this would break the direct correspondence of states in our abstract and concrete models. We encode this information in the node labels of our concrete structure (α node_labelAA). Any edge label that appears in the field true_labels corresponds to an edge with the empty successor set in our abstract model.

> α edge_labelGBA = <| pos_lab : α list; neg_lab : α list; acc_set : α ltl_frml list |>
> α node_labelGBA = <| frmls : α ltl_frml list |>
> α concrGBA = <|
>   graph : (α node_labelGBA, α edge_labelGBA) gfg;
>   init : num list;
>   all_acc_frmls : α ltl_frml list;
>   atomic_prop : α list
> |>

Fig. 4. The types used to encode the concrete representation of GBAs.

Using these two types we define our concrete alternating automata by combining the graph with a list of atomic propositions and an init field corresponding to the set of sets given by I in the abstract automaton.

As the GBA transition structure corresponds to an ordinary graph, we can define it in the natural way (see Fig. 4). By Definition 7, the states of the GBA are sets of states of the VWAA, which are LTL formulae in our case, so we label the GBA states by lists of LTL formulae. The acc_set field is a list of formulae for which the edge is accepting. So rather than grouping all the accepting edges in a set T_f, every edge that is accepting for f should contain f in the field acc_set. Additionally the field all_acc_frmls declares all acceptance sets that exist in the GBA.

4.3   Abstraction Functions

To establish the correspondence between our concrete and abstract automata we define abstraction functions that take a concrete automaton and return its abstract counterpart. These abstraction functions can be seen as defining the semantics of the concrete structure.

To abstract the states of the automaton we visit all nodes in the graph and read their labels. For the transitions we introduce the following function:

transform_label AP pos neg $=$
    FOLDR (λ a $sofar$. char ($\mathscr{P}(AP)$) a $\cap$ $sofar$)
      (FOLDR (λ a $sofar$. char_neg ($\mathscr{P}(AP)$) a $\cap$ $sofar$) ($\mathscr{P}(AP)$)
      neg) pos

The functions char and char_neg are defined exactly as the sets Σ_p and $\Sigma_{\neg p}$ in Definition 6, where $\Sigma = \mathscr{P}(AP)$ in this case. The function transform_label defines how the fields pos_lab and neg_lab of the concrete edge labels should be interpreted. It computes all subsets of Σ that contain all atomic propositions in pos and do not contain any atomic proposition in neg.

Note that different values of pos and neg can lead to the same abstract interpretation by transform_label. One reason is that the order of the lists does not matter, the other is that whenever some atomic proposition appears in both lists, the value of transform_label is the empty set.

To abstract the transition function we have to compute a set of abstract transitions given a formula φ. We do this by finding the node labeled by φ in the graph, grouping its outgoing edges by the value of edge_grp, looking up all the identifiers of the successor states and computing the first components of the transitions using transform_label. If there is no such node in the graph, the function returns the empty set. We call this function abstr_transAA. The procedure for the abstract GBA follows the same idea but does not have to bother with conjunctive edges.

The final states are abstracted by collecting all states of the concrete VWAA that have is_final set to *true*. From the concrete GBA we get the sets T_f by collecting all transitions where the acc_set field in the edge label contains the formula f.

4.4   Concrete Translations

Concrete LTL to VWAA. First we describe our concrete algorithm for the first part of the translation, from LTL formulae to VWAA, now encoded with the concrete graph types described in Sect. 4.2. We reimplement the core functions $\overline{\varphi}$ and $\otimes$ and a concrete version of δ, called concr_trans, using lists, and show that when abstracted with transform_label, concr_trans corresponds to δ.

Theorem 3

$$\vdash \mathsf{set}\ (\mathsf{MAP}\ (\mathsf{abstr\_edge}\ AP)\ (\mathsf{concr\_trans}\ \phi))\ =\ \mathsf{trans}\ (\mathscr{P}(AP))\ \phi$$

Here trans is δ, computed for a specific alphabet, and abstr_edge applies transform_label to the lists of positive and negative atomic propositions of a concrete edge, and transforms the list of successors into a set.

Additionally we specify functions for adding nodes and edges to the graph representing the alternating automaton, add_state and add_edge. The function add_state is a wrapper around the generic function of the graph type for adding nodes that additionally decides whether or not a state should be final by checking if the formula is an "until"-formula. The function add_edge decides whether to add the edge to the true_labels field of the node, which it does if the set of successors is empty, or by using real edges in the graph. Because add_edge may be called for a node that is not in the graph, its return value uses the option type.

Using these auxiliaries, we define a recursive function called expand_graph (see Fig. 5). It maintains a list of nodes to process and the current state of the graph. In every iteration the first element of the list is processed by computing its outgoing transitions with concr_trans and adding the successors and the edges to the graph. The list of nodes that still need to be processed is extended by the new successors if they have not been processed already. For a given formula φ, expand_graph is initially called with the list of formulae in $\overline{\varphi}$ (the set of initial states by Definition 6), and, as its first parameter, the graph containing only these formulae and no edges.

To show termination of expand_graph we use the fact that in the list of nodes to be processed we always remove one element f and replace it with its sucessors, all of which are subformulae of f. As the "subformulae of" relation is a partial order, this lets us use the multiset ordering to define a wellfounded order on the second argument of expand_graph that decreases in every iteration.

Concrete VWAA to GBA. The second part of the concrete translation, from VWAA to GBA, takes a concrete alternating automaton as input and computes a concrete GBA. The states of the GBA are labeled by lists of states of the VWAA. As the set of outgoing transitions of a GBA state depends on the transitions of the VWAA states in its label, we need to compute these from the input VWAA. We do this by defining a concrete version of the function abstr_transAA called get_concr_transAA.

```
expand_graph g [] = SOME g
expand_graph g₁ (f :: fs) =
  (let
     trans = concr_trans f ;
     sucs = nub (FOLDR (λ e pr. e.sucs ⧺ pr) [] trans) ;
     g₂ = FOLDR (λ p g. add_state g p) g₁ sucs ;
     g₃ =
       FOLDR (λ e g?. monad_bind g? (add_edge f e))
         (SOME g₂) trans ;
     new_to_process =
       FILTER
         (λ s. ¬MEM s (graph_states g₁) ∧ s ≠ f ∧ ¬MEM s fs)
         sucs
   in
   case g₃ of
   | NONE ⇒ NONE
   | SOME g ⇒ expand_graph g (new_to_process ⧺ fs))
```

Fig. 5. Concrete function implementing LTL to VWAA. The first argument is the graph of an alternating automaton and the second argument is the list of nodes that still need to be processed.

To compute the transition of a GBA state labeled by a list of VWAA states L, we compute get_concr_transAA for every q in L and then apply a fold with our concrete version of $\otimes$ to the list of transitions. For every edge we then need to check for which of the final states f of the VWAA the conditions of T_f, given in Definition 7, apply. Remember that this includes a check whether there is a transition in $\delta(f)$ that does not contain f in its successor set. To perform these checks more efficiently, we precompute the transitions for all final states of the VWAA.

Finally we need to remove all transitions that are not $\preceq$-minimal. To do this we define a concrete counterpart of $\preceq$. Having defined this relation, we find the minimal elements by comparing all the computed transitions of a state pairwise. We then add the successor states and the edges to the graph and extend the list of nodes to be processed by the new nodes.

Showing termination of this function is more involved than for the first part. The reason is that there is no partial order on the states of the GBA in general, indeed it can have non trivial cycles. To show termination we use the following insight: either the statespace of the graph grows, or it stays the same and the list of nodes to be processed becomes shorter. The first part is a wellfounded relation, as there is an upper bound on the total number of possible states, namely the powerset of the states of the alternating automaton, $\mathscr{P}(Q)$. Here we need to show that all new states computed by concr_trans are really in $\mathscr{P}(Q)$. If the statespace of the graph does not grow in some iteration of expand_graph, we know that all successors of the currently processed node must already have been processed. Thus the list of nodes to be processed gets shorter by one, as the current node is removed. Combining these two orders lexicographically leads to a wellfounded relation. The same approach to prove termination of a graph expansion algorithm was adopted in Schimpf *et al.* [15].

4.5   Verifying the Concrete Functions

After having defined our concrete automata types and concrete functions that implement the translations we show two things. First, they never return NONE on any reasonable input. For the VWAA to GBA translation we require a concrete alternating automaton as produced by the concrete LTL to VWAA translation. Second, applying the abstraction functions gives us exactly the abstract automata that we get by chaining the abstract translation function with the restriction to reachable states. For the LTL to VWAA translation we prove the following theorem, which essentially corresponds to the left hand side of Fig. 2. The function concr_ltl2vwaa computes the list of initial states and calls expand_graph.

$$\vdash \forall \varphi.$$
$$\exists c_{AA}.$$
$$\text{concr\_ltl2vwaa } \varphi = \text{SOME } c_{AA} \wedge$$
$$\text{abstr\_AA } c_{AA} = \text{restr\_states } (\text{ltl2vwaa } \varphi)$$

To show the first part we need to show that we do not call add_edge for a node that is not in the graph, since this is the only possibility for expand_graph to return NONE (see Fig. 5). We do this by showing that all nodes in the list that still have to be processed must have been added to the graph already.

The second part amounts to showing that, after applying the abstraction functions, the states, the transition function, the initial and the final states are equal to the corresponding fields in the result of the abstract translation.

Using Theorem 3 we show that for every state q that has already been processed it holds that all states that are one step reachable from q are either already in the graph, or in the list of nodes to be processed. Reachability here means the reflexive and transitive closure of δ. From this lemma follows that we will eventually include all reachable states of the abstract automaton. To show that only such states are included we again use Theorem 3 and show the invariant that every state in the graph is indeed reachable. In these two steps we use the assumption that the initial states are computed correctly, which we prove independently.

For the transition function we need to show that add_edge adds the edges computed by concr_trans in the intended way. To show this we show that for all nodes in the graph g that have been processed already, abstr_transAA g q is equal to $\delta(q)$.

The proofs for the abstractions of final and initial states amount to showing that the concrete computation of $\overline{\varphi}$ corresponds to the abstract function, and that exactly the nodes labeled by an "until" formula have is_final set to *true*.

For the second part of the translation, from concrete VWAA to concrete GBA, we prove the following theorem, which corresponds to the right hand side of Fig. 2:

$$\vdash \mathsf{concr\_ltl2vwaa} \; \varphi \; = \; \mathsf{SOME} \; c_{AA} \; \wedge \; a_{AA} \; = \; \mathsf{abstr\_AA} \; c_{AA} \; \Rightarrow$$
$$\exists \, c_{GB} \, .$$
$$\mathsf{concr\_vwaa2gba} \; c_{AA} \; = \; \mathsf{SOME} \; c_{GB} \; \wedge$$
$$\mathsf{abstr\_GBA} \; c_{GB} \; = \; \mathsf{restr\_states} \; (\mathsf{vwaa2gba} \; a_{AA})$$

We have similar proof obligations here as in the previous case, we need to show that states, transition function and initial states are correctly abstracted. For the acceptance condition we show that for all f in all_acc_frml of the concrete automaton: if we collect all transitions in the concrete graph labeled by f, we get exactly the set T_f. Additionally, for every f in all_acc_frml we show that $T_f \in \mathcal{T}$, and for the other direction if $T_f \in \mathcal{T}$, then f is in all_acc_frml.

The states are handled by showing that the concrete computation of the transition function corresponds to the abstract definition and then using the same ideas as in the first translation step. For the transition function we have the advantage that it is more directly encoded in the edges of the graph. On the other hand we need to compute the transition functions of the VWAA states, that the GBA state is labeled by, correctly, and handle the minimization by $\preceq$. For the minimization we need to show that two concrete transitions are related by our concrete version of $\preceq$ if and only if their abstract counterparts are related by $\preceq$. This implies that we are removing the right transitions in the concrete function.

Translation to CakeML. The CakeML ecosystem includes a general mechanism for translating (a subset of) HOL functions into provably equivalent CakeML ASTs (Myreen and Owens [13]). We use this technology to transform our concrete algorithm into CakeML syntax, to which we can then apply the CakeML compiler, generating assembly code. Under minimal assumptions (including: CakeML's model of the hardware corresponds to that of the chip that actually executes the code, and the correctness of the assembler and linker used to generate the final executable), the correctness of the CakeML compiler lets us conclude that this machine code will implement the algorithm exactly as written in the HOL formulation. In turn, the abstraction proofs described earlier then give us a high-assurance connection between the machine code that executes and the mathematical results of G&O.

At this stage, we embody our algorithms in a simple tool that parses an LTL formula on standard input, and prints out the two translated automata as (typically rather large) S-expressions. We have not benchmarked our executable's performance to any degree. Certainly, we are confident that CakeML-compiled code and a naïve representation of graphs/automata will not perform as well as hand-tuned C tools that have had extensive development. On the other hand, the development in HOL4 and CakeML gives us extremely high assurance that our tool is correct.

5 Related Work

The most complete verification effort of algorithms in the context of LTL model checking was done by Esparza *et al.* [6]. They describe a fully verified implementation of an LTL model checker in the Isabelle theorem prover. The work builds on a previously described verification [15] of the LTL to generalized Büchi automata translation which was introduced by Gerth *et al.* [8]. The algorithm uses a tableau construction and is more amenable to a direct verification as it does not include the intermediate step of alternating automata. The work has been extended to use Promela as input language to describe systems [14] and to use partial order reductions [1]. Additional optimization techniques for Büchi automata have been verified as independent functions in Schimpf and Smaus [16]. Another mechanization of a translation algorithm from LTL to automata was reported on in Esparza *et al.* [5]. The authors introduce a new algorithm targeting deterministic automata and emphasize the importance of interactive theorem provers, which allowed them to uncover errors in their original proofs.

One approach that has been developed to refine abstract definitions into efficient code is the Isabelle Refinement Framework [9,10]. Both powerful and generic, it allows the refinement of abstract types into more efficient data structures. We believe that our rather custom abstraction would have been hard to achieve in this framework, as the structure of the abstract automata are quite different to the concrete ones, and multiple abstract details are encoded in the same concrete types. For example, consider the accepting edges of the GBA. While the abstract automaton provides all these edges in a set of sets, in the concrete world they are embedded in the graph using the edge labels.

Alternating automata in the context of interactive theorem proving were previously addressed by Merz [12]. This work mechanizes a proof of the closure of weak alternating automata under complementation, using winning strategies of logical games. As an application, Merz presents a translation from LTL into very weak alternating automata. The translation mechanized by Merz generates more states than G&O (all sub-formulæ and negations *vs.* only temporal subformulae), and he does not address the second, exponential, translation to GBAs. This work also remains completely abstract, without mentioning concrete algorithms.

6 Conclusion

In this paper we have presented a formalization of the algorithm for translating LTL formulae into generalized Büchi automata presented in G&O, which uses very weak alternating automata as an intermediate representation.

We introduce an encoding of both alternating automata and generalized Büchi automata in a compilable, generic graph type that uses an efficient lookup structure. This is especially interesting for alternating automata, as they are a powerful computational model leading to elegant algorithms, *e.g.*, Vardi [18].

To cope with the complexity of the algorithm, we divide the formalization into an abstract and a concrete part. In the abstract part we mechanize the proofs and show correctness of the translation as it is presented in G&O. The correspondence between the abstract and concrete models is established using abstraction functions that map concrete automata to abstract ones. We implement the algorithm on our concrete types and show that applying the abstractions to the resulting automata leads to the automata given by the abstract translation.

This approach turned out to be fruitful: we were able to reproduce the abstract correctness results fairly quickly. Not having to additionally cope with arguments about concrete data structures, termination and details concerning our graph type, made a big difference. We would like to extend our ideas to include optimization steps in the translation, by showing independent correctness in the abstract world and efficiently embedding them in the expansion algorithm. So far, efficiency has not been a big concern for us; rather we have focused on producing verified code for the algorithm in G&O. In future work we would like to optimize the code and provide an empirical comparison to existing tools.

Finally we use the CakeML compiler to produce fully verified code implementing our concrete functions. This step significantly strengthens the confidence we can have in the machine code, as we do not have to trust a standard compiler. Translation of LTL formulae into automata is only one part of a complete model checker, but our experience suggests that an extremely high assurance model checker embodying sophisticated optimizations is entirely feasible.

References

1. Brunner, J., Lammich, P.: Formal verification of an executable LTL model checker with partial order reduction. In: Rayadurgam, S., Tkachuk, O. (eds.) NFM 2016. LNCS, vol. 9690, pp. 307–321. Springer, Cham (2016). https://doi.org/10.1007/978-3-319-40648-0_23
2. Courcoubetis, C., Vardi, M., Wolper, P., Yannakakis, M.: Memory-efficient algorithms for the verification of temporal properties. Form. Meth. Syst. Des. $\mathbf{1}(2)$, 275–288 (1992)
3. Couvreur, J.-M., Duret-Lutz, A., Poitrenaud, D.: On-the-fly emptiness checks for generalized Büchi automata. In: Godefroid, P. (ed.) SPIN 2005. LNCS, vol. 3639, pp. 169–184. Springer, Heidelberg (2005). https://doi.org/10.1007/11537328_15
4. Erwig, M.: Functional programming with graphs. In: Simon, L., Jones, P., Tofte, M., Berman, A.M. (eds.) Proceedings of the 1997 ACM SIGPLAN International Conference on Functional Programming (ICFP 1997), Amsterdam, The Netherlands, 9–11 June 1997, pp. 52–65. ACM (1997)
5. Esparza, J., Křetínský, J., Sickert, S.: From LTL to deterministic automata - a safraless compositional approach. Form. Meth. Syst. Des. $\mathbf{49}(3)$, 219–271 (2016)
6. Esparza, J., Lammich, P., Neumann, R., Nipkow, T., Schimpf, A., Smaus, J.-G.: A fully verified executable LTL model checker. In: Sharygina, N., Veith, H. (eds.) CAV 2013. LNCS, vol. 8044, pp. 463–478. Springer, Heidelberg (2013). https://doi.org/10.1007/978-3-642-39799-8_31

7. Gastin, P., Oddoux, D.: Fast LTL to Büchi automata translation. In: Berry, G., Comon, H., Finkel, A. (eds.) CAV 2001. LNCS, vol. 2102, pp. 53–65. Springer, Heidelberg (2001). https://doi.org/10.1007/3-540-44585-4_6

8. Gerth, R., Peled, D., Vardi, M.Y., Wolper, P.: Simple on-the-fly automatic verification of linear temporal logic. In: Dembiński, P., Średniawa, M. (eds.) Protocol Specification, Testing and Verification XV, PSTV 1995. IFIP Advances in Information and Communication Technology, pp. 3–18. Springer, Boston (1996). https://doi.org/10.1007/978-0-387-34892-6_1

9. Lammich, P.: Automatic data refinement. In: Blazy, S., Paulin-Mohring, C., Pichardie, D. (eds.) ITP 2013. LNCS, vol. 7998, pp. 84–99. Springer, Heidelberg (2013). https://doi.org/10.1007/978-3-642-39634-2_9

10. Lammich, P., Tuerk, T.: Applying data refinement for monadic programs to Hopcroft's algorithm. In: Beringer, L., Felty, A. (eds.) ITP 2012. LNCS, vol. 7406, pp. 166–182. Springer, Heidelberg (2012). https://doi.org/10.1007/978-3-642-32347-8_12

11. Loding, C., Thomas, W.: Alternating automata and logics over infinite words. In: van Leeuwen, J., Watanabe, O., Hagiya, M., Mosses, P.D., Ito, T. (eds.) TCS 2000. LNCS, vol. 1872, pp. 521–535. Springer, Heidelberg (2000). https://doi.org/10.1007/3-540-44929-9_36

12. Merz, S.: Weak alternating automata in Isabelle/HOL. In: Aagaard, M., Harrison, J. (eds.) TPHOLs 2000. LNCS, vol. 1869, pp. 424–441. Springer, Heidelberg (2000). https://doi.org/10.1007/3-540-44659-1_26

13. Myreen, M.O., Owens, S.: Proof-producing synthesis of ML from higher-order logic. In: Thiemann, P., Findler, R.B. (eds.) ACM SIGPLAN International Conference on Functional Programming, ICFP 2012, Copenhagen, Denmark, 9–15 September 2012, pp. 115–126. ACM (2012)

14. Neumann, R.: Using promela in a fully verified executable LTL model checker. In: Giannakopoulou, D., Kroening, D. (eds.) VSTTE 2014. LNCS, vol. 8471, pp. 105–114. Springer, Cham (2014). https://doi.org/10.1007/978-3-319-12154-3_7

15. Schimpf, A., Merz, S., Smaus, J.-G.: Construction of Büchi automata for LTL model checking verified in Isabelle/HOL. In: Berghofer, S., Nipkow, T., Urban, C., Wenzel, M. (eds.) Theorem Proving in Higher Order Logics: Proceedings of 22nd International Conference, TPHOLs 2009, Munich, Germany, 17–20 August 2009, pp. 424–439. Berlin, Heidelberg (2009). https://doi.org/10.1007/978-3-642-03359-9

16. Schimpf, A., Smaus, J.-G.: Büchi automata optimisations formalised in Isabelle/HOL. In: Banerjee, M., Krishna, S.N. (eds.) ICLA 2015. LNCS, vol. 8923, pp. 158–169. Springer, Heidelberg (2015). https://doi.org/10.1007/978-3-662-45824-2_11

17. Vardi, M.Y.: Nontraditional applications of automata theory. In: Hagiya, M., Mitchell, J.C. (eds.) TACS 1994. LNCS, vol. 789, pp. 575–597. Springer, Heidelberg (1994). https://doi.org/10.1007/3-540-57887-0_116

18. Vardi, M.Y.: Alternating automata: unifying truth and validity checking for temporal logics. In: McCune, W. (ed.) CADE 1997. LNCS, vol. 1249, pp. 191–206. Springer, Heidelberg (1997). https://doi.org/10.1007/3-540-63104-6_19

CALCCHECK: A Proof Checker for Teaching the "Logical Approach to Discrete Math"

Wolfram Kahl[✉]

McMaster University, Hamilton, ON, Canada
kahl@cas.mcmaster.ca

Abstract. For calculational proofs as they are propagated by Gries and Schneider's textbook classic "A Logical Approach to Discrete Math" (LADM), automated proof checking is feasible, and can provide useful feedback to students acquiring and practicing basic proof skills. We report on the CALCCHECK system which implements a proof checker for a mathematical language that resembles the rigorous but informal mathematical style of LADM so closely that students very quickly recognise the system, which provides them immediate feed-back, as not an obstacle, but as an aid, and realise that the problem is finding proofs.

Students interact with this proof checker trough the "web application" front-end CALCCHECK_Web which provides some assistance for proof entry, but intentionally no assistance for proof finding. Upon request, the system displays, side-by-side with the student input, a version of that input annotated with the results of checking each step for correctness.

CALCCHECK_Web has now been used twice for teaching an LADM-based second-year discrete mathematics course, and students have been solving exercises and submitting assignments, midterms, and final exams on the system — for examinations, there is the option to disable proof checking and leave only syntax checking enabled. CALCCHECK also performed the grading, with very limited human overriding necessary.

1 Introduction

The textbook "A Logical Approach to Discrete Math" (referred to as "LADM") by Gries and Schneider (1993) is a classic introduction to reasoning in the *calculational style*, which allows for rigorous-yet-readable proofs. Gries and Schneider (1995) establish a precise logical foundation for such calculations in propositional logic, and Gries (1997) expands this also to predicate logic, so that we do not need to dwell on these aspects in the current paper.

We present a mechanised theory language that has been designed to be as close to the "informal" but rigorous language of LADM, and the proof checker CALCCHECK designed for supporting teaching based on LADM. A predecessor system (Kahl 2011) using LATEX-based interaction in the style of *f*UZZ (Spivey 2008) only supported checking isolated calculations in a hard-coded LADM-like expression language, and recognised only hard-coded theorem numbers in unstructured hints; the current version of CALCCHECK admits user-defined

© Springer International Publishing AG, part of Springer Nature 2018
J. Avigad and A. Mahboubi (Eds.): ITP 2018, LNCS 10895, pp. 324–341, 2018.
https://doi.org/10.1007/978-3-319-94821-8_19

operators, and has a completely new language for theories, structured proofs, and structured calculation hints.

Since students still need to learn "what a proof is" and "how different proofs can be", we consciously do not offer any assistance in proof finding, but we turned the proof checker into a web application, so that students can obtain instant feedback for their proof attempts, all while writing proofs that are recognisably in the style of the textbook.

For example, on p. 55 of LADM we find the following calculation (with relatively detailed hints), reproduced here almost exactly (only with slightly different spacing):

As an example, we prove theorem (3.44a): $p \wedge (\neg p \vee q) \equiv p \wedge q$:

$$p \wedge (\neg p \vee q)$$
$=$ ⟨ Golden rule (3.35), with $q := \neg p \vee q$ ⟩
$$p \equiv \neg p \vee q \equiv p \vee \neg p \vee q$$
$=$ ⟨ Excluded middle (3.28) ⟩
$$p \equiv \neg p \vee q \equiv true \vee q$$
$=$ ⟨ (3.29), $true \vee p \equiv true$ ⟩
$$p \equiv \neg p \vee q \equiv true$$
$=$ ⟨ Identity of $\equiv$ (3.3) ⟩
$$p \equiv \neg p \vee q$$
$=$ ⟨ (3.32), $p \vee \neg q \equiv p \vee q \equiv p$,
         with $p, q := q, p$ — to eliminate operator $\neg$ ⟩
$$p \equiv p \vee q \equiv q$$
$=$ ⟨ Golden rule (3.35) ⟩
$$p \wedge q$$

In CALCCHECK, this theorem together with this proof can be entered as follows in plain Unicode text:

```
Theorem (3.44) (3.44a) "Absorption": p ∧ (¬ p ∨ q) ≡ p ∧ q
Proof:
    p ∧ (¬ p ∨ q)
  ≡( "Golden rule" (3.35) with `q = ¬ p ∨ q` )
    p ≡ ¬ p ∨ q ≡ p ∨ ¬ p ∨ q
  ≡( "Excluded middle" (3.28) )
    p ≡ ¬ p ∨ q ≡ true ∨ q
  ≡( (3.29) `true ∨ p ≡ true` )
    p ≡ ¬ p ∨ q ≡ true
  ≡( "Identity of ≡" (3.3) with `q = p ≡ ¬ p ∨ q` )
    p ≡ ¬ p ∨ q
  ≡( (3.32) `p ∨ q ≡ p ∨ ¬ q ≡ p`
       with `p, q = q, p` — to eliminate operator ¬ )
    p ≡ p ∨ q ≡ q
  ≡( "Golden rule" (3.35) )
    p ∧ q
```

Except for the comment "— to eliminate operator $\neg$", everything here is formal content, and checked by the system. We will explain some of the details in Sect. 2. It should however be obvious that the correspondence is very close, with the small differences mostly due to either the fact that we are using a plain text format, or to the requirement that the language needs to be unambiguously parse-able for automatic checking to become feasible.

A student encountering only the theorem statement of this in their homework might, if allowed to use two theorem references per hint, write the variant shown below to the left in the CALCCHECK$_{\text{Web}}$ interface in their web browser:

```
Theorem (3.44) (3.44a) "Absorption":
    p ∧ (¬ p ∨ q) ≡ p ∧ q
Proof:
|   p ∧ (¬ p ∨ q)
  ≡( "Golden rule" )
    p ≡ ¬ p ∨ q ≡ p ∨ ¬ p ∨ q
  ≡( "Excluded middle", "Zero of ∨" )
    p ≡ ¬ p ∨ q ≡ true
  ≡( "Identity of ≡", (3.30) )
    p ≡ p ∨ q ≡ q
  ≡( "Golden rule" )
    p ∧ q
```

Theorem (3.44) (3.44a) "Absorption": p ∧ (¬ p ∨ q) ≡ p ∧ q
Proof:
  Proving `p ∧ (¬ p ∨ q) ≡ p ∧ q`:
    p ∧ (¬ p ∨ q)
  ≡("Golden rule")
    — CalcCheck: Found (3.35) "Golden rule"
    — CalcCheck: — OK
    p ≡ (¬ p ∨ q ≡ p ∨ (¬ p ∨ q))
  ≡("Excluded middle", "Zero of ∨")
    — CalcCheck: Found (3.28) "Excluded middle"
    — CalcCheck: Found (3.29) "Zero of ∨"
    — CalcCheck: — OK
    p ≡ (¬ p ∨ q ≡ true)
  ≡("Identity of ≡", (3.30))
    — CalcCheck: Found (3.3) "Identity of ≡"
    — CalcCheck: Found (3.30) "Identity of ∨"
    — CalcCheck: Could not justify this step!
    p ≡ (p ∨ q ≡ q)
  ≡("Golden rule")
    — CalcCheck: Found (3.35) "Golden rule"
    — CalcCheck: — OK
    p ∧ q
— CalcCheck: 1 out of 4 steps not justified
— CalcCheck: Calculation matches goal — OK

After sending this to the server for checking, the box to the right will be filled in by the system as shown in the screen-shot above, and the student will likely notice that they mis-typed the number of theorem (3.32), one of the few nameless theorems in LADM that are emphasised as worth remembering the number of.

We proceed with explaining the basics of the CALCCHECK theory language in Sect. 2. In Sect. 3 we strive to give an idea of how interaction with such theories works in practice, before proceeding to more advanced language features: In Sect. 4 we present the main hard-coded proof structuring principles, and in Sect. 5 we discuss our treatment of quantification, substitution, and metavariables. More complicated hints are covered in Sect. 6, and mechanisms for selectively making reasoning features available in Sect. 7. Finally, we highlight some aspects of the implementation in Sect. 8 and discuss some related work in Sect. 9. Some additional documentation is available at the CALCCHECK home page at http://CalcCheck.mcmaster.ca/.

2   The Basic CALCCHECK Language

A CALCCHECK module consists of a sequence of *top-level items* (TLIs), which include declarations, axioms, theorems, and several kinds of administrative items, as for example precedence declarations.

```
Precedence 40 for: _∧_
Associating to the right: _∧_
Declaration: _∧_ : 𝔹 → 𝔹 → 𝔹
Axiom (3.35) "Golden rule": p ∧ q ≡ p ≡ q ≡ p ∨ q
```

The language is layout-sensitive: Everything after the first line in a top-level item, or inside other high-level syntactic constructs, needs to be indented at least two spaces farther than the first line. The only exception to this is the "Proof:" for a theorem, which starts in the same column as the theorem.

Instead of the word Theorem, one may alternatively use Lemma, Corollary, Proposition, or Fact without any actual differences. (Technically, Axioms are theorems that are just not allowed to have proof.) A theorem may have any number of *theorem numbers* (always in parentheses and without spaces, such as (3.35) above and (3.44a) in Sect. 1) and *theorem names* (always in pretty double quotes — as opposed to the plain double quotation mark character " " " — such as ""Absorption"" in Sect. 1). The same names and numbers may be given to several theorems, which implements the way LADM uses "Absorption (3.44)" to refer to uses of either (3.44a) or (3.44b) or both.

A *calculation*, such as the proof body in Sect. 1, consists of a sequence of *expressions* interleaved with *calculation operators* (in Sect. 1 only "≡") attached to a pair of *hint brackets* "⟨ ... ⟩" enclosing a *hint*. A hint is a sequence of *hint items* separated by commas or "and" or both; so far, we only have seen *theorem references* as hint items. (Comments, such as "— to eliminate operator ¬" in Sect. 1, are currently only supported inside hints.)

A theorem reference can be either a theorem name in pretty double quotes, or a theorem number in parentheses, or an expression in back-ticks ('... '), or several theorem references separated by white-space. In Sect. 1 we have seen a few examples of the latter in the LADM calculation; they refer to the intersection of the sets of theorems referred to by the constituent atomic theorem references. It is configurable whether theorem references in the shape of expressions can be used alone; this is forbidden by default: Learning the theorem names is, for the most part, learning the vocabulary of the language of discrete math, and therefore part of the learning objectives. (With this setting, a theorem with no names nor numbers cannot be referred to, but may still be useful for documentation, for example as a Fact.)

For the expression syntax, almost arbitrary sequences of printable Unicode characters are legal identifiers, as in Agda (Norell 2007), so that almost all lexemes need to be separated by spaces. (Parentheses need no spaces.)

CALCCHECK follows LADM in supporting conjunctional operators: The expression $1 < 2 \in S \subseteq T$ is considered shorthand for $(1 < 2) \wedge (2 \in S) \wedge (S \subseteq T)$. The set of conjunctional operators and their precedence is not hard-coded; for emulating LADM we write in our "prelude":

```
Precedence 50 for:
  _=_ , _≠_ , _<_ , _>_ , _≤_ , _≥_ , _≮_ , _≰_ , _≯_ , _≱_ , _∤_ ,
  _∈_ , _∉_ , _∋_ , _∌_ , _⊂_ , _⊆_ , _⊄_ , _⊈_ , _⊃_ , _⊅_ , _⊇_ , _⊉_
Conjunctional:
  _=_ , _≠_ , _<_ , _>_ , _≤_ , _≥_ , _≮_ , _≰_ , _≯_ , _≱_ , _∤_ ,
  _∈_ , _∉_ , _∋_ , _∌_ , _⊂_ , _⊆_ , _⊄_ , _⊈_ , _⊃_ , _⊅_ , _⊇_ , _⊉_
```

As could be seen already in the example at the beginning of this section, such declarations of operator precedence and associating behaviour come before the actual declarations of the operator: Like LADM, CALCCHECK supports operator overloading. Since operator precedence and associating behaviour have to be declared before the actual Declarations, it is easy to enforce coherent precedences also in larger developments. (In LADM, this declaration-independent precedence table can be found on the inside cover.)

Underscores denote argument positions of mixfix operators. Arbitrary binary infix operators can be used as calculation operators, that is, preceding hint brackets ⟨ ... ⟩. The calculation notation as such is considered conjunctional, which enables us to use the non-conjunctional associative operator ≡ as calculation operator in the examples in Sect. 1, or, later, also implication.

Two of the steps in the first calculation specify substitutions to variables in the referenced theorem with expressions (containing variables of the currently-proven theorem), for example "with $p, q := q, p$" in the second-last step. This is also allowed in CALCCHECK, except that the substitution is delimited by back-ticks. We choose back-ticks because they are also used in MarkDown for embedding code in prose — CALCCHECK allows MarkDown blocks as top-level items for "literate theories" documentation. We us backticks for embedding expressions, and other expression-level material such as substitutions, inside "higher-level structures" in many places. In theorems and proofs, essentially the only places where backticks are not used are after the colon in the theorem statement, and outside the hints in calculations.

3   The CALCCHECK$_{\text{Web}}$ Front-End

Since CALCCHECK source files are just plain Unicode text files, editing them using any editor is certainly possible. However, the preferred way to edit CALCCHECK source, and the only way currently offered to students, is via the "web application" CALCCHECK$_{\text{Web}}$, which can be accessed via websocket-capable web browsers.

A "notebook" style view is presented with a vertical sequence of "cells". Mark-Down TLIs are shown in cells containing a single box, and "code cells" with a horizontal split into two boxes (as already shown in the screenshot at the end of Sect. 1). The left box is for code entry, and the right box is populated with feedback from the server, which performs, upon request, syntax checking, or syntax and proof checking combined. (For exams, proof checking can be disabled.)

For text and code entry, CALCCHECK$_{\text{Web}}$ provides symbol input via mostly LATEX-like escape sequences; typing a backslash triggers a pop-up displaying the escape sequence typed so far, and the possible next characters. In experienced use, this pop-up is irrelevant, and disappears when characters beyond the completion of the escape sequence are entered. Alternatively, upon a TAB key press, this pop-up also displays a menu, as in the following screenshot:

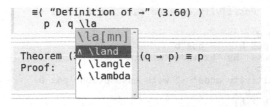

Similar completion is provided for theorem names, after typing a prefix (of at least length three) of a theorem name preceded by either pretty opening double quotes "" or the simple double-quote character '"', hitting the TAB key brings up a theorem name completion menu containing only theorem names currently in scope, but intentionally not filtered in any other way.

Support for indentation currently provided includes toggling display of initial spaces as "visible space" "␣", and key bindings for increasing or decreasing the indentation of whole blocks.

4   Structured Proofs

Calculations, as shown in Sect. 1, are just one kind of proof supported by CALCCHECK. LADM emphasises the use of axioms (and theorems) in calculations over other inference rules, so not many other proof structures are needed. Besides calculations, the other options for proof in CALCCHECK (explained in more detail below) are:

- "By *hint*" for discharging simple proof obligations,
- "Assuming '*expression*':" corresponding to implication introduction,
- "By cases: '*expression*₁',…,'*expression*ₙ'" for proofs by case analysis,
- "By induction on '*var* : *type*':" for proofs by induction,
- "For any '*var* : *type*':" corresponding to ∀-introduction,
- "Using *hint*:" for turning theorems into inference rules, see Sect. 6.3.

With these (nestable) proof structures, we essentially formalise the slightly more informal practices of LADM, which, in Chapt. 4, introduces what appears to be formal syntax for proofs by cases, and for proofs of implications by assuming the antecedents. However, in actual later LADM proofs, this syntax is typically not used. For example, on p. 305 we find some cases listed in a way that does not easily correspond to the pattern in LADM Chapt. 4, and the assumption of the antecedent is almost hidden in the surrounding prose that replaces the explicit proof structure. We can emulate the calculation there very closely again, and we embed it into a fully formal proof that is, in our opinion, at least as clear and readable as the arrangement in LADM:

```
Theorem (15.34) "Positivity of squares": b ≠ 0 ⇒ pos (b · b)
Proof:
  Assuming `b ≠ 0`:
    By cases: `pos b`, `¬ pos b`
      Completeness: By "Excluded middle"
      Case `pos b`:
        By "Positivity under ·" with assumption `pos b`
      Case `¬ pos b`:
          pos (b · b)
      ≡( (15.23) `- a · - b = a · b` )
          pos ((- b) · (- b))
      ⇐( "Positivity under ·" (15.31) )
          pos (- b) ∧ pos (- b)
      ≡( "Idempotency of ∧", "Double negation" )
          ¬ ¬ pos (- b)
      ≡( "Positivity under unary minus" (15.33) with assumption `b ≠ 0` )
          ¬ pos b        — This is Assumption `¬ pos b`
```

Our syntax for assuming the antecedent should be self-explaining — the keyword assumption for producing hint items referring to an assumption (which may also be given a local theorem name in double quotes) may also be written Assumption. The assumed expression is again delimited by backticks.

For proof by cases, we follow the pattern proposed in LADM Chap. 4, except that we insist on a proof of Completeness of the list of patterns to be explicitly supplied. In the case above, we discharge this proof obligation via By "Excluded middle" — this is another variant of proofs, where just a hint (that is, a sequence of hint items) is provided after the keyword By. The expression of the current Case is available in the proof via the Assumption keyword.

At the end of the calculation above, we have "— This is ..."; this is used in LADM without the words "This is" as a "formal comment" indicating that the last expression in the calculation is the indicated assumption, or, more frequently, an instance of the indicated theorem. Later, Gries and Schneider (1995) explain this via the inference rule "Equanimity". For CALCCHECK, such "— This is ..." clauses are not considered comments at all, but are part of the calculation syntax, and require exactly this phrasing. As in LADM, this can be used at either end of a calculation. Several further details of the above proof of "Positivity of squares" will be explained below in Sects. 6 and 7.

The first proof structure beyond calculations that is introduced in the course is actually successor-based natural induction, where natural numbers have been introduced inductively from zero "0" and the successor operator "S_", and the inductive definitions for operations have been provided as sequences of axioms, as the following for subtraction:

```
Declaration:  _-_  : ℕ → ℕ → ℕ
Axiom "Subtraction from zero":                              0 - n     = 0
Axiom "Subtraction of zero from successor":      (S m) - 0       = S m
Axiom "Subtraction of successor from successor": (S m) - (S n) = m - n
```

With this, even nested induction proofs such as the following become easy to produce for the students:

```
Theorem "Subtraction after addition": (m + n) - n = m
Proof:
  By induction on `m : N`:
    Base case:
        (0 + n) - n
      =( "Identity of +" )
        n - n
      =( "Self-cancellation of subtraction" )
        0
    Induction step `(S m + n) - n = S m`:
      By induction on `n : N`:
        Base case:
            (S m + 0) - 0
          =( "Identity of +" )
            S m - 0
          =( "Subtraction of zero from successor" )
            S m
        Induction step:
            (S m + S n) - S n
          =( "Definition of +" )
            S (m + S n) - S n
          =( "Subtraction of successor from successor" )
            (m + S n) - n
          =( "Adding the successor", "Definition of +" )
            (S m + n) - n
          =( Induction hypothesis `(S m + n) - n = S m` )
            S m
```

The proof goals for base case and induction step may optionally be made explicit — we show this here only for the outer induction step. In nested induction steps where several induction hypotheses are available, the system currently requires the keyword phrase **Induction hypothesis** to be accompanied by the chosen induction hypothesis, but only for pedagogical reasons.

Currently, besides natural induction, also induction on sequences is supported by this hard-coded By induction on proof format; the 'm : $\mathbb{N}$' after this keyword phrase above indicates the induction variable and its type, which selects the induction principle, if one is implemented and activated for that type.

5   Quantification, Substitution, Metavariables

For quantification, CALCCHECK follows the spirit of LADM, but in the concrete syntax is closer to the Z notation (Spivey 1989): The general pattern of quantified expressions is "*bigOp varDecls | rangePredicate • body*", and we have, for example:

$$(\textstyle\sum i \mid 0 \le i < 5 \bullet i\,!) \;=\; 0\,! + 1\,! + 2\,! + 3\,! + 4\,!$$
$$(\forall\, k, n : \mathbb{N} \mid k < n < 3 \bullet k \cdot n < 5) \;\equiv\; 0 \cdot 1 < 5 \wedge 0 \cdot 2 < 5 \wedge 1 \cdot 2 < 5$$

The range predicate, when omitted together with the " | ", defaults to *true*. As in Z, parentheses around quantifications can be omitted, and the scope of the variable binding then extends "as far as syntactically possible". (This a conscious notational departure from LADM, where parentheses around quantifications are compulsory, and ":" is used instead of "•".) In another notational departure, we denote function application by (typically space-separated) juxtaposition, "$f\ x$", instead of "$f.x$" for atomic arguments in LADM.

The following proof is for a stronger variant of the LADM theorem (8.22) "Change of dummy", which both LADM and Gries (1997) show without the range predicate R in the assumption (but when LADM refers to (8.22) later, in Chaps. 12 and 17, it actually always would have to use our variant). Here, as in LADM, "$\star$" is used as a metavariable for a quantification operator, that is, a symmetric and associative binary operator (usually equipped with an identity).

```
Theorem (8.22a) "Change of restricted dummy":
   (∀ x | R • (∀ y • x = f y  ≡  y = g x))
 ⇒ (⋆ x | R • P) = (⋆ y | R[x = f y] • P[x = f y])
Proof:
   Assuming "Inverse" `∀ x | R • ∀ y • x = f y  ≡  y = g x`:
      (⋆ y | R[x = f y] • P[x = f y])
   =( "One-point rule for ⋆" )
      (⋆ y | R[x = f y] • (⋆ x | x = f y • P))
   =( "Nesting for ⋆" )
      (⋆ y, x | R[x = f y] ∧ x = f y • P)
   =( Substitution )
      (⋆ y, x | R[x = z][z = f y] ∧ x = f y • P)
   =( "Replacement" )
      (⋆ y, x | R[x = z][z = x] ∧ x = f y • P)
   =( Substitution )
      (⋆ y, x | R ∧ x = f y • P)
   =( "Dummy list permutation for ⋆" )
      (⋆ x, y | R ∧ x = f y • P)
   =( "Nesting for ⋆" )
      (⋆ x | R • (⋆ y | x = f y • P))
   =( "Range replacement in nested ⋆" with assumption "Inverse" )
      (⋆ x | R • (⋆ y | y = g x • P))
   =( "One-point rule for ⋆" )
      (⋆ x | R • P[y = g x])
   =( Substitution )
      (⋆ x | R • P)
```

LADM and Gries (1997) both refrain from formalising the assumption "f has an inverse" as part of the theorem statement, since they present *all* general quantification theorems before introducing universal quantification. With a different theory organisation, we introduce universal quantification as instance of a restricted theory of general quantification, and then use universal quantification to state and prove theorems like this about general quantification which mention universal quantification.

This "Change of restricted dummy" theorem is really a metatheorem: Its statement contains metavariables x and y for *different* variables, and P and R for expressions that may have free occurrences of x, and it also contains explicit substitutions. Gries (1997) calls such proofs of metatheorems using metatheorems "schematic proofs". [1] The fact that P and R must not have free occurrences of y is expressed by Gries and Schneider as the *proviso* "$\neg occurs('y', 'P, R')$" in the metalanguage.

CALCCHECK takes a slightly different approach to metavariables: For consistency with LADM, we keep the inference rule substitution, and use only the substitution notation $E[v := G]$. Once quantification is introduced, we emphasise that substitution binds variables, too (where only occurrences of v in E are bound in $E[v := G]$), and application of substitution may need to rename bound variables (in E) to avoid capture of free (in G) variables. Expression equality in CALCCHECK is only up to renaming of bound variables; students are encouraged to use "Reflexivity of =" calculation steps to document such renaming.

When introducing quantification and variable binding, we (re-)explain axiom schemas, and emphasise that metavariables are *instantiated* (and not substituted), but do not provide notation for that. (Instantiation of metavariables does not rename binders and therefore can capture variables that are free in the instantiating expression. Such capture is the point of metavariables — in (8.22a) above, R is meant to be instantiated with expression containing free occurences of x.)

In a theorem statement, metavariables for expressions are defined (and recognised by CALCCHECK) as looking like free variables in the scope of a variable binder. Metavariables with occurrences in scope of different sets of variable binders may only be instantiated with expressions in which only the intersection of all these binders occurs free. Bound variables that are allowed to occur in metavariables for expressions have to be considered metavariables for variables, and matched consistently. Thus, the "$\neg occurs$" provisos can be derived from the theorem statement; for the theorem above, if metavariable reporting (by default disabled) and proviso reporting are both enabled, CALCCHECK generates the following output:

Theorem (8.22a) "Change of restricted dummy": $(\forall\, x \mid R \bullet (\forall\, y \bullet x = f\,y \equiv y = g\,x)$
$) \Rightarrow (\bigstar\, x \mid R \bullet P) = (\bigstar\, y \mid R[x := f\,y] \bullet P[x := f\,y])$
— CalcCheck: Metavariables: $P = P[\![\, x\,]\!]$, $R = R[\![\, x\,]\!]$, $f = f[\![\, y\,]\!]$, $g = g[\![\, x, y\,]\!]$
— CalcCheck: Proviso: $\neg occurs('x', 'f')$, $\neg occurs('y', 'P, R')$

The proof above contains three steps where the hint is the keyword Substitution; this hint item is used for performing substitutions. For both Substitution steps here, it is necessary that $\neg occurs('z', 'R')$; for such new variable binders, this is handled automatically by remembering also which variables *are* allowed to occur in R, as shown in the "**Metavariables**" information report above.

Above we used "Replacement" (3.84a): $e = f \wedge P[z := e] \equiv e = f \wedge P[z := f]$ (called "Substitution" in LADM). This is another example for a metatheorem;

---

[1] Gries (1997) restricts metavariables to be named by single upper-case letters, (non-meta-)variables by single lower-case letters. Gries (1997) then distinguishes between "uniform substitution" written $E[V := G]$ for metavariables V, and "textual substitution" written E_G^v for variables v, where only the latter renames variable binders to avoid capture of free variables of G. However, the use of "$R[x := f\,y]$" in the statement and proof of (8.22) there is then unclear — it will have to be understood as "textual substitution" since otherwise y might be captured by binders in R.

its statement involves substitution, and the metavariables z for variables and P for expressions that may have free occurrences of z. Since CALCCHECK currently does not use second-order matching, the reverse Substitution step preceding the application of "Replacement" (3.84a) is necessary for establishing the matching of the metavariables z and P, here to the variable z and the expression $R[x := z]$ respectively. The second Substitution could be merged with the "Replacement" (3.84a) step, but has been left separate here for readability.

Note that "Dummy list permutation" is a quantification axiom missing from LADM and also not mentioned by Gries (1997), but used implicitly in the proof of (8.22) in both places.

The proof above is almost identical to the proof for (8.22) of LADM, except for the step using the assumption "Inverse", where the proof for (8.22) only has to invoke that assumption. In the proof above, we use the following lemma:

```
Theorem "Range replacement in nested ⋆":
    (∀ x | R • (∀ y • Q₁ ≡ Q₂))
    ⇒ (⋆ x | R • (⋆ y | Q₁ • P)) = (⋆ x | R • (⋆ y | Q₂ • P))
Proof:
      ∀ x | R • (∀ y • Q₁ ≡ Q₂)
    ≡( "Nesting for ∀" )
      ∀ x • ∀ y | R • Q₁ ≡ Q₂
    ≡( "Trading for ∀" )
      ∀ x • ∀ y • R ⇒ (Q₁ ≡ Q₂)
    ≡( (3.62) )
      ∀ x • ∀ y • R ∧ Q₁ ≡ R ∧ Q₂
    ⇒( Subproof:
          Assuming "A" `∀ x • ∀ y • R ∧ Q₁ ≡ R ∧ Q₂`:
              (⋆ x | R • (⋆ y | Q₁ • P))
            =( "Nesting for ⋆" )
              (⋆ x, y | R ∧ Q₁ • P)
            =( Assumption "A" )
              (⋆ x, y | R ∧ Q₂ • P)
            =( "Nesting for ⋆" )
              (⋆ x | R • (⋆ y | Q₂ • P))
      )
      (⋆ x | R • (⋆ y | Q₁ • P)) = (⋆ x | R • (⋆ y | Q₂ • P))
```

In the first step here, two different rules that are both called "Nesting for ∀" are applied in sequence, and in opposite directions. The last hint here contains a single Subproof hint item; inside such a subproof, any kind of proof can be written.

The necessity to distinguish metavariables for variables becomes most obvious from considering theorem (11.7) of LADM (Gries (1997) does not cover set comprehension):

Theorem (11.7) (11.7x) "Simple Membership": $x \in \{ x \mid P \} \equiv P$
— CalcCheck: Metavariables: $P = P[\![x]\!]$

If one were to consider the left-most x here as a normal free variable, then the rule for deriving provisos given above would imply that x must not occur free in P, since the right-most P does not occur in the scope of a binder for x.

It is useful to consider (11.7) in the context of metatheorem (9.16) of LADM and Gries (1997): "P is a theorem iff ($\forall x \bullet P$) is a theorem." In the universally quantified version, both occurrences of P are within scope of a binder for x, so no proviso is derived:

Theorem (11.7) (11.7∀) "Simple Membership": ($\forall$ x $\bullet$ x$\in$ { x $\mid$ P } $\equiv$ P)
— CalcCheck: Metavariables: P = P⟦ x ⟧

This is really a theorem — in our development, we actually prove this version first, and then obtain (11.7x) via instantiation.

By classifying x in (11.7x) as a metavariable for variables, we identify the "free-looking" occurrence of x as a binder in the scope of which the right-most P occurs. The effect of this approach is to let CALCCHECK derive the same metavariable occurrence and ¬*occurs* provisos for (11.7x) as for (11.7∀), compatible with (9.16).

Note, however, that (9.16) talks about *theorems*, not metatheorems (or theorem schemas). A version that would make sense for metatheorems would need to add the meta-proviso that the same provisos are derived. As a case in point, consider the one-point rule:

Axiom (8.14) "One-point rule" "One-point rule for ∀": ($\forall$ x $\mid$ x = E $\bullet$ P) $\equiv$ P[x ≔ E]
— CalcCheck: Metavariables: E = E⟦ ⟧ , P = P⟦ x ⟧
— CalcCheck: Proviso: ¬*occurs*(`x`, `E`)

Naïvely applying (9.16) to that would yield the following, where E always occurs in scope of a binder for x:

Axiom "Spuriously-quantified one-point rule for ∀": ($\forall$ x $\bullet$ ($\forall$ x $\mid$ x =E $\bullet$ P) $\equiv$ P[x ≔ E])
— CalcCheck: Metavariables: E = E⟦ x ⟧ , P = P⟦ x ⟧

This "axiom"-schema however is invalid for instantiations where x occurs free in E — just try to instantiate P with ($x < 4$) and E with ($5 \cdot x$).

6   Combined Hint Items

While in Sect. 1, the keyword "with" appeared followed by substitutions, in "Positivity of squares" in Sect. 4 as well as in "Change of restricted dummy" in Sect. 5 there are occurrences of the shape "hi_1 with hi_2" for two hint items hi_1 and hi_2. This is the simplest case of the following pattern:

$$hi_1 \text{ with } hi_2 \text{ and } \ldots \text{ and } hi_n$$

In CALCCHECK, this pattern has the two formal interpretations explained in Sects. 6.1 and 6.2, together covering probably most of the informal uses of the word "with" in LADM.

6.1 Conditional Rewriting

If among the theorems, assumptions, and induction hypotheses referred to by hi_1 there is one that can be seen as an implication with an equality (or equivalence) as consequent,

$$A_1 \Rightarrow \cdots \Rightarrow A_m \Rightarrow L = R,$$

then this is used as a conditional rewrite rule: If rewriting using $L \longrightarrow R$ succeeds with substitution σ, then CALCCHECK attempts to prove the antecedents $A_1\sigma, \ldots,$ $A_m\sigma$ using the hint items $hi_2, \ldots, hi_n$.

The with uses in Sects. 4 and 5 all are of this kind.

6.2 Rule Transformation

A different way the hint item construct above can be used is by extracting rewriting rules from hi_2 to hi_n and using these to rewrite the theorems referenced by hi_1. The results of that rewriting are then used to prove the goal of the hint. The following proof contains two such cases:

```
Theorem "Positivity": pos a ≡ a ≠ 0 ∧ ¬ pos (- a)
Proof:
    a ≠ 0 ∧ ¬ pos (- a)
  ≡( "Positivity under unary minus" (15.33) with (3.62) )
    a ≠ 0 ∧ pos a
  ≡( "Positive implies non-zero" with (3.60) )
    pos a
```

The two instances of hi_1 here are:

```
    Axiom (15.33) "Positivity under unary minus":
                      b ≠ 0  ⇒  (pos b ≡ ¬ pos (- b))
    Theorem "Positive implies non-zero": pos a ⇒ a ≠ 0
```

These are rewritten using:

```
    Theorem (3.60) "Definition of ⇒":  p ⇒ q ≡ p ∧ q ≡ p
    Theorem (3.62):  p ⇒ (q ≡ r) ≡ p ∧ q ≡ p ∧ r
```

In both cases, this rewriting produces precisely what is needed for the respective calculation step.

6.3 Theorems as Proof Methods — "Using"

LADM contains, on p. 80, an example for the "proof method" *proof by contrapositive*, almost completely in prose, with only a two-step calculation corresponding to the third and fourth steps in the calculation part of our fully formal proof:

```
Theorem "Example for use of Contrapositive":
    x + y ≥ 2  ⇒  x ≥ 1 ∨ y ≥ 1
Proof:
  Using "Contrapositive":
    Subproof for `¬ (x ≥ 1 ∨ y ≥ 1)  ⇒  ¬ (x + y ≥ 2)`:
        ¬ (x ≥ 1 ∨ y ≥ 1)
      ≡( "De Morgan" )
        ¬ (x ≥ 1) ∧ ¬ (y ≥ 1)
      ≡( "Complement of <" with (3.14) )
        x < 1 ∧ y < 1
      ⇒( "<-Monotonicity of +" )
        x + y < 1 + 1
      ≡( Evaluation )
        x + y < 2
      ≡( "Complement of <" with (3.14) )
        ¬ (x + y ≥ 2)
```

The general pattern for keyword Using is with a hint item and followed by an indented sequence of subproofs:

Using hi_1:

 sp_1

 $\vdots$

 sp_n

Technically, this is considered as syntactic sugar for a single-hint-item proof using a combined hint item in the pattern explained above:

By hi_1 with sp_1 and … and sp_n

However, using the By shape would be quite awkward to write for larger subproofs.

Pragmatically, one rather tends to consider "Using" as a *proof method generator* — mutual implication, antisymmetry laws, set extensionality, indirect equality, etc. all are frequently used to produce readable proofs in this way. Since hi_1 can again be a combined hint item, the "Using" proof pattern introduces considerable flexibility.

"Using" also liberates the user from the restriction to the induction principles hard-coded for "By induction on": Given, for example the induction principle for sequences with empty sequence ϵ and list "cons" operator $\_\vartriangleleft\_$ (as in LADM):

```
Axiom "Induction over sequences":
    P[xs ≔ ε]
  ⇒ (∀ xs : Seq A | P • (∀ x : A • P[xs ≔ x ◁ xs]))
  ⇒ (∀ xs : Seq A • P)
```

The example proof below Using this induction principle also is the first proof we show containing our construct for ∀-introduction: "For any 'vs': *proof-for-P*" proves ∀ vs • P, and "For any 'vs' satisfying 'R': *proof-for-P*" proves ∀ vs | R • P while *proof-for-P* may use assumption R.

```
Theorem (13.7) "Tail is different":
                    ∀ xs : Seq A • ∀ x : A • x ◄ xs ≠ xs
Proof:
  Using "Induction over sequences":
    Subproof for `∀ x : A • x ◄ ϵ ≠ ϵ`:
      For any `x : A`: By "Cons is not empty"
    Subproof for `∀ xs : Seq A | (∀ x : A • x ◄ xs ≠ xs)
                    • (∀ z : A • (∀ x : A • x ◄ z ◄ xs ≠ z ◄ xs))`:
      For any `xs : Seq A`
          satisfying "Ind. Hyp." `(∀ x : A • x ◄ xs ≠ xs)`:
        For any `z : A`, `x : A`:
          x ◄ z ◄ xs ≠ z ◄ xs
          ≡( "Definition of ≠", "Injectivity of ◄" )
          ¬ (x = z ∧ z ◄ xs = xs)
          ⇐( "De Morgan", "Weakening", "Definition of ≠" )
          z ◄ xs ≠ xs    — This is Assumption "Ind. Hyp."
```

7 Activation of Features

The CALCCHECK language has actually no hard-coded operators — everything can be introduced by the user via "Declaration" TLIs.

To make available functionality of the proof checker that depends on certain language elements, it is necessary to "Register" operators for built-in operators, and to "Activate" theorems on which built-in functionality relies. For example:

- Equality _=_ and equivalence _≡_ need to be registered to become available for extraction of equations for rewriting.
- *true* needs to be registered in particular for making it possible to omit "— This is (3.4)" at the end of an equivalence calculation ending in true.
- Activation of associativity and symmetry (commutativity) properties is necessary for using the internal AC representation and AC matching for the respective operators, which enables the reasoning up to symmetry and associativity that LADM also adapts throughout.

These first three items are already required for LADM Chap. 3, but only these — to force students to produce proofs conforming to the setting of Chap. 3, the remaining features need to be turned off.

LADM Chap. 4 "Relaxing the Proof Style" introduces the structured proof mechanisms described in Sect. 4 together with a number of other relaxations, that are all justified in terms of Chapter-3-style proofs. Correspondingly, CALCCHECK needs to be made aware of these justifications:

- Implication needs to be registered for Assuming and conditional rewriting (Sect. 6.1) to become available.
- Registration of conjunction is required in particular for implicit use of "Shunting" in conditional rewriting, and, as the operator underlying universal quantification, also for implicit use of "Instantiation" (i.e., ∀-elimination) in rule extraction from hint items.
- Transitivity of equality and equivalence is built-in, and also transitivity of equality with other operators, as an instance of Leibniz. For two or more

non-equality operators to be accepted as calculation operators in the same calculation, the corresponding transitivity law needs to be activated.
- For equality (or equivalence) calculations to be accepted for example when proving an implication, the relevant reflexivity law needs to be activated.
- Activation of converse laws, such as (3.58) "Consequence": $p \Leftarrow q \; \equiv \; p \Rightarrow q$, makes mentioning their use superfluous.
- Activation of monotonicity and antitonicity laws makes it possible to use a style similar to that explained by (Gries 1997, Sect. 4.1), but not restricted to formulae: Writing "Monotonicity with ..." respectively "Antitonicity with ..." then replaces the deeply-nested with-cascades of monotonicity laws that otherwise are frequently necessary.

Beyond LADM Chap. 4, some further features also depend on declared correspondence of user-defined operators with built-in constructors:

- Disjunction is required for representing set enumerations $\{1, 2\}$ as set comprehensions $\{x \mid x = 1 \lor x = 2\}$.
- Arithmetic operators like $\_+\_$, $\_-\_$, $\_\cdot\_$ and Boolean operators including also $\neg\_$ need to be registered for the keyword hint item Evaluation, seen in the first proof in Sect. 6.3, to be able to evaluate ground expressions.
- The built-in induction mechanisms also require registration of the respective operators.

8   Implementation Aspects

CALCCHECK is implemented in Haskell, with CALCCHECK$_{\text{Web}}$ using Haste by Ekblad (2016) to compile the client part from Haskell to JavaScript running in the user's web browser, and to generate the client server communication.

The core of proof checking in CALCCHECK consists in translating hints into rewrite rules, and attempting to confirm the correctness of individual proof steps by rewriting. For a calculation step "$e_1 \; op\langle \; hint \; \rangle \; e_2$", the system will use the rewriting rules derived from $hint$ to search for a common reduct of e_1 and e_2 if op is an equality operator, and otherwise (respectively alternatively) attempt to rewrite "$e_1 \; op \; e_2$" to $true$. Rewriting is mainly performed in depth-limited breadth-first search.

Since the previous, LaTeX-based version of CALCCHECK (Kahl 2011), the term datastructure used in the AC-enabled rewriting engine has seen the addition of binding structures essentially along the lines of Talcott (1993), and also a separate representation of metavariables. As mentioned before, both syntax checking and proof checking run on the server; each time a user triggers checking from a cell, all preceding cells are sent along, since they might contain changes that affect even parsing, and also changes in the theorem names they provide. For each code cell, the theorem names it provides are sent back to the client in addition to the visible feedback, and used for theorem name completion.

For typical use, in particular in the teaching context, CALCCHECK "notebooks" consist of two parts: A "prefix" that is preloaded once by the server process, and contains all the theory imports, declarations, local theorems, activations, etc., that should be available everywhere in the user view, and a "suffix" that is displayed in the user's browser as described in Sect. 3. In suffixes, import declarations and certain other features (configurable) are not available, so that the only interaction with the server file system is saving the user state of the suffix into files with server-generated names; saving is restricted to users registered via the local learning management system.

Each attempt to use a hint for justifying some goal (in particular calculation steps) is guarded by a time-out, and for grading, longer time-outs are used. During the recent final exam written on 12 CALCCHECK$_{Web}$ notebook server processes by 199 students, the 6-core machine acting as server has been observed to occasionally reach loads beyond the equivalent of one core being 100% busy, peaking at 1.4 cores.

9   Discussion of Related Work

A system with apparently quite similar goals is Lurch (Carter and Monks 2017), which lets users use conventional mathematical prose for the top-level structure of proofs, with embedded mathematical formulae marked up (unobtrusively for the prose reader) with their rôles in the mathematical development. Although this may in a certain sense be perceived to be "nicer", it is mainly nicer in the sense of supporting students of mathematics who will be expected to confidently write mathematical prose that will not normally be expected to be subjected to mechanised checking. The goal of CALCCHECK however is different: It is targeting future computer scientists and software professionals, who will need to be ready to productively use formal specification languages and automated proof systems of many different kinds, whether these are full-fledged proof assistants like Coq or Isabelle, or model checkers or automated provers like Spin or Prover9, or "modelling languages" like JML. For use of all these systems, precise understanding of issues of scope and variable binding is needed; this is frequently "hand-waived" in conventional mathematical prose. By offering a precise concept of what a proof is, and by being able to force students to produce proofs with varying levels of detail, CALCCHECK also strives to equip students with a mindset from which understanding the limitations of other verification systems will be easier, so that they will be better positioned to use them productively.

A flavour of calculational proof presentation that is slightly different from LADM are the "structured derivations" of Back (2010). These share with CALCCHECK the goal of readable fully formal, mechanically checkable proofs; MathEdit by Back et al. (2007) appears to have been a first attempt to provide tool support for this.

10   Conclusion

The proofs we arrive at are perhaps not always the ultimate in the elegance the calculational style is famous for, but they are coming close, and by virtue of providing formal syntax for useful kinds of structured proofs, frequently it is actually easier to achieve elegance in CalcCheck than in the calculational style embedded in conventional mathematical prose for larger-scale proof structure. Many students showed significant skills in finding quite elegant and widely different proofs even in exam settings, and student feedback about CalcCheck has been almost unanimously positive.

References

Back, R.-J.: Structured derivations: A unified proof style for teaching mathematics. Formal Aspects Comput. **22**(5), 629–661 (2010). https://doi.org/10.1007/s00165-009-0136-5

Back, R.-J., Bos, V., Eriksson, J.: MathEdit: Tool support for structured calculational proofs. TUCS Technical report 854, Turku Centre for Computer Science (2007)

Carter, N.C., Monks, K.G.: A web-based toolkit for mathematical word processing applications with semantics. In: Geuvers, H., England, M., Hasan, O., Rabe, F., Teschke, O. (eds.) CICM 2017. LNCS (LNAI), vol. 10383, pp. 272–291. Springer, Cham (2017). https://doi.org/10.1007/978-3-319-62075-6_19

Ekblad, A.: High-performance client-side web applications through Haskell EDSLs. In: Mainland, G., (ed) Proceedings 9th International Symposium on Haskell, Haskell 2016, pp. 62–73. ACM (2016). https://doi.org/10.1145/2976002.2976015

Gries, D.: Foundations for calculational logic. In: Broy, M., Schieder, B. (eds.) Mathematical Methods in Program Development, pp. 83–126. Springer, Heidelberg (1997). https://doi.org/10.1007/978-3-642-60858-2_16

Gries, D., Schneider, F.B.: A Logical Approach to Discrete Math. Monographs in Computer Science. Springer, New York (1993). https://doi.org/10.1007/978-1-4757-3837-7

Gries, D., Schneider, F.B.: Equational propositional logic. Inform. Process. Lett. **53**, 145–152 (1995). https://doi.org/10.1016/0020-0190(94)00198-8

Kahl, W.: The teaching tool CalcCheck a proof-checker for Gries and Schneider's "Logical Approach to Discrete Math". In: Jouannaud, J.-P., Shao, Z. (eds.) CPP 2011. LNCS, vol. 7086, pp. 216–230. Springer, Heidelberg (2011). https://doi.org/10.1007/978-3-642-25379-9_17

Norell, U.: Towards a practical programming language based on dependent type theory. Ph.D. thesis, Department of Computer Science and Engineering, Chalmers University of Technology (2007). http://wiki.portal.chalmers.se/agda/pmwiki.php

Spivey, J.M.: The Z Notation: A Reference Manual. Prentice Hall International Series in Computer Science. Prentice Hall (1989)

Spivey, M.: The fuzz type-checker for Z, Version 3.4.1, and the fuzz Manual, 2nd edn. (2008). http://spivey.oriel.ox.ac.uk/corner/Fuzz. Accessed 15 April 2018

Talcott, C.L.: A theory of binding structures and applications to rewriting. Theoret. Comput. Sci. **112**, 68–81 (1993). https://doi.org/10.1016/0304-3975(93)90240-T

Understanding Parameters of Deductive Verification: An Empirical Investigation of KeY

Alexander Knüppel[(✉)], Thomas Thüm, Carsten Immanuel Pardylla, and Ina Schaefer

TU Braunschweig, Braunschweig, Germany
{a.knueppel,t.thuem,c.burmeister,i.schaefer}@tu-bs.de

Abstract. As formal verification of software systems is a complex task comprising many algorithms and heuristics, modern theorem provers offer numerous parameters that are to be selected by a user to control how a piece of software is verified. Evidently, the number of parameters even increases with each new release. One challenge is that default parameters are often insufficient to close proofs automatically and are not optimal in terms of verification effort. The verification phase becomes hardly accessible for non-experts, who typically must follow a time-consuming trial-and-error strategy to choose the right parameters for even trivial pieces of software. To aid users of deductive verification, we apply machine learning techniques to empirically investigate which parameters and combinations thereof impair or improve provability and verification effort. We exemplify our procedure on the deductive verification system KeY 2.6.1 and specified extracts of OpenJDK, and formulate 53 hypotheses of which only three have been rejected. We identified parameters that represent a trade-off between high provability and low verification effort, enabling the possibility to prioritize the selection of a parameter for either direction. Our insights give tool builders a better understanding of their control parameters and constitute a stepping stone towards automated deductive verification and better applicability of verification tools for non-experts.

Keywords: Deductive verification · Design by contract
Formal methods · Theorem proving · KeY · Control parameters
Automated reasoning

1 Introduction

Formal methods are intended to provide adequate solutions for software developers to rigorously *prove* that a piece of software is in line with a given specification [6,9,40,41]. Besides light-weight methods intended to uncover the majority of defects early, such as code reviews and testing, there is need for advanced strategies to find the last defects. For instance, model checking is an automatic

© Springer International Publishing AG, part of Springer Nature 2018
J. Avigad and A. Mahboubi (Eds.): ITP 2018, LNCS 10895, pp. 342–361, 2018.
https://doi.org/10.1007/978-3-319-94821-8_20

technique verifying that a given formal model (e.g., state machines) adheres to its specification [8,39]. Although we expect our considerations to be more generally applicable to other formal verification techniques, we focus on deductive verification, which is another technique that targets program verification directly on source code [1,2,7,21,43]. Essentially, an implementation together with its formal specification is translated into a logical formula and validity is proved by a theorem prover [43].

Despite considerable advances over the last decades and the advantage to be directly applied to source code, deductive verification is still only hesitantly applied in industrial software projects. Reasons are manifold. For example, there are doubts about the cost-effectiveness of formal methods [29]. In particular, most legacy systems are not designed with formal verification in mind, which makes post-hoc specification and verification expensive for industrial projects [4]. Moreover, developing sufficient formal specifications is error-prone and tedious [2,3], and typically requires high expertise of the underlying proof theory. Even worse, full automation is not always possible because of the undecidability of the halting problem.

However, when full automation is feasible, a subsequent and often overlooked hurdle for inexperienced users is to parameterize the verification tool. Different implementations and specifications have different needs and modern verification tools provide parameters that are set by a user to control the verification process. For example, the parameter *loop treatment* can be used to decide whether loops are always unrolled or specified loop invariants are used. Consequently, successful verification often depends on those parameters.

Understanding all parameters requires a considerable amount of knowledge. Non-expert users may face problems when a piece of software cannot be verified even when implementation and specification are seemly correct. Furthermore, minimizing verification effort is important for industrial software. Complex software systems are frequently changed and formal specifications are adapted accordingly. In this process, past proof results may become invalid. A naive solution is then to follow a trial-and-error strategy by applying different parameter configurations, after which verification is restarted. This strategy, however, wastes a considerable amount of resources, making it less applied in industry.

We argue that a better understanding of parameterization allows tool builders to better support users of deductive verification. In particular, we investigate whether specific parameters have a larger influence on automated provability and whether specific options increase or decrease the verification effort. We focus on deductive verification following the *Design by Contract* paradigm [35]. Contracts are an extension to Hoare triples [23] and constitute a methodology to specify methods of imperative languages (i.e., Java or C) with *preconditions* and *postconditions*, and classes with *class invariants*. Callers of a contract-specified method have the obligation to fulfill the precondition and may therefore rely on the postcondition. Class invariants have to hold before and after method execution.

There exist numerous languages with support for contracts, such as Eiffel [36], Spec# [2], and the Java modeling language (JML) [31]. For the purpose of this paper, we analyze parameters of the state-of-the-art verification system KeY 2.6.1 [1]. KeY is a modern theorem prover with a large community intended to verify JML-specified Java programs. To empricially investigate KeY's parameters, we formulate a total of 53 hypotheses in terms of provability and verification effort derived from the literature and documentation of KeY. Morever, we construct parameter-influence models based on our measurements to reason about which options influence the verification effort the most. To this end, we employ SPLConqueror [45], a framework which incorporates machine-learning techniques to measure the influence of parameters on non-functional properties. In summary, our contributions are the following.

- We formulate and empirically validate 53 hypotheses about parameterization in KeY, which provide clear recommendations for users who aim to verify pieces of software automatically.
- We empirically evaluate the influence of KeY's parameters with respect to provability and verification effort with machine learning.
- We identify parameters that depict a trade-off between higher provability and lower verification effort and discuss consequences for users and tool builders.

2   Problem Statement

With formal verification, our goal is to identify the last remaining defects. When automatic software verification fails, users are confronted with a diverse set of reasons. Typically, most common reasons consist of (a) a wrong implementation, (b) a wrong or insufficient specification (e.g., loop invariants are missing or too weak), (c) insufficient heuristics of the verification tool (e.g., when automatically inferring loop invariants or instantiating quantifiers), or (d) the verification task times out after the maximum number of proof steps or heap memory is exceeded.

As if these hurdles are not enough, a subsequent challenge is that parameterization has also a great effect on the outcome and the default values are oftentimes not sufficient. Getting the parameters right from the beginning makes deductive verification significantly more successful and cost-effective.

We divide parameters of deductive verification broadly into two categories. The first category describes qualitative parameters that explicitly change *what to prove*. For instance, there is an option in KeY to ignore *integer overflows*. Consequently, implementations that cause an integer overflow are not verifiable with this setting. Depending on the context and how the implementation is facilitated, however, verifying the absence of integer overflows is crucial. Those parameters must be set by users or have at least a well-chosen default value.

The second category describes parameters that only influence provability and verification effort (i.e., *how to prove*). For instance, there is a parameter in KeY for method call treatment; a method call is either always replaced with an existing contract (i.e., contracting), or its implementation is always inlined (i.e., method expand). Typically, contracting is faster and results in lower verification

```
/*@ public normal_behavior
  @ requires T > 0;
  @ ensures \result < T;
  @*/
public /*@ pure @*/ int modT(int input, int T) {
  return input % T;
}
```

Listing 1. Method Computing the Modulo of Integer Values

```
/*@ public normal_behavior
  @ requires e != null;
  @ ensures contains(e);
  @ ensures collectionSize == \old(collectionSize) + 1;
  @ ensures \result;
  @ assignable elements;
  @*/
boolean add(/*@nullable@*/ Object e);
```

Listing 2. Method ArrayList.add(Object) Specified with Contracts in JML

effort. However, in case of missing or insufficient contracts, a method can only be proved correct with method inlining.

To verify a piece of software automatically, a user must first identify what to prove and has to set respective parameters accordingly (e.g., enabling detection of integer overflows). In a second step, a user typically starts the verification process with default parameters or the last used configuration to check whether they suffice. In case of failure, oftentimes parameterization is changed and verification is restarted in a trial-and-error manner. Moreover, as frequent changes to software systems are the common case, reducing the verification effort is another important requirement. Hence, having a better understanding of the control parameters and providing better tool support would tremendously help inexperienced users to apply deductive verification more successfully.

In the following, we depict two examples, where default parameters are either insufficient or result in an increased verification effort. In Listing 1, we illustrate a small example of a formally specified method modT(int, int) that gets two integer values as input and computes the modulo between both. The precondition is denoted by keyword requires and states that input parameter T must be greater than 0. The postcondition is denoted by keyword ensures and states that the return value will always be less than T given the precondition. Keyword \result represents the return value. Notably, method modT(int, int) is not automatically verifiable with KeY's default parameters. As the example is small, implementation and specification are readily comprehensible and seemly fit together. In particular, the reason is a parameter called *Arithmetic treatment*. The set of possible options is {Basic, DefOps, Model Search}, where Basic is the default value. However, unlike DefOps, Basic is incapable of evaluating the modulo operator.

Starting from the default parameters, choosing DefOps as value for *Arithmetic treatment* suffices to verify method modT(int, int). Although it is possible to modify the postcondition to \result== input%T and verify it indeed with the default parameters, such modifications are hard to find for real-world software systems.

In Listing 2, we depict another example, where we specified method add(Object) of class ArrayList in JML. Class ArrayList is part of the Collection-API and implements the interface Collection. The precondition states that callers of add(Object) can only rely on the postcondition when they provide an instantiated object. The postcondition states that (a) the input object is indeed part of the list after successful method execution, (b) the list's size is incremented by one, and (c) the return value is true. In a postcondition, keyword \old evaluates the expression before method execution. Keyword assignable represents the framing condition (i.e., a set of locations). Implicitly, locations excluded from the frame are not allowed to be modified. The example also contains *queries*, which are side-effect free methods that can be called in specifications (e.g., method contains(Object)). Internally, method add(Object) calls method ensureCapacity(Object), which increases the capacity of the respective list by one, if necessary. While the example depicts a trivial contract, the verification effort with KeY's default parameters can be reduced from 33,748 proof steps to 13,328 proof steps when changing the parameter *Quantifier treatment* from No Splits with Progs to No Splits.

Ideally, with assistance of extended tooling, a developer is capable of understanding the influence of various control parameters on provability and verification effort. For instance, a recommendation system may suggest to change the option of parameter *Arithmetic treatment* for the example depicted in Listing 1, as only after carefully studying the tool tips in KeY it becomes apparent that *Arithmetic treatment*::Basic cannot evaluate the modulo operator.

3 Parameters of Deductive Verification with KeY

As already mentioned, KeY provides numerous parameters to control *what* and *how* a piece of software is verified. In particular, there exist three categories of parameters in KeY, namely *search strategy options*, *taclet options*, and *general options*. Search strategy options control to what extent and in which order KeY applies inference rules to automatically verify a method. Taclet options control rather *what* to prove (e.g., integer overflow) and, thus, are typically fixed for a verification target. General options enable the employment of an SMT solver or allow for one step simplification, which combines single inference rules into one.

Although KeY's automated proof strategy algorithm is configurable, the difficulty here is that it has grown over many years, with participation of many different researchers and universities. The effect of some parameters on provability and verification effort is therefore challenging to anticipate. We aim to provide a better understanding on how specific parameters improve or impair provability and verification effort. Based on our experience and observations combined with studying the online documentation[1], numerous publications [13, 15, 17, 18, 26, 27, 32, 42], all tool tips in KeY, and the KeY book [1], we formulated a total of 38 assumptions that we empirically evaluate by deriving 51 statistical hypotheses in the next section. Related to Listings 1 and 2, two examples for assumptions about the parameters *Arithmetic treatment* and *Quantifier treatment* with respect to provability and verification effort are the following.

Assumption 19 (Arithmetic Treatment - Basic and DefOps). *If a specification case is provable with option* Basic, *it is also provable with option* DefOps.

Assumption 23 (Quantifier Treatment - Verification Effort). *The verification effort with option* Free *is at least as great as with option* No Splits with Progs. *The verification effort with option* No Splits with Progs *is at least as great as with option* No Splits. *The verification effort with option* No Splits *is at least as great as with option* None.

In Table 1, we give an overview on all our assumptions in a short from. The first column denoted by *Assumption* represents an identifier for the respective assumption and the second column denoted by *Parameter* represents the parameter which is subject to the assumption. For our significance tests, we only relate two values of a parameter with each other (i.e., fourth and fifth column). That is why there exist more experiments (i.e., third column) than assumptions. For instance, Assumption 23 relates four values with each other in terms of verification effort. The type of the *depended variable* is represented in the sixth column, where provability is denoted by P and verification effort is denoted by VE. In particular, for reasoning about the verification effort we always assume that a verification target is provable with both subjected options. The last column represents whether one option improves, impairs, or does not influence the outcome. For provability, $O_a \leq O_b$ means that O_b provides a higher chance of provability that O_a. For verification effort, $O_a \leq O_b$ means than O_b leads to a greater effort than O_a. The opposite meaning for each depended variable is denoted by $\geq$ and no significant influence is denoted by $<>$.

We considered Assumption 27 separately. The reason is that this assumption does not compare options, but states that specification cases are not provable based on particular conditions:

Assumption 27 (Class Axiom Rule). *If a method writes onto a location on the heap and there exists at least one class invariant that refers to this location, then option* Off *is not sufficient to verify this method.*

To summarize, we formulated a total of 38 assumptions on 47% of all available parameters in KeY. In Fig. 1, we depict all assumptions about parameters for

---

[1] https://www.key-project.org/applications/program-verification/ and http://i12 www.ira.uka.de/key/download/quicktour/quicktour-2.0.zip.

Table 1. Investigated parameters and formulated hypotheses

Assumption	Parameter	Hypothesis	First option	Second option	Requirement	Dependency
1	Stop at	1	Default	Unclosable	P	<>
2	Stop at	2	Default	Unclosable	VE	<>
3	One step simplification	3	Enabled	Disabled	P	<>
4	One step simplification	4	Enabled	Disabled	VE	≤
5	Proof splitting	5	Delayed	Free	P	≤
6	Proof splitting	6	Delayed	Free	VE	≤
7	Proof splitting	7	Off	Free	P	≤
		8	Off	Delayed	P	≤
8	Proof splitting	9	Off	Free	VE	≤
		10	Off	Delayed	VE	≤
9	Loop treatment	11	Invariant	Loop scope invariant	P	≤
10	Loop treatment	12	Invariant	Loop scope invariant	VE	≥
11	Dependency contracts without accessible-clauses	13	On	Off	P	<>
12	Dependency contracts without accessible-clauses	14	On	Off	VE	<>
13	Query treatment without queries	15	On	Restricted	P	<>
		16	On	Off	P	<>
14	Query treatment without queries	17	On	Restricted	VE	<>
		18	On	Off	VE	<>
15	Query treatment	19	Off	Restricted	P	≤
		20	Restricted	On	P	≤
16	Query treatment	21	Restricted	On	VE	<>
17	Expand local queries	22	On	Off	VE	≥
18	Expand local queries	23	On	Off	P	≥
19	Arithmetic treatment	24	Basic	DefOps	P	≤
20	Arithmetic treatment	25	DefOps	ModelSearch	P	<>
21	Quantifier treatment without quantifiers	26	None	No splits	P	<>
		27	None	No splits with progs	P	<>
		28	None	Free	P	<>
22	Quantifier treatment without quantifiers	29	None	No splits	VE	<>
		30	None	No splits with progs	VE	<>
		31	None	Free	VE	<>
23	Quantifier treatment	32	None	No splits	P	≤
		33	No splits	No splits with progs	P	≤
		34	No splits with progs	Free	P	≤
24	Quantifier treatment	35	Free	No splits with progs	VE	≥
		36	No splits with progs	No splits	VE	≥
		37	No Splits	None	VE	≥

(continued)

Table 1. (*continued*)

Assumption	Parameter	Hypothesis	First option	Second option	Requirement	Dependency
25	Class axiom rule without axioms	38	Free	Delayed	P	<>
		39	Free	Off	P	<>
26	Class axiom rule without axioms	40	Free	Delayed	VE	<>
		41	Free	Off	VE	<>
27	Class axiom rule	52	Considered separately			
28	Class axiom rule	53	Off	Delayed	P	<>
29	Strings	42	On	Off	P	≥
30	Strings	43	On	Off	VE	<>
31	BigInt	44	On	Off	P	≥
32	BigInt	45	On	Off	VE	<>
33	IntegerSimplificationRules	46	Full	Minimal	P	≥
34	IntegerSimplificationRules	47	Full	Minimal	VE	<>
35	Sequences	48	On	Off	P	≥
36	Sequences	49	On	Off	VE	<>
37	MoreSeqRules	50	On	Off	P	≥
38	MoreSeqRules	51	On	Off	VE	<>

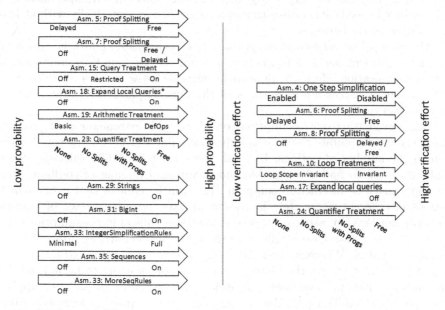

* Queries must be verifiable with *Method Treatment: Expand*

Fig. 1. Overview on assumptions for provability and verification effort

which we identified an order of their options with respect to higher and lower provability and verification effort. Confirmation of these assumptions helps in prioritizing parameters for fine-tuning in terms of provability and verification effort.

4   Empirical Evaluation of KeY's Parameters

In a series of experiments, we evaluate the assumptions that we formulated in the last section. All formulated assumptions, evaluation artifacts, results, and the verification target can be found online.[2]

4.1   Experimental Setup

For our verification target, we formally specified parts of OpenJDK's Collection-API with JML. Reasons to focus on OpenJDK are threefold, namely (a) it represents a widespread and highly applied real-world software, (b) there exists already an informal specification in the JavaDocs comments that we utilize for our formal specification, and (c) it is open-source and the only Java distribution that allows us to add contracts and freely distribute it. A method can have more than one contract (e.g., when different preconditions are connected with distinct postconditions), which we refer to as *specification cases* in the following. In total, our test study comprises 27 specification cases distributed over the interface Collection and classes ArrayList, LinkList, Arrays, and Math, which we specified in a complementary study [30]. To reduce bias in our experiments, we only included specification cases that can be verified automatically with at least one parameter configuration.

All assumptions refer either to *provability* or *verification effort*. For provability, the dependent variable is whether a proof can be found automatically or not. For verification effort, the dependent variable is the number of proof steps. Independent variables are in both cases the current parameterization and the specification cases.

While evaluating our assumptions on all possible parameter configurations yields the most accurate result, it is impractical due to the combinatorial explosion in the number of options. In total, there exist 1,990,656 valid parameter configurations. To find a reasonable and meaningful number of configurations, we apply pairwise interaction sampling [11], which requires that every pair of options of different parameters is present at least once in the set of parameter configurations. In essence, 1,084 configurations suffice. For assumptions that compare at least two options related to the verification effort, we apply the non-parametric paired Wilcoxon-Test [50]. The rationale for a non-parametric test is that we cannot expect the distribution of the proof effort to be normal. For assumptions that only consider one option (i.e., Assumption 27), we apply a 1-sample Wilcoxon-Test [50]. For assumptions that compare at least two values related to provability, we apply a McNemar-Test [34]. For each experiment, we define a significance level of 5% and we set the maximal number of proof nodes to 500,000, after which a verification task times out.

---

[2] http://github.com/AlexanderKnueppel/UnderstandingParametersInKeY.

4.2   Empricial Evaluation of Assumptions

In Table 2, we summarize all experiments together with their statistical hypotheses, respective p-value, and outcome. Depending on the statistical test and formulation of the assumption, we need to define and evaluate different kind of statistical hypotheses. If the assumption states that *there is no significant difference*, we cannot formulate a null hypothesis H_0 that we would like to reject in favor of the assumption. In this case, we use the assumption itself as the null hypothesis and denote the *hypothesis type* as H_0. The consequence is that we can only reject or not reject our assumption, but never accept it. The preferred outcome is *not rejected*, as otherwise our assumption would be indeed wrong. If we can formulate a null hypothesis, we use the hypothesis type $H_\leq$ and $H_\geq$ for assumptions that paraphrase *at least* or *at most* relationships between options (e.g., Assumption 23) or simply H_A otherwise. In this case, the preferred outcome is *accepted*. The result is always *not rejected*, if we could not reject the null hypothesis.

Based on our results, we had to neglect two hypotheses, namely Assumption 21 and Assumption 22 stating that *Quantifier treatment* does not effect provability and verification effort if the logical formula under verification is quantifier-free. We discovered that each verification of a program in KeY works with quantification internally as soon as assignable-clauses are used, even when no quantifiers are used in the contracts. Furthermore, we had to reject Assumption 2 (i.e., parameter *Stop at* does not influence the verification effort), as the statistical result was significant (p-value: $2,488 * 10^{-2}$).

Four hypotheses about how the verification effort is influenced were accepted (i.e., Hypothesis 4, 6, 9, and 12). Six additional assumptions about how provability is influenced were accepted after an additional manual inspection (Hypothesis 7, 8, 23, 24, 32, 46). The reason for manual inspection is that the employed McNemar-Test is always two-sided. This means that the direction of difference (i.e., positive or negative) is not directly apparent. For the manual inspection, we use contingency tables to decide whether the null hypothesis can indeed be rejected. In Table 3, we depict the significant hypotheses, which were tested in the McNemar-Test, with their contingency tables. A hypothesis can be accepted if the sum of the first row is unequal to the sum of the first column, and analogously for the second row and second column. *Closed* and *Open* refer to whether a verification task was solved automatically or not. For instance, for Hypothesis 7, a total of 114 verification tasks of all verification tasks performed were solved automatically with *Proof splitting*::Free and *Proof splitting*::Off, whereas 142 verification tasks could not be solved automatically with either option.

One oddity is Assumption 37. 100% of the data correlates with its statement, which is why we could not compute the p-value but accepted the hypothesis nonetheless. In summary, three hypotheses were rejected and eleven hypotheses were accepted. The remaining hypotheses could neither be rejected nor confirmed and may have to be investigated in more detail and with additional verification targets in future studies.

Table 2. Experimental results with accepted and rejected assumptions

Parameter	Assumption	Experiment	Hypothesis type	p-value	Result
Stop at	1	1	H_0	NA	Not rejected
	2	2	H_0	$2,488 * 10^{-2}$	**Rejected**
One step simplification	3	3	H_0	NA	Not rejected
	4	4	H_A	$< 2,2 * 10^{-16}$	**Accepted**
Proof splitting	5	5	$H_<$	NA	Not rejected
	6	6	H_A	$7,7 * 10^{-9}$	**Accepted**
Proof splitting	7	7	$H_<$	$3,252 * 10^{-9}$	**Accepted***
		8	$H_<$	$3,252 * 10^{-9}$	**Accepted***
	8	9	H_A	$4,147 * 10^{-2}$	**Accepted**
		10	H_A	$6,063 * 10^{-1}$	Not rejected
Loop treatment	9	11	H_0	NA	Not rejected
	10	12	H_A	$9,186 * 10^{-3}$	**Accepted**
Dependency contracts	11	13	H_0	NA	Not rejected
	12	14	H_0	1	Not rejected
Query treatment	13	15	H_0	NA	Not rejected
		16	H_0	NA	Not rejected
	14	17	H_0	$1,422 * 10^{-1}$	Not rejected
		18	H_0	1	Not rejected
	15	19	$H_<$	$1,573 * 10^{-1}$	Not rejected
		20	$H_<$	NA	Not rejected
	16	21	H_A	$1,706 * 10^{-1}$	Not rejected
Expand local queries	17	22	H_A	$6,601 * 10^{-2}$	Not rejected
	18	23	$H_<$	$4,55 * 10^{-2}$	**Accepted***
Arithmetic treatment	19	24	$H_>$	$9,237 * 10^{-13}$	**Accepted***
	20	25	H_A	$3,173 * 10^{-1}$	Not rejected
Quantifier treatment	21	26–28	-	-	**Rejected**
	22	29–31	-	-	**Rejected**
	23	32	$H_<$	$4,55 * 10^{-2}$	**Accepted***
		33	$H_<$	$1,573 * 10^{-1}$	Not rejected
		34	$H_<$	NA	Not rejected
		35	H_A	$7,186 * 10^{-1}$	Not rejected
	24	36	H_A	$2,869 * 10^{-1}$	Not rejected
		37	H_A	$1,562 * 10^{-1}$	Not rejected
Class axiom rules	25	38	H_0	NA	Not rejected
		39	H_0	NA	Not rejected
	26	40	H_0	NA	Not rejected
		41	H_0	NA	Not rejected
	27	52	H_0	NA	Not rejected
	28	53	H_0	NA	Not rejected
Strings	29	42	$H_\geq$	NA	Not rejected
	30	43	H_0	1	Not rejected
BigInt	31	44	$H_\geq$	NA	Not rejected
	32	45	H_0	1	Not rejected
IntegerSimplificationRules	33	46	$H_\geq$	$5,32 * 10^{-4}$	**Accepted***
	34	47	H_0	$8,783 * 10^{-1}$	Not rejected
Sequences	35	48	$H_\geq$	NA	Not rejected
	36	49	H_0	$2,61 * 10^{-1}$	Not rejecte d
MoreSeqRules	37	50	$H_\geq$	NA	**Accepted***
	38	51	H_0	$3,458 * 10^{-1}$	Not rejected

* After manual inspection

Table 3. Contingency tables of manually inspected assumptions

Hypothesis 7 Proof Splitting		Off	
		Closed	Open
Free	Closed	114	37
	Open	0	142

Hypothesis 24 Query Treatment		Basic	
		Closed	Open
DefOps	Closed	119	51
	Open	0	164

Hypothesis 8 Proof Splitting		Off	
		Closed	Open
Delayed	Closed	116	37
	Open	0	143

Hypothesis 32 Quantifier Treatment		None	
		Closed	Open
No Splits	Closed	132	4
	Open	0	107

Hypothesis 23 Expand local queries		On	
		Closed	Open
Off	Closed	82	0
	Open	4	105

Hypothesis 46 IntegerSimplificationRules		Full	
		Closed	Open
Minimal	Closed	122	0
	Open	12	151

4.3 Learning a Parameter-Influence Model

Our assumptions only state which options do have or do not have an effect on provability or verification effort. However, in terms of verification effort, it is also interesting to know which options have a larger effect than others. While previous assumptions help to exclude some options when optimizing the verification effort, we are even interested in prioritizing options according to their impact.

Based on our experiments, we collected a considerable amount of data points, which depict the verification effort with respect to different options. In a nutshell, we decided to learn a parameter-influence model based on our data to draw these relevant conclusions about which options have a larger impact on the verification effort. To derive a general model, we use all of our specification cases as input for machine-learning techniques. To this end, we employ SPLConqueror [45], a framework which incorporates machine-learning techniques to measure the influence of parameters on non-functional properties.

Our samples comprise only specification cases that are automatically verified, as the number of proof steps in unclosed proofs is not meaningful. Moreover, to increase our confidence in the performance-influence model, we applied cross-validation to learn a total of ten models. To this end, we partitioned all verified specification cases into ten subsets and used nine randomly chosen subsets to train each of the ten models.

In Fig. 2, we depict the results of our ten prediction models using boxplots that relate numerous options with verification effort. Each boxplot presents a factor that indicates whether an option improves (negative value) or impairs (positive value) the verification effort. Notably, there exist numerous options that were discarded in the training process of all ten models, as too few verification tasks could be closed with them automatically. Therefore, we also omitted them in Fig. 2.

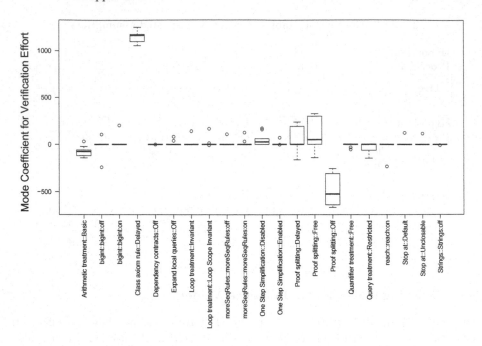

Fig. 2. Influence of various options on verification effort

For most options, the median is close to zero, which is why we cannot reason about their influence. Exceptions are *Arithmetic treatment*::Basic, *Class axiom*::Delayed, *One Step Simplification*::Disabled, *Proof splitting*::Delayed, *Proof splitting*::Free, *Proof splitting*::Off, and *Query treatment*::Restricted. With respect to the median, we achieve the largest improvement with *Proof splitting*::Off followed by *Arithmetic treatment*::Basic. *Class axiom*::Delayed leads to the largest deterioration of the verification effort.

While we did not make any assumptions about options *Arithmetic treatment*::Basic and *Class axiom*::Delayed with respect to verification effort, Fig. 2 reveals that *Arithmetic treatment*::Basic almost always reduces the effort, whereas *Class axiom*::Delayed always and significantly impairs it. Hence, we can derive two new assumptions from our parameter-influence models. Nonetheless, these assumptions are not derived from the literature, but based on our exploratory study and have to be evaluated in future studies.

To briefly summarize, we identified seven options that reasonably impact the verification effort. Our results allow us to prioritize these options when provability is already ensured. However, we also discovered that *Arithmetic treatment*::Basic only insignificantly reduces the verification effort based on our specified extract of OpenJDK and we previously confirmed that what is provable with *Arithmetic treatment*::Basic is also provable with *Arithmetic treatment*::DefOps (cf. Assumption 19). Hence, *Arithmetic treatment*::DefOps may be the better choice for all verification tasks. Notably, numerous of our assumptions coincides with the models' predictions (i.e., Assumption 4, 6, 8, and 16).

4.4 Threats to Validity

The measured verification effort may not be representative, as we decided to count the proof nodes that KeY produces internally, which may vary in complexity and execution time. One alternative is to measure the overall execution time needed to verify a method. We decided against it, as execution timing depends on numerous external factors, such as computing power, parallel processes, and even the currently active virtual machine, whereas the number of proof nodes is a reproducible measurement.

Our verification target (i.e., OpenJDK's Collection API) may not comprise enough representative specification cases, as we did not specify many loop invariants or complex algorithms. Nevertheless, we specified real-world Java code, for which the specification effort was already tremendously high (i.e., it took us numerous iterations and months to be amenable for automatic verification). For all employed specification cases there exist at least one parameter configuration, which suffices to automatically verify it. Moreover, we computed all results on high-end servers over a period of two months. Specifying and verifying more specification cases would take considerably longer.

We only formulated assumptions about parameters that are used in KeY 2.6.1. It is thus questionable whether our considerations can be generalized to other verification systems. However, parameterization for non-expert users is also challenging for other techniques and tools, such as model checking with Java Pathfinder [46]. Furthermore, the chosen dependent variables (i.e., provability and verification effort) are typically most meaningful for users and tool builders of other verification systems, too.

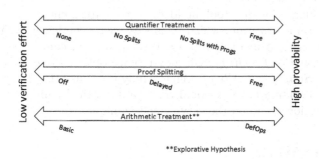

Fig. 3. Trade-off between lower verification effort and higher provability

5   Suggestions for Users and Tool Builders

Our assumptions depict numerous options that a user should prioritize for tuning to increase provability and verification effort. For instance, parameters that need to be changed (i.e., differ from the default option) are *Quantifier Treatment* and *Proof splitting*, which can be set to Free, and *Arithmetic Treatment*, which can be set to DefOps. Moreover, *Stop At* should stay at Default and *Expand Local Queries* should stay at On.

If provability is ensured, verification effort can be tweaked. Our hypotheses state that *One Step Simplification* should be set to free,

Proof Splitting should be set to off, and *Loop Treatment* to Loop Scope Invariant. In particular, the parameter-influence models illustrated that *Proof Splitting::*Off decreases verification effort the most compared to all options. Moreover, the models revealed that *Arithmetic Treatment* and *Class Axiom Rule* have also an impact. *Arithmetic Treatment::*Basic is preferred to *Arithmetic Treatment::*DefOps, whereas *Class Axiom Rule::*Delayed should be avoided when possible. Nevertheless, our suggestion is to always start with *Arithmetic Treatment::*DefOps, as it is often needed for provability and the gain in terms of verification effort seems to be insignificant.

Based on our results, we can also identify parameters, whose options represent a trade-off between higher provability and lower verification effort. In essence, these parameters are *Quantifier treatment*, *Proof splitting*, and *Arithmetic treatment*, which are illustrated in Fig. 3.

Whenever provability is ensured, these options allow a user to decrease the verification effort.

Derived from our assumptions, some options have no measured impact on the verification effort but influence provability. It is thus questionable why a user is confronted with these options. A solution would be to provide different modes for different requirements, such as a *simple view* for inexperienced users that hides specific options. In particular, such a view may discard parameters *BigInt*, *IntegerSimplificationRules*, *Sequences*, *MoreSeqRules*, *One Step Simplification*, and *Stop At* for the mentioned reason.

Another suggestion for tool builders is to implement a recommendation system for parameterization, which enhances user experience. KeY could provide hints to users to increase provability if a method cannot be verified. For instance, option *Proof Splitting::*Off may replace option *Proof Splitting::*Free or option *IntegerSimplificationRules::*Minimal may replace option *IntegerSimplificationRules::*Full. Moreover, KeY could try to automatically fine-tune parameters during verification based on the very same technique.

6   Related Work

A survey on different languages for behavioral contracts was done by Hatcliff et al. [21]. Besides KeY with its specialization on Java source code, there exist alternative tools for deductive program verification of other languages, such as Spec# [2], VCC [10] for verifying concurrent C, and the Why platform [49], which comprises tools for the verification of WhyML, Java, and C programs [12, 16,33]. For the purpose of this paper, we concentrated on KeY, as (a) it provides numerous parameters, (b) it has an active community, and (c) we already gained ample and practical experiences with it [24,46–48].

Gouw et al. [14] investigated the correctness of OpenJDK's TimSort with KeY and discovered an exploitable bug in its implementation. They changed parameterization even *during* the search for proofs, which is difficult as it requires to find meaningful interruption points. This is an indicator that an advanced understanding of the parameters is indispensable to verify real-world software with deductive verification.

Another formal verification technique requiring an understanding about its parameters is model checking. SPIN [25] is a software model checker that focuses on finite state machines and provides numerous configurable options and optimizations, such as partial order reduction, state compressions, and bitstate hashing. Java Pathfinder (JPF) [22] is a software model checker focusing on Java source code. JPF can be parameterized and extended in a variety of ways and is build upon a general and uniform configuration management. Configuring JPF for *efficiently* finding defects for a given verification task needs a considerable amount of knowledge about model checking.

Optimizing the selection of parameters of configurable programs is a widely researched area. Benavides et al. [5] analyzed the performance of CSP, SAT, and BDD solvers in finding a valid configuration. Ochoa et al. [37] transform a set of configurations into a CSP solver to find a non-conflicting set of configurations that adhere to particular business objectives, such as costs, time, and human resources, of multiple stakeholders. Siegmund et al. [45] proposed SPL Conqueror, which we used to learn our prediction models. Despite its initial connection to software product lines, SPL Conqueror is used by various researchers to learn models that predict the influence of non-functional properties in configurable software [19,20,28,38,44]. We provide an additional use case for SPL Conqueror, as we learned parameter-influence models to argue about how the selection of particular options influence the verification effort.

7   Conclusion

Our long-term goal is to make deductive verification accessible for mainstream software developers. Although formal methods improved significantly over the last decades, software developers still struggle to specify and verify even trivial pieces of software. One often overlooked hurdle that inexperienced users face is parameterization. While parameterization of formal method tools comes with the promise to ease the process of automatic verification, we exhibited that setting the right values for the ever growing amount of parameters is challenging.

In particular, our focus is on parameters of deductive verification, where we used the verification system KeY 2.6.1 as an example.

We formulated a total of 38 assumptions how options in KeY improve or impair provability and verification effort. We derived a total of 51 statistical hypotheses and empirically measured the effect of different parameter configurations by employing significance tests and machine-learning techniques.

Our empirical investigation is a stepping stone towards automated deductive verification and better applicability for non-experts. Only three of our initial assumptions had been invalidated. We identified options that should be prioritized according to their impact on verification effort when provability is ensured. Moreover, we identified three parameters (i.e., *Quantifier Treatment, Proof Splitting*, and *Arithmetic Treatment*), whose options represent a trade-off between provability and verification effort. Our insights provide valuable recommendations to users on which parameters to prioritize given a verification requirement. Moreover, tool builders can utilize our insights to improve on the user experience. For instance, implementing a recommendation system for parameters based on our investigation would help users to verify software more easily. Furthermore, KeY may hide insignificant parameters in specific verification scenarios or fine-tune parameters automatically during proofs.

For future work, it is necessary to employ more verification targets to investigate the assumptions that could not be accepted. Moreover, it would be interesting to implement a system for parameter recommendations that provides even more fine-grained recommendations based on contracts and methods. Furthermore, we only measured the effect of single values of parameters on provability and verification effort. However, specific values of different parameters may interact with each other and therefore have a larger or even reversed effect when used in combination. Finally, investigating parameterization of other verification tools is indispensable to help more industrial software developers to integrate formal methods in their everyday software development tasks.

Acknowledgments. This work was supported by the DFG (German Research Foundation) under the Researcher Unit FOR1800: Controlling Concurrent Change (CCC). We acknowledge Richard Bubel, Reiner Hähnle, Dominik Steinhöfel, Norber Siegmund, Alexander Grebhahn, Christian Kästner, Sven Apel, and Stefan Krüger for fruitful discussion and valuable feedback throughout this work. We also thank all reviewers for their valuable feedback and corrections.

References

1. Ahrendt, W., Beckert, B., Bubel, R., Hähnle, R., Schmitt, P.H., Ulbrich, M.: Deductive Software Verification-The KeY Book: From Theory to Practice. Springer, Heidelberg (2016). https://doi.org/10.1007/978-3-319-49812-6
2. Barnett, M., Fähndrich, M., Leino, K.R.M., Müller, P., Schulte, W., Venter, H.: Specification and verification: the Spec# experience. Commun. ACM **54**, 81–91 (2011)
3. Baumann, C., Beckert, B., Blasum, H., Bormer, T.: Lessons learned from microkernel verification-specification is the new bottleneck. arXiv preprint arXiv:1211.6186 (2012)
4. Beckert, B., Bormer, T., Grahl, D.: Deductive verification of legacy code. In: Margaria, T., Steffen, B. (eds.) ISoLA 2016. LNCS, vol. 9952, pp. 749–765. Springer, Cham (2016). https://doi.org/10.1007/978-3-319-47166-2_53
5. Benavides, D., Trinidad, P., Ruiz-Cortés, A.: Using constraint programming to reason on feature models. In: Proceedings of the International Conference on Software Engineering and Knowledge Engineering (SEKE), pp. 677–682 (2005)

6. Bowen, J., Stavridou, V.: Safety-critical systems, formal methods and standards. Softw. Eng. J. **8**(4), 189–209 (1993)
7. Burdy, L., Cheon, Y., Cok, D.R., Ernst, M.D., Kiniry, J., Leavens, G.T., Leino, K.R.M., Poll, E.: An overview of JML tools and applications. Int. J. Softw. Tools Technol. Transf. (STTT) **7**(3), 212–232 (2005)
8. Clarke, E.M., Grumberg, O., Peled, D.A.: Model Checking. MIT Press, Cambridge (1999)
9. Clarke, E.M., Wing, J.M.: Formal methods: state of the art and future directions. ACM Comput. Surv. (CSUR) **28**(4), 626–643 (1996)
10. Cohen, E., Dahlweid, M., Hillebrand, M., Leinenbach, D., Moskal, M., Santen, T., Schulte, W., Tobies, S.: VCC: a practical system for verifying concurrent C. In: Berghofer, S., Nipkow, T., Urban, C., Wenzel, M. (eds.) TPHOLs 2009. LNCS, vol. 5674, pp. 23–42. Springer, Heidelberg (2009). https://doi.org/10.1007/978-3-642-03359-9_2
11. Cohen, M.B., Dwyer, M.B., Shi, J.: Interaction testing of highly-configurable systems in the presence of constraints. In: Proceedings of the 2007 International Symposium on Software Testing and Analysis, pp. 129–139. ACM (2007)
12. Cuoq, P., Kirchner, F., Kosmatov, N., Prevosto, V., Signoles, J., Yakobowski, B.: Frama-C. In: Eleftherakis, G., Hinchey, M., Holcombe, M. (eds.) SEFM 2012. LNCS, vol. 7504, pp. 233–247. Springer, Heidelberg (2012). https://doi.org/10.1007/978-3-642-33826-7_16
13. Darvas, Á., Mehta, F., Rudich, A.: Efficient well-definedness checking. In: Armando, A., Baumgartner, P., Dowek, G. (eds.) IJCAR 2008. LNCS (LNAI), vol. 5195, pp. 100–115. Springer, Heidelberg (2008). https://doi.org/10.1007/978-3-540-71070-7_8
14. de Gouw, S., Rot, J., de Boer, F.S., Bubel, R., Hähnle, R.: OpenJDK's Java.utils.Collection.sort() is broken: the good, the bad and the worst case. In: Kroening, D., Păsăreanu, C.S. (eds.) CAV 2015. LNCS, vol. 9206, pp. 273–289. Springer, Cham (2015). https://doi.org/10.1007/978-3-319-21690-4_16
15. de Moura, L., Bjørner, N.: Z3: an efficient SMT solver. In: Ramakrishnan, C.R., Rehof, J. (eds.) TACAS 2008. LNCS, vol. 4963, pp. 337–340. Springer, Heidelberg (2008). https://doi.org/10.1007/978-3-540-78800-3_24
16. Filliâtre, J.-C., Marché, C.: The Why/Krakatoa/Caduceus platform for deductive program verification. In: Damm, W., Hermanns, H. (eds.) CAV 2007. LNCS, vol. 4590, pp. 173–177. Springer, Heidelberg (2007). https://doi.org/10.1007/978-3-540-73368-3_21
17. Gladisch, C.D.: Model generation for quantified formulas with application to test data generation. Proc. Int. J. Softw. Tools Technol. Transfer **14**(4), 439–459 (2012)
18. Gosling, J.: The Java Language Specification. Addison-Wesley Professional, Boston (2000)
19. Grebhahn, A., Siegmund, N., Apel, S., Kuckuk, S., Schmitt, C., Köstler, H.: Optimizing performance of stencil code with SPL conqueror. In: Proceedings of the 1st International Workshop on High-Performance Stencil Computations (HiStencils), pp. 7–14 (2014)
20. Guo, J., Czarnecki, K., Apely, S., Siegmundy, N., Wasowski, A.: Variability-aware performance prediction: a statistical learning approach. In: Proceedings of the 28th IEEE/ACM International Conference on Automated Software Engineering, pp. 301–311. IEEE Press (2013)
21. Hatcliff, J., Leavens, G.T., Leino, K.R.M., Müller, P., Parkinson, M.: Behavioral interface specification languages. ACM Comput. Surv. **44**(3), 16:1–16:58 (2012)

22. Havelund, K., Pressburger, T.: Model checking Java programs using Java PathFinder. J. Softw. Tools Technol. Transfer **2**(4), 366–381 (2000)
23. Hoare, C.A.R.: Proof of correctness of data representations. Acta Informatica **1**(4), 271–281 (1972)
24. Holthusen, S., Nieke, M., Thüm, T., Schaefer, I.: Proof-carrying apps: contract-based deployment-time verification. In: Margaria, T., Steffen, B. (eds.) ISoLA 2016. LNCS, vol. 9952, pp. 839–855. Springer, Cham (2016). https://doi.org/10.1007/978-3-319-47166-2_58
25. Holzmann, G.J.: The model checker SPIN. IEEE Trans. Softw. Eng. (TSE) **23**(5), 279–295 (1997)
26. Hubbers, E., Poll, E.: Reasoning about card tears and transactions in Java Card. In: Wermelinger, M., Margaria-Steffen, T. (eds.) FASE 2004. LNCS, vol. 2984, pp. 114–128. Springer, Heidelberg (2004). https://doi.org/10.1007/978-3-540-24721-0_8
27. Huisman, M., Mostowski, W.: A symbolic approach to permission accounting for concurrent reasoning. In: 2015 14th International Symposium on Proceedings of the Parallel and Distributed Computing (ISPDC), pp. 165–174. IEEE (2015)
28. Kienzle, J., Mussbacher, G., Collet, P., Alam, O.: Delaying decisions in variable concern hierarchies. ACM SIGPLAN Not. **52**, 93–103 (2016)
29. Knight, J.C., DeJong, C.L., Gibble, M.S., Nakano, L.G.: Why are formal methods not used more widely? In: Proceedings of the Fourth NASA Formal Methods Workshop. Citeseer (1997)
30. Knüppel, A., Pardylla, C.I., Thüm, T., Schaefer, I.: Experience report on formally verifying parts of openJDK's API with KeY. In: Proceedings of the Fourth Workshop on Formal Integrated Development Environment. Springer, Heidelberg (2018)
31. Leavens, G.T., Cheon, Y.: Design by Contract with JML, September 2006
32. Leavens, G.T., Poll, E., Clifton, C., Cheon, Y., Ruby, C., Cok, D., Müller, P., Kiniry, J., Chalin, P., Zimmerman, D.M., Dietl, W.: JML Reference Manual, May 2013
33. Marché, C., Moy, Y.: The Jessie Plugin for Deductive Verification in Frama-C. INRIA Saclay Île-de-France and LRI, CNRS UMR (2012)
34. McNemar, Q.: Note on the sampling error of the difference between correlated proportions or percentages. Psychometrika **12**(2), 153–157 (1947)
35. Meyer, B.: Object-Oriented Software Construction, 1st edn. Prentice-Hall Inc., Upper Saddle River (1988)
36. Meyer, B.: Applying design by contract. IEEE Comput. **25**(10), 40–51 (1992)
37. Ochoa, L., González-Rojas, O., Thüm, T.: Using decision rules for solving conflicts in extended feature models. In: Proceedings of the International Conference on Software Language Engineering (SLE), pp. 149–160. ACM, October 2015
38. Olaechea, R., Stewart, S., Czarnecki, K., Rayside, D.: Modelling and multi-objective optimization of quality attributes in variability-rich software. In: Proceedings of the Fourth International Workshop on Nonfunctional System Properties in Domain Specific Modeling Languages, p. 2. ACM (2012)
39. Robby, Rodríguez, E., Dwyer, M.B., Hatcliff, J.: Checking JML specifications using an extensible software model checking. Framework **8**(3), 280–299 (2006)
40. Rushby, J.: Formal methods and their role in the certification of critical systems. In: Shaw R. (ed.) Safety and Reliability of Software Based Systems, pp. 1–42. Springer, London (1997). https://doi.org/10.1007/978-1-4471-0921-1_1
41. Sannella, D.: A survey of formal software development methods. Department of Computer Science, Laboratory for Foundations of Computer Science, University of Edinburgh (1988)

42. Scheurer, D., Hähnle, R., Bubel, R.: A general lattice model for merging symbolic execution branches. In: Ogata, K., Lawford, M., Liu, S. (eds.) ICFEM 2016. LNCS, vol. 10009, pp. 57–73. Springer, Cham (2016). https://doi.org/10.1007/978-3-319-47846-3_5

43. Schumann, J.M.: Automated Theorem Proving in Software Engineering. Springer, Heiedelberg (2001). https://doi.org/10.1007/978-3-662-22646-9

44. Siegmund, N., Grebhahn, A., Apel, S., Kästner, C.: Performance-influence models for highly configurable systems. In: Proceedings of the 2015 10th Joint Meeting on Foundations of Software Engineering, pp. 284–294. ACM (2015)

45. Siegmund, N., Rosenmüller, M., Kuhlemann, M., Kästner, C., Apel, S., Saake, G.: SPL conqueror: toward optimization of non-functional properties in software product lines. Softw. Qual. J. **20**(3–4), 487–517 (2012)

46. Thüm, T., Meinicke, J., Benduhn, F., Hentschel, M., von Rhein, A., Saake, G.: Potential synergies of theorem proving and model checking for software product lines. In: Proceedings of the International Software Product Line Conference (SPLC), pp. 177–186. ACM (2014)

47. Thüm, T., Schaefer, I., Apel, S., Hentschel, M.: Family-based deductive verification of software product lines. In: Proceedings of the International Conference on Generative Programming and Component Engineering (GPCE), pp. 11–20. ACM, September 2012

48. Thüm, T., Winkelmann, T., Schröter, R., Hentschel, M., Krüger, S.: Variability hiding in contracts for dependent software product lines. In: Proceedings of the Workshop on Variability Modelling of Software-intensive Systems (VaMoS), pp. 97–104. ACM (2016)

49. Why Development Team: Why: a software verification platform. http://why.lri.fr/. Accessed 16 Dec 2010

50. Wohlin, C., Runeson, P., Höst, M., Ohlsson, M.C., Regnell, B.: Experimentation in Software Engineering. Springer, Heidelberg (2012). https://doi.org/10.1007/978-3-642-29044-2

Software Verification with ITPs Should Use Binary Code Extraction to Reduce the TCB
(Short Paper)

Ramana Kumar[1,2]([⊠]), Eric Mullen[3], Zachary Tatlock[3],
and Magnus O. Myreen[4]

[1] Data61, CSIRO, Sydney, Australia
ramana.kumar@cl.cam.ac.uk
[2] UNSW, Sydney, Australia
[3] University of Washington, Seattle, WA, USA
[4] Chalmers University of Technology, Gothenburg, Sweden

Abstract. LCF-style provers emphasise that all results are secured by logical inference, and yet their current facilities for code extraction or code generation fall short of this high standard. This paper argues that extraction mechanisms with a small trusted computing base (TCB) ought to be used instead, pointing out that the recent CakeML and Œuf projects show that this is possible in HOL and within reach in Coq.

1 Introduction

Software verification is a primary use of interactive theorem provers (ITPs). To verify a system, one uses the logic of the prover to model the system, specify its desired properties, and prove that the model satisfies the specification. To run the verified system, many provers facilitate *extracting code* from the model, to be compiled and executed in a mainstream functional language. While widely used, this approach leads to an unsettlingly large *trusted computing base* (TCB) – the unverified components a system depends on for correct construction and execution. The TCB of conventional extraction includes pretty printers that "massage" code into the target language (e.g., OCaml or Haskell) without proof, as well as the target language implementation (unverified compiler and runtime).

Recently, two verification projects, CakeML and Œuf, have shown that code extraction with a small TCB is possible without a significant increase in verification effort. They enable ordinary software verification in a prover, and (mostly automated) compilation of a verified system *within the prover* to extract *binary code* for execution. The TCB of this approach no longer includes sophisticated compilation or extraction machinery; binary "extraction" simply prints the literal bytes of verified machine code to a file outside of the ITP.

© Springer International Publishing AG, part of Springer Nature 2018
J. Avigad and A. Mahboubi (Eds.): ITP 2018, LNCS 10895, pp. 362–369, 2018.
https://doi.org/10.1007/978-3-319-94821-8_21

In this paper, our first contribution is to explain the principles of the binary extraction approach taken by CakeML and Œuf to reduce the TCB. For CakeML, this is the first demonstration of the toolchain outside of compiler bootstrapping. Our second contribution is a detailed account of what remains in the TCB after applying binary code extraction, including sketches of how to build a validation story for the remaining assumptions. We argue that, given the feasibility of these ideas, all ITP-based software verification projects should strive for a small TCB.

2   Binary Code Extraction Workflow

The main idea of binary extraction is to *stay inside the prover* until you have binary code. We show how this workflow naturally extends the conventional approach with a simple example: computing the frequencies of words in a file.

First, we specify correctness by defining valid_wordfreq_output, a predicate that holds when *output* represents the word frequencies of *file_contents*:

valid_wordfreq_output *file_contents output* $\iff$
$\exists\, ws.$
    set ws = set (words_of *file_contents*) $\wedge$ sorted ($\lambda x\, y.\, x < y$) ws $\wedge$
    output = concat (map ($\lambda w.$ format_output (w,frequency *file_contents* w)) ws)

An *output* is correct if there is a list of words ws such that the set of words in ws is the same as the set of words in *file_contents*; each element in ws is strictly less than the next one; and *output* is a concatenation of lines, each of which corresponds to an element of ws and contains the frequency of that word in *file_contents* formatted according to a format_output function.

Next, we implement the specification as a functional program inside the logic. The main function for our example, compute_wordfreq_output, is defined below using helpers: insert_line inserts each word from a line into an ordered binary tree; to_list flattens an ordered binary tree into a sorted list; and map and foldl are the usual functions over lists. The format_output and words_of functions are the same as in the specification.

    compute_wordfreq_output *input_lines* =
    map format_output (to_list (foldl insert_line empty_tree *input_lines*))

    insert_line t *line* = foldl insert_word t (words_of *line*)

After defining the functional implementation, we prove that it computes the desired result. This is where most of the manual proof effort associated with ITP-based verification is applied. However, by sticking with shallowly embedded logical functions, we avoid the explosion of details that would arise in a program logic over a deep embedding. For the word frequency example, we prove the following theorem in about 100 lines of tactic-based proof script.

$$\vdash \text{valid\_wordfreq\_output } \textit{file\_contents} \tag{1}$$
$$\text{(concat (compute\_wordfreq\_output (lines\_of } \textit{file\_contents})))}$$

Conventional Approach. One would now use code extraction to pretty print the verified program into a mainstream functional language like Haskell or OCaml — crucially without any proof relating a formal semantics of the target language with either the ITP-generated code or the functions in the logic.

Binary Extraction Approach. The proposed *binary extraction* approach takes a different route, which stays in the logic longer and produces proved guarantees about the behaviour of the generated code. This route requires some infrastructure — a synthesis tool and a verified compiler — but these need only be built once. Here are the steps of binary extraction:

S1: *Use proof-producing or verified synthesis* to translate the shallowly embedded logical functions, such as compute_wordfreq_output, into a deep embedding in a programming language with a formal semantics. The result is a pure program that is proved to perform the desired computation.

S2: *Add some verified wrapper I/O code* so that the program from S1 can interact with its environment. The result is a complete standalone program. The previous step (S1) is automatic, but this step (S2) might be interactive or automatic depending on the desired I/O interaction.

S3: *Compile the verified standalone program in the logic* so that the compiler's correctness theorem can be applied to a theorem about its evaluation, i.e., ⊢compile source = compiler_output for a particular source and compiler_output.

These steps yield a theorem stating that compiler_output is a machine-code program that performs I/O according to S2 and implements the computation from S1, which will have been verified against its specification using typical ITP methods. For our example, the computation is compute_wordfreq_output, so we connect the behaviour of compiler_output to the valid_wordfreq_output property via the algorithm-level theorem (1). Compared with the conventional approach, the only extra manual effort in S1–S3 is the verification of the I/O wrapper in S2.

Below, we show how S1–S3 are realised in HOL4 by CakeML, and how they are almost realised in Coq by Œuf. That these ideas are supported (or very nearly supported) in both provers demonstrates cross-prover applicability.

Binary Extraction in CakeML. S1–S3 are supported by different parts of the CakeML ecosystem and the underlying HOL4 theorem prover. (S1:) Functional programs written in HOL are translated to pure CakeML functions by an automatic proof-producing synthesis tool [21]; (S2:) the wrapper code is added to the code from S1 manually and verified using characteristic formulae (CF) for CakeML [8]; and (S3:) the verified CakeML compiler's backend [23] is evaluated in the logic using HOL4's evaluation engine by Barras [2]. S1 and S3 are automatic, while S2 currently requires some expertise from the user.

For our example, S2 involves writing a wrapper like the code shown below and proving a CF separation-logic-style correctness theorem for it.

```
val _ = (append_prog o process_topdecs) '
  fun wordfreq u =
    case TextIO.inputLinesFrom (List.hd (CommandLine.arguments()))
    of SOME lines =>
      TextIO.print_list (compute_wordfreq_output lines) '
```

The first line above instructs HOL4 to add code to the CakeML program being constructed. The code between the quotation marks, ' ... ', is CakeML concrete syntax for the top-level CakeML function. compute_-wordfreq_output is the synthesised CakeML function corresponding to the compute_wordfreq_output HOL function. Other names refer to functions in the CakeML basis library.

Once S1–S3 have been completed, the theorems from each step are easily composed to produce an end-to-end correctness theorem[1], which we explain below.

$$
\begin{aligned}
\vdash\ &\mathsf{wfCL}\ [pname;\ fname] \land \mathsf{wfFS}\ fs \land \mathsf{hasFreeFD}\ fs \land \\
&\mathsf{get\_file\_contents}\ fs\ fname = \mathsf{Some}\ file\_contents \land \\
&\mathsf{x64\_installed}\ \mathsf{compiler\_output}\ (\mathsf{basis\_ffi}\ [pname;\ fname]\ fs)\ mc\ ms \Rightarrow \\
&\exists\, io\_events\ ascii\_output. \\
&\quad \mathsf{machine\_sem}\ mc\ (\mathsf{basis\_ffi}\ [pname;\ fname]\ fs)\ ms \subseteq \\
&\quad \mathsf{extend\_with\_resource\_limit}\ \{\ \mathsf{Terminate\ Success}\ io\_events\ \} \land \\
&\quad \mathsf{extract\_fs}\ fs\ io\_events = \mathsf{Some}\ (\mathsf{add\_stdout}\ fs\ ascii\_output) \land \\
&\quad \mathsf{valid\_wordfreq\_output}\ file\_contents\ ascii\_output
\end{aligned}
\tag{2}
$$

The first two lines make assumptions about the environment, namely: the command line must consist of two well-formed (wfCL) words, *pname* and *fname*; the file system *fs* must be well-formed (wfFS) with a free file descriptor (hasFreeFD); and *fname* must exist in *fs* with contents *file_contents*. The third line is more interesting and concerns the initial machine state *ms*. We assume that *ms* is an x86-64 machine state where compiler_output has been installed into memory and is ready to go; we also assume that CakeML's foreign-function interface (basis_ffi) behaves according to our model of the file system and standard streams.

If all these assumptions are true, then the machine-code level execution (machine_sem) will terminate. During execution the machine will perform some *io_events* (or some prefix of them, if it runs out of memory). The extract_fs line states that running the file system model *fs* through the *io_events* has the effect of adding some *ascii_output* to standard output. The last line states that this *ascii_output* is correct according to our specification valid_wordfreq_output.

Binary Extraction in Œuf. Œuf accomplishes S1–S3 similarly to CakeML, but in Coq and building on CompCert [16]. (S1:) A Gallina program is translated to Œuf functions by an untrusted Coq plugin and translation validated to ensure equivalence [20]. Œuf then compiles the code to CompCert's Cminor IR. (S2:) I/O wrapper code is written in C, compiled to Cminor via CompCert, and linked

---
[1] https://code.cakeml.org/tree/master/tutorial/solutions.

to the code from S1. (S3:) The combined stand-alone program is compiled to assembly using CompCert, relying on tools like Valex to formally validate assembling [13]. An extracted version of the Œuf compiler is still currently used, since CompCert is not yet fully executable within Coq [17].

For our example, a user would write an I/O wrapper in C similar to:

```
int main(void) { union list* input = to_coq_str(read_stdin());
                 union list* freqs = OEUF_CALL(wordfreq, input);
                 write_stdout(of_coq_str(freqs)); return 0; }
```

The `OEUF_CALL` macro constructs a closure and passes arguments to Œuf extracted code while `to_coq_str` and `of_coq_str` translate between the Œuf string representation (lists of Boolean 8-tuples) and the standard C representation (`char*`). Proving S2 requires showing that C data conversions and system calls adhere to the Œuf ABI [20], which we have specified for this example.[2] Assuming these specifications, composing theorems for S1–S3 yields an end-to-end guarantee:

$$\begin{aligned}
&\mathsf{Oeuf.compile(wordfreq')} = \mathsf{OK}\ c \wedge \mathsf{Oeuf.link(c, shim)} = \mathsf{OK}\ p \wedge \\
&\mathsf{CompCert.compile(p)} = \mathsf{OK}\ b \wedge \mathsf{initSt}(b, s_1) \Rightarrow \\
&\exists\ w\ \tau\ s_2.\ s_1 \xrightarrow{\tau} s_2 \wedge \mathsf{finalSt}(b, s_2) \wedge \mathsf{stdIn}(\tau) = w \wedge \mathsf{stdOut}(\tau) = \mathsf{wordfreq}(w)
\end{aligned} \tag{3}$$

The first two lines relate the original Gallina function, shim (wrapper), and compiled output for the whole program; and require that state s_1 is a valid initial state, (i.e., that b has been correctly loaded). The final line guarantees that, under these assumptions, the program will safely execute and terminate in a final state[3] (as CompCert assembly semantics are deterministic) while generating trace τ of I/O events and that this trace corresponds to reading string w from standard input and writing wordfreq(w) to standard output. stdIn and stdOut filter the trace and relate low-level values to Gallina values.

3   Trusted Computing Base

The binary extraction approach yields both a proved result (theorem (2) or (3)), and the verified binary executable itself (compiler_output or b) printed into a file. To show what remains in the TCB for correct execution, we analyse the CakeML version of the word frequency example.

The ITP: Its Logic and Implementation. We trust our theorem prover: that classical higher-order logic is consistent [10,14,22], and that the HOL4 kernel implements this logic correctly. Trusting the ITP implementation means trusting the ~4000 lines of Standard ML code in the kernel, the rationale underlying HOL4's LCF-based design [19], and the compiler (Poly/ML), OS, and hardware on which HOL4 runs. It is possible (though we have not done it here) to obtain

---

[2] https://github.com/uwplse/oeuf/tree/master/demos/word_freq.
[3] No resource limits are assumed since CompCert semantics model infinite memory.

externally checkable proof certificates from HOL4 (via OpenTheory [12]), mitigating the need to trust any specific ITP implementation. These kinds of trust are intrinsic to any ITP-based approach.

The Specification. We need to correctly formalise the desired behaviour of our program (valid_wordfreq_output), because it is not checked by any proofs. However, a specification can be tested by proving sanity-checking theorems about it and evaluating (the executable parts of) its definition on concrete examples. Trust across this *specification gap* is intrinsic to any kind of formal verification.

The Extraction Procedure. To execute verified code, it must at some point exit the theorem prover and appear in memory associated with a running process. We trust the function — a very simple one for binary extraction — that reads the code (as a term in logic) and prints it into an executable image template. We trust that this file is not tampered with, and that the linker (next paragraph) and OS loader operate correctly. These assumptions are captured in the x64_installed predicate, which specifies the expected state of the machine after the executable is loaded. In addition to carefully defining x64_installed, we could validate this assumption using runtime checks on startup: e.g., that the registers pointing to the ends of the CakeML heap are valid and aligned. Trusting something between formal models and reality is unavoidable.

The Execution Environment. The final theorem is about execution of a formal machine model (machine_sem). We trust the hardware to behave according to this model, and that the OS and other processes do not interfere with the CakeML process. (We model interference and assume it avoids the CakeML process's memory [7].) Machine models, like Fox's L3 models that we use, can be validated by systematic testing against the hardware [4,6]. Our verified program interacts with its environment, and we model how we expect the environment to behave, with functions like basis_ffi and extract_fs. The I/O facilities (command line, files, and standard streams) available in CakeML's basis library are supported by a small C interface to the underlying system calls (e.g., open). We trust our implementation of this interface, and the C compiler (on the interface code only) and linker. The verified program may exit prematurely if it runs out of memory: we eliminate this occurrence only by observation.

4 Broader Context and Vision

The ITP community is pursuing several approaches relevant to reducing the TCB of code extraction. Important aspects of extraction have been proven correct for Coq [18] and Isabelle/HOL [3,9]; the CertiCoq [1] team and Hupel and Nipkow [11] are working toward verified code generators for Coq and Isabelle/HOL respectively; and frameworks like the Isabelle Refinement Framework [15] and Fiat [5] are exploring other approaches to proof-producing code extraction. CakeML and Œuf are distinguished by striving to be a natural replacement for conventional extraction, using conventional programming languages for synthesis, and aiming to completely eliminate the compiler from the TCB by proving results about the behaviour of the whole program binary including effectful wrapper code.

Given the advances from throughout the community and the fact that similar results are supported across different ITPs, we feel that extraction with a small TCB is on the cusp of wide-scale feasibility for verified systems. Much is left to study and build before these approaches achieve the convenience and performance of conventional extraction techniques, but we have demonstrated that it is already possible to rigorously connect facts established in the logic of an ITP to binary executable code under a substantially smaller TCB, and without substantial increase in verification effort. We enthusiastically urge the rest of the ITP community to adopt and advance the ideas behind binary code extraction.

References

1. Anand, A., Appel, A., Morrisett, G., Paraskevopoulou, Z., Pollack, R., Belanger, O.S., Sozeau, M., Weaver, M.: CertiCoq: A verified compiler for Coq. In: CoqPL (2017)
2. Barras, B.: Programming and computing in HOL. In: Aagaard, M., Harrison, J. (eds.) TPHOLs 2000. LNCS, vol. 1869, pp. 17–37. Springer, Heidelberg (2000). https://doi.org/10.1007/3-540-44659-1_2
3. Berghofer, S., Nipkow, T.: Executing higher order logic. In: Callaghan, P., Luo, Z., McKinna, J., Pollack, R., Pollack, R. (eds.) TYPES 2000. LNCS, vol. 2277, pp. 24–40. Springer, Heidelberg (2002). https://doi.org/10.1007/3-540-45842-5_2
4. Campbell, B., Stark, I.: Randomised testing of a microprocessor model using SMT-solver state generation. SCP **118**, 60–76 (2016)
5. Delaware, B., Pit-Claudel, C., Gross, J., Chlipala, A.: Fiat: deductive synthesis of abstract data types in a proof assistant. In: POPL, pp. 689–700 (2015)
6. Fox, A., Myreen, M.O.: A trustworthy monadic formalization of the ARMv7 instruction set architecture. In: Kaufmann, M., Paulson, L.C. (eds.) ITP 2010. LNCS, vol. 6172, pp. 243–258. Springer, Heidelberg (2010). https://doi.org/10.1007/978-3-642-14052-5_18
7. Fox, A.C.J., Myreen, M.O., Tan, Y.K., Kumar, R.: Verified compilation of CakeML to multiple machine-code targets. In: CPP, pp. 125–137 (2017)
8. Guéneau, A., Myreen, M.O., Kumar, R., Norrish, M.: Verified characteristic formulae for CakeML. In: Yang, H. (ed.) ESOP 2017. LNCS, vol. 10201, pp. 584–610. Springer, Heidelberg (2017). https://doi.org/10.1007/978-3-662-54434-1_22
9. Haftmann, F., Nipkow, T.: A code generator framework for Isabelle/HOL. In: TPHOLs (2007)
10. Harrison, J.: Towards self-verification of HOL light. In: Furbach, U., Shankar, N. (eds.) IJCAR 2006. LNCS (LNAI), vol. 4130, pp. 177–191. Springer, Heidelberg (2006). https://doi.org/10.1007/11814771_17
11. Hupel, L., Nipkow, T.: A verified compiler from Isabelle/HOL to CakeML. In: Ahmed, A. (ed.) ESOP 2018. LNCS, vol. 10801, pp. 999–1026. Springer, Cham (2018). https://doi.org/10.1007/978-3-319-89884-1_35
12. Hurd, J.: The opentheory standard theory library. In: Bobaru, M., Havelund, K., Holzmann, G.J., Joshi, R. (eds.) NFM 2011. LNCS, vol. 6617, pp. 177–191. Springer, Heidelberg (2011). https://doi.org/10.1007/978-3-642-20398-5_14

13. Kästner, D., Leroy, X., Blazy, S., Schommer, B., Schmidt, M., Ferdinand, C.: Closing the gap - the formally verified optimizing compiler CompCert. In: Safety-critical Systems Symposium 2017, SSS 2017, pp. 163–180. Developments in System Safety Engineering: Proceedings of the Twenty-fifth Safety-critical Systems Symposium, CreateSpace, Bristol, United Kingdom, February 2017. https://hal.inria.fr/hal-01399482
14. Kumar, R., Arthan, R., Myreen, M.O., Owens, S.: Self-formalisation of higher-order logic - semantics, soundness, and a verified implementation. JAR **56**(3), 221–259 (2016)
15. Lammich, P.: Refinement to imperative/HOL. ITP (2015)
16. Leroy, X.: Formal certification of a compiler back-end, or: programming a compiler with a proof assistant. In: 33rd ACM Symposium on Principles of Programming Languages, pp. 42–54. ACM Press (2006)
17. Leroy, X.: Using coq's evaluation mechanisms in anger (2015). http://gallium.inria.fr/blog/coq-eval/
18. Letouzey, P.: Extraction in Coq: an overview. In: Beckmann, A., Dimitracopoulos, C., Löwe, B. (eds.) CiE 2008. LNCS, vol. 5028, pp. 359–369. Springer, Heidelberg (2008). https://doi.org/10.1007/978-3-540-69407-6_39
19. Milner, R.: LCF: a way of doing proofs with a machine. In: Bečvář, J. (ed.) MFCS 1979. LNCS, vol. 74, pp. 146–159. Springer, Heidelberg (1979). https://doi.org/10.1007/3-540-09526-8_11
20. Mullen, E., Pernsteiner, S., Wilcox, J.R., Tatlock, Z., Grossman, D.: Œuf: minimizing the Coq extraction TCB. In: CPP 2018, pp. 172–185 (2018)
21. Myreen, M.O., Owens, S.: Proof-producing translation of higher-order logic into pure and stateful ML. JFP **24**(2–3), 284–315 (2014)
22. Pitts, A.M.: The HOL System: Logic, 3rd edn. https://hol-theorem-prover.org#doc
23. Tan, Y.K., Myreen, M.O., Kumar, R., Fox, A.C.J., Owens, S., Norrish, M.: A new verified compiler backend for CakeML. In: ICFP, pp. 60–73 (2016)

Proof Pearl: Constructive Extraction of Cycle Finding Algorithms

Dominique Larchey-Wendling$^{(\boxtimes)}$

Université de Lorraine, CNRS, LORIA, Nancy, France
dominique.larchey-wendling@loria.fr

Abstract. We present a short implementation of the well-known Tortoise and Hare cycle finding algorithm in the constructive setting of Coq. This algorithm is interesting from a constructive perspective because it is both very simple and potentially non-terminating (depending on the input). To overcome potential non-termination, we encode the given termination argument (there exists a cycle) into a bar inductive predicate that we use as termination certificate. From this development, we extract the standard OCaml implementation of this algorithm. We generalize the method to the full Floyd's algorithm that computes the entry point and the period of the cycle in the iterated sequence, and to the more efficient Brent's algorithm for computing the period only, again with accurate extractions of their respective standard OCaml implementations.

Keywords: Cycle finding · Bar inductive predicates
Partial algorithms in Coq · Correctness by extraction

1 Introduction

The Tortoise and the Hare (T&H for short) in particular and cycle detection [1] in general are standard algorithms that will very likely cross the path of any would-be computer scientist. They aim at detecting cycles in *deterministic* sequences of values, i.e. when the next value depends only on the current value. They have many applications, from pseudorandom number strength measurement, integer factorization through Pollard's rho algorithm [18] or more generally cryptography, etc., even celestial mechanics. But our interest with those algorithms lies more in the framework in which we want to implement and certify them:

- first we want to prove the partial correction of cycle detection algorithms without assuming their termination. Hence, we do not restrict our study to finite domains. In the finitary case indeed, the *pigeon hole principle* ensures that there is always a cycle to detect and termination can be certified by cardinality considerations [9];

Work partially supported by the TICAMORE project ANR grant 16-CE91-0002.

J. Avigad and A. Mahboubi (Eds.): ITP 2018, LNCS 10895, pp. 370–387, 2018.
https://doi.org/10.1007/978-3-319-94821-8_22

- inductive/constructive type theories constitute challenging contexts for these algorithms because they are inherently partial. The reason for that is the undecidability of the existence of a cycle in an arbitrary given sequence. Hence, we need to work with partial recursive functions;
- Coq could wrongfully be considered being limited to total functions. To work around this, Hilbert's ϵ-operator (a non-constructive form of the axiom of choice) is sometimes postulated as a convenient way to deal with partial functions [7]. Why not, HOL is based on it? We argue that we can stay fully constructive and we will show that axiom-free Coq can work with such partial recursive functions provided they are precisely specified.

T&H was attributed to Robert W. Floyd by Donald E. Knuth [15] but may in fact be a folk theorem (see [2], footnote 8 on page 21). The idea is to launch from a starting point x_0 both a slow tortoise (the tortoise steps once at each iteration) and a quick hare (the hare steps twice at each iteration). Then the hare will recapture the tortoise if and only if there is a cycle in the sequence from x_0. We refer to [1] for further visual explanations about the origin and intuition behind this very well-known algorithm. The T&H algorithm computes only a meeting point for the two fabulous animals. We will call *Floyd's cycle finding* the algorithm that builds on this technique to compute the entry point and period of the cycle. We also consider *Brent's period only finding algorithm* [6] that proceeds with a slow hare and a so-called "teleporting tortoise."

Because these algorithms do not always terminate, defining the corresponding fully specified Coq fixpoints might be considered challenging. The folklore *fuel* trick could be used to simulate general recursion in Coq. The idea is to use a term of the type $X \rightarrow (\text{fuel} : \text{nat}) \rightarrow \text{option } Y$ to represent a partial recursive function $f : X \rightarrow Y$. The fuel argument ensures termination, as e.g. a bound on the number of recursive sub-calls. This fuel trick has several problems: computing a big-enough fuel value from the input $x : X$ might be as complicated as showing the termination of $f(x)$ itself; but also, the fuel argument is *informative* and is thus preserved by extraction, both as a parasitic argument and as a companion program for computing the input value of fuel from the value of x; and finally, the output type is now option Y instead of Y so an extra match construct is necessary. We will show how to replace the informative fuel argument with a *non-informative* bar inductive predicate to ensure termination. As such, it is erased by extraction, getting us rid of parasitic arguments. In particular, we obtain an OCaml extraction of T&H certified by less than 80 lines of Coq code.

2 Formalization of the Problem

Given a set X and a function $f : X \rightarrow X$, we define the n-th iterate $f^n : X \rightarrow X$ of f by induction on n: $f^0 = x \mapsto x$ and $f^{n+1} = f \circ f^n$. In Coq, this definition corresponds to the code of the iterator (with a convenient compact f^n notation):

```
Fixpoint iter {X : Type} (f : X → X) n x : X :=
   match n with 0 ↦ x | S n ↦ f (fⁿ x) end
where " fⁿ " := (iter f n).
```

We get the identity $f^{a+b}(x) = f^a(f^b(x))$ by induction on a. Given a starting point $x_0 \in X$, we consider the infinite sequence $x_0, f(x_0), f^2(x_0), \ldots, f^n(x_0), \ldots$ of iterates of f on x_0, i.e. the map $n \mapsto f^n(x_0)$. From a classical logic perspective, two mutually exclusive alternatives are possible:

A1 the sequence $n \mapsto f^n(x_0)$ is injective, i.e. $f^i(x_0) \neq f^j(x_0)$ holds unless $i = j$. In this case, there is no cycle in the iterated sequence from x_0;

A2 there exist $i \neq j$ such that $f^i(x_0) = f^j(x_0)$ and in this case, there is a cycle in the iterated sequence from x_0.

It is however not possible to computationally distinguish those two cases: no cycle finding algorithm can be both correct and always terminating. To show this undecidability result, one can reduce the Halting problem to the cycle detection problem (see file cycle_undec.v).

The T&H algorithm terminates exactly when Alternative A2 above holds and loops forever when Alternative A1 holds. There are many equivalent characterizations of the existence of a cycle, which we call the *cyclicity property*.

Proposition 1 (Cyclicity). *For any set X, any function $f : X \to X$ and any $x_0 \in X$, the four following conditions are equivalent:*

1. *there exist $i, j \in \mathbb{N}$ such that $i \neq j$ and $f^i(x_0) = f^j(x_0)$;*
2. *there exist $\lambda, \mu \in \mathbb{N}$ such that $0 < \mu$ and $f^\lambda(x_0) = f^{\lambda+\mu}(x_0)$;*
3. *there exist $\lambda, \mu \in \mathbb{N}$ s.t. $0 < \mu$ and for any $i, j \in \mathbb{N}$, $f^{i+\lambda}(x_0) = f^{i+\lambda+j\mu}(x_0)$;*
4. *there exists $\tau \in \mathbb{N}$ such that $0 < \tau$ and $f^\tau(x_0) = f^{2\tau}(x_0)$.*

Proof. For $1 \Rightarrow 2$, if $i < j$ then choose $\lambda = i$ and $\mu = j - i$ (exchange i and j if otherwise $j < i$). For $2 \Rightarrow 3$, first show $f^\lambda(x_0) = f^{\lambda+j\mu}(x_0)$ by induction on j. Then $f^{i+\lambda}(x_0) = f^i(f^\lambda(x_0)) = f^i(f^{\lambda+j\mu}(x_0)) = f^{i+\lambda+j\mu}(x_0)$. For $3 \Rightarrow 4$, choose $\tau = (1 + \lambda)\mu$ and derive $f^\tau(x_0) = f^{2\tau}(x_0)$ using $i = (1 + \lambda)\mu - \lambda$ and $j = 1 + \lambda$. For $4 \Rightarrow 1$, choose $i = \tau$ and $j = 2\tau$. That proof is mechanized as Proposition cyclicity_prop in file utils.v. $\square$

The functional specification of T&H is to compute a meeting index $\tau \in \mathbb{N}$ such that $0 < \tau$ and $f^\tau(x_0) = f^{2\tau}(x_0)$ (corresponding to Item 4 of Proposition 1), provided such a value exists. Operationally, the algorithm consists in enumerating the sequence of pairs $(f(x_0), f^2(x_0)), (f^2(x_0), f^4(x_0)), \ldots, (f^n(x_0), f^{2n}(x_0)), \ldots$ in an efficient way until the two values $f^n(x_0)$ and $f^{2n}(x_0)$ are equal.

2.1   An OCaml Account of the Tortoise and the Hare

The T&H algorithm can be expressed in OCaml as the following two functions:

```
tort_hare_rec : (f : α → α) → (x : α) → (y : α) → int
tortoise_hare : (f : α → α) → (x₀ : α) → int
let rec tort_hare_rec f x y =
    if x = y then 0 else 1 + tort_hare_rec f (f x) (f (f y))
let tortoise_hare f x₀ = 1 + tort_hare_rec f (f x₀) (f (f x₀))
```

In general, the tail-recursive version is preferred because tail-recursive functions can be compiled into loops without the help of a stack. The code of the function `tortoise_hare_tail` contains a sub-function `loop` where the first argument f of `tortoise_hare_tail` is fixed and second argument x_0 is unused.

$$\texttt{tortoise\_hare\_tail} : (f : \alpha \to \alpha) \to (x_0 : \alpha) \to \texttt{int}$$
$$\quad \texttt{loop} : (n : \texttt{int}) \to (x : \alpha) \to (y : \alpha) \to \texttt{int}$$

```
let tortoise_hare_tail f x0 =
    let rec loop n x y = if x = y then n else loop (1 + n) (f x) (f (f y))
    in loop 1 (f x0) (f (f x0))
```

Notice that the pre-condition of cyclicity (any item of Proposition 1) is necessary otherwise the above OCaml code does not terminate and is thus incorrect. Any correctness proof must include that cyclicity pre-condition, or a stronger one.

2.2   Goals and Contributions

The goal of this work is double:

Goal 1: functional correctness. Using purely constructive means, build fully specified Coq terms that compute a meeting point for the tortoise and the hare, with the sole pre-condition of cyclicity. Reiterate this for Floyd's and Brent's cycle finding algorithms;

Goal 2: operational correctness. Ensure that the extraction of the previous Coq terms give the corresponding standard OCaml implementations. In particular, derive the above implementations of `tortoise_hare` and `tortoise_hare_tail` by extraction.

From these two goals, trusting Coq extraction mechanism, we get the functional correctness of the standard OCaml implementations for free.

For T&H, solving Goal 1 can be viewed as constructing a term of type

$$\texttt{th\_coq} : (\exists \tau, 0 < \tau \wedge f^\tau x_0 = f^{2\tau} x_0) \to \{\tau \mid 0 < \tau \wedge f^\tau x_0 = f^{2\tau} x_0\}$$

from the assumptions of a type $X : \texttt{Type}$, a procedure $=^?_X$ for deciding equality over X, and a sequence given by $f : X \to X$ and $x_0 : X$. Notice the assumption $=^?_X : \forall x\, y : X, \{x = y\} + \{x \neq y\}$ of an *equality decider* for X that is necessary in Coq. Indeed, unlike OCaml which has a built-in polymorphic equality decider,[1] Coq does (and can) not have equality deciders for every possible type.

Solving Goal 2 means that after extraction of OCaml code from `th_coq`, we get the same function code as `tortoise_hare` (resp. `tortoise_hare_tail`).

This paper is a companion for Coq implementations of cycle finding algorithms. The corresponding source code can be found at

https://github.com/DmxLarchey/The-Tortoise-and-the-Hare

---

[1] OCaml equality decider is partially correct, e.g. it throws exceptions on functions.

The implementation involves around 3000 lines of Coq code but this does not reflect the compactness of our implementation of T&H. Indeed, it contains Floyd's and Brent's algorithms as well, and there are several accompanying files illustrating certified recursion through bar inductive predicates. To witness the conciseness of our approach, we give a standalone tail-recursive implementation of T&H of less than 80 lines, not counting comments (see th_alone.v). This project compiles under Coq 8.6 and is available under a Free Software license.

The designs of the cycle finding algorithms that we propose are all based on bar inductive predicates used as termination certificates for Coq fixpoint recursion. In Scct. 3, we give a brief introduction to these predicates from a programmer's point of view and show why they are suited for solving termination problems. The corresponding Coq source code can be found in file bar.v.

In Sect. 4, we present two fully specified implementations of T&H, one non-tail recursive and one tail-recursive. We give a detailed account of the algorithmic part of the implementation that we isolate from *logical obligations*. We explain how bar inductive predicates are used to separate/postpone termination proofs from algorithmic considerations. The corresponding file is tortoise_hare.v.

In Sect. 5, we give an overview of the implementation of the full Floyd cycle finding algorithm that computes the characteristic index and period of an iterated sequence. The corresponding Coq file is floyd.v. In Sect. 6, we give a brief account of our implementation of Brent's period finding algorithm, in fact two implementations: one suited for binary numbers and one suited for unary numbers. The corresponding source code files are brent_bin.v and brent_una.v.

The T&H has already been the subject of implementations in Coq [9,11], but under different requirements. In Sect. 7, we compare our development with those alternative approaches. From a constructive point of view, we analyse the pre-conditions under which correctness is established in each case.

3   Termination Using Bar Inductive Predicates

In this section, we explain how to use *bar inductive predicates* [12] — a constructive and axiom-free form of bar induction[2] — as termination certificates.

As explained in Sect. 3.2, in the context we use them (decidable terminated cases), these predicates have the same expressive power as the *accessibility predicates* used for well-founded recursion in Coq in the modules Wf and Wellfounded from the standard library (see also left part of Fig. 1). But we think that bar inductive predicates have several advantages over accessibility predicates:

- compared to the general accessibility predicates of [4], they do not need the simultaneous induction/recursion schemes of Dybjer [10] (not integrated in Coq so far) in case of nested/mutual recursion [17];

---

[2] Conventional bar induction often requires *Brouwer's thesis* which precisely postulates that bar predicates are inductive.

$$\frac{\forall y, R\ y\ x \to \mathsf{Acc}\ y}{\mathsf{Acc}\ x} \qquad \qquad \frac{T\ x}{\mathsf{bar}\ x} \qquad \frac{\forall y, R\ x\ y \to \mathsf{bar}\ y}{\mathsf{bar}\ x}$$

Fig. 1. Inductive rules for Acc and bar termination certificates.

– unlike standard accessibility predicates (module Wf) which involve thinking about termination before implementing the algorithm, or *inductively defined domain predicates* (see [3] pp. 427–432) which involve thinking about termination together with the algorithm, bar inductive predicates focuses on *terminated cases* and *recursive sub-calls* so termination proofs can be separated from the algorithm.

We argue that separating/postponing the proof of termination makes the use of bar inductive predicates more versatile. At least, we hope that we illustrate our case here. Of course, a comprehensive comparison with [5] would be necessary to complete our case. The reader could be interested in recent developments that show that the method of bar inductive predicates scales well to more complicated nested/mutual recursive schemes [17].

We do not really introduce new concepts in the section. But we want to stress the links between the notion of *cover-induction* [8] and the notion of *bar inductive predicate* (e.g. inductive bars [12]). We insist on these notions because we will specialize the following generic implementation to get an "extraction friendly" Coq definition of cycle finding algorithms.

3.1 Dependently Typed Recursion for Bar Inductive Predicates

Let us consider a type X, a unary relation $T : X \to \mathsf{Prop}$ and a binary relation $R : X \to X \to \mathsf{Prop}$. Here are some possible intuitive interpretations of T and R:

T x**:** the computation at point x is terminated (no recursive sub-call);
R x y**:** a call at point x may trigger a recursive sub-call at point y.

We define the inductive predicate bar $: X \to \mathsf{Prop}$ which covers points where computation is warrantied to terminate, by the two rules on the right of Fig. 1:

```
Variables  (X : Type) (T : X → Prop) (R : X → X → Prop).
Inductive  bar (x : X) : Prop :=
  | in_bar_0 :  T x                      → bar x
  | in_bar_1 :  (∀y, R x y → bar y)  → bar x.
```

The first rule in_bar_0 states that a terminated computation terminates. The second rule in_bar_1 states that if every recursive sub-call y of x terminates then so is the call at x. Notice that the predicate bar x : Prop carries no computational content and thus cannot be used to perform computational choices. Termination is only warrantied by the bar x predicate, it is not performed by it.

Hence we assume a decider term $T_{\text{dec}} : \forall x, \{T\,x\} + \{\neg T\,x\}$ for terminated points. We then define `bar_rect`, a dependently typed recursion principle for bar x. For this, we need the following inductions hypotheses:

Hypothesis $(T_{\text{dec}} : \forall x, \{T\,x\} + \{\neg T\,x\})$.
Variable $(P : X \to \mathtt{Type})$.
Hypothesis $(H_T : \forall x, T\,x \to P\,x)$ $(H_{\text{bar}} : \forall x, (\forall y, R\,x\,y \to P\,y) \to P\,x)$.

where H_T gives the value for terminated points and H_{bar} combines the values of the recursive sub-calls into a value for the call itself. With these assumptions, we get the following dependently typed induction principle:

```
Fixpoint bar_rect x (H : bar x) {struct H} : P x :=
   match T_dec x with
     | left  H_x  ↦ H_T _ H_x
     | right H_x  ↦ H_bar _ (fun y H_y ↦ bar_rect y G₁?)
   end.
```

where $\mathbb{G}_1^?$ is a proof term for a *logical obligation*:

$$\mathbb{G}_1^? \;\;/\!/\;\; \dots, x : X, H : \mathtt{bar}\ x, H_x : \neg T\,x, y : X, H_y : R\,x\,y \vdash \mathtt{bar}\ y$$

Notice that for Coq to accept such a `Fixpoint` definition as well-typed, one must ensure that the given proof of goal $\mathbb{G}_1^?$ is a sub-term of the term $H : \mathtt{bar}\ x$ because H is declared as the *structurally decreasing argument* of this fixpoint. Hence, the first step in the proof of $\mathbb{G}_1^?$ is `destruct H`.

The above implementation of `bar_rect` expects the proof term of $\mathbb{G}_1^?$ to be given *before* the actual `Fixpoint` definition of `bar_rect`. This can be mitigated with the use of the very handy `refine` tactic that can delay the proof obligations after the *incomplete proof term* is given (see bar.v for details). As a final remark concerning the term `bar_rect`, there are two ways of stopping a chain of recursive sub-calls: the first is obviously to reach a terminated point (i.e. $T\,x$) but the chain can also stop when there is zero recursive sub-calls (i.e. $R\,x\,y$ holds for no y). While the first condition of terminated points is decidable, the second condition of the nonexistence of recursive sub-calls is usually not decidable. Hence when using the accessibility predicate Acc $(\mathtt{fun}\ u\ v \mapsto R\,v\,u \wedge \neg T v)\ x$ which mixes both T and R (see Theorem `bar_Acc_eq_dec` below), detecting the first termination condition is less natural.

3.2   Accessibility vs. Bar Inductive Predicates

We show that bar inductive predicates generalize accessibility predicates defined in the Coq standard library module `Wf`,

Theorem `bar_empty_Acc_eq` $(X : \mathtt{Type})$ $(R : X \to X \to \mathtt{Prop})$ $(x : X)$:
$$\mathtt{bar}\ (\mathtt{fun}\ \_ \mapsto \mathtt{False})\ R\ x \Longleftrightarrow \mathtt{Acc}\ R^{-1}\ x$$

which is obvious from the rules of Fig. 1 because when $T = \texttt{fun}\ \_ \mapsto \texttt{False}$ is empty, one cannot use rule in_bar_0. Then, we show that when $T : X \to \texttt{Prop}$ is (logically) decidable, then $\texttt{bar}\ T\ R$ can be encoded as an accessibility predicate:

Theorem bar_Acc_eq_dec X $(T : X \to \texttt{Prop})$ $(R : X \to X \to \texttt{Prop})$:

$$(\forall x, T\, x \lor \neg T\, x) \to \forall x, \texttt{bar}\ T\ R\ x \iff \texttt{Acc}\ (\texttt{fun}\ u\, v \mapsto R\, v\, u \land \neg T\, v)\ x$$

From our point of view, the advantage of **bar** over **Acc** is that they keep the two forms of termination separate (x is terminated by T vs. x generates no recursive sub-call), making them easier to reason or compute with. Moreover, using **Acc** incites at using only well-founded relations (or even decreasing measures) whereas **bar** focuses on terminated points/recursive calls and thus can be used more freely as exemplified in Sects. 4 and 5.

3.3 Constructive Epsilon via Bar Inductive Predicates

As a first illustration of using bar inductive predicates, we show how to implement *Constructive Indefinite Ground Description* defined in the standard library module **ConstructiveEpsilon**.

Theorem Constructive_Epsilon $(Q : \texttt{nat} \to \texttt{Prop})$:

$$\big(\forall n, \{Q\, n\} + \{\neg Q\, n\}\big) \to (\exists n, Q\, n) \to \{n : \texttt{nat} \mid Q\, n\}.$$

We instantiate bar_rect with $(T := Q)$, $(R\, x\, y := \texttt{S}\, x = y)$ and $(P\_ := \{x \mid Q\, x\})$. We only have to transform the termination certificate $\exists n, Q\, n$ into a bar inductive predicate at the purely logical/**Prop** level. For this, we show $(\exists n, Q\, n) \to \texttt{bar}\ Q\ R\ 0$: from $Q\, n$ deduce $\texttt{bar}\ Q\ R\ n$ using in_bar_0 and then $\texttt{bar}\ Q\ R\ (n-1)$,... down to $\texttt{bar}\ Q\ R\ 0$ by descending induction[3] using in_bar_1.

Notice that using the previous development, we can already implement the functional specification of the T&H algorithm:

$$\texttt{th\_min} : (\exists \tau, 0 < \tau \land f^\tau\, x_0 = f^{2\tau}\, x_0) \to \{\tau \mid 0 < \tau \land f^\tau\, x_0 = f^{2\tau}\, x_0\}$$

by application of Constructive_Epsilon with $(Q\, n := 0 < n \land f^n\, x_0 = f^{2n}\, x_0)$. Indeed, such $Q : \texttt{nat} \to \texttt{Prop}$ is computationally decidable as both $< : \texttt{nat} \to \texttt{nat} \to \texttt{Prop}$ and $=_X : X \to X \to \texttt{Prop}$ are computationally decidable.[4]

However, this approach will not give us the operational specification of T&H because this implementation of th_min using Constructive_Epsilon extracts into the inefficient unbounded minimization algorithm on the right-hand side (see also th_min.v). There, μmin corresponds to unbounded minimiza-

```
let th_min f x_0 =
  let rec μmin n =
    if (n = 0) or (f^n x_0 ≠ f^{2n} x_0)
    then μmin (1 + n)
    else n
  in μmin 0
```

tion. This th_min program recomputes $f^n\, x_0$ and $f^{2n}\, x_0$ for each value of n before a cycle is detected, making it really inefficient.

---

[3] Descending induction is implemented by nat_rev_ind in file utils.v.

[4] For $=_X$, this is precisely the assumption of the $=^?_X$ equality decider.

4   The Tortoise and the Hare via Bar Inductive Predicates

In this section, we use the methodology of Sect. 3 (i.e. termination via bar inductive predicates) to design a fully specified implementation of the T&H algorithm
that satisfies Goals 1 and 2 of Sect. 2.2. We could use bar_rect to implement
this algorithm but we do not use it directly. Indeed, we want to finely control the
computational content of our terms so that we can extract the expected OCaml
code accurately. However, we will mimic the implementation of bar_rect several
times. The corresponding file for this section is tortoise_hare.v.

The T&H detects potential cycles in the iterated values of an endo-function
$f : X \to X$. As explained in Sect. 2.2, for the remaining of this section, we assume
the following pre-conditions for the hare to recapture the tortoise:

Variables $(X : \texttt{Type})$ $(=^?_X : \forall x\, y : X, \{x = y\} + \{x \neq y\})$
$(f : X \to X)$ $(x_0 : X)$ $(H_0 : \exists \tau,\, 0 < \tau \wedge f^\tau\, x_0 = f^{2\tau}\, x_0)$.

that is a type X with an equality decider $=^?_X$, a sequence f starting at point x_0
satisfying a cyclicity assumption H_0 (see Proposition 1). These pre-conditions
are not minimal for establishing the correctness of T&H[5] but we do think they
are general enough to accommodate most use cases of T&H.

4.1   A Non-tail Recursive Implementation

Let us start with the non-tail recursive implementation of T&H, as is done
in the OCaml code of tortoise_hare (see Sect. 2.1). We define a bar inductive predicate which will be used as termination certificate for the main loop
tort_hare_rec. Compared to the generic inductive definition of bar of Sect. 3,
$\mathsf{bar_{th}}$ is a binary (instead of unary) bar predicate specialized with $(T\, x\, y :=
x = y)$ and $(R\, x\, y\, u\, v := u = f\, x \wedge v = f\, (f\, y))$:

```
Inductive bar_th (x y : X) : Prop :=
  | in_bar_th_0 :  x = y                          → bar_th x y
  | in_bar_th_1 :  bar_th (f x) (f (f y))  → bar_th x y
```

This definition matches the inductive rules of Fig. 2 (left part). We define a fully
specified Coq term tort_hare_rec mimicking both the OCaml code of Sect. 2.1
(with the addition of a termination certificate of type $\mathsf{bar_{th}}\, x\, y$) and the code of
bar_rect of Sect. 3:

```
Fixpoint tort_hare_rec x y (H : bar_th x y) : {k | f^k x = f^{2k} y} :=
  match x =^?_X y with
    | left E   ↦ exist _ 0 G₁?
    | right C  ↦ match tort_hare_rec (f x) (f (f y)) G₂? with
                  | exist _ k H_k ↦ exist _ (S k) G₃?
                end
  end.
```

---

[5] See th_rel.v where an arbitrary decidable relation $R : X \to X \to \texttt{Prop}$ replaces $=_X$.

$$\frac{x = y}{\mathrm{bar_{th}}\ x\ y} \qquad \frac{\mathrm{bar_{th}}\ (f\,x)\ (f\,(f\,y))}{\mathrm{bar_{th}}\ x\ y} \qquad \Bigg| \qquad \frac{x = y}{\mathrm{bar_{tl}}\ i\ x\ y} \qquad \frac{\mathrm{bar_{tl}}\ (\mathsf{S}\,i)\ (f\,x)\ (f\,(f\,y))}{\mathrm{bar_{tl}}\ i\ x\ y}$$

Fig. 2. Inductive rules for $\mathrm{bar_{th}} : X \to X \to \mathsf{Prop}$ and $\mathrm{bar_{tl}} : \mathsf{nat} \to X \to X \to \mathsf{Prop}$.

where $\mathbb{G}_1^?$, $\mathbb{G}_2^?$ and $\mathbb{G}_3^?$ are three proof terms of the following types:

$$\mathbb{G}_1^?\ /\!/\ \ldots, E : x = y \vdash f^0\ x = f^{2.0}\ y$$
$$\mathbb{G}_2^?\ /\!/\ \ldots, C : x \neq y, H : \mathrm{bar_{th}}\ x\ y \vdash \mathrm{bar_{th}}\ (f\,x)\ (f\,(f\,y))$$
$$\mathbb{G}_3^?\ /\!/\ \ldots, H_k : f^k\ (f\,x) = f^{2k}\ (f\,(f\,y)) \vdash f^{\mathsf{S}\,k}\ x = f^{2(\mathsf{S}\,k)}\ y$$

These can be established before the Fixpoint definition of tort_hare_rec or else (preferably), using the Coq refine tactic, after the statement of the computational part of tort_hare_rec, as remaining logical obligations (see tortoise_hare.v for exact Coq code). Recall that the termination certificate H must structurally decrease, i.e. the proof term for $\mathbb{G}_2^?$ must be a sub-term of H.

We can now define tortoise_hare by calling tort_hare_rec but we need to provide a termination certificate:

```
Definition  tortoise_hare : {τ | 0 < τ ∧ f^τ x₀ = f^{2τ} x₀} :=
    match tort_hare_rec (f x₀) (f (f x₀)) G₁? with
    | exist _ k H_k ↦ exist _ (S k) G₂?
    end.
```

There are two remaining logical obligations, $\mathbb{G}_1^?$ being the termination certificate:

$$\mathbb{G}_1^?\ /\!/\ \ldots, H_0 : \exists \tau, 0 < \tau \wedge f^\tau x_0 = f^{2\tau} x_0 \vdash \mathrm{bar\_th}\ (f\,x_0)\ (f\,(f\,x_0))$$
$$\mathbb{G}_2^?\ /\!/\ \ldots, H_k : f^k\ (f\,x_0) = f^{2k}\ (f\,(f\,x_0)) \vdash 0 < \mathsf{S}\,k \wedge f^{\mathsf{S}\,k}\ x_0 = f^{2(\mathsf{S}\,k)}\ x_0$$

We prove $\mathbb{G}_1^?$ as follows: from H_0, we (non-computationally) deduce m such that $0 < m$ and $f^m x_0 = f^{2m} x_0$. Using in_bar_th_0 we immediately get $\mathrm{bar_{th}}\ (f^m x_0)\ (f^{2m} x_0)$. Then using in_bar_th_1 repeatedly from $m, m - 1, \ldots$ down to 1 we get $\mathrm{bar_{th}}\ (f^1 x_0)\ (f^2 x_0)$. $\mathbb{G}_2^?$ is obtained by trivial computations over nat using the f_equal/omega tactics.

The Coq command Recursive Extraction tortoise_hare produces the corresponding OCaml code of Sect. 2.1 except that the OCaml type int is replaced with nat and the OCaml built-in equality decider is replaced with (a to be provided implementation of) $=_X^?$.

4.2 A Tail-Recursive Implementation

Now we proceed with the tail-recursive implementation of T&H. We define a ternary bar inductive predicate corresponding to the recursive call of the loop in the OCaml code of tort_hare_tail in Sect. 2.1:

```
Inductive  bar_tl (i : nat) (x y : X) : Prop :=
    | in_bar_tl_0 :  x = y                            → bar_tl i x y
    | in_bar_tl_1 :  bar_tl (S i) (f x) (f (f y))  → bar_tl i x y
```

the corresponding inductive rules being described in Fig. 2 (right part). Then we can define the internal `loop` of `tort_hare_tail` by a (local) fixpoint over the fourth argument of type $\mathbf{bar}_{t1}\ i\ x\ y$:

```
Fixpoint loop i x y (H : bart1 i x y) : {k | i ⩽ k ∧ f^(k−i) x = f^(2(k−i)) y} :=
    match x =?_X y with
      | left E  ↦ exist _ i G₁?
      | right C ↦ match loop (S i) (f x) (f (f y)) G₂? with
                    | exist _ k H_k ↦ exist _ k G₃?
                  end
    end.
```

where $\mathbb{G}_1^?, \mathbb{G}_2^?$ and $\mathbb{G}_3^?$ are three proof terms of the following types:

$$\mathbb{G}_1^?\ //\ \dots, E : x = y \vdash i \leqslant i \wedge f^{i-i}\ x = f^{2(i-i)}\ y$$
$$\mathbb{G}_2^?\ //\ \dots, C : x \neq y, H : \mathbf{bar}_{t1}\ i\ x\ y \vdash \mathbf{bar}_{t1}\ (S\,i)\ (f\,x)\ (f\,(f\,y))$$
$$\mathbb{G}_3^?\ //\ \dots, H_k : S\,i \leqslant k \wedge f^{(k-S\,i)}\ (f\,x) = f^{2(k-S\,i)}\ (f^2\,y) \vdash i \leqslant k \wedge f^{k-i}\ x = f^{2(k-i)}\ y$$

and $\mathbb{G}_2^?$ is the termination certificate and must be a sub-term of H. Then we proceed with the implementation of `tortoise_hare_tail` which calls `loop`:

```
Definition tortoise_hare_tail : {τ | 0 < τ ∧ f^τ x₀ = f^(2τ) x₀} :=
    match loop 1 (f x₀) (f (f x₀)) G₁? with
      | exist _ k H_k ↦ exist _ k G₂?
    end.
```

We must provide a termination certificate $\mathbb{G}_1^?$ and establish the specification $\mathbb{G}_2^?$:

$$\mathbb{G}_1^?\ //\ \dots, H_0 : \exists \tau, 0 < \tau \wedge f^\tau\ x_0 = f^{2\tau}\ x_0 \vdash \mathbf{bar}_{t1}\ 1\ (f\,x_0)\ (f\,(f\,x_0))$$
$$\mathbb{G}_2^?\ //\ \dots, H_k : 1 \leqslant k \wedge f^{k-1}\ (f\,x_0) = f^{2(k-1)}\ (f\,(f\,x_0)) \vdash 0 < k \wedge f^k\ x_0 = f^{2k}\ x_0$$

$\mathbb{G}_1^?$ is proved by descending induction much like what is done in the non-tail recursive case and $\mathbb{G}_2^?$ is quite trivial to obtain using the `f_equal`/`omega` tactics.

The extracted OCaml code corresponds to the `tortoise_hare_tail` implementation of Sect. 2.1. The file `th_alone.v` contains a standalone implementation of `tortoise_hare_tail` in less than 80 lines, not counting comments.

5 Floyd's Cycle Finding Algorithm in Coq

In this section, we give an overview of Floyd's index and period finding algorithm as implemented in the file `floyd.v`. It has the same pre-conditions as the T&H algorithms of Sect. 4:

```
Variables (X : Type) (=?_X : ∀x y : X, {x = y} + {x ≠ y})
          (f : X → X) (x₀ : X) (H₀ : ∃τ, 0 < τ ∧ f^τ x₀ = f^(2τ) x₀).
```

It does not only finds a meeting point for the tortoise and the hare but computes the characteristic pair of values (λ, μ) of the cycle which satisfy the predicate cycle_spec with the following body:

Definition cycle_spec $(\lambda\, \mu : \mathtt{nat}) : \mathtt{Prop} :=$

$0 < \mu \wedge f^{\lambda} x_0 = f^{\lambda+\mu} x_0 \wedge \forall i\, j,\, i < j \to f^{i} x_0 = f^{j} x_0 \to \lambda \leqslant i \wedge \mu\ \mathtt{div}\ (j - i).$

where div represents the *divisibility order* over nat (i.e. d div n means $\exists q, n = qd$). The *index* λ is such that $f^{\lambda}(x_0)$ is the entry point of the cycle and $\mu > 0$ is the *period* (or length) of the cycle. The third conjunct $(\forall i\, j,\, i < j \to \ldots)$ states that any (non-empty) cycle $i \rightsquigarrow j$ occurs after λ and has a length divisible by μ.

Hence under the above pre-conditions, Floyd's algorithm has the functional specification floyd_find_cycle : $\{\lambda : \mathtt{nat}\ \&\ \{\mu : \mathtt{nat}\ |\ \mathtt{cycle\_spec}\ \lambda\ \mu\}\}$. The operational specification is simply that the Coq term extracts to a standard OCaml implementation derived from [1]. floyd_find_cycle is implemented as the combination of three sub-terms, floyd_meeting_pt which first computes a meeting point for the tortoise and the hare, then floyd_index that computes the index and finally floyd_period that computes the period. We describe these three sub-terms in specific sub-sections, each sub-section potentially having its own set of additional pre-conditions, mimicking Coq sectioning mechanism. For each of these terms, we use a tailored bar inductive predicate to ensure termination under the corresponding pre-conditions.

5.1   Computing a Meeting Point

The term floyd_meeting_pt needs no further pre-conditions. We use the inductive $\mathtt{bar_{th}} : X \to X \to \mathtt{Prop}$ of Fig. 2 as termination certificate, the same that we used for the non-tail recursive tortoise_hare implementation. However, we do get a tail-recursive term here because we only compute a meeting point, not its index in the sequence as in the $\mathtt{bar_{t1}}$/tortoise_hare_tail case.

Let bar_th_meet : $\forall x\, y,\, \mathtt{bar_{th}}\, x\, y \to \{c : X \mid \exists k,\, c = f^{k} x \wedge c = f^{2k} y\}$.
Definition floyd_meeting_pt : $\{c \mid \exists \tau,\, 0 < \tau \wedge c = f^{\tau} x_0 \wedge c = f^{2\tau} x_0\}$.

We define bar_th_meet as a local fixpoint using the same technique as in the $\mathtt{bar_{th}}$/tortoise_hare case. Then, we show H_0' : $\mathtt{bar_{th}}\, (f\, x_0)\, (f\, (f\, x_0))$ as a consequence of H_0 and we derive floyd_meeting_pt from the following instance bar_th_meet $(f\, x_0)\, (f\, (f\, x_0))\, H_0'$.

5.2   Computing the Index

The term floyd_index uses the post-condition of floyd_meeting_pt as a further pre-condition, i.e. a meeting point c for the tortoise and the hare. We use the predicate $\mathtt{bar_{in}} : \mathtt{nat} \to X \to X \to \mathtt{Prop}$ of Fig. 3 as termination certificate.

Variables $(c : X)\ (H_c : \exists \tau,\, 0 < \tau \wedge c = f^{\tau} x_0 \wedge c = f^{2\tau} x_0)$.
Let bar_in_inv $(i : \mathtt{nat})\ (x\, y : X)$:
         $\mathtt{bar_{in}}\, i\, x\, y \to \mathtt{least\_le}\ (\mathtt{fun}\ n \mapsto i \leqslant n \wedge f^{n-i} x = f^{n-i} y)$.
Definition floyd_index : least_le $(\mathtt{fun}\ l \mapsto \exists k,\, 0 < k \wedge f^{l} x_0 = f^{k+l} x_0)$.

$$\frac{x = y}{\text{bar}_{\text{in}}\ i\ x\ y} \qquad \frac{\text{bar}_{\text{in}}\ (S\,i)\ (f\,x)\ (f\,y)}{\text{bar}_{\text{in}}\ i\ x\ y} \qquad \Bigg| \qquad \frac{c = y}{\text{bar}_{\text{pe}}\ i\ y} \qquad \frac{\text{bar}_{\text{pe}}\ (S\,i)\ (f\,y)}{\text{bar}_{\text{pe}}\ i\ y}$$

Fig. 3. Inductive rules for $\text{bar}_{\text{in}} : \text{nat} \to X \to X \to \text{Prop}$ and $\text{bar}_{\text{pe}} : \text{nat} \to X \to \text{Prop}$.

We define bar_in_inv as a local fixpoint where least_le P is the least[6] n : nat which satisfies $P\,n$. We show $H'_c : \text{bar}_{\text{in}}\ 0\ x_0\ c$ as a consequence of H_c and we derive floyd_index from the following instance bar_in_inv $0\ x_0\ c\ H'_c$.

5.3   Computing the Period

The further pre-condition of floyd_period is a point c which belongs to a (non-empty) cycle, a direct consequence of the post-condition of floyd_meeting_pt. Termination is certified by the predicate $\text{bar}_{\text{pe}} : \text{nat} \to X \to \text{Prop}$ of Fig. 3.

Variables $(c : X)$ $(H_c : \exists k,\ 0 < k \wedge c = f^k\,c)$.
Let bar_pe_inv $i\ x : \text{bar}_{\text{pe}}\ i\ x \to$ least_le $(\text{fun}\ n \mapsto i \leqslant n \wedge x = f^{n-i}\,y)$.
Definition floyd_period : least_div $(\text{fun}\ n \mapsto 0 < n \wedge c = f^n\,c)$.

We define bar_pe_inv as a local fixpoint. We prove $H'_c : \text{bar}_{\text{pe}}\ 1\ (f\,c)$ using H_c and we get floyd_period from the following instance bar_pe_inv $1\ (f\,c)\ H'_c$. Here, least_div P is the least n s.t. $P\,n$ for the divisibility order div.

5.4   Gluing all Together

We finish with the term floyd_find_cycle:

Definition floyd_find_cycle : $\{\lambda : \text{nat}\ \&\ \{\mu : \text{nat} \mid \text{cycle\_spec}\ \lambda\ \mu\}\}$.

It needs no further pre-conditions than those given at the beginning Sect. 5. It is composed of the successive applications of floyd_meeting_pt, floyd_index and floyd_period where the post-condition of floyd_meeting_pt serves as input for the extra pre-conditions of floyd_index and floyd_period. We conclude with a short proof that the computed index and period satisfy cycle_spec. The extracted OCaml program corresponds to the following code with the same remarks regarding $=^?_X$ and nat/int as with tortoise_hare from Sect. 4.1.

```
let floyd_meeting_pt f x_0 =
    let rec loop x y = if x = y then x else loop (f x) (f² y) in loop (f x_0) (f² x_0)
let floyd_index f x_0 c =
    let rec loop i x y = if x = y then i else loop (1 + i) (f x) (f y) in loop 0 x_0 c
let floyd_period f c =
    let rec loop i y = if c = y then i else loop (1 + i) (f y) in loop 1 (f c)
let floyd_find_cycle f x_0 =
    let c = floyd_meeting_pt f x_0 in (floyd_index f x_0 c, floyd_period f c)
```

---

[6] Least for the natural order $\leqslant$ over nat.

6   Brent's Period Finding Algorithm

In the file brent_bin.v, we propose a correctness proof of Brent's algorithm in the same spirit as what was done for Floyd's cycle finding algorithm of Sect. 5. Brent's algorithm [6] only computes the period μ of the cycle. The index λ can be computed afterwards by using two tortoises separated by μ steps. Brent's algorithm is more efficient than T&H: it can be proved that a run of Brent's algorithm on (f, x_0) always generates less calls to f than a run of the T&H (or Floyd's cycle finding) algorithm on the same input (see [14], Sect. 7.1.2).

In this section, we just describe the functional specification and the operational specification (i.e. the extracted OCaml code) of Brent's algorithm of which we propose two implementations. The first one in file brent_bin.v is suited for a binary representation of natural numbers, but it is not efficient with unary natural numbers. The other implementation in file brent_una.v is also efficient on unary natural numbers such as those of type nat.

Given a type X : Type, an equality decider $=^?_X$: $\forall x\, y$: $X, \{x = y\} + \{x \neq y\}$, input values $f : X \to X$ and $x_0 : X$, and a cycle existence certificate (see Proposition 1), Brent's algorithm computes a μ satisfying the specification period_spec : $\forall \mu$: nat, Prop with the following body

$$0 < \mu \wedge (\exists \lambda, f^\lambda\, x_0 = f^{\lambda+\mu}\, x_0) \wedge \forall i\, j, i < j \to f^i\, x_0 = f^j\, x_0 \to \mu \text{ div } (j - i)$$

that is, it computes the period of the cycle. The term brent_bin extracts to something close to the following OCaml code:

```
let brent_bin f x0 =
   let rec loop p l x y =
      if        x = y then l
      else if p = l then loop 2p 1 y (f y)
      else              loop p (1 + l) x (f y)
   in loop 1 1 x0 (f x0)
```

where int is replaced with nat, $(x = y)$ with $(x =^?_X y)$ and $(p = l)$ with (eq_nat_dec $x\, y$). However this code is not optimal with a unary representation of numbers such as nat: in particular $2p$ and $p = l$ are slow (linear) to compute.

To get a more efficient implementation, one should either use a binary representation of numbers or switch to brent_una which has the same specification as the binary version but extracts to the following OCaml code:

```
let brent_una f x0 =
   let rec loop l p m x y =
      if        x = y then l
      else if m = 0 then loop 1 (1 + p) p y (f y)
      else               loop (1 + l) (1 + p) (m − 1) x (f y)
   in loop 1 1 0 x0 (f x0)
```

This code is much better suited for unary numbers. In particular, $m = 0$ and $m - 1$ are computed via pattern-matching on m in constant time.

7   Correctness by Extraction and Related Works

Correctness is a property of programs *with respect to a given specification*. As trivial as this remark may seem, it is important to keep it in mind because the purpose of extraction is to erase the logical content of Coq programs and to keep only their computational content: *specifications are erased by extraction*. One cannot claim that a program extract(t) is correct just because it has been extracted from a Coq term $t : T$. The correctness property is only ensured with respect to the particular specification T that (by the way) had just been erased.

7.1   Correctness of the Tortoise and the Hare

We illustrate this critical aspect of extraction on the non-tail recursive OCaml implementation of T&H. The type of `tortoise_hare` is

Definition tortoise_hare $\{X\}$ $(=^?_X : \forall x\, y : X, \{x = y\} + \{x \neq y\})\ f\ x_0 :$
$(H_0 : \exists \tau, 0 < \tau \wedge f^\tau\, x_0 = f^{2\tau}\, x_0) \to \{\tau : \mathbf{nat} \mid 0 < \tau \wedge f^\tau\, x_0 = f^{2\tau}\, x_0\}.$

as reported from Sect. 4. The pre-conditions of this specification are all the logical properties of the input parameters, i.e. the fact that $=^?_X$ is an equality decider for X and the cyclicity property H_0. The post-condition is the fact that τ is a meeting index for the fabulous animals. By extraction, the OCaml code `tortoise_hare` of Sect. 2.1 is correct w.r.t. to this specification. In the extracted program however, the pre-conditions on $=^?_X$ and of cyclicity, and the post-conditions $0 < \tau$ and $f^\tau\, x_0 = f^{2\tau}\, x_0$ have disappeared.

Now consider this "alternative" (and cheating) implementation of T&H:

Variables $(X : \mathbf{Type})\ (=^?_X : \ldots)\ (f : X \to X)\ (x_0 : X)\ (H_0 : \mathbf{False}).$
Fixpoint th_false_rec $x\ y\ (H : \mathbf{False}) : \mathbf{nat} :=$
   match $x =^?_X y$ with
    | left $E \mapsto 0$
    | right $C \mapsto$ S (th_false_rec $(f\, x)\ (f\, (f\, y))$ match H with end)
   end.
Definition th_false := S (th_false_rec $(f\, x_0)\ (f\, (f\, x_0))\ H_0$).

The pre-conditions for `th_false` are the same as those of `tortoise_hare` except that cyclicity has been replaced with absurdity ($H_0 : \mathbf{False}$). The post-condition has been erased. Yet, up to renaming, extraction of OCaml code from `th_false` and from `tortoise_hare` yields the *very same program*. So the OCaml program `tortoise_hare`/`th_false` is correct w.r.t. two very different specifications: one is useful (cyclicity) and one is useless (absurdity).

The file infinite_loop.ml contains `th_false` and gives other illustrations of abuses of extraction in absurd contexts. As a conclusion, before using or running an extracted algorithm, one should first check for assumptions in the specification using the Coq command `Print Assumptions` which lists all the potentially hidden pre-conditions such as axioms or parameters which are usually not displayed in the types of terms in Coq to avoid bloating them.

7.2 Comparison with Related Works

Given the stunning simplicity of T&H, our implementation is hardly the first attempt at certifying this algorithm. As for Coq, may be there are others, but we are aware of two previous developments. One (unpublished) proof by J.C. Filliâtre [11] and another, more recent, by J.F. Dufourd [9].

Illustrating how Hilbert's ϵ-operator could be used to manage partial functions within Coq was the main goal of his implementation [11] (private communication with J.C. Filliâtre); and the correctness of T&H was not really a goal of that project. The use of `epsilon/epsilon_spec` to *axiomatize* Hilbert's ϵ-operator is, from a constructive point of view, our main criticism against that implementation. Hence, while it is true that the term `find_cycle` of fillia_orig.v extracts to some OCaml code very similar to `tortoise_hare_tail` of Sect. 2.1, the corresponding specification has stronger pre-conditions than our own. Following the observations on correctness of Sect. 7.1, even if erased by extraction, Hilbert's ϵ-operator is still a pre-condition and a particularly strong form of the axiom of choice. Anything that exists can be reified with this operator. While not necessarily contradictory in itself, that kind of axiom is incompatible with several extensions of Coq [7]. We do not think it can be accepted from a constructive point of view because admitting the ϵ-operator allows to write non-recursive functions in Coq. We suggest the interested reader to consult the file collatz.v to see how the ϵ-operator "solves" the Halting problem or the Collatz problem [16]. It is our claim that although `find_cycle` implements some correctness property of T&H, the pre-conditions under which this correctness is achieved cannot be constructively or computationally satisfied. In the file fillia_modif.v, we propose a modified version of [11] where the ϵ-operator is replaced with `Constructive_Epsilon` from Sect. 3.3. This requires in-depth changes, in particular, on the induction principle used to ensure termination. We additionally mention a more recent Isabelle/HOL proof of P. Gammie [13] which seems to reuse the technique of J.C. Filliâtre.

The work of J.F. Dufourd [9] is based on different assumptions and the correctness proof of T&H that he obtains derives from a quite large library on functional orbits over finite domains of around 20 000 lines of code. The pre-conditions do not assume cyclicity. Instead there is a stronger assumption of finiteness of the domain, from which cyclicity can be derived using the pigeon hole principle (PHP). Admittedly, this finiteness assumption is not unreasonable, even constructively: most use cases of cycle finding algorithms occur over finite domains. However, it is our understanding that Pollard's rho algorithm [18] is run on a finite domain of *unkown (i.e. non-informative) cardinality*. As far as we can understand J.F. Dufourd's code, his inductive proofs are cardinality based. Hence an informative bound on the cardinal of the domain is likely a pre-condition for correctness. Thus, we think that his correctness proof might not be applicable to Pollard's rho algorithm (non-informative finiteness, see below).

On the other hand, modifying our specification of `tortoise_hare_tail` to replace cyclicity by finiteness involves the PHP (see php.v). In that case, finiteness could be expressed as the predicate $(\exists l : \text{list}\, X, \forall x : X, \text{In}\, x\, l)$ which postulates the *non-informative existence* of a list covering all the type X. Hence we would obtain a specification compatible with the context of Pollard's rho algorithm. This short development can be found in th_finite.v.

References

1. Cycle detection – Wikipedia, The Free Encyclopedia. https://en.wikipedia.org/wiki/Cycle_detection
2. Aumasson, J.-P., Meier, W., Phan, R.C.-W., Henzen, L.: The Hash Function BLAKE. ISC. Springer, Heidelberg (2014). https://doi.org/10.1007/978-3-662-44757-4
3. Bertot, Y., Castéran, P.: Interactive Theorem Proving and Program Development - Coq'Art: The Calculus of Inductive Constructions. Texts in Theoretical Computer Science. An EATCS Series. Springer, Heidelberg (2004). https://doi.org/10.1007/978-3-662-07964-5
4. Bove, A., Capretta, V.: Modelling general recursion in type theory. Math. Structures Comput. Sci. **15**(4), 671–708 (2005). https://doi.org/10.1017/S0960129505004822
5. Bove, A., Krauss, A., Sozeau, M.: Partiality and recursion in interactive theorem provers – an overview. Math. Structures Comput. Sci. **26**(1), 38–88 (2016). https://doi.org/10.1017/S0960129514000115
6. Brent, R.P.: An improved Monte Carlo factorization algorithm. BIT Numer. Math. **20**(2), 176–184 (1980). https://doi.org/10.1007/BF01933190
7. Castéran, P.: Utilisation en Coq de l'opérateur de description (2007). http://jfla.inria.fr/2007/actes/PDF/03_casteran.pdf
8. Coen, C.S., Valentini, S.: General Recursion and Formal Topology. In: PAR-2010, Partiality and Recursion in Interactive Theorem Provers. EPiC Series in Computing, vol. 5, pp. 72–83. EasyChair (2012). https://doi.org/10.29007/hl75
9. Dufourd, J.F.: Formal study of functional orbits in finite domains. Theoret. Comput. Sci. **564**, 63–88 (2015). https://doi.org/10.1016/j.tcs.2014.10.041
10. Dybjer, P.: A general formulation of simultaneous inductive-recursive definitions in type theory. J. Symb. Log. **65**(2), 525–549 (2000). https://doi.org/10.2307/2586554
11. Filliâtre, J.C.: Tortoise and the hare algorithm (2007). https://github.com/coq-contribs/tortoise-hare-algorithm
12. Fridlender, D.: An interpretation of the fan theorem in type theory. In: Altenkirch, T., Reus, B., Naraschewski, W. (eds.) TYPES 1998. LNCS, vol. 1657, pp. 93–105. Springer, Heidelberg (1999). https://doi.org/10.1007/3-540-48167-2_7
13. Gammie, P.: The Tortoise and Hare Algorithm (2015). https://www.isa-afp.org/entries/TortoiseHare.html
14. Joux, A.: Algorithmic Cryptanalysis. Cryptography and Network Security. Chapman & Hall/CRC (2009)

15. Knuth, D.E.: The Art of Computer Programming, Volume 2: Seminumerical Algorithms. Addison-Wesley Longman Publishing Co., Inc., Boston (1997)
16. Lagarias, J.: The Ultimate Challenge: The $3x+1$ Problem. American Mathematical Society (2010). http://bookstore.ams.org/mbk-78
17. Larchey-Wendling, D., Monin, J.F.: Simulating Induction-Recursion for Partial Algorithms. In: TYPES 2018, Braga, Portugal (2018). https://members.loria.fr/DLarchey/files/papers/TYPES_2018_paper_19.pdf
18. Pollard, J.M.: A monte carlo method for factorization. BIT Numer. Math. **15**(3), 331–334 (1975). https://doi.org/10.1007/BF01933667

Fast Machine Words in Isabelle/HOL

Andreas Lochbihler[(✉)]

Institute of Information Security, Department of Computer Science, ETH Zurich,
Zurich, Switzerland
`andreas.lochbihler@inf.ethz.ch`

Abstract. Code generated from a verified formalisation typically runs faster when it uses machine words instead of a syntactic representation of integers. This paper presents a library for Isabelle/HOL that links the existing formalisation of words to the machine words that the four target languages of Isabelle/HOL's code generator provide. Our design ensures that (i) Isabelle/HOL machine words can be mapped soundly and efficiently to all target languages despite the differences in the APIs; (ii) they can be used uniformly with the three evaluation engines in Isabelle/HOL, namely code generation, normalisation by evaluation, and term rewriting; and (iii) they blend in with the existing formalisations of machine words. Several large-scale formalisation projects use our library to speed up their generated code. To validate the unverified link between machine words in the logic and those in the target languages, we extended Isabelle/HOL with a general-purpose testing facility that compiles test cases expressed within Isabelle/HOL to the four target languages and runs them with the most common implementations of each language. When we applied this to our library of machine words, we discovered miscomputations in the 64-bit word library of one of the target-language implementations.

1 Introduction

Nowadays, algorithms are routinely verified formally using proof assistants and many proof assistants support the generation of executable code from the formal specification. The generated code is used for animating the formal specification [10,38,41,45], validating the formal models [16,18,39], proving properties by evaluation [1,21,23,48], and to obtain actual tools with formal guarantees such as CompCERT [37], CakeML [30], CeTA [49], CAVA [15], Cocon [28], DRAT-trim [25], and GRAT [34].

Usability of the generated code requires that it be efficient. This is mainly achieved by using (i) optimised data structures, which have been verified in the proof assistant, and (ii) hardware support for computing with data, in particular integers and arrays. To that end, the code generators of many proof assistants can be configured to map types and their operations to those provided by the target language rather than to implement them according to their construction in the logic. For example, integers can use optimised libraries like GMP instead

© Springer International Publishing AG, part of Springer Nature 2018
J. Avigad and A. Mahboubi (Eds.): ITP 2018, LNCS 10895, pp. 388–410, 2018.
https://doi.org/10.1007/978-3-319-94821-8_23

of being implemented as lists of binary digits and arrays are translated to read-only arrays with constant-time access instead of lists with linear-time access. In today's practice, such mappings are often unverified (an exception is CakeML's verified bignum library [30]) and are therefore part of the trusted code base (TCB). As we discuss below, verified code generation could shrink the TCB, but it has not yet reached maturity.

Apart from efficiency, such mappings bridge the gap between formal logic and the real world. The mapped data types are used to exchange data with non-verified code, e.g., drivers, application interfaces, test harnesses, and foreign function interfaces (FFI) in general [42, 44]. The proof assistant Isabelle/HOL in version Isabelle2017, e.g., provides the necessary mappings for arbitrary-precision integers, booleans, lists, and strings.[1]

In this paper, we extend this list for Isabelle/HOL with machine words of 8, 16, 32, and 64 bits (Sect. 3), and with machine words of unspecified size (Sect. 4). By reusing Isabelle/HOL's formalisation of fixed-size words [11,12], our library inherits the infrastructure for reasoning about machine words and integrates smoothly with existing formalisations. The key challenge was to simultaneously support all target languages of Isabelle/HOL's code generator (Standard ML, OCaml, Haskell, and Scala) with their varying APIs and all evaluation mechanisms (code generation, normalisation by evaluation, and term rewriting). Supporting all target languages and all evaluators is crucial to obtain a usable and versatile library that works together with many other Isabelle/HOL libraries.

We have validated our unverified mappings by running many test cases. To that end, we have developed a general-purpose framework for Isabelle/HOL to run and test the generated code (Sect. 5.1). After we had fixed the initial mistakes in our mappings, our test cases even found a bug in the implementation of 64-bit words in PolyML 5.6.1 in 64-bit mode, which is the Standard ML implementation that runs Isabelle2017 (Sect. 5.2).

Our library is available on the Archive of Formal Proofs [40], which includes a user guide as documentation. Several projects and tools use it already. Users report significant performance improvements over using arbitrary-precision integers (Sect. 6). The testing framework is distributed with Isabelle2017 (file `HOL-Library.Code_Test`).

Contributions. The main contributions of this paper are the following:

1. We describe the design of an Isabelle/HOL library for fixed-size words that are mapped to machine words in different target languages. By using our library, users can generate faster code from their formalisations.
2. We analyse the pitfalls and subtleties of code adaptations and show how to ensure that code adaptations work for all target languages and all evaluators. Our library demonstrates the feasibility of our approach. This analysis is of

---

[1] Immutable arrays are supported for Standard ML and Haskell, but not the other target languages of Isabelle/HOL's code generator. In the version for Isabelle2017, the Collections framework by Lammich [33] provides mutable arrays for Standard ML, Haskell, and Scala, but not OCaml.

interest even to other proof assistants that provide code extraction: libraries similar to ours suffer from such subtle soundness bugs although they target just one language, not four (Sect. 7).

3. To justify the soundness of our mapping, we generalise the correctness notion for code generation such that logical underspecification can be refined during code generation. We argue that the new notion is meaningful and identify conditions under which it coincides with the existing correctness notion. Such refinements can also be used in other contexts where abstract types are implemented by concrete data structures.

4. We develop a general-purpose framework for running and testing the generated code, which can be used independently of our machine word library. For example, it can compute with infinite codatatype values using Haskell's built-in laziness. The existing ML-based evaluation mechanism does not terminate for such computations.

Design choices. Our goal is to develop a practical and efficient library suitable for large-scale projects, not a fragile research prototype. Thus, it must work with the technology that is already mature. In our case, this is Isabelle's existing code generator with the four target languages and its unverified mappings, which inflate the trusted code base. Although we cannot obtain formal guarantees on the generated code itself, it *is* generated from a verified formalisation in a systematic way supported by a sound theory. So our library merely adds the correctness of the (validated) mappings to the TCB, which already includes the compiler and library of the target language anyway.

The alternative would be to target the ongoing work on verified code generation such as CertiCoq [2], Œuf [43], and CakeML [30] with its Isabelle link [27]. Our mappings could then be verified down to assembly language or machine code and would thus not enlarge the TCB. Given the present state of these projects, such a library would be less versatile than ours. For example, the Isabelle link to CakeML lacks abstract datatypes, which many Isabelle/HOL projects use for code generation. Moreover, even CakeML, the most mature of the three, produces machine code that is often slower than the output of unverified compilers, although the run times' orders of magnitude are about equal [47].

More importantly, our approach will still be relevant when such mappings will be verified in the future, as the key challenge of fitting different APIs under one hood will persist. The reason is that there will be several verified compilation chains, e.g., CertiCoq and Œuf for Coq, and a versatile library should support code generation with all of them. Clearly, careful API design can avoid some of the differences, e.g., signed vs. unsigned words. But others like the varying word sizes will remain as they reflect crucial design choices in the compilation chain. In Standard ML, e.g., the word size varies by compiler precisely because each compiler organises the heap in its own way, stealing some bits of every word for memory management. A performant library must deal with such compiler-specific issues, as library users should not have to care about these details.

2   Background on Isabelle/HOL

This section introduces aspects of the proof assistant Isabelle/HOL that are relevant for this paper. Isabelle/HOL implements an extension of classical higher-order logic (HOL) [46] with Haskell-style type classes for overloading [22]. Its standard library formalises machine words [11,12] as a HOL type α word where the type parameter α determines the number of bits via a type-class operation len-of$_\alpha$ :: nat. More precisely, α word is defined as a copy of the integers from 0 to $2^{\text{len-of}_\alpha} - 1$. For example, 32 word and 64 word denote the type of 32-bit and 64-bit words, respectively, using an encoding of numbers as types. So, the arithmetic and bit-wise operations on words are derived from those on the integers, i.e., the results are truncated by taking the remainder w.r.t. $2^{\text{len-of}_\alpha}$. Technically, these operations are overloaded for integers and words using type classes.

The Lifting and Transfer tools [26] can lift definitions and transfer theorems across quotients. We use them for the special case of subtypes (typedefs in HOL), e.g., from integers to words. In this case, a lifted definition is executable if the original term is.

The code generator [20,21] generates code from a fragment of HOL to functional programming languages, mapping HOL types and functions to datatypes and functions in the target language. Four languages are supported: Standard ML, OCaml, Haskell, and Scala. The code generator ensures partial correctness of the generated code. That is, if the code terminates successfully, then the result satisfies the properties that have been proven about the HOL functions. This guarantee relies on the code generator's assumption that the generated functional program behaves according to a higher-order rewrite system (HORS). In this view, datatype constructors are uninterpreted function symbols and the equations of a function yield a set of rewrite equations. Executing the generated program in the target language corresponds to performing rewrite steps with the corresponding equations on the term representation. Since the equations have counterparts in HOL, all these steps could also have been taken in the logic, so the result is derivable in HOL. Conversely, nothing can be said if the execution raises an exception or does not terminate.[2]

Moreover, this approach decouples the logical definitions from the extracted code, as the HORS does not attach logical meaning to the function symbols themselves. Any HOL function of the right type can thus serve as a datatype constructor and any HOL equation can be used to implement a function if the constraints of the target language are met. They are therefore called code constructors and code equations. For example, one can change the implementation of nat to a binary representation without changing the definition in the logic or the proofs. This corresponds to data refinement [20].

Isabelle/HOL's code generator also provides a minimalistic foreign function interface (FFI) via **code-printing** declarations [19, Sect. 7]. These declarations instruct the code generator to output a specified string instead of what it would

---

[2] Non-termination does not affect logical soundness as the function definitions' consistency in HOL must have been established independently of the code generator.

normally generate for a HOL type or HOL function in the specified target language. As they act on the concrete syntax, such declarations are called code adaptations. They are used to map integers,[3] booleans, lists, unit, and strings to their counterparts in the target language. Code adaptations lack a formal semantics and are therefore part of the TCB.

The code equations can also be used to evaluate HOL terms and to prove theorems by execution. Isabelle/HOL has three different mechanisms to do so: (i) generating and running Standard ML code for ground terms and propositions; (ii) symbolic normalisation by evaluation (NBE) [1]; and (iii) term rewriting within Isabelle. The first mechanism uses the full power of the code generator, mapping HOL types to Standard ML data types and functions to SML functions using the code equations and the code adaptations. NBE represents HOL values as a term data type in Standard ML and HOL functions as Standard ML functions that manipulate terms according to code equations; no code adaptations are used. Term rewriting uses only the code equations in a call-by-value strategy. Note that the same set of code equations is used for all target languages and for all mechanisms.

Only term rewriting is checked by Isabelle's kernel and can thus be trusted. When the other two evaluation mechanisms are used, code generation and possibly the code adaptations become part of the TCB. In return, they are much faster than term rewriting. When proving theorems, Isabelle tags all theorems whose derivation involved some step outside of the kernel, such as NBE or code generation. So everyone can easily check whether Isabelle's kernel has completely checked all steps of a theorem's derivation.

With the existing Isabelle/HOL setup for α word, the generated code represents words as arbitrary-precision integers and all operations take the remainder modulo $2^{\text{len-of}_\alpha}$. This sets the efficiency baseline for evaluating our library.

3   Fixed-Size Machine Words

We now introduce HOL types for words of 8, 16, 32, and 64 bits (Sect. 3.1), present the code adaptations for all target languages (Sects. 3.2 and 3.3), and argue why they are sound (Sect. 3.4).

3.1   Types of Unsigned Words

Recall that the type α word of Isabelle/HOL words is polymorphic in the number of bits. Yet, code adaptations can only be given for type constructors such as word, not compound types like 32 word. As target languages provide only monomorphic word types, we must map 32 word to a different target language type than say 64 word. We therefore first introduce new HOL type constructors

---

[3] Isabelle/HOL provides two types of integers: int and integer. The latter is always mapped to target-language integers and the former can be implemented using the latter. Here, we ignore this distinction and always assume that integers are implemented by target-language integers.

for (unsigned) machine words of 8, 16, 32, and 64 bits. In detail, we define types uint8, uint16, uint32, and uint64 as type copies of the existing unsigned word formalisation. As the construction is identical for all bit lengths, we only show the one for uint32 and use uint* to refer to all four types.

All arithmetic and bit-wise operations are lifted from 32 word to uint32 using the Lifting tool [26]. Cast operations between all uint* types are also available. Here, we give just two examples: the overloaded addition operation (+) and the conversion function word-of-int from integers.

lift-definition (+) :: uint32 ⇒ uint32 ⇒ uint32
　　is (+) :: 32 word ⇒ 32 word ⇒ 32 word .

lift-definition uint32-of-int :: int ⇒ uint32 is word-of-int .

In principle, we could easily transfer all the existing theorems about these operations, too. But our library does not do so as we consider uint* primarily as types for code generation, not for proving theorems. Instead, whenever we must prove a theorem about uint*, we first transfer the statement to α word (for the appropriate choice of α) using Transfer and then use the existing, well-engineered proof automation. This approach avoids duplicating theorems and tactics and thus saves the subsequent maintenance efforts.

3.2   Setting up Code Generation

With the uint* types and their operations in place, we can now design the code constructors, code equations, and code adaptations. Our design should achieve three goals:

1. It should work simultaneously for all four target languages, all three evaluation mechanisms, and all strategies of Quickcheck [5]. Recall that the code constructors and equations are shared by all target languages and evaluation mechanisms. So we must find constructors and equations that are suitable for all of them.
2. The code adaptations for the uint* operations should yield very efficient code.
3. The adaptations should be as small as possible to reduce the chance of errors.

In case of conflicting goals, we will value the efficiency of the target language mapping higher than the efficiency of the other evaluators (normalisation and term rewriting). The evaluators are typically used only for small HOL programs, where efficiency is not as crucial as in generated applications.

To support evaluation by normalisation and term rewriting, we design our code equations such that they implement uint* in terms of α word, which in turn is implemented using arbitrary-precision integers. In detail, we declare uint* as abstract datatypes to the code generator [20], such that code equations cannot pattern match on the code constructor for uint*. This ensures that code equations respect the abstraction barrier of uint*, so that we can later change the generated code using adaptations without worrying that users of our library might have declared code equations that look into the internal construction of uint*. In fact, all this setup is already in place by the way we have defined uint* and their operations using the Lifting tool. Moreover, the conversion function uint32-of-int from integers to uint32 acts as the "smart constructor" to create values of type uint32.

Next, we describe the code adaptations that map the uint* types and functions to the target language primitives. Yet, the provided word types vary across the target languages and even across different implementations of the same language. Table 1 lists the available word sizes for the most common implementations of the four target languages (marked with

Table 1. Bounded integers in the standard library by target language. Supported fixed sizes are marked with √. The last row lists the bit size of the default type. Grey cells indicate that only signed operations are available.

bits	PolyML 32	PolyML 64	SMLNJ	mlton	OCaml 32	OCaml 64	GHC	Scala
8	√	√	√	√			√	√
16				√			√	√
32	√	√	√	√	√	√	√	√
64		√	√	√	√	√	√	√
?	31	63	31	32	31	63	≥ 30	32

√). As can be seen, the support varies widely: only 32-bit words are provided by all implementations. PolyML provides 64-bit words only when run in 64-bit mode. For OCaml and Scala, most word types provide only signed operations, which interpret the most significant bit as a sign (marked as grey cells). Following α word, our library provides unsigned words, so extra effort will be needed in these cases. The last row shows the bit widths of the languages' standard word type. We will look at this row in more detail in Sect. 4.

The code adaptations for the types and most operations are straightforward as the libraries provide suitable functions. The type uint32, e.g., is mapped as follows. In the remainder of this section, we discuss the non-trivial cases.

code-printing type-constructor uint32 →
    (SML) `Word32.word` (OCaml) `int32` (Haskell) `Data.Word.Word32` (Scala) `Int`

If a target language does not provide a particular bit width (e.g., 8 and 16 bits in OCaml), we omit the code adaptations. The generated code will thus follow the code equations that the evaluators use. So, 8- and 16-bit words are implemented in OCaml using arbitrary-precision `Big_ints`, taking the remainder w.r.t. 2^8 or 2^{16} after every operation. With some more effort, they could also be implemented using 32-bit words.

Division and remainder require a more elaborate design of the code equations, which the drawing below illustrates. We define a cascade of constants div, uint32-div, uint32-sdiv, ... that model the division operators of the different target languages. Code equations (→) implement each constant using the next one. Code adaptations (⟹) map the constants to right target languages; they thereby terminate the cascade early.

$$\text{div}_{\text{uint32}} \longrightarrow \text{uint32-div} \longrightarrow \text{uint32-sdiv} \longrightarrow \text{div}_{32\ \text{word}} \longrightarrow \text{div}_{\text{int}} \longrightarrow \ldots$$
$$\quad\quad\quad \Downarrow \quad\quad\quad\quad\quad\quad \Downarrow \quad\quad\quad\quad\quad\quad\quad \Downarrow$$
$$\quad\quad\text{SML, Haskell} \quad\quad \text{OCaml, Scala} \quad\quad\quad \text{SML, Haskell, OCaml, Scala}$$

We now look at the different constants implementing division. As is customary in HOL, division by 0 yields 0 and taking the remainder w.r.t. 0 is the identity function [24], but the target languages typically raise exceptions. To avoid the exceptions and thus make the generated code fail less often, we define a new constant uint32-div that is unspecified for 0 and add 0 as a special case to div's code equation (and the remainder's):

definition uint32-div x y = (if $y = 0$ then undefined (div) x 0 else x div y)

lemma [code]: $(x$ div $y) = ($if $y = 0$ then 0 else uint32-div x $y)$

Here, undefined is an unspecified, polymorphic HOL constant. By applying it to the div function and the arguments x and 0, we get a fresh, unspecified formal value $x/0$ for every x. This way, mapping uint32-div to target language operations remains sound even if these return different results for dividing different x by 0—provided that the same value is consistently returned for the same x, if any.[4] For example,

code-printing constant uint32-div $\rightarrow$
  (SML) Word32.div (_, _)   (Haskell) Prelude.div

where (_, _) expresses that Standard ML's Word32.div takes both arguments as a tuple.

Unfortunately, mapping uint32-div to OCaml's and Scala's division operations directly would be unsound, as OCaml's int32 and Scala's Int are signed. Therefore, we define another division operation uint32-sdiv on uint32 that interprets uint32 as signed words and coincides with uint32-div when a division by zero occurs. Next, we prove a code equation that implements uint32-div using uint32-sdiv. The following equation expresses the algorithm adapted from Hacker's Delight [50, Sect. 9.3] on α word, where $\ll$ and $\gg$ denote unsigned bit shifts to the left and right, and sdiv denotes signed division. We prove the equation for all x and y of type α word with $y \neq 0$, and then lift it to all the uint* types. This is possible thanks to the polymorphic α word.

$$(x \text{ div } y, x \text{ mod } y) = (\text{ if } 1 \ll (\text{len-of}_\alpha - 1) \leq y \text{ then if } x < y \text{ then } (0, x) \text{ else } (1, x - y)$$
$$\text{else let } q = ((x \gg 1) \text{ sdiv } y) \ll 1; \ r = x - q * y \text{ in}$$
$$\text{if } r \geq y \text{ then } (q + 1, r - y) \text{ else } (q, r))$$

---

[4] Alternatively, we could have (under-)specified uint32-div with a conditional definition like

definition uint32-div where $y \neq 0 \longrightarrow$ uint32-div x y = x div y

lemma [code]: uint32-div x y = (if $y = 0$ then Code.abort "Div0" (λ_{-}. uint32-div x y) else x div y)

As the precondition makes the defining equation unsuitable for code generation, we would have to manually state and derive an unconditional code equation like the one shown, with which division by zero would make the normalisation evaluator fail to terminate. The definition with undefined requires no further setup for code generation and does not cause non-termination.

Thus, we get the following OCaml and Scala code adaptations for division. Note that there are no code adaptations for uint32-div for OCaml and Scala.

code-printing constant uint32-sdiv → (OCaml) Int32.div   (Scala) _ / _

The cascade of constants also applies to evaluation by normalisation and term rewriting, as they use the same code equations. Since there are no code adaptations, they follow the cascade until the end, i.e., arithmetic on integers. That is, they perform division and remainder on uint32 by testing for the zero divisor and only then performing a *signed* division according to the given algorithm, which is implemented via 32 word and the arbitrary-precision integers. This is an example of where we accept inefficiencies in the evaluators in favour of better generated code. Accordingly, the same roundabout way of implementing division also applies for uint* types that are not supported natively by the target language. In OCaml, e.g., 8- and 16-bit words follow the cascade until the code adaptations for arbitrary-precision integers branch off to OCaml's Big_int library.

The other operations affected by the signed interpretation are dealt with in a similar way. The smart constructor uint32-of-int :: int ⇒ uint32, in particular, requires adjusting the integer range from HOL's 0 to $2^{32} - 1$ to OCaml's and Scala's -2^{31} to $2^{31} - 1$. Like for division and remainder, we state and verify a conversion algorithm for arbitrary bit lengths as a lemma on α word and lift it to uint* using the Transfer package.

This simple idea of a cascade of constants with selective code adaptations yields more efficient code than what Isabelle code generation experts had come up with previously. Traditionally, code adaptations identified a domain on which the implementations in all target languages behave the same. The division and remainder operations on arbitrary-precision integers in Isabelle/HOL's standard library illustrate this approach. They are not directly mapped to the target language operations because they differ on negative numbers: dividing -5 by 3, e.g., yields -1 in Scala and OCaml whereas it results in -2 in Haskell and Standard ML (and Isabelle/HOL). Isabelle's standard library instead defines a special division-modulo operation divmod-abs that first takes absolute values and serialises it to target-language expressions that do the same.

definition divmod-abs m $n = (|m| \text{ div } |n|, |m| \text{ mod } |n|)$

code-printing divmod-abs →
    (SML) `IntInf.divMod (IntInf.abs _, IntInf.abs _)`
    (Haskell) `divMod (abs _) (abs _)`

and similarly for OCaml and Scala. The original division and remainder operations are implemented using divmod-abs where signs and values for negative numbers are adjusted as necessary. This approach clearly is not optimal with respect to efficiency, as some computations such as taking the absolute value are performed twice, once in the code equation for div (and mod) and once again in the code adaptation. In particular, those operations are computed even if the target language's operations exactly fit Isabelle's (like in the case of Haskell and Standard ML). For PolyML 5.6.1, we measured that the overhead of these checks and additional operations is about 100%, i.e., a division operation takes twice as long as it would have to. So users have to pay the performance penalty even if they are not interested in generating code in languages with mismatching operations.

In contrast, our cascading approach has no overhead for languages with perfectly matching operations (Standard ML and Haskell) and much less overhead for the others, where we have precisely modelled the target language operations in the logic and verified the implementation. The same could be done for division and remainder on integers.

3.3   Dealing with Underspecification

The bit shift operations $\ll$, $\gg$, and $\ggg$ (right shift with sign extension) are not affected by the signed interpretation, but they behave differently in different target languages. In Scala, they only take the lower bits of the shift into account. For example, shifting 1 by 65 bits to the left as a uint32 yields 2, as the lower $5 = \log_2 32$ bits of 65 denote the value 1. In Haskell and OCaml, the result of these operations is unspecified when the shift is negative or exceeds the word size. In Standard ML, the bit shift operations correctly honour all bits of the shift, but the shift must be given as a Word, whose size varies with the implementation (as shown in the last row in Table 1). In Isabelle, however, the shifts are specified as (unbounded) natural numbers, so we must take overflows into account.

Given the underspecification in Haskell and OCaml, we cannot model the target language's bit shifts exactly in HOL, as we do for sdiv. Instead, we resort to underspecification in HOL, too. For each shift operation, we define a version which is specified only for the bit shifts that do not exceed the word size. For $\ll$ on uint32, e.g., we define

$$\text{uint32-shiftl } x\ i = (\text{if } i < 0 \lor i \geq 32 \text{ then undefined } (\ll) \ x\ i \text{ else } x \ll \text{nat } i)$$

where we model the underspecification using undefined as we did for uint32-div in Sect. 3.2. We prove a code equation for $\ll$ (and one for uint32-shiftl for the evaluators)

$$x \ll n = (\text{if } n < 32 \text{ then uint32-shiftl } x \ (\text{int } n) \text{ else } 0),$$

where nat and int convert between integers and natural numbers, and map uint32-shiftl directly to the target languages.

3.4   Soundness of Code Adaptations for Underspecified HOL Functions

Recall from Sect. 2 that the HORS view on code generation assumes that the successful execution steps of the generated program corresponds to rewrite steps in HOL. This guarantees partial correctness of the generated code. Clearly, code adaptations violate this invariant. Fortunately, we can generalise the reasoning to code adaptations for fully specified HOL functions, by assuming that there is a HOL proof tactic that can justify the result of a successful execution of the mapped code. This assumption can either be validated using tests (Sect. 5) or by giving a formal semantics to the generated code and verifying the translation [30]. For our library, this approach works for all the arithmetic operations, as the only such underspecified operations are division and remainder, where in the underspecified cases the generated code fails with an exception.

Unfortunately, this argument does not carry over to the bit shifts described in Sect. 3.3. Clearly, evaluating $1 \ll 65$ in, say, Scala does return a specific value—namely 2—and there is no way to prove that the unspecified HOL value undefined $(\ll)$ 1 65 equals 2. The code adaptations thus tighten the specification, i.e., they correspond to a kind of refinement. We now describe the correctness guarantees obtained by such an implicit refinement and identify the necessary assumptions on the target language operations.

The set-theoretic semantics of HOL assigns arbitrary values of the right type to unspecified constants, i.e., constants that have been declared, but not (yet) defined [32]. We can therefore consider the underspecification of a function as picking sufficiently many freshly declared constants and returning one of them for each argument where the underspecification occurs. Skolemizing over all the arguments and even the intended HOL function, we end up with an equivalent specification, e.g., the family $\lambda x\ i.$ undefined $(\ll)$ $x\ i$ of unspecified uint32 values. We can view this underspecification as model-theoretic non-determinism, which code adaptations can refine. Like deferred Isabelle/HOL definitions of constants that have been declared earlier, a code adaptation conceptually defines the family of unspecified values as the values that the target language implementation will compute. Clearly, these definitions are only conceptual, because they never manifest as a definitional theorem that Isabelle's kernel could check. Moreover, the chosen values depend on the particular target language implementation that will run the generated code. In this view, code adaptations constitute a deferred definition mechanism that executes when code is generated and whose effect is revoked at the end of code generation (as these definitions are not recorded in the logic).

This interpretation shows that any result computed by the generated code must be a possible value in *some* HOL model. Assuming that the formalisation is consistent, we obtain a (weaker) version of partial correctness, namely every theorem provable in HOL applies to the result. This is because the theorems hold in all HOL models and the result lives in one of them. Yet, we can no longer argue that the result is derivable from the HOL definitions, i.e., that *all* HOL models enforce this result. In other words, the generated code can only produce results which are *consistent* with the formalisation, but not necessarily *enforced* by it. In summary, we obtain the guarantee that it is impossible to prove in HOL that the result violates any provable property of the formalisation.

Our correctness argument hinges on three requirements, which our library meets:

1. The unspecified values are indeed logically unspecified. Otherwise, the refinement can lead to inconsistencies.
2. The function computed by the code adaptation in the target language implementation must be definable in HOL. In particular, the function must be pure, i.e., consistently return the same result for the same arguments, independent of the calling context, and its HOL definition must not introduce cyclic dependencies [31]. Obviously, it must also coincide with the mapped HOL function on the domain where it is specified.[5]
3. The code must not be used to prove theorems in the logic. Theorems proven by the refined code could silently introduce the implicit refinements as axioms into the logic. That is, some theorems might actually not be derivable from the stated axioms.

The last requirement means that implicit refinement via code adaptations must not be used when we prove theorems by code generation. The proofs of the code equations for the bit shift operations show that their results do not depend on the unspecified behaviour of the auxiliary functions like uint32-shiftl, i.e., we can use these operations in proofs by evaluation. However, users might directly call these auxiliary functions with unintended arguments (e.g., uint32-shiftl 1 2^{32}). To be safe, we ensure that in the code target Eval, which is used for proving theorems, code adaptations never cause implicit refinements. We achieve this by explicitly checking whether the arguments lie in the specified domain and otherwise raise an exception. For example,

```
code-printing constant uint32-shiftl → (Eval)
  (fn x => fn i => if i < 0 orelse i >= 32 then raise (Fail "<<")
                   else Word32.<<(x, Word.fromLargeInt(IntInf.toLarge i)))
```

Admittedly, it might have been easier to include the range checks for the shift operations in the code adaptations of all targets, not just Eval. This would have saved us from implicit refinements and their implications on soundness, at the cost of two more integer comparisons per executed bit shift. But in the next section, we take underspecification to the level of types, where we cannot avoid it any more.

4   Machine Words of Unspecified Length

Words of 8, 16, 32, and 64 bits are not optimally efficient for all target languages. Some implementations offer words of 31 and 63 bits, which are implemented

---

[5] The bit shifts are underspecified only in Haskell and OCaml. In Haskell, this assumption is satisfied as the bit shift operations belong to the Safe Haskell subset where pure functions cannot have side effects, i.e., referential transparency holds. As OCaml maps bit shifts directly to C, the interpretation of undefined behaviour would allow to the compiler to violate this assumption. However, to our knowledge, none of the state-of-the-art compilers exploits such technically undefined bit shifts badly. They all map it consistently to some bit shift instructions on the hardware, which does meet our requirements. The compilation strategy can change in the future though.

more efficiently as they need not boxing in memory. They use the missing bit to distinguish between primitive values and pointers, exploiting that the lowest bit of a pointer is always 0 due to memory alignment constraints. Accordingly, the bit length also depends on whether the runtime runs in 32-bit or 64-bit mode. The last row in Table 1 shows these bit widths by implementation. The Haskell API specifies only a lower bound of 30 bits; GHC in version 7.6.3 provides 32 bits in 32-bit mode and 64 bits in 64-bit mode. This may change in future versions, e.g., if the memory management starts to use some bits of a processor words for tagging like PolyML and OCaml do. The table therefore shows only the API constraint.

In this section, we introduce a type uint that maps to these machine words in the target languages. Generated code can thus benefit from unboxing, i.e., run faster with less memory. As the exact bit width varies across target language, implementation, and architecture, we again resort to underspecification in HOL to achieve sound code adaptations. That is, uint denotes the type of all machine words of a given non-zero length, but we do not specify the length in HOL. Formally, we introduce an uninterpreted type default-size and specify that len-of$_{default-size}$ be some positive number.[6] Then, uint denotes the type of all words of length len-of$_{default-size}$, which the code generator maps to Word.word in Standard ML, Data.Word.Word in Haskell, int in OCaml, and Int in Scala.

typedecl default-size

specification len-of$_{default-size}$ > 0 **by** auto

typedef uint = UNIV :: default-size word set ..

code-printing type-constructor uint →
    (SML) Word.word   (Haskell) Data.Word.Word (OCaml) int (Scala) Int

The operations and code adaptations for uint are analogous to uint∗, as described in Sect. 3. Signed and underspecified operations are handled in the same way, too. We map len-of$_{default-size}$ to the target language's bit width, e.g., Word.wordSize in Standard ML.

The underspecification for uint is much more invasive than uint∗'s. For the latter, only a few auxiliary operations like uint32-shiftl are underspecified, but all of the official operations are fully specified. On uint, in contrast, we do not even know what number $3 * 5$ denotes. For example, $3 * 5 = 7$ holds in HOL models where len-of$_{default-size}$ = 3. Evaluation by code generation therefore does not make sense for uint and our code adaptations ensure that all uint operations always raise exceptions in the evaluation target Eval. It might be possible to configure the other evaluators (normalisation and term rewriting) such that they

---

[6] Technically, the command **specification** defines the constant using Hilbert choice ε and derives the given property, after the specification has been shown to be satisfiable (**by** auto). So some unintended equations about len-of$_{default-size}$ are provable, e.g., len-of$_{default-size}$ = ($\varepsilon x.\ x > 0$). To avoid violating requirement 3.4 from Sect. 3.4, we hide the defining equation and only work with the specification. Arthan [4] discusses the problem of unintended identities for underspecified constants in detail.

evaluate uint expressions symbolically, but we have not succeeded in doing so yet. Therefore, evaluation and proving theorems by execution is currently not supported for uint.

So, what can be done with those unspecified uint? Here are three useful applications. First, Lammich [33] has implemented bit vectors as a list of uint. He formalises bit vectors on polymorphic words α word, making no assumptions about α. For example, the n-th bit of the bit vector is stored in the $(n \bmod \text{len-of}_\alpha)$-th bit of the $(n \operatorname{div} \text{len-of}_\alpha)$-th list element. So, $v \, ! \, (n \operatorname{div} \text{len-of}_\alpha) \, !! \, (n \bmod \text{len-of}_\alpha)$ looks up the n-th bit in the bit vector v, where $l \, ! \, i$ returns the i-th element of the list or array l. Then, he lifts his formalisation to uint using the Transfer tool. Thus, the generated code adapts the size of the list to the target language implementation.

Second, hashing does not rely on the exact size of the values. Algorithms based on hashing deal with clashes anyway, so their correctness does not depend on the exact hash values. Yet, hashing must be fast. Taking uint for hash values enables such fast hashing.

Third, finite rings $\mathbb{Z}/p\mathbb{Z}$ can be implemented via uint if $p^2 < 2^{\text{len-of}_{\text{default-size}}}$, which can be tested dynamically. We evaluate such an implementation in Sect. 6.

5 Validation

The code adaptations in our library are rather complicated, with many subtleties and corner cases. It is therefore imperative to validate the code adaptations. In theory, as all word types are finite, we could certify the code adaptations by running the generated code for all possible argument values and checking that the mapped HOL term evaluates to the same result (unless it is unspecified). In practice, this might be feasible for uint8 and uint16, but the argument space for 32- and 64-bit words is too large. Therefore, we content ourselves by running selected test cases.

In this section, we present a generic-purpose testing framework in Isabelle/HOL (Sect. 5.1) and the design and results of our validation (Sect. 5.2).

5.1   Automating Regression Tests for Code Generation

To automate the testing, we have developed a general-purpose testing tool for Isabelle/HOL's code generator, which is distributed with Isabelle2017 (theory HOL-Library.Code_Test). Our tool provides a new command **code-test** that takes a list of test cases and a list of target language implementations. A test case is any boolean HOL term. The supported target language implementations are PolyML, MLton, SMLNJ, GHC, OCaml, and Scala. For each target language implementation, the command performs five steps:

1. It generates code for all the test cases in the corresponding target language.
2. It produces a test harness tailored to the target language implementation.
3. If necessary, it compiles the generated code and the test harness.

4. It executes all test cases by running the (compiled) program.
5. It reports which test cases have succeeded or failed, and for the failed ones, it outputs the evaluation result for selected subterms, e.g., the two sides of (in)equalities.

If code generation or any of the test cases fails, the command raises an error in Isabelle/HOL, which makes it suitable for regression testing.

For example, the following invocation tests that our code adaptations correctly use Scala's signed division on bytes for computing the unsigned fraction:

code-test $251 \operatorname{div} 3 = (83 :: \operatorname{uint8})$ **in** Scala

In case of a failure, **code-test** outputs to what the left and right-hand side have evaluated in Scala. To that end, **code-test** also generates code for reifying the result value as a HOL term in the target language. This HOL term is then serialised as a YXML string in the same format that Isabelle/PIDE uses to communicate with the prover process [51]. Term reification is shared with the counter-example generator Quickcheck [8, Sect. 3.3.4], so it automatically works for most user-defined types, in particular all (co)datatypes.

The different target language implementations are modularly supported by drivers. A driver gets as input (i) the directory for the code, (ii) the names of the generated files, and (iii) the name of the generated function that executes all test cases. The driver outputs (i) the names and contents of its test harness files, and (ii) bash commands for compiling and running the code and the test harness.

Drivers must be registered with our tool under an identifier, e.g., PolyML and MLton, and with an associated code target, e.g., SML. The tool then takes care of all the rest, such as parsing the user's input, invoking Isabelle/HOL's code generator, writing all files to a fresh temporary directory, compiling and running the program, and showing the pretty-printed result to the user. Thus, users can easily write and register their own drivers when they want to test other implementations.

5.2   Test Case Selection and Validation Results

As is common practice, we partition the argument values into equivalence classes and select only one representative from each equivalence class. For uint∗, we consider the three classes $\{0, \ldots, 2^{l-1} - 1\}$, $\{2^{l-1}, \ldots, 2^l - 1\}$, and $\{2^l, \ldots\}$, where l denotes the bit length of the word type. For bit indices, we choose the classes $\{0, \ldots, l - 2\}$, $\{l - 1\}$, and $\{l, \ldots\}$. The most significant bit $l - 1$ has its own class because of the signed operations.

We have run all these test cases with all implementations and all evaluators. In fact, the test cases are routinely run by the regression test system of the Archive of Formal Proofs. This ensures that incompatible changes in Isabelle/HOL's code generator configuration are quickly detected.

During the development of our library, the test cases revealed many errors in the code adaptations, both syntactic and semantic errors, e.g., forgetting appropriate casts in Scala to counter the automatic promotion to Int. Of course, we have addressed all the errors and now all test cases pass. This indicates that our test cases are reasonable.

Surprisingly, the tests did not only reveal errors in our code adaptations. For PolyML 5.6.1, which Isabelle2017 runs on, one of our tests on 64-bit words failed when PolyML runs in 64-bit mode. The problem is that PolyML's Word64 structure does not correctly implement division. For example, Word64.div(0wxFFFFFFFFFFFFFFFB, 0wx3) evaluates to 0wx55555553 instead of 0wx5555555555555553. The error occurs only in 64-bit mode because PolyML does not provide a Word64 structure in 32-bit mode. Meanwhile, Matthews has implemented the Word64 structure differently in PolyML 5.7, thereby eliminating the bug. Isabelle2017 itself is not affected by the error because its implementation does not use 64-bit words. To support evaluation of uint64 terms in Isabelle2017, our library tests at load time whether the underlying 64-bit PolyML system provides the incorrect Word64 structure and—if so—generates a replacement based on arbitrary-precision integers.

6   Evaluation

We have been developing our library of machine words since 2013. Meanwhile, it has been picked up by several other users in their projects. This shows that our library is usable. Moreover, we can evaluate the performance by looking at real-world use cases instead of unrealistic micro-benchmarks. In this section, we describe how the projects used our library and comment on the performance impact we are aware of. For one project, we also ran the benchmarks to measure the performance impact of our library ourselves.

Divason et al. [14] have verified the Berlekamp-Zassenhaus algorithm for factoring polynomials over the integers. The algorithm factors a given polynomial over the finite rings $\mathbb{Z}/p^k\mathbb{Z}$ for $k = 1, 2, 4, 8, \ldots$ using Berlekamp's algorithm and Hensel's lifting lemma. Zassenhaus' algorithm then reconstructs the factorisation over the integers. Divason et al. have parametrised the factorisation algorithm and the Hensel lifting over the arithmetic operations. So they can choose the most efficient implementation dynamically according to the following strategy. If $p^k < 2^{16}$, all computations are done in uint32 as multiplying two 16-bit numbers stays below 2^{32}. If $2^{16} \leq p^k < 2^{32}$, their implementation uses uint64. Otherwise, arbitrary-precision integers are used.

To quantify the performance gain by using uint∗, we ran three versions of the factorisation algorithm (generated in Haskell from AFP version 70d9faada9d0). The first version omits the range checks for p^k and always uses arbitrary-precision integers. This establishes the baseline. The second version chooses the implementation type according to the above strategy. The third version uses uint if $p^k < \sqrt{2^{\text{len-of}_{\text{default-size}}}}$ and arbitrary-precision integers otherwise. We used the benchmarks by Divason et al. [14]: 400 randomly generated polynomials with 100 to 500 coefficients. The measurements were performed on an Intel i7 quad core at 2.4 GHz with 16 GB RAM running Ubuntu 14.04 LTS. The generated code was compiled with GHC 7.6.3 with option -O2.

Factoring all 400 polynomials using arbitrary-precision integers took 33.70 min in total. Using uint32/uint64 reduces the time to 27.42 min, i.e., a reduction by 18.6%. Per polynomial, the time reduction ranged between 10.0% and 39.9% with median 19.9% and relative standard deviation 5.9%. This shows that our library consistently provides better efficiency than computing with GMP integers, despite the additional range checks. The difference between uint32/uint64 and uint was insignificant. This is because GHC 7.6.3 always boxes machine words (Data.Word.Word). To measure the effect of boxing, we also ran the second and third version with PolyML 5.7.1 in 64-bit mode, where uint is only 63 bits, but unboxed: on average, uint is 4.0% faster than uint32/uint64.

Fleury et al. [17] are developing a verified SAT solver using Isabelle/HOL. For efficiency reasons, uint32 words are used for propositional variables and literals, where the positive and negative literals of a variable v are given by $2 \cdot v$ and $2 \cdot v + 1$, respectively. Both literals of a variable can thus be computed efficiently using bit operations. Fleury told us in personal communication that switching from GMP integers to uint32 improved performance of the generated Standard ML code considerably. Bit shifts on GMP integers are apparently significantly slower, even if the values fit in GMP's small integers.

The Isabelle Collections framework and the Monadic Refinement framework [33, 35, 36] use our library for implementing hash functions, on which verified hash sets and hash arrays build. Using these frameworks, Esparza et al. [15] have generated an LTL model checker in Standard ML from their formalisation. They observed a speed-up of one order of magnitude when they changed hashing from arbitrary-precision integers to our library.

Lochbihler and Züst [42] obtain a Haskell implementation of the TLS protocol generated from Isabelle/HOL. Unlike in the other projects, they use uint∗ not for efficiency reasons, but for exchanging data with foreign Haskell functions and for constructing the protocol messages. The socket API functions take arguments that are machine words of 8, 16, or 32 bits, and some fields in the protocol messages also have such bit lengths.

7   Related Work

Many proof assistants provide libraries for fixed-size words. Those that support code generation to machine integers are all tailored to one particular target language, usually the language the prover is implemented in. In contrast, our library shows how to fit the varying APIs of four target languages into one library while retaining efficiency.

The coq-bits library [6, 29] by Blot et al. models signed 8-, 16-, and 32-bit words in Coq. Using Coq's code adaptation command **Extract Inlined Constant**, the library maps all word types to OCaml's **int** type. They program exhaustive test cases in Coq and prove that the test cases suffice to establish that the translation is correct. But they run the test cases only for 8- and 16-bit words, as exhaustively testing 32- or 64-bit words is impractical. Thereby, they have missed that their mapping is unsound for 32-bit integers when Ocaml runs in 32-bit mode as **int** has only 31 significant bits then.

Armand et al. [3] added OCaml's 31-bit machine integers to Coq's evaluation engine, which is comparable to Isabelle/HOL's normalisation evaluator [1]. Théry [48] relies on them to establish by evaluation that the Mini-Rubik cube can always be solved in at most 11 steps. While we have made sure that our library supports evaluation, the normalisation evaluator uses the symbolic representation for the uint* types. Changing this representation to Standard ML machine words would require a complete re-design of the evaluator since it does not support any form of code adaptation. Our mappings therefore need not be trusted for normalisation. Regarding execution times, code extraction is much faster than normalisation in Isabelle anyway and even more so with our library.

Maude provides fixed-size words similar to Isabelle's Word library [9, Sect. 9.5]. Yet, they are not mapped to machine words, but emulated using arbitrary-precision integers.

Greve et al. [18] describe how ACL2 code can be written in such a way that the underlying LISP compiler uses unboxed machine words (fixnum) instead of arbitrary-precision integers. They annotate their code with many declarations that restrict the allowed integer range to signed 32 bit words. ACL2's guard checker accordingly demands a proof that the range is respected. Divason et al. [14] had to prove similar respectfulness theorems when they implemented the $GF(p^k)$ operations on the uint* types. Most proofs were automatic using the Transfer package and the existing theorems for α word. Like for Haskell, the exact range of fixnum in LISP is implementation-defined; at least 16 bits are required. Greve et al. ignore this issue and assume that at least 32 bits are provided.

PVS's ground evaluator generates LISP code from PVS specifications. It also supports unchecked code adaptations, which are called semantic attachments [10]. Muñoz' library PVSio [44] provides semantic attachments for, among others, floating point arithmetic, which replaces exact arithmetic on reals. Semantic attachments cannot be used to prove theorems by ground evaluation to prevent inconsistencies, e.g., due to rounding errors. Isabelle/HOL's code generator allows code adaptations for proofs. We therefore carefully craft the adaptations for the target Eval and raise exceptions in underspecified cases.

The problem of refining underspecified functions for code generation is also addressed by the Isabelle Monadic Refinement framework [36] and its Coq counterpart Fiat [13]. In both frameworks, programs must be written in a non-determinism monad. They can then be refined within the logic towards a deterministic implementation. This refinement approach could be used to model the non-determinism due to the different bit sizes in the various target language implementations. Users would however have to write all their functions in the monad and refine the non-determinism way before code generation. This would severely impair the usability of our library. We therefore opted for model-theoretic refinement and accepted that this refinement is unverified.

8 Conclusion and Future Work

We have presented a library for efficiently computing with machine words of 8, 16, 32, and 64 bits in Isabelle/HOL. It distinguishes itself from other such libraries in that it simultaneously supports all four target languages of Isabelle's code generator and all of Isabelle's evaluation mechanisms. Thus, formalisations based on our library do not have to commit to a particular language and can instead be used in any Isabelle context. We achieve this flexibility using a model-theoretic refinement semantics for code adaptations. To validate our library, we have developed a general-purpose regression test framework for Isabelle/HOL and tested the correctness of our code adaptations. Our library has successfully boosted the performance of the generated code in several projects.

We have also used the test framework to obtain HOL evaluators in Haskell, OCaml, and Scala. Haskell in particular is useful as its lazy evaluation semantics handles infinite codatatype values, on which the existing call-by-value evaluators do not terminate.

Our code adaptations are unverified—like all code adaptations for Isabelle/HOL. The adaptations and the machine word implementations in the target languages are therefore in the trusted code base whenever our library is used for code generation. This applies to (i) tools obtained by code generation and (ii) proofs by evaluation. As Isabelle tags all theorems whose proof has not been checked by the kernel, users can always check whether a theorem has gone through the kernel. If they do not want to trust the adaptations, they can always prove their theorems by term rewriting (or normalisation).

We will add more word types, e.g., signed words, on demand. While their formalisation is very easy thanks to the length-polymorphic Word library, getting the code adaptations right requires a careful study of the language specifications.

When the projects on verified code generation reach maturity, we hope to formally verify our mappings to reduce the TCB. In the meantime, it would be interesting to systematize the test case generation, e.g., by model-driven testing as implemented in HOL-Testgen [7]. We could validate the code adaptations further and check whether target language implementations correctly implement the operations. In this scenario, our library is only the starting point. Other libraries like Yu's formalisation of IEEE floating point numbers [52] could also benefit from validation. Although testing can never formally establish the correctness of code adaptations, it is a very practical approach to ensuring soundness.

Acknowledgements. Peter Lammich contributed an initial formalisation of machine words of unspecified length. Rafael Häuselmann helped to implement the **code_test** command. René Thiemann and Mathias Fleury encouraged us to develop the library further. The author was supported by the Swiss National Science Fund under grant 153217.

References

1. Aehlig, K., Haftmann, F., Nipkow, T.: A compiled implementation of normalisation by evaluation. J. Funct. Program. **22**(1), 9–30 (2012)
2. A. Anand, A. Appel, G. Morrisett, Z. Paraskevopoulou, R. Pollack, O. Savary Belanger, M. Sozeau, and M. Weaver. CertiCoq: A verified compiler for Coq. In: CoqPL 2017 (2017)
3. Armand, M., Grégoire, B., Spiwack, A., Théry, L.: Extending COQ with imperative features and its application to SAT verification. In: Kaufmann, M., Paulson, L.C. (eds.) ITP 2010. LNCS, vol. 6172, pp. 83–98. Springer, Heidelberg (2010). https://doi.org/10.1007/978-3-642-14052-5_8
4. Arthan, R.: On definitions of constants and types in HOL. J. Autom. Reason. **56**(3), 205–219 (2016)
5. Blanchette, J.C., Bulwahn, L., Nipkow, T.: Automatic proof and disproof in Isabelle/HOL. In: Tinelli, C., Sofronie-Stokkermans, V. (eds.) FroCoS 2011. LNCS (LNAI), vol. 6989, pp. 12–27. Springer, Heidelberg (2011). https://doi.org/10.1007/978-3-642-24364-6_2
6. Blot, A., Dagand, P.É., Lawall, J.: From sets to bits in Coq. In: Kiselyov, O., King, A. (eds.) FLOPS 2016. LNCS, vol. 9613, pp. 12–28. Springer, Cham (2016). https://doi.org/10.1007/978-3-319-29604-3_2
7. Brucker, A.D., Wolff, B.: HOL-TESTGEN: an interactive test-case generation framework. In: Chechik, M., Wirsing, M. (eds.) FASE 2009. LNCS, vol. 5503, pp. 417–420. Springer, Heidelberg (2009). https://doi.org/10.1007/978-3-642-00593-0_28
8. Bulwahn, L.: Counterexample Generation for Higher-Order Logic Using Functional and Logic Programming. Ph.D. thesis, Fakultät für Informatik, Technische Universität München (2013)
9. Clavel, M., et al.: All About Maude - A High-Performance Logical Framework. LNCS, vol. 4350. Springer, Heidelberg (2007). https://doi.org/10.1007/978-3-540-71999-1
10. Crow, J., Owre, S., Rushby, J., Shankar, N., Stringer-Calvert, D.: Evaluating, testing, and animating PVS specifications. Technical report, Computer Science Laboratory. SRI International, Menlo Park, CA (2001)
11. Dawson, J.: Isabelle theories for machine words. In: Goldsmith, M., Roscoe, B. (eds.) AVOCS 2007, vol. 250(1). ENTCS, pp. 55–70. Elsevier (2009)
12. Dawson, J., Graunke, P., Huffman, B., Klein, G., Matthews, J.: Machine words in Isabelle/HOL (2017). http://isabelle.in.tum.de/dist/library/HOL/HOL-Word/document.pdf
13. Delaware, B., Pit-Claudel, C., Gross, J., Chlipala, A.: Fiat: Deductive synthesis of abstract data types in a proof assistant. In: POPL 2015, pp. 689–700. ACM, New York (2015)
14. Divasón, J., Joosten, S., Thiemann, R., Yamada, A.: A formalization of the Berlekamp-Zassenhaus factorization algorithm. In: CPP 2017, pp. 17–29. ACM, New York (2017)
15. Esparza, J., et al.: A fully verified executable LTL model checker. In: Sharygina, N., Veith, H. (eds.) CAV 2013. LNCS, vol. 8044, pp. 463–478. Springer, Heidelberg (2013). https://doi.org/10.1007/978-3-642-39799-8_31
16. Farzan, A., Meseguer, J., Roşu, G.: Formal JVM code analysis in JavaFAN. In: Rattray, C., Maharaj, S., Shankland, C. (eds.) AMAST 2004. LNCS, vol. 3116, pp. 132–147. Springer, Heidelberg (2004). https://doi.org/10.1007/978-3-540-27815-3_14

17. Fleury, M., Blanchette, J.C., Lammich, P.: A verified SAT solver with watched literals using imperative HOL. In: CPP 2018, pp. 158–171. ACM (2018)
18. Greve, D., Wilding, M., Hardin, D.: High-speed, analyzable simulators. In: Kaufmann, M., Manolios, P., Strother Moore, J. (eds.) Computer-Aided Reasoning: ACL2 Case Studies. Advances in Formal Methods, vol. 4, pp. 113–135. Springer, Boston (2000). https://doi.org/10.1007/978-1-4757-3188-0_8
19. Haftmann, F.: Code generation from Isabelle/HOL theories (2017). http://isabelle. in.tum.de/dist/Isabelle2017/doc/codegen.pdf
20. Haftmann, F., Krauss, A., Kunčar, O., Nipkow, T.: Data refinement in Isabelle/HOL. In: Blazy, S., Paulin-Mohring, C., Pichardie, D. (eds.) ITP 2013. LNCS, vol. 7998, pp. 100–115. Springer, Heidelberg (2013). https://doi.org/10.1007/978-3-642-39634-2_10
21. Haftmann, F., Nipkow, T.: Code generation via higher-order rewrite systems. In: Blume, M., Kobayashi, N., Vidal, G. (eds.) FLOPS 2010. LNCS, vol. 6009, pp. 103–117. Springer, Heidelberg (2010). https://doi.org/10.1007/978-3-642-12251-4_9
22. Haftmann, F., Wenzel, M.: Constructive type classes in Isabelle. In: Altenkirch, T., McBride, C. (eds.) TYPES 2006. LNCS, vol. 4502, pp. 160–174. Springer, Heidelberg (2007). https://doi.org/10.1007/978-3-540-74464-1_11
23. Hales, T.C., Harrison, J., McLaughlin, S., Nipkow, T., Obua, S., Zumkeller, R.: A revision of the proof of the Kepler conjecture. Disc. Comput. Geom. 44(1), 1–34 (2010)
24. Harrison, J.: Theorem Proving with the Real Numbers. Springer, London (1998). https://doi.org/10.1007/978-1-4471-1591-5
25. Heule, M., Hunt, W., Kaufmann, M., Wetzler, N.: Efficient, verified checking of propositional proofs. In: Ayala-Rincón, M., Muñoz, C.A. (eds.) ITP 2017. LNCS, vol. 10499, pp. 269–284. Springer, Cham (2017). https://doi.org/10.1007/978-3-319-66107-0_18
26. Huffman, B., Kunčar, O.: Lifting and Transfer: a modular design for quotients in Isabelle/HOL. In: Gonthier, G., Norrish, M. (eds.) CPP 2013. LNCS, vol. 8307, pp. 131–146. Springer, Cham (2013). https://doi.org/10.1007/978-3-319-03545-1_9
27. Hupel, L., Nipkow, T.: A verified compiler from Isabelle/HOL to CakeML. In: Ahmed, A. (ed.) ESOP 2018. LNCS, vol. 10801, pp. 999–1026. Springer, Cham (2018). https://doi.org/10.1007/978-3-319-89884-1_35
28. Kanav, S., Lammich, P., Popescu, A.: A conference management system with verified document confidentiality. In: Biere, A., Bloem, R. (eds.) CAV 2014. LNCS, vol. 8559, pp. 167–183. Springer, Cham (2014). https://doi.org/10.1007/978-3-319-08867-9_11
29. Kennedy, A., Benton, N., Jensen, J.B., Dagand, P.-E.: Coq: the world's best macro assembler? In: PPDP 2013, pp. 13–24. ACM, New York (2013)
30. Kumar, R., Myreen, M.O., Norrish, M., Owens, S.: CakeML: a verified implementation of ML. In: POPL 2014, pp. 179–191. ACM, New York (2014)
31. Kunčar, O.: Correctness of Isabelle's cyclicity checker: implementability of overloading in proof assistants. In: CPP 2015, pp. 85–94. ACM, New York (2015)
32. Kunčar, O., Popescu, A.: A consistent foundation for Isabelle/HOL. In: Urban, C., Zhang, X. (eds.) ITP 2015. LNCS, vol. 9236, pp. 234–252. Springer, Cham (2015). https://doi.org/10.1007/978-3-319-22102-1_16
33. Lammich, P.: Collections framework. Archive of Formal Proofs (2009). http://isa-afp.org/entries/Collections.html, Formal proof development

34. Lammich, P.: The GRAT tool chain. In: Gaspers, S., Walsh, T. (eds.) SAT 2017. LNCS, vol. 10491, pp. 457–463. Springer, Cham (2017). https://doi.org/10.1007/978-3-319-66263-3_29

35. Lammich, P., Lochbihler, A.: The Isabelle collections framework. In: Kaufmann, M., Paulson, L.C. (eds.) ITP 2010. LNCS, vol. 6172, pp. 339–354. Springer, Heidelberg (2010). https://doi.org/10.1007/978-3-642-14052-5_24

36. Lammich, P., Tuerk, T.: Applying data refinement for monadic programs to Hopcroft's algorithm. In: Beringer, L., Felty, A. (eds.) ITP 2012. LNCS, vol. 7406, pp. 166–182. Springer, Heidelberg (2012). https://doi.org/10.1007/978-3-642-32347-8_12

37. Leroy, X.: A formally verified compiler back-end. J. Autom. Reason. **43**(4), 363–446 (2009)

38. Liu, H., Moore, J.S.: Executable JVM model for analytical reasoning: a study. In: IVME 2003, pp. 15–23. ACM (2003)

39. Lochbihler, A.: A Machine-Checked, Type-Safe Model of Java Concurrency : Language, Virtual Machine, Memory Model, and Verified Compiler. Ph.D. thesis, Karlsruher Institut für Technologie, Fakultät für Informatik, July 2012

40. Lochbihler, A.: Native word. Archive of Formal Proofs (2017). http://devel.isa-afp.org/entries/Native_Word.html, Formal proof development

41. Lochbihler, A., Bulwahn, L.: Animating the formalised semantics of a Java-like language. In: van Eekelen, M., Geuvers, H., Schmaltz, J., Wiedijk, F. (eds.) ITP 2011. LNCS, vol. 6898, pp. 216–232. Springer, Heidelberg (2011). https://doi.org/10.1007/978-3-642-22863-6_17

42. Lochbihler, A., Züst, M.: Programming TLS in Isabelle/HOL. Isabelle Workshop (2014). http://www.andreas-lochbihler.de/pub/lochbihler14iw.pdf

43. Mullen, E., Pernsteiner, S., Wilcox, J.R., Tatlock, Z., Grossman, D.: Œuf: Minimizing the Coq extraction TCB. In: CPP 2018, pp. 172–185. ACM (2018)

44. Muñoz, C.: Rapid prototyping in PVS. Contractor Report NASA/CR-2003-212418, NASA, Langley Research Center, Hampton VA 23681–2199, USA (2003)

45. Nipkow, T.: Teaching semantics with a proof assistant: no more LSD trip proofs. In: Kuncak, V., Rybalchenko, A. (eds.) VMCAI 2012. LNCS, vol. 7148, pp. 24–38. Springer, Heidelberg (2012). https://doi.org/10.1007/978-3-642-27940-9_3

46. Nipkow, T., Wenzel, M., Paulson, L.C. (eds.): Isabelle/HOL. LNCS, vol. 2283. Springer, Heidelberg (2002). https://doi.org/10.1007/3-540-45949-9

47. Owens, S., Norrish, M., Kumar, R., Myreen, M.O., Tan, Y.K.: Verifying efficient function calls in CakeML. In: ICFP 2017, Proc. ACM Program. Lang., vol. 1, pp. 18:1–18:27. ACM (2017)

48. Théry, L.: Proof pearl: revisiting the Mini-Rubik in Coq. In: Mohamed, O.A., Muñoz, C., Tahar, S. (eds.) TPHOLs 2008. LNCS, vol. 5170, pp. 310–319. Springer, Heidelberg (2008). https://doi.org/10.1007/978-3-540-71067-7_25

49. Thiemann, R., Sternagel, C.: Certification of termination proofs using CeTA. In: Berghofer, S., Nipkow, T., Urban, C., Wenzel, M. (eds.) TPHOLs 2009. LNCS, vol. 5674, pp. 452–468. Springer, Heidelberg (2009). https://doi.org/10.1007/978-3-642-03359-9_31

50. Warren, H.S.: Hacker's Delight, 2 edn. Addison-Wesley (2012)

51. Wenzel, M.: Isabelle as document-oriented proof assistant. In: Davenport, J.H., Farmer, W.M., Urban, J., Rabe, F. (eds.) CICM 2011. LNCS (LNAI), vol. 6824, pp. 244–259. Springer, Heidelberg (2011). https://doi.org/10.1007/978-3-642-22673-1_17
52. Yu, L.: A formal model of IEEE floating point arithmetic. Archive of Formal Proofs (2013). http://isa-afp.org/entries/IEEE_Floating_Point.html, Formal proof development

Relational Parametricity and Quotient Preservation for Modular (Co)datatypes

Andreas Lochbihler[(✉)] and Joshua Schneider[(✉)]

Institute of Information Security, Department of Computer Science, ETH Zürich,
Zürich, Switzerland
{andreas.lochbihler,joshua.schneider}@inf.ethz.ch

Abstract. Bounded natural functors (BNFs) provide a modular framework for the construction of (co)datatypes in higher-order logic. Their functorial operations, the mapper and relator, are restricted to a subset of the parameters, namely those where recursion can take place. For certain applications, such as free theorems, data refinement, quotients, and generalised rewriting, it is desirable that these operations do not ignore the other parameters. In this paper, we generalise BNFs such that the mapper and relator act on both covariant and contravariant parameters. Our generalisation, $\mathrm{BNF_{CC}}$, is closed under functor composition and least and greatest fixpoints. In particular, every (co)datatype is a $\mathrm{BNF_{CC}}$. We prove that subtypes inherit the $\mathrm{BNF_{CC}}$ structure under conditions that generalise those for the BNF case. We also identify sufficient conditions under which a $\mathrm{BNF_{CC}}$ preserves quotients. Our development is formalised abstractly in Isabelle/HOL in such a way that it integrates seamlessly with the existing parametricity infrastructure.

1 Introduction

Datatypes and codatatypes are a fundamental tool in functional programming and proof assistants. Proof assistants based on type theory usually provide (co)datatypes as a built-in concept (e.g. Coq [31], Agda [29], Lean [28]), whereas other tools defer the construction to definitional (Isabelle [7], HOL [36]) or axiomatic packages (PVS [30], Dafny [23]). Traytel et al. [38] proposed *bounded natural functors* (BNFs) as a semantic criterion for (co)datatypes that are constructible in higher-order logic, which was subsequently implemented as a definitional package in Isabelle/HOL [7]. Notably, the BNF class includes important non-free types such as finite sets and discrete probability distributions. The package allows a modular approach: Once a type constructor has been proved to be a BNF, it can be used to define new (co)datatypes.

Electronic supplementary material The online version of this chapter (https://doi.org/10.1007/978-3-319-94821-8_24) contains supplementary material, which is available to authorized users.

© Springer International Publishing AG, part of Springer Nature 2018
J. Avigad and A. Mahboubi (Eds.): ITP 2018, LNCS 10895, pp. 411–431, 2018.
https://doi.org/10.1007/978-3-319-94821-8_24

For example, following the coalgebraic theory of systems [34], deterministic discrete systems are modelled as elements of a codatatype (a, b) dds, where a is the type of inputs, and b is the type of outputs of the system. In Isabelle/HOL, the following command defines this type with constructor Dds and destructor run.

$$\text{codatatype } (a, b) \text{ dds} = \text{Dds } (\text{run} \colon a \Rightarrow b \times (a, b) \text{ dds})$$

Note that (a, b) dds on the right-hand side occurs inside a product $(b \times \natural)$ and a function type $(a \Rightarrow \natural)$, which both are BNFs. Yet, not all recursive specifications produce valid HOL (co)datatypes [13]. For example, a datatype must not recurse through the domain of predicates $\natural \Rightarrow$ bool. Otherwise, HOL's set-theoretic semantics would have to contain an injection into a non-empty set from its powerset. To avoid such inconsistencies, BNFs distinguish *live* type parameters from *dead* ones, and (co)recursion is limited to live parameters. For the function space $a \Rightarrow b$, b is live and a is dead, and the same holds for (a, b) dds.

Many type constructors come with a map operation (mapper) that lifts unary functions on the type parameters to the whole type. For lists, e.g., the mapper $\text{map}_{\text{list}} \colon\colon (a \Rightarrow b) \Rightarrow a \text{ list} \Rightarrow b \text{ list}$ applies the given function to all elements in the list. The function space's mapper $\text{map}_\Rightarrow g \ h \ f = h \circ f \circ g$ transforms both the domain and the range of f. Every BNF has a mapper, but it acts only on the live parameters. For the function space, the BNF mapper $\text{map}_\Rightarrow \text{id } h$ therefore transforms only the range, but not the domain. Similarly, the BNF mapper map_{dds} for DDS's has type $(b \Rightarrow b') \Rightarrow (a, b) \text{ dds} \Rightarrow (a, b') \text{ dds}$, i.e., it can transform a system's outputs. But it is useless if we want to transform the inputs. For example, consider a system S turning integers into booleans (e.g., testing whether the partial sum of inputs is even). Then, we cannot easily use it on natural numbers. In contrast, if the mapper acted also on the contravariant type parameter a, i.e., $\text{map}_{\text{dds}} \colon\colon (a' \Rightarrow a) \Rightarrow (b \Rightarrow b') \Rightarrow (a, b) \text{ dds} \Rightarrow (a', b') \text{ dds}$, then the new system could be written as $\text{map}_{\text{dds}} \text{ int id } S$, where int $\colon\colon$ nat $\Rightarrow$ int embeds the natural numbers in the integers.

This limitation of the BNF mapper is pervasive. First of all, it also affects the derived relator, which lifts relations rather than functions. For example, the list relator $\text{rel}_{\text{list}} R$ relates two lists iff they have the same length and the elements at the same indices are related by R. The function space relator $A \mapsto B$ takes a relation A on the domain and a relation B on the codomain. It relates two functions if they map A-related inputs to B-related outputs. But when seen as a BNF, A is always fixed to the identity relation $(=)$. Accordingly, due to the modular construction of (co)datatypes, the DDS relator lifts only a relation on outputs, and the input's relation is fixed to $(=)$.

Mappers and relators are used by many reasoning tools built on relational parametricity [33]. A polymorphic term is relationally parametric if its instances for related types are related, too. This requires an interpretation of types as relations and, thus, relators. The BNF restriction to live parameters hampers many applications of the interpretation: Quotients cannot be lifted through dead parameters [17], data refinement cannot happen in dead parameter positions [9,21], rewriting with equivalence relations must not affect dead parameters [37], free theorems can talk only about live parameters [39], and so on. Whenever—today in Isabelle/HOL—any of this is needed for dead parameters, too, one has to manually define more general mappers and relators ad hoc for the affected types.

In this paper, we generalise the BNF notion to BNF_{CC}, where dead parameters are refined into covariant, contravariant, and fixed parameters—"CC" stands for covariance and contravariance. While live parameters are technically covariant, we reserve the latter term for non-live parameters. For example, the type of second-order functions $(a \Rightarrow b) \Rightarrow c$ is a BNF where only c is live and a and b are dead. Considered as a BNF_{CC}, c is live, b is contravariant because it occurs at a negative position with respect to the function space, and a is covariant as it occurs at a positive, but not strictly positive position. The BNF_{CC} mapper and relator act on all type parameters but the fixed ones. For dds, e.g., we do obtain the desired mapper that lets us transform both the inputs and the outputs. The BNF_{CC} notion coincides with the BNF notion when there are only live and fixed parameters.

The key feature of BNFs is that they are closed under composition and least and greatest fixpoints, i.e., (co)datatypes. BNF_{CC}s also enjoy these properties. So, they are as modular as the BNFs and can be integrated into Isabelle's (co)datatype package. We emphasise that BNF_{CC}s do not allow users to define more (co)datatypes than BNFs do. The difference is that BNF_{CC}s yield more versatile mappers and relators with useful properties. Moreover, they integrate nicely with the rest of the functor-based infrastructure.

The main contributions of this paper are the following:

- We introduce the notion of BNF_{CC} as a generalisation of BNF (Sect. 2). BNF_{CC}s are equipped with more general relators and mappers than what the underlying natural functor provides. These operations are useful for various reasoning tasks (Sect. 1.2).
- We prove that BNF_{CC}s are closed under composition (Sect. 3) and least and greatest fixpoints (Sect. 4). This makes BNF_{CC}s modular and avoids syntactic conditions on definitions. In particular, every (co)datatype defined with Isabelle/HOL's (co)datatype package [7,8] is a BNF_{CC}.
- We prove that subtypes preserve the BNF_{CC} structure under certain conditions (Sect. 5). Consequently, non-uniform (co)datatypes [8] are BNF_{CC}s, too. If there are no covariant and contravariant parameters, our conditions are equivalent to those for BNFs.
- We prove that BNF_{CC}s lift quotients unconditionally through live parameters and under mild conditions through covariant and contravariant parameters (Sect. 6). This makes Isabelle's Lifting package more powerful, as the BNF theory only proves lifting for live parameters.

We formalised all our constructions and proofs in Isabelle/HOL [25]. Since reasoning about abstract functors is impossible in HOL, we axiomatised two generic BNF_{CC}s with sufficiently many parameters of each kind, and used them for the concrete constructions. The formalisation includes the examples from Sects. 1 and 2. In addition, we give informal proof sketches for most propositions and theorems in Appendix A (Online Resource). The implementation of BNF_{CCS} as an extension of the existing packages is left as future work (Sect. 8).

1.1 Background: Bounded Natural Functors

A bounded natural functor (BNF) [38] is a type constructor F of some arity equipped with a mapper, conversions to sets, and a cardinality bound on those sets. A type parameter of F is either *live* or *dead*; dead parameters are ignored by the BNF operations. The mapper is given by the polymorphic operation $\mathsf{map_F} :: \overline{(l \Rightarrow l')} \Rightarrow (\bar{l}, \bar{\mathfrak{d}})\, \mathsf{F} \Rightarrow (\bar{l'}, \bar{\mathfrak{d}})\, \mathsf{F}$ on the live parameters $\bar{l}$, whereas the dead parameters $\bar{\mathfrak{d}}$ remain fixed.[1] We assume without loss of generality that all live parameters precede the dead ones in the parameter list. For each live type parameter l_i, a BNF comes with a polymorphic setter $\mathsf{set}_\mathsf{F}^i :: (\bar{l}, \bar{\mathfrak{d}})\, \mathsf{F} \Rightarrow l_i\ \mathsf{set}$. The cardinality bound $\mathsf{bd_F}$ is assumed to be infinite and may depend only on non-live parameters. The BNF operations must satisfy the following laws [7]:

$$\mathsf{map_F}\ \overline{\mathsf{id}} = \mathsf{id} \qquad \mathsf{map_F}\ \overline{(f \circ g)} = \mathsf{map_F}\ \bar{f} \circ \mathsf{map_F}\ \bar{g} \qquad \forall i.\ |\mathsf{set}_\mathsf{F}^i| \le \mathsf{bd_F}\ (1)$$

$$\forall i.\ \mathsf{set}_\mathsf{F}^i\ (\mathsf{map_F}\ \bar{f}\ x) = f_i\ `\ \mathsf{set}_\mathsf{F}^i\ x \qquad \frac{\forall i.\ \forall y \in \mathsf{set}_\mathsf{F}^i\ x.\ f_i\ y = g_i\ y}{\mathsf{map_F}\ \bar{f}\ x = \mathsf{map_F}\ \bar{g}\ x}\ (2)$$

$$\mathsf{rel_F}\ \bar{R} \circ \mathsf{rel_F}\ \bar{S} \sqsubseteq \mathsf{rel_F}\ \overline{(R \circ S)}\ (3)$$

Here, $f\, `\, A = \{y \mid \exists x \in A.\ y = f\ x\}$ denotes A's image under f, $|A|$ is A's cardinality, $\circ$ is relation composition, and $\sqsubseteq$ is relation containment. Relations are represented as binary predicates of type $\mathfrak{a} \otimes \mathfrak{b} = (\mathfrak{a} \Rightarrow \mathfrak{b} \Rightarrow \mathsf{bool})$. The relator $\mathsf{rel_F}$ is defined as

$$\mathsf{rel_F}\ \bar{R}\ x\ y = (\exists z.\ (\forall i.\, \mathsf{set}_\mathsf{F}^i\ z \subseteq \{(a,b) \mid R_i\ a\ b\}) \land \mathsf{map_F}\ \overline{\pi_1}\ z = x \land \mathsf{map_F}\ \overline{\pi_2}\ z = y)\ (4)$$

where π_1 and π_2 project a pair to its components. The relator extends $\mathsf{map_F}$ to relations, interpreting functions f by their graphs $\mathsf{Gr}\ f = (\lambda x\ y.\ y = f\ x)$:

Lemma 1. *If F is a BNF, then* $\mathsf{Gr}\ (\mathsf{map_F}\ \bar{f}) = \mathsf{rel_F}\ \overline{(\mathsf{Gr}\ f)}$.

BNFs are closed under various operations: functor composition, "killing" of live type parameters, and least and greatest fixpoints. Examples of basic BNFs are the identity functor, products ($\times$), sums ($+$), and function spaces ($\Rightarrow$), where the domain is dead. Finite lists $\mathfrak{a}$ list are a datatype and hence a BNF, too, with mapper $\mathsf{map_{list}}$, relator $\mathsf{rel_{list}}$, and bound $\aleph_0$. The setter $\mathsf{set_{list}}$ returns the set of elements in the list.

---

[1] The notation $\bar{x}$ stands for a meta-syntactic list of formal entities $x_1, x_2, \ldots, x_n$. We use this notation quite liberally, such that the expanded type of $\mathsf{map_F}$ reads

$$(l_1 \Rightarrow l_1') \Rightarrow (l_2 \Rightarrow l_2') \Rightarrow \ldots \Rightarrow (l_m \Rightarrow l_m') \Rightarrow (l_1, \ldots, l_m, \mathfrak{d}_1, \ldots, \mathfrak{d}_n)\, \mathsf{F} \Rightarrow (l_1', \ldots, l_m', \mathfrak{d}_1, \ldots, \mathfrak{d}_n)\, \mathsf{F}.$$

Similarly, we write $\forall i.\ \varphi$ for the conjunction of all instances of φ over the index i. Superscripts select a subsequence, e.g., $\bar{x}^{>2}$ represents $x_3, x_4, \ldots, x_n$.

1.2   Examples and Applications

We now illustrate the benefits of parametricity-based reasoning using small examples, which all require the generalised mappers and relators. Although all our examples revolve around the DDS codatatype, parametricity-based reasoning is not restricted to coalgebraic system models. It can equally be used for all the other (co)datatypes, and whenever a type parameter is covariant or contravariant (e.g., a in (a, b) tree $=$ Leaf b | Node $(a \Rightarrow (a, b)$ tree$))$, the BNF$_{CC}$ theory makes the reasoning more powerful than the BNF theory.

Free Theorems. Wadler [39] showed how certain theorems can be derived from parametricity by instantiating the relations with the graphs of functions and using Lemma 1, which we generalise to BNF$_{CC}$s in Sect. 2. As shown in the introduction, the inputs and outputs of a DDS can be transformed with the mapper map$_{dds}$. Parallel || and sequential • composition for DDS's, e.g., are defined corecursively by

primcorec (||) :: (a, b) dds $\Rightarrow (c, \eth)$ dds $\Rightarrow (a + c, b + \eth)$ dds where
  run $(S_1 || S_2) = (\lambda x.$ case x of
    Inl $a \Rightarrow$ let $(b, S_1') =$ run S_1 a in (Inl $b, S_1' || S_2)$
    | Inr $c \Rightarrow$ let $(d, S_2') =$ run S_2 c in (Inr $d, S_1 || S_2'))$
primcorec (•) :: (a, b) dds $\Rightarrow (b, c)$ dds $\Rightarrow (a, c)$ dds where
  run $(S_1 • S_2) = (\lambda a.$ let $(b, S_1') =$ run S_1 a; $(c, S_2') =$ run S_2 b in $(c, S_1' • S_2'))$

where Inl and Inr denote the injections into the sum type.

The following "free" theorems are derived from the parametricity laws by rewriting only; no coinduction is needed. Note that the BNF mapper on live parameters only would not be any good for • as the function g occurs both in the live and dead positions.

$$\text{map}_{dds}\ f\ h\ S_1 || \text{map}_{dds}\ g\ k\ S_2 = \text{map}_{dds}\ (\text{map}_+\ f\ g)\ (\text{map}_+\ h\ k)\ (S_1 || S_2)$$
$$\text{map}_{dds}\ f\ g\ S_1 • S_2 = \text{map}_{dds}\ f\ \text{id}\ (S_1 • \text{map}_{dds}\ g\ \text{id}\ S_2)$$
$$S_1 • \text{map}_{dds}\ g\ h\ S_2 = \text{map}_{dds}\ \text{id}\ h\ (\text{map}_{dds}\ \text{id}\ g\ S_1 • S_2)$$

Reasoning with parametricity is especially useful in larger applications. The first author formalised a cryptographic algebra based on sub-probabilistic discrete systems (sPDS) similar to Maurer's random systems [26]. Deriving the free theorems from parametricity pays off particularly for transformers of sPDS, which are formalised as a codatatype that recurses through another codatatype of probabilistic resumptions. Proofs by coinduction would require substantially more effort even for such simple theorems.

Data Refinement. Data refinement changes the representation of data in a program. It offers a convenient way to go from abstract data structures like sets to efficient ones like red-black trees, which are the key to generate efficient code from a formalisation. Several tools automate the data refinement and synthesise an implementation from an abstract specification in this way [9,10,14,21]. As these tools are based on parametricity, (nested) data refinement is only possible in type parameters on which the relators act. A more general relator thus increases the refinement capabilities.

For example, consider a DDS traverse G parametrised by a finite graph G. Upon input of a node set A, it returns all successor nodes $G[A]$ of A that have not yet been visited. Such a DDS can be used to implement a breadth-first or depth-first search traversal of a graph. Suppose that the correctness proof works with abstract graphs, say, represented by a finite set of edges (type $(a \times a)$ fset), whereas the refinement traverse_i represents the graph as a list of edges and the inputs and outputs as lists (we use Haskell-style list comprehension syntax). Using the canonical DDS coiterator dds-of and the refinement relation fset-as-list $:: a$ list $\otimes a$ fset for implementing finite sets by lists, we get the following refinement theorem. Note that we need the general relator rel$_{\text{dds}}$ to lift the refinement relations on the inputs and outputs. (Recall that $\mapsto$ is the function space relator.)

primcorec dds-of $:: (s \Rightarrow a \Rightarrow b \times s) \Rightarrow s \Rightarrow (a, b)$ dds where
   run (dds-of $f\ s$) = map$_\times$ id (dds-of f) $\circ f\ s$

definition traverse $:: (a \times a)$ fset $\Rightarrow (a$ fset, a fset) dds where
   traverse G = dds-of $(\lambda \mathcal{V}\ A.\ (G[A] - \mathcal{V}, \mathcal{V} \cup A))\ \emptyset$

definition traverse$_i$ $:: (a \times a)$ list $\Rightarrow (a$ list, a list) dds where
   traverse$_i$ E = dds-of $(\lambda \mathcal{V}\ A.\ [y \mid (x,y) \leftarrow E, x \in \text{set}_{\text{list}}\ A, y \notin \mathcal{V}], \mathcal{V} \cup \text{set}_{\text{list}}\ A))\ \emptyset$

lemma REFINEMENT: (fset-as-list $\mapsto$ rel$_{\text{dds}}$ fset-as-list fset-as-list) traverse$_i$ traverse

Quotients. Quotient preservation theorems are used to modularly construct quotient types and to lift functions and lemmas to them [16,17,20]. For example, the type of finite sets fset is a quotient of lists where the order and multiplicity of the elements are ignored. Given the quotient preservation theorems for $\Rightarrow$ and dds, Isabelle's Lifting package can lift this fset-list quotient to traverse's type. It can thus synthesise a definition for traverse using traverse$_i$ and prove the REFINEMENT lemma automatically given a proof that traverse$_i$ respects the quotient.

The refinement relation fset-as-list can additionally be parametrised by a refinement relation R on the elements [20]: fset-as-list$'$ R = rel$_{\text{list}}$ $R \circ$ fset-as-list. Combining traverse$_i$'s parametricity with REFINEMENT using some BNF$_{\text{CC}}$ relator properties, one can then automatically derive a stronger refinement rule, where the node type can simultaneously be refined; the assumption expresses that R must preserve the identity of nodes, as expected from traverse$_i$'s implicit dependence on the equality operation.

$$\frac{(R \mapsto R \mapsto (=))\ (=)\ (=)}{(\text{fset-as-list}'\ R \mapsto \text{rel}_{\text{dds}}\ (\text{fset-as-list}'\ R)\ (\text{fset-as-list}'\ R))\ \text{traverse}_i\ \text{traverse}}$$

Generalised Rewriting. Rewriting replaces subterms with equal terms. In generalised rewriting, relations other than equality are considered, and the context in which rewriting takes place must have an appropriate congruence property [37]. For example, the DDS seen outputs all the elements in the current input set that it has seen before. It is a monotone system with respect to the subset relation, which we express using the DDS relator. The graph traversal traverse is also monotone in the underlying graph provided that the input sets remain the same.

definition seen :: (a fset, a fset) dds **where** seen = dds-of ($\lambda S\ A.\ (S \cap A, S \cup A)$) $\emptyset$

lemma SEEN-MONO: rel$_{dds}$ ($\subseteq$) ($\subseteq$) seen seen
lemma TRAVERSE-MONO: rel$_{dds}$ ($=$) ($\subseteq$) (traverse G) (traverse H) **if** $G \subseteq H$

Now suppose that H is a supergraph of G, or equivalently $G \subseteq H$. Using the parametricity of sequential composition, we can thus rewrite traverse $G \bullet$ seen to traverse $H \bullet$ seen, where the systems are related by rel$_{dds}$ ($=$) ($\subseteq$).

2   Bounded Natural Functors with Co- and Contravariance

The operations specified by a BNF act only on live type parameters. As discussed in the introduction, many types admit more general operations, for example the function space's mapper map$_\Rightarrow$ $g\ h\ f = h \circ f \circ g$. Yet, the BNF structure is restricted to the mapper map$_\Rightarrow$ id h, which targets only the range, but not the domain.

In this section, we define bounded natural functors with covariance and contravariance (BNF$_{CC}$) as a generalization of BNFs. A BNF$_{CC}$ has a mapper and relator which take additional covariant and contravariant arguments corresponding to (a subset of) the dead parameters $\bar{\mathfrak{d}}$. Thus $\bar{\mathfrak{d}}$ is refined into three disjoint sequences: $\bar{\mathfrak{c}}$ for covariant, $\bar{\mathfrak{k}}$ for contravariant, and $\bar{\mathfrak{f}}$ for the remaining fixed parameters which are ignored by the generalised operations. The names covariant and contravariant indicate whether the mapper preserves the order of composition or swaps it, and whether the relator is monotone or anti-monotone in the corresponding argument, respectively. (Live parameters behave like covariant parameters in this regard. We use "covariant" only for parameters that are not live.) For example, the function space $\mathfrak{k} \Rightarrow \mathfrak{l}$ is a BNF$_{CC}$ that is live in $\mathfrak{l}$ and contravariant in $\mathfrak{k}$, as map$_\Rightarrow$'s type $(\mathfrak{k}' \Rightarrow \mathfrak{k}) \Rightarrow (\mathfrak{l} \Rightarrow \mathfrak{l}') \Rightarrow (\mathfrak{k} \Rightarrow \mathfrak{l}) \Rightarrow (\mathfrak{k}' \Rightarrow \mathfrak{l}')$ indicates. Similarly, the BNF$_{CC}$ $(\mathfrak{c} \Rightarrow \mathfrak{k}) \Rightarrow \mathfrak{l}$ is live in $\mathfrak{l}$, covariant in $\mathfrak{c}$, and contravariant in $\mathfrak{k}$.

Definition 1 (BNF$_{CC}$). *A BNF$_{CC}$ is a type constructor F with operations*

$$map_F :: \overline{(\mathfrak{l} \Rightarrow \mathfrak{l}')} \Rightarrow \overline{(\mathfrak{c} \Rightarrow \mathfrak{c}')} \Rightarrow \overline{(\mathfrak{k}' \Rightarrow \mathfrak{k})} \Rightarrow (\bar{\mathfrak{l}}, \bar{\mathfrak{c}}, \bar{\mathfrak{k}}, \bar{\mathfrak{f}})\ F \Rightarrow (\bar{\mathfrak{l}'}, \bar{\mathfrak{c}'}, \bar{\mathfrak{k}'}, \bar{\mathfrak{f}})\ F$$
$$rel_F :: \overline{\mathfrak{l} \otimes \mathfrak{l}'} \Rightarrow \overline{\mathfrak{c} \otimes \mathfrak{c}'} \Rightarrow \overline{\mathfrak{k} \otimes \mathfrak{k}'} \Rightarrow (\bar{\mathfrak{l}}, \bar{\mathfrak{c}}, \bar{\mathfrak{k}}, \bar{\mathfrak{f}})\ F \otimes (\bar{\mathfrak{l}'}, \bar{\mathfrak{c}'}, \bar{\mathfrak{k}'}, \bar{\mathfrak{f}})\ F$$

and, like for plain BNFs, a cardinality bound bd_F and set functions set^i_F for all live parameters l_i. The cardinality bound may depend on $\bar{c}, \bar{t}$, and $\bar{f}$, but not on $\bar{l}$. We define two conditions pos_F, neg_F for the relator rel_F subdistributing over relation composition:[2]

$$pos_F, neg_F :: \overline{(c \otimes c') \times (c' \otimes c'')} \Rightarrow \overline{(t \otimes t') \times (t' \otimes t'')} \Rightarrow bool$$

$pos_F \ \overline{(C, C')} \ \overline{(K, K')} \longleftrightarrow$

$$(\forall \overline{L} \ \overline{L'}. \ rel_F \ \overline{L} \ \overline{C} \ \overline{K} \bigcirc rel_F \ \overline{L'} \ \overline{C'} \ \overline{K'} \sqsubseteq rel_F \ \overline{(L \bigcirc L')} \ \overline{(C \bigcirc C')} \ \overline{(K \bigcirc K')}) \quad (5)$$

$neg_F \ \overline{(C, C')} \ \overline{(K, K')} \longleftrightarrow$

$$(\forall \overline{L} \ \overline{L'}. \ rel_F \ \overline{(L \bigcirc L')} \ \overline{(C \bigcirc C')} \ \overline{(K \bigcirc K')} \sqsubseteq rel_F \ \overline{L} \ \overline{C} \ \overline{K} \bigcirc rel_F \ \overline{L'} \ \overline{C'} \ \overline{K'}) \quad (6)$$

The BNF_{CC} operations must satisfy the conditions shown in Fig. 1:

1. *The mapper map_F is functorial with respect to all non-fixed parameters (7) and relationally parametric (8).*
2. *The BNF laws about the setters (the cardinality bound, naturality, and congruence) are satisfied for the mapper $map^\star_F \ \bar{l} = map_F \ \bar{l} \ \overline{id} \ \overline{id}$ restricted to live arguments (9).*
3. *The relator rel_F is monotone in live and covariant arguments, and antimonotone in contravariant arguments; the relator $rel^\star_F \ \overline{L} = rel^\star_F \ \overline{L} \ \overline{(=)} \ \overline{(=)}$ restricted to live arguments is strongly monotone (10).*[3]
4. *The relator preserves equality and distributes over converses $\_^{-1}$ (11).*
5. *The relator distributes over relation composition if the relations for covariant and contravariant parameters are equality (12).*

In comparison to plain BNFs, the BNF_{CC} relator is a primitive operation because it is not obvious how to generalise the characterisation (4) in terms of the mapper and setters to covariant and contravariant arguments. We therefore require several properties of the relator. Note that strong monotonicity (10) and negative composition subdistributivity (12) on live arguments are equivalent to the characterisation of $rel^\star_F$, given the other axioms.

Distributivity over relation composition is split into two directions (positive and negative) because concrete functors satisfy the directions under different conditions and some theorems only need one of the two directions. The names positive and negative stem from Isabelle's Lifting package, which needs the appropriate direction for positive or negative positions in types. In this paper, we often

---

[2] In our formalisation, pos_F and neg_F take type tokens to avoid issues with hidden polymorphism in the live and fixed type parameters. We omit this detail in the paper to simplify the notation.

[3] When $pos_F \ \overline{((=), C)} \ \overline{((=), K)} = neg_F \ \overline{((=), C)} \ \overline{((=), K)} = \text{True}$ for all $\overline{C}$ and $\overline{K}$, then the two monotonicity rules (10) are equivalent to the following combined rule:

$$\frac{\forall i. \ \forall a \in set^i_F \ x. \ \forall b \in set^i_F \ y. \ L_i \ a \ b \longrightarrow L'_i \ a \ b \qquad \forall i. \ C_i \sqsubseteq C'_i \qquad \forall i. \ K'_i \sqsubseteq K_i}{rel_F \ \overline{L} \ \overline{C} \ \overline{K} \ x \ y \longrightarrow rel_F \ \overline{L'} \ \overline{C'} \ \overline{K'} \ x \ y}.$$

$$\mathsf{map_F}\ \overline{\mathsf{id}}\ \overline{\mathsf{id}}\ \overline{\mathsf{id}} = \mathsf{id} \qquad \mathsf{map_F}\ (\overline{\ell \circ \ell'})\ (\overline{c \circ c'})\ (\overline{k' \circ k}) = \mathsf{map_F}\ \overline{\ell}\ \overline{c}\ \overline{k} \circ \mathsf{map_F}\ \overline{\ell'}\ \overline{c'}\ \overline{k'} \quad (7)$$

$$\overline{((L \mapsto L')} \mapsto \overline{(C \mapsto C')} \mapsto \overline{(K' \mapsto K)} \mapsto \mathsf{rel_F}\ \overline{L}\ \overline{C}\ \overline{K} \mapsto \mathsf{rel_F}\ \overline{L'}\ \overline{C'}\ \overline{K'})\ \mathsf{map_F}\ \mathsf{map_F} \quad (8)$$

$$\forall i.\ |\mathsf{set_F^i}| \leq \mathsf{bd_F} \qquad \forall i.\ \mathsf{set_F^i}\ (\mathsf{map_F^\star}\ \overline{\ell}\ x) = \ell_i\ {}^\backprime \mathsf{set_F^i}\ x \qquad \dfrac{\forall i.\ \forall y \in \mathsf{set_F^i}\ x.\ \ell_i\ y = \ell_i'\ y}{\mathsf{map_F^\star}\ \overline{\ell}\ x = \mathsf{map_F^\star}\ \overline{\ell'}\ x} \quad (9)$$

$$\dfrac{\forall i.\ L_i \sqsubseteq L_i' \quad \forall i.\ C_i \sqsubseteq C_i' \quad \forall i.\ K_i' \sqsubseteq K_i}{\mathsf{rel_F}\ \overline{L}\ \overline{C}\ \overline{K} \sqsubseteq \mathsf{rel_F}\ \overline{L'}\ \overline{C'}\ \overline{K'}} \qquad \dfrac{\forall i.\ \forall a \in \mathsf{set_F^i}\ x.\ \forall b \in \mathsf{set_F^i}\ y.\ L_i\ a\ b \longrightarrow L_i'\ a\ b}{\mathsf{rel_F^\star}\ \overline{L}\ x\ y \longrightarrow \mathsf{rel_F^\star}\ \overline{L'}\ x\ y} \quad (10)$$

$$\mathsf{rel_F}\ \overline{(=)}\ \overline{(=)}\ \overline{(=)} = (=) \qquad (\mathsf{rel_F}\ \overline{L}\ \overline{C}\ \overline{K})^{-1} = \mathsf{rel_F}\ \overline{L^{-1}}\ \overline{C^{-1}}\ \overline{K^{-1}} \quad (11)$$

$$\mathsf{pos_F}\ \overline{((=),(=))}\ \overline{((=),(=))} \qquad \mathsf{neg_F}\ \overline{((=),(=))}\ \overline{((=),(=))} \quad (12)$$

Fig. 1. Conditions on the operations of a $\mathrm{BNF_{CC}}$

derive sufficient criteria for each direction, for concrete functors and $\mathrm{BNF_{CC}}$ constructions. For example, the function space $\mathfrak{k} \Rightarrow \mathfrak{l}$ satisfies the positive direction unconditionally, i.e., $\mathsf{pos}_{\Rightarrow}\ {}_- = \mathsf{True}$. In contrast, the negative direction does not always hold. But it does if the contravariant relations are functional, i.e., graphs of functions:

$$\dfrac{\text{left-unique } K \quad \text{right-total } K \quad \text{right-unique } K' \quad \text{left-total } K'}{\mathsf{neg}_{\Rightarrow}\ (K, K')}, \quad (13)$$

where left-unique $R = (\forall x\ z\ y.\ R\ x\ z \land R\ y\ z \longrightarrow x = y)$ and left-total $R = (\forall x.\ \exists y.\ R\ x\ y)$, and right-unique and right-total are defined analogously.

The precise relationship between BNFs and $\mathrm{BNF_{CC}}$s is as follows:

Proposition 1. *1. Every BNF $(\overline{\mathsf{l}}, \overline{\mathsf{d}})$ F is a BNF_{CC} where $\overline{\mathsf{l}}$ are live, $\overline{\mathsf{d}}$ are fixed, and $\mathsf{map_F}$, $\overline{\mathsf{set_F}}$, $\mathsf{bd_F}$, and $\mathsf{rel_F}$ are inherited from the BNF. So $\mathsf{pos_F} = \mathsf{neg_F} = \mathsf{True}$'.*

2. Every BNF_{CC} $(\overline{\mathsf{l}}, \overline{\mathsf{c}}, \overline{\mathsf{k}}, \overline{\mathsf{f}})$ F is a BNF with live parameters $\overline{\mathsf{l}}$ and dead parameters $\overline{\mathsf{c}}, \overline{\mathsf{k}}, \overline{\mathsf{f}}$ for the mapper $\mathsf{map_F^\star}$, setters $\overline{\mathsf{set_F}}$, bound $\mathsf{bd_F}$, and relator $\mathsf{rel_F^\star}$.

The $\mathrm{BNF_{CC}}$ axioms are either BNF axioms or routinely proved from them, and vice versa. The only exception is $\mathsf{rel_F^\star}$'s equational characterisation (4) for a $\mathrm{BNF_{CC}}$, which implies, e.g., that $\mathsf{neg_F} = \mathsf{True}$ [7]. To show the characterisation, we use the following property, which generalises Lemma 1 to the $\mathrm{BNF_{CC}}$ mapper and relator. It follows from the functor laws (7), parametricity (8), and equality preservation (11).

Lemma 2. *For a BNF_{CC} F, the graph of $\mathsf{map_F}\ \overline{\ell}\ \overline{c}\ \overline{k}$ is the relator applied to the graphs of $\overline{\ell}, \overline{c}$, and the converse graphs of $\overline{k}$: $\mathsf{Gr}\ (\mathsf{map_F}\ \overline{\ell}\ \overline{c}\ \overline{k}) = \mathsf{rel_F}\ \overline{(\mathsf{Gr}\ \ell)}\ \overline{(\mathsf{Gr}\ c)}\ \overline{(\mathsf{Gr}\ k)^{-1}}$.*

We now give some examples of $\mathrm{BNF_{CC}}$s. Every BNF without dead parameters is also a $\mathrm{BNF_{CC}}$ with all parameters being live by Proposition 1. This includes all sums-of-product (co)datatypes, which are also known as polynomial (co)datatypes. Many other BNFs such as distinct lists, finite and countable sets, and discrete probability distributions fall into this class, too. For these, our BNF_{CC} generalisation would not have been necessary. But there are other types where BNF_{CC}s do make a difference:

(a) We previously mentioned the function type $\ell \Rightarrow \mathfrak{l}$ with mapper $\mathsf{map}_{\Rightarrow}$ and relator $\mapsto$, where $\mathfrak{l}$ is live and ℓ is contravariant.

(b) The powerset functor $\mathfrak{c}$ set has the image operation as the mapper and the relator

$$\mathsf{rel}_{\mathsf{set}} \; C \; X \; Y = (\forall x \in X. \; \exists y \in Y. \; C \; x \; y) \wedge (\forall y \in Y. \; \exists x \in X. \; C \; x \; y).$$

The parameter c is covariant and not live only because there is no bound on the cardinality. We have $\mathsf{pos}_{\mathsf{set}} \; _- = \mathsf{neg}_{\mathsf{set}} \; _- = \mathsf{True}$.

(c) Sets $\mathfrak{c}$ bset_b with a finite cardinality bound $b \in \mathbb{N}$ are a subtype of the powerset functor $\mathfrak{c}$ set. For $b > 2$, bset_b is not a BNF in $\mathfrak{c}$ [15]. We will see in Sect. 5 that we obtain the BNF_{CC} properties by composition and subtyping. We have $\mathsf{pos}_{\mathsf{bset}_b -} = \mathsf{True}$, and right-unique $C \vee$ left-unique C' implies $\mathsf{neg}_{\mathsf{bset}_b} \; (C, C')$.

(d) Predicates ℓ pred $= \ell \Rightarrow$ bool are the contravariant powerset functor with mapper $\mathsf{map}_{\Rightarrow} \; k$ id and relator $K \mapsto (=)$. Interestingly, the negative subdistributivity condition $\mathsf{neg}_{\mathsf{pred}}$ is weaker than $\mathsf{neg}_{\Rightarrow}$ because the live parameter of $\Rightarrow$ has been instantiated to bool. We thus get that $\mathsf{neg}_{\mathsf{pred}} \; (K, K')$ is implied by left-unique $K \wedge$ right-total $K \vee$ right-unique $K' \wedge$ left-total K', i.e., only one of the two relations must be functional, not both as in (13). Clearly, $\mathsf{pos}_{\mathsf{pred}} \; _- = \mathsf{True}$.

(e) Filters $\mathfrak{c}$ filter (sets of sets closed under finite intersections and supersets) can be viewed as a semantic subtype of $\mathfrak{c}$ pred pred $= (\mathfrak{c} \Rightarrow \mathsf{bool}) \Rightarrow \mathsf{bool}$. Here, $\mathfrak{c}$ is covariant because we go twice through $\Rightarrow$'s left-hand side.

These examples propagate: whenever one of these types occurs inside a larger type, this type also benefits from BNF_{CC}'s greater generality over BNF's.

3   Simple Operations on BNF_{CC}s

We now show that BNF_{CC}s are closed under functor composition, like BNFs are. This property is crucial for a modular construction of (co)datatypes. It allows us to construct arbitrarily complex signatures from simple building blocks, because the BNF_{CC} properties follow by construction. For example, the type ℓ option $\Rightarrow (\mathfrak{c}_1 \times \mathfrak{c}_2)$ set is a composition of the type constructors $\Rightarrow$, option, set, and $\times$. For BNF_{CC}s, we distinguish three kinds of composition depending on whether the composition occurs in a live (set in $\ell \Rightarrow \boxtimes$), covariant ($\times$ in $\boxtimes$ set), or contravariant parameter (option in $\boxtimes \Rightarrow \mathfrak{l}$).

Before we turn to composition, we discuss two technical issues: *demoting* and *merging* parameters. For BNFs, demotion is known as killing, which transforms a live parameter into a dead. For BNF_{CC}s, there are three kinds of demotion ($\longrightarrow$):

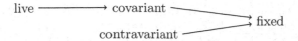

Demotion is a preparatory step for composition: If composition happens in a covariant or contravariant position, the live parameters of the inner functor are no longer live. Demotion first transforms all live parameters into covariant ones. During composition in a covariant or contravariant parameter, we can thus assume that the inner functor has no live parameters.

Merging unifies two type parameters of a $\mathrm{BNF_{CC}}$. Both type parameters must be of the same kind (live, covariant, contravariant, or fixed)—otherwise, they must be demoted first. For example, we can merge c_1 and c_2 in $(c_1 \times c_2)$ set directly to obtain the unary covariant functor $(c \times c)$ set. In contrast, before merging ℓ and $\mathfrak{l}$ in $\ell \Rightarrow \mathfrak{l}$, we must demote the live parameter $\mathfrak{l}$ and the contravariant parameter ℓ to fixed. Treating merging as a separate operation simplifies the composition theorem (Theorem 1 below) as we can assume without loss of generality that the two functors do not share any parameters.

Proposition 2. *BNF$_{CC}$s are closed under all kinds of demotion and merging.*

Demoting a live parameter adds an argument to the conditions for composition distribution, i.e., it removes the corresponding relations from the universal quantifiers in (5 and 6). So the conditions become weaker. It may therefore be useful to associate one type constructor with several $\mathrm{BNF_{CC}}$ instances that differ in the live parameters. In $\ell \Rightarrow \mathfrak{l}$, e.g., demoting $\mathfrak{l}$ to c allows us to relax the conditions on ℓ's relations by imposing some on c's. In the covariant case, negative distributivity $\mathrm{neg}_\Rightarrow (C, C')\,(K, K')$ holds if right-unique K', left-total K', left-unique C', and right-total C'. But in the live case, $\mathrm{neg}_\Rightarrow (K, K')$ does not hold for right-unique K', left-total K' in general. This difference will be crucial for quotient preservation (Sect. 6).

We now return to composition and show that the class $\mathrm{BNF_{CC}}$ of functors is closed under composition. We only discuss composition in a single parameter. This is not a restriction because composing with multiple functors simultaneously is equivalent to a sequence of single compositions, independent of the order. We also assume that the two functors do not share any parameters. A subsequent merge step can always introduce the sharing. We distinguish four different kinds of composition depending on which parameter the inner functor instantiates. For each case, we obtain different sufficient criteria for relator subdistributivity, as shown in the next theorem.

Theorem 1. *BNF$_{CC}$s are closed under composition in all kinds of parameters. Formally, let $(\overline{\mathfrak{l}_F}, \overline{c_F}, \overline{\ell_F}, \overline{f_F})$ F and $(\overline{\mathfrak{l}_G}, \overline{c_G}, \overline{\ell_G}, \overline{f_G})$ G be BNF$_{CC}$s such that no parameter is shared between F and G. We consider four kinds of composing F with G into a new functor FG, where i denotes the position of the composition in F's corresponding parameter list:[4]*

---

[4] For example, if we instantiate the third covariant parameter of F with G, then $i = 3$.

Live. $(\overline{l_F^{\leq i}}, (\overline{l_G}, \overline{c_G}, \overline{\ell_G}, \overline{f_G})\,G, \overline{l_F^{\geq i}}, \overline{c_F}, \overline{\ell_F}, \overline{f_F})\,F$ *is a* BNF_{CC} *with* $\overline{l_F^{\neq i}}, \overline{l_G}$ *live,*
$\overline{c_F}, \overline{c_G}$ *covariant,* $\overline{\ell_F}, \overline{\ell_G}$ *contravariant, and* $\overline{f_F}, \overline{f_G}$ *fixed.*
$\mathsf{pos}_F\ \overline{(C_F, C_F')}\ \overline{(K_F, K_F')}$ *and* $\mathsf{pos}_G\ \overline{(C_G, C_G')}\ \overline{(K_G, K_G')}$ *are sufficient for*
$\mathsf{pos}_{FG}\ \overline{(C_F, C_F')}\ \overline{(C_G, C_G')}\ \overline{(K_F, K_F')}\ \overline{(K_G, K_G')}$; *it is the same for* neg_{FG}.

Covariant. *If* $\overline{l_G}$ *is empty, then* $(\overline{l_F}, \overline{c_F^{\leq i}}, (\overline{c_G}, \overline{\ell_G}, \overline{f_G})\,G, \overline{c_F^{\geq i}}, \overline{\ell_F}, \overline{f_F})\,F$ *is a*
BNF_{CC} *with* $\overline{l_F}$ *live,* $\overline{c_F^{\neq i}}, \overline{c_G}$ *covariant,* $\overline{\ell_F}, \overline{\ell_G}$ *contravariant, and* $\overline{f_F}, \overline{f_G}$
fixed. $\mathsf{pos}_G\ \overline{(C_G, C_G')}\ \overline{(K_G, K_G')}$ *and* $\mathsf{pos}_F\ \overline{(C_F, C_F')}^{<i}\ (\mathsf{rel}_G\ \overline{C_G}\ \overline{K_G}, \mathsf{rel}_G\ \overline{C_G'}\ \overline{K_G'})$
$\overline{(C_F, C_F')}^{>i}\ \overline{(K_F, K_F')}$ *are sufficient for* $\mathsf{pos}_{FG}\ \overline{(C_F, C_F')}^{\neq i}\ \overline{(C_G, C_G')}\ \overline{(K_F, K_F')}$
$\overline{(K_G, K_G')}$; *it is the same for* neg_{FG}.

Contravariant. *If* $\overline{l_G}$ *is empty, then* $(\overline{l_F}, \overline{c_F}, \overline{\ell_F^{\leq i}}, (\overline{c_G}, \overline{\ell_G}, \overline{f_G})\,G, \overline{\ell_F^{\geq i}}, \overline{f_F})\,F$ *is a*
BNF_{CC} *with* $\overline{l_F}$ *live,* $\overline{c_F}, \overline{\ell_G}$ *covariant,* $\overline{\ell_F^{\neq i}}, \overline{c_G}$ *contravariant, and* $\overline{f_F}, \overline{f_G}$
fixed. $\mathsf{neg}_G\ \overline{(C_G, C_G')}\,\overline{(K_G, K_G')}$ *and* $\mathsf{pos}_F\ \overline{(C_F, C_F')}\ \overline{(K_F, K_F')}^{<i}\ (\mathsf{rel}_G\ \overline{C_G}\ \overline{K_G},$
$\mathsf{rel}_G\ \overline{C_G'}\ \overline{K_G'})\ \overline{(K_F, K_F')}^{>i}$ *are sufficient for* $\mathsf{pos}_{FG}\ \overline{(C_F, C_F')}\ \overline{(K_G, K_G')}\ \overline{(K_F,}$
$\overline{K_F')}^{\neq i}\ \overline{(C_G, C_G')}$; *it is the same for* neg_{FG}. *(Note that in the new functor,*
$\overline{C_G}, \overline{C_G'}$ *are now contravariant and* $\overline{K_G}, \overline{K_G'}$ *covariant.)*

Fixed. *If* $\overline{l_G}, \overline{c_G}, \overline{\ell_G}$ *are all empty, then* $(\overline{l_F}, \overline{c_F}, \overline{\ell_F}, \overline{f_F^{\leq i}}, \overline{f_G}\,G, \overline{f_F^{\geq i}})\,F$ *is a*
BNF_{CC} *with* $\overline{l_F}$ *live,* $\overline{c_F}$ *covariant,* $\overline{\ell_F}$ *contravariant, and* $\overline{f_F^{\neq i}}, \overline{f_G}$ *fixed.*
$\mathsf{pos}_F\ \overline{(C_F, C_F')}\ \overline{(K_F, K_F')}$ *is sufficient for* $\mathsf{pos}_{FG}\ \overline{(C_F, C_F')}\ \overline{(K_F, K_F')}$; *it is the*
same for neg_{FG}.

For example, consider the composition of $(l_1, \ell_1)\,F = \ell_1 \Rightarrow l_1$ with
$(c_1, \ell_2)\,G = \ell_2 \Rightarrow c_1$ in the contravariant parameter ℓ_1. In G, the range c_1 is normally live, but it has already been demoted such that there are no more live
parameters. We obtain the BNF_{CC} $(\ell_2 \Rightarrow c_1) \Rightarrow l_1$, where ℓ_2 is now covariant and
c_1 is contravariant, while l_1 remains live. The conditions $\mathsf{pos}_\Rightarrow\ (K_2, K_2) = \mathsf{True}$
and $\mathsf{neg}_\Rightarrow\ (K_2 \mapsto C_1, K_2' \mapsto C_1')$ are sufficient for negative subdistributivity of the
composed relator $(K_2 \mapsto C_1) \mapsto L_1$, i.e., they imply $\mathsf{neg}_{(\Rightarrow)\Rightarrow}\ (K_2, K_2')\ (C_1, C_1')$.

4 Least and Greatest Fixpoints

Bounded natural functors have been introduced mainly to construct
(co)datatypes modularly in HOL. A (co)datatype $\overline{a}$ T defined by the command

$$(\mathsf{co})\mathsf{datatype}\ \overline{a}\ \mathsf{T} = \mathsf{ctor_T}\ (\mathsf{dtor_T} : (\overline{a}\ \mathsf{T}, \overline{a})\ F)$$

corresponds to the least (greatest) solution X of the fixpoint equation
$\overline{a}\ X \cong (\overline{a}\ X, \overline{a})\ F$, up to the (co)algebra isomorphism given by the constructor
$\mathsf{ctor_T}$ and destructor $\mathsf{dtor_T}$. Whenever the (co)recursion goes through a live type
parameter of F, the fixpoint exists and it is again a BNF for the remaining live
parameters—this is the closure property under fixpoints.[5]

---

[5] For mutually recursive (co)datatypes, the solutions are taken over a system of equations instead of a single fixpoint equation. The BNF_{CC} theory generalises to systems
of equations in the same way as the BNF theory does.

In this section, we show that every (co)datatype defined over a BNF_{CC} can be extended to a BNF_{CC} in a meaningful way, namely such that the following primitive (co)datatype operations are parametric with respect to the generalised relator: the constructor $ctor_T$, the destructor $dtor_T$, and a (co)recursor, which witnesses initiality or finality of the (co)algebra. In the following, we consider a BNF_{CC} F and its least fixpoint T taken over the first live parameter. We define T's generalised mapper by primitive (co)recursion according to the fixpoint equation

$$\mathsf{map_T}\ \overline{\ell}\ \overline{c}\ \overline{k}\ (\mathsf{ctor_T}\ x) = \mathsf{ctor_T}\ (\mathsf{map_F}\ (\mathsf{map_T}\ \overline{\ell}\ \overline{c}\ \overline{k})\ \overline{\ell}\ \overline{c}\ \overline{k}\ x),$$

and T's generalised relator (co)inductively as the least or greatest predicate closed under

$$\frac{\mathsf{rel_F}\ (\mathsf{rel_T}\ \overline{L}\ \overline{C}\ \overline{K})\ \overline{L}\ \overline{C}\ \overline{K}\ x\ y}{\mathsf{rel_T}\ \overline{L}\ \overline{C}\ \overline{K}\ (\mathsf{ctor_T}\ x)\ (\mathsf{ctor_T}\ y)}.$$

Note that $\mathsf{rel_T}$ is well-defined since $\mathsf{rel_F}$ is monotone in the live arguments. This choice of the relator (and therefore of the mapper, due to Lemma 2) is intuitively correct as we obtain a general form of parametricity to the extent permitted by $\mathsf{rel_F}$:

Proposition 3. *The constructor, destructor, and (co)recursor for T are parametric with respect to $\mathsf{rel_T}$.*

The canonical BNF map function for T, which acts only on T's live parameters, is equal to $\mathsf{map_T^*}$ by definition. Similarly, the restricted relator $\mathsf{rel_T^*}$ satisfies the BNF characterisation (4). The setters $\overline{\mathsf{set_T}}$ satisfy only the restricted parametricity law $(\mathsf{rel_T^*}\ \overline{L} \mapsto \mathsf{rel_{set}}\ L_i)\ \mathsf{set_T^i}\ \mathsf{set_T^i}$. As they ignore the covariant and contravariant parameters, the general parametricity law $(\mathsf{rel_T}\ \overline{L}\ \overline{C}\ \overline{R} \mapsto \mathsf{rel_{set}}\ L_i)\ \mathsf{set_T^i}\ \mathsf{set_T^i}$ does not make sense and does not hold in general either. For example, the setter for the function space $\ell \Rightarrow \mathfrak{l}$ takes the range of the function. Choosing $K = \bot$, where $\bot$ is the empty relation, and $L = (=)$, then $(K \mapsto L)\ (\lambda\_.\ \mathsf{True})\ (\lambda\_.\ \mathsf{False})$, but clearly not $\mathsf{rel_{set}}\ (=)\ (\mathsf{range}\ (\lambda\_.\ \mathsf{True}))\ (\mathsf{range}\ (\lambda\_.\ \mathsf{False}))$.

Theorem 2. *BNF_{CC}s are closed under least and greatest fixpoints through live parameters. In particular, if T is the least or greatest fixpoint through one of F's live parameters, then $\mathsf{pos_F}\ \overline{(C,C')}\ \overline{(K,K')}$ implies $\mathsf{pos_T}\ \overline{(C,C')}\ \overline{(K,K')}$, and the same for $\mathsf{neg_F}$ and $\mathsf{neg_T}$.*

5 Subtypes

In HOL, a new type $\overline{a}\ T$ is defined by carving out a non-empty subset S of an already existing type $\overline{a}\ F$. Such a type definition creates an embedding isomorphism $\mathsf{Rep_T}\ ::\ \overline{a}\ T \Rightarrow \overline{a}\ F$ between $\overline{a}\ T$ and S with inverse $\mathsf{Abs_T}\ ::\ \overline{a}\ F \Rightarrow \overline{a}\ T$, where $\mathsf{Abs_T}$ is unspecified outside of S. If F is a BNF, then the new type T can inherit F's BNF structure provided that S is "well-behaved." Biendarra [6] identified the following two conditions on S, from which his Isabelle/HOL command `lift-bnf` derives the BNF properties.

- *Closed under the BNF mapper:* whenever $x \in S$, then $\mathsf{map}_F^\star \, \bar{\ell} \, x \in S$; and
- *Reflects projections:* if $\mathsf{map}_F^\star \, \overline{\pi_1} \, z \in S$ and $\mathsf{map}_F^\star \, \overline{\pi_2} \, z \in S$, then $z \in S$.

Meanwhile, Popescu [32] weakened the second condition as follows: whenever $\mathsf{map}_F^\star \, \overline{\pi_1} \, z \in S$ and $\mathsf{map}_F^\star \, \overline{\pi_2} \, z \in S$, then there exists $y \in S$ such that $\mathsf{set}_F^i \, y \subseteq \mathsf{set}_F^i \, z$ for all i, $\mathsf{map}_F^\star \, \overline{\pi_1} \, y = \mathsf{map}_F^\star \, \overline{\pi_1} \, z$, and $\mathsf{map}_F^\star \, \overline{\pi_2} \, y = \mathsf{map}_F^\star \, \overline{\pi_2} \, z$.

In this section, we generalise Biendarra's and Popescu's conditions to BNF_{CCS}:

Theorem 3 (BNF_{CC} inheritance for subtypes). *Let $(\bar{\mathsf{l}}, \bar{\mathsf{c}}, \bar{\mathsf{t}}, \bar{\mathsf{f}})$ F be a BNF_{CC} and let $(\bar{\mathsf{l}}, \bar{\mathsf{c}}, \bar{\mathsf{t}}, \bar{\mathsf{f}})$ T be isomorphic to the non-empty set $S :: (\bar{\mathsf{l}}, \bar{\mathsf{c}}, \bar{\mathsf{t}}, \bar{\mathsf{f}})$ F set via the morphisms Rep_T and Abs_T. The type T inherits the BNF_{CC} structure from F via*

$$\mathsf{map}_T \, \bar{\ell} \, \bar{c} \, \bar{k} = \mathsf{Abs}_T \circ \mathsf{map}_F \, \bar{\ell} \, \bar{c} \, \bar{k} \circ \mathsf{Rep}_T \qquad \mathsf{set}_T^i = \mathsf{set}_F^i \circ \mathsf{Rep}_T \qquad \mathsf{bd}_T = \mathsf{bd}_F$$
$$\mathsf{rel}_T \, \overline{L} \, \overline{C} \, \overline{K} \, x \, y = \mathsf{rel}_F \, \overline{L} \, \overline{C} \, \overline{K} \, (\mathsf{Rep}_T \, x) \, (\mathsf{Rep}_T \, y)$$

if $\mathsf{rel}_T \, \overline{(L \bigcirc L')} \, \overline{(=)} \, \overline{(=)} \sqsubseteq \mathsf{rel}_T \, \overline{L} \, \overline{(=)} \, \overline{(=)} \bigcirc \mathsf{rel}_T \, \overline{L'} \, \overline{(=)} \, \overline{(=)}$ for all $\overline{L}, \overline{L'}$, and $x \in S$ implies $\mathsf{map}_F \, \bar{\ell} \, \bar{c} \, \bar{k} \, x \in S$. Moreover, $\mathsf{pos}_F \, \overline{(C, C')} \, \overline{(K, K')}$ implies $\mathsf{pos}_T \, \overline{(C, C')} \, \overline{(K, K')}$.

Negative subdistributivity can often be reduced to proving closedness under zippings, which generalises reflection of projections in the BNF case. We allow a condition neg_T' that is stronger than neg_F, assuming that $\mathsf{neg}_T' \, \overline{((=), (=))} \, \overline{((=), (=))}$ still holds. The set S is *closed under zippings for* neg_T' iff

$$\cfrac{x \in S \qquad y \in S \qquad \mathsf{rel}_F \, \overline{L} \, \overline{(C \bigcirc C')} \, \overline{(K \bigcirc K')} \, x \, y}{\mathsf{rel}_F \, \overline{(\lambda a \, (a', b). \, a' = a \wedge L \, a \, b)} \, \overline{C} \, \overline{K} \, x \, z \qquad \mathsf{rel}_F \, \overline{(\lambda(a, b'). \, b. \, b' = b \wedge L \, a \, b)} \, \overline{C'} \, \overline{K'} \, z \, y}{z \in S}$$

for all x, y, z and all $\overline{L}, \overline{C}, \overline{C'}, \overline{K}, \overline{K'}$ such that $\mathsf{neg}_T' \, \overline{(C, C')} \, \overline{(K, K')}$.

Lemma 3. *Let S be closed under zippings for neg_T'. Then $\mathsf{neg}_T' \, \overline{(C, C')} \, \overline{(K, K')}$ implies $\mathsf{neg}_T \, \overline{(C, C')} \, \overline{(K, K')}$.*

Corollary 1. *BNF_{CCS} are closed under subtypes that are closed under the BNF_{CC} mapper and zippings (for some condition on negative subdistributivity).*

Non-uniform (co)datatypes are therefore also BNF_{CCS}, as they are defined as subtypes of ordinary (co)datatypes [8], and the subtype predicate is invariant under the mapper.

The assumptions on S in Theorem 3 and Corollary 1 are indeed generalisations of Popescu's and Biendarra's conditions, respectively. For when there are neither covariant nor contravariant parameters, the assumptions on S in Theorem 3 are equivalent to Popescu's conditions, given the BNF relator characterisation (4). Similarly, closure under zippings is equivalent to Biendarra's reflecting projections in that case.

Note that closure under zippings strictly implies negative subdistributivity. For example, sets of cardinality at most two are a BNF and a subtype of the finite powerset BNF fset. Yet, the cardinality restriction to at most two does not reflect projections (take $z = \{a, b\} \times \{0, 1\}$). Our Theorem 3 handles this case, but Lemma 3 cannot be used as closedness under zippings is not provable. The current implementation of lift-bnf cannot handle this case either.

Since BNF$_{CC}$s do not require the relator distributing unconditionally over relation composition, there can be several relators that extend the mapper in the sense of Lemma 2. For example, filters (Sect. 2) are a subtype of the BNF$_{CC}$ obtained by composing the contravariant powerset functor with itself. This view yields the mapper map$_{filter}$ c $F = \{X \mid c^{-1}(X) \in F\}$ from the literature. (We omit the conversions between sets and predicates for clarity). Yet, there are two relator candidates for filter: First, the construction in Theorem 3 gives rel$^1_{filter}$ R F $G =$ rel$_{pred}$ (rel$_{pred}$ R) F G. Second, the canonical categorical extension of a functor on SET to REL [12,34] gives

$$\text{rel}^2_{filter}\ R\ F\ G = (\exists Z.\ R \in Z \wedge F = \{U \mid \pi_1^{-1}(U) \cap R \in Z\} \wedge G = \{V \mid \pi_1^{-1}(V) \cap R \in Z\})$$

where $f^{-1}(V)$ denotes the preimage of V under f. The latter relator is strictly stronger than the former. For example, the drawing on the right shows a filter $F = \{\{a_1, a_2, a_3\}\}$ on a three-element type, a filter $G = \{\{b_1\}, \{b_1, b_2\}\}$ on a two-element type, and a relation R between the elements. We have

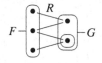

rel$^1_{filter}$ R F G, but not rel$^2_{filter}$ R F G. In this case, rel$^2_{filter}$ is the right choice as it gives pos$_{filter\ \_}$ = neg$_{filter\ \_}$ = True [12]. But sometimes the relator definition from Theorem 3 is better. Probability distributions with a finite cardinality bound on the support, e.g., preserve quotients only with the relator from Theorem 3 (Sect. 6).

6 Quotient Preservation

We now consider quotient relationships between types and how BNF$_{CC}$s preserve such relationships. This allows a modular construction of quotients by composing BNF$_{CC}$s.

A type $\mathfrak{a}$ is a *quotient* of another type $\mathfrak{r}$ under a partial equivalence relation R on $\mathfrak{r}$ iff $\mathfrak{a}$ is isomorphic to $\mathfrak{r}$'s equivalence classes. A quotient $\mathfrak{a}$ can thus be viewed as an abstraction of $\mathfrak{r}$, and, conversely, $\mathfrak{r}$ as a refinement of $\mathfrak{a}$. (This definition subsumes both subtypes and total quotients. One must consider partial equivalence relations in a higher-order setting for reasons similar to why parametricity uses relations instead of functions [16].) A type constructor $\overline{\mathfrak{b}}$ F *preserves quotients* in the type parameters $\overline{\mathfrak{b}^{\in I}} = \mathfrak{b}_{i_1}, \ldots, \mathfrak{b}_{i_m}$ iff $\overline{\mathfrak{a}}$ F is a quotient of $\overline{\mathfrak{r}}$ F whenever $\overline{\mathfrak{a}^{\in I}}$ are quotients of $\overline{\mathfrak{r}^{\in I}}$ and $\overline{\mathfrak{a}^{\notin I}} = \overline{\mathfrak{r}^{\notin I}}$ (for some construction of the equivalence relation; we provide the details below). For lists, e.g., a quotient between element types $\mathfrak{a}$ and $\mathfrak{r}$ yields a quotient between lists of such elements, $\mathfrak{a}$ list and $\mathfrak{r}$ list. Note that quotient preservation is different from the construction of a quotient type or subtype from a BNF_{CC}. The former, which we discuss in this section, deals with type instantiation, while the latter produces a truly new type.

In HOL, a quotient between types is described by a relation $Q :: \tau \otimes \alpha$ that is right-total and right-unique. Such a relation induces (i) an embedding morphism $rep :: \alpha \Rightarrow \tau$, (ii) an abstraction morphism $abs :: \tau \Rightarrow \alpha$, and (iii) the underlying partial equivalence relation $R = Q \circ Q^{-1}$.

The embedding rep picks an unspecified element in the equivalence class, which may require the axiom of choice, and $abs\ r$ is unspecified if no equivalence class contains r. Due to this underspecification, it is useful to keep track of rep and abs as primitive operations, e.g., for code generation. Similarly, Isabelle's Lifting package [17] maintains the explicit characterisation of the equivalence relation R to simplify the respectfulness proof obligations presented to the user. The predicate Quot formalises these relationships:

$$\text{Quot } R \text{ } abs \text{ } rep \text{ } Q \longleftrightarrow (Q \leq \text{Gr } abs \wedge \text{Gr } rep \leq Q^{-1} \wedge R = Q \circ Q^{-1}). \quad (14)$$

Quotient preservation can thus be expressed as an implication. For lists, e.g., we have that Quot R abs rep Q implies Quot $(\text{rel}_{\text{list}} R)$ $(\text{map}_{\text{list}} abs)$ $(\text{map}_{\text{list}} rep)$ $(\text{rel}_{\text{list}} Q)$. Note how the relator and mapper lift the relations and morphisms from elements to lists. BNFs preserve quotients in all live parameters; this is an easy consequence of relator monotonicity and distributivity.

Theorem 4 ([20, Sect. 4.7]). *BNFs preserve quotients in live parameters, in the following sense:* Quot $(\text{rel}_F \overline{R})$ $(\text{map}_F \overline{abs})$ $(\text{map}_F \overline{rep})$ $(\text{rel}_F \overline{Q})$ *holds whenever* $(\overline{l}, \overline{o})$ F *is a BNF and* $\forall i.$ Quot R_i abs_i rep_i Q_i.

This theorem does not fully generalise to BNF_{CC}s with covariant and contravariant parameters, as the counterexample in Appendix B (Online Resource) shows. We obtain the following result, however, which shows that positive subdistributivity of the relator over the quotient relations and their converses is a sufficient condition for quotient preservation.

Theorem 5. *Let* $(\overline{l}, \overline{c}, \overline{e}, \overline{f})$ F *be a* BNF_{CC}. *Assume that* $\forall i.$ Quot R_χ^i abs_χ^i rep_χ^i T_χ^i *for all* $\chi \in \{L, C, K\}$. *If* pos_F $\overline{(Q_C, Q_C^{-1})}$ $\overline{(Q_K, Q_K^{-1})}$, *then*

$$\text{Quot } (\text{rel}_F \overline{R_L} \text{ } \overline{R_C} \text{ } \overline{R_K}) (\text{map}_F \overline{abs_L} \text{ } \overline{abs_C} \text{ } \overline{rep_K}) (\text{map}_F \overline{rep_L} \text{ } \overline{rep_C} \text{ } \overline{abs_K}) (\text{rel}_F \overline{Q_L} \text{ } \overline{Q_C} \text{ } \overline{Q_K}).$$

We now illustrate how this theorem applies to different BNF_{CC}s. Note that it applies to all the BNF_{CC}s mentioned at the end of Sect. 2, as their relators all positively distribute over all relation compositions (if we use the right relator for filters as dicussed in Sect. 5). For a BNF_{CC} F constructed from these primitives, pos_F _ = True need not hold, though, as BNF_{CC} composition in negative positions swaps the positive and negative conditions. Nevertheless, we can derive pos_F (Q, Q^{-1}) for quotient relations Q by using our composition theorems, as the following two examples illustrate. First, predicates over predicates c $pp = (c \Rightarrow \text{bool}) \Rightarrow \text{bool}$ do preserve quotients. By the contravariant case of Theorem 1, $\mathsf{pos}_{\mathsf{pp}}\,(Q, Q^{-1})$ follows from $\mathsf{pos}_{\mathsf{pred}}\,(Q \mapsto (=), Q^{-1} \mapsto (=))$ and $\mathsf{neg}_{\mathsf{pred}}\,(Q, Q^{-1})$. The former is trivial as $\mathsf{pos}_{\mathsf{pred}}\,\_ = \mathsf{True}$. For the latter, observe that predicates $\mathfrak{k}$ pred are obtained from the function space $\mathfrak{k} \Rightarrow \mathfrak{c}$ by instantiating $\mathfrak{c}$ with the nullary $\mathsf{BNF}_{\mathsf{CC}}$ bool. So, by Theorem 1 (the covariant case), $\mathsf{neg}_{\mathsf{pred}}\,(Q, Q^{-1})$ follows from $\mathsf{neg}_{\mathsf{bool}} = \mathsf{True}$ and $\mathsf{neg}_{\Rightarrow}\,((=), (=))\,(Q, Q^{-1})$, which is easily proved using Q being a quotient relation. In this reasoning, it is essential that we do not use the function space $\mathsf{BNF}_{\mathsf{CC}}$ with the live codomain. Instead, we first demote the codomain to a covariant parameter (fixed would also do). For in the live case, Theorem 1 gives us only the implication from $\mathsf{neg}_{\Rightarrow}\,(Q, Q^{-1})$ (without the live parameter relations as arguments) to $\mathsf{neg}_{\mathsf{pred}}\,(Q, Q^{-1})$, but $\mathsf{neg}_{\Rightarrow}\,(Q, Q^{-1})$ does not hold as it quantifies over all live parameter relations. This illustrates the weakening by demotion that we discussed below Proposition 2.

The second example shows that it is important to associate several $\mathsf{BNF}_{\mathsf{CC}}$s with one type constructor, even in a single type expression. The codatatype

$$\texttt{codatatype}\ (\mathfrak{c}, \mathfrak{k})\ \mathsf{T} = \mathsf{ctor}_{\mathsf{T}}\ ((\mathfrak{c} \Rightarrow \mathfrak{k}) \Rightarrow (\mathfrak{c}, \mathfrak{k})\ \mathsf{T})$$

is the final coalgebra of the functor $(\mathfrak{l}, \mathfrak{c}, \mathfrak{k})\ \mathsf{F} = (\mathfrak{c} \Rightarrow \mathfrak{k}) \Rightarrow \mathfrak{l}$ and it preserves quotients. To derive $\mathsf{pos}_{\mathsf{T}}\,(C, C')\,(K, K')$ modularly from the construction, we must treat F's outer function space with live codomain (as the corecursion goes through this parameter) and F's inner function space with covariant codomain (for the same reason as in the pp case).

7 Related Work

We have already discussed the related work on bounded natural functors [6–8,20,38] in the previous sections. Here, we discuss how $\mathsf{BNF}_{\mathsf{CC}}$s fit into the Isabelle ecosystem, and compare our approach to previous work for other theorem provers.

The Transfer package by Huffman and Kunčar [17] implements Mitchell's representation independence [27] using a database of parametricity theorems and (conditional) respectfulness theorems for equality and quantifiers. $\mathsf{BNF}_{\mathsf{CC}}$ relators can be directly used in the parametricity rules, making them more versatile than BNF relators thanks to the generalisation to covariant and contravariant arguments. The respectfulness theorems follow from monotonicity and positive or negative relator distributivity, whose preconditions our composition theorems carefully track. Moreover, Gilcher's automatic derivation of parametricity theorems [11] also benefits from the generalised relators.

The Lifting package [17] lifts constants over quotients and derives appropriate transfer rules using databases of quotient preservation theorems and relator monotonicity and distributivity. Like for Transfer, our theorems can be fed directly into these databases, making the Lifting package more useful.

Lammich's Autoref tool [21,22] performs data refinement based on parametricity. Currently, Lammich must manually derive relators for (co)datatypes.

BNF$_{CC}$s offer a systematic way to define relators and to derive their fundamental properties.

Apart from HOL, parametricity has recently received a lot of attention in dependent type theories as implemented in Coq, Agda, and Lean. In these rich logics, it is possible to internalise Reynolds' relational interpretation of types [5]. So, the parametricity theorem is just a syntactic translation of a type and its proof can be systematically programmed. Various such translations have been studied for different subsets of the logics [2,3,19]; Anand and Morrisett provide a good overview [2]. These works prove (by induction over the syntax of the logic) that all functions defineable in the logic are parametric and then implement this proof as a tool such as ParamCoq [19] and ParamCoq-iff [2]. As HOL lacks the syntactic nature of type theories and its classical axioms forbid a general parametricity result, we follow a semantic approach using BNF$_{CC}$s instead. This has the advantage that our approach is modular: only semantic properties matter, but not the particular way that something was defined in.

Moreover, most of the syntax-directed type-theoretic works hardly study how the relational interpretation can be used. At best, free theorems are derived (e.g., Anand and Morrisett derive respectfulness of α-equivalence of λ-terms from an operational semantics being parametric). Parametricity is also the foundation for two data refinement frameworks in Coq, Fiat [10] and CoqEAL [9], similar to Autoref [22] in Isabelle/HOL. They define the relators manually in an ad hoc way and it is unclear whether the syntax-directed works could be used instead. In contrast, BNF$_{CC}$s provide a framework to systematically define mappers and relators and to derive their rich properties. They thus directly lead to a wealth of applications, including free theorems, data refinement, and type abstraction through quotients.

8   Conclusion and Future Work

BNF$_{CC}$s generalise the concept of bounded natural functors, which are motivated by the construction of (co)datatypes in HOL. They equip both covariant and contravariant type parameters with a functorial structure, even when they do not meet the requirements of bounded naturality. Hence, the mapper and relator of a BNF$_{CC}$ act on these type parameters, too. We have shown that BNF$_{CC}$s are closed under the most important type construction mechanisms in HOL: composition, datatypes, codatatypes, and subtypes. This way, we obtain canonical definitions of the mapper and the relator for these constructions, together with proofs of some useful properties. For (co)datatypes, it is crucial that we stay compatible with the BNF restrictions, which motivates our unified view on the functorial structure of types. Applications of parametricity, such as data refinement, quotients, and generalised rewriting, benefit from the extended operations.

We have not yet automated the BNF$_{CC}$ construction in Isabelle/HOL, but we have formalised the constructions and proofs in an abstract setting. Moreover, we applied the BNF$_{CC}$ theory manually in a few applications. In the CryptHOL framework [4,24], e.g., the first author manually defined the generalised mapper and relator for the codatatype

codatatype $(\mathfrak{a}, \mathfrak{b}, \mathfrak{c})$ gpv = GPV $((\mathfrak{a} + (\mathfrak{b} \times (\mathfrak{c} \Rightarrow (\mathfrak{a}, \mathfrak{b}, \mathfrak{c})$ gpv$)))$ option pmf)

which models sub-probabilistic discrete systems, and proved properties like relator monotonicity and distributivity. Following the BNF_{CC} theory, we have refactored the definitions and proofs. By exploiting the modularity, they became cleaner, simpler, and shorter.

BNF_{CC}s are functors on the category of sets, but for covariant and contravariant parameters, they need not be functors on the category of relations, as the relator need not distribute unconditionally over relation composition [17]. This is a necessary consequence of dealing with the full function space. Therefore, the relator is not uniquely determined by the mapper, either, and one must choose the relator that fits one's needs best.

There are now four groups of type parameters: live, covariant, contravariant, and fixed. Are they enough or do we need further refinements? In the category of sets, this is as far as we can possibly get while retaining the functorial structure. But in some cases, we would like to go beyond. For example, the state $\mathfrak{s}$ in a state monad $(\mathfrak{s}, \mathfrak{a})$ stateM = $\mathfrak{s} \Rightarrow \mathfrak{a} \times \mathfrak{s}$ occurs in a positive and a negative position, so demotion makes $\mathfrak{s}$ fixed. The BNF_{CC} mapper and relator therefore ignore it. One could generalise the mapper to $\mathfrak{s}$ if we restrict the morphisms to bijections, i.e., change the underlying category to bijections. Similarly, if a type parameter has a type class constraint, only type class homomorphisms can be mapped in general. Extending BNF_{CC}s into this direction is left as future work.

Moreover, we have not studied whether quotient types [17,18] can be equipped with a BNF_{CC} structure in general. We are still working on identifying the conditions under which a quotient inherits the BNF structure from the raw type. For the extension to BNF_{CC}s, we conjecture that we must first generalise the setter concept from live to covariant and contravariant parameters, as unsound (set) functors seem to require repair even in the BNF case [1]. Furthermore, we are interested in lifting a family of quotient relations between two BNF_{CC}s to a quotient relation between their fixpoints. This is necessary for refining a whole collection of types that is closed under (co)datatype formation, as needed, e.g., in [35].

Acknowledgements. The authors thank Dmitriy Traytel, Andrei Popescu, and the anonymous reviewers for inspiring discussions and suggestions how to improve the presentation. The authors are listed alphabetically.

References

1. Adámek, J., Gumm, H.P., Trnková, V.: Presentation of set functors: a coalgebraic perspective. J. Log. Comput. **20**, 991–1015 (2010)
2. Anand, A., Morrisett, G.: Revisiting parametricity: inductives and uniformity of propositions. CoRR abs/1705.01163 (2017). http://arxiv.org/abs/1705.01163
3. Atkey, R., Ghani, N., Johann, P.: A relationally parametric model of dependent type theory. In: POPL 2014, pp. 503–515. ACM (2014)

4. Basin, D., Lochbihler, A., Sefidgar, S.R.: CryptHOL: game-based proofs in higher-order logic. Cryptology ePrint Archive: Report 2017/753 (2017). https://eprint.iacr.org/2017/753

5. Bernardy, J.P., Jansson, P., Paterson, R.: Proofs for free: parametricity for dependent types. J. Funct. Program. **22**(2), 107–152 (2012)

6. Biendarra, J.: Functor-preserving type definitions in Isabelle/HOL. Bachelor thesis, Fakultät für Informatik, Technische Universität München (2015)

7. Blanchette, J.C., et al.: Truly modular (Co)datatypes for Isabelle/HOL. In: Klein, G., Gamboa, R. (eds.) ITP 2014. LNCS, vol. 8558, pp. 93–110. Springer, Cham (2014). https://doi.org/10.1007/978-3-319-08970-6_7

8. Blanchette, J.C., Meier, F., Popescu, A., Traytel, D.: Foundational nonuniform (co)datatypes for higher-order logic. In: LICS 2017, pp. 1–12. IEEE (2017)

9. Cohen, C., Dénès, M., Mörtberg, A.: Refinements for free!. In: Gonthier, G., Norrish, M. (eds.) CPP 2013. LNCS, vol. 8307, pp. 147–162. Springer, Cham (2013). https://doi.org/10.1007/978-3-319-03545-1_10

10. Delaware, B., Pit-Claudel, C., Gross, J., Chlipala, A.: Fiat: deductive synthesis of abstract data types in a proof assistant. In: POPL 2015, pp. 689–700. ACM (2015)

11. Gilcher, J., Lochbihler, A., Traytel, D.: Conditional parametricity in Isabelle/HOL (extended abstract). Poster at TABLEAU/FroCoS/ITP 2017 (2017). http://www.andreas-lochbihler.de/pub/gilcher2017ITP.pdf

12. Gumm, H.P.: Functors for coalgebras. Algebra Univ. **45**, 135–147 (2001)

13. Gunter, E.L.: Why we can't have SML-style datatype declarations in HOL. In: TPHOLs 1992. IFIP Transactions, vol. A-20, pp. 561–568. Elsevier, North-Holland (1992)

14. Haftmann, F., Krauss, A., Kunčar, O., Nipkow, T.: Data refinement in Isabelle/HOL. In: Blazy, S., Paulin-Mohring, C., Pichardie, D. (eds.) ITP 2013. LNCS, vol. 7998, pp. 100–115. Springer, Heidelberg (2013). https://doi.org/10.1007/978-3-642-39634-2_10

15. Hölzl, J., Lochbihler, A., Traytel, D.: A formalized hierarchy of probabilistic system types. In: Urban, C., Zhang, X. (eds.) ITP 2015. LNCS, vol. 9236, pp. 203–220. Springer, Cham (2015). https://doi.org/10.1007/978-3-319-22102-1_13

16. Homeier, P.V.: A design structure for higher order quotients. In: Hurd, J., Melham, T. (eds.) TPHOLs 2005. LNCS, vol. 3603, pp. 130–146. Springer, Heidelberg (2005). https://doi.org/10.1007/11541868_9

17. Huffman, B., Kunčar, O.: Lifting and Transfer: a modular design for quotients in Isabelle/HOL. In: Gonthier, G., Norrish, M. (eds.) CPP 2013. LNCS, vol. 8307, pp. 131–146. Springer, Cham (2013). https://doi.org/10.1007/978-3-319-03545-1_9

18. Kaliszyk, C., Urban, C.: Quotients revisited for Isabelle/HOL. In: SAC 2011, pp. 1639–1644. ACM (2011)

19. Keller, C., Lasson, M.: Parametricity in an impredicative sort. CoRR abs/1209.6336 (2012). http://arxiv.org/abs/1209.6336

20. Kunčar, O.: Types, abstraction and parametric polymorphism in higher-order logic. Ph.D. thesis, Fakultät für Informatik, Technische Universität München (2016)

21. Lammich, P.: Automatic data refinement. In: Blazy, S., Paulin-Mohring, C., Pichardie, D. (eds.) ITP 2013. LNCS, vol. 7998, pp. 84–99. Springer, Heidelberg (2013). https://doi.org/10.1007/978-3-642-39634-2_9

22. Lammich, P., Lochbihler, A.: Automatic refinement to efficient data structures: a comparison of two approaches. J. Autom. Reasoning (2018). https://doi.org/10.1007/s10817-018-9461-9

23. Leino, K.R.M.: Dafny: an automatic program verifier for functional correctness. In: Clarke, E.M., Voronkov, A. (eds.) LPAR 2010. LNCS (LNAI), vol. 6355, pp. 348–370. Springer, Heidelberg (2010). https://doi.org/10.1007/978-3-642-17511-4_20
24. Lochbihler, A.: CryptHOL. Archive of Formal Proofs (2017). http://isa-afp.org/entries/CryptHOL.html, Formal proof development
25. Lochbihler, A., Schneider, J.: Bounded natural functors with covariance and contravariance. Archive of Formal Proofs (2018). http://isa-afp.org/entries/BNF_CC.html, Formal proof development
26. Maurer, U.: Indistinguishability of random systems. In: Knudsen, L.R. (ed.) EUROCRYPT 2002. LNCS, vol. 2332, pp. 110–132. Springer, Heidelberg (2002). https://doi.org/10.1007/3-540-46035-7_8
27. Mitchell, J.C.: Representation independence and data abstraction. In: POPL 1986, pp. 263–276. ACM (1986)
28. de Moura, L., Kong, S., Avigad, J., van Doorn, F., von Raumer, J.: The Lean theorem prover (System Description). In: Felty, A.P., Middeldorp, A. (eds.) CADE 2015. LNCS (LNAI), vol. 9195, pp. 378–388. Springer, Cham (2015). https://doi.org/10.1007/978-3-319-21401-6_26
29. Norell, U.: Towards a practical programming language based on dependent type theory. Ph.D. thesis, Department of Computer Science and Engineering, Chalmers University of Technology (2007)
30. Owre, S., Shankar, N.: Abstract datatypes in PVS. Technical Report CSL-93-9R, Computer Science Laboratory, SRI International (1993)
31. Paulin-Mohring, C.: Inductive definitions in the system Coq – rules and properties. In: Bezem, M., Groote, J.F. (eds.) TLCA 1993. LNCS, vol. 664, pp. 328–345. Springer, Heidelberg (1993). https://doi.org/10.1007/BFb0037116
32. Popescu, A.: Personal communication (2017)
33. Reynolds, J.C.: Types, abstraction and parametric polymorphism. In: IFIP 1983. Information Processing, vol. 83, pp. 513–523. IFIP, North-Holland (1983)
34. Rutten, J.J.M.M.: Universal coalgebra: a theory of systems. Theor. Comput. Sci. **249**(1), 3–80 (2000)
35. Schneider, J.: Formalising the run-time costs of HOL programs. Master's thesis, Department of Computer Science, ETH Zurich (2017)
36. Slind, K., Norrish, M.: A brief overview of HOL4. In: Mohamed, O.A., Muñoz, C., Tahar, S. (eds.) TPHOLs 2008. LNCS, vol. 5170, pp. 28–32. Springer, Heidelberg (2008). https://doi.org/10.1007/978-3-540-71067-7_6
37. Sozeau, M.: A new look at generalized rewriting in type theory. J. Formalized Reasoning **2**(1), 41–62 (2009)
38. Traytel, D., Popescu, A., Blanchette, J.C.: Foundational, compositional (co)datatypes for higher-order logic. In: LICS 2012, pp. 596–605. IEEE (2012)
39. Wadler, P.: Theorems for free! In: FPCA 1989, pp. 347–359. ACM (1989)

Towards Verified Handwritten Calculational Proofs
(Short Paper)

Alexandra Mendes[1,2](✉) and João F. Ferreira[1,3]

[1] School of Computing, Teesside University, Middlesbrough, UK
alexandra@archimendes.com
[2] HASLab/INESC TEC, Universidade do Minho, Braga, Portugal
[3] INESC-ID/IST, University of Lisbon, Lisbon, Portugal

Abstract. Despite great advances in computer-assisted proof systems, writing formal proofs using a traditional computer is still challenging due to mouse-and-keyboard interaction. This leads to scientists often resorting to pen and paper to write their proofs. However, when handwriting a proof, there is no formal guarantee that the proof is correct. In this paper we address this issue and present the initial steps towards a system that allows users to handwrite proofs using a pen-based device and that communicates with an external theorem prover to support the users throughout the proof writing process. We focus on calculational proofs, whereby a theorem is proved by a chain of formulae, each transformed in some way into the next. We present the implementation of a proof-of-concept prototype that can formally verify handwritten calculational proofs without the need to learn the specific syntax of theorem provers.

Keywords: Handwritten mathematics · Interactive theorem proving
Mathematical proof · Calculational method · Handwriting

1 Introduction

Mathematical proof is at the core of many scientific disciplines, but the development of correct mathematical proofs is still a challenging activity. In recent years, there have been great advances in computer-assisted proof systems that support the development of formally verified proofs (e.g. Isabelle/HOL [25] and Coq [8]). However, writing proofs using a traditional computer poses difficulties due to mouse-and-keyboard interaction. That is why scientists often resort to pen and paper to support them in their thinking process and to record their proofs. The problem is that when handwriting a proof, there is no formal guarantee that the proof is correct.

To formally verify a handwritten proof, one has to translate it into a theorem prover's language. This process takes considerable time and effort and requires a good knowledge of the theorem prover's syntax. This makes the writing of verified proofs feel unnatural and difficult, further encouraging scientists to use the pen and paper approach.

To the best of our knowledge, the problem of formally verifying handwritten proofs is still open. In this paper, we present a proof-of-concept research prototype that attempts to bridge the gap between the natural mathematical practice of handwriting proofs and their mechanical verification. This is the first step towards a system for pen-based devices that allows users to handwrite proofs and that communicates with an external theorem prover to support the users throughout the proof writing process. We focus on calculational proofs [3], whereby a theorem is proved by a chain of formulae, each transformed in some way into the next. We used the method of rapid prototyping [29] to demonstrate the feasibility of the system. Since innovations communicated verbally can be difficult to imagine, a prototype can give the prospective users a better sense of what can be achieved as well as giving us proof that our goal is attainable.

In the next sections, we present some background and related work (Sect. 2), summarise requirements taken from existing literature (Sect. 3), present a proof-of-concept prototype (Sect. 4), and conclude by discussing the next steps (Sect. 5).

2   Background and Related Work

This work is being developed in the context of teaching and research on correct-by-construction program design. Starting with the pioneering work of Dijkstra and Gries [9,14], a calculational method emerged, emphasising the use of systematic mathematical calculation in the design of algorithms. Proofs written in the calculational format consist of a chain of formulae, each transformed in some way into the next, with each step optionally accompanied by a hint justifying the validity of that step (see Fig. 1(c) for an example).

Calculational proofs are known for their readability and for helping to avoid mistakes, but errors can still occur. Indeed, the need for mechanical verification of calculational proofs has been widely recognised. For example, in [22], the authors point out errors in some of Dijkstra's calculations and send a clear message to the calculational community: *"If your proofs are so rigorous and so amenable to mechanization, stop just saying so and do it"*. However, they question how hard it would be to learn and use a proof checker and whether transforming proofs for mechanical checking would make them ugly and hard to understand.

Some work has been done towards mechanised calculational proofs. For example, Leino and Polikarpova [20] extended Dafny [19] to support proof calculations. The authors state that *"It would be wonderful if we could just take a pen-and-paper calculational proof and get it machine-checked completely automatically"*, further supporting the need for verified handwritten proofs. Also, Tesson et al. [28] design and implement a set of tactics for the Coq proof assistant to help writing proofs in calculational form.

With the advent of pen-input devices the possibilities to improve on the interaction limitations of traditional computers are enormous, in particular when it comes to mathematical input. These devices enable software tools such as MathBrush [18] and Microsoft's ink math assistant [26], which allow the recognition, evaluation, and manipulation of handwritten mathematical input (for an extended list of pen-based mathematical tools, see [23,24]). Our work differs from these tools on the emphasis and domain of application: while these emphasise the recognition and evaluation of expressions, our focus is on supporting the handwriting, manipulation, and verification of calculational proofs. As far as we know, there is only one system that supports the manipulation of handwritten calculational proofs using pen-based devices: the MST editor [23,24]. This editor provides structured manipulation of handwritten expressions and provides features to enable flexible and interactive presentations. A limitation, however, is that proofs remain unverified. We attempt to address this limitation by supporting computer-assisted verification of handwritten proofs. Proof assistants such as Isabelle/HOL [25] and Coq [8] can be used to achieve this, but the use of these requires knowledge of the theorem prover's syntax and its intricacies, which has the reputation of being a demanding task. Our work attempts to overcome this by providing a system to interface with a theorem prover without any specific knowledge of the backend proof assistant.

The availability of several different IDEs for existing theorem provers indicates that the human-prover interaction is a concern. For Isabelle/HOL alone there are several IDEs available, including Proof General [2], Isabelle/jEdit [31], and Isabelle/Clide [21]. All of these require the use of keyboard and/or mouse and the knowledge of the, often idiosyncratic, theorem prover's syntax. In fact, the community is still investigating how to improve current IDEs, as demonstrated by the development of PIDE [31–33], a framework for prover interaction and integration, and by Company-Coq [27], an extension of Proof General's Coq mode. Moreover, the support provided by these IDEs for auto-completion of mathematical symbols suggests that typing these does not come naturally.

3   Requirements

The scope of the requirements presented in this section is limited to the context of our work (teaching and research on calculational methods for correct-by-construction program design) and is mostly based on comments found on research papers written by exponents of the calculational method.

R1: Support for calculational mathematics. Given the context of our work, the system should allow the user to write calculational proofs. It should be easy to input mathematical symbols and unconventional mathematical formulae (e.g. the Eindhoven quantifier notation [5]).

R2: Support for structure editing. Similarly to Math/pad and MST [6,24], the system should provide structure editing operations to assist the user in effectively writing handwritten calculational proofs and to ensure that human errors are less likely to be introduced. Frequently used structural operations like selection and copy of expressions and sub-expressions, group/ungroup of sub-expressions, and distributivity operations should be supported.

R3: Support for handwritten input. It should be possible to handwrite calculational proofs as one would normally do when using pen and paper. Moreover, as identified in [24], the result of structure editing rules on hand written expressions should remain handwritten, since it is undesirable to mix different writing and font styles, as doing so, can make presentations confusing.

R4: Support for learning, teaching and research. The system should support learning, teaching, and research on calculational methods for correct-by-construction program design. It has already been argued that calculational proofs offer some pedagogic advantages over conventional informal proofs [12,13,15] and that students prefer or understand better calculational proofs [10]. Moreover, a structure editor can assist students and teachers in learning and explaining how certain rules are applied [23]. A structure editor can also assist researchers who use the calculational method, since they usually write calculations that involve a great deal of syntactic manipulations of uninterpreted and unconventional mathematical formulae (e.g. [4,11,16]). It is desirable for the system to reduce the cognitive load of its users by providing intelligent visual hints throughout the proof writing process.

R5: Support for formal verification. The system should allow mechanical verification of handwritten calculational proofs. The need for mechanical verification of calculational proofs has been identified many years ago [22,30]. More recently, there has been work on mechanisation of calculational proofs [7,17,20], but the problem of verifying handwritten calculational proofs remains open. Moreover, any provers should be used transparently, i.e. the system should hide all the knowledge required to translate handwritten input into syntax accepted by provers. It is important to note that this includes usability aspects other than just syntax—for example, the system should be able to represent mathematical objects in a way that is appropriate in the context of the proof and in the context of the theorem prover. Similar to the idea put forward by Verhoeven and Backhouse [30], our system could be seen as a user interface to a theorem prover. Ideally, the system should allow expert users to define new interaction methods with the backend provers.

4   Proof-of-Concept Prototype

Our proof-of-concept research prototype is implemented in C# and is based on an extension of Classroom Presenter (CP) [1] that uses the library MST [24]. Reusing these existing tools provides us with a structured editor of handwritten mathematics that immediately meets the requirements **R1**, **R2**, **R3**, and, to a certain extent, **R4** (for example, intelligent visual hints are still missing).

The novelty of our prototype is that it adds preliminary support for verification (**R5**): we added a new tool to CP's toolbar that transforms individual steps of a calculation into lemmas that can be proved by Isabelle/HOL. An example of a proof that can be verified by our system is shown in Fig. 1(**c**). The user can select the new tool by clicking on the new toolbar button, and after clicking

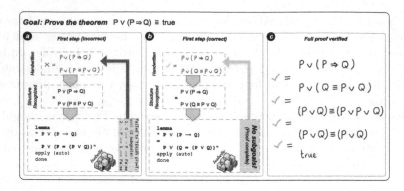

Fig. 1. System overview. Users handwrite calculational proofs and proof steps are translated into Isabelle/HOL to be verified. Invalid steps will be flagged (**a**), allowing users to fix them (**b**). Full proofs can be verified on a step-by-step basis (**c**). (Color figure online)

on a step relation (e.g. equality), the system performs verification of the corresponding individual step. Depending on the validity of the step, the system either shows a green check-mark or a red cross (as illustrated). The proof shown depends on the definition of implication, which can be defined either as

$$P \Rightarrow Q \; \equiv \; P \equiv P \wedge Q \qquad \text{or as} \qquad P \Rightarrow Q \; \equiv \; Q \equiv P \vee Q$$

This means that $P \Rightarrow Q$ can be replaced by either $P \equiv P \wedge Q$ or by $Q \equiv P \vee Q$. However, a common mistake done by students is to swap the conjunction by the disjunction (and vice-versa). This common mistake is illustrated in Fig. 1(a), where we can also see an overview of the steps taken by our prototype to verify a proof step. Once the structure of the handwritten input is created (using the features available in the MST library), our system converts the recognised structure of the step into a lemma that can be interpreted by Isabelle/HOL. We also attach a proof to each generated lemma. In the current version of the system, we simply instruct Isabelle/HOL to try and prove the goal automatically (achieved by `apply(auto)`). In Fig. 1(a), Isabelle/HOL fails to prove the step automatically and the user is informed. In Fig. 1(b), Isabelle/HOL succeeds in verifying the step.

More specifically, the first step of the calculation shown in Fig. 1(b) would be translated into the following:

```
lemma
 " P \<or> (P \<longrightarrow> Q)
 =
    P \<or> (Q = (P \<or> Q))"
apply (auto)
done
```

Currently, this translation only supports calculational propositional logic operators [9,14], but our code can easily be extended to include a larger mathematical domain (however, other proof tactics may be needed). Other translations are possible and will be explored in future iterations of this work. Communication with Isabelle/HOL is currently performed via an external process that we programmed. This process accepts plain-text input encoding a lemma and its proof. It then embeds the input into a generated Isabelle/HOL theory file created on the fly, attempts to verify the theory, and returns a modified version of Isabelle/HOL's output to the user interface. The handwritten step is then annotated with either a green check-mark or a red cross, depending on the result of the verification. Figure 2 shows a screenshot of our current prototype in use when writing the proof discussed above. It shows that the first step has already been verified and the user is currently applying the distributivity rule using a gesture (this is one of MST's features that we imported into our system).

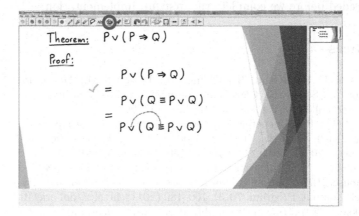

Fig. 2. Screenshot of our prototype in action. The first step of the calculation was verified using the new tool (highlighted button). The user is currently applying the distributivity rule using a gesture. (Color figure online)

5   Conclusion and Future Work

We have described our first steps towards a system that allows users to verify handwritten calculational proofs. Our proof-of-concept prototype supports propositional logic proofs and uses Isabelle/HOL as the backend prover. The major novelty of this work lies on the implementation of the requirement for formal verification (**R5**): the prototype can formally verify handwritten calculational proofs without the need to learn how to use a theorem prover. The implementation of the prototype shows that the system we envisage is feasible and that it has the potential to assist its users in writing correct proofs.

Having a prototype will now allow us to demonstrate the potential of such a tool to target users. Our next step will be to demonstrate the prototype in learning, teaching, and research environments to obtain user feedback. This will enable us to understand further requirements of likely users of this tool. Once this step is completed, we will implement a more complete system that will improve the interaction with the backend prover (e.g. feedback from the background proof assistant to include hints for completing proof steps or counter-examples). For this, we plan to use PIDE [31–33]. We also intend to use hints handwritten by users to justify proof steps in the verification process, allowing the detection of inconsistent justifications. The feedback from users will inform the way in which hints will be dealt with. We further plan to support multiple backend provers and to link proofs of programs with the code generation mechanisms available in some theorem provers, such as Isabelle/HOL and Coq. We will take into account the feedback received from users and adapt the system to meet any further requirements that arise. We plan to continue using rapid prototyping to demonstrate any new features to users before providing complete implementations. We anticipate that several iterations of rapid prototyping and evaluations will be needed before we complete the first full implementation.

References

1. Anderson, R., Anderson, R., Chung, O., Davis, K.M., Davis, P., Prince, C., Razmov, V., Simon, B.: Classroom presenter - a classroom interaction system for active and collaborative learning. In: WIPTE (2006)
2. Aspinall, D.: Proof general: a generic tool for proof development. In: Graf, S., Schwartzbach, M. (eds.) TACAS 2000. LNCS, vol. 1785, pp. 38–43. Springer, Heidelberg (2000). https://doi.org/10.1007/3-540-46419-0_3
3. Backhouse, R.: The calculational method. Inf. Process. Lett. **53**(3), 121 (1995)
4. Backhouse, R., Ferreira, J.F.: On Euclid's algorithm and elementary number theory. Sci. Comput. Program. **76**(3), 160–180 (2011). https://doi.org/10.1016/j.scico.2010.05.006. http://joaoff.com/publications/2010/euclid-alg
5. Backhouse, R., Michaelis, D.: Exercises in quantifier manipulation. In: Uustalu, T. (ed.) MPC 2006. LNCS, vol. 4014, pp. 69–81. Springer, Heidelberg (2006). https://doi.org/10.1007/11783596_7
6. Backhouse, R., Verhoeven, R.: Math∫pad: a system for on-line preparation of mathematical documents. Softw. Concepts Tools **18**, 80–89 (1997)
7. Bauer, G., Wenzel, M.: Calculational reasoning revisited an Isabelle/Isar experience. In: Boulton, R.J., Jackson, P.B. (eds.) TPHOLs 2001. LNCS, vol. 2152, pp. 75–90. Springer, Heidelberg (2001). https://doi.org/10.1007/3-540-44755-5_7
8. Bertot, Y., Castran, P.: Interactive Theorem Proving and Program Development: Coq'Art The Calculus of Inductive Constructions. Springer, Heidelberg (2010). https://doi.org/10.1007/978-3-662-07964-5
9. Dijkstra, E.W., Scholten, C.S.: Predicate Calculus and Program Semantics. Springer, New York (1990). https://doi.org/10.1007/978-1-4612-3228-5
10. Ferreira, J.F., Mendes, A.: Students' feedback on teaching mathematics through the calculational method. In: 39th ASEE/IEEE Frontiers in Education Conference. IEEE (2009)

11. Ferreira, J.F., Mendes, A.: A calculational approach to path-based properties of the Eisenstein-Stern and Stern-Brocot trees via matrix algebra. J. Log. Algebraic Methods Program. **85**(5, Part 2), 906–920 (2016). https://doi.org/10.1016/j.jlamp.2015.11.004. http://www.sciencedirect.com/science/article/pii/S2352220815001418. Articles dedicated to Prof. J. N. Oliveira on the occasion of his 60th birthday
12. Ferreira, J.F., Mendes, A., Backhouse, R., Barbosa, L.S.: Which mathematics for the information society? In: Gibbons, J., Oliveira, J.N. (eds.) TFM 2009. LNCS, vol. 5846, pp. 39–56. Springer, Heidelberg (2009). https://doi.org/10.1007/978-3-642-04912-5_4
13. Ferreira, J.F., Mendes, A., Cunha, A., Baquero, C., Silva, P., Barbosa, L.S., Oliveira, J.N.: Logic training through algorithmic problem solving. In: Blackburn, P., van Ditmarsch, H., Manzano, M., Soler-Toscano, F. (eds.) TICTTL 2011. LNCS (LNAI), vol. 6680, pp. 62–69. Springer, Heidelberg (2011). https://doi.org/10.1007/978-3-642-21350-2_8
14. Gries, D., Schneider, F.B.: A Logical Approach to Discrete Math. Springer, New York (1993)
15. Gries, D., Schneider, F.B.: Teaching math more effectively, through calculational proofs. Am. Math. Mon. **102**(8), 691–697 (1995)
16. Hinze, R.: Scans and convolutions—a calculational proof of Moessner's theorem. In: Scholz, S.-B., Chitil, O. (eds.) IFL 2008. LNCS, vol. 5836, pp. 1–24. Springer, Heidelberg (2011). https://doi.org/10.1007/978-3-642-24452-0_1
17. Kahl, W.: Calculational relation-algebraic proofs in Isabelle/Isar. In: Berghammer, R., Möller, B., Struth, G. (eds.) RelMiCS 2003. LNCS, vol. 3051, pp. 178–190. Springer, Heidelberg (2004). https://doi.org/10.1007/978-3-540-24771-5_16
18. Labahn, G., Lank, E., MacLean, S., Marzouk, M., Tausky, D.: Mathbrush: a system for doing math on pen-based devices. In: Proceedings of the 2008 The Eighth IAPR International Workshop on Document Analysis Systems, DAS 2008, pp. 599–606. IEEE Computer Society, Washington, DC (2008). https://doi.org/10.1109/DAS.2008.21
19. Leino, K.R.M.: Dafny: an automatic program verifier for functional correctness. In: Clarke, E.M., Voronkov, A. (eds.) LPAR 2010. LNCS (LNAI), vol. 6355, pp. 348–370. Springer, Heidelberg (2010). https://doi.org/10.1007/978-3-642-17511-4_20
20. Leino, K.R.M., Polikarpova, N.: Verified calculations. In: Cohen, E., Rybalchenko, A. (eds.) VSTTE 2013. LNCS, vol. 8164, pp. 170–190. Springer, Heidelberg (2014). https://doi.org/10.1007/978-3-642-54108-7_9
21. Lüth, C., Ring, M.: A web interface for Isabelle: the next generation. In: Carette, J., Aspinall, D., Lange, C., Sojka, P., Windsteiger, W. (eds.) CICM 2013. LNCS (LNAI), vol. 7961, pp. 326–329. Springer, Heidelberg (2013). https://doi.org/10.1007/978-3-642-39320-4_22
22. Manolios, P., Moore, J.S.: On the desirability of mechanizing calculational proofs. Inf. Process. Lett. **77**(2–4), 173–179 (2001). https://doi.org/10.1016/S0020-0190(00)00200-3
23. Mendes, A.: Structured editing of handwritten mathematics. Ph.D. thesis, School of Computer Science, University of Nottingham (2012)
24. Mendes, A., Backhouse, R., Ferreira, J.F.: Structure editing of handwritten mathematics: improving the computer support for the calculational method. In: Proceedings of the Ninth ACM International Conference on Interactive Tabletops and Surfaces, ITS 2014, pp. 139–148. ACM, New York (2014). https://doi.org/10.1145/2669485.2669495

25. Nipkow, T., Wenzel, M., Paulson, L.C.: Isabelle/HOL: A Proof Assistant for Higher-Order Logic. Springer, Heidelberg (2016). https://doi.org/10.1007/3-540-45949-9. https://isabelle.in.tum.de/doc/prog-prove.pdf
26. Microsoft OneNote. https://www.onenote.com. Accessed 02 Feb 2018
27. Pit-Claudel, C., Courtieu, P.: Company-Coq: taking proof general one step closer to a real IDE. In: CoqPL 2016: International Workshop on Coq for Programming Languages (2016)
28. Tesson, J., Hashimoto, H., Hu, Z., Loulergue, F., Takeichi, M.: Program calculation in Coq. In: Johnson, M., Pavlovic, D. (eds.) AMAST 2010. LNCS, vol. 6486, pp. 163–179. Springer, Heidelberg (2011). https://doi.org/10.1007/978-3-642-17796-5_10
29. Tripp, S.D., Bichelmeyer, B.: Rapid prototyping: an alternative instructional design strategy. Educ. Tech. Res. Dev. **38**(1), 31–44 (1990)
30. Verhoeven, R., Backhouse, R.: Interfacing program construction and verification. In: Wing, J.M., Woodcock, J., Davies, J. (eds.) FM 1999. LNCS, vol. 1709, pp. 1128–1146. Springer, Heidelberg (1999). https://doi.org/10.1007/3-540-48118-4_10. http://dl.acm.org/citation.cfm?id=647545.730778
31. Wenzel, M.: Isabelle/jEdit – a prover IDE within the PIDE framework. In: Jeuring, J., Campbell, J.A., Carette, J., Dos Reis, G., Sojka, P., Wenzel, M., Sorge, V. (eds.) CICM 2012. LNCS (LNAI), vol. 7362, pp. 468–471. Springer, Heidelberg (2012). https://doi.org/10.1007/978-3-642-31374-5_38
32. Wenzel, M.: Asynchronous user interaction and tool integration in Isabelle/PIDE. In: Klein, G., Gamboa, R. (eds.) ITP 2014. LNCS, vol. 8558, pp. 515–530. Springer, Cham (2014). https://doi.org/10.1007/978-3-319-08970-6_33
33. Wenzel, M., Wolff, B.: Isabelle/PIDE as platform for educational tools. arXiv preprint arXiv:1202.4835 (2012)

A Formally Verified Solver
for Homogeneous Linear Diophantine
Equations

Florian Meßner, Julian Parsert, Jonas Schöpf, and Christian Sternagel[(✉)]

University of Innsbruck, Innsbruck, Austria
{florian.g.messner,julian.parsert,jonas.schoepf,
christian.sternagel}@uibk.ac.at

Abstract. In this work we are interested in minimal complete sets of solutions for homogeneous linear diophantine equations. Such equations naturally arise during AC-unification—that is, unification in the presence of associative and commutative symbols. Minimal complete sets of solutions are for example required to compute AC-critical pairs. We present a verified solver for homogeneous linear diophantine equations that we formalized in Isabelle/HOL. Our work provides the basis for formalizing AC-unification and will eventually enable the certification of automated AC-confluence and AC-completion tools.

Keywords: Homogeneous linear diophantine equations
Code generation · Mechanized mathematics · Verified code
Isabelle/HOL

1 Introduction

(Syntactic) unification of two terms s and t, is the problem of finding a substitution σ that, applied to both terms, makes them syntactically equal: $s\sigma = t\sigma$. For example, it is easily verified that $\sigma = \{x \mapsto z, y \mapsto z\}$ is a solution to the unification problem $\mathsf{f}(x, y) \approx^? \mathsf{f}(z, z)$. Several syntactic unification algorithms are known, some of which have even been formalized in proof assistants.

By throwing a set of equations E into the mix, we arrive at *equational* or *E-unification*, where we are interested in substitutions σ that make two given terms equivalent with respect to the equations in E, written $s\sigma \approx_E t\sigma$. While for syntactic unification most general solutions, called *most general unifiers*, are unique, E-unification is distinctly more complex: depending on the specific set of equations, E-unification might be undecidable, have unique solutions, have minimal complete sets of solutions, etc.

For AC-unification we instantiate E from above to a set AC of associativity and commutativity equations for certain function symbols. For example, by taking $\mathsf{AC} = \{(x \cdot y) \cdot z \approx x \cdot (y \cdot z), x \cdot y \approx y \cdot x\}$, we express that $\cdot$ (which we write

---

This work is supported by the Austrian Science Fund (FWF): project P27502.

J. Avigad and A. Mahboubi (Eds.): ITP 2018, LNCS 10895, pp. 441–458, 2018.
https://doi.org/10.1007/978-3-319-94821-8_26

Table 1. An example HLDE and its minimal complete set of solutions

	solution	x y z
$x + y = 2z$	z_1	2 0 1
	z_2	0 2 1
	z_3	1 1 1

infix, for convenience) is the only associative and commutative function symbol. Obviously, the substitution σ from above is also a solution to the AC-unification problem $x \cdot y \approx_{AC}^? z \cdot z$ (since trivially $z \cdot z \approx_{AC} z \cdot z$). You might ask: *is it the only one?* It turns out that it is not. More specifically, there is a minimal complete set (see Sect. 2 for a formal definition) consisting of the five AC-unifiers:

$$\{x \mapsto z_3, \qquad y \mapsto z_3, \qquad z \mapsto z_3\}$$
$$\{x \mapsto z_1 \cdot z_1, \qquad y \mapsto z_2 \cdot z_2, \qquad z \mapsto z_1 \cdot z_2\}$$
$$\{x \mapsto z_1 \cdot z_1 \cdot z_3, \; y \mapsto z_3, \qquad z \mapsto z_1 \cdot z_3\}$$
$$\{x \mapsto z_3, \qquad y \mapsto z_2 \cdot z_2 \cdot z_3, \; z \mapsto z_2 \cdot z_3\}$$
$$\{x \mapsto z_1 \cdot z_1 \cdot z_3, \; y \mapsto z_2 \cdot z_2 \cdot z_3, \; z \mapsto z_1 \cdot z_2 \cdot z_3\}$$

But how can we compute it? The answer involves minimal complete sets of solutions for homogeneous linear diophantine equations (HLDEs for short). From the initial AC-unification problem $x \cdot y \approx_{AC}^? z \cdot z$ we derive the equation in Table 1, which basically tells us that, no matter what we substitute for x, y, and z, there have to be exactly twice as many occurrences of the AC-symbol $\cdot$ in the substitutes for x and y than there are in the substitute for z.

The minimal complete set of solutions to this equation, labeled by fresh variables, is depicted in Table 1, where the numbers indicate how many occurrences of the corresponding fresh variable are contributed to the substitute for the variable in the respective column. The AC-symbol $\cdot$ is used to combine fresh variables occurring more than once. For example, the solution labeled by z_1 contributes two occurrences of z_1 to the substitute for x and one occurrence of z_1 to the substitute for z, while not touching the substitute for y at all.

Now each combination of solutions for which x, y, and z are all nonzero[1] gives rise to an independent minimal AC-unifier (in general, given n solutions, there are 2^n combinations, one for each subset of solutions). The unifiers above correspond to the combinations: $\{z_3\}$, $\{z_1, z_2\}$, $\{z_1, z_3\}$, $\{z_2, z_3\}$, $\{z_1, z_2, z_3\}$. We refer to the literature for details on how exactly we obtain unifiers from sets of solutions to HLDEs and why this works [1,12]. Suffice it to say that minimal complete sets of solutions to HLDEs give rise to minimal complete sets of AC-unifiers. [2] The main application we have in mind, relying on minimal complete sets of AC-unifiers, is computing AC-critical pairs. This is for example useful for

---

[1] The "nonzero" condition naturally arises from the fact that substitutions cannot replace variables by nothing.

[2] Actually, this only holds for elementary AC-unification problems, which are those consisting only of variables and one specific AC-symbol. However, arbitrary AC-unification problems can be reduced to sets of elementary AC-unification problems.

proving confluence of rewrite systems with and without AC-symbols [6,10,11] and required for normalized completion [8,14].

In this paper we investigate how to compute minimal complete sets of solutions of HLDEs, with our focus on formal verification using a proof assistant. In other words, we are only interested in *verified* algorithms (that is, algorithms whose correctness has been machine-checked). More specifically, our contributions are as follows:

- We give an Isabelle/HOL formalization of HLDEs and their minimal complete sets of solutions (Sect. 3).
- We describe a simple algorithm that computes such minimal complete sets of solutions (Sect. 2) and discuss an easy correctness proof that we formalized in Isabelle/HOL (Sect. 4).
- After several rounds of program transformations, making use of standard optimization techniques and improved bounds from the literature (Sect. 5), we obtain a more efficient solver (Sect. 6)—to the best of our knowledge, the first formally verified solver for HLDEs.

Our formalization is available in the *Archive of Formal Proofs* [9] (development version, changeset d5fabf1037f8). Through Isabelle's code generation feature [4] a verified solver can be obtained from our formalization.

2 Main Ideas

For any formalization challenge it is a good idea to start from as simple a grounding as possible: trying to reduce the number of involved concepts to a bare minimum *and* to keep the complexity of involved proofs in check.

When formalizing an algorithm, once we have a provably correct implementation, we might still want to make it more efficient. Instead of doing all the (potentially hard) proofs again for a more efficient (and probably more involved) variant, we can often prove that the two variants are equivalent and thus carry over the correctness result from a simple implementation to an efficient one. This is also the general plan we follow for our formalized HLDE solver.

To make things simpler when computing minimal complete sets of solutions for an HLDE $a \bullet x = b \bullet y$ (where a and b are lists of coefficients and $v \bullet w$ denotes the *dot product* of two lists $v = [v_1, \ldots, v_k]$ and $w = [w_1, \ldots, w_k]$ defined by $v_1 w_1 + \cdots + v_k w_k$), we split the task into three separate phases:

- *generate* a finite search-space that covers all potentially minimal solutions
- *check* necessary criteria for minimal solutions (throwing away the rest)
- *minimize* the remaining collection of candidates

Generate. For the first phase we make use of the fact that for every minimal solution (x, y) the entries of x are bounded by the maximal coefficient in b, while the entries of y are bounded by the maximal coefficient in a (which we will prove in Sect. 3).

Moreover, we generate the search-space in *reverse lexicographic* order, where for arbitrary lists of numbers $u = [u_1, \ldots, u_k]$ and $v = [v_1, \ldots, v_k]$ we have $u <_{\mathsf{rlex}} v$ iff there is an $i \leq k$ such that $u_i < v_i$ and $u_j = v_j$ for all $i < j \leq k$. This allows for a simple recursive implementation and can be exploited in the minimization phase.

Assuming that x-entries of solutions are bounded by A and y-entries are bounded by B, we can implement the generate-phase by the function

```
generate A B m n = tl [(x, y). y ← gen B n, x ← gen A m]
```

where we use Haskell-like list comprehension and tl is the standard *tail* function on lists dropping the first element—which in this case is the trivial (and non-minimal) solution consisting only of zeroes—and gen B n computes all lists of natural numbers of length n whose entries are bounded by B, in reverse lexicographic order.

```
gen B 0 = [[]]
gen B (Suc n) = [x#xs. xs ← gen B n, x ← [0..B]]
```

Our initial example $x + y = 2z$ can be represented by the two lists of coefficients [1,1] and [2] and the corresponding search-space is generated by generate 2 1 2 1, resulting in

```
[([1,0],[0]),([2,0],[0]),([0,1],[0]),([1,1],[0]),
 ([2,1],[0]),([0,2],[0]),([1,2],[0]),([2,2],[0]),
 ([0,0],[1]),([1,0],[1]),([2,0],[1]),([0,1],[1]),
 ([1,1],[1]),([2,1],[1]),([0,2],[1]),([1,2],[1]),([2,2],[1])]
```

Check. Probably the most obvious necessary condition for (x, y) to be a minimal solution is that it is actually a solution, that is, $a \bullet x = b \bullet y$ (taking the later minimization phase into account, it is in fact also a sufficient condition). We can implement the check-phase, given two lists of coefficients a and b, by

```
check a b = filter (λ(x, y). a • x = b • y)
```

using the standard *filter* function on lists that only preserves elements satisfying the given predicate.

For our initial example check [1,1] [2] (generate 2 1 2 1) computes the first two phases, resulting in [([2,0],[1]),([1,1],[1]),([0,2],[1])].

Minimize. It is high time that we specify in what sense minimal solutions are to be minimal. To this end, we use the *pointwise less-than-or-equal* order $\leq_{\mathsf{v}}$ on lists (whose strict part $<_{\mathsf{v}}$ is defined by $x <_{\mathsf{v}} y$ iff $x \leq_{\mathsf{v}} y$ but not $y \leq_{\mathsf{v}} x$). Now minimization can be implemented by the function

```
minimize [] = []
minimize ((x,y)#xs) =
    (x,y) # filter (λ(u,v). x@y ≮ᵥ u@v) (minimize xs)
```

where @ is Isabelle/HOL's list concatenation. This is also where we exploit the fact that the input to minimize is sorted in reverse lexicographic order: then, since (x,y) is up front, we know that all elements of xs are strictly greater with respect to $<_{rlex}$; moreover, $u <_v v$ implies $u <_{rlex} v$ for all u and v; and thus, x@y is not $<_v$-greater than any element of xs, warranting that we put it in the resulting minimized list without further check.

A Simple Algorithm. Putting all three phases together we obtain a straightforward algorithm for computing all minimal solutions of an HLDE given by its lists of coefficients a and b

```
solutions a b =
  let A = max b; B = max a; m = length a; n = length b in
  minimize (check a b (generate A B m n))
```

where length xs—which we sometimes write $|xs|$—computes the length of a list xs. We will prove the correctness of solutions in Sect. 4.

Performance Tuning. There are several potential performance improvements over the simple algorithm from above. In a first preparatory step, we categorize solutions into *special* and *non-special* solutions (Sect. 5). The former are minimal by construction and can thus be excluded from the minimization phase. For the latter, several necessary conditions are known that are monotone in the sense that all prefixes and suffixes of a list satisfy them whenever the list itself does. Now merging the generate and check phases by "pushing in" these conditions as far as possible has the potential to drastically cut down the explored searchspace. We will discuss the details in Sect. 6.

3 An Isabelle/HOL Theory of HLDEs and Their Solutions

In this section, after putting our understanding of HLDEs and their solutions on firmer grounds, we obtain bounds on minimal solutions that serve as a basis for the two algorithms we present in later sections.

A *homogeneous linear diophantine equation* is an equation of the form

$$a_1 x_1 + a_2 x_2 + \cdots + a_m x_m = b_1 y_1 + b_2 y_2 + \cdots + b_n y_n$$

where coefficients a_i and b_j are fixed natural numbers. Moreover, we are only interested in solutions (x, y) over the naturals.

That means that all the required information can be encoded into two lists of natural numbers $a = [a_1, \ldots, a_m]$ and $b = [b_1, \ldots, b_n]$. From now on, let a and b be fixed, which is achieved by Isabelle's locale mechanism in our formalization:[3]

locale *hlde* = **fixes** $a\ b$ :: *nat list* **assumes** $0 \notin$ set a **and** $0 \notin$ set b

---

[3] For technical reasons (regarding *code generation*) we actually have the two locales *hlde-ops* and *hlde* in our formalization.

In the locale, we also assume that a and b do not have any zero entries (which is useful for some proofs; note that arbitrary HLDEs can be transformed into equivalent HLDEs satisfying this assumption by dropping all zero-coefficients).

Solutions of the HLDE represented by a and b are those pairs of lists (x, y) that satisfy $a \bullet x = b \bullet y$. Formally, the *set of solutions* $\mathcal{S}(a, b)$ is given by

$$\mathcal{S}(a, b) = \{(x, y) \mid a \bullet x = b \bullet y \wedge |x| = m \wedge |y| = n\}$$

A solution is *(pointwise) minimal* iff there is no nonzero solution that is pointwise strictly smaller. The *set of (pointwise) minimal* solutions is given by

$$\mathcal{M}(a, b) = \{(x, y) \in \mathcal{S}(a, b) \mid x \neq 0 \wedge \nexists (u, v) \in \mathcal{S}(a, b). \ u \neq 0 \wedge u \, @ \, v <_{\mathsf{v}} x \, @ \, y\}$$

where we use the notation $v \neq 0$ to state that a list v is *nonzero*, that is, does not exclusively consist of zeroes. While the above definition might look asymmetric, since we only require x and u to be nonzero, we actually also have that y and v are nonzero, because (x, y) and (u, v) are both solutions and a and b do not contain any zeroes.

Huet [5, Lemma 1] has shown that, given a minimal solution (x, y), the entries of x and y are bounded by $\max b$ and $\max a$, respectively. In preparation for the proof of this result, we prove the following auxiliary fact.

Lemma 1. *If x is a list of natural numbers of length n, then either*

(1) $x_i \equiv 0 \pmod{n}$ for some $1 \leq i \leq n$, or
(2) $x_i \equiv x_j \pmod{n}$ for some $1 \leq i < j \leq n$.

Proof. Let X be the set of elements of x and $M = \{y \bmod n \mid y \in X\}$. If $|M| < |X|$ then property (2) follows by the pigeonhole principle. Otherwise, $|M| = |X|$ and either x contains already duplicates and we are done (again by establishing property (2)), or the elements of x are pairwise disjoint. In the latter case, we know that $|M| = n$. Since all elements of M are less than n by construction, we obtain $M = \{0, \ldots, n - 1\}$. This, in turn, means that property (1) is satisfied.   □

Now we are in a position to prove a variant of Huet's Lemma 1 for improved bounds (which were, to the best of our knowledge, first mentioned by Clausen and Fortenbacher [2]), where, given two lists u and v of same length, we use $\max_v^{\neq 0}(u)$ to denote $\max(\{0\} \cup \{u_i \mid 1 \leq i \leq |v| \wedge v_i \neq 0\})$, that is, the maximum of those u-elements whose corresponding v-elements are nonzero.

Lemma 2. *Let (x, y) be a minimal solution. Then we have $x_i \leq \max_y^{\neq 0}(b)$ for all $1 \leq i \leq m$ and $y_j \leq \max_x^{\neq 0}(a)$ for all $1 \leq j \leq n$.*

Proof. Since the two statements above are symmetric, we concentrate on the first one. Let $M = \max_y^{\neq 0}(b)$ and assume that there is $x_k > M$ with $1 \leq k \leq m$. We will show that this contradicts the minimality of (x, y). We have

$$M \cdot \sum_{j=1}^{n} y_j \geq b \bullet y = a \bullet x \geq a_k x_k > a_k \cdot M$$

and thus $\sum_{j=1}^{n} y_j > a_k$.

At this point we give an explicit construction for a corresponding existential statement in Huet's original proof. The goal is to construct a pointwise increasing sequence of lists $\boldsymbol{u} = \boldsymbol{u}^1, \ldots, \boldsymbol{u}^{a_k}$ such that for all $v \in$ set $\boldsymbol{u}$ we have (1) $v \leq_v y$ and also (2) $0 < \sum_{i=1}^{n} v_i \leq a_k$. This is achieved by taking $\boldsymbol{u}^i = (\text{inc } y \, 0)^i \, 0_{|y|}$ where 0_n denotes a list of n zeroes and we employ the auxiliary function

```
inc y i v =
  if i < length y then
    if v ! i < y ! i then v[i := v ! i + 1]
    else inc y (Suc i) v
  else v
```

that, given two lists y and v, increments v at the smallest position $j \geq i$ such that $v_j < y_j$ (if this is not possible, the result is v). Here $x \, ! \, i$ denotes the ith element of list x and $x[i := v]$ a variant of list x, where the ith element is v.

As long as there is "enough space" (as guaranteed by $\sum_{j=1}^{n} y_j > a_k$), $\boldsymbol{u}^i$ is pointwise smaller than y and the sum of its elements is i for all $1 \leq i \leq a_k$, thereby satisfying both of the above properties.

Now we obtain a list u that in addition to (1) and (2) also satisfies (3) $b \bullet u \equiv 0$ (mod a_k). This is achieved by applying Lemma 1 to the list of natural numbers $\text{map} (\lambda x. \, b \bullet x) \, \boldsymbol{u}$, and analyzing the resulting cases. Either such a list is already in $\boldsymbol{u}$ and we are done, or $\boldsymbol{u}$ contains two lists $\boldsymbol{u}^i$ and $\boldsymbol{u}^j$ with $i < j$, for which $b \bullet \boldsymbol{u}^i \equiv b \bullet \boldsymbol{u}^j$ (mod a_k) holds. In the latter case, the pointwise subtraction $\boldsymbol{u}^j -_v \boldsymbol{u}^i$ satisfies properties (1) to (3).

Remember that $x_k > M$. Together with properties (1) and (2) we know

$$b \bullet u \leq M \cdot \sum_{j=1}^{n} u_j \leq M \cdot a_k < a_k x_k$$

By (3), we further have $b \bullet u = a_k c$ for some $0 < c < x_k$, showing that (x, y) is strictly greater than the nonzero solution $(0_{|m|}[k := c], u)$. Finally, a contradiction to the minimality of (x, y). □

As a corollary, we obtain Huet's result, namely that all x_i are bounded by $\max b$ and all y_j are bounded by $\max a$, since $\max_v^{\neq 0}(c) \leq \max c$ for all lists v and c.

4   Certified Minimal Complete Sets of Solutions

Before we prove our algorithm from Sect. 2 correct, let us have a look at a characterization of the elements of minimize that we require in the process (where $<_{\text{rlex}}$ as well as $<_v$ are extended to pairs of lists by taking their concatenation).

Lemma 3. set (minimize xs) = $\{x \in$ set $xs \mid \nexists y \in$ set $xs. \, y <_v x\}$ *whenever* xs *is sorted with respect to* $<_{\text{rlex}}$.

Proof. An easy induction over xs shows the direction from right to left. For the other direction, let x be an arbitrary but fixed element of minimize xs. Another easy induction over xs shows that then x is also in xs. Thus it remains to show that there is no y in xs which is $<_\mathsf{v}$-smaller than x. Assume that there is such a y for the sake of a contradiction and proceed by induction over xs. If $xs = []$ we are trivially done. Otherwise, $xs = z \mathbin{\#} zs$ and when x is in minimize zs and y is in zs, the result follows by IH. In the remaining cases either $z = x$ or $z = y$, but not both (since this would yield $z <_\mathsf{v} z$). For the former we have $x \leq_\mathsf{rlex} y$ by sortedness and for the latter we obtain $y \not<_\mathsf{v} x$ by the definition of minimize (since x is in minimize zs), both contradicting $y <_\mathsf{v} x$.   □

In the remainder of this section, we will prove completeness (all minimal solutions are generated) and soundness (only minimal solutions are generated) of solutions.

Lemma 4 (Completeness). $\mathcal{M}(a, b) \subseteq$ set (solutions $a\ b$)

Proof. Let (x, y) be a minimal solution. We use the abbreviations $A = \max b$, $B = \max a$, and $C =$ set (check $a\ b$ (generate $A\ B\ m\ n$)). Then, by Lemma 3, we have set (solutions $a\ b$) $= \{x \in C \mid \nexists y \in C.\ y <_\mathsf{v} x\}$. Note that (x, y) is in C (which contains all solutions within the bounds provided by A and B, by construction) due to Lemma 2. Moreover, $y \not<_\mathsf{v} x \mathbin{@} y$ for all $y \in C$ follows from the minimality of (x, y), since C is clearly a subset of $\mathcal{S}(a, b)$. Together, the previous two statements conclude the proof.   □

Lemma 5 (Soundness). set (solutions $a\ b$) $\subseteq \mathcal{M}(a, b)$

Proof. Let (x, y) be in solutions $a\ b$. According to the definition of $\mathcal{M}(a, b)$ we have to show that (x, y) is in $\mathcal{S}(a, b)$ (which is trivial), x is nonzero, and that there is no $<_\mathsf{v}$-smaller solution (u, v) with nonzero u. Incidentally, the last part can be narrowed down to: there is no $<_\mathsf{v}$-smaller *minimal* solution (u, v) (since for every solution we can find a $\leq_\mathsf{v}$-smaller minimal solution by well-foundedness of $<_\mathsf{v}$, and the left component of minimal solutions is nonzero by definition).

We start by showing that x is nonzero. Since there are no zeroes in a and b, and (x, y) is a solution, x can only be a zero-list if also y is. However, the elements of solutions $a\ b$ are sorted in strictly increasing order with respect to $<_\mathsf{rlex}$ and the first one is already not the pair of zero-lists, by construction.

Now, for the sake of a contradiction, assume that there is a minimal solution $(u, v) <_\mathsf{v} (x, y)$. By Lemma 4, we obtain that (u, v) is also in solutions $a\ b$. But then, due to its minimality, (u, v) is also in C (the same set we already used in the proof of Lemma 4). Moreover, (x, y) is in C by construction. Together with Lemma 3 and $(u, v) <_\mathsf{v} (x, y)$, this results in the desired contradiction.   □

As a corollary of the previous two results, we obtain that solutions computes exactly all minimal solutions, that is set (solutions $a\ b$) $= \mathcal{M}(a, b)$.

5   Special and Non-special Solutions

For each pair of variable positions i and j, there is exactly one minimal solution such that only the x-entry at position i and the y-entry at position j are nonzero. Since all other entries are 0, the equation collapses to $a_i x_i = b_j y_j$. Taking the minimal solutions (by employing the least common multiple) of this equation, we solve for x_i and then for y_j and obtain the nonzero x-entry $d_{ij} = \text{lcm}(a_i, b_j)/a_i$ and the nonzero y-entry $e_{ij} = \text{lcm}(a_i, b_j)/b_j$, respectively. Given i and j, we obtain the *special solution* (x, y) where x is $[0, \ldots, d_{ij}, \ldots, 0]$ and y is $[0, \ldots, e_{ij}, \ldots, 0]$.

All special solutions can be computed in advance and outside of our minimization phase, since special solutions are minimal (the only entries where a special solution could decrease are d_{ij} and e_{ij}, but those are minimal due to the properties of least common multiples). We compute all special solutions by the following function

```
special_solutions a b =
    [sij a b i j. i ← [1..length a], j ← [1..length b]]
```

where

```
sij a b i j = ((replicate (length a) 0)[i := dij a b i j],
               (replicate (length b) 0)[j := eij a b i j])
dij a b i j = lcm (a ! i) (b ! j) div (a ! i)
eij a b i j = lcm (a ! i) (b ! j) div (b ! j)
```

We have already seen a relatively crude bound on minimal solutions in Sect. 3. A further bound, this time for minimal non-special solutions, follows.

Lemma 6. *Let (x, y) be a non-special solution such that $x_i \geq d_{ij}$ and $y_j \geq e_{ij}$ for some $1 \leq i \leq m$ and $1 \leq j \leq n$. Then (x, y) is not minimal.*

Proof. Assume that (x, y) is a minimal solution and consider the special solution $(u, v) = ([0, \ldots, d_{ij}, \ldots, 0], [0, \ldots, e_{ij}, \ldots, 0])$. Due to $x_i \geq d_{ij}$ and $y_j \geq e_{ij}$ we obviously have $u @ v \leq_\mathsf{v} x @ y$. Since (x, y) is not special itself, we further obtain $u @ v <_\mathsf{v} x @ y$, contradicting the supposed minimality of (x, y). □

This result allows us to avoid all candidates that are pointwise greater than or equal to some special solution during our generation phase, which is the motivation for the following functions for bounding the elements of non-special minimal solutions. The function max_y, bounding entries of y, is directly taken from Huet [5]. Moreover, max_x is our counterpart to max_y bounding entries of x. As max_x is symmetric to max_y, we only give details for the latter, which is

```
max_y x j =
    if j < n ∧ Eⱼ x ≠ ∅ then  min (Eⱼ x)
    else max a
```

where E_j is defined by

$$E_j\ x = \{e_{ij} - 1 \mid i < |x| \wedge x_i \geq d_{ij}\}$$

from which we can show that all minimal solutions satisfy the following bounds

$$\text{boundr}\ x\ y \longleftrightarrow (\forall 1 \leq j \leq n.\ y_j \leq \text{max\_y}\ x\ j)$$
$$\text{subdprodl}\ x\ y \longleftrightarrow (\forall k \leq m.\ [a]^k \bullet [x]^k \leq b \bullet y)$$
$$\text{subdprodr}\ y \longleftrightarrow (\forall l \leq n.\ [b]^l \bullet [y]^l \leq a \bullet \text{map}\ (\text{max\_x}\ [y]^l)\ [1..m])$$

where boundr, subdprodl, and subdprodr are mnemonic for *bound on entries of right component*, *bound on sub dot product of left component*, and *bound on sub dot product of right component*, respectively.

Lemma 7. *Let $(x, y) \in \mathcal{M}(a, b)$ be a non-special minimal solution. Then, all of the following hold:*

(1) boundr $x\ y$,
(2) subdprodl $x\ y$, and
(3) subdprodr $x\ y$.

Proof. Property (1) directly corresponds to condition (c) of Huet. Thus, we refer to our formalization for details but note that this is where Lemma 6 is employed (apart from motivating the definitions of max_x and max_y in the first place).

Property (2), which is based on Huet's condition (d), follows from (x, y) being a solution and the fact that the dot product cannot get larger by dropping (same length) suffixes from both operands.

The last property (3) is based on condition (b) from Huet's paper. Again, we refer to our formalization for details.   □

Given a bound B and a list of coefficients as, the function alls computes all pairs whose first component is a list xs of length $|as|$ with entries at most B and whose second component is $as \bullet xs$. Note that the resulting list is sorted in reverse lexicographic order with respect to first components of pairs.[4]

```
alls B []     = [([], 0)]
alls B (a#as) = [(x # xs, s + a * x).
                 (xs, s) ← alls B as, x ← [0..B]]
```

Example 1. For $a = [1,1]$ (corresponding to the left-hand side coefficients of our initial example) and $B = 2$ the list computed by allsB a is

```
[(([0,0],0),([1,0],1),([2,0],2),([0,1],1),([1,1],2),([2,1],3),
  ([0,2],2),([1,2],3),([2,2],4)]
```

---

[4] Also, in case you are wondering, the second component of the pairs will only play a role in Sect. 6, where it will avoid unnecessary recomputations of sub dot products. However, including these components already for alls serves the purpose of enabling later proofs of program transformations (or *code equations* as they are called in Isabelle).

Since for a potential solution (x, y) elements of x and of y have different bounds, we employ

```
generate A B a b =
    tl (map (λ(x, y). (fst x, fst y)) (alls2 A B a b))
```

where

```
alls2 A B a b = [(xs, ys). ys ← alls B b, xs ← alls A a]
```

Note that the result of generate is sorted with respect to $<_{\mathsf{rlex}}$. If we use max b and max a as bounds for x and y, respectively, then generate takes care of the new generate phase.

The static bounds on individual candidate solutions we obtain from Lemma 2 can be checked by the predicate

$$\text{static\_bounds } x \ y \longleftrightarrow$$
$$(\forall 1 \leq i \leq m. \ x_i \leq \max^{\neq 0}_y(b)) \wedge (\forall 1 \leq j \leq n. \ y_j \leq \max^{\neq 0}_x(a))$$

The new check phase is based on the following predicate, which is a combination of these static bounds, the fact that we are only interested in solutions, and the three further bounds from Lemma 7

```
check_cond (x, y) = static_bounds x y ∧ a • x = b • y ∧
    boundr x y ∧ subdprodl x y ∧ subdprodr y)
```

and implemented by check' = filter check_cond.

The new minimization phase finally, is still implemented by minimize, only that this time its input will often be a shorter list.

Combining all three phases, non-special solutions are computed by

```
non_special_solutions =
    let A = max b; B = max a in
    minimize (check' (generate' A B a b))
```

By including all special solutions we arrive at the intermediate algorithm solve, which already separates special from non-special solutions, but still requires further optimization:

```
solve a b = special_solutions a b @ non_special_solutions a b
```

The proof that solve a b correctly computes the set of minimal solutions, that is set (solve a b) $= \mathcal{M}(a, b)$, is somewhat complicated by the additional bounds, but structurally similar enough to the corresponding proof of solutions that we refer the interested reader to our formalization.

Having covered the correctness of our algorithm, it is high time to turn towards performance issues.

6   A More Efficient Algorithm for Code Generation

While the list of non-special solutions computed in Sect. 5 lends itself to formalization (due to its separation of concerns regarding the generate and check phases), it may waste a lot of time on generating lists that will not pass the later checks.

Example 2. Recall our initial example with coefficients $a = [1,1]$ and $b = [2]$. Let $A = \max b = 2$ and $B = \max a = 1$. Then, the list generated by allsB b contains for example a y-entry $([0],0)$. This is combined with all nine elements of allsA a (listed in Example 1) before filtering takes place, even though only a single x-entry, namely $([0,0],0)$, will survive the check phase (since all others exceed the bound $\max_{[0]}^{\neq 0}(b) = 0$ for some entry).

We now proceed to a more efficient variant of non_special_solutions which computes the same results (alas, we cannot hope for better asymptotic behavior, since computing minimal complete sets of solutions of HLDEs is NP-complete). While all of the following has been formalized, we will not give any proofs here, due to their rather technical nature and a lack of further insights. We start with the locale

> **locale** *bounded_gen_check* =
> **fixes** C **and** B
> **assumes** $C\ (x \mathbin{\#} xs)\ s = \textit{False}$ **if** $x > B$
> **and** $C\ (x' \mathbin{\#} xs)\ s'$ **if** $C\ (x \mathbin{\#} xs)\ s,\ x' \leq x,\ s' \leq s$

which takes a condition C, a bound B, and defines a function gen_check that combines (to a certain extent) generate' and check' from the previous section.

```
gen_check [] = [([], 0)]
gen_check (a # as) = concat (map (incs a 0) (gen_check as))
```

Here, the auxiliary function incs is defined by (note that termination of this function relies on the fact that there is an upper bound—namely B, as ensured by the first assumption of the locale—on the entries of the generated lists):

```
incs a x (xs, s) =
  let t = s + a*x in
  if C (x#xs) t then (x#xs, t) # incs a (x+1) (xs, s) else []
```

The idea of gen_check is to length-incrementally (starting with rightmost elements) generates all lists whose elements are bounded by B, such that only intermediate results that satisfy C are computed.

For us, the crucial property of gen_check is its connection to alls, which is covered by the following result (for which we need the second locale assumption).

Lemma 8. gen_check a = filter (suffs C a) (alls B a)

Where suffs C a (x,s) ensures that $|x| = |a|$, $s = a \bullet x$ and all non-empty
suffixes of the list x (including x itself) satisfy condition C.

Now we can define generate_check in terms of two instantiations of the
locale *bounded_gen_check* (meaning that each time the locale parameters C
and B are replaced by terms for which all assumptions of the locale are satisfied), using appropriate conditions C_1, C_2 and bounds B_1, B_2, respectively.
This results in the two instances gen_check$_1$ and gen_check$_2$ of gen_check,
where gen_check$_1$ receives a further parameter y, which stands for a fixed
y-entry against which we are trying to generate x-entries.

To be more precise, we use the following instantiations

$$B_1 = \lambda b.\ \max b$$
$$B_2 = \max a$$
$$C_1\ b\ y\ x\ s \longleftrightarrow x = [\ ] \vee s \le b \bullet y \wedge x \le \max_y^{\neq 0}(b)$$
$$C_2\ y\ s \longleftrightarrow y = [\ ] \vee (y \le \max a\ \wedge s \le a \bullet \mathrm{map}\ (\mathrm{max\_x}\ y)\ [1..|a|])$$

Combining gen_check$_1$ and gen_check$_2$ we obtain a function that computes
candidate solutions as follows:

generate_check a b = $[(x,y) \mid y \leftarrow$ gen_check$_2$ $b, x \leftarrow$ gen_check$_1$ y $a]$

Using Lemma 8 it can be shown that generate_check behaves exactly the
same way as first generating candidates using alls2 and then filtering them
according to conditions C_1 and C_2.

generate_check a b =
$\quad [(x,y) \leftarrow$ alls2 $(B_1\ b)\ B_2\ a\ b.$ suffs $(C_1\ b\ (\mathrm{fst}\ y))\ a\ x \wedge$ suffs $C_2\ b\ y]$

We further filter this list of candidate solutions in order to get rid of superfluous
entries, resulting in the function fast_filter defined by

filter P (map $(\lambda(x,y).\ (\mathrm{fst}\ x, \mathrm{fst}\ y))$ (tl (generate_check a b)))

where P $(x,y) =$ static_bounds x $y \wedge a \bullet x = b \bullet y \wedge$ boundr x y.

Extensionally fast_filter is equivalent to what non_special_solutions of
our intermediate algorithm above does before minimization.

Lemma 9. *Let* $A = \max b$ *and* $B = \max a$. *Then*

fast_filter a b = check' a b (generate' A B a b)

This finally allows us to use the following more efficient definition of solve for
code generation (of course all results on solve carry over, since extensionally the
two versions of solve are the same, as shown by Lemma 9).

solve a b = special_solutions a b @ minimize (fast_filter a b)

Generating the Solver. At this point we generate Haskell code for solve (and
also for the library functions integer_of_nat and nat_of_integer, which will
be used in our main file) by

export-code solve integer_of_nat nat_of_integer
in *Haskell* **module-name** *HLDE* **file** *"generated/"*

(For this step a working Isabelle installation is required.)

The only missing part is the (hand written) main entry point to our program in Main.hs (it takes an HLDE as command line argument in Haskell syntax, makes sure that the coefficients are all nonzero, hands the input over to solve, and prints the result):

```
main = getArgs >>= parse

parse [s] = start s parse _   = do
  hPutStrLn stderr usage
  exitWith (ExitFailure 1)

start input = do
  let (a, b) = read input :: ([Integer], [Integer])
  if 0 `elem` a || 0 `elem` b then do
    hPutStrLn stderr "0-coefficients are not allowed"
    exitWith (ExitFailure 2)
  else if null a || null b then do
    hPutStrLn stderr "empty lists coefficients are not allowed"
    exitWith (ExitFailure 3)
  else
    mapM_ (putStrLn . show . (\(x, y) ->
      (map integer_of_nat x, map integer_of_nat y))) (
      solve (map nat_of_integer a) (map nat_of_integer b))

usage = {- ... -}
```

A corresponding binary hlde can be compiled using the command (provided of course that our AFP entry and a Haskell compiler are both installed):

```
isabelle afp_build HLDE
```

We conclude this section by an example run (joining output lines to save space):

```
$ ./hlde "([2,1],[1,1,2])"
([1,0],[2,0,0]) ([1,0],[0,2,0]) ([1,0],[0,0,1]) ([0,1],[1,0,0])
([0,1],[0,1,0]) ([0,2],[0,0,1]) ([1,0],[1,1,0])
```

7   Evaluation

We compare our verified algorithms—the *simple* algorithm (S) of Sect. 4, the *intermediate* algorithm of Sect. 5 (I), and the *efficient* algorithm of Sect. 6 (E)—with the fastest unverified implementation we are aware of: a *graph* algorithm (G) due to Clausen and Fortenbacher [2].

Table 2. Comparing runtimes of verified algorithms and fastest known algorithm

HLDE with coefficients			verified algorithms			
a	b	#sols	S time (s)	I time (s)	E time (s)	G time (s)
[1,1]	[2]	3	0.001	0.001	0.001	n/a
[1,1]	[3]	4	0.001	0.001	0.001	n/a
[1,1,1]	[3]	10	0.001	0.002	0.001	n/a
[1,1,1]	[3,3,2]	26	0.002	0.003	0.002	n/a
[1,2,5]	[1,2,3,4]	39	0.2	0.3	0.07	0.012
[1,1,1,2,3]	[1,1,2,2]	44	0.2	0.01	0.01	0.006
[2,5,9]	[1,2,3,7,8]	119	188.00	212.00	21.00	0.081
[2,2,2,3,3,3]	[2,2,2,3,3,3]	138	262.00	49.00	0.07	0.012
[1,4,4,8,12]	[3,6,9,12,20]	232	-	-	221.00	0.180

In Table 2 we give the resulting runtimes (in seconds) for computing minimal complete sets of solutions of a small set of benchmark HLDEs (in increasing order of number of solutions; column #sols): the first four lines cover our initial example and three slight modifications, while the remaining examples are taken from Clausen and Fortenbacher).

However, there are two caveats: on the one hand, the runtimes for G are direct transcriptions from Clausen and Fortenbacher (hence also the missing entries for the first four examples), that is, they where generated on hardware from more than two decades ago; on the other hand, G uses improved bounds for the search-space of potential solutions, which are not formalized and thus out of reach for our verified implementations.

Anyway, our initial motivation was to certify minimal complete sets of AC-unifiers. Which is, why we want to stress the following: already for the first four examples of Table 2 the number of AC-unifiers goes from five, over 13, then 981, up to 65 926 605. For the remaining examples we were not even able to compute the number of minimal AC-unifiers (running out of memory on 20 GB of RAM); remember that in the worst case for an elementary unification problem whose corresponding HLDE has n minimal solutions, the number of minimal AC unifiers is in the order of 2^n. Thus, applications that rely on minimal complete sets of AC-unifiers will most likely not succeed on examples that are much bigger than the one in line three of Table 2, rendering certification moot.

On the upside, we expect HLDEs arising from realistic examples involving AC-unification to be quite small, since the nesting level of AC-symbols restricts the length of a and b and the multiplicity of variables restricts individual entries.

8   Related Work

In the literature, there are basically three approaches for solving HLDEs: lexicographic algorithms, completion procedures, and graph theory based algorithms.

Already in the 1970s Huet devised *an algorithm to generate the basis of solutions to homogeneous linear diophantine equations* in a paper of the same title [5], the first instance of a lexicographic algorithm. Our formalization of HLDEs and bounds on minimal solutions is inspired by Huet's elegant and short proofs. We also took up the idea of separating special and non-special solutions from Huet's work. Moreover, the structure of our algorithm mostly corresponds to Huet's informal description of his lexicographic algorithm: a striking difference is that we use a reverse lexicographic order. This facilities a construction relying on recursive list functions without the need of accumulating parameters. Compared to the beginning of our work, where we tried to stay with the standard lexicographic order, this turned out to lead to greatly simplified proofs.

In 1989, Lankford [7] proposed the first completion procedure solving HLDEs.

Fortenbacher and Clausen [2] give an accessible survey of these earlier approaches and in addition present the first graph theory based algorithm. They conclude that any of the existing algorithms is suitable for AC-unification: on the one hand there are huge performance differences for some big HLDEs; on the other hand AC-unification typically requires only relatively small instances; moreover, if the involved HLDEs grow too big the number of minimal AC-unifiers explodes massively, dwarfing the resource requirements for solving those HLDEs.

Later, Contejean and Devie [3] gave the first algorithm that was able to solve *systems* of linear diophantine equations (and is inspired by a geometric interpretation of the algorithm due to Fortenbacher and Clausen).

In contrast to our purely functional algorithm, all of the above approaches have a distinctively imperative flavor, and to the best of our knowledge, none of them have been formalized using a proof assistant.

9   Conclusions and Further Work

We had two main reasons for choosing a lexicographic algorithm (also keeping in mind that the problem being NP-complete, all approaches are asymptotically equivalent): (1) our ultimate goal is AC-unification and as Fortenbacher and Clausen [2] put it *"How important are efficient algorithms which solve [HLDEs] for [AC-unification]? [...] any of the algorithms presented [...] might be chosen [...],"* and (2) Huet's lexicographic algorithm facilitates a simple purely functional implementation that is amenable to formalization.

Structure and Statistics. Our formalization comprises 3353 lines of code. These include 73 definitions and functions as well as 281 lemmas and theorems, most of which are proven using Isabelle's *Intelligible Semi-Automated Reasoning* language Isar [13]. The formalization is structured into the following theory files:

List_Vector covering facts (about dot products, pointwise subtraction, several orderings, etc.) concerning vectors represented as lists of natural numbers.
Linear_Diophantine_Equations covering the abstract results on HLDEs discussed in Sect. 3.

Sorted_Wrt, Minimize_Wrt covering some facts about sortedness and minimization with respect to a given binary predicate.

Simple_Algorithm containing the simple algorithm of Sect. 2 and its correctness proof (Sect. 4).

Algorithm containing an intermediate algorithm (Sect. 5) that separates special from non-special solutions, as well as a more efficient variant (Sect. 6).

Solver_Code issuing a single command to generate Haskell code for solve and compiling it into a program hlde.

Future Work. Our ultimate goal is of course to reuse the verified algorithm in an Isabelle/HOL formalization of AC-unification.

Another direction for future work is to further improve our algorithm. For example, the improved bounds $\sum_{i=1}^{m} x_i \leq \max b$ and $\sum_{j=1}^{n} y_j \leq \max a$ are discussed by Clausen and Fortenbacher [2]. Moreover, already Huet [5] mentions the optimization of explicitly computing x_1 after ($[x_2, \ldots, x_m], y$) is fixed (which potentially divides the number of generated lists by the maximum value in b).

References

1. Baader, F., Nipkow, T.: Term Rewriting and All That. Cambridge University Press, New York (1998)
2. Clausen, M., Fortenbacher, A.: Efficient solution of linear diophantine equations. J. Symbolic Comput. **8**(1), 201–216 (1989). https://doi.org/10.1016/S0747-7171(89)80025-2
3. Contejean, É., Devie, H.: An efficient incremental algorithm for solving systems of linear diophantine equations. Inf. Comput. **113**(1), 143–172 (1994). https://doi.org/10.1006/inco.1994.1067
4. Haftmann, F., Nipkow, T.: Code generation via higher-order rewrite systems. In: Blume, M., Kobayashi, N., Vidal, G. (eds.) FLOPS 2010. LNCS, vol. 6009, pp. 103–117. Springer, Heidelberg (2010). https://doi.org/10.1007/978-3-642-12251-4_9
5. Huet, G.: An algorithm to generate the basis of solutions to homogeneous linear diophantine equations. Inf. Process. Lett. **7**(3), 144–147 (1978). https://doi.org/10.1016/0020-0190(78)90078-9
6. Klein, D., Hirokawa, N.: Confluence of non-left-linear TRSs via relative termination. In: Bjørner, N., Voronkov, A. (eds.) LPAR 2012. LNCS, vol. 7180, pp. 258–273. Springer, Heidelberg (2012). https://doi.org/10.1007/978-3-642-28717-6_21
7. Lankford, D.: Non-negative integer basis algorithms for linear equations with integer coefficients. J. Autom. Reasoning **5**(1), 25–35 (1989). https://doi.org/10.1007/BF00245019
8. Marché, C.: Normalized rewriting: an alternative to rewriting modulo a set of equations. J. Symbolic Comput. **21**(3), 253–288 (1996). https://doi.org/10.1006/jsco.1996.0011
9. Meßner, F., Parsert, J., Schöpf, J., Sternagel, C.: Homogeneous Linear Diophantine Equations. The Archive of Formal Proofs, October 2017. https://devel.isa-afp.org/entries/Diophantine_Eqns_Lin_Hom.shtml, Formal proof development

10. Nagele, J., Felgenhauer, B., Middeldorp, A.: CSI: new evidence – a progress report. In: de Moura, L. (ed.) CADE 2017. LNCS (LNAI), vol. 10395, pp. 385–397. Springer, Cham (2017). https://doi.org/10.1007/978-3-319-63046-5_24

11. Shintani, K., Hirokawa, N.: CoLL: a confluence tool for left-linear term rewrite systems. In: Felty, A.P., Middeldorp, A. (eds.) CADE 2015. LNCS (LNAI), vol. 9195, pp. 127–136. Springer, Cham (2015). https://doi.org/10.1007/978-3-319-21401-6_8

12. Stickel, M.: A unification algorithm for associative-commutative functions. J. ACM **28**(3), 423–434 (1981). https://doi.org/10.1145/322261.322262

13. Wenzel, M.: Isabelle/Isar - a versatile environment for human-readable formal proof documents. Ph.D. thesis, Institut für Informatik (2002)

14. Winkler, S., Middeldorp, A.: Normalized completion revisited. In: Proceedings of the 24th International Conference on Rewriting Techniques and Applications (RTA). Leibniz International Proceedings in Informatics, vol. 21, pp. 319–334. Schloss Dagstuhl (2013). https://doi.org/10.4230/LIPIcs.RTA.2013.319

Formalizing Implicative Algebras in Coq

Étienne Miquey[✉]

Équipe Gallinette Inria, LS2N (CNRS), Nantes, France
etienne.miquey@inria.fr

Abstract. We present a Coq formalization of Alexandre Miquel's *implicative algebras* [18], which aim at providing a general algebraic framework for the study of classical realizability models. We first give a self-contained presentation of the underlying *implicative structures*, which roughly consist of a complete lattice equipped with a binary law representing the implication. We then explain how these structures can be turned into models by adding separators, giving rise to the so-called implicative algebras. Additionally, we show how they generalize Boolean and Heyting algebras as well as the usual algebraic structures used in the analysis of classical realizability.

1 Introduction

Krivine Classical Realizability. It is well-known since Griffin's seminal work [10] that a classical Curry-Howard correspondence can be obtained by adding control operators to the λ-calculus. Several calculi were born from this idea, amongst which Krivine λ_c-calculus [13], defined as the λ-calculus extended with Scheme's `call/cc` operator (for *call-with-current-continuation*). Elaborating on this calculus, Krivine's developed in the late 90s the theory of *classical realizability* [13], which is a complete reformulation of its intuitionistic twin. Originally introduced to analyze the computational content of classical programs, it turned out that classical realizability also provides interesting semantics for classical theories. While it was first tailored to Peano second-order arithmetic (*i.e.* second-order type systems), classical realizability actually scales to more complex classical theories, *e.g.* ZF [14], and gives rise to surprisingly new models. In particular, its generalizes Cohen's forcing [14,17] and allows for the direct definition of a model in which neither the continuum hypothesis nor the axiom of choice hold [16].

Algebraization of Classical Realizability. During the last decade, the study of the algebraic structure of the models that classical realizability induces have been an active research topic. This line of work was first initiated by Streicher, who proposed the concept of *abstract Krivine structure* [24], followed by Ferrer,

Electronic supplementary material The online version of this chapter (https://doi.org/10.1007/978-3-319-94821-8_27) contains supplementary material, which is available to authorized users.

© Springer International Publishing AG, part of Springer Nature 2018
J. Avigad and A. Mahboubi (Eds.): ITP 2018, LNCS 10895, pp. 459–476, 2018.
https://doi.org/10.1007/978-3-319-94821-8_27

Frey, Guillermo, Malherbe and Miquel who introduced other structures peculiar to classical realizability [5–9]. In addition to the algebraic study of classical realizability models, these works had the interest of building the bridge with the algebraic structures arising from intuitionistic realizability. In particular, Streicher showed in [24] how classical realizability could be analyzed in terms of *triposes* [21], the categorical framework arising from intuitionistic realizability models, while the later work of Ferrer *et al.* [6,7] connected it to Hofstra and Van Oosten's notion of *ordered combinatory algebras* [12]. More recently, Alexandre Miquel introduced the elegant concepts of *implicative structure* and *implicative algebra*[18][1], which appear to encompass the previous approaches and which we present in this paper.

Implicative Structures. In addition to providing an algebraic framework conducive to the analysis of classical realizability, an important feature of implicative structures is that they allow us to identify *realizers* (*i.e.* λ-terms) and *truth values* (*i.e.* formulas). Concretely, implicative structures are complete lattices equipped with a binary operation $a \to b$ satisfying properties coming from the logical implication. As we will see, they indeed allow us to interpret both the formulas and the terms in the same structure. For instance, the ordering relation $a \le b$ will encompass different intuitions depending on whether we regard a and b as formulas or as terms. Namely, $a \le b$ will be given the following meanings:

- the formula a is a *subtype* of the formula b;
- the term a is a *realizer* of the formula b;
- the realizer a is *more defined* than the realizer b.

The last item corresponds to the intuition that if a is a realizer of all the formulas of which b is a realizer, a is more precise than b, or more powerful as a realizer.

In terms of the Curry-Howard correspondence, this means that not only do we identify types with formulas and proofs with programs, but *we also identify types and programs*.

Implicative Algebras. Because we consider formulas as realizers, any formula will be at least realized by itself. In particular, the lowest formula ⊥ is realized. While this can be dazzling at first sight, it merely reflects the fact that implicative structures do not come with an intrinsic criterion of consistency. To overcome this, we will introduce the notion of *separator*, which is similar to the usual notion of filter for Boolean algebras. *Implicative algebras* will be defined as implicative structures equipped with a separator. As we shall see, they capture the algebraic essence of classical realizability models. In particular, we will embed both the λ_c-calculus and its second-order type system in such a way that the adequacy is preserved. Implicative algebras therefore appear to be the adequate algebraic structure to study classical realizability and the models it induces.

Coq Formalization. The formalization of implicative algebras that we present in this paper has been written using the Coq proof assistant. It was written

---

[1] Independently, very similar structures can be found in Frédéric Ruyer's Ph.D. thesis [22] under the name of *applicative lattices*.

during the author's PhD, as a way of (1) checking the correctness of implicative algebras properties (which, at the time, were neither published nor formally written with their proofs), and (2) easing the further study of similar structures[2].

Technically, it relies on Charguéraud's *locally nameless representation* of λ-terms [2]. and the corresponding LN library[3], which was developed at the occasion of the POPLmark challenge [1]. As for the different algebraic structures evoked in the paper, we systematically represent them as classes using Sozeau-Oury's `Class` mechanism [23]. Interestingly, apart from the technical details mentioned above to define terms (and types), the formalization of the different results mostly follows the corresponding pen and paper proofs.

Outline of the paper. We begin by briefly recalling the structures of classical realizability models in Sect. 2. We then present in Sect. 3 the concept of implicative structures and explain how it generalizes well-known algebraic structures[4]. We then show in Sect. 4 how λ_c-terms and second-order types can be adequately embedded within implicative structures. Finally, we introduce implicative algebras in Sect. 5. We study their internal logic and finally explain how they give rise to models. *It should be clear to the reader that the notion of implicative algebra and its properties are due to Alexandre Miquel [18].*

The theorems in the paper are hyperlinked with their formalizations in the Coq development[5]. Detailed proofs can be found in [19, Chapter 10] from which this paper is partially taken.

2   Krivine Classical Realizability

Due to the lack of space, it is not possible to fully introduce here Krivine classical realizability and its models defined using the machinery of the λ_c-calculus[6]. Rather than that, we choose to present it through the lenses of Streicher's *abstract Krivine structures* (AKS), which are merely an axiomatization of the Krivine abstract machine for the λ_c-calculus viewed as an algebraic structure:

Definition 1 (AKS). *An* abstract Krivine structure *is given by a septuple* $(\Lambda, \Pi, app, push, k_-, k, s, \text{cc}, \mathbf{PL}, \perp\!\!\!\perp)$ *where:*

1. Λ *and* Π *are non-empty sets, called the* terms *and the* stacks *of the AKS;*
2. $app : t, u \mapsto tu$ *is a function (called* application*) from* $\Lambda \times \Lambda$ *to* Λ*;*
3. $push : t, \pi \mapsto t \cdot \pi$ *is a function (called* push*) from* $\Lambda \times \Pi$ *to* Π*;*
4. $k_- : \pi \mapsto k_\pi$ *is a function from* Π *to* Λ *(k_π is called a continuation);*

---

[2] Namely, one goal of the author's PhD work was to define similar algebras based on the decomposition of the implication as $\neg A \vee B$ and $\neg(A \wedge \neg B)$ (see [19]).

[3] In doing so, our development implicitly relies on assumptions of functional and propositional extensionnality, which we do not need nor use.

[4] We will not recall the definition of lattices, Heyting algebras and so on, for a more detailed introduction we refer the reader to [19, Chapter 9].

[5] Available at https://gitlab.com/emiquey/ImplicativeAlgebras/.

[6] For a detailed introduction on this topic, we refer the reader to [13] or [19].

5. $\boldsymbol{k}, \boldsymbol{s}$ and cc *are three distinguished terms of* Λ;
6. $\bot\!\!\!\bot \subseteq \Lambda \times \Pi$ *(called the* pole*) is a relation between terms and stacks, also written* $t \star \pi \in \bot\!\!\!\bot$*. This relation fulfills the following axioms for all terms* $t, u, v \in \Lambda$ *and all stacks* $\pi, \pi' \in \Lambda$:

$$
\begin{aligned}
t \star u \cdot \pi \in \bot\!\!\!\bot &\Rightarrow tu \star \pi \in \bot\!\!\!\bot \\
t \star \pi \in \bot\!\!\!\bot &\Rightarrow \boldsymbol{k} \star t \cdot u \cdot \pi \in \bot\!\!\!\bot \\
tv(uv) \star \pi \in \bot\!\!\!\bot &\Rightarrow \boldsymbol{s} \star t \cdot u \cdot v \cdot \pi \in \bot\!\!\!\bot
\end{aligned}
\qquad
\begin{aligned}
t \star \boldsymbol{k}_\pi \cdot \pi \in \bot\!\!\!\bot &\Rightarrow \mathsf{cc} \star t \cdot \pi \in \bot\!\!\!\bot \\
t \star \pi \in \bot\!\!\!\bot &\Rightarrow \boldsymbol{k}_\pi \star t \cdot \pi' \in \bot\!\!\!\bot
\end{aligned}
$$

7. $\mathbf{PL} \subset \Lambda$ *is a subset of* Λ *(whose elements are called the* proof-like *terms), which contains* $\boldsymbol{k}, \boldsymbol{s}, \mathsf{cc}$ *and is closed under application.*

Given any subset of stacks $X \subseteq \Pi$ (which we call a *falsity value*), we write $X^{\bot\!\!\!\bot}$ for its orthogonal set with respect to the pole:

$$
X^{\bot\!\!\!\bot} \triangleq \{t \in \Lambda : \forall \pi \in X, t \star \pi \in \bot\!\!\!\bot\}
$$

Orthogonality for subsets $X \subseteq \Lambda$ (*i.e.* a *truth value*) is defined identically. Intuitively, classical realizability models are mainly given by the choice of the sets $\bot\!\!\!\bot$ and $\mathbf{PL}$ together with the interpretation of formulas as falsity values. Valid formulas are the one admitting a proof-like *realizer*, that is to say a term $t \in \mathbf{PL}$ such that $t \in \|A\|^{\bot\!\!\!\bot}$ where $\|A\| \in \mathcal{P}(\Pi)$ is the falsity value of A.

3   Implicative Structures

3.1   Definition

Intuitively, *implicative structures* are tailored to represent both the formulas of second-order logic and realizers arising from Krivine's λ_c-calculus. We shall see in the sequel how they indeed allow us to define λ-terms, but let us introduce them by focusing on their logical facet. We are interested in formulas of second-order logic, that is to say of system F, which are defined by a simple grammar:

$$
A, B ::= X \mid A \Rightarrow B \mid \forall X.A
$$

Implicative structures are therefore defined as meet-complete lattices (for the universal quantification) with an internal binary operation satisfying the properties of the implication:

Definition 2. *An* implicative structure *is a complete meet-semilattice* $(\mathcal{A}, \preccurlyeq)$ *equipped with a binary operation* $(a, b) \mapsto (a \to b)$*, called the* implication *of* $\mathcal{A}$*, that fulfills the following axioms:*

1. *Implication is anti-monotonic with respect to its first operand and monotonic with respect to its second operand, in the sense that for all* $a, a_0, b, b_0 \in \mathcal{A}$:

(*Variance*)   *If* $a_0 \preccurlyeq a$ *and* $b \preccurlyeq b_0$ *then* $(a \to b) \preccurlyeq (a_0 \to b_0)$.

2. *Arbitrary meets distribute over the second operand of implication, in the sense that for all $a \in \mathcal{A}$ and for all subsets $B \subseteq \mathcal{A}$:*

$$(\text{Distributivity}) \qquad\qquad \bigwedge_{b \in B} (a \to b) = a \to \bigwedge_{b \in B} b$$

Remark 3. *In the particular case where $B = \emptyset$, the axiom of distributivity states that $a \to \top = \top$ for all $a \in \mathcal{A}$.*

3.2   Examples of Implicative Structures

Complete Heyting Algebras. The first example of implicative structures is given by complete Heyting algebras. Indeed, the axioms of implicative structures are intuitionistic tautologies verified by any complete Heyting algebra. Therefore, every complete Heyting algebra induces an implicative structure with the same arrow:

Proposition 4. *Every complete Heyting algebra is an implicative structure.*

Proof. Since $\mathcal{H}$ is complete, by definition we have $a \to b = \bigvee \{x \in \mathcal{H} : a \wedge x \preccurlyeq b\}$, from which we deduce that $a \curlywedge c \preccurlyeq b \Leftrightarrow a \preccurlyeq c \to b$. The axioms defining implicative structures are straightforward to prove using these observations.

The converse is obviously false, since the implication of an implicative structure $\mathcal{A}$ is in general not determined by the lattice structure of $\mathcal{A}$. Besides, since any (complete) Boolean algebra is in particular a (complete) Heyting algebra, *a fortiori* any complete Boolean algebra induces an implicative structure:

Proposition 5. *If $\mathcal{B}$ is a complete Boolean algebra, then $\mathcal{B}$ induces an implicative structure where the implication is defined for all $a, b \in \mathcal{B}$ by $a \to b \triangleq \neg a \curlyvee b$.*

Dummy Structures. Given a complete lattice $\mathcal{L}$, it is easy to check that the following definitions induce dummy implicative structures:

Proposition 6. *If $\mathcal{L}$ is a complete lattice, the following definitions give rise to implicative structures: 1. $a \to b \triangleq \top$ 2. $a \to b \triangleq b$ (for all $a, b \in \mathcal{L}$)*

Both definitions lead to implicative structures which are meaningless from the point of view of logic. Nonetheless, they will provide us with useful counter-examples.

Ordered Combinatory Algebras. We recall the notion of *ordered combinatory algebra*, abbreviated in OCA, which is a variant[7] of Hofstra and Van Oosten's notion of ordered partial combinatory algebras [12]. Ferrer *et al.* structures to represent Krivine realizability, called $^\mathcal{I}$OCA or $^\mathcal{K}$OCA, are particular cases of OCA [5–7].

---

[7] In partial combinatory algebras, the application is defined as a partial function.

Definition 7 (OCA). *An* ordered combinatory algebra *is given by a quintuple* $(\mathcal{A}, \leq, \boldsymbol{app}, \boldsymbol{k}, \boldsymbol{s})$, *where:*

- $\leq$ *is a partial order over* $\mathcal{A}$,
- $\boldsymbol{app} : (a, b) \mapsto ab$ *is a monotonic function*[8] *from* $\mathcal{A} \times \mathcal{A}$ *to* $\mathcal{A}$,
- $\boldsymbol{k} \in \mathcal{A}$ *is such that* $\boldsymbol{k}ab \leq a$ *for all* $a, b \in \mathcal{A}$,
- $\boldsymbol{s} \in \mathcal{A}$ *is such that* $\boldsymbol{s}abc \leq ac(bc)$ *for all* $a, b, c \in \mathcal{A}$.

Given any ordered combinatory algebra, we can define an implication on the complete lattice $\mathcal{P}(\mathcal{A})$ which give rise to an implicative structure:

Proposition 8. *If* $\mathcal{A}$ *is an ordered combinatory algebra, then the complete lattice* $\mathcal{P}(\mathcal{A})$ *equipped with the implication*[9]:

$$A \to B \; \triangleq \; \{r \in \mathcal{A} : \forall a \in A.ra \in B\} \qquad\qquad (\forall A, B \subseteq \mathcal{A})$$

is an implicative structure.

Proof. Both conditions (variance/distributivity) are trivial from the definition.

Implicative Structure of Classical Realizability. Our final example of implicative structure—which is the main motivation of this work—is given by classical realizability. As we saw in Sect. 2, the construction of classical realizability models, whether it be from Krivine's realizability algebras [14–16] in a set-theoretic like fashion or in Streicher's AKS [24], takes place in a structure of the form $(\Lambda, \Pi, \cdot, \bot\!\!\!\bot)$ where Λ is the set of realizers; Π is the set of stacks; $(\cdot) : \Lambda \times \Pi \to \Pi$ is a binary operation for pushing a realizer onto a stack and $\bot\!\!\!\bot \subseteq \Lambda \times \Pi$ is the pole. Given such a quadruple, we can define for all $a, b \in \mathcal{A}$:

$$\mathcal{A} \triangleq \mathcal{P}(\Pi) \qquad a \preccurlyeq b \triangleq a \supseteq b \qquad a \to b \triangleq a^{\bot\!\!\!\bot} \cdot b = \{t \cdot \pi : t \in a^{\bot\!\!\!\bot}, \pi \in b\}$$

where as usual $a^{\bot\!\!\!\bot}$ is $\{t \in \Lambda : \forall \pi \in a, (t, \pi) \in \bot\!\!\!\bot\} \in \mathcal{P}(\Lambda)$, the orthogonal set of $a \in \mathcal{P}(\Pi)$ with respect to the pole $\bot\!\!\!\bot$. It is easy to verify that:

Proposition 9. *The triple* $(\mathcal{A}, \preccurlyeq, \to)$ *is an implicative structure.*

Proof. The proof is again trivial. Variance conditions correspond to the usual monotonicity of truth and falsity values in Krivine realizability [13], while the distributivity follows directly by unfolding the definitions.

---

[8] Observe that the application, which is written as a product, is neither commutative nor associative in general.

[9] This definition is related with the consttruction of a realizability tripos from an OCA $\mathcal{A}$. Indeed, given a set X, the ordering on predicates of $\mathcal{P}(\mathcal{A})^X$ is defined by:

$$\varphi \vdash_X \psi \; \triangleq \; \exists r \in \mathcal{A}.\forall x \in X.\forall a \in \mathcal{A}.(a \in \varphi(x) \Rightarrow ra \in \psi(x))$$

where r is broadly a *realizer* of $\forall x \in X.\varphi(x) \Rightarrow \psi(x)$. See [12] for further details.

4 Interpreting the λ-calculus

4.1 Interpretation of λ-terms

We motivated the definition of implicative structures with the aim of obtaining a common framework for the interpretation both of types and programs. We shall now see how λ-terms can indeed be defined in implicative structures.

From now on, let $\mathcal{A} = (\mathcal{A}, \preccurlyeq, \rightarrow)$ denotes an arbitrary implicative structure.

Definition 10 (Application). *Given two elements* $a, b \in \mathcal{A}$*, we call the* application *of* a *to* b *and write* ab *the element of* $\mathcal{A}$ *that is defined by:*

$$ab \;\triangleq\; \bigwedge \{c \in \mathcal{A} : a \preccurlyeq (b \rightarrow c)\}.$$

If we think of the order relation $a \preccurlyeq b$ as "*a is more precise than b*", the above definition actually defines the application ab as the meet of all the elements c such that $b \rightarrow c$ is an approximation of a. This definition fulfills the usual properties of the λ-calculus:

Proposition 11 (Properties of application). *For all* $a, a', b, b', c \in \mathcal{A}$ *:*

1. If $a \preccurlyeq a'$ and $b \preccurlyeq b'$, then $ab \preccurlyeq a'b'$ *(Monotonicity)*
2. $(a \rightarrow b)a \preccurlyeq b$ *(β-reduction)*
3. $a \preccurlyeq (b \rightarrow ab)$ *(η-expansion)*
4. $ab = \min\{c \in \mathcal{A} : a \preccurlyeq (b \rightarrow c)\}$ *(Minimum)*
5. $ab \preccurlyeq c \;\Leftrightarrow\; a \preccurlyeq (b \rightarrow c)$ *(Adjunction)*

Proof. Simple lattice manipulations using the properties of the arrow.

Remark 12 (Galois connection). *The adjunction* $ab \preccurlyeq c \Leftrightarrow a \preccurlyeq (b \rightarrow c)$ *expresses the existence of a family of Galois connections* $f_b \dashv g_b$ *indexed by all* $b \in \mathcal{A}$*, where the left and right adjoints* $f_b, g_b : \mathcal{A} \rightarrow \mathcal{A}$ *are defined by:*

$$f_b : a \mapsto ab \qquad and \qquad g_b : c \mapsto (b \rightarrow c) \qquad (for\ all\ a, b, c \in \mathcal{A})$$

Recall that in a Galois connection, the left adjoint is fully determined by the right one (and vice-versa). In the particular case of a complete Heyting algebra $(\mathcal{H}, \preccurlyeq, \rightarrow)$*, this implies that the application is characterized by* $ab = a \curlywedge b$ *for all* $a, b \in \mathcal{H}$*. Indeed, in any Heyting algebra, the adjunction* $a \curlywedge b \preccurlyeq c \Leftrightarrow a \preccurlyeq (b \rightarrow c)$ *holds for all* $a, b, c \in \mathcal{H}$*, by uniqueness of the left adjoint,* ab *and* $a \curlywedge b$ *are thus equal.*

Definition 13 (Abstraction). *Given a function* $f : \mathcal{A} \rightarrow \mathcal{A}$*, we call* abstraction *of* f *and write* λf *the element of* $\mathcal{A}$ *defined by:*

$$\lambda f \;\triangleq\; \bigwedge_{a \in \mathcal{A}} (a \rightarrow f(a))$$

Once again, if we think of the order relation $a \preccurlyeq b$ as "*a is more precise than b*", the meet of the elements of a set S is an element containing the union of all the informations given by the elements of S. With this in mind, the above definition sets λf as the union of all the step functions $a \to f(a)$. This definition, together with the definition of the application, fulfills again properties expected from the λ-calculus:

Proposition 14 (Properties of the abstraction). *The following holds for any* $f, g : \mathcal{A} \to \mathcal{A}$:

1. *If for all* $a \in \mathcal{A}$, $f(a) \preccurlyeq g(a)$, *then* $\lambda f \preccurlyeq \lambda g$.   (*Monotonicity*)
2. *For all* $a \in \mathcal{A}$, $(\lambda f)a \preccurlyeq f(a)$.   (*$\beta$-reduction*)
3. *For all* $a \in \mathcal{A}$, $a \preccurlyeq \lambda(x \mapsto ax)$.   (*$\eta$-expansion*)

Proof. Again, the proof consists in easy lattices manipulations.

We call *λ-term with parameters (in $\mathcal{A}$)* any term defined from the following grammar (see Footnote 11):

$$t, u ::= x \mid a \mid \lambda x.t \mid tu$$

where x is a variable and a is an element of $\mathcal{A}$. We can thus associate to each closed λ-term with parameters t an element $t^{\mathcal{A}}$ of $\mathcal{A}$, defined by induction on the size of t as follows (where $a \in \mathcal{A}$):

$$a^{\mathcal{A}} \triangleq a \qquad (tu)^{\mathcal{A}} \triangleq t^{\mathcal{A}} u^{\mathcal{A}} \qquad (\lambda x.t)^{\mathcal{A}} \triangleq \lambda(a \mapsto (t[a/x])^{\mathcal{A}})$$

Thanks to the properties of the application and of the abstraction in implicative structures that we proved, we can check that the embedding of λ-term is sound with respect to the β-reduction:

Proposition 15. *For all closed λ-terms t and u with parameters, if* $t \longrightarrow_{\beta} u$, *then* $t^{\mathcal{A}} \preccurlyeq u^{\mathcal{A}}$.

Proof. By induction on the reduction $t \longrightarrow_{\beta} u$ using Propositions 11 and 14.

Again, if we think of the order relation $a \preccurlyeq b$ as "*a is more precise than b*", it makes sense that the β-reduction $t \longrightarrow_{\beta} u$ is reflected in the ordering $t^{\mathcal{A}} \preccurlyeq u^{\mathcal{A}}$: the result of a computation contains indeed less information than the computation itself[10].

4.2   Adequacy

We now dispose of a structure in which we can interpret types and λ-terms. We saw that the interpretation of terms was intuitively sound with respect to the β-reduction. We shall now prove that the typing rules of System F are adequate with respect to the interpretation of terms, that is to say that if t is a closed λ-term of type T, then $t^{\mathcal{A}} \preccurlyeq T^{\mathcal{A}}$. The last statement can again be understood as the fact that a term (*i.e.* a computation) carries more information than its type, just like a realizer of a formula is more informative about the formula than the formula itself.

---

[10] For instance, 0 contains less information than $15 - (3 \times 5)$ or than $\mathbb{1}_{\mathbb{Q}}(\sqrt{(2)})$.

Adequacy of the Interpretation. We shall now sketch the formalization of the former result. First, we extend the usual formulas of System F by defining second-order formulas with parameters as:

$$A, B ::= a \mid X \mid A \Rightarrow B \mid \forall X.A \qquad\qquad (a \in \mathcal{A})$$

We can then embed closed formulas with parameters into the implicative structure $\mathcal{A}$. The embedding is trivially defined by:

$$a^{\mathcal{A}} \triangleq a \qquad (A \Rightarrow B)^{\mathcal{A}} \triangleq A^{\mathcal{A}} \to B^{\mathcal{A}} \qquad (\forall X.A)^{\mathcal{A}} \triangleq \curlywedge_{a \in \mathcal{A}} (A\{X := a\})^{\mathcal{A}}$$

where $a \in \mathcal{A}$. We define a type system for the λ_c-calculus with parameters[11] (that is λ-terms with parameter plus an instruction cc). Typing contexts are defined as usual by finite lists of hypotheses of the shape $(x : A)$ where x is a variable and A a formula with parameters. The inference rules, given in Fig. 1, are the same as in System F (with the extended syntaxes of terms and formulas with parameters), plus the additional rules for cc.

In order to prove the adequacy of the type system with respect to the embedding, we define substitutions, which we write σ, as functions mapping variables (of terms and types) to element of $\mathcal{A}$:

$$\sigma ::= \varepsilon \mid \sigma[x \mapsto a] \mid \sigma[X \mapsto a] \qquad\qquad (a \in \mathcal{A},\, x, X \text{ variables})$$

In the spirit of the proof of adequacy in classical realizability, we say that a substitution σ realizes a typing context Γ, which we write $\sigma \Vdash \Gamma$, if for all bindings $(x : A) \in \Gamma$ we have $\sigma(x) \preccurlyeq (A[\sigma])^{\mathcal{A}}$.

$$
\frac{(x : A) \in \Gamma}{\Gamma \vdash x : A}
\qquad
\frac{\Gamma, x : A \vdash t : B}{\Gamma \vdash \lambda x.t : A \to B}
\qquad
\frac{\Gamma \vdash t : A \to B \qquad \Gamma \vdash t : A}{\Gamma \vdash tu : B}
$$

$$
\frac{\Gamma \vdash t : A}{\Gamma \vdash t : \forall X.A}\,(X \notin FV(\Gamma))
\qquad
\frac{\Gamma \vdash t : \forall X.A}{\Gamma \vdash t : A\{X := B\}}
\qquad
\frac{}{\Gamma \vdash \mathbf{cc} : ((A \to B) \to A) \to A}
$$

Fig. 1. Second-order type system for the λ_c-calculus

Theorem 16 (Adequacy). *The typing rules of Fig. 1 are adequate with respect to the interpretation of terms and formulas: if t is a λ_c-term with parameters, A a formula with parameters and Γ a typing context such that $\Gamma \vdash t : A$ then for all substitutions $\sigma \Vdash \Gamma$, we have $(t[\sigma])^{\mathcal{A}} \preccurlyeq (A[\sigma])^{\mathcal{A}}$.*

---

[11] In practice, we use Charguéraud's locally nameless representation [2] for terms and formulas. Without giving too much details, we actually define pre-terms and pre-types which allow both for names (for free variables) and De Bruijn indices (for bounded variables). Terms and types are then defined as pre-terms and pre-types without free De Bruijn indices. As for the embedding from pre-terms (resp. pre-types) into an implicative structure, we define them by means of inductive predicates: Inductive **translated** : trm $\to$ X $\to$ Prop := ... for which we proved the expected properties.

Proof. The proof resembles the usual proof of adequacy in classical realizability (see [13,19]), namely by induction on typing derivations.

Corollary 17. *For all λ-terms t, if $\vdash t : A$, then $t^A \preccurlyeq A^A$.*

4.3   Combinators

The previous results indicate that any closed λ-term is, through the interpretation, lower than the interpretation of its principal type. We give here some examples of closed λ-terms which are in fact equal to their principal types through the interpretation in $\mathcal{A}$. Let us now consider the following combinators:

$$\mathbf{i} \triangleq \lambda x.x \qquad \mathbf{k} \triangleq \lambda xy.x \qquad \mathbf{s} \triangleq \lambda xyz.xz(yz) \qquad \mathbf{w} \triangleq \lambda xy.xyy$$

It is well-known that these combinators can be given the following polymorphic types:

$$
\begin{array}{l|l}
\mathbf{i} : \forall X.X \Rightarrow X & \mathbf{s} : \forall XYZ.(X \Rightarrow Y \Rightarrow Z) \Rightarrow (X \Rightarrow Y) \Rightarrow X \Rightarrow Z \\
\mathbf{k} : \forall XY.X \Rightarrow Y \Rightarrow X & \mathbf{w} : \forall XY.(X \Rightarrow X \Rightarrow Y) \Rightarrow X \Rightarrow Y
\end{array}
$$

Through the interpretation these combinators are identified with their types:

Proposition 18. *The following equalities hold in any implicative structure $\mathcal{A}$:*

1. *$\mathbf{i}^A = \bigwedge_{a \in \mathcal{A}} (a \to a)$*
2. *$\mathbf{k}^A = \bigwedge_{a,b \in \mathcal{A}} (a \Rightarrow b \Rightarrow a)$*
3. *$\mathbf{s}^A = \bigwedge_{a,b,c \in \mathcal{A}} ((a \to b \to c) \to (a \to b) \to a \to c)$*
4. *$\mathbf{w}^A = \bigwedge_{a,b,c \in \mathcal{A}} ((a \to a \to b) \to a \to b)$*

Proof. The inequality from left to right are consequences of the adequacy. The converse inequalities are proved by hands, using the properties of application and abstraction in implicative structures (Propositions 11 and 14).

Finally, in the spirit of the previous equalities, we define the interpretation of cc by the interpretation of its principal type, that is:

$$\mathbf{cc}^A \triangleq \bigwedge_{a,b} (((a \to b) \to a) \to a)$$

Remark 19. *It is not always the case that a term is equal to its principal type. Consider for instance a dummy implicative structure $\mathcal{A}$ where $a \to b = \top$ for all elements $a, b \in \mathcal{A}$. Suppose in addition that $\mathcal{A}$ has at least two distinct elements, so that $\bot \neq \top$. Then the following holds:*

1. *For any $a, b \in \mathcal{A}$, we have $ab = \bigwedge \{c : a \preccurlyeq b \to c\} = \bigwedge \mathcal{A} = \bot$.*
2. *For any $f : \mathcal{A} \to \mathcal{A}$, we have $\lambda f = \bigwedge_{a \in \mathcal{A}} (a \to f(a)) = \bigwedge_{a \in \mathcal{A}} \top = \top$.*
3. *$\mathbf{ii} : \forall X.X \to X$, yet $(\mathbf{ii})^A = \bot \neq \top = (\forall X.X \to X)^A$.*
4. *$\mathbf{i}^A = \top \neq \bot = (\mathbf{skk})^A$.*

4.4   The Problem of Consistency

The last remark shows us that not all implicative structures are suitable for interpreting intuitionistic or classical logic. We thus need to introduce a criterion of consistency.

Definition 20 (Consistency). *We say that an implicative structure is:*

- intuitionistically consistent *if $t^A \neq \bot$ for all closed λ-terms;*
- classically consistent *if $t^A \neq \bot$ for all closed λ_c-terms.*

We shall now relate the previous definition to the usual definition of consistency in classical realizability models. Recall that any abstract Krivine structure $\mathcal{K} = (\Lambda, \Pi, \mathsf{app}, \mathsf{push}, \mathsf{k}_{-}, \mathbf{k}, \mathbf{s}, \mathsf{cc}, \mathbf{PL}, \bot\!\!\!\bot)$ induces an implicative structure $(\mathcal{A}, \preccurlyeq, \rightarrow)$ where $\mathcal{A} = \mathcal{P}(\Pi)$, $a \preccurlyeq b \Leftrightarrow a \supseteq b$ and $a \rightarrow b = a^{\bot\!\!\!\bot} \cdot b$. A realizability model is said to be consistent when there is no proof-like term realizing $\bot$. In terms of abstract Krivine structures, the consistency can then be expressed by this simple criterion:

$$\mathcal{K} \text{ is consistent if and only if } \{\bot\}^{\bot\!\!\!\bot} \cap \mathbf{PL} = \Pi^{\bot\!\!\!\bot} \cap \mathbf{PL} = \emptyset$$

We thus need to check that this criterion of consistency for the AKS implies the consistency of the induced implicative structure, *i.e.* that if t is a closed λ_c-term, then $t^A \neq \bot$. By definition of the implicative structure $\mathcal{A}$ induced by $\mathcal{K}$, we have that $t^A \in \mathcal{A} = \mathcal{P}(\Pi)$. Therefore, t^A is a falsity value from the point of view of the AKS. To ensure that it is not equal to $\bot$ (*i.e.* Π), it is enough to find a realizer of t^A in $\mathcal{K}$. The consistency of $\mathcal{K}$ precisely states that $\bot$ does not have any realizer.

Our strategy to find a realizer for t^A in $\mathcal{K}$ is to use t itself. First, we reduce the problem to the set of terms that are identifiable with the combinatory terms of $\mathcal{K}$. We call a *combinatory term* any term that is obtained by combination of the combinators $(\mathbf{k}, \mathbf{s}, \mathsf{cc})$. To each combinatory term t we associate a term t^A in Λ, whose definition by induction is trivial:

$$\mathbf{k}^A \triangleq \mathbf{k} \qquad \mathbf{s}^A \triangleq \mathbf{s} \qquad \mathsf{cc}^A \triangleq \mathsf{cc} \qquad (tu)^A \triangleq \mathsf{app}(t^A, u^A)$$

Since the set $\mathbf{PL}$ is closed under application, for any combinatory term t, its interpretation t^A is in $\mathbf{PL}$. The combinatory completeness of $(\mathbf{k}, \mathbf{s}, \mathsf{cc})$ with respect to closed λ_c-terms ensures us that there exists a combinatory term t_0 (viewed as a λ-term) such that $t_0 \longrightarrow_\beta t$. By Proposition 15, we thus have $t_0^A \preccurlyeq t^A$. It is thus enough to show that $t_0^A \neq \bot$: we reduced the original problem for closed λ_c-terms to combinatory terms.

It only remains to show that for any combinatory term t_0, its interpretation t_0^A is not $\bot$. For the reasons detailed above, it is sufficient to prove that t_0^A is realized. We prove that t_0^A is in fact realized by t_0^A:

Lemma 21. *For any combinatory term t, t^A realizes t^A, i.e. $t^A \Vdash t^A$.*

Proof. By induction on the structure of t, by combining usual results of classical realizability and properties of implicative structures.

We can thus conclude that the consistency of $\mathcal{K}$ induces the one (in the sense of Definition 20) of the associated implicative structure:

Proposition 22. *If $\mathcal{K}$ is a consistent abstract Krivine structure, then the implicative structure it induces is classically consistent.*

Proof. Let t be any closed λ_c-term. We want to show that $t^{\mathcal{A}} \neq \bot = \boldsymbol{\Pi}$. We show that $t^{\mathcal{A}}$, which belongs to $\mathcal{P}(\boldsymbol{\Pi})$ is realized by a proof-like term.

It is worth noting that the criterion of consistency is defined with respect to the set **PL** together with the pole. These sets are already at the heart of the definition of Krivine's realizability models, where valid formulas are precisely the formulas realized by a proof-like term. We shall then introduce the corresponding ingredient for implicative structures.

5   Implicative Algebras

5.1   Separation

Definition 23 (Separator). *Let $(\mathcal{A}, \preccurlyeq, \to)$ be an implicative structure. We call a separator over $\mathcal{A}$ any set $\mathcal{S} \subseteq \mathcal{A}$ such that for all $a, b \in \mathcal{A}$, the following conditions hold:*

1. $\mathbf{k}^{\mathcal{A}} \in \mathcal{S}$, and $\mathbf{s}^{\mathcal{A}} \in \mathcal{S}$.   *(Combinators)*
2. *If $a \in \mathcal{S}$ and $a \preccurlyeq b$, then $b \in \mathcal{S}$.*   *(Upwards closure)*
3. *If $(a \to b) \in \mathcal{S}$ and $a \in \mathcal{S}$, then $b \in \mathcal{S}$.*   *(Closure under modus ponens)*

A separator $\mathcal{S}$ is said to be classical *if $\mathbf{cc}^{\mathcal{A}} \in \mathcal{S}$ and* consistent *if $\bot \notin \mathcal{S}$.*

Remark 24 (Alternative definition) . *In presence of condition (2), it is easy to show that condition (3) is equivalent to the following condition:*

(3') *If $a \in \mathcal{S}$ and $b \in \mathcal{S}$ then $ab \in \mathcal{S}$.*   *(Closure under application)*

Intuitively, thinking of elements of an implicative structure as truth values, a separator should be understood as the set which distinguishes the valid formulas. Considering the elements as terms, it should rather be viewed as the set of valid realizers. Indeed, conditions (1) and (3') ensure that all closed λ-terms are in any separator. Reading $a \preccurlyeq b$ as "*the formula a is a subtype of the formula b*", condition (2) ensures the validity of semantic subtyping. Thinking of the ordering as "*a is a realizer of the formula b*", condition (2) states that if a formula is realized, then it is in the separator.

Definition 25 (Implicative algebra). *We call* implicative algebra *any quadruple $(\mathcal{A}, \preccurlyeq, \to, \mathcal{S})$ where $(\mathcal{A}, \preccurlyeq, \to)$ is an implicative structure and $\mathcal{S}$ is a separator over $\mathcal{A}$. We say that an implicative algebra is* classical *if its separator is.*

Example 26 (Complete Boolean algebras). *It is easy to verify that for any complete Boolean algebra $\mathcal{B}$, combinators are interpreted by the maximal element in the induced implicative structure: $\mathbf{k}^{\mathcal{B}} = \mathbf{s}^{\mathcal{B}} = \mathbf{cc}^{\mathcal{B}} = \top$. Therefore, the singleton $\{\top\}$ is a classical separator for the induced implicative structure. Any non-degenerated complete Boolean algebras thus induces a classically consistent implicative algebra.*

Example 27 (Abstract Krivine structure). *Recall that any AKS induces an implicative structure $(\mathcal{A}, \preccurlyeq, \rightarrow)$ where $\mathcal{A} = \mathcal{P}(\mathbf{\Pi})$, $a \preccurlyeq b \Leftrightarrow a \supseteq b$ and $a \rightarrow b = a^{\perp\!\perp} \cdot b$. The set of realized formulas, namely $\mathcal{S} = \{a \in \mathcal{A} : a^{\perp\!\perp} \cap \mathbf{PL} \neq \emptyset\}$, defines a valid separator.*

5.2 λ_c-terms

The first property that we shall state about classical separators is that they contain the interpretation of all closed λ_c-terms. This follows again from the combinatory completeness of the basis $(\mathbf{k}, \mathbf{s}, \mathbf{cc})$ for the λ_c-calculus[12]. Indeed, if $\mathcal{S}$ is a classical separator over an implicative structure $(\mathcal{A}, \preccurlyeq, \rightarrow)$, it is clear that any combinatory term is in the separator. Again, by combinatory completeness, if t is a closed λ_c-term, there exists a combinatory term t_0 such that $t_0 \longrightarrow_\beta t$, and therefore $t_0^{\mathcal{A}} \preccurlyeq t^{\mathcal{A}}$ (by Proposition 15). By upward closure of separators, we deduce that:

Proposition 28. *If $(\mathcal{A}, \preccurlyeq, \rightarrow, \mathcal{S})$ is a (classical) implicative algebra and t is a closed λ-term (resp. λ_c-term), then $t^{\mathcal{A}} \in \mathcal{S}$.*

From the previous proposition and the adequacy of second-order typing rules for the λ_c-calculus (Theorem 16), we obtain that:

Corollary 29. *If $(\mathcal{A}, \preccurlyeq, \rightarrow, \mathcal{S})$ is a (classical) implicative algebra, t is a closed λ-term (resp. λ_c-term) and A is a formula such that $\vdash t : A$, then $A^{\mathcal{A}} \in \mathcal{S}$.*

Remark 30. *The latter corollary provides us with a methodology for proving that an element of a given implicative algebra is in the separator. In the spirit of realizability, where the standard methodology to prove that a formula is realized consists in using typed terms and adequacy as much as possible, we can use typed terms to prove automatically that the corresponding formulas belongs to the separator. We shall use this methodology abundantly in the sequel. In the Coq development, this corresponds to a tactic called* `realizer` *which allows us to prove that an element belongs to the separator simply by furnishing a realizer:*

Lemma composition: $\forall$ *a b c, (a $\mapsto$ b) $\mapsto$ (b $\mapsto$ c) $\mapsto$ a $\mapsto$ c $\in \mathcal{S}$*
Proof. intros. `realizer` *((λ^+ λ^+ λ^+([\$1] ([\$2] \$0)))). Qed. (** $\lambda xyz.y(xz)$ *)*

---

[12] In order to avoid the certification of the corresponding compilation function, we state this well-known fact as an axiom (the only one) in our development.

5.3 Internal Logic

In order to be able to define triposes from implicative algebras, the first step is to equip them with a structure of Heyting algebra. To this end, we begin with defining an entailment relation in the spirit of filtered OCAs [12]. We then define quantifiers and connectives as usual in classical realizability (see [13]), and we verify that they satisfy the usual logical rules. In the rest of this section, we work within a fixed implicative algebra $(\mathcal{A}, \preccurlyeq, \rightarrow, \mathcal{S})$.

Definition 31 (Entailment). *For all $a, b \in \mathcal{A}$, we say that a entails b and write $a \vdash_{\mathcal{S}} b$ if $a \rightarrow b \in \mathcal{S}$. We say that a and b are equivalent and write $a \cong_{\mathcal{S}} b$ if $a \vdash_{\mathcal{S}} b$ and $b \vdash_{\mathcal{S}} a$.*

Proposition 32 (Properties of $\vdash_{\mathcal{S}}$). *For any $a, b, c \in \mathcal{A}$, the following holds:*

    1. $a \vdash_{\mathcal{S}} a$ *(Reflexivity)*
    2. If $a \vdash_{\mathcal{S}} b$ and $b \vdash_{\mathcal{S}} c$ then $a \vdash_{\mathcal{S}} c$.   *(Transitivity)*
    3. If $a \preccurlyeq b$ then $a \vdash_{\mathcal{S}} b$.   *(Subtyping)*
    4. If $a \cong_{\mathcal{S}} b$, then $a \in \mathcal{S}$ if and only if $b \in \mathcal{S}$.   *(Closure under $\cong_{\mathcal{S}}$)*
    5. If $a \vdash_{\mathcal{S}} b \rightarrow c$ then $a \curlywedge b \vdash_{\mathcal{S}} c$.   *(Half-adjunction property)*
    6. $\bot \vdash_{\mathcal{S}} a$ *(Ex falso quod libet)*
    7. $a \vdash_{\mathcal{S}} \top$ *(Maximal element)*

Proof. Straightforward from the definitions, using $\lambda xyz.y(xz)$ to realize the second item and $\lambda xy.xyy$ to realize the fifth.

Besides, the entailment relation behaves like Heyting's arrow with respect to the preorder relation $\vdash_{\mathcal{S}}$ in terms of monotonicity:

Proposition 33 (Compatibility with $\rightarrow$). *For all $a, a', b, b' \in \mathcal{A}$, we have:*

    1. If $b \vdash b'$ then $a \rightarrow b \vdash a \rightarrow b'$.   *2. If $a \vdash a'$ then $a' \rightarrow b \vdash a \rightarrow b$.*

Proof. Direct using $\lambda xyz.x(yz)$ and $\lambda xyz.y(xz)$ as realizers.

Negation. We define the negation by $\neg a \triangleq a \rightarrow \bot$. If the separator is classical, we can prove that for any $a \in \mathcal{A}$, we have:

Proposition 34 (Double negation). *If $\mathcal{S}$ is a classical separator, for any $a \in \mathcal{A}$ we have:*   *1. $a \vdash_{\mathcal{S}} \neg\neg a$*   *2. $\neg\neg a \vdash_{\mathcal{S}} a$*

Proof. The first item is realized by $\lambda xk.kx$, while the second follows from the inequality $((a \rightarrow \bot) \rightarrow a) \rightarrow a \preccurlyeq ((a \rightarrow \bot) \rightarrow \bot) \rightarrow a$, whose left member is realized by **cc**.

Quantifiers. Following the usual definitions in classical realizability (see [13, 19]), the universal quantification of a family of truth values is naturally defined as its meet while the existential quantification is defined through a negative encoding:

$$\bigvee_{i \in I} a_i \triangleq \bigwedge_{i \in I} a_i \qquad\qquad \exists_{i \in I} a_i \triangleq \bigwedge_{c \in \mathcal{A}} \left(\bigwedge_{i \in I} (a_i \to c) \to c \right)$$

While it could have seemed more natural to define existential quantifiers through joins, we should recall that the arrow does not commute with joins in general[13]. It is clear that these definitions are compatible with the expected semantic rules:

Proposition 35. (Universal quantifier). *The following semantic typing rules are valid in any implicative structures:*

$$\frac{\Gamma \vdash t : a_i \quad \text{for all } i \in I}{\Gamma \vdash t : \forall_{i \in I}\, a_i} \qquad\qquad \frac{\Gamma \vdash t : \forall_{i \in I}\, a_i \quad i_0 \in I}{\Gamma \vdash t : a_{i_0}}$$

Proposition 36. (Existential quantifier). *The following semantic typing rules are valid in any implicative structures:*

$$\frac{\Gamma \vdash t : a_{i_0} \quad i_0 \in I}{\Gamma \vdash \lambda x.xt : \exists_{i \in I}\, a_i} \qquad\qquad \frac{\Gamma \vdash t : \exists_{i \in I}\, a_i \quad \Gamma, x : a_i \vdash u : c \quad (\text{for all } i \in I)}{\Gamma \vdash t(\lambda x.u) : c}$$

Proof. Straightforward using the adjunction of the application (Proposition 11).

Sum and Product. We define it by the usual encodings in System F:

$$a \times b \triangleq \bigwedge_{c \in \mathcal{A}} ((a \to b \to c) \to c) \qquad\qquad a + b \triangleq \bigwedge_{c \in \mathcal{A}} ((a \to c) \to (b \to c) \to c)$$

Recall that the pair $\langle a, b \rangle$ is encoded by the λ-term $\lambda x.xab$, while first and second projections are respectively defined by $\pi_1 \triangleq \lambda xy.x$ and $\pi_2 \triangleq \lambda xy.y$. We can check that the expected semantic typing rules are valid.

Proposition 37 (Product). *The following semantic typing rules are valid:*

$$\frac{\Gamma \vdash t : a \quad \Gamma \vdash u : b}{\Gamma \vdash \lambda z.ztu : a \times b} \qquad \frac{\Gamma \vdash t : a \times b}{\Gamma \vdash t\pi_1 : a} \qquad \frac{\Gamma \vdash t : a \times b}{\Gamma \vdash t\pi_2 : b}$$

Proposition 38 (Sum). *The following semantic typing rules are valid*

$$\frac{\Gamma \vdash t : a}{\Gamma \vdash \lambda lr.lt : a + b} \qquad \frac{\Gamma \vdash t : b}{\Gamma \vdash \lambda lr.rt : a + b} \qquad \frac{\Gamma \vdash t : a_1 + a_2 \quad \Gamma, x_i : a_i \vdash u_i : c}{\Gamma \vdash t(\lambda x_1.u_1)(\lambda x_2.u_2) : c}$$

---

[13] When it does, the realizability tripos actually collapses to a forcing tripos, see [18, 19].

Proof. Straightforward lattices manipulations, similar to the proof for the existential quantifier.

The natural candidate to computationally represents a "meet" of a and b is the product type $a \times b$. We can verify that it satisfies the expected property (in Heyting algebras) with respect to the arrow:

Proposition 39 (Adjunction). *For any $a, b, c \in \mathcal{A}$, we have $a \vdash_S b \to c$ if and only if $a \times b \vdash_S c$.*

Proof. Both directions are proved using the expected realizer and subtyping: from left to right, we use $\lambda xy.yx$ to realize $(a \to b \to c) \to a \times b \to c$; from right to left, we realize $(a \times b \to c) \to a \to b \to c$ with $\lambda pxy.p(\lambda z.zxy)$.

5.4   Implicative Tripos

It is clear from the properties of implicative algebras presented in the last sections that the entailment relation together with the sum and products induce a structure of Heyting prealgebra (indeed, the entailment relation only defines a preorder). By considering the quotient $\mathcal{A}/{\cong_S}$ of the former Heyting prealgebra by the relation $\cong_S$, and lifting the previous definitions of connectives and quantifiers to equivalence classes, we thus obtain a Heyting algebra[14]. This construction is actually the main step towards the definition of the implicative tripos [18,19], which allows us to recover the usual categorical interpretation of realizability models. In particular, it provides us with a framework in which simple criteria allows us to compare classical realizability and forcing models.

6   Conclusion and Future Work

6.1   Conclusion

We presented in this paper Miquel's concept of *implicative algebra* [18], that relies on the primitive notion of implicative structure. These structures are defined as a particular class of meet-complete lattices equipped with an arrow, where this arrow satisfies commutations with arbitrary meets which are the counterpart of the logical commutation between the universal quantification and the implication. We showed that implicative algebras are a generalization Streicher's AKSs [24] and Ferrer *et al.*'s ^{K}OCAs [6,7]. Besides, they provide us with a framework in which both λ_c-terms and their types can be interpreted. This has the nice consequence that we really consider the elements of the implicative structure as λ_c-terms and that we can compute with truth values. Through the formalization, this is reflected by a tactic allowing us to prove that elements belong to the separator simply by furnishing realizers.

---

[14] If the implicative algebra is classical, for all $a \in \mathcal{A}$ we saw that $\neg\neg a \cong_S a$. Through the same quotient, this implies that $\neg\neg[a] = [a]$ for all $a \in \mathcal{A}$, and that the induced Heyting algebra is actually a Boolean algebra.

6.2   Future Work

For future work, it would be interesting to push the formalization further to be able to represent implicative triposes. However, this poses the challenge of manipulating quotients and equivalent classes. The safe definition of quotients within CIC (and thus Coq) is indeed a tricky question [3,4,11], and as for now, we do not know which solution (reasoning modulo setoids, quotient as types classes, etc.) would be the more adapted to our situation.

In a more theoretic perspective, implicative algebras take position on a presentation of logic through universal quantification and the implication. The computational counterpart of this choice is that the presentation relies on the call-by-name λ_c-calculus. This raises the question of knowing whether it is possible to have alternative presentations with similar structures based on different connectives (and thus different calculi). We partially undertook this investigation in [19] by studying different presentations based on disjunctive and conjunctive connectives and related to Munch-Maccagnoni's system L [20]. Yet, the equivalence between all presentations still remains to prove.

Acknowledgments. The author wishes to thank Assia Mahboubi for pushing him to write the current paper.

References

1. The POPLmark Challenge. https://www.seas.upenn.edu/~plclub/poplmark/
2. Charguéraud, A.: The locally nameless representation. J. Autom. Reason. **49**(3), 363–408 (2012). https://doi.org/10.1007/s10817-011-9225-2
3. Chicli, L., Pottier, L., Simpson, C.: Mathematical quotients and quotient types in Coq. In: Geuvers, H., Wiedijk, F. (eds.) TYPES 2002. LNCS, vol. 2646, pp. 95–107. Springer, Heidelberg (2003). https://doi.org/10.1007/3-540-39185-1_6
4. Cohen, C.: Types quotients en Coq. In: Hermann (ed.) Actes des 21éme journées francophones des langages applicatifs (JFLA 2010). INRIA, Vieux-Port La Ciotat, France (2010). http://jfla.inria.fr/2010/actes/PDF/cyrilcohen.pdf
5. Ferrer, W., Malherbe, O.: The category of implicative algebras and realizability. ArXiv e-prints, December 2017. https://arxiv.org/abs/1712.06043
6. Ferrer Santos, W., Guillermo, M., Malherbe, O.: Realizability in OCAs and AKSs. ArXiv e-prints (2015). https://arxiv.org/abs/1512.07879
7. Ferrer Santos, W., Frey, J., Guillermo, M., Malherbe, O., Miquel, A.: Ordered combinatory algebras and realizability. Math. Struct. Comput. Sci. **27**(3), 428-458 (2017). https://doi.org/10.1017/S0960129515000432
8. Frey, J.: Realizability toposes from specifications. In: Altenkirch, T. (ed.) 13th International Conference on Typed Lambda Calculi and Applications (TLCA 2015). Leibniz International Proceedings in Informatics (LIPIcs), vol. 38, pp. 196–210. Schloss Dagstuhl-Leibniz-Zentrum fuer Informatik, Dagstuhl, Germany (2015). https://doi.org/10.4230/LIPIcs.TLCA.2015.196
9. Frey, J.: Classical realizability in the CPS target language. Electron. Notes Theor. Comput. Sci. **325**(Suppl. C), 111–126 (2016), The Thirty-second Conference on the Mathematical Foundations of Programming Semantics (MFPS XXXII). https://doi.org/10.1016/j.entcs.2016.09.034

10. Griffin, T.G.: A formulae-as-type notion of control. In: Proceedings of the 17th ACM SIGPLAN-SIGACT Symposium on Principles of Programming Languages, POPL 1990, pp. 47–58. ACM, New York (1990). https://doi.org/10.1145/96709.96714

11. Hofmann, M.: Extensional Concepts in Intensional Type Theory. Ph.D. thesis, University of Edinburgh (1995)

12. Hofstra, P., Van Oosten, J.: Ordered partial combinatory algebras. Math. Proc. Cambridge Philos. Soc. **134**(3), 445-463 (2003). https://doi.org/10.1017/S0305004102006424

13. Krivine, J.L.: Realizability in classical logic. In: Interactive Models of Computation and Program Behaviour. Panoramas et synthèses **27** (2009)

14. Krivine, J.L.: Realizability algebras: a program to well order r. Log. Methods Comput. Sci. **7**(3) (2011). https://doi.org/10.2168/LMCS-7(3:2)2011

15. Krivine, J.L.: Realizability algebras II : new models of ZF + DC. Log. Methods Comput. Sci. **8**(1), 10, 28 p. (2012). https://doi.org/10.2168/LMCS-8(1:10)2012

16. Krivine, J.L.: Quelques propriétés des modèles de réalisabilité de ZF, February 2014. http://hal.archives-ouvertes.fr/hal-00940254

17. Miquel, A.: Existential witness extraction in classical realizability and via a negative translation. Log. Methods Comput. Sci. **7**(2), 188–202 (2011). https://doi.org/10.2168/LMCS-7(2:2)2011

18. Miquel, A.: Implicative algebras: a new foundation for realizability and forcing. ArXiv e-prints (2018). https://arxiv.org/abs/1802.00528

19. Miquey, É.: Classical realizability and side-effects. Ph.D. thesis, Université Paris Diderot; Universidad de la República, Uruguay, November 2017. https://hal.inria.fr/tel-01653733

20. Munch-Maccagnoni, G.: Focalisation and classical realisability. In: Grädel, E., Kahle, R. (eds.) CSL 2009. LNCS, vol. 5771, pp. 409–423. Springer, Heidelberg (2009). https://doi.org/10.1007/978-3-642-04027-6_30

21. Pitts, A.M.: Tripos theory in retrospect. Math. Struct. Comput. Sci. **12**(3), 265–279 (2002). https://doi.org/10.1017/S096012950200364X

22. Ruyer, F.: Proofs, Types and Subtypes. Ph.D. thesis, Université de Savoie, November 2006. https://tel.archives-ouvertes.fr/tel-00140046

23. Sozeau, M., Oury, N.: First-class type classes. In: Mohamed, O.A., Muñoz, C., Tahar, S. (eds.) TPHOLs 2008. LNCS, vol. 5170, pp. 278–293. Springer, Heidelberg (2008). https://doi.org/10.1007/978-3-540-71067-7_23

24. Streicher, T.: Krivine's classical realisability from a categorical perspective. Math. Struct. Comput. Sci. **23**(6), 1234–1256 (2013). https://doi.org/10.1017/S0960129512000989

Boosting the Reuse of Formal Specifications

Mariano M. Moscato[1]([⊠]), Carlos G. Lopez Pombo[2]([⊠]), César A. Muñoz[3]([⊠]), and Marco A. Feliú[1]([⊠])

[1] National Institute of Aerospace, Hampton, VA, USA
{mariano.moscato,marco.feliu}@nianet.org
[2] Instituto de Investigación En Ciencias de la Computación (ICC),
CONICET–Universidad de Buenos Aires, Buenos Aires, Argentina
[3] NASA Langley Research Center, Hampton, VA, USA
cesar.a.munoz@nasa.gov

Abstract. Advances in theorem proving have enabled the emergence of a variety of formal developments that, over the years, have resulted in large corpuses of formalizations. For example, the NASA PVS Library is a collection of 55 formal developments written in the Prototype Verification System (PVS) over a period of almost 30 years and containing more than 28000 proofs. Unfortunately, the simple accumulation of formal developments does not guarantee their reusability. In fact, in formal systems with very expressive specification languages, it is often the case that a particular conceptual object is defined in different ways. This paper presents a technique to establish sound connections between formal definitions. Such connections support the possibility of (partial) borrowing of proved results from one formal description into another, improving the reusability of formal developments. The technique is described using concepts from the field of universal algebra and algebraic specification. The technique is illustrated with concrete examples taken from formalizations available in the NASA PVS Library.

1 Introduction

Proof assistants have been actively used for decades now. Advances in formal verification techniques have enabled their use in the development cycle of critical systems. The routine use of proof assistants in some domains has resulted in the generation of a large number of formalizations. This generation of content can be seen as an unguided collective effort, since it includes the work of very different actors, from purely academic environments to industrial organizations. Thus, each formalization is biased by the particular background, goals, and style of its creators and the subtleties of each theorem prover.

© Springer International Publishing AG, part of Springer Nature 2018
J. Avigad and A. Mahboubi (Eds.): ITP 2018, LNCS 10895, pp. 477–494, 2018.
https://doi.org/10.1007/978-3-319-94821-8_28

With the aim of promoting the reuse of existing efforts, large corpuses of formal models have been created and augmented by accumulating individual endeavors. For instance, the NASA PVS Library[1] is a collection of formal models written in the Prototype Verification System (PVS) [1] and maintained by the NASA Langley Formal Methods Team. While this library is extensively used at NASA and other places, it falls short in reusability. This is often the case of formal systems featuring powerful formalisms such as higher-order logic. In such settings, the same conceptual object can be described in different and, sometimes, incompatible ways. While some of these differences can be avoided, they are often intentional. For example, a particular way of stating some definition could help the construction of a proof for a specific property, but could make the definition not suitable for computation. Examples of this phenomenon arise naturally when working with structured data types. Depending on the objectives of a particular project, it may be more convenient to represent a graph with a set of nodes and a set of edges, while in a different context, it may be preferable to use an adjacency list instead. Several examples of multiple definitions for the same conceptual element can be found in the NASA PVS Library.

This paper proposes a technique for *(1)* stating a formal connection between (parts of) different specifications, which provide the same functionality, and *(2)* supporting the transference of properties (and their proofs) between these specifications. From a practical point of view, the proposed technique improves the possibility of reuse of existing developments. This technique has the following distinguishing features: it is *explicitly verifiable*, since it provides a formal proof of the correctness of the link between definitions that can be checked in the same environment (PVS), it is *nonintrusive*, since its use does not require any modification of existing developments, it is *automatable*, since most of the steps of the proposed technique are suitable for automation, and it is *general* enough to deal with cases that could not be addressed by similar techniques.

The rest of the paper is devoted to the description of the representation technique and the illustration of its use by means of a practical case study. In Sect. 2, basic definitions are presented in order to state the notation to be used in the following sections. Section 3 describes formally the datatype connection technique. A real case study with significative practical consequences is detailed in Sect. 4. Section 5 discusses related work. Finally, Sect. 6 summarizes the results and discusses further work.

2   Preliminary Definitions

Most of the definitions and results presented in this section are taken and adapted from [2,3]. The formalism used throughout the rest of the paper is based on higher-order logic and its syntax and semantics follows mainly [4].

A concrete PVS specification is used to illustrate the notions and concepts introduced in this and the next section. PVS supports a strongly-typed higher-order language with several additional features intended to provide the user

---

[1] https://shemesh.larc.nasa.gov/fm/ftp/larc/PVS-library/.

```
fseqs [T: TYPE+]: THEORY BEGIN          list [T: TYPE]: DATATYPE BEGIN
  [...]                                   null: null?
  default: T = choose({t:T | TRUE})       cons (car: T, cdr:list):cons?
                                        END list
  barray(n: nat): TYPE = {f: [nat->T] |
  FORALL (i: nat): i >= n => f(i) = default}

  fseq:TYPE = [#length:nat, seq:barray(length)#]

  empty?_fseq(f): bool = (f'length = 0)
  [...]
END fseqs
```

Fig. 1. Fragments of different PVS theories representing finite-sized containers. Left: finite sequences (NASA PVS Library). Right: finite lists (PVS Prelude).

with a rich set of tools for the formalization of concepts. Regarding datatype definitions, PVS includes built-in support for structured, algebraic, and enumeration types, among other characteristics. As in the vast majority of modern computerized deduction systems, specifications in PVS must be grouped in syntactic constructions called *theories*. A comprehensive battery of theories containing fundamental definitions and properties is provided under the name of *PVS prelude*. There, notions for basic datatypes such as booleans, numbers, and characters, and well-known data structures, such as arrays and lists, can be found. All the definitions in the prelude are implicitly available to any user-defined theory. Additionally, the NASA PVS Library is an ongoing effort that collects a significant amount of both elemental and specialized PVS developments. In order to use definitions from user-defined theories, they must be explicitly imported. To improve the presentation of the notions needed to describe the representation technique, practical aspects such as the structured nature of PVS specifications are not reflected in the theoretical development. Nevertheless, comments on how to bridge that gap are provided when it is considered adequate.

Figure 1 presents the running example for this and the next section[2]. On its left-hand side, an excerpt from a formalization of finite sequences by Butler and Maddalon is depicted. This theory, which is part of the NASA PVS Library, takes as parameter the type (T) of the elements being contained in the sequence. The definition of the type of the finite sequences (fseq) is based on an auxiliary type named **barray**, for *bounded array*. The arrays in PVS are formalized as functions from natural numbers to T. The **barray** datatype depends on a natural number, which represents the bound of the array; queries for an index beyond the bound are defined to result in an arbitrary, but constant, value (**default**). The existence of such value is guaranteed by the declaration of T as TYPE+, which forces T to be a nonempty type. The right side of the figure shows the formalization of an algebraic datatype representing finite lists of a generic type T. In this case, the **DATATYPE** construct is used as an abbreviation for the definition of a regular PVS

---

[2] Keywords in PVS are not case sensitive. Uppercase is used here to differentiate them from the rest of the tokens.

theory, containing a type (list), a constant (null), three functions (cons, car, and cdr) and two predicates (null? and cons?). The constant null denotes the empty list and the function cons can be used to construct a list from an element and another list. The predicates null? and cons? determine whether a given list is empty or not, respectively. The function car returns the first element of a list. The function cdr returns the rest of its elements.

Definition 1 (Type). *Let T be a set of sort symbols. Type(T) is defined as the smallest set satisfying: (1) $T \subseteq$ Type(T), and (2) if $t_1, \cdots, t_k \in$ Type(T), then the list $[t_1, \cdots, t_k] \in$ Type(T).*

Definition 2 (Signature of Symbols). *Let T be a set of sort symbols and $s, t \in$ Type(T). The signature (or type) of a function symbol f is of the form $[s, t]$ and is also written as $[s \rightarrow t]$, while the signature (or type) of a predicate symbol p is of the form $[t]$.*

Definition 3 (Signature of a Language). *Let T be a set of sort symbols. The signature of a language is a structure $\langle \mathcal{C}, \mathcal{F}, \mathcal{P} \rangle$ where $\mathcal{C}$ is a set of constant symbols and $\mathcal{F}$ (resp. $\mathcal{P}$) is an indexed set of function (resp. predicate) symbols, each one with its corresponding type over Type(T) accessible through the function type : $\mathcal{C} \cup \mathcal{F} \cup \mathcal{P} \rightarrow$ Type(T).*

Henceforth, whenever a signature Π is used, its sets of symbols will be referred to as $\mathcal{C}_\Pi$, $\mathcal{F}_\Pi$ and $\mathcal{P}_\Pi$. The notion of language signature refers to the available symbols in the context of a given PVS theory. Thus, for example, the signature Π_{fseqs} includes all the symbols defined in the PVS prelude, the constant symbol default, and the predicate symbol empty?_fseq. The set T_{fseqs} of sort symbols, on which the signature Π_{fseqs} stands, contains all the sort symbols defined in the prelude plus barray and fseq. Because list is part of the PVS prelude, its corresponding set of sort symbols T_{list} and signature Π_{list} contain the sort and language symbols previously defined, plus the sort symbol list and the constant, function, and predicate symbols mentioned above. Thus, $\Pi_{\text{list}} \subset \Pi_{\text{fseqs}}$.

Definition 4 (The Language of Higher-Order Logic with Equality). *Let T be a set of sort symbols and Π a language signature over Type(T). Let $\mathcal{X}$ be a set of flexible symbols for which the function type : $\mathcal{X} \rightarrow$ Type(T) reports the type of each symbol. Term$(\Pi, \mathcal{X})$ is defined as the smallest set satisfying:*

- *$\mathcal{X} \subseteq$ Term$(\Pi, \mathcal{X})$, $\mathcal{C}_\Pi \subseteq$ Term$(\Pi, \mathcal{X})$, and $\mathcal{F}_\Pi \subseteq$ Term$(\Pi, \mathcal{X})$,*
- *for any function symbol f in $\mathcal{F}_\Pi$ s.t. type$(f) = [t \rightarrow t_{k+1}]$ and $t = [t_1, \cdots, t_k]$, for all $r_i \in$ Term$(\Pi, \mathcal{X})$ with $1 \leq i < k$ s.t. type$(r_i) = t_i$, $f(r_1, \cdots, r_k) \in$ Term$(\Pi, \mathcal{X})$ and type$(f(r_1, \cdots, r_k)) = t_{k+1}$, and*
- *for any flexible symbol x in $\mathcal{X}$ s.t. type$(x) = [t \rightarrow t_{k+1}]$ and $t = [t_1, \cdots, t_k]$, for all $r_i \in$ Term$(\Pi, \mathcal{X})$ with $1 \leq i < k$ s.t. type$(r_i) = t_i$, $x(r_1, \cdots, r_k) \in$ Term$(\Pi, \mathcal{X})$ and type$(x(r_1, \cdots, r_k)) = t_{k+1}$.*

In its turn, $Form(\Pi, \mathcal{X})$ is the smallest set satisfying:

- *for any type t in $Type(T)$, $\{r, r'\} \subseteq Term(\Pi, \mathcal{X})$, and $type(r) = type(r') = t$, $r = r' \in Form(\Pi, \mathcal{X})$, and*
- *for any predicate symbol p in $\mathcal{P}_\Pi$ s.t. $type(p) = [t_1, \cdots, t_k]$, for all $r_i \in Term(\Pi, \mathcal{X})$ with $1 \le i < k$ s.t. $type(r_i) = t_i$, $p(r_1, \cdots, r_k) \in Form(\Pi, \mathcal{X})$*
- *for any formulas φ and ψ in $Form(\Pi, \mathcal{X})$, $t \in Type(T)$ and $x \in \mathcal{X}$, $\{\neg\varphi, \varphi \lor \psi, (\exists x : t)\varphi\} \subseteq Form(\Pi, \mathcal{X})$.*

The language of higher-order logic over signature Π and a set of flexible symbols $\mathcal{X}$ will be denoted $\mathcal{L}_{\Pi, \mathcal{X}}$.

Due to space limitations, and because it is not central to the proposed work, a formal definition of model theory for the language of higher-order logic of Definition 4 is not explicitly presented. The next definition presents the classical proof calculus for higher-order logics which mainly follows [5, Chap. 11].

Definition 5 (Proof Theory). *Let T be a set of sort symbols and Π be a language signature over $Type(T)$. The notion of syntactic deduction is expressed by the relation $\vdash \subseteq 2^{Form(\Pi, \mathcal{X})} \times Form(\Pi, \mathcal{X})$ and defined by the following set of rules: (1) the usual rules for first-order logic [5, Sect. 2.1.1] and (2) rules for the introduction and elimination of higher-order quantifiers: see [5, Sect. 11.1.1] for a description of the rules and its interpretation in terms of substitution using lambda expressions and their application [6, Sect. 3.1] or [7, Chap. 4].*

The calculus implemented in PVS is a version of sequent calculus for higher-order logic, which fulfill the requisites stated by the Definition 5.

Definition 6 (Theory Presentation). *Let T a set of sort symbols and Π a language signature over $Type(T)$. A theory presentation over flexible symbols $\mathcal{X}$ is a structure $\langle T, \Pi, \Gamma, \Delta \rangle$, where $\Gamma \cup \Delta \subseteq \mathcal{L}_{\Pi, \mathcal{X}}$ and $\{\Gamma \vdash \varphi\}_{\varphi \in \Delta}$.*

A theory presentation in the context of PVS involves all the symbols defined in a given theory as well as all the symbols available from other (imported) theories, along with all the explicit and implicit axioms, introduced for example by defining types, constants, predicates, or functions, and the properties stated as theorems.

3 Representation Technique

Both types from the example of Fig. 1, `list` and `fseq`, can be seen as formalizations of the same ideal object: an ordered finite-sized container. Nevertheless, they are fairly different from a practical point of view. PVS supports the evaluation of ground expressions through a translation to Common Lisp, which enables useful features such as rapid prototyping and computational reflection [8]. Expressions involving `fseq` cannot be evaluated since they are based on total functions over the infinite domain of natural numbers. On the other hand, expressions involving `list` are completely amenable to evaluation because of its recursive definition. However, proving properties of list often requires induction, while the same properties in fseq are proven by straightforward instantiations of existential or universal quantifications. Hence, it is useful to have a connection between fseq and list in order to transfer the functions and properties defined on fseq to list with minimum human interaction.

The representation of theory presentations, as stated in this paper, finds its theoretical foundations in the field of algebraic specification [9,10] and more specifically in its subsequent development through the use of category theory [11,12]. The definitions shown below constitute the basic elements used in constructing representations between theory presentations.

Definition 7 (Type Map). *Let T and T' be sets of sort symbols. The function $\tau : T \to Type(T')$ is a type map if it is a total function mapping sort symbols in T to types in $Type(T')$. Given a type map τ, its homomorphic extension to lists of types will be denoted by $\widehat{\tau_T^{T'}} : Type(T) \to Type(T')$.*

Continuing with the example of Fig. 1, the following equations should hold: *(1)* $\widehat{\tau_{fseq}^{list}}(\text{fseq}) = \text{list}$ and *(2)* $\widehat{\tau_{fseq}^{list}}(t) = t$ for any other $t \in T_{fseqs}$. The extension of $\widehat{\tau_{fseq}^{list}}$ to lists of types is done positionwise.

Definition 8 (Language Signature Map). *Let T and T' be sets of sort symbols and $\Pi = \langle C, F, P \rangle$ and $\Pi' = \langle C', F', P' \rangle$ be language signatures over $Type(T)$ and $Type(T')$ respectively. Let X and X' be sets of flexible symbols and $\tau : T \to Type(T')$ be a type map.*

A total function $\langle \sigma^C, \sigma^F, \sigma^P \rangle_\tau : \Pi \to Term(\Pi', X') \cup Form(\Pi', X')$ is a language signature map if it satisfies: (1) $\sigma^C : C \to Term(\Pi', X')$ is a total function s.t. $\widehat{\tau_T^{T'}}(type(c)) = type(r')$, whenever $\sigma^C(c) = r'$, (2) $\sigma^F : F \to Term(\Pi', X')$ is a total function s.t. $\widehat{\tau_T^{T'}}(type(f)) = type(r')$, whenever $\sigma^F(f) = r'$, and (3) $\sigma^P : P \to Form(\Pi', X')$ is a total function s.t. $\widehat{\tau_T^{T'}}(type(p)) = [t'_1, \ldots, t'_k]$, whenever $\sigma^P(p) = \varphi'$, $fv(\varphi') = \{x'_1, \ldots, x'_k\}$, where $fv : Form(\Pi', X') \to X'$ is a function that yields the free variables of a formula and $type(x'_i) = t'_i$.

Language signature maps show how constants, functions and predicates from the source theory presentation are interpreted in the target one. As the target theory presentation is not guaranteed to have the exact same signature, constant and function symbols from the source theory can be represented by terms in the target. The same consideration applies to predicate symbols from the source theory. In the example, the only predicate defined in the source theory (fseqs) is empty?_fseq and holds when the sequence is empty. This symbol is represented by the term null?(l), whose free variable l is of type list, as stated by the type map $\widehat{\tau_{fseq}^{list}}$. The notion of term representation, defined below, relates a term in the source language with its intended representation in the target language.

Definition 9 (Term Representation). *Let T and T' be two sets of sort symbols and $\Pi = \langle C, F, P \rangle$ and $\Pi' = \langle C', F', P' \rangle$ be language signatures over $Type(T)$ and $Type(T')$ respectively. Let $\tau : T \to Type(T')$ be a type map, $\sigma = \langle \sigma^C, \sigma^F, \sigma^P \rangle_\tau$ be a language signature map, and X and X' be sets of flexible* *symbols. The* term representation relation $Repr_\sigma \subseteq Term(\Pi, \mathcal{X}) \times Term(\Pi', \mathcal{X}')$ *is the smallest relation such that:*

(1) it relates flexible symbols in $\mathcal{X}$ to flexible symbols in $\mathcal{X}'$ preserving:

    (a) typing, i.e., for all $x \in \mathcal{X}$ and $x' \in \mathcal{X}'$, if $Repr_\sigma(x, x')$ then $\widehat{\tau_T^{T'}}(type(x)) = type(x')$, and

    (b) free occurrences of symbols, i.e., if $Repr_\sigma(t, t')$ each free flexible symbol in t is related to the same flexible symbol from $\mathcal{X}'$ via $Repr_\sigma$,

(2) it is homomorphic with respect to language signature representation, i.e.,

    – *for all function symbol f in $\mathcal{F}$ such that $type(f) = [t_1, \cdots, t_k \to t_{k+1}]$, and for all $r_i \in Term(\Pi, \mathcal{X})$ and $r'_i \in Term(\Pi', \mathcal{X}')$ s.t. $type(r_i) = t_i$, $type(r'_i) = t'_i$, and $Repr_\sigma(r_i, r'_i)$ holds for each $i \in \mathbb{N}$ such that $1 \leq i \leq k$,*

$$Repr_\sigma(f(r_1, \cdots, r_k), (\lambda x'_1 \cdots x'_k.\sigma^\mathcal{F}(f))(r'_1, \cdots, r'_k)) \text{ holds,}$$

being $x'_1, \cdots, x'_{k-1}$ the free flexible symbols occurring in $\sigma^\mathcal{F}(f)$.

    – *for all predicate symbol p in $\mathcal{P}$ such that $type(p) = [t_1, \cdots, t_k]$, and for all $r_i \in Term(\Pi, \mathcal{X})$ and $r'_i \in Term(\Pi', \mathcal{X}')$ s.t. $type(r_i) = t_i$, $type(r'_i) = t'_i$, and $Repr_\sigma(r_i, r'_i)$ holds for each $i \in \mathbb{N}$ such that $1 \leq i \leq k$,*

$$p(r_1, \cdots, r_k) \text{ iff } \sigma^\mathcal{P}(p) \Big|_{[x'_1, \cdots, x'_k]}^{[r'_1, \cdots, r'_k]}$$

being $x'_1, \cdots, x'_k$ the free flexible symbols occurring in $\sigma^\mathcal{P}(p)$.

A term representation relation $Repr_\sigma$ relies on the signature map σ on which it is constructed. In fact, $Repr_\sigma$ inherits several of the properties fulfilled by σ. For example, since signature maps are type-preserving (see Definition 8) and condition (1) above assures type-preservation in the case of representation of flexible symbols, it can be assured that $Repr_\sigma$ preserves typing as well, i.e., if $Repr_\sigma(t, t')$ then $\widehat{\tau_T^{T'}}(t) = t'$. Additionally, properties such as totality, injectivity and surjectivity can be inherited from signature maps to term representation relations. In the following, the case of surjective term-representation relations is addressed first in order to ease the reading. A more general case is explained later.

The representation of formulas is just a translation between two different theory presentations but within the same logical language. The only nontrivial cases require the application of language signature maps, for the case of predicate symbols, and term representation as it was defined above.

Definition 10 (Formula Representation). *Let T and T' be sets of sort symbols, $\Pi = \langle \mathcal{C}, \mathcal{F}, \mathcal{P} \rangle$ and $\Pi' = \langle \mathcal{C}', \mathcal{F}', \mathcal{P}' \rangle$ be language signatures over $Type(T)$ and $Type(T')$ respectively, $\tau : T \to Type(T')$ be a type map, $\sigma = \langle \sigma^\mathcal{C}, \sigma^\mathcal{F}, \sigma^\mathcal{P} \rangle_\tau$ be a surjective language signature map, and $\mathcal{X}$ and $\mathcal{X}'$ be sets of flexible symbols. A formula representation $Tr_\sigma : \mathcal{L}(\Pi, \mathcal{X}) \to \mathcal{L}(\Pi', \mathcal{X}')$ is defined as follows:*

$$Tr_\sigma(r_1 = r_2) = r'_1 = r'_2 \qquad\qquad , \text{ s.t. } Repr_\sigma(r_i, r'_i) \text{ for } i \in \{1, 2\}.$$

$$Tr_\sigma(p(r_1, \cdots, r_k)) = \sigma^\mathcal{P}(p) \Big|_{[x'_1, \cdots, x'_k]}^{[r'_1, \cdots, r'_k]} \text{ , for all } p \in \mathcal{P} \text{ and}$$

$$Repr_\sigma(r_i, r'_i) \text{ for } i \in \{1, \cdots, k\}.$$

$$Tr_\sigma(\varphi \vee \psi) = Tr_\sigma(\varphi) \vee Tr_\sigma(\psi)$$

$$Tr_\sigma((\exists x : t)\varphi) = (\exists \, x' : \widehat{\tau_T^{T'}}(t)) \, Tr_\sigma(\varphi), \text{ for all } x \in \mathcal{X}.$$

Tr_σ *extends to sets of formulas as* $Tr_\sigma(\Delta) = \{\, Tr_\sigma(\varphi) \mid \varphi \in \Delta \,\}.$

The next definition provides the means for connecting two theory presentations. This type of map was introduced in [13] under the name *map of entailment system*. More recently, a similar type of map used to connect the semantics of two theory presentations, was given the name of *theoroidal co-morphisms of institutions* by Goguen and Roçu [14].

Definition 11 (Theoroidal Representation). *Let T and T' be sets of sort symbols, Π and Π' be language signatures over $Type(T)$ and $Type(T')$ respectively, $\tau : T \to Type(T')$ be a type map, σ be a language signature map, and $\mathcal{T} = \langle T, \Pi, \Gamma, \Delta \rangle$ and $\mathcal{T}' = \langle T', \Pi', \Gamma', \Delta' \rangle$ two theory presentations. Then, a formula representation Tr_σ is said to be a* theoroidal representation *if $\Gamma' \vdash Tr_\sigma(\Gamma)$. It is said to be* axiom preserving *if $Tr_\sigma(\Gamma) \subseteq \Gamma'$.*

Note that if σ is a surjective signature map, Tr_σ is axiom preserving. It can be shown that theoroidal representations compose in a very smooth component-wise form yielding theoroidal representations (see [13, p. 24] for details). Theorem 1 states that theoroidal representations preserve deductibility, i.e., the existence of a proof, while axiom preserving theoroidal representations are stronger by providing proof representation.

Theorem 1 (Deductibility Preservation). *Let T and T' be sets of sort symbols, Π and Π' be language signatures over $Type(T)$ and $Type(T')$ respectively, $\tau : T \to Type(T')$ be a type map, σ be a language signature map, and $\mathcal{T} = \langle T, \Pi, \Gamma, \Delta \rangle$ and $\mathcal{T}' = \langle T', \Pi', \Gamma', \Delta' \rangle$ two theory presentations related by the theoroidal representation Tr_σ. Then,*

$$\Gamma \vdash \varphi \quad \text{implies} \quad \Gamma' \vdash Tr_\sigma(\varphi).$$

Proof. First observe that $\Gamma' \vdash Tr_\sigma(\Gamma)$ holds by Definition 11 and then, $Tr_\sigma(\Gamma) \vdash Tr_\sigma(\varphi)$ follows by induction on the structure of the proof. The base case is when $\varphi \in \Gamma$, which holds by definition of $Tr_\sigma(\Gamma)$ in Definition 10. The inductive steps follow by considering each of the rules for introducing and eliminating logical symbols (i.e., $\neg$, $\vee$ and $\exists$). The inductive hypothesis guarantees that the hypothesis of the rule follow from $Tr_\sigma(\Gamma)$. After applying the same rule, and considering that Tr_σ preserve the logical structure of the formula, its definition can be fold to obtain the translation of the original formula. $\square$

Corollary 1 (Theorem Preservation). *Let T and T' be sets of sort symbols and Π and Π' be language signatures over $Type(T)$ and $Type(T')$ respectively, $\tau : T \to Type(T')$ be a type map, σ be a language signature map, and $\mathcal{T} = \langle T, \Pi, \Gamma, \Delta \rangle$ and $\mathcal{T}' = \langle T', \Pi', \Gamma', \Delta' \rangle$ two theory presentations related by the axiom preserving theoroidal representation Tr_σ. Then,*

$$\varphi \in \Delta \quad \text{implies} \quad Tr_\sigma(\varphi) \in \Delta'.$$

Proof. The proof follows trivially from Theorem 1 and by definition of axiom preserving theoroidal representations in Definition 11. $\square$

A mechanizable method for transferring theorems between theory presentations can be inferred from the structure of the proof of Theorem 1. The mechanization is particularly easier to achieve when dealing with conditions compatible with the hypothesis of Corollary 1.

Up to this point, some strong restrictions were posed on mapping notions presented in this section, i.e., type maps, language signature maps, term representations, formula representations, and theoroidal representations, in order to ease the presentation of the technique. These restrictions are totality and surjectivity. Establishing a theorem preserving representation from one theory to another should be more flexible.

Relaxing the restriction about surjectivity of representations implies that no every element in the target domain is required to represent an element in the source domain. To allow such relaxation, the way in which formulas are translated (Definition 10) needs to be modified. In particular, quantified variables appearing in the representation of a quantified formula must range only over those elements from $\widehat{\tau_T^{T'}}(t)$ that in fact represent elements from t. Then, a way to refer to representability of elements must be included at the logical level of the language of the signatures. While in higher-order settings such as the one provided by PVS is possible to define the notions needed at that level (symbol, term, formula, term representation, etc.), an alternative way is proposed in order to reduce the amount of practical effort needed to apply the technique. This alternative approach is based on the semantic counterpart of the *Repr* relation. Given a source and a target theory representation, Π and Π' respectively, the semantic version of the term representation relation (*Repr*) can be stated as a relation $\overline{repr} \subseteq \overline{Type(T)} \times \overline{Type(T')}$ such that:

(1) for every pair of predicates $\overline{p} \in \{\overline{p}\}_{p \in \mathcal{P}}$ and $\overline{p'} \in \{\overline{p'}\}_{p' \in \mathcal{P}'}$ s.t. $\sigma^{\mathcal{P}}(p) = p'$, and for all $x_i \in \overline{Type(T)}$, with $1 \leq i \leq n$ and $n = arity(p)$, and $x'_i \in \overline{Type(T')}$ s.t. $\overline{repr}(x_i, x'_i)$,

$$\overline{p}(x_1, \cdots, x_n) \;\text{iff}\; \overline{p'}(x'_1, \cdots, x'_n) \qquad \text{and}$$

(2) for every pair of functions $\overline{f} \in \{\overline{f}\}_{f \in \mathcal{F}}$ and $\overline{f'} \in \{\overline{f'}\}_{f' \in \mathcal{F}'}$ s.t. $\sigma^{\mathcal{F}}(f) = f'$, and for all $x_i \in \overline{Type(T)}$ and $x'_i \in \overline{Type(T')}$ s.t. $\overline{repr}(x_i, x'_i)$ with $1 \leq i \leq n$ and $n = arity(f)$,

$$\overline{repr}(\overline{f}(x_1, \cdots, x_n), \overline{f'}(x'_1, \cdots, x'_n)).$$

The similarity between the conditions on $\overline{repr}$ and *Repr* is not casual and can be used to prove its equivalence. The PVS definition shown in the left side of Fig. 2 can be proposed to denote $\overline{repr}$ for the example. The lemma depicted on the right side is to be proved to assure that it is in fact a good candidate. Then, the representation of a quantified formula $(\exists x : t)\varphi$ in Definition 10 can be stated as $(\exists \; x' : \widehat{\tau_T^{T'}}(t))(\,((\exists x : t)\overline{repr}(x, x')) \wedge Tr_\sigma(\varphi)\,)$.

Another restriction that needs to be relaxed in the definitions above, is the totality of the maps. It is not unusual that different formalisations, responding to different needs, only specify the portion of the language signature, i.e., types, constants, functions and predicates, required by the context where that particular specification is used. This observation means that the approach presented before should be able to cope with partial language signature maps.

```
repr(f:fseq,l:list): bool =           empty?_fseq_homomorphic: LEMMA
  length(l) = f'length AND               FORALL(f:fseq,l:list|repr(f,l)):
    FORALL(i:nat): i < length(l)           empty?_fseq(f) IFF null?(l)
      IMPLIES nth(l,i) = f'seq(i)
```

Fig. 2. PVS implementation of the $\overline{repr}$ for the example of Fig. 1 (*left*) and the lemma about its homomorphism w.r.t. the `empty?_fseq` function (*right*).

From a practical point of view, a possible way to support this feature with minimum impact in the definitions above is to restrict the sets of syntactic symbols taken into consideration in the representation process. First, note that not every type in the source theory is needed in the context of the target theory. For example, when constructing the representation of `fseqs` using `list` elements, the type `barray` is just an auxiliary concept that needs no counterpart on the `list` side. Then, the domain of the intended type map $\tau_{\mathrm{fseq}}^{\mathrm{list}}$ should be $Type(T_{\mathrm{fseqs}}) \backslash \mathtt{barray}$. Secondly, the domain of the total functions $\sigma^{\mathcal{C}}$, $\sigma^{\mathcal{F}}$ and $\sigma^{\mathcal{P}}$ in the language signature map (Definition 8) would be a subset of the whole set of constant (resp. function and predicate) symbols of the source theory presentation. In the example, the domain of $\sigma^{\mathcal{C}}$ should be $\mathcal{C}_{\mathrm{fseqs}} \backslash \{\mathtt{default}\}$. Finally, the term, formula, and theoroidal representation relation is now restricted to accept those terms and formulas that can be constructed using only the symbols of interest. It is important to note that this restriction on the symbols could provoke that some proofs from the source theory can not be preserved in the target theory. This occurs when terms or formulas that can only be constructed using discarded symbols are explicitly provided during the application of a proof step, such as in the introduction of a new hypothesis (cut rule, for example) or in the instantiation of existential-strength quantifiers. While there is no automatic way to solve this kind of problem, it is easily and even mechanically discoverable.

When language signatures are analysed in detail, it is possible to recognize that function and predicate symbols play different roles depending on their logical definition. Some symbols may be axiomatized defining the result of their application to terms and performing observations of their properties that cannot be obtained by other means. Some other symbols may be defined as the composition of other functions (resp. predicates) leading to conservative extensions [15, Sect. 2.3.3] of a theory presentation where such a symbol does not exists. The representation of a symbol in the latter group can be automatically generated whenever the representation of those symbols in the former group have already been provided. In such case, the result of the language signature map applied to the constituent parts of the symbol's definition can be used to extend the language signature map to apply to both sets of symbols. Additionally, the proof that the representation of the symbol does not invalidate the correctness of the term representation relation (second item in Definition 9) can be automatically constructed. It should proceed as a proof by induction on the complexity of the term. This technique is particularly useful when defining representations for algebraic datatypes, since once representations for the constructors, selectors, and recognizers are defined, the rest of the definitions are stated in terms of them or some other symbol whose definition relies on them. Thus, the representation of such symbols can be mostly automated.

Like most modern proof-assistance environments, PVS provides a variety of features intended to help the user in writing complex formalizations. Some of these features include the ability of structuring specifications through the use of specific clauses (EXPORTING and IMPORTING) and the definition of theory schemas through the use of theory parameters, among others. While specific uses for both characteristics are mentioned above, they can also be used for different purpose. For instance, the IMPORTING clause can be used to extend a PVS theory for which a representation is already defined. In such case, the theoretical notion of *theory extension* needs to be further studied in order to establish how the theoroidal representation is affected by this relationship between theories. Theory extensions can be formalized by considering a special kind of (axiom preserving) theoroidal representations relying on language signature maps whose components are injective functions, and term representation relations that are total and one-to-one. These conditions force the formula representation to be analogue to an injective translation, forcing the target theory to extend the source one. *Conservativity* can also be posed in terms of conditions imposed to language signature maps and term representation relations.

The use of theory parameters to define theory schema is one of the most useful PVS features regarding the development of formalizations. When the theory parameters are not part of the representation, i.e., when they are trivially represented, no further consideration is needed. Such is the case of the fseqs and list example presented in the previous section. However, the theory parameters can also be represented in a non-trivial way in the target theory, as illustrated by the case study presented in the next Section. To formally cope with the impact that both features could impose on the proposed technique, a complete study of the ways theory presentations can be related is necessary, but left as further work.

4   Case Study

Polynomials are widely used to provide smooth approximations of non-linear functions. At the beginning of the last century, Bernstein developed a novel way to represent polynomials in his proof of the StoneWeierstrass approximation theorem [16]. This representation has proved to be specially useful in the field of computerized graphics. Muñoz and Narkawicz [17] developed a formalization based on Bernstein polynomials that can be effectively used to find minimum and maximum values for arbitrary polynomial expressions. Such formalization, provided as a PVS specification available as part of the NASA PVS Library, relied on the fragment of PVS that can be soundly evaluated [8].

```
Polynomial: TYPE = [nat->Coefficient]
Polyproduct: TYPE = [nat->Polynomial]
MultiPolynomial : TYPE = [nat->Polyproduct]

mpoly_eval(bspoly,degmono,cf,m,n)(X) : real =
  sigma(0,n-1,LAMBDA(i:nat):cf(i)*pprod_eval(bspoly(i),degmono,m)(X))
```

Fig. 3. Excerpt from the original multipolynomial formalization part of the Bernstein development.

As part of the upgrade from the version 5 to 6 of PVS, the internal implementation of some PVS data structures was changed. While this change improved the overall performance of the system, it also affected the ground evaluation used in proof strategies developed as part of the Bernstein development. To overcome this problem, it was necessary to change the way in which polynomials were modeled in the formalization. Because of this, the Bernstein development and its strategies were not originally ported from PVS 5 to PVS 6. Recently, the technique proposed in this paper was applied and a new version of the Bernstein algorithms were developed in just a fraction of the time originally estimated. In this section, the case study of the representation of Bernstein polynomials is presented.

The Bernstein development was built around a formalization of multivariate polynomials, or *multipolynomials* for short. Any multipolynomial in m variables (denoted by P below) can be seen as a sum of a finite number, say n, of products between a real coefficient c_i and a so-called *polyproduct*, which is a product of univariate polynomials ($p_{i,j}$).

$$P(x_1, \cdots, x_m) = \sum_{i=1}^{n} c_i \prod_{j=1}^{m} p_{i,j}(x_j) \tag{1}$$

In the first version of the Bernstein development, multipolynomials were modeled using arrays, i.e., functions from natural numbers into a type, as shown in Fig. 3. Every index k of a `Polynomial` array p provides access to the coefficient of the k-th power in the univariate polynomial p. Indices i and j in Eq. 1 would be used to access `Multipolynomial` and `Polyproduct` arrays respectively. The type `Coefficient` is a renaming for `real`, the PVS type denoting real numbers. Figure 3 also shows the formalization of the evaluation function `mpoly_eval`. Its parameters are respectively: the multipolynomial to be evaluated, the degree of the polynomials in each polyproduct, the coefficients c_i and the values m and n from Eq. 1; X represents the variables $x_1, \cdots, x_m$. The function `pprod_eval`, omitted for brevity, is the function used to evaluate polyproducts and it is formalized similarly to `mpoly_eval`.

The way in which arrays are used in the algorithms defined as part of the Bernstein development are not amenable for evaluation in PVS. However, only a finite prefix of every array is really used. Therefore, a new formalization is proposed where the array prefixes are represented as finite lists. The additional difficulty of this case with respect to the example of Fig. 1 is that MultiPolynomial is in fact formalized by a nesting of arrays. Thus, the proposed representation is to be applied at each level of this nesting in order to mimic every type in Fig. 3 with the types depicted in Fig. 4.

Consequently, this representation of arrays as finite lists is formalized in PVS as shown in Fig. 5. The theory arrays_into_lists, in fact, uses finite lists of elements of a given unrestricted type (T2) to represent arrays containing elements of a possibly different type (T1). The relation between the type of contained elements T2 and T1 needs to be explicitly stated by a corresponding representing function, inner_repr in the figure. Once these three formal elements (T1,T2, and inner_repr) are provided to arrays_into_lists as theory parameters, the representation relation between arrays and finite lists is stated in the repr relation: a list l represents an array if every element of the list represents the element in the corresponding position of the array.

Note that the proposed representation, in contrast to the example of the previous section, is not injective. Each list l of elements of type T2 can be used to represent any of an infinite number of arrays starting with T1 elements in the order induced by l. For instance, the empty list can be used to represent any array. This coarse-grain property of the representation has to be taken care of by the theory using it to establish the representation relation between MultiPolynomial and MultiPolynomialList.

Figure 6 shows such a theory. The theory parameters are, respectively, two natural numbers representing the number of terms in the sum and the degree of the polynomial (n and m in Eq. 1) and the degree of the polynomials in each polyproduct. Note how every importing of the arrays_into_lists theory is adequately instantiated using, at each nesting level, the representation of the nested type stated by the previous importing clause. Then, the representation relation states that a MultipolynomialList represents a MultiPolynomial if for every list in MultipolynomialList at every nesting level: *(1)* it represents the corresponding array in MultiPolynomial according to the arrays_into_list theory and *(2)* its length is correct according to the theory parameters. This latter condition assures that the lists being used to represent the MultiPolynomial are exactly those that have to be used. Finally, the equivalence between evaluation functions of both formalizations is stated and proved. The proof proceeds by exploring symmetrically the structure of both functions and leveraging the equivalence lemmas from the auxiliary functions.

```
PolynomialList : TYPE = list[Coefficient]
PolyproductList : TYPE = list[PolynomialList]
MultiPolynomialList : TYPE = list[PolyproductList]

a2l_mpoly_eval(bsplist,degmono,cf,m,n)(X) : real =
  sigma(0,n-1,LAMBDA(i:nat):cf(i)*pprod_eval(nth(bsplist,i),degmono,m)(X))
```

Fig. 4. Excerpt from the new formalization for multipolynomials based on lists.

```
arrays_into_lists[T1,T2: TYPE, inner_repr: [T1,T2->bool]]: THEORY BEGIN
  repr(A:[nat->T1], l:list[T2]): bool =
    FORALL(i:below(length(l))): inner_repr(A(i), nth(l,i))
END arrays_into_lists
```

Fig. 5. PVS theory for the representation of arrays using finite lists.

```
multipoly_into_polylist[n,m:posnat, degmono: [nat->nat]]: THEORY BEGIN
  IMPORTING
    arrays_into_lists[Coefficient,Coefficient,=] AS polynomial,
    arrays_into_lists[Polynomial,PolynomialList,polynomial.repr]
    AS polyproduct,
    arrays_into_lists[Polyproduct,PolyproductList,polyproduct.repr]
    AS multipolynomial

  repr(mp: MultiPolynomial, pl: MultipolynomialList): bool =
    multipolynomial.represents(pl,mp) AND
    n = length(pl) AND
    (FORALL (pp_i: below(n)): m = length(nth(pl,pp_i))) AND
    (FORALL (pp_i: below(n), var_i: below(m)):
     length(nth(nth(pl,pp_i),var_i)) = degmono(var_i) + 1)

  a21_multibs_eval_equivalence: LEMMA
    FORALL(mp: MultiPolynomial, pl: MultiPolynomialList):
    repr(mp,pl) IMPLIES
      FORALL(X: Vars, cf: [nat->real]):
        multibs_eval(mp,dm,cf,m,n)(X) = a21_multibs_eval(pl,dm,cf,m,n)(X)

END multipoly into polylist
```

Fig. 6. Representation of multipolynomials on arrays using multipolynomials as lists.

New versions of the Bernstein algorithms and proof strategies were also defined, changing only the few places where explicit references to the datatype representing multipolynomials were found. As the correctness theorems for such algorithms were expressed in terms of the evaluation of the multipolynomial, the equivalence lemma a21_multibs_eval_equivalence was used to easily show that all the algorithms on MultipolynomialList are also correct. The proposed translation of multipolynomials had the direct impact of providing an executable and proven correct version of the Bernstein algorithms in a small fraction of the time it was estimated just for fixing the previous version. Moreover, the new version outperformed the previous one. This is the case because lists are translated to Lisp in a much more efficient way than arrays, and because the original definition also performed some internal conversions to lists that are unnecessary in the new version.

5   Related Work

The idea of establishing formal connections between datatypes has been approached from different flanks and support for related features has been added to a variety of formal systems. Notably, much of the effort has been posed on the connection between isomorphic datatypes. The goal of the technique presented in this paper is more general. Since its main motivation emerged from the practical problem of facilitate the reuse of existent formalizations, limiting the scope of application to isomorphic types would be too restrictive. The downside of such design decision is that automatization is harder to achieve.

For the sake of brevity, the universe of comparison explored in the following is restricted to the context of higher-order logic environments, such as Isabelle [18] and Coq [19]. In such context, the technique presented in this paper is closely related to the formalization of quotients in Isabelle [20]. There, the connection between a so-called *raw* type and a more abstract type is established through the *lifting* of terms from the raw to the abstract type and the *transfer* of theorems between them. Besides the similarities, there are some important differences. First, conceptually the work on quotients appears to be designed with a directionality in mind. While this directionality is not explicit in the transfer-related features, since theorems can be transfered from raw to abstract and vice versa, such notion is still present in the lifting functionalities. On the other hand, in the technique presented in this paper the roles of abstract and raw type are not forced explicitly. Indeed, the representation presented in this paper is not just about abstraction (or concretization) relations, but it allows for more general relationships between types of different nature. More concretely, the lifting of terms described in [20] is based on the existence of two functions, *Abs* and *Rep*, that relate an abstract instance with its raw counterpart. Meanwhile, the corresponding relations in the technique proposed in this paper are not necessarily functional. This feature makes it easy to apply the proposed technique to the example presented in the Sect. 4. Nevertheless, there is a cost associated: the level of automation described in [20] is not reachable, in general, for the proposed technique. For particular cases, as those identified in the NASA PVS Library, automation seems feasible and it is work in progress.

Also addressing the problem of transfering theorems along isomorphic types, Zimmermann and Herbelin [21] have proposed a technique for Coq that shares much of the spirit of the one presented here. Nevertheless, similarly to [20], their work also relies in a functional notion to relate elements from different datatypes. Again, while automation is easier to achieve, in fact a simple algorithm to translate proofs is explicitly presented in the mentioned paper, the use of a relational representation mechanism makes the technique presented here applicable to cases that can not be addressed in functional representation settings. Additionally, the transfer mechanism in [21] does not support some features already supported in this paper, such as the compositionality of the technique showed in the case study. Other similar approaches have been proposed in the context of the Coq system. For example, the concept of signature for higher-order functions by Sozeau [22] resembles the idea of transfer of theorems proposed in this paper. Furthermore, the work of Magaud [23] on translation of proof terms for Coq also addressed the problem of minimizing the effort in sharing theorems of related datatypes. Both approaches are less general than the work presented in this paper.

Regarding PVS, the system provides native support for theory interpretations [24]. This feature allows for the instantiation of uninterpreted symbols such as constant, function, and even type symbols. The necessary proof obligations are automatically generated by the type-checker in order to ensure the validity of the interpretation. While this feature greatly improves reusability of theories, the examples described in this paper fall out of its scope of application because interpretations are limited to refinement of uninterpreted symbols.

A typical instantiation of the problem of formal connections is the refinement of abstract into concrete datatypes. In this practical but more restricted case, elaborated features can be developed and a higher level of automation can be achieved. Notable examples of refinement in higher-order logic systems were developed by Lammich [25] for Isabelle/HOL and Cohen et al. [26] for Coq, among others.

There are also similar techniques that were specially developed with the aim to connect different dependent types. From the observation that when working with dependent types is not unusual to find cases were the only difference between them is given by its *logical* description, while the structural definition is almost or directly the same, McBride have developed the notion of *Ornaments* [27], which allows to handle type declarations as first-order citizens of the formal setting and establish relationships between them. These relationships are particularly aimed at qualifying one of the types as more informative than the other. Ornaments are specially designed to relate inductive structures, for instance, natural numbers (defined inductively using zero and successor) and finite-length lists. On the contrary, the technique presented in this paper does not suffer from such restriction: the (possible dissonant) nature of the datatypes being connected does not limit the applicability of the connection process. Somehow in the same line of Ornaments but motivated by the extraction of code from a formally verified description, Dagand et al. have presented a technique grounded on an application of the notion of Galois connections that outperforms Ornaments in that it supports the connection between more general datatypes [28]. The constructive setting in which such technique is developed makes it very different from the approach explained in these pages.

6   Conclusion

This paper presents a formal study of the concept of connections between theory presentations in a higher-order logic setting. The concept is approached in a way as general as possible in order to maximize the range of application of the representation technique. In particular, the proposed approach does not only apply to refinement of abstract into concrete types, but also to more general relationships between types. A non-trivial case study is presented to illustrate the usefulness of the proposed technique.

While the technique is not implemented as an automatic procedure yet, the systematic nature of the approach hints that automation can be achieved for a considerable part of the process. For specific cases, such as when dealing with structured and algebraic datatypes, the automation of a significant part of the technique is planned to be undertaken as future work. On the theoretical side, further research is needed to understand how the technique presented in this paper is affected by the various theory extension mechanisms provided by modern proof assistants.

Acknowledgments. Research by the first, second, and forth author was supported by the National Aeronautics and Space Administration under NASA/NIA Cooperative Agreement NNL09AA00A. Research by the second author was also supported by the *Agencia Nacional de Promoción Científica y Tecnológica* (ANPCyT) under grant PICT 2013-2129 and by the *Consejo Nacional de Investigaciones Científicas y Técnicas* (CONICET) under grant PIP 11220130100148CO.

References

1. Owre, S., Rushby, J.M., Shankar, N.: PVS: a prototype verification system. In: Kapur, D. (ed.) CADE 1992. LNCS, vol. 607, pp. 748–752. Springer, Heidelberg (1992). https://doi.org/10.1007/3-540-55602-8_217
2. Burris, S., Sankappanavar, H.P.: A Course in Universal Algebra. Graduate Texts in Mathematics. Springer, Berlin (1981)
3. Enderton, H.B.: A Mathematical Introduction to Logic. Academic Press, New York (1972)
4. van Benthem, J., Doets, K.: Higher-order logic. In: Gabbay, D., Guenthner, F. (eds.) Handbook of Philosophical Logic, 2nd edn., vol. 1, pp. 189–243. Kluwer Academic Publishers (2001)
5. Troelstra, A.S., Schwichtenberg, H.: Basic Proof Theory. Number 43 in Cambridge Tracts in Theoretical Computer Science. Cambridge University Press, Cambridge (1996)
6. Girard, J.Y., Lafont, Y., Taylor, P.: Proofs and Types. Number 7 in Cambridge Tracts in Theoretical Computer Science. Cambridge University Press, Cambridge (1989)
7. Barendregt, H.P.: Lambda calculi with types. In: Abramsky, S., Gabbay, D., Maibaum, T.S.E. (eds.) Handbook of Logic in Computer Science, Volume II. Oxford University Press (1999)
8. Muñoz, C.: Rapid prototyping in PVS. Contractor report NASA/CR-2003-212418, NASA, Langley Research Center, Hampton VA 23681-2199, USA, May 2003
9. Ehrig, H., Mahr, B., Orejas, F.: Introduction to algebraic specification. Part 1: formal methods for software development. Comput. J. **35**(5), 468–477 (1992)
10. Ehrig, H., Mahr, B., Orejas, F.: Introduction to algebraic specification. Part 2: from classical view to foundations of system specifications. Comput. J. **35**(5), 468–477 (1992)
11. McLane, S.: Categories for Working Mathematician. Graduate Texts in Mathematics. Springer, Berlin (1971)
12. Pierce, B.C.: Basic Category Theory for Computer Scientists. MIT Press, Cambridge (1991)

13. Meseguer, J.: General logics. In: Ebbinghaus, H.D., Fernandez-Prida, J., Garrido, M., Lascar, D., Artalejo, M.R. (eds.) Proceedings of the Logic Colloquium 1987, Granada, Spain, North Holland, vol. 129, pp. 275–329 (1989)
14. Goguen, J.A., Roşu, G.: Institution morphisms. Formal Aspects Comput. 13(3–5), 274–307 (2002)
15. Turski, W.M., Maibaum, T.S.E.: The Specification of Computer Programs. International Computer Science Series. Addison-Wesley Publishing Co., Inc., Boston (1987)
16. Bernstein, S.: Démonstration du théorème de weierstrass fondée sur le calcul des probabilités. Commun. Kharkov Math. Soc. 13(1), 1–2 (1912)
17. Muñoz, C., Narkawicz, A.: Formalization of a representation of Bernstein polynomials and applications to global optimization. J. Autom. Reasoning 51(2), 151–196 (2013)
18. Nipkow, T., Wenzel, M., Paulson, L.C. (eds.): Isabelle/HOL – A Proof Assistant for Higher-Order Logic. LNCS, vol. 2283. Springer, Heidelberg (2002). https://doi.org/10.1007/3-540-45949-9
19. Bertot, Y., Castéran, P.: Interactive Theorem Proving and Program Development: CoqArt: The Calculus of Inductive Constructions. Springer, Heidelberg (2013). https://doi.org/10.1007/978-3-662-07964-5
20. Huffman, B., Kunčar, O.: Lifting and transfer: a modular design for quotients in Isabelle/HOL. In: Gonthier, G., Norrish, M. (eds.) CPP 2013. LNCS, vol. 8307, pp. 131–146. Springer, Cham (2013). https://doi.org/10.1007/978-3-319-03545-1_9
21. Zimmermann, T., Herbelin, H.: Automatic and transparent transfer of theorems along isomorphisms in the Coq proof assistant. arXiv preprint arXiv:1505.05028 (2015)
22. Sozeau, M.: A new look at generalized rewriting in type theory. J. Formalized Reasoning 2(1), 41–62 (2010)
23. Magaud, N.: Changing data representation within the Coq system. In: Basin, D., Wolff, B. (eds.) TPHOLs 2003. LNCS, vol. 2758, pp. 87–102. Springer, Heidelberg (2003). https://doi.org/10.1007/10930755_6
24. Owre, S., Shankar, N.: Theory interpretations in PVS. Technical report SRI-CSL-01-01, Computer Science Laboratory, SRI International, Menlo Park, CA (2001)
25. Lammich, P.: Refinement based verification of imperative data structures. In: Proceedings of the 5th ACM SIGPLAN Conference on Certified Programs and Proofs, CPP 2016, pp. 27–36. ACM, New York (2016)
26. Cohen, C., Dénès, M., Mörtberg, A.: Refinements for free!. In: Gonthier, G., Norrish, M. (eds.) CPP 2013. LNCS, vol. 8307, pp. 147–162. Springer, Cham (2013). https://doi.org/10.1007/978-3-319-03545-1_10
27. McBride, C.: Ornamental algebras, algebraic ornaments (2010). Unpublished
28. Dagand, P.É., Tabareau, N., Tanter, É.: Foundations of dependent interoperability. J. Funct. Program. 28 (2018)

Towards Formal Foundations
for Game Theory

Julian Parsert[✉] and Cezary Kaliszyk

Department of Computer Science, University of Innsbruck, Innsbruck, Austria
{julian.parsert,cezary.kaliszyk}@uibk.ac.at

Abstract. Utility functions form an essential part of game theory and economics. In order to guarantee the existence of these utility functions sufficient properties are assumed in an axiomatic manner. In this paper we discuss these axioms and the von-Neumann-Morgenstern Utility Theorem, which names precise assumptions under which expected utility functions exist. We formalize these results in Isabelle/HOL. The formalization includes formal definitions of the underlying concepts including continuity and independence of preferences. We make the dependencies more precise and highlight some consequences for a formalization of game theory.

1 Introduction

Utility theory seeks to describe how humans evaluate and compare alternatives or outcomes using mathematical tools. This theory forms the basis of game theory and therefore several fields in economics. Hence, we believe that formalizations in either of those areas require a solid base in utility theory.

In their pioneering work "Theory of Games and Economic Behavior" von Neumann and Morgenstern axiomatically describe, how actors evaluate *uncertain outcomes* [22]. They developed the theory of expected utility, which describes a scheme based on the expected value of outcomes. Utility functions allow the use of many mathematical tools for optimization etc. Hence, much effort is put into precisely specifying properties which guarantee the existence of such functions. To this end, von Neumann and Morgenstern dedicate the first chapters of [22] to specifying the assumptions necessary (and sufficient) for preference relations to admit expected utility representation. This is now known as the von-Neumann-Morgenstern Utility Theorem. These assumptions are introduced as *axioms* upon which the entire book is based. Kahneman and others criticized [20] the theory of expected utility and developed alternatives [7]. Moreover, impossibility results were proven [19]. Nevertheless, it still remains the standard theory in game theory [14] and the most common tool in economic reasoning [9].

Our goal is to provide a solid foundation of utility theory upon which further work in both economics and game theory can be conducted. We do so by introducing formal definitions in Isabelle/HOL and deriving results that not

© The Author(s) 2018
J. Avigad and A. Mahboubi (Eds.): ITP 2018, LNCS 10895, pp. 495–503, 2018.
https://doi.org/10.1007/978-3-319-94821-8_29

only support the intuition of expected utility, but also help automated theorem provers in proving subsequent results. With that we prove the von-Neumann-Morgenstern Expected Utility Theorem.

Related Work. Arrow's impossibility theorem has been formalized by Wiedijk [25] and Nipkow [13]. Gammie has formalized some results in social choice theory, as well as stable matching [4,5]. Kuhn's theorem has been formalized by Vestergaard [21] and generalized by Le Roux [17]. The same author later worked on a formalization of Nash equilibria for two player games [18]. Recently, Martin-Dorel and Soloviev formalized boolean games with non-deterministic aspects. In addition, algorithmic game theory results have been formalized in Coq [1].

The concepts we discuss are also relevant for the formalization of economic concepts. Related work includes the verification of financial systems [16] and binomial pricing models [3]. As part of the ForMaRE project [10] VCG-Auctions [8] have been formalized. In microeconomics we discussed a formalization of two economic models and the First Welfare Theorem [15].

To our knowledge the only work that uses expected utility theory is that of Eberl [2]. The focus there is not the underlying utility theory, but rather its use in social decision schemes. Since our focus is the this underlying theory and in particular the von Neumann-Morgenstern Utility Theorem, we found that there is only little overlap.

2   Isabelle/HOL, Probability, and Notations

Isabelle/HOL [24] is an *Interactive Theorem Prover* based on higher-order logic. Due to space limitations, we refer the reader to the Isar reference manual [23] for Isabelle's foundations and notations. We introduce a few reoccurring notions of HOL-Probability, but we refer to [6] for a more detailed explanation.

It is common to denote the composition of probability mass functions (pmfs) p and q with a probability α as follows $\alpha\, p + (1 - \alpha)\, q$. This notation corresponds to the following Isabelle definition:

definition mix_pmf :: *real* $\Rightarrow$ *'a pmf* $\Rightarrow$ *'a pmf* $\Rightarrow$ *'a pmf* **where**
   mix_pmf $a\ p\ q\ =\ (bernoulli - pmf\ a) \gg= (\lambda b.$ **if** b **then** p **else** $q)$

In particular, we compose a Bernoulli distribution that returns either *True* or *False* with probability a, with a function that returns p if the random variable is *True* or q otherwise. We use Isabelle's standard definition for the support of a pmf, set_pmf, while return_pmf applied to x returns a pmf yielding x with the probability 1.

A preference relation is a transitive and reflexive binary relation (i.e. a preorder). The notations $x \succeq y$, $x \succeq [R]\ y$, and $R(x, y)$ are equivalent and denote a preference relation where x is weakly preferred to y. Despite its potential ambiguity, we will be using the first alternative if the specific relation can be inferred from context. Similarly, the symbols $x \succ y$ and $x \succ[R]$ y denote the strict preference relation where $x \succ y$ iff $x \succeq y \land \neg y \succeq x$, whereas $x \approx y$ and $x \approx[R]$ y denote the indifference relation where $x \approx y$ iff $x \succeq y \land y \succeq x$.

We will use the terms "pmf" and "lottery" interchangeably. In economic and game theoretic literature the latter is more common, while the former is used in probability theory and Isabelle/HOL.

3   Preference Relations and Their Properties

We present and discuss important definitions which we will use in subsequent sections.

First we briefly introduce rational preferences and Utility functions. However, since both have been thoroughly discussed and formalized in the authors' previous work [15] we will not go into detail or mention results involving these.

Definition 1 (Rational Preferences). *A binary relation R over a carrier set C is called a rational preference relation, if R is a total preorder on C. Hence R is total, transitive, and reflexive.*

We refer to [15] or the sources for a more detailed account of the Definitions 1 and 2 as well as derived results.

Definition 2 (Utility function). *A function $u : C \mapsto \mathbb{R}$ is said to represent a rational preference relation R over C, if*

$$\forall x\, y \in C. \ x \succeq[R]\ y \iff u(x) \geq u(y).$$

The function u is called utility function.

Based on these two definitions we continue with the new additions. Firstly, we consider continuous preferences. Definition 3 is sometimes also called the *Archimedean axiom.*

Definition 3 (Continuous Preferences). *A binary relation R over a carrier set C, is said to be continuous if, $\forall\ p\ q\ r \in C$,*

$$p \succeq[R]\ q \land q \succeq[R]\ r \longrightarrow \exists \alpha \in [0\ldots 1].(\texttt{mix\_pmf}\ \alpha\ p\ r) \approx[R]\ q.$$

Intuitively this means that if $p \succ q$, then lotteries that are *close* to p are also preferred to q. An alternative interpretation would be, that if preferences are continuous, there are no outcomes that are so bad (not preferred with respect to R) that no probability is small enough to "redeem" them by composing with a better alternative.

Next, we define independence of preferences. Informally, we want independence to entail that the (preference) relation between two elements p and q only depends on the parts where p and q differ.

Definition 4 (Independence of Preferences). *A binary relation R over a carrier set C, is independent if, $\forall p\, q\, x \in C.\ \forall \alpha \in (0\ldots 1]$,*

$$p \succeq [R]\ q \longleftrightarrow (\texttt{mix\_pmf}\ \alpha\ p\ x) \succeq [R]\ (\texttt{mix\_pmf}\ \alpha\ q\ x).$$

Independence implies that the relation between $\alpha\, p + (1-\alpha)\, x$ and $\alpha\, q + (1-\alpha)\, x$ only depends on the relation of p and q rather than their combination with x.

Even though utility functions have been defined, the special case of *expected* utility functions has not been discussed. We will do so now.

Definition 5 (Expected Utility Form[1]). *Given a set P of probability mass functions over a set of outcomes $\mathcal{O}$ and a preference relation R over P, a utility function $\mathcal{U} : P \mapsto \mathbb{R}$ representing R has expected utility form, if there exists a utility function $u^2 : \mathcal{O} \mapsto \mathbb{R}$ such that for all $p \in P$,*

$$\mathcal{U}(p) := \sum_{x \in \mathcal{O}} p(x) * u(x).$$

Notice that Definition 5 introduces two kinds of utility functions, the expected utility function $\mathcal{U}$ and the Bernoulli utility function u. The function $\mathcal{U}$ assigns a utility value to lotteries/pmfs that *range over* outcomes, while u assigns a utility value to outcomes themselves. The utility of a lottery p which equals $\mathcal{U}(p)$ is then defined to be the expected value of the utility function u with the lottery p.

4   The Setup

In this section we introduce notations that we use and discuss further concepts and assumptions.

First, we assume the set of outcomes $\mathcal{O}$ to be a non-empty finite set[3]. Next, we define the carrier set $\mathcal{P}$ to be the set of all probability mass functions (pmf) over the finite set of outcomes $\mathcal{O}$, $\mathcal{P} := \{l \mid support\ l \subseteq \mathcal{O}\}$. This set can be visualized using a probability simplex. Figure 1 shows such a simplex with three outcomes. Note, that if $|\mathcal{O}| > 1$ then the set P is uncountable. Now, we can define *degenerate lotteries* to be all lotteries that yield one outcome with the probability 1. In Fig. 1 these are simply the corner points (i.e., the points $\mathcal{O}_{1-3}$). A rational preference relation over $\mathcal{P}$ is denoted with $\mathcal{R}$. Since the final result requires $\mathcal{R}$ to be continuous and independent (cf. Definitions 3 and 4) most literature assumes these from the get go. We found that not all assumptions were necessary for the results. Therefore, in the formalization we chose to introduce assumptions only when necessary. Nevertheless, for the sake of readability we assume $\mathcal{R}$ to be rational (1), continuous (3), and independent (4) in the subsequent sections. For more detail on the necessity of assumptions we refer to the formalization.

With this setup, we can state the theorem we are aiming for, the von-Neumann-Morgenstern Utility Theorem (Theorem 1).

---

[1] This form is also known as the von-Neumann-Morgenstern utility function.

[2] This function is sometimes referred to as Bernoulli utility function.

[3] The discussed theorem also holds for infinite sets [9]. However, this has not been formalized.

Theorem 1 (von-Neumann-Morgenstern Utility Theorem). *The prefer-ence relation $\mathcal{R}$ over the carrier set $\mathcal{P}$ can be represented by a utility function of expected utility form (Definition 5) if and only if $\mathcal{R}$ is rational (1), continuous (3), and satisfies independence (4). More formally, $\mathcal{R}$ satisfies (1), (3), (4), if and only if, $\exists u : \mathcal{O} \mapsto \mathbb{R}$ such that $\forall p\, q \in \mathcal{P}$,*

$$p \succeq q \iff \sum_{x \in \mathcal{O}} p(x) * u(x) \geq \sum_{x \in \mathcal{O}} q(x) * u(x).$$

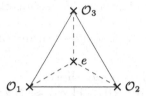

Fig. 1. This is the probability simplex for the case where $|\mathcal{O}| = 3$. The set $\{l \mid \text{support } l \subseteq \mathcal{O}\}$ is exactly the set of all points on this simplex. The point e is the pmf with the probability $\frac{1}{3}$ for all three outcomes $(\frac{1}{3}\mathcal{O}_1 + \frac{1}{3}\mathcal{O}_2 + \frac{1}{3}\mathcal{O}_3)$.

5   The Proof Outline

We will present the key insights and ideas leading to a proof of Theorem 1. All the definitions and proofs can be found in the formalization. Since we use the setup introduced in the previous section all assumptions and notations carry over. In particular $\succeq$ will denote the previously introduced relation $\mathcal{R}$.

Theorem 1 is proved by showing two implications. Both directions can be found in the formalization. However, we will discuss the more difficult direction. That is, a preference relation satisfying (1), (3), and (4) admits expected utility representation.

The set of degenerate lotteries is finite, trivially there exists at least one most preferred element (with respect to $\mathcal{R}$). Moreover, we can prove Lemma 1.

Lemma 1. *Every best[4] degenerate lottery B_{deg} is at least as good as any other lottery in $\mathcal{P}$.*

$$\forall y \in \mathcal{P}. B_{deg} \succeq y$$

The same can be shown for the worst (least preferred) elements. Thus proving that there exists at least one best $\mathcal{B}$ and one worst $\mathcal{W}$ element in $\mathcal{P}$ such that

$$\forall x \in \mathcal{P}. \mathcal{B} \succeq x \wedge x \succeq \mathcal{W}. \tag{1}$$

---

[4] We will use the "best" and "worst" to denote most and least preferred with respect to $\mathcal{R}$.

If $\mathcal{B} \approx \mathcal{W}$ any constant function would represent the preference relation $\mathcal{R}$, thus proving Theorem 1 for this special case. Hence, we will assume $\mathcal{B} \succ \mathcal{W}$.

From the assumption of continuity and Property 1, we know that $\forall p \in \mathcal{P}$,

$$\exists \alpha. \ \alpha \, \mathcal{B} + (1 - \alpha) \, \mathcal{W} \approx p.$$

Moreover, we can show that such an α is unique. If it was not, we could create two distinct lotteries $p = \alpha \, \mathcal{B} + (1 - \alpha) \, \mathcal{W}$ and $q = \beta \, \mathcal{B} + (1 - \beta) \, \mathcal{W}$ with $\alpha > \beta$ and $p \approx q$. However, since $\mathcal{B} \succ \mathcal{W}$ and p has a higher chance of the best outcome than q, we deduce $p \succ q$, a contradiction. This shows that for all lotteries $p \in \mathcal{P}$, there exists a unique *calibration probability* α, such that, $\alpha \mathcal{B} + (1 - \alpha)\mathcal{W} \approx p$.

The key idea is to define a function that assigns the unique calibration probability to every lottery in $\mathcal{P}$. This is realised with the utility function `util`. Given a pmf p its unique calibration α is obtained (using the indefinite choice operator `SOME`) and returned.

definition `util` :: *'a pmf $\Rightarrow$ real* **where**
    `util` $p \ = \ (SOME \ \alpha. \ \alpha \in \{0 \ldots 1\} \wedge p \approx [\mathcal{R}] \ \text{mix\_pmf} \ \alpha \ \mathcal{B} \ \mathcal{W})$

The next lemma shows that `util` indeed is a utility function as per Definition 2.

Lemma 2. *For all p and q in $\mathcal{P}$,*

$$p \succeq q \iff \text{util}(p) \geq \text{util}(q)$$

Lemma 2 is already an important result. However, since we are not only interested in general utility functions, but utility functions that adhere to expected utility form (Definition 5), we also need to prove the following Lemma.

Lemma 3. `util` *is linear. That is, for all p, q in $\mathcal{P}$,*

$$\text{util}(\alpha \, p + (1 - \alpha) \, q) = \alpha \, \text{util}(p) + (1 - \alpha) \, \text{util}(q)$$

Proof Outline. First, we generate two lotteries that have the same preference as p and q using `util`, $\mathcal{B}$, and $\mathcal{W}$. After substituting these generated lotteries in the left hand side of the equation, we can distribute α, rearrange the terms and apply the definition of `util` to derive the right hand side. For a detailed account of this lemma, we refer to the formalization. □

One of the most prominent modern books on game theory [11] defines von-Neumann-Morgenstern utility functions simply as linear functions which `util` indeed is (Lemma 3). Since linearity is the defining property of expected utility functions Lemma 4 can be proven. Note, that `util` has the wrong type *'a pmf $\Rightarrow$ real*. Therefore, we simply define the Bernoulli utility function u with the following lambda abstraction $(\lambda x. \ \text{util}(\text{return\_pmf} \ x))$ of type *'a $\Rightarrow$ real*.

Lemma 4. *Given a $p \in \mathcal{P}$*

$$\mathcal{U}(p) = \sum_{x \in \mathcal{O}} p(x) * u(x)$$

This shows the existence of an expected utility function assuming (1), (3), and (4), thus proving one direction of Theorem 1.

6 Conclusions

As mentioned in Sect. 1 multiple prominent books including [11,22], introduce the theory of expected utility as a set of axioms upon which their work is based. Thus, a formalization of utility theory is crucial for further development in game theory and economics. The presented formalization amounts to almost 2400 lines of code including over 120 lemmas. These can be used for future work such as Nash's theorem [12] on the existence of mixed strategy equilibria.

Acknowledgments. We thank Manuel Eberl for his help with Isabelle's HOL-Probability. This work is supported by the European Research Council (ERC) grant no 714034 *SMART* and the Austrian Science Fund (FWF) project P26201.

References

1. Bagnall, A., Merten, S., Stewart, G.: A library for algorithmic game theory in Ssreflect/Coq. J. Formalized Reasoning **10**(1), 67–95 (2017). https://jfr.unibo.it/article/view/7235
2. Eberl, M.: Randomised social choice theory. Archive of Formal Proofs, May 2016. http://isa-afp.org/entries/Randomised_Social_Choice.shtml. Formal proof development
3. Echenim, M., Peltier, N.: The binomial pricing model in finance: a formalization in Isabelle. In: de Moura, L. (ed.) CADE 2017. LNCS (LNAI), vol. 10395, pp. 546–562. Springer, Cham (2017). https://doi.org/10.1007/978-3-319-63046-5_33
4. Gammie, P.: Some classical results in social choice theory. Archive of Formal Proofs, November 2008. http://isa-afp.org/entries/SenSocialChoice.html. Formal proof development
5. Gammie, P.: Stable matching. Archive of Formal Proofs, October 2016. http://isa-afp.org/entries/Stable_Matching.html. Formal proof development
6. Hölzl, J.: Construction and stochastic applications of measure spaces in higher-order logic. Ph.D. thesis, Technical University Munich (2013). http://nbn-resolving.de/urn:nbn:de:bvb:91-diss-20130219-1116512-0-6
7. Kahneman, D., Tversky, A.: Prospect theory: an analysis of decision under risk. Econometrica **47**(2), 263–291 (1979). http://www.jstor.org/stable/1914185
8. Kerber, M., Lange, C., Rowat, C., Windsteiger, W.: Developing an auction theory toolbox. AISB, pp. 1–4 (2013)
9. Kreps, D.: Notes on the Theory of Choice. Underground Classics in Economics. Avalon Publishing (1988). https://books.google.at/books?id=9D0Oljs5GrQC
10. Lange, C., Rowat, C., Kerber, M.: The ForMaRE project – formal mathematical reasoning in economics. In: Carette, J., Aspinall, D., Lange, C., Sojka, P., Windsteiger, W. (eds.) CICM 2013. LNCS (LNAI), vol. 7961, pp. 330–334. Springer, Heidelberg (2013). https://doi.org/10.1007/978-3-642-39320-4_23
11. Maschler, M., Solan, E., Zamir, S.: Game Theory. Cambridge University Press, New York (2013)
12. Nash, J.F.: Equilibrium points in *n*-person games. Proc. Nat. Acad. Sci. U.S.A. **36**, 48–49 (1950)
13. Nipkow, T.: Arrow and Gibbard-Satterthwaite. Archive of Formal Proofs (2008). https://www.isa-afp.org/entries/ArrowImpossibilityGS.shtml

14. Nisan, N., Roughgarden, T., Tardos, E., Vazirani, V.V.: Algorithmic Game Theory. Cambridge University Press, New York (2007)
15. Parsert, J., Kaliszyk, C.: Formal microeconomic foundations and the first welfare theorem. In: Proceedings of the 7th ACM SIGPLAN International Conference on Certified Programs and Proofs, CPP 2018, pp. 91–101. ACM (2018). https://doi.org/10.1145/3167100
16. Passmore, G.O., Ignatovich, D.: Formal verification of financial algorithms. In: de Moura, L. (ed.) CADE 2017. LNCS (LNAI), vol. 10395, pp. 26–41. Springer, Cham (2017). https://doi.org/10.1007/978-3-319-63046-5_3
17. Roux, S.: Acyclic preferences and existence of sequential nash equilibria: a formal and constructive equivalence. In: Berghofer, S., Nipkow, T., Urban, C., Wenzel, M. (eds.) TPHOLs 2009. LNCS, vol. 5674, pp. 293–309. Springer, Heidelberg (2009). https://doi.org/10.1007/978-3-642-03359-9_21
18. Roux, S.L., Martin-Dorel, É., Smaus, J.: An existence theorem of Nash equilibrium in Coq and Isabelle. In: Bouyer, P., Orlandini, A., Pietro, P.S. (eds.) Proceedings Eighth International Symposium on Games, Automata, Logics and Formal Verification, GandALF 2017, EPTCS, vol. 256, Roma, Italy, 20–22 September 2017, pp. 46–60 (2017). https://doi.org/10.4204/EPTCS.256.4
19. Roux, S.L., Pauly, A.: Extending finite-memory determinacy to multi-player games. Inf. Comput. (2018). http://www.sciencedirect.com/science/article/pii/S0890540118300270
20. Tversky, A., Kahneman, D.: Judgment under uncertainty: heuristics and biases. Science 185(4157), 1124–1131 (1974). http://science.sciencemag.org/content/185/4157/1124
21. Vestergaard, R.: A constructive approach to sequential nash equilibria. Inf. Process. Lett. 97(2), 46–51 (2006). https://doi.org/10.1016/j.ipl.2005.09.010
22. von Neumann, J., Morgenstern, O.: Theory of Games and Economic Behavior. Princeton University Press (1947). https://books.google.at/books?id=AUDPAAAAMAAJ
23. Wenzel, M.: The Isabelle/Isar Reference Manual (2017)
24. Wenzel, M., Paulson, L.C., Nipkow, T.: The Isabelle framework. In: Mohamed, O.A., Muñoz, C., Tahar, S. (eds.) TPHOLs 2008. LNCS, vol. 5170, pp. 33–38. Springer, Heidelberg (2008). https://doi.org/10.1007/978-3-540-71067-7_7
25. Wiedijk, F.: Formalizing Arrow's theorem. Sadhana 34(1), 193–220 (2009). https://doi.org/10.1007/s12046-009-0005-1

Verified Timing Transformations in Synchronous Circuits with λπ -Ware

João Paulo Pizani Flor[✉] and Wouter Swierstra

Utrecht University, Utrecht, The Netherlands
{J.P.PizaniFlor,W.S.Swierstra}@uu.nl

Abstract. We define a DSL for hardware description, called λπ-Ware, embedded in the dependently-typed language Agda, which makes the DSL well-scoped and well-typed by construction. Other advantages of dependent types are that circuit models can be simulated and verified in the same language, and properties can be proven not only of specific circuits, but of circuit generators describing (infinite) families of circuits. This paper focuses on the relations between circuits computing the same values, but with different levels of statefulness. We define common recursion schemes, in combinational and sequential versions, and express known circuits using these recursion patterns. Finally, we define a notion of convertibility between circuits with different levels of statefulness, and prove the core convertibility property between the combinational and sequential versions of our vector iteration primitive. Circuits defined using the recursion schemes can thus have different architectures with a guarantee of functional equivalence up to timing.

1 Introduction

Modelling electronic circuits has been a fertile ground for functional programming (Sheeran 2005) and theorem proving (Hanna and Daeche 1992). There have been numerous efforts to describe, simulate, and verify circuits using functional languages such as MuFP (Sheeran 1984) and more recently CλaSH (Baaij 2015) and ForSyDe (Sander and Jantsch 2004).

Functional languages have also been used to *host* an Embedded Domain-Specific Language (EDSL) for hardware description. Some of these EDSLs, such as Wired (Axelsson et al. 2005), capture low-level information about the layout of a circuit; others aim to use the host language to provide a higher-level of abstraction to describe the circuit's intended behaviour. A notable example of the latter approach is Lava (Bjesse et al. 1999) and its several variants (Gill et al. 2009; Singh 2004).

Also interactive theorem proving and programming with dependent types have been fruitfully used to support hardware verification efforts, with some based on HOL (Melham 1993; Boulton et al. 1992), some on Coq (Braibant 2011; Braibant and Chlipala 2013) and some on Martin-Löf Type Theory (Brady et al. 2007)

© The Author(s) 2018
J. Avigad and A. Mahboubi (Eds.): ITP 2018, LNCS 10895, pp. 504–522, 2018.
https://doi.org/10.1007/978-3-319-94821-8_30

Following this line of research, we utilize a dependently-typed programming language (Agda) as the *host* of our hardware EDSL, for its proving capabilities and convenience of embedding.

In particular, this paper focuses on verification related to *timing*, that is, the behaviour of a circuit in terms of its inputs *over time*. When designing hardware, a compromise must be made between the *area* occupied by a circuit and the *number of clock cycles* it takes to produce its results.

A *combinational* (stateless) architecture better harnesses potential parallelism but might negatively influence other constraints such as frequency and power consumption. A more *sequential* circuit (stateful), on the other hand, will occupy less area but might be a bottleneck in computational throughput and impact other parts of the design that depend on its outputs.

There are many different ways to implement any specific functional behaviour, and it can be difficult to find the right spot in the design space upfront. Timing-related circuit transformations are quite invasive and error-prone – making it difficult to correct bad design decisions *a posteriori*. With this paper, we attenuate some of these issues by defining a language for circuit description that facilitates the exploration of different points in the timing design space. More concretely, this paper makes the following contributions:

- We show how to embed a typed hardware DSL, $\lambda\pi$-Ware, in the general purpose dependently typed programming language Agda (Sect. 3), together with an executable semantics based on state transitions (Sect. 4).
- Next, we define common recursion patterns to build circuits in both combinational and sequential architectures (Sect. 5). We show how some well-known circuits can be expressed in terms of these recursion patterns.
- Finally, we define a precise relation between the combinational and sequential versions of circuits that *exhibit equivalent behaviour* (Sect. 5.1). By proving that different versions of our recursion schemes are convertible, we allow hardware designers to enable different levels of parallelism while being certain that semantics are being preserved up to timing.

Altogether, these contributions help to *separate the concerns* between the *values* a circuit must produce and the *timing* with which they are produced. In this way, timing decisions can more easily be modified later in the design process.

The codebase in which the ideas exposed in this paper are developed is available online.[1] For the sake of presentation, code excerpts in this paper may differ slightly from the corresponding ones in the repository.

2   Overview

We begin by shortly demonstrating the usage of $\lambda\pi$-Ware. Although inspired by our previous work (Π - Ware (Pizani Flor et al. 2016)), $\lambda\pi$-Ware uses variable binding for sharing and loops, instead of pointfree combinators. Furthermore,

---

[1] https://gitlab.com/joaopizani/lambda1-hdl/tree/paper-2017-comb-seq.

$\lambda\pi$-Ware has a universe of (simply-)structured types, whereas the types of Π-Ware were vectors only. In this section, we illustrate the language by means of two variations on a simple circuit. Later sections cover the syntax and semantics of $\lambda\pi$-Ware in greater detail.

Example: Horner's Method. We look at two circuits for calculating the value of a polynomial at a given point, one with a combinational architecture and another sequential, both based on Horner's method.

For any coefficients $a_0, \ldots, a_n$ in $\mathbb{N}$, we can define a polynomial as follows:

$$p(x) = \sum_{i=0}^{n} a_i x^i = a_0 + a_1 x + a_2 x^2 + a_3 x^3 + \cdots + a_n x^n,$$

In order to compute the value of the polynomial at a specific point x_0 of its domain, Horner's method proceeds by using the following sequence of values:

$$b_n := a_n$$
$$b_{n-1} := a_{n-1} + b_n x_0$$
$$\vdots$$
$$b_0 := a_0 + b_1 x_0.$$

Then b_0 is the value of our polynomial at x_0, that is, $p(x_0)$. By iteratively expanding definitions for each of the b_i in the equations above, one arrives at a factorized form of the polynomial clearly equivalent to the usual series of powers.

Combinational Version. Horner's method is easily expressed as a *fold*, and in $\lambda\pi$-Ware we can build a combinational (stateless) circuit to compute this fold, for any given degree n. When reading the signature of the horner-comb definition below, one must note that *only* the parameters with the type former λH are circuit inputs, and the others are synthesis parameters.

horner-comb : $\forall$ n (x_0 : λH $\mathbb{N}$) (a_n : λH $\mathbb{N}$) (as : Vec (λH $\mathbb{N}$) n) $\rightarrow$ λH $\mathbb{N}$
horner-comb x_0 = foldl-comb (λ s a $\rightarrow$ a :+: x_0 :*: s)

This circuit computes the value of a polynomial of degree n at a given point. It has three inputs: the point at which to evaluate the polynomial (x_0), the coefficient of highest degree (a_n) and the remaining coefficients (as). Later in Sect. 4 we present the detailed semantics of circuits, but for now we can say that horner-comb n behaves similarly to foldl from Agda's standard library.

Figure 1 shows the architecture of horner-comb, where we can clearly see that the circuit contains no loops nor memory cells and that the *body* of the foldl is *replicated* n times. In the horner-comb model, area is linearly proportional to the degree of the polynomial, and if we want to reduce area occupation, we need to introduce *state* into the picture somehow.

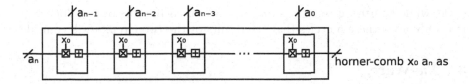

Fig. 1. Block diagram of the horner-comb circuit.

Sequential Version. Next, we describe a fully *sequential* circuit to do the same calculation, using internal state to produce a *sequence of outputs*. With this architecture the area is constant (independent of the degree of the polynomial). The output value of the circuit at clock cycle i corresponds to the sum of all polynomial terms with degree smaller than or equal to i, evaluated at point x_0.

horner-seq : $\forall\, (x_0 : \lambda H\ \mathbb{N})\ (a : \lambda H\ \mathbb{N}) \to \lambda H\ \mathbb{N}$
horner-seq x_0 = foldl-seq $(\lambda\, s\, a \to a :+: x_0 :*: s)$

The circuit takes two inputs: x_0, the point at which we desire to evaluate the polynomial; and a, a *single* input containing the n- (i+1)-th coefficient at the i-th clock cycle. The circuit is defined using the foldl-seq combinator, that iterates its argument function. This function corresponds to the loop body, mapping the current approximation, s, and the current value of the input a to a new approximation. As we shall see, to execute this *sequential* circuit, we will need to provide an initial value for the state, s.

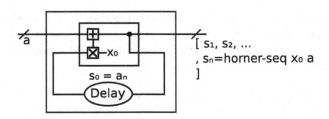

Fig. 2. Block diagram of the horner-seq circuit.

Figure 2 shows the architecture of horner-seq, where we see that the body of the foldl is the same as in the combinational version. But now instead of n instances of the body we have a single instance, with one of its outputs *tied back* in a loop with a memory cell (shift register).

We have seen that the combinational and sequential definitions are syntactically similar, but have very different timing behaviour and generate very different architectures. First of all, the coefficients input of horner-comb is a vector (a *bus* in hardware parlance), while the corresponding input of horner-seq is a single number. Also, all the coefficients are consumed by horner-comb in a single clock cycle, while horner-seq consumes the sequence of coefficients over n clock cycles. It is only after these n cycles that the results of the two circuits will coincide.

3 λπ-Ware

We begin by fixing the universe of types, U, for the elements that circuits may produce or consume. This type is parameterized by the type of data carried over the circuit's wires (B). A typical choice of B would be bits or booleans, with other choices possible when modelling a higher-level circuit, such as integers or a datatype representing assembly instructions for a microprocessor.

```
data U (B : Set) : Set where
   unit   : U B
   ι      : U B
   _⇒_    : (σ τ : U B) → U B
   _⊗_    : (σ τ : U B) → U B
   _⊕_    : (σ τ : U B) → U B
   vec    : (τ : U B) (n : ℕ) → U B
```

The collection of type codes consists of a unit ($\mathbb{1}$) and base (ι) types, closed under function space ($\Rightarrow$), products ($\otimes$), coproducts ($\oplus$) and homogeneous arrays of fixed size (vec). Each element of U B is mapped to the corresponding Agda type, in particular the code ι is mapped to B, the base type in our type universe.

Core Datatype. As mentioned before, our language is a deep-embedding in Agda, and circuits are elements of the λB datatype. Let us start by discussing the most fundamental constructors of λB, shown below. Additional constructors are discussed further ahead.

```
data λB : (Γ : Ctxt B) (τ : U B) → Set where
   ⟨_⟩  : (g : Gate τ) → λB Γ τ
   var  : (i : Γ ∋ τ)   → λB Γ τ
   _$_  : (f : λB Γ (σ ⇒ τ)) (x : λB Γ σ) → λB Γ τ
   let' : (x : λB Γ σ) (b : λB (σ :: Γ) τ) → λB Γ τ
   loop : (c : λB (σ :: Γ) (σ ⊗ τ))        → λB Γ τ
```

We use *typed De Bruijn* indices for variable binding, however, there is a convenience layer on top of λB, called λH, as seen in the overview section. Definitions using λH are essentially a shallow embedding of circuits into Agda (using Higher-Order Abstract Syntax (HOAS)), offering a more convenient programming interface by having named variables. The *unembedding* technique (Atkey et al. 2009) guarantees that it is always possible to go from a circuit definition using λH to an equivalent one using λB.

Returning to the λB datatype itself, it is indexed by a context (Γ : Ctxt B) representing the arguments to the circuit or any free variables currently in scope. The datatype is also indexed by the circuit's output type, (τ : U B).

The whole development is parameterized by a type of primitive gates, Gate : U B → Set, and the ⟨_⟩ constructor creates a circuit from such a fundamental gate. One example of such type of gates is the usual triple ({NOT, AND, OR}) with Bool as the chosen base type; circuit designers, however, are free to choose the fundamental gates that best fit their domain.

Our language does have an eliminator (_$_) for arrow types, but no introduction form. Arrow types can only be introduced by using gates, and this is by design, as we target synthesizability and circuits must be first-order to be synthesized. Using arrow types for gates allows for convenient partial application, while for general abstraction we use host language definitions as metaprograms.

While the constructors shown above form the heart of the λB datatype, there are also constructors for products, coproducts and vectors:

$$
\begin{array}{ll}
\_,\_ & : \lambda B\ \Gamma\ \tau_1\ \rightarrow\ \lambda B\ \Gamma\ \tau_2 \rightarrow \lambda B\ \Gamma\ (\tau_1 \otimes \tau_2) \\
\text{case} \otimes \_ \text{of} \_ & : \lambda B\ \Gamma\ (\sigma_1 \otimes \sigma_2)\ \rightarrow\ \lambda B\ (\sigma_1 :: \sigma_2 :: \Gamma)\ \tau \rightarrow \lambda B\ \Gamma\ \tau \\
\text{inl} & : \lambda B\ \Gamma\ \tau_1 \rightarrow \lambda B\ \Gamma\ (\tau_1 \oplus \tau_2) \\
\text{inr} & : \lambda B\ \Gamma\ \tau_2 \rightarrow \lambda B\ \Gamma\ (\tau_1 \oplus \tau_2) \\
\text{case} \oplus \_ \text{either} \_ \text{or} \_ & : \lambda B\ \Gamma\ (\sigma_1 \oplus \sigma_2)\ \rightarrow\ \lambda B\ (\sigma_1 :: \Gamma)\ \tau\ \rightarrow\ \lambda B\ (\sigma_2 :: \Gamma)\ \tau \\
& \rightarrow \lambda B\ \Gamma\ \tau \\
\text{nil} & : \lambda B\ \Gamma\ (\text{vec}\ \tau\ \text{zero}) \\
\text{cons} & : \lambda B\ \Gamma\ \tau \rightarrow \lambda B\ \Gamma\ (\text{vec}\ \tau\ n) \rightarrow \lambda B\ \Gamma\ (\text{vec}\ \tau\ (\text{suc}\ n)) \\
\text{mapAccumL-comb} & : \lambda B\ (\sigma :: \rho :: \Gamma)\ (\sigma \otimes \tau)\ \rightarrow\ \lambda B\ \Gamma\ \sigma\ \rightarrow\ \lambda B\ \Gamma\ (\text{vec}\ \rho\ n) \\
& \rightarrow \lambda B\ \Gamma\ (\sigma \otimes \text{vec}\ \tau\ n)
\end{array}
$$

We give the elimination forms for both products and coproducts uniformly as case constructs, instead of projections that matches on its argument and introduces newly bound variables to the context. For vectors, λB has the two usual introduction forms: one to produce an empty vector of any type (nil) and to extend an existing vector with a new element (cons). Finally, the *accumulating map*, mapAccumL-comb, performs a combination of map and foldl: The input vector with elements of type ρ is pointwise transformed into one with elements of type τ, all the while threading an accumulating parameter of type σ from left to right.

This eliminator is less general than the usual type theoretic elimination principle for vectors; embedding this more general eliminator would require dependent types and higher-order functions in our circuit language. To keep our object language simple, however, we chose a more simple elimination principle capable of expressing the most common hardware constructs.

4   Semantics and Properties

Where the previous section defined the *syntax* of our circuit language, we now turn our attention to its *semantics*. Although there are many different interpretations that we could assign to our circuits, for the purpose of this paper we will focus on describing a circuit's input/output behaviour.

State Transition Semantics. Circuits defined in λB can be classified in two ways. *Combinational* circuits do not have any loops; *sequential* circuits may contain loops. To define the semantics of sequential circuits, we will need to define the type of state associated with a particular circuit. To do so, we define the inductive family λs:

```
data λs  :  (c : λB Γ τ) → Set where
   _s,_   : (sx : λs x) (sy : λs y)     → λs (x , y)
   sLoop  : {c : λB (σ :: Γ) (σ ⊗ τ)} → (si : El σ) → (sc : λs c) → λs (loop c)
   ...
```

This family has a constructor for each constructor of λB. Most of these constructors either contain no significant information, or simply follow the structure of the circuit, like in the clause for pairs, _s,_, shown above. The most interesting case is sLoop, in which the state required to simulate a circuit of the form loop c consists of a value of type El σ — where σ is the type of the state that the circuit produces — together with any additional state that may arise from the loop body.

One other constructor of λs deserves special attention: sMapAccumL-comb. A circuit built with `mapAccumL-comb` consists of n *copies* of a subcircuit f connected in a row. Hence, the state of such a circuit consists of a *vector* of states, one for each of the copies of f. Correspondingly, we define the state associated with such an accumulating map as follows:

```
sMapAccumL-comb  :  (sf : Vec (λs f) n) (se : λs e) (sxs : λs xs)
                    → λs (mapAccumL-comb f e xs)
```

With this definition of state in place, we turn our attention to the semantics of our circuits. We will sketch the definition of our single step semantics, $[\![\_|]\!]$s, mapping a circuit, initial state and environment to a new state and the value produced by the circuit.

$$[\![\_|]\!]\text{s} : (\text{c} : \lambda\text{B } \Gamma \tau) (\text{m} : \lambda\text{s c}) (\gamma : \text{Env El } \Gamma) \to \lambda\text{s c} \times \text{El } \tau$$

The environment γ assigns values to any free variables in our circuit definition. The base cases for our semantics are as follows:

```
[[ ⟨ g ⟩ |]]s m γ  =  m , ([-]g g)
[[ var i |]]s m γ  =  m , lookup i γ
```

In the case for gates, we apply the semantics of our atomic gates, described by the auxiliary function $[\![-]\!]g$; in the case for variables, we lookup the corresponding value from the environment. Both these cases do not refer to the circuit's state. This state becomes important when simulating loops. In the clauses for application, let′ and loop, shown in Listing 1, we do need to consider the circuit's state.

$$[\![\; f \$ x \;|\!]\!]s \; (mf \; s\$ \; mx) \; \gamma \;=\; \textbf{let} \; (mx' , rx) \;=\; [\![\; x \;|\!]\!]s \; mx \; \gamma$$
$$(mf' , rf) \;=\; [\![\; f \;|\!]\!]s \; mf \; \gamma$$
$$\textbf{in} \; ((mf' \; s\$ \; mx') , (rf \; rx))$$

$$[\![\; let' \; x \; b \;|\!]\!]s \; (sLet \; mx \; mb) \; \gamma \;=\; \textbf{let} \; (mx' , rx) \;=\; [\![\; x \;|\!]\!]s \; mx \; \gamma$$
$$(mb' , rb) \;=\; [\![\; b \;|\!]\!]s \; mb \; (rx :: \gamma)$$
$$\textbf{in} \; ((sLet \; mx' \; mb') , rb)$$

$$[\![\; loop \; f \;|\!]\!]s \; (sLoop \; ml \; mf) \; \gamma \;=\; \textbf{let} \; (mf' , (ml' , rl)) \;=\; [\![\; f \;|\!]\!]s \; mf \; (ml :: \gamma)$$
$$\textbf{in} \; ((sLoop \; ml' \; mf') , rl)$$

Listing 1: State-combining clauses of the single-step state transition semantics.

In the cases of application and **let**, each subcircuit simply "takes a step" independently and the next state of the whole circuit is a combination of the next states of each subcircuit. The case for loop is slightly more interesting: the loop body,f, takes an additional input, namely the current state given by the ml parameter of sLoop constructor.

The further clauses of the transition function handle the introduction and elimination forms of products, coproducts and vectors. They are all defined simply by recursive evaluation of the subcircuits, and are straightforward enough to omit from the presentation here. For example, the clause for coproduct elimination is shown below:

$$[\![\; case\oplus \; x\lor y \; either \; f \; or \; g \;|\!]\!]s \; (sCase\oplus \; mxy \; mf \; mg) \; \gamma \;=\;$$
$$\textbf{let} \; (mxy' , rx\lor ry) \;=\; [\![\; x\lor y \;|\!]\!]s \; mxy \; \gamma$$
$$\textbf{in} \; [\; map\times \; (flip \; (sCase\oplus \; mxy') \; mg) \; id \circ ([\![\; f \;|\!]\!]s1 \; mf \; \gamma)$$
$$, \; map\times \; (\quad (sCase\oplus \; mxy') \; mf) \; id \circ ([\![\; g \;|\!]\!]s1 \; mg \; \gamma)$$
$$] \; rx\lor ry$$

First the coproduct value $(x\lor y)$ is evaluated, computing a result value and its next state. The result of the evaluation $(rx\lor ry)$ is then fed to Agda's coproduct eliminator $([\_,\_])$; the functions that process the left and right injections proceed accordingly. In either case, the value is fed into evaluation of the appropriate body (either f or g), and the result is then used as the result of the whole coproduct evaluation.

Similarly our elimination principle for vectors, mapAccumL-comb, is worth highlighting:

$\llbracket$ mapAccumL-comb f e xs $\rrbracket$s (sMapAccumL-comb mfs me mxs) γ =
   let (me' , re)   = $\llbracket$ e $\rrbracket$s me γ
        (mxs' , rxs)   = $\llbracket$ xs $\rrbracket$s mxs γ
        (rz , mfs' , rys) = mapAccumL2 (transformF $\llbracket$ f $\rrbracket$s2 γ) re mfs rxs
   in (sMapAccumL-comb mfs' me' mxs' , (rz , rys))

The above clause is key in the relation that we later establish (Sect. 5.1) between combinational and sequential versions of circuits. The three key sub-steps involved in this clause are: evaluation of the left identity element (e), the evaluation of the row of inputs (xs) and the row of step function copies (f).

The first two steps are as expected: both the identity and row of inputs take a step, and we thus obtain the next state and result values of each. The core step is then evaluating the row of copies of f, and its semantics are given using the auxiliary function mapAccumL2.

The mapAccumL2 function is simply a two-input version of an accumulating map, which works by simply zipping the pair of input vectors and calling the mapAccumL function from Agda's standard library.

mapAccumL : $(\sigma \rightarrow \alpha \rightarrow (\sigma \times \beta)) \rightarrow \sigma \rightarrow$ Vec α n $\rightarrow \sigma \times$ Vec β n
mapAccumL f s []   = s , []
mapAccumL f s (x :: xs) = **let** s' , y   = f s x
                          s'' , ys = mapAccumL f s' xs
                  in s'' , (y :: ys)
mapAccumL2 : $(\sigma \rightarrow \alpha \rightarrow \gamma \rightarrow (\sigma \times \beta \times \delta)) \rightarrow \sigma \rightarrow$ Vec α n $\rightarrow$ Vec γ n
              $\rightarrow \sigma \times$ Vec β n $\times$ Vec δ n
mapAccumL2 f s xs ys
         = map$\times$ id unzip \$ mapAccumL (uncurry $\circ$ f) s (zip xs ys)

In the semantics of mapAccumL-comb, we apply mapAccumL2 to the vector with the result of xs (called rxs) as well as the vector with states for the copies of f (called mfs). Then, as the result of the application we obtain the final accumulator value and vector of result values, *together with the vector of next state values* (mfs').

Multi-step Semantics. To describe the behaviour of a circuit *over time*, we need to define another semantics. More specifically, in this work we consider only discrete-time synchronous circuits, and thus we will show how to use $\llbracket \_ \rrbracket$s to define a multi-step state-transition semantics.

$\llbracket \_ \rrbracket$n : (c : λB Γ τ) (m : λs c) n (γ : Vec (Env El Γ) n) $\rightarrow \lambda$s c $\times$ Vec (El τ) n
$\llbracket \_ \rrbracket$n c m n = mapAccumL $\llbracket$ c $\rrbracket$s m

When simulating a circuit for n cycles, we need to take not *one* input environment but n, and instead of producing a single value, the simulation returns a vector of n values. Just as we saw for mapAccumL-comb, we ensure that the newly computed state is threaded from one simulation cycle to the next.

This is exactly the behaviour of an accumulating map, thus the use of mapAccumL here. The use of mapAccumL here is the key to the connection between the multi-cycle of circuits using loop and the single-cycle behavior of circuits using mapAccumL-comb.

5   Combinational and Sequential Combinators

With λπ-Ware we intend to give a hardware developer more freedom to explore the trade-offs between area, frequency and number of cycles that a circuit might take to complete a computation. This freedom comes from the proven guarantees of convertibility between combinational and sequential versions of circuits.

To make it easier to explore this design space, we provide some *circuit combinators* for common patterns. Each of these patterns comes in a pair of sequential and combinational versions, with a lemma relating the two. If a circuit is defined using one of these combinators, changing between architectures is as easy as changing the combinator version used. The associated lemma guarantees the relation between the functional behaviour of the versions.

All combinators in this section are derived from the two primitive constructors loop and mapAccumL-comb. By appropriate partial application and the use of "wrappers" to create the loop body, all sequential combinators are derived from loop. Similarly, using the same wrappers but with mapAccumL-comb, we derive all combinational combinators.

Of notice is also the fact that, in this section, we present the combinators in *De Bruijn* style, as this is the most useful representation to use when evaluating circuit (generators), which is covered in 5.1.

The map *Combinators.* For example, we might want to easily build circuits that *map* a certain function over its inputs. We will define both the sequential and combinational map combinators in terms of a third circuit, mapper. The sequential version is given by map-seq:

$$\text{mapper} \;:\; (f \;:\; \lambda B \;(\rho :: \Gamma)\; \tau) \rightarrow \lambda B \;(\sigma :: \rho :: \Gamma)\; (\sigma \otimes \tau)$$
$$\text{mapper } f \;=\; \#_0 \,,\, K_1\, f$$
$$\text{map-seq} \;:\; (f \;:\; \lambda B \;(\rho :: \Gamma)\; \tau) \rightarrow \lambda B \;(\rho :: \Gamma)\; \tau$$
$$\text{map-seq } f \;=\; \text{loop} \,\{\sigma = \mathbb{1}\}\; (\text{mapper } f)$$

We define map-seq by applying loop to the mapper f circuit. In mapper, the next state (first projection of the pair) is a copy of its first input ($\#_0$), whereas the second projection is made by the *weakened* f, which discards its first input.

The combinational version of the same combinator (map-comb) is defined in terms of mapAccumL-comb and mapper:

$$\text{map-comb} \;:\; (f \;:\; \lambda B \;(\rho :: \Gamma)\; \tau) \;(xs \;:\; \lambda B\; \Gamma\; (\text{vec } \rho\; n)) \rightarrow \lambda B\; \Gamma\; (\text{vec } \tau\; n)$$
$$\text{map-comb } f\; xs \;=\; \text{snd}\, (\text{mapAccumL-comb}\, (\text{mapper } f)\; \text{unit } xs)$$

In the above definition we note that we are free to choose the type of the "initial element" (2nd argument), but we use $\mathbb{1}$ (value unit), as units can always be used regardless of the base type chosen in the development. Furthermore, we use snd to extract only the second element of the pair (the output vector), and discard the "final element" outputted.

The foldl-scanl Combinators. Perhaps even more useful than mapping is *scanning* and *folding* over a vector of inputs. To obtain the sequential and combinational versions of such combinators, we again apply the loop and mapAccumL-comb primitives to a special body which wraps the binary operation (f) of the scan/fold.

$$\text{folder}\ :\ (f\ :\ \lambda B\ (\sigma :: \rho :: \Gamma)\ \sigma) \to \lambda B\ (\sigma :: \rho :: \Gamma)\ (\sigma \otimes \sigma)$$
$$\text{folder}\ f\ =\ \#_0\ ,\ f$$

$$\text{foldl-scanl-seq}\ :\ (f\ :\ \lambda B\ (\sigma :: \rho :: \Gamma)\ \sigma) \to \lambda B\ (\rho :: \Gamma)\ \sigma$$
$$\text{foldl-scanl-seq}\ f\ =\ \text{loop}\ (\text{folder}\ f)$$

The wrapper called folder makes the next state equal to the first input of the binary operator, and the output be the result of applying the binary operator. In the above definition of foldl-scanl-seq, we get the behaviour of scanl and foldl *combined*: The circuit outputs from clock cycle 0 to n form the result of the scanl operation, and the last one at cycle n+1 is the value of the foldl.

The combinational version also has such a combined behaviour:

$$\text{foldl-scanl-comb}\ :\ (f\ :\ \lambda B\ (\sigma :: \rho :: \Gamma)\ \sigma)\ (e\ :\ \lambda B\ \Gamma\ \sigma)\ (xs\ :\ \lambda B\ \Gamma\ (\text{vec}\ \rho\ n))$$
$$\to \lambda B\ \Gamma\ (\sigma \otimes \text{vec}\ \sigma\ n)$$
$$\text{foldl-scanl-comb}\ f\ e\ xs\ =\ \text{mapAccumL-comb}\ (\text{folder}\ f)\ e\ xs$$

In foldl-scanl-comb, we obtain a pair as output, of which the first element is the foldl component, and the second element is the scanl (vector) component. Thus by simply applying the fst and snd functions we can obtain the usual foldl and scanl.

Whereas these combinators capture some common patterns in hardware design, their usefulness also depends on lemmas relating their combinational and sequential versions.

5.1   Convertibility of Combinational and Sequential Versions

In this section we make precise the relation between circuits with different levels of statefulness. For conciseness, only the extreme cases are handled: completely stateless (combinational) versus completely sequential. However, nothing in the following treatment precludes it from being used for *partial unrolling*.

We will show that when two circuits are deemed "convertible up to timing", they can be substituted for one another with minor interface changes in the surrounding context but no alteration of the values ultimately produced.

The relation of convertibility relies on the fact that any sequential circuit will have an occurrence of the loop constructor. As such, a less stateful variant of such a circuit can be obtained by substituting the occurrence of loop with one of mapAccumL-comb, thereby unrolling the loop. The fundamental relation between loop and mapAccumL-comb is what we now establish. First, recall the types of the single- and multi-step semantic functions:

$$[\![\_]\!]s \; : \; (c \; : \; \lambda B \; \Gamma \; \tau) \; (m \; : \; \lambda s \; c) \quad (\gamma \; : \; \mathsf{Env} \; \mathsf{El} \; \Gamma) \qquad \to \lambda s \; c \times \mathsf{El} \; \tau$$
$$[\![\_]\!]n \; : \; (c \; : \; \lambda B \; \Gamma \; \tau) \; (m \; : \; \lambda s \; c) \; n \; (\gamma \; : \; \mathsf{Vec} \; (\mathsf{Env} \; \mathsf{El} \; \Gamma) \; n) \to \lambda s \; c \times \mathsf{Vec} \; (\mathsf{El} \; \tau) \; n$$

Now, to establish the desired relation, we apply both the single and multi-cycle semantics. The *step function* subcircuit (called f) is equal in both cases, and the mapAccumL-comb case takes 2 extra parameters besides f.

$$[\![\; \mathsf{mapAccumL\text{-}comb} \; f \; (\mathsf{val} \; m) \; xs \;]\!]s \; (\mathsf{sMapAccumL} \; (\mathsf{replicate} \; mf)) \; \gamma \; : \; \mathsf{Tpar}$$
$$[\![\; \mathsf{loop} \; f \qquad\qquad\qquad]\!]n \; (\mathsf{sLoop} \; m \; mf) \; n \; (\mathsf{map} \; (\_ :: \gamma) \; xs) \; : \; \mathsf{Tloop}$$
$$\textbf{where} \; \mathsf{Tpar} \;\; = \; \lambda s \; (\ldots) \times \mathsf{Tp} \; \sigma \times \mathsf{Vec} \; (\mathsf{Tp} \; \tau) \; n$$
$$\mathsf{Tloop} \; = \; \lambda s \; (\ldots) \times \qquad\quad \mathsf{Vec} \; (\mathsf{Tp} \; \tau) \; n$$

The second parameter of mapAccumL-comb must be a circuit whose value is the same as the first parameter of sLoop, and we use here the simplest possible such circuit: (val m). The third parameter (xs) is the input vector of size n, and is used to build the vector of environments used by the multi-cycle semantics (map $(\_ :: \gamma)$ xs).

Finally, the state of the mapAccumL-comb case is built by simply replicating one state of f by n times. Stating the convertibility property in this way makes it be valid only for a state-independent f, that is, when the input/output semantics of f is independent of the state.

$$\mathsf{state\text{-}independent} \; : \; \forall \; (c \; : \; \lambda B \; \Gamma \; \tau) \to \mathsf{Set}$$
$$\mathsf{state\text{-}independent} \; c \; = \; \forall \; sa \; sb \; \gamma \to [\![\; c \;]\!]s \; sa \; \gamma \equiv [\![\; c \;]\!]s \; sb \; \gamma$$

This restriction on f could be somewhat further loosened (as is discussed in Sect. 6.2), but we work here with state-independent loop bodies to simplify the presentation.

As we have seen, the results from applying each semantic function have different types (Tpar and Tloop), so the relation comparing these results is more subtle than just equality. We define this relation, called $\_=^*\_$, as follows:

$$\_=^*\_ \; : \; \mathsf{Tpar} \to \mathsf{Tloop} \to \mathsf{Set}$$
$$(\_ , s' , xs') =^* (sm'' , xs'') \; = \; (s' \equiv \mathsf{gets}_0 \; sm'') \times (xs' \equiv xs'')$$

Both sides of $(\_=^*\_)$ consist of a pair of next state and circuit outputs. In the mapAccumL-comb case, the next state can be ignored in the comparison, but in the loop case, the *value* stored in the loop state (obtained by gets_0) must be equal to the first output of evaluating mapAccumL-comb. With the comparison function defined, we can finally completely express the relation we desire:

$$\begin{aligned}
&\llbracket \text{ mapAccumL-comb f (val m) xs} \rrbracket s \text{ (sMapAccumL-comb (replicate mf))} \ \gamma \\
=^* &\llbracket \text{ loop f} \qquad\qquad\qquad\qquad \rrbracket n \text{ (sLoop m mf) n (map (\_:: } \gamma\text{) xs)}
\end{aligned}$$

Proof of the Basic Relation. The proof of the basic convertibility relation between mapAccumL-comb and loop proceeds by induction on the input vector xs. Due to the deliberate choice of semantics for both constructors involved, and the choice of the right parameters for the application of each, a considerable part of the proof is achieved by just the built-in reduction behaviour of the proof assistant (Agda).

The only key lemma involved is shown below. Namely, the state-independence principle is shown to hold for a whole vector, assuming that it holds for the body circuit f.

$$\begin{aligned}
\text{state-independent-vec} \ : \ &\forall \text{ (mfas mfbs : Vec (} \lambda s \text{ f) n) (p : state-independent f)} \\
\rightarrow \quad &\text{mapAccumL2 (transformF } \llbracket \text{ f } \rrbracket s2 \ \gamma\text{) e mfas xs} \\
\equiv \ &\text{mapAccumL2 (transformF } \llbracket \text{ f } \rrbracket s2 \ \gamma\text{) e mfbs xs}
\end{aligned}$$

This lemma is useful because both left-hand side and right-hand side of the convertibility relation can be transformed into applications of mapAccumL2 simply by reduction, but with different state vector parameters. Thus the lemma is used to bring the sub-goals to a state where they can be closed by using the induction hypothesis.

```
mapAccumL-comb-seq :
        ⟦ mapAccumL-comb f (val m) xs ⟧s (sMapAccumL-comb (replicate mf)) γ
    =* ⟦ loop f                        ⟧n (sLoop m mf) n (map (_:: γ) xs)
mapAccumL-comb-seq f mf m (x :: xs) γ p  = g:m , g:ys where
    m' , mf' , y  = (transformF ⟦ f ⟧s2 γ) m mf x   -- take one step
    ih:m , ih:ys  = mapAccumL-comb-seq f mf' m' xs γ p   -- ind. hyp.
    lemma = state-independent-vec f xs (replicate mf) (replicate mf') p
    g:m  : p₁ (p₂ (⟦ mapAccumL-comb f ... ⟧s ...)) ≡ gets₀ (p₁ (⟦ loop f ⟧n ...))
    g:ys : p₂ (p₂ (⟦ mapAccumL-comb f ... ⟧s ...)) ≡        p₂ (⟦ loop f ⟧n ...)

    g:m  =            (cong ... lemma) ⟨ trans ⟩ ih:m
    g:ys = cong₂ ... ((cong ... lemma) ⟨ trans ⟩ ih:ys)
```

Convertibility of Derived Combinators. When building circuits using the derived combinators (map, foldl-scanl, etc.), the convertibility between different (more or less stateful) variants of such circuits rely on the convertibility between the different variants of the combinators themselves.

The basic convertibility principle shown above between mapAccumL-comb and loop is the most general one, and can be directly applied to the derived combinators as well, as they are all just a specialized instance of mapAccumL-comb or loop. However, for the derived combinators, some more specific properties are useful.

With regards to the map combinators, for example, we wish that the vectors produced by the combinational and sequential versions be equal, without any regard for initial or final states. This can be succinctly expressed as:

$$
\begin{aligned}
&\text{snd} (\llbracket \text{ map-comb f xs } \rrbracket \text{s units } \gamma) \\
\equiv\ &\text{snd} (\llbracket \text{ map-seq f } \qquad \rrbracket \text{n units}' \ (\text{map } (\_:: \gamma)\ \text{xs}))
\end{aligned}
$$

Where units and units′ are simply the states (composed of units) that need to be passed to the semantic function but are irrelevant for the computed vectors.

On the other hand, when comparing foldl-comb to foldl-seq, the intermediate values produced in the output of foldl-seq are disregarded, and only the final state matters.

$$
\begin{aligned}
&\text{fst} (\llbracket \text{ foldl-comb f (val e) xs } \rrbracket \text{s m} \qquad \gamma) \\
\equiv\ &\text{fst} (\llbracket \text{ foldl-seq f } \qquad\qquad \rrbracket \text{n (sFoldl e m) (map } (\_:: \gamma)\ \text{xs}))
\end{aligned}
$$

Both of these properties (for map and for foldl) can simply be proven by application of the general property shown above for mapAccumL-comb and loop. This is because the definition of the derived combinators is just a partial application of mapAccumL-comb and loop, along with projections.

5.2 Applications of the Combinational and Sequential Combinators

In this section we describe several variants of circuit families that compute matrix multiplication, as a commonly used application of the aforementioned techniques.

The first design choice involved in this example application is *how to represent matrices*, i.e., the choice of the *matrix type*. Traditionally in computing contexts, matrices are mostly represented in two ways: *row major* (vector of rows) and *column major* (vector of columns). As it turns out, *both* representations are useful for our purposes, so we show both here:

$$
\begin{aligned}
&\text{RMat CMat : (r c : } \mathbb{N}) \rightarrow \mathbb{U}\ \mathbb{N} \\
&\text{RMat r c } = \text{ vec (vec } \mathbb{N} \text{ c) r} \\
&\text{CMat r c } = \text{ vec (vec } \mathbb{N} \text{ r) c}
\end{aligned}
$$

Here, RMat r c and CMat r c both represent matrices with r rows and c columns, the difference being only whether they are row- or column-major. Going further with the example, we need to define the basic ingredient of matrix multiplication: the *dot product* of two equally-sized vectors.

dp : λH (vec ℕ n) → λH (vec ℕ n) → λH ℕ
dp xs ys = foldl-comb _:+:_ (val 0) (zipWith-comb _:*:_ xs ys)

The dot product is simply defined as element-wise multiplication of the vectors and summing up the results. We can then use the dot product m times in order to multiply a vector by a compatibly-sized matrix.

vec×mat-comb : λH (vec ℕ n) → λH (CMat n m) → λH (vec ℕ m)
vec×mat-comb v m = map-comb (dp v) m

Here an important detail resides: as the dot product is done for *each column* of the matrix, the matrix argument of vec×mat must be in *column-major* representation. Also, here we start having choices: we may either have the computation done combinationally as above, or sequentially as below:

vec×mat-seq : λH (vec ℕ n) → λH (vec ℕ n) → λH ℕ
vec×mat-seq v m = map-seq (dp v) m

With the multi-step semantics in mind, we know that each of the m columns of the matrix will be present on the circuit's second input, one per clock cycle, and that collecting the output values for m cycles gives the same vector of results as the one from the combinational version.

For defining the multiplication of two matrices, we simply use vec×mat on each row of the left matrix. If using vec×mat-comb, we obtain a matrix multiplication circuit with area proportional to r * c, whereas by using vec×mat-seq the area is proportional to r * 1.

mat×mat-comb : λH (RMat n m) → λH (CMat m p) → λH (RMat n p)
mat×mat-comb mr mc = map-comb (flip vec×mat-comb mc) mr

mat×mat-seq : λH (RMat n m) → λH (vec ℕ m) → λH (vec ℕ n)
mat×mat-seq mr mc = map-comb (flip vec×mat-seq mc) mr

In the combinational version (mat×mat-comb), all the rows in the resulting matrix are computed in parallel, with the column-positioned values inside each row computed also in parallel. In the sequential version, at each clock cycle one whole *column* is produced, with the row-positioned values inside each column computed in parallel.

Matrix multiplication as defined here has two nested recursion blocks, and thus four ways in which it could be sequentialized. Above we have shown two possible such choices, and the other two can simply be obtained by swapping map-comb for map-seq.

6   Discussion

6.1   Related Work

There is a rich tradition of using functional programming languages to model and verify hardware circuits, Sheeran (2005) gives a good overview – we restrict ourselves to the most closely related languages here. Languages embedded in Haskell, such as Lava and Wired, typically rely on automated theorem provers and testing using QuickCheck for verification. In $\lambda\pi$-Ware, however, we can perform *inductive* verification of our circuits. Existing embeddings in most theorem provers, such as Coquet (Braibant 2011) and Π - Ware (Pizani Flor et al. 2016), have a more limited treatment of variable scoping and types. More recent work by Choi et al. (2017) is higher level, but sacrifices the ability to be simulated directly (using denotational semantics) in the theorem prover.

6.2   Future Work

Other Timing Transformations. While our language easily lets you explore possible designs, trading time and space, there are several alternative transformations, such as *pipelining* that we have not yet tried to describe in this setting.

While we have a number of combinators for transforming between combinational and sequential circuits, these are mostly aimed at linear, list-like data. Even though these structures are the most prevalent in hardware design, we would like to explore related timing transformations on tree-structured circuits. To this end, it would be interesting to look into the formalization and verification of *flattening* transformations, and of the work done in the field of *nested data parallelism*.

Relaxed Unrolling Restriction. In Sect. 5.1 we mention that the proof of semantics preservation for loop unrolling relies on the premise that the loop body is *state-independent*, that is, it has the same input/output behaviour for any given state. This premise can be relaxed somewhat, and proving that loop unrolling still preserves semantics under this relaxed premise is (near-)future work.

The relaxed restriction on the body f of a loop to be unrolled is as follows:

$$\mathsf{state\text{-}input\text{-}independent} : \forall\,(\mathsf{c} : \lambda\mathsf{B}\,\Gamma\,\tau) \to \mathsf{Set}$$
$$\mathsf{state\text{-}input\text{-}independent}\ \mathsf{c} = \mathsf{fst}\,(\llbracket\,\mathsf{c}\,\rrbracket\mathsf{s}\,\mathsf{sa}\,\gamma) \equiv \mathsf{fst}\,(\llbracket\,\mathsf{c}\,\rrbracket\mathsf{s}\,\mathsf{sa}\,\delta)$$

That is, the next state (fst projection) is equal even with evaluation taking different input environments. This condition is necessary because when writing the combinational version of a loop construct we must give each copy of f its own initial state. As the desired initial state for each such copy must be known at verification time, it cannot depend on input.

Using the definitions from Sect. 5.1 along with the relaxed hypotheses above, we can show that not only total, but also *partial unrolling* preserves semantics up to timing.

7   Conclusion

There are several advantages to be gained by embedding a hardware design DSL in a host language with dependent types, such as Agda. Among these advantages are the easy enforcement of some well-formedness characteristics of circuits, the power given by the host's type system to express object language types and design constraints. The crucial advantage though, is the ability to have *modelling, simulation, synthesis and theorem proving* in the same language.

By using the host language's theorem-proving abilities, we are able not only to show properties of individual circuits, but of (infinite) classes of circuits, defined by using *circuit generators*. Particularly interesting is the ability to have *verified transformations*, preserving some semantics.

The focus of this paper lies on timing-related transformations, but we also recognize the promise of theorem proving for the formalization of other non-functional aspects of circuit design, such as power consumption, error correction, fault-tolerance and so forth. The formal study of all these aspects of circuit construction and program construction could benefit from mechanized verification.

Acknowledgments. We would like to thank the very helpful feedback gathered during the visit to Chalmers University of Technology, funded by COST Action EUTypes CA15123. Especially valuable were the meetings and discussions with Mary Sheeran, whose deep knowledge of the field oriented this work in its beginning stage. Also, we are very thankful to the feedback given during our presentation on this topic at the TFP2017 conference in Canterbury.

This work was supported by the Netherlands Organization for Scientific Research (NWO) project on *A Dependently Typed Language for Verified Hardware*.

References

Atkey, R., Lindley, S., Yallop, J.: Unembedding domain-specific languages. In: Proceedings of the 2nd ACM SIGPLAN Symposium on Haskell, Haskell 2009, pp. 37–48. ACM, New York (2009). https://doi.org/10.1145/1596638.1596644. ISBN 978-1-60558-508-6

Axelsson, E., Claessen, K., Sheeran, M.: Wired: wire-aware circuit design. In: Borrione, D., Paul, W. (eds.) CHARME 2005. LNCS, vol. 3725, pp. 5–19. Springer, Heidelberg (2005). https://doi.org/10.1007/11560548_4

Baaij, C.P.R.: Digital circuits in CλaSH: functional specifications and type-directed synthesis. info:eu-repo/semantics/doctoralThesis. University of Twente, Enschede, January 2015. https://doi.org/10.3990/1.9789036538039

Bjesse, P., Claessen, K., Sheeran, M., Singh, S.: Lava: hardware design in Haskell. ACM SIGPLAN Not. 34(1), 174–184 (1999). https://doi.org/10.1145/291251.289440. ISSN 03621340

Boulton, R.J., Gordon, A.D., Gordon, M.J.C., Harrison, J., Herbert, J., Van Tassel, J.: Experience with embedding hardware description languages in HOL. In: TPCD, vol. 10, pp. 129–156 (1992)

Brady, E., Mckinna, J., Hammond, K.: Constructing correct circuits: verification of functional aspects of hardware specifications with dependent types. In: Trends in Functional Programming 2007 (2007)

Braibant, T.: Coquet: a Coq library for verifying hardware. In: Jouannaud, J.-P., Shao, Z. (eds.) CPP 2011. LNCS, vol. 7086, pp. 330–345. Springer, Heidelberg (2011). https://doi.org/10.1007/978-3-642-25379-9_24

Braibant, T., Chlipala, A.: Formal verification of hardware synthesis. In: Sharygina, N., Veith, H. (eds.) CAV 2013. LNCS, vol. 8044, pp. 213–228. Springer, Heidelberg (2013). https://doi.org/10.1007/978-3-642-39799-8_14

Choi, J., Vijayaraghavan, M., Sherman, B., Chlipala, A., Arvind: Kami: a platform for high-level parametric hardware specification and its modular verification. Proc. ACM Program. Lang. 1(ICFP), 24:1–24:30 (2017). https://doi.org/10.1145/3110268.. ISSN 2475-1421

Gill, A., Bull, T., Kimmell, G., Perrins, E., Komp, E., Werling, B.: Introducing Kansas Lava. In: Morazán, M.T., Scholz, S.-B. (eds.) IFL 2009. LNCS, vol. 6041, pp. 18–35. Springer, Heidelberg (2010). https://doi.org/10.1007/978-3-642-16478-1_2

Hanna, F.K., Daeche, N.: Dependent types and formal synthesis. Philos. Trans. Phys. Sci. Eng. 339(1652), 121–135 (1992). http://www.jstor.org/stable/54016. ISSN 0962-8428

Melham, T.: Higher Order Logic and Hardware Verification. Cambridge Tracts in Theoretical Computer Science, vol. 31. Cambridge University Press, Cambridge (1993). https://doi.org/10.1017/CBO9780511569845. http://www.cs.ox.ac.uk/tom.melham/pub/Melham-1993-HOL.html. ISBN 0-521-41718-X

Pizani Flor, J.P., Sijsling, Y., Swierstra, W.: π-ware: hardware description and verification in Agda. In: Uustalu, T. (ed.) 21th International Conference on Types for Proofs and Programs (TYPES 2015). Leibniz International Proceedings in Informatics (LIPIcs) (2016)

Sander, I., Jantsch, A.: System modeling and transformational design refinement in ForSyDe [formal system design]. IEEE Trans. Comput. Aided Des. Integr. Circuits Syst. 23(1), 17–32 (2004). https://doi.org/10.1109/TCAD.2003.819898. ISSN 0278-0070

Sheeran, M.: Hardware design and functional programming: a perfect match (2005). http://www.jucs.org/jucs_11_7/hardware_design_and_functional/jucs_11_7_1135_1158_sheeran.pdf

Sheeran, M.: muFP, a language for VLSI design. In: Proceedings of the 1984 ACM Symposium on LISP and Functional Programming, pp. 104–112. ACM Press (1984). https://doi.org/10.1145/800055.802026. ISBN 0897911423

Singh, S.: Designing reconfigurable systems in lava. In: Proceedings 17th International Conference on VLSI Design 2004, pp. 299–306 (2004). https://doi.org/10.1109/ICVD.2004.1260941

A Formal Equational Theory
for Call-By-Push-Value

Christine Rizkallah[1]([✉]), Dmitri Garbuzov[2], and Steve Zdancewic[2]

[1] University of New South Wales, Sydney, Australia
c.rizkallah@unsw.edu.au
[2] University of Pennsylvania, Philadelphia, USA
{dmitri,stevez}@cis.upenn.edu

Abstract. Establishing that two programs are contextually equivalent
is hard, yet essential for reasoning about semantics preserving program
transformations such as compiler optimizations. We adapt Lassen's normal form bisimulations technique to establish the soundness of equational
theories for both an untyped call-by-value λ-calculus and a variant of
Levy's call-by-push-value language. We demonstrate that our equational
theory significantly simplifies the verification of optimizations.

1 Introduction

Establishing program equivalence is a well-known and long-studied problem [15].
Programmers informally think about equivalences when coding: For example, to
convince themselves that some refactoring doesn't change the meaning of the
program. Tools such as compilers rely on program equivalences when optimizing
and transforming code.

Contextual equivalence is the gold standard of what it means for two (potentially open) program terms M_1 and M_2 to be equal. Although the exact technical
definition varies from language to language, the intuition is that M_1 is contextually equivalent to M_2 if for every closing context $C[-]$, $C[M_1]$ "behaves the same"
as $C[M_2]$. Such contextual equivalences justify program optimizations where we
can replace a less-optimal program M_1 by a better M_2 in the program context
$C[-]$, without affecting the intended behavior of the program.

While the literature is full of powerful and general techniques for establishing program equivalences using pen-and-paper proofs, not many of these general
techniques have been mechanically verified (with some notable exceptions [4]).
Moreover, some of these techniques are difficult to apply in practice. For example,
complete methods, such as applicative bisimulation [1,2], environmental bisimulation [17], and "closed-instances of uses" (CIU) techniques [13], typically require

C. Rizkallah—Work done while at University of Pennsylvania
This work is supported by NSF grant 1521539. Any opinions, findings, and conclusions or recommendations expressed in this material are those of the authors and do
not necessarily reflect the views of the NSF.

© Springer International Publishing AG, part of Springer Nature 2018
J. Avigad and A. Mahboubi (Eds.): ITP 2018, LNCS 10895, pp. 523–541, 2018.
https://doi.org/10.1007/978-3-319-94821-8_31

quantification over *all* closed function arguments or closing contexts, which significantly complicates the proofs, especially in the presence of mutually recursive function definitions.

Building on his earlier work [9], Lassen introduced the notion of *eager normal form bisimulations* [10], which is applicable to the call-by-value (CBV) λ-calculus and yet avoids quantification over all arguments or contexts. Lassen's normal form bisimulation $\mathcal{N}$ is *sound* with respect to contextual equivalence but it is not complete—this is the price to be paid for the easier-to-establish equivalences. However, $\mathcal{N}$ is still useful for many equivalence proofs because the relation includes reduction and is a congruence. Lassen defines an equational theory for CBV that is included in $\mathcal{N}$ and hence, it is also sound. This approach is appealing for formal verification because $\mathcal{N}$ is a simple co-inductive relation defined in terms of the operational semantics—mathematical objects that are relatively straightforward to work with in theorem provers.

Here, our aim is to use a similar technique to verify typical compiler optimizations. We introduce a proof structure based on $\mathcal{N}$ and use it to establish the correctness of suitable equational theories. Our development makes a key technical simplification compared to Lassen's work [10]: it does not rely on establishing that $\mathcal{N}$ is itself sound. Instead, we use a variant of $\mathcal{N}$ to prove that two related terms co-terminate. We then directly prove that the equational theory is a congruence, and is therefore sound—we explain this in precise detail below.

We demonstrate our development in two settings. First, we use the untyped call-by-value (CBV) λ-calculus as a familiar vehicle for explaining the ideas in a way that allows for comparison with other approaches. Next, we *scale up* the development to the more complex setting of an untyped variant of Levy's call-by-push-value (CBPV) λ-calculus [12] that includes an explicit letrec construct—this formalization is much more challenging and is our central contribution.

CBPV is a well-known formalism whose metatheory is well-behaved with respect to many extensions. We choose this particular CBPV variant because its features and semantics are closely related to the intermediate languages used by modern compilers [6,7,14]. This makes it attractive for formal verification, since the results can be made applicable to compiler intermediate representations without much extra effort. Our CBPV equational theory is not complete but it includes β-reduction, the operational semantics, and it is a congruence. We show that it nevertheless *trivializes* verifying many typical compiler optimizations.

To summarize, we present sound equational theories for CBV (Sect. 2) and for a "lower-level" CBPV calculus that includes mutually recursive definitions (Sect. 3). Our CBPV theory makes it trivial to verify various standard compiler optimizations (Sect. 4). All of our results are formalized [16] in Coq.

2   The Pure Untyped Call-by-Value λ-calculus

In this section, we first demonstrate our proof technique in the case of the untyped call-by-value λ-calculus, which serves as a way to introduce the ideas and as a model of how to proceed in more complex languages.

$$\frac{P\,t}{\mathcal{E}\,P\,t} \qquad \frac{\mathcal{E}\,P\,s}{\mathcal{E}\,P\,(s\,t)} \qquad \frac{\mathcal{E}\,P\,t}{\mathcal{E}\,P\,(v\,t)} \qquad \frac{}{\mathsf{canon}\ x} \qquad \frac{}{\mathsf{canon}\ (\lambda x.\,t)} \qquad \frac{\mathcal{E}\,(\lambda s.\,\exists\,x\,v:\,s = x\,v)\,t}{\mathsf{canon}\ t}$$

Fig. 1. Evaluation context closure and canonical forms

The set *Term* of λ-terms are variables (x, y, z), applications, and λ-abstractions:

$$s, t ::= x \mid s\,t \mid \lambda x.\,t$$

We identify terms up to α-equivalence. Variables and λ-abstractions are values $\mathcal{V}$ (represented by u and v) and applications are not. If s and t are terms then $s[x := v]$ is defined to be the result of substituting a value v for x in s via capture-avoiding substitution. A term of the form $(\lambda x.s)\,t$ is called a β-*redex*, and has β-*reduct* $s[x := t]$. We say s β-*reduces to* t, written $s \to_\beta t$, if a subterm of s is a β-redex such that t is the result of replacing this subterm by its β-reduct. We define $s \sim_\beta t$ to be the least equivalence relation containing $\to_\beta$. When $s \sim_\beta t$ holds, we say s and t are β-*equivalent*.

Given s and t, these rules define the small step operational semantics $s \longrightarrow t$:

$$\frac{}{(\lambda x.\,t)\,v \longrightarrow t[x := v]} \qquad \frac{s_1 \longrightarrow s_2}{s_1\,t \longrightarrow s_2\,t} \qquad \frac{t_1 \longrightarrow t_2}{v\,t_1 \longrightarrow v\,t_2}$$

We prove that, as expected, values do not step and that the operational semantics is deterministic. A term t is in *normal form*, written nf t, iff $\not\exists t' : t \longrightarrow t'$.

2.1 Progress and Canonical Forms

The bisimulation relation that we will construct relies on relating programs in normal form. We therefore define the predicate canon that holds for terms that are "canonical" in the sense that they are either values or stuck computations. To identify stuck computations, the intuitive idea is to identify terms whose evaluation is blocked because a free variable is in active position.

The set of evaluation contexts for λ-calculus is typically given by the following grammar, where [] is a "hole" indicating where the next evaluation step will occur:

$$E ::= [] \mid E\,t \mid v\,E$$

For formalization purposes, we will need to work with terms t of the form $E[s]$, where s (typically a redex) replaces the hole in E. However, rather than reifying evaluation contexts as a datatype and defining corresponding "plugging" and "unique decomposition" operations, we find it more convenient to work with a relational definition of the same concepts.

The higher-order predicate $\mathcal{E}$, whose type is $(\textit{Term} \to \mathbb{P}) \to \textit{Term} \to \mathbb{P}$ (where $\mathbb{P}$ is the type of propositions), helps us identify the evaluation context of terms that satisfy a certain property. The $\mathcal{E}$ predicate is defined inductively, as shown in Fig. 1. It is parameterized by a proposition P and its structure mirrors

$$\frac{R\,s\,t}{\mathcal{P}\,R\,s\,t} \qquad \frac{\mathcal{P}\,R\,s\,t}{\mathcal{P}\,R\,(\lambda x.\,s)\,(\lambda x.\,t)} \qquad \frac{\mathcal{P}\,R\,s\,s'}{\mathcal{P}\,R\,(s\,t)\,(s'\,t)} \qquad \frac{\mathcal{P}\,R\,t\,t'}{\mathcal{P}\,R\,(s\,t)\,(s\,t')}$$

Fig. 2. Linearly compatible closure $\mathcal{P}$ of a (binary) relation R.

that of the grammar of evaluation contexts. Intuitively, the predicate $\mathcal{E}\,P\,t$ holds if the term t is equal to a term of the form $E[s]$ and $P\,s$ holds. We call $\mathcal{E}\,P$ the *evaluation context closure* of the predicate P.

Using $\mathcal{E}$, it is easy to define stuck computations as the evaluation context closure of an appropriate predicate. The canonical forms predicate canon is defined to hold for values or stuck computations as shown in Fig. 1. Note that the expression $\lambda s.\,\exists x\,v:\ s = x\,v$ that appears in the third rule is a *meta-level* abstraction: it is a proposition that holds of s when there exists some x and v such that s is of the form $x\,v$ (i.e. s is a variable applied to some value).

We prove that canon is the predicate we want, by showing that it captures (all and only) terms that are in normal form. More formally, we show that for any term t, nf $t \leftrightarrow$ canon t. The proof directly follows from the progress theorem which states that for any term t, either canon t holds or $\exists t' : t \longrightarrow t'$ holds.

2.2   Contextual Equivalence

Contextual equivalence is the standard way of defining program equality. Two programs s and t are contextually equivalent if $C[s]$ and $C[t]$ either both terminate or both diverge for every closing context $C[-]$. Note that for the untyped λ-calculus in which divergence is the only effect, it is not necessary to explicitly check that they always compute the same result. The intuition is that if there is any situation where they can return different results, one can craft a context in which one terminates and the other diverges. Hence from co-termination in any context one can conclude that they can always be used interchangeably [15].

We define contextual equivalence for CBV. To start, we define a predicate $\mathcal{P}$ that lifts a binary relation R on terms into a relation $\mathcal{P}\,R$ linearly compatible with the λ-calculus syntax, as shown in Figure 2. A relation R that is closed under such a lifting is called *linearly compatible* i.e., if $\forall s\,t : \mathcal{P}\,R\,s\,t \to R\,s\,t$.

We define what it means for terms s and t to appear in the same context $C[-]$ by taking $\mathcal{C}\,s\,t\,s'\,t' \leftrightarrow \mathcal{P}(\lambda u\,v.\ u = s \wedge v = t)\,s'\,t'$. We call $(\mathcal{C}\,s\,t)$ the *contextual closure* of s and t. Intuitively, $\mathcal{C}\,s\,t\,s'\,t'$ holds exactly when there exists a context $C[-]$ such that $s' = C[s]$ and $t' = C[t]$. Note that quantifying over all such s' and t' precisely captures the idea of quantifying over all contexts.

A term s *terminates at* t, written $s \Downarrow t$, if $s \longrightarrow^* t \wedge$ nf t. Two terms s and t *co-terminate*, written co-terminate $s\,t$, if $(\exists s' : s \Downarrow s') \leftrightarrow (\exists t' : t \Downarrow t')$. Two terms s and t are *contextually equivalent* when for all contexts $C[-]$, $C[s]$ halts if and only if $C[t]$ does too. More formally, terms s and t are contextually equivalent, written $s \equiv t$, if $\forall s'\,t' : \mathcal{C}\,s\,t\,s'\,t' \to$ co-terminate $s'\,t'$. It will be useful to have a separate notion of when a relation R implies co-termination: A relation R is *adequate* if $\forall a\,b : R\,a\,b \to$ co-terminate $a\,b$.

$$\frac{R\,s\,t}{Eq\,R\,s\,t} \qquad \frac{}{Eq\,R\,t\,t} \qquad \frac{Eq\,R\,s\,t}{Eq\,R\,t\,s} \qquad \frac{Eq\,R\,s\,s' \quad Eq\,R\,s'\,t}{Eq\,R\,s\,t}$$

Fig. 3. Equivalence closure of a relation R

$$\frac{}{\dot{\mathcal{P}}\,R\,x\,x} \qquad \frac{R\,s\,s' \quad \dot{\mathcal{P}}\,R\,s'\,t}{\dot{\mathcal{P}}\,R\,s\,t} \qquad \frac{\dot{\mathcal{P}}\,R\,s\,t}{\dot{\mathcal{P}}\,R\,(\lambda x.\,s)\,(\lambda x.\,t)} \qquad \frac{\dot{\mathcal{P}}\,R\,s\,s' \quad \dot{\mathcal{P}}\,R\,t\,t'}{\dot{\mathcal{P}}\,R\,(s\,t)\,(s'\,t')}$$

Fig. 4. Alternative formulation of compatible closure $\dot{\mathcal{P}}$.

2.3   Equational Theory

With these definitions in mind, we can now define what it means for a program transformation, or more generally for an arbitrary relation, to be sound with respect to contextual equivalence. A term transformation function $f : Term \rightarrow Term$ is *sound* if $\forall t : t \equiv (f\,t)$. A relation R is *sound* if $\forall s\,t : R\,s\,t \rightarrow s \equiv t$.

To reach our goal of easily verifying various program optimizations, which are often reduction or β-reduction in context, we next develop a sound equational theory. We want an equational theory that is sound (with respect to contextual equivalence) and that includes the operational semantics. Moreover, the equational theory should be a *congruence relation* (i.e. a linearly compatible equivalence relation). Given an equational theory that is congruent and that includes β-reduction, we can conclude that it also includes β-equivalence. Hence, even though it is not complete, it still relates a sufficient number of terms.

Our equational theory simplifies verifying optimizations. We just need to prove that the optimization is included in the equational theory (rather than directly in contextual equivalence). Such a proof can be obtained by verifying and contextually lifting simple transformations—Sect. 4 gives several examples.

We intuitively want our equational theory to be the congruence closure of the operational step reduction. This way the theory includes the semantics and is a congruence. For terms s and t, a *single-step reduction*, written $s \Rightarrow t$, holds if $\mathcal{P}(\longrightarrow)\,s\,t$. The *equivalence closure* of a relation R, defined using the rules in Fig. 3, is the reflexive, symmetric, and transitive closure over R. The equivalence closure of the single-step reduction relation defines the *equational theory for λ-calculus*. Two terms s and t are equal according to the *equational theory for λ-calculus*, written $s \Leftrightarrow t$, if $Eq\,(\Rightarrow)\,s\,t$. It is straightforward to see that our equational theory includes the operational semantics. Proving soundness is significantly more involved and is explained in the next section.

Although the definition of $\mathcal{P}$ given in Fig. 2 is a good way to understand the concept of closing a relation under contexts, working with it directly in proofs can be cumbersome. However, because we only care about the equivalence closure of single-step reduction, we can refactor the definitions to build in reflexivity and a bit of transitivity in a way that simplifies the proofs. The resulting variation of $\mathcal{P}$, called $\dot{\mathcal{P}}$, is shown in Fig. 4. The equivalence closure of $\dot{\mathcal{P}}$ is the same as that of $\mathcal{P}$ (i.e., $\forall R\,s\,t : Eq\,(\mathcal{P}\,R)\,s\,t \leftrightarrow Eq\,(\dot{\mathcal{P}}\,R)\,s\,t$), but we no longer need to deal with the nested proof structure needed for the reflexive-transitive closure of $\mathcal{P}$—instead we can work directly by induction on $\hat{\mathcal{P}}$, which is already reflexive and transitive. We prove that both versions yield the same equational theory.

2.4   Soundness of the Equational Theory for CBV

Recall that proving that a relation R is *sound* involves proving that terms related by R are contextually equivalent. This is done by proving that R is a linearly compatible relation and by proving that terms related by R co-terminate. Lassen defines a normal form bisimulation relation and uses it to assist in proving soundness of an equational theory for λ-calculus. Similar to Lassen, we also make use of normal form bisimulations in our proof technique. Our proof structure, however, is different than Lassen's. We prove that our equational theory is linearly compatible directly rather than proving that the normal form bisimulation is linearly compatible—we expand on this comparison at the end of this section. This proof follows directly due to the way we define the equational theory, and is, hence, simpler to extend to the more complex call-by-push-value language.

Normal Form Bisimulation. A normal form bisimulation is a bisimulation between executions of terms that either terminate at related normal forms, or diverge. Since program executions can be infinite, normal form bisimulation is defined co-inductively. We first define $\mathcal{N}_s$, which defines one step in the bisimulation, and then use it to define our normal form bisimulation $\mathcal{N}$.

$$\frac{s \longrightarrow^* x \quad t \longrightarrow^* x}{\mathcal{N}_s\, R\, s\, t} \qquad\qquad \frac{s \longrightarrow^* \lambda x.\, s' \quad t \longrightarrow^* \lambda x.\, t' \quad R\, s'\, t'}{\mathcal{N}_s\, R\, s\, t}$$

$$\frac{s \longrightarrow^* s' \quad t \longrightarrow^* t' \quad R\, v\, v'}{\mathcal{E}\,(\lambda t.\, t = x\, v)\, s' \quad \mathcal{E}\,(\lambda t.\, t = x\, v')\, t'}{\mathcal{N}_s\, R\, s\, t}(*) \qquad\qquad \frac{s \longrightarrow^+ s' \quad t \longrightarrow^+ t' \quad R\, s'\, t'}{\mathcal{N}_s\, R\, s\, t}$$

Fig. 5. Normal form bisimulation steps

Figure 5 defines the *normal form bisimulation step* $\mathcal{N}_s$ of a relation R. The normal form bisimulation $\mathcal{N}$ is the greatest relation such that if $\mathcal{N}\, s\, t$ then there exists a relation R such that $\forall s'\, t' : R\, s'\, t' \rightarrow \mathcal{N}\, s'\, t'$ and $\mathcal{N}_s\, R\, s\, t$.

The *co-induction* lemma for $\mathcal{N}$ intuitively states that to establish $\mathcal{N}$, we need to find a bisimulation relation R that is preserved by $\mathcal{N}_s$. We prove that $\mathcal{N}$ is an equivalence, as we need this in our soundness proof.

Theorem 1. $\mathcal{N}$ *is adequate.* $\forall s\, t : \mathcal{N}\, s\, t \rightarrow$ co-terminate $s\, t$.

Congruence of Equational Theory. A relation is a *congruence* if it is a linearly compatible equivalence. By definition, our equational theory is an equivalence.

Theorem 2. *The relation $\Leftrightarrow$ is linearly compatible.*

Proof. By induction on $\mathcal{P}$. The proof heavily relies on the way $\mathcal{P}$ is defined. $\square$

Soundness of the Equational Theory for CBV

Theorem 3. $\mathcal{N}$ *includes reduction.* $\forall s\, t:\ s \Rightarrow t \to \mathcal{N}\, s\, t.$

Theorem 4. $\mathcal{N}$ *includes the equational theory.* $\forall s\, t:\ s \Leftrightarrow t \to \mathcal{N}\, s\, t.$

Theorem 5. *The equational theory is sound.* $\forall s\, t:\ s \Leftrightarrow t \to s \equiv t.$

Proof. Given $s \Leftrightarrow t$, we want to show that $\forall s'\, t' : \mathcal{C}\, s\, t\, s'\, t' \to$ co-terminate $s'\, t'$. For any terms s' and t' such that $\mathcal{C}\, s\, t\, s'\, t'$, we know $\mathcal{P}(\lambda u\, v.\ u = s \wedge v = t)\, s'\, t'$ by definition of $\mathcal{C}$. From $s \Leftrightarrow t$ and $\mathcal{P}(\lambda u\, v.\ u = s \wedge v = t)\ s'\ t'$ we can infer $\mathcal{P}(\Leftrightarrow)\ s'\ t'$. Since $\Leftrightarrow$ is linearly compatible (Theorem 2) we know $s' \Leftrightarrow t'$. Since $\mathcal{N}$ includes $\Leftrightarrow$ (Theorem 4) it follows that $\mathcal{N}\, s'\, t'$, and by adequacy (Theorem 1), we can conclude co-terminate $s'\, t'$. $\square$

Comparison. Similar to Lassen we make use of $\mathcal{N}$ but the structure of our soundness proof differs. Lassen defines $\mathcal{N}$ and proves that it is sound. He proves that his equational theory is included in $\mathcal{N}$ and is, therefore, also sound.

We show that our equational theory is sound by directly showing it is linearly compatible and only using $\mathcal{N}$ to assist in proving that the equational theory is adequate. In fact, our CBV version of $\mathcal{N}$ is adequate yet unsound for contextual equivalence. It relates $(\lambda y.y)(x\lambda z.z)$ and $(\lambda y.\Omega)(x\lambda z.z)$ which are both stuck on x (as $\mathcal{N}_s$'s $(*)$ rule allows the evaluation contexts on each side to be unrelated), and the context $(\lambda x.[-])\lambda u.u$ distinguishes them. Our proof method does not rely on $\mathcal{N}$ being sound and this simpler definition suffices for proving adequacy.

We decided to take this approach because the way we define our equational theory makes proving that it is a linearly compatible relation entirely straightforward. The definition of $\mathcal{N}$ is designed to ease the adequacy proof.

3   A Call-by-Push-Value Language

We show how the same essential proof structure can be applied to a more complex language to establish the soundness of its equational theory. We choose a variant of Levy's call-by-push-value (CBPV) language as our target because it is a "low-level" language with a rich equational theory. We redefine and reuse some of the notation that was introduced for CBV in the context of CBPV.

3.1   Syntax

The top of Fig. 6 shows the syntax for our variant of Levy's call-by-push-value calculus (CBPV) [12], which serves as the basis for our definitions. CBPV is a

$$\begin{aligned}
\text{Values} \ni V \quad &::= \quad x \ \mid \ n \ \mid \ \mathsf{thunk}\, M \\
\text{Terms} \ni M, N &::= \quad \mathsf{force}\, V \quad \mid \quad \mathsf{letrec}\, x_1 = M_1, ..\,, x_n = M_n \,\mathsf{in}\, N \\
& \quad \mid \ \mathsf{prd}\, V \quad \mid \quad M \,\mathsf{to}\, x \,\mathsf{in}\, N \\
& \quad \mid \ V \!\cdot\! M \quad \mid \quad \lambda x.M \\
& \quad \mid \ V_1 \oplus V_2 \quad \mid \quad \mathsf{if0}\, V\ M_1\ M_2 \\
\text{Sorts} \ni S \quad &::= \quad \mathsf{V} \ \mid \ \mathsf{C}
\end{aligned}$$

$$\frac{}{M \rightsquigarrow M} \qquad \frac{\{\,\mathsf{thunk}\,(\mathsf{letrec}\,\overline{x_i = M_i}^{\,i}\,\mathsf{in}\, M_i)/x_i\,\}^{\,i}\, N \rightsquigarrow N'}{\mathsf{letrec}\,\overline{x_i = M_i}^{\,i}\,\mathsf{in}\, N \rightsquigarrow N'}$$

$$\frac{}{\mathsf{force}\,(\mathsf{thunk}\, M) \to M} \qquad \frac{}{\mathsf{if0}\, 0\ M_1\ M_2 \to M_1} \qquad \frac{}{\mathsf{if0}\, n\ M_1\ M_2 \to M_2}\ (n \neq 0)$$

$$\frac{M \rightsquigarrow \mathsf{prd}\, V}{M \,\mathsf{to}\, x \,\mathsf{in}\, N \to \{V/x\}\, N} \qquad \frac{M \rightsquigarrow \lambda x.N}{V \!\cdot\! M \to \{V/x\}\, N} \qquad \frac{M \to M'}{V \!\cdot\! M \to V \!\cdot\! M'}$$

$$\frac{M \to M'}{M \,\mathsf{to}\, x \,\mathsf{in}\, N \to M' \,\mathsf{to}\, x \,\mathsf{in}\, N} \qquad \frac{M \rightsquigarrow (n_1 \oplus n_2)}{M \,\mathsf{to}\, x \,\mathsf{in}\, N \to \{n_1[\![\oplus]\!]n_2/x\}\, N}$$

$$\frac{N \rightsquigarrow N' \quad N' \to M}{N \to M} \qquad \frac{\{\,\mathsf{thunk}\,(\mathsf{letrec}\,\overline{x_i = M_i}^{\,i}\,\mathsf{in}\, M_i)/x_i\,\}^{\,i}\, N \longrightarrow N'}{\mathsf{letrec}\,\overline{x_i = M_i}^{\,i}\,\mathsf{in}\, N \longrightarrow N'}$$

$$\frac{}{\mathsf{wf_V}\, x} \quad \frac{}{\mathsf{wf_V}\, n} \quad \frac{\mathsf{wf_C}\, M}{\mathsf{wf_V}\,(\mathsf{thunk}\, M)} \quad \frac{\mathsf{wf_V}\, V}{\mathsf{wf_C}\,(\mathsf{prd}\, V)} \quad \frac{\mathsf{wf_C}\, M \quad \mathsf{wf_C}\, N}{\mathsf{wf_C}\,(M \,\mathsf{to}\, x \,\mathsf{in}\, N)}$$

$$\frac{\overline{\mathsf{wf_C}\, M}^{\,i} \quad \mathsf{wf_C}\, N}{\mathsf{wf_C}\,(\mathsf{letrec}\,\overline{x_i = M_i}^{\,i}\,\mathsf{in}\, N)} \quad \frac{\mathsf{wf_C}\, M}{\mathsf{wf_C}\,(\lambda x.M)} \quad \frac{\mathsf{wf_V}\, V}{\mathsf{wf_C}\,(\mathsf{force}\, V)}$$

$$\frac{\mathsf{wf_V}\, V \quad \mathsf{wf_C}\, M}{\mathsf{wf_C}\,(V \!\cdot\! M)} \quad \frac{\mathsf{wf_V}\, V_1 \quad \mathsf{wf_V}\, V_2}{\mathsf{wf_C}\,(V_1 \oplus V_2)} \quad \frac{\mathsf{wf_V}\, V \quad \mathsf{wf_C}\, M_1 \quad \mathsf{wf_C}\, M_2}{\mathsf{wf_C}\,(\mathsf{if0}\, V\ M_1\ M_2)}$$

Fig. 6. Syntax, operational semantics, and wellformedness for the CBPV language.

somewhat lower level and more structured functional language than the ordinary λ-calculus. The key feature is that it distinguishes *values* from *computations*. Values include variables x, natural numbers n, and suspended computations $\mathsf{thunk}\, M$, whereas computations include: $\mathsf{force}\, V$, which runs a suspended thunk; mutually recursive definitions, $\mathsf{letrec}\,\overline{x_i = M_i}^{\,i}\,\mathsf{in}\, M$; monadically-structured sequences of computations, written $M \,\mathsf{to}\, x \,\mathsf{in}\, N$, which runs M to produce a computation of the form $\mathsf{prd}\, V$ and then binds V as x in N; λ-abstraction $\lambda x.M$ and application $V \!\cdot\! M$ (V is the *argument* and M is the *function*); binary arithmetic operations $V_1 \oplus V_2$, where $\oplus$ is addition, subtraction, or less-than; and finally, conditional statements $\mathsf{if0}\, V\ M_1\ M_2$ that run M_1 if V is 0, otherwise, run M_2.

The most difficult aspect of formalizing this language is dealing with the $\mathsf{letrec}\,\overline{x_i = M_i}^{\,i}\,\mathsf{in}\, M$ form. The intended semantics of this term is that each of the x_i's is bound in all of the M_i's and in M. As a consequence, we have to define a multiway substitution operation, which we denote $\{\,\overline{V_i/x_i}^{\,i}\,\}\, M$. It means the simultaneous substitution of each V_i for x_i in M. Our Coq code uses de Bruijn indices, but we present the language with named variables for better readability.

3.2   Wellformedness

In Levy's presentation, the value and computation terms are separated *syntactically*, with distinct grammars for each. In our Coq formulation, we have found it simpler to combine both syntaxes into one recursive definition, which avoids combining nested recursion (for the lists of bindings found in letrec) with mutual recursion. As a consequence, we separately define (mutually recursive) wellformedness predicates that distinguish values from computations; we say values have sort V and computations have sort C.

A term M is wellformed when there exists a sort S such that $\mathsf{wf}_S\ M$ according to the rules in Fig. 6. Thanks to this separation, many of our definitions later on are adapted to account for an extra condition parameter in order to only account for wellformed terms. As a matter of notation, we use the metavariable V to mean values and M, M', N, etc. to mean computations.

In another departure from Levy's original presentation of CBPV, our version is *untyped*. Because most of our results pertain to the dynamic semantics of the language, we have eschewed types here; however, incorporating a type system should be a fairly straightforward adaptation of our formalism. In particular, the parameters necessary to account for wellformedness can simply be instantiated with a typing predicate instead.

3.3   Structural Operational Semantics

The value-computation distinction is a key feature of the CBPV design: its evaluation order is completely determined thanks to restrictions that ensure there is never a choice between a substitution step and a congruence rule. The middle portion of Fig. 6 gives the details of the operational semantics, whose small-step evaluation relation is denoted by $M \longrightarrow M'$.

Once again, the real challenge for our formalization is how to deal with letrec. The usual approach is to allow the structural operational semantics to unroll a letrec as a step of computation, which, in our setting would amount to using the following rule: $\mathsf{letrec}\ \overline{x_i = M_i}^{\,i}\ \mathsf{in}\ M \longrightarrow \{\,\mathsf{thunk}\,(\mathsf{letrec}\ \overline{x_i = M_i}^{\,i}\ \mathsf{in}\ M_i)/x_i\,\}^{\,i}\ M$. Note that because the variables x_i range over values but the right-hand sides of the letrec bindings are computations M_i, we have to wrap each computation in a thunk during the unrolling. However, rather than using this rule directly, we have opted to construct the operational semantics in such a way that letrec unrolling doesn't "count" as an operational step. Therefore the operational semantics rules in Fig. 6 rely on an auxiliary relation $\rightsquigarrow$ that unrolls a letrec to expose either a prd or a λ term. The operational semantics rules are otherwise straightforward and consist of three "real" steps of computation: forcing a thunk, sequencing, and β-substitution, and three congruence rules that search for the next redex. Our choice to handle letrec in this way is not strictly necessary (our techniques would apply with the "standard" interpretation above); however, we prefer this formulation, despite its slight cost in complexity, because we anticipate that treating letrec as having no runtime cost is more consistent with the semantics of low-level compiler intermediate representations [7].

Given the definitions of the wellformedness relation and operational semantics, it is easy to establish some basic facts. A term M is in *normal form*, written nf M, if $\not\exists M' : M \longrightarrow M'$. We prove that the step relation is deterministic and hence each term has at most one normal form:

- For terms M, N, and N', if $M \longrightarrow N$ and $M \longrightarrow N'$, then $N = N'$.
- For terms M, N, and N', if $M \longrightarrow^* N$, $M \longrightarrow^* N'$, nf N, and nf N', then $N = N'$.

The step relation also preserves wellformedness (note that values do not step):

- For terms M and M', if $\mathsf{wf_C}\ M$ and $M \longrightarrow M'$, then $\mathsf{wf_C}\ M'$.
- For terms M and M', if $\mathsf{wf_C}\ M$ and $M \longrightarrow^* M'$, then $\mathsf{wf_C}\ M'$.

3.4 Progress, and Canonical and Error Forms

Similar to CBV, the bisimulation relation for CBPV also relies on relating terms in normal form However, for CBPV there are also certain *erroneous terms*, that are not considered to be "good" CBPV programs and that cannot step. In this pure setting, such programs could be ruled out by using a type system, but similar issues arise if we extended the language with nontrivial constants and partial operations (such as division or array lookup). We therefore define two predicates error and canon, which partition normal terms into two sets: those that we consider "erroneous" and those that are "canonical" in the sense that they are good CBPV terms that are nevertheless stuck. Canonical terms arise because we want to be able to relate *open* terms. For example, (force x) to x' in M cannot step because it is blocked on trying to evaluate the term force x.

To define these predicates, the intuitive idea would be to identify terms whose evaluation is blocked because the next step of computation must force a variable (or, perform an ill-typed action). Setting aside letrec for the moment, the evaluation contexts for our CBPV language are given by the following grammar:

$$E ::= [\,] \quad | \quad E \text{ to } x \text{ in } N \quad | \quad V \cdot E$$

We want terms of the form $E[\text{force } x]$ to be canonical, and, similarly, terms like $E[(\lambda x.M) \text{ to } x' \text{ in } N]$ to be erroneous. However, as our operational semantics treats letrec as transparent, we must account for unrolling of recursive definitions.

We introduce a higher-order predicate $\mathcal{E}$ whose type is $(\textit{Term} \to \mathbb{P}) \to \textit{Term} \to \mathbb{P}$ (similar to the one we defined for the CBV λ-calculus). $\mathcal{E}$ is defined inductively, as shown in Fig. 7. It is parameterized by a proposition P and its structure mirrors that of the grammar of evaluation contexts, except that it also builds in a case for unrolling letrec. Intuitively, the predicate $\mathcal{E}\ P\ M$ holds if the term M unrolls to a term of the form $E[M']$ and $P(M')$ holds. We call $\mathcal{E}\ P$ the *evaluation context closure* of the predicate P.

$$\frac{P\,(M)}{\mathcal{E}\,P\,M} \qquad \frac{\mathcal{E}\,P\,M}{\mathcal{E}\,P\,(M\,\mathsf{to}\,x\,\mathsf{in}\,N)} \qquad \frac{\mathcal{E}\,P\,M}{\mathcal{E}\,P\,(V\!\cdot\!M)}$$

$$\frac{\mathcal{E}\,P\,(\{\,\mathsf{thunk}\,(\mathsf{letrec}\,\overline{x_i = M_i}^{\,i}\,\mathsf{in}\,M_i)/x_i\,\}^{\,i}\,N)}{\mathcal{E}\,P\,(\mathsf{letrec}\,\overline{x_i = M_i}^{\,i}\,\mathsf{in}\,N)}$$

$$\frac{M \rightsquigarrow \lambda x.M'}{\mathsf{errorP}\,(M\,\mathsf{to}\,x\,\mathsf{in}\,N)} \qquad \frac{M \rightsquigarrow \mathsf{prd}\,V'}{\mathsf{errorP}\,(V\!\cdot\!M)} \qquad \mathsf{error}\,M \Leftrightarrow \mathcal{E}\,\mathsf{errorP}\,M$$

$$\frac{M \rightsquigarrow \mathsf{prd}\,V}{\mathsf{canon}\,M} \quad \frac{M \rightsquigarrow \lambda x.M'}{\mathsf{canon}\,M} \quad \frac{\mathcal{E}\,\mathsf{forcevar}\,M}{\mathsf{canon}\,M} \quad \mathsf{forcevar}\,M \leftrightarrow (\exists x.\,M = \mathsf{force}\,x)$$

Fig. 7. Evaluation context closure, and canonical and error forms

Using $\mathcal{E}$, it is straightforward to define error M as the evaluation context closure of a predicate errorP that picks out the ill-formed terms. Similarly, canon M instantiates $\mathcal{E}$ with the predicate forcevar, which holds of a term M exactly when M is force x for some variable x. We prove that these predicates form a partition of wellformed computations (note that values do not step):

- For a term M, if $\mathsf{wf_C}\,M$ and canon M, then nf $M \wedge \neg\mathsf{error}\,M$.
- For a term M, if $\mathsf{wf_C}\,M$ and error M, then nf $M \wedge \neg\mathsf{canon}\,M$.
- For a term M, if $\mathsf{wf_C}\,M$, then canon M or error M or $\exists N : M \longrightarrow N$.

3.5   Contextual Equivalence

Recall that contextual equivalence equates two terms if their behavior is the same in all program contexts. In the context of CBPV, we are only interested in reasoning about wellformed terms. We therefore define a *conditional linearly compatible closure* that restricts the context to that of terms which respect a condition. The condition is later used to restrict the context to wellformed terms.

We define *conditional linearly compatible closure*, $\mathcal{P}$, similar to that of CBV but with an additional condition P. It lifts a relation R into a relation $\mathcal{P}\,R\,P$ linearly compatible with the CBPV syntax and that relates terms for which P holds (defined in Appendix A). A relation that is closed under such lifting for wellformed terms is called *wellformed linearly compatible*. Formally, a relation R on terms is *wellformed linearly compatible* if $\forall M\,N : \mathcal{P}\,R\,(\lambda x.\,\exists S : \mathsf{wf}_S\,x)\,M\,N \to R\,M\,N$. The conditional contextual closure $\mathcal{C}$ is defined as follows:

$$\mathcal{C}\,P\,M\,N\,M'\,N' \leftrightarrow \mathcal{P}(\lambda u v.\,u = M \wedge v = N)\,P\,M'\,N'$$

Intuitively, $\mathcal{C}\,P\,M\,N\,M'\,N'$ holds exactly when there exists a context $C[-]$ such that $M' = C[M]$, $N' = C[N]$, $P\,(C[M])$, and $P\,(C[N])$. Instantiating P with $(\lambda x.\,\exists S : \mathsf{wf}_S\,x)$ and quantifying over all M' and N' captures the idea of quantifying over all contexts for wellformed terms.

$$\frac{P\,M}{Eq\,R\,P\,M\,M} \qquad \frac{P\,M \quad (R\,M\,M' \lor R\,M'\,M) \quad Eq\,R\,P\,M'\,N}{Eq\,R\,P\,M\,N}$$

Fig. 8. Conditional equivalence closure of a relation R over a predicate P

A term M *terminates to* N, written $M \Downarrow N$, if $M \longrightarrow^* N \land \mathsf{nf}\,N$. Two terms M and N *co-terminate*, written $\mathsf{co\text{-}terminate}\,M\,N$, if $(\exists M' : M \Downarrow M') \leftrightarrow (\exists N' : N \Downarrow N')$. Two terms M and N are *contextually equivalent*, written $M \equiv N$, if $\forall M'\,N' : \mathcal{C}\,(\lambda x.\,\exists\,S : \mathsf{wf}_S\,x)\,M\,N\,M'\,N' \to \mathsf{co\text{-}terminate}\,M'\,N'$. The definitions of soundness and adequacy are identical to the ones in Sect. 2 except that they operate on CBPV terms.

3.6   Equational Theory

The equational theory for CBPV is defined in a similar fashion to that for CBV. Once again, we are interested in a theory that is sound and that includes the operational semantics. As mentioned earlier, also ensuring that the equational theory is a congruence relation results in a theory that includes β-equivalence. Once again all our definitions in the CBPV context are parameterized with a condition, which is in turn used to limit our scope to wellformed terms.

A relation R is *conditionally reflexive* on a predicate P if $\forall x : P\,x \to R\,x\,x$. A relation R is *conditionally symmetric* on a predicate P if $\forall x\,y : P\,x \to P\,y \to R\,x\,y \to R\,y\,x$. A relation R is *conditionally transitive* on a predicate P if $\forall x\,y\,z : P\,x \to P\,y \to P\,z \to R\,x\,y \to R\,y\,z \to R\,x\,z$. A relation R is a *conditional equivalence* on a predicate P if it is conditionally reflexive, conditionally symmetric, and conditionally transitive on P.

Our CBPV operational semantics does not explicitly unroll letrec as a step of computation. Nevertheless, we would like our equational theory to equate letrec terms to their unrolled version. Therefore, we first define a reduction relation.

Definition 1. *Given two terms M and N, the reduction relation $M \overset{\cdot}{\longrightarrow} N$ is defined using the following two rules:*

$$\frac{M \longrightarrow N}{M \overset{\cdot}{\longrightarrow} N} \qquad \frac{}{\mathsf{letrec}\,\overline{x_i = M_i}^{\,i}\,\mathsf{in}\,M \overset{\cdot}{\longrightarrow} \{\,\mathsf{thunk}\,(\mathsf{letrec}\,\overline{x_i = M_i}^{\,i}\,\mathsf{in}\,M_i)/x_i\,\}^{\,i}\,M}$$

Definition 2. *For wellformed terms M and N, the* parallel reduction *relation, written $M \Rightarrow N$, holds if $\mathcal{P}(\overset{\cdot}{\longrightarrow})\,M\,N$.*

The conditional equivalence closure of a relation R over a predicate P is the reflexive, symmetric, and transitive closure of R where P holds. It is defined using the rules in Fig. 8 where the second rule combines symmetry and transitivity.

Lemma 1. *For any two elements x and y related by the conditional equivalence closure of a relation R on P, $P\,x$ holds and $P\,y$ holds.*

Lemma 2. *Given two relations R and R', and given a predicate P, if $\forall x'\,y'$: $P\,x' \to P\,y' \to R\,x'\,y' \to R'\,x'\,y'$ and R' is a conditional equivalence on P, then for any two elements x and y related using the conditional equivalence closure of R on P, $R'\,x\,y$ holds.*

Two terms M and N are equal according to the *equational theory for CBPV*, written $M \Leftrightarrow_S N$, if $Eq\,(\Rightarrow)\,\mathsf{wf}_S\,M\,N$. Once again, it is straightforward to see that our equational theory includes the operational semantics. We present the soundness proof of the equational theory for CBPV in the next section.

3.7   Soundness of the Equational Theory for CBPV

The soundness proof of the equational theory for CBPV follows the same structure as that for CBV. The proof development is restricted to wellformed terms.

Normal Form Bisimulation. Similar to CBV, $\mathcal{N}_s$ defines one step in the bisimulation for CBPV and is used to co-inductively define our bisimulation $\mathcal{N}$ for CBPV. One difference to the CBV definition is that $\mathcal{N}$ also takes a sort as input and relates two terms at that sort. Moreover, the definition relies on $\mathcal{E}_C$, which lifts a binary relation R across two terms and ensures that they share an evaluation context up to R. Namely, $\mathcal{E}_C\,R\,M\,N\,M'\,N'$ holds if $R\,M'\,N'$ and if $M'\,N'$ are evaluation contexts over M and N, respectively.

Figure 9 defines the *normal form bisimulation step* $\mathcal{N}_s$ of a relation R (that given a sort relates terms). The normal form bisimulation $\mathcal{N}$ is the greatest relation such that if $\mathcal{N}\,S\,M\,N$ then there exists a relation R such that $\forall S'\,M'\,N'$: $R\,S'\,M'\,N' \to \mathcal{N}\,S'\,M'\,N'$ and $\mathcal{N}_s\,R\,S\,M\,N$.

Theorem 6. *$\mathcal{N}$ is a conditional equivalence on wellformed terms.*

Congruence of Equational Theory. A relation is called a *congruence* if it is both an equivalence over wellformed terms and wellformed linearly compatible. By definition, our equational theory is an equivalence over wellformed terms.

Theorem 7. *The relation $(\lambda M\,N : \exists S : M \Leftrightarrow_S N)$ is wellformed linearly compatible.*

Soundness of Equational Theory for CBPV

Theorem 8. *$\mathcal{N}$ includes parallel reduction*

$$\forall\,S\,M\,N : \mathsf{wf}_S\,M \to \mathsf{wf}_S\,N \to M \Rightarrow N \to \mathcal{N}\,S\,M\,N$$

Theorem 9. *$\mathcal{N}$ includes the equational theory*

$$\forall\,S\,M\,N : M \Leftrightarrow_S N \to \mathcal{N}\,S\,M\,N$$

Proof. Follows directly from Lemma 2 along with Theorems 6 and 8.   □

$$\frac{}{\mathcal{N}_s\, R\, \mathsf{V}\, x\, x} \quad \frac{}{\mathcal{N}_s\, R\, \mathsf{V}\, n\, n} \quad \frac{R\, \mathsf{C}\, M\, N}{\mathcal{N}_s\, R\, \mathsf{V}\, (\mathsf{thunk}\, M)\, (\mathsf{thunk}\, N)} \quad \frac{M \longrightarrow^* M' \quad N \longrightarrow^* N' \quad \mathsf{error}\, M' \quad \mathsf{error}\, N'}{\mathcal{N}_s\, R\, \mathsf{C}\, M\, N}$$

$$\frac{M \longrightarrow^* M' \quad N \longrightarrow^* N' \quad M' \rightsquigarrow \mathsf{prd}\, V}{N' \rightsquigarrow \mathsf{prd}\, V' \quad R\, \mathsf{V}\, V\, V'}{\mathcal{N}_s\, R\, \mathsf{C}\, M\, N} \quad \frac{M \longrightarrow^* M' \quad N \longrightarrow^* N' \quad M' \rightsquigarrow \lambda x.M''}{N' \rightsquigarrow \lambda x.N'' \quad R\, \mathsf{C}\, M''\, N''}{\mathcal{N}_s\, R\, \mathsf{C}\, M\, N}$$

$$\frac{M \longrightarrow^* M' \quad N \longrightarrow^* N'}{\mathcal{E}_C\, R\, (\mathsf{force}\, x)\, (\mathsf{force}\, x)\, M'\, N'}{\mathcal{N}_s\, R\, \mathsf{C}\, M\, N} \quad \frac{M \longrightarrow^* M' \quad N \longrightarrow^* N'}{\mathcal{E}_C\, R\, (\mathsf{if0}\, x\, M_1\, M_2)\, (\mathsf{if0}\, x\, M_1'\, M_2')\, M'\, N'}{R\, \mathsf{C}\, M_1\, M_1' \quad R\, \mathsf{C}\, M_2\, M_2'}{\mathcal{N}_s\, R\, \mathsf{C}\, M\, N}$$

$$\frac{M \longrightarrow^* M' \quad N \longrightarrow^* N'}{M' \rightsquigarrow x \oplus V \quad N' \rightsquigarrow x \oplus V'}{R\, \mathsf{V}\, V\, V' \quad (\nexists M'' : V = \mathsf{thunk}\, M'')}{\mathcal{N}_s\, R\, \mathsf{C}\, M\, N} \quad \frac{M \longrightarrow^* M' \quad N \longrightarrow^* N'}{M' \rightsquigarrow n \oplus V \quad N' \rightsquigarrow n \oplus V'}{R\, \mathsf{V}\, V\, V' \quad (\nexists M'' : V = \mathsf{thunk}\, M'')}{\mathcal{N}_s\, R\, \mathsf{C}\, M\, N}$$

$$\frac{M \longrightarrow^* M' \quad N \longrightarrow^* N' \quad \mathcal{E}_C\, R\, (M_1\, \mathsf{to}\, x\, \mathsf{in}\, M_2)\, (N_1\, \mathsf{to}\, x\, \mathsf{in}\, N_2)\, M'\, N'}{M_1 \rightsquigarrow V_1 \oplus V_2 \quad N_1 \rightsquigarrow V_1' \oplus V_2' \quad R\, \mathsf{V}\, V_1\, V_1' \quad R\, \mathsf{V}\, V_2\, V_2' \quad R\, \mathsf{C}\, M_2\, N_2}{((V_1 = x \wedge (\nexists M : V_2 = \mathsf{thunk}\, M)) \vee (V_2 = x \wedge (\nexists M : V_1 = \mathsf{thunk}\, M)))}{\mathcal{N}_s\, R\, \mathsf{C}\, M\, N}$$

$$\frac{M \longrightarrow^+ M' \quad N \longrightarrow^+ N' \quad R\, \mathsf{C}\, M'\, N'}{\mathcal{N}_s\, R\, \mathsf{C}\, M\, N}$$

Fig. 9. Normal form bisimulation steps

Theorem 10. $\mathcal{N}$ *is adequate*

$$\forall S\, M\, N : \mathsf{wf}_S\, M \to \mathsf{wf}_S\, N \to \mathcal{N}\, S\, M\, N \to \mathsf{co\text{-}terminate}\, M\, N$$

We only prove and rely on the adequacy and equivalence of $\mathcal{N}$. Although unnecessary, we suspect our CBPV $\mathcal{N}$ is a congruence, thus, is sound. Unlike CBV, $\mathcal{N}$ here uses $\mathcal{E}_C$ which, unlike $\mathcal{E}$, only relates terms that share an evaluation context.

Theorem 11. *The equational theory is sound.* $\forall S\, M\, N : M \Leftrightarrow_S N \to M \equiv N.$

Proof. Given $M \Leftrightarrow_S N$, we want to show

$$\forall M'\, N' : \mathcal{C}\, (\lambda x.\, \exists S : \mathsf{wf}_S\, x)\, M\, N\, M'\, N' \to \mathsf{co\text{-}terminate}\, M'\, N'.$$

For any terms M' and N' such that $\mathcal{C}\, (\lambda x.\, \exists S : \mathsf{wf}_S\, x)\, M\, N\, M'\, N'$, we know $\mathcal{P}(\lambda u v.\, u = M \wedge v = N)\, (\lambda x.\, \exists S : \mathsf{wf}_S\, x)\, M'\, N'$ by definition of $\mathcal{C}$. From $M \Leftrightarrow_S N$ and $\mathcal{P}(\lambda u v.\, u = M \wedge v = N)\, (\lambda x.\, \exists S : \mathsf{wf}_S\, x)\, M'\, N'$ we can infer $\mathcal{P}\, (\lambda M\, N : \exists S : M \Leftrightarrow_S N)\, (\lambda x.\, \exists S : \mathsf{wf}_S\, x)\, M'\, N'$. Since $(\lambda M\, N : \exists S : M \Leftrightarrow_S N)$ is linearly compatible (Theorem 7) we know $M' \Leftrightarrow_{S'} N'$ for some sort S'. From Lemma 1 we know $\mathsf{wf}_{S'}\, M'$ and $\mathsf{wf}_{S'}\, N'$. Since $\mathcal{N}$ includes the equational theory (Theorem 9) it follows that $\mathcal{N}\, S'\, M'\, N'$. Since $\mathcal{N}$ is adequate (Theorem 10) we conclude that $\mathsf{co\text{-}terminate}\, M'\, N'$. $\square$

4   Verifying Optimizations Using Our Equational Theory

Compilers rely on program equivalences when optimizing and transforming programs. In the context of verified compilation, as found in the CompCert [11] and CakeML [8] projects, formal verification of particular program equivalences is crucial: the correctness of classic optimizations like constant folding, code inlining, loop unrolling, etc., hinges on such proofs. Therefore, techniques that facilitate such machine-checked proofs have the potential for a broad impact.

We demonstrate how our CBPV equational theory makes it trivial to verify many typical compiler optimizations and indicate which imperative optimizations correspond to our results.

Connection to Compilers for Imperative Languages. In separate work [7], we prove an equivalence between our functional CBPV language and *control flow graphs* (CFG), which many compilers for imperative languages use to represent low-level programs. Control flow graphs are suited to program analysis and optimizations. However, formalizing the behavior and metatheory of CFG programs is non-trivial: CFG programs don't compose well, their semantics depends on auxiliary state, and, as a consequence, they do not enjoy a simple equational theory that can be used for reasoning about the correctness of program transformations. The equational theory developed in this paper can also be used to reason about CFG optimizations.

Optimizations. Figure 10 summarizes various desirable CBPV compiler optimizations that are easily proven sound using our equational theory by term rewriting. We give a short C description of the optimizations before explaining how we verified them in Coq.

CBPV equation	optimization
$\mathsf{force}\,(\mathsf{thunk}\,M) \equiv M$	block merging "direct jump case"
$V \cdot \lambda x.M \equiv \{V/x\}\,M$	block merging "phi case"
$\mathsf{prd}\,V\,\mathsf{to}\,x\,\mathsf{in}\,M \equiv \{V/x\}\,M$	move elimination
$(n_1 \oplus n_2)\,\mathsf{to}\,x\,\mathsf{in}\,M \equiv \{n_1[\![\oplus]\!]n_2/x\}\,M$	constant folding
$\mathsf{thunk}\,(\lambda y.M) \cdot \lambda x.N \equiv \{\mathsf{thunk}\,(\lambda y.M)/x\}\,N$	function inlining
$\mathsf{if0}\,0\,M_1\,M_2 \equiv M_1$	dead branch elimination "true branch"
$\mathsf{if0}\,n\,M_1\,M_2 \equiv M_2$ where $(n \neq 0)$	dead branch elimination "false branch"
$\mathsf{if0}\,n\,M\,M \equiv M$	branch elimination

Fig. 10. CBPV equations and corresponding imperative optimizations

Block merging "direct jump case": merges two blocks of the control-flow-graph if the first is the unique predecessor of the second and jumps directly.

Block merging "phi case": replaces functions applied to arguments with their result (does one step of β-reduction). At the CFG level, this optimization corresponds to eliminating a jump to a block containing a "phi" node.

Move elimination: eliminates a move by substituting the target register's value.

Function inlining: replaces function calls in the program with the bodies of the called functions. Note that the function inlining CBPV equation is just an instance of the block merging "phi case" equation.

Dead branch elimination: removes dead branches in conditional statements.

Branch elimination: replaces conditional statements that have identical branches with the code of one of the branches. Unlike the three previous optimizations, this one does not follow directly from one step in the operational semantics. However, it still falls in our equational theory and it was easy to verify it using an additional case analysis on n in the proof.

Proof Technique. We define a function mk-cmp that takes an optimization function $f : Term \rightarrow Term$ and a program M and returns an optimized program M' where the optimization function f has been applied recursively throughout the program (first on all subprograms then again recursively on the result). The equations in Fig. 10 are examples of such functions f. Given the term on the left-hand side of the equation as input, f returns the right-hand side; given any other term, f acts as the identity. Applying mk-cmp to these functions applies them throughout the program repeatedly and results in the actual optimizations that we verify.

We first prove two general theorems about mk-cmp, one about preserving wellformedness and the other about preserving soundness.

Theorem 12. mk-cmp *preserves wellformedness*

$$\forall f\, S\, M : (\forall S'\, N : \mathsf{wf}_{S'}\, N \rightarrow \mathsf{wf}_{S'}\, (f\, N)) \rightarrow$$
$$\mathsf{wf}_S\, M \rightarrow \mathsf{wf}_S\, (\mathsf{mk\text{-}cmp}\, f\, M)$$

This theorem intuitively states that if f preserves wellformedness then so does mk-cmp f.

Theorem 13. mk-cmp *preserves soundness*

$$\forall f\, S\, M : (\forall S'\, N : \mathsf{wf}_{S'}\, N \rightarrow \mathsf{wf}_{S'}\, (f\, N)) \rightarrow$$
$$(\forall S'\, N : \mathsf{wf}_{S'}\, N \rightarrow N \equiv (f\, N)) \rightarrow$$
$$\mathsf{wf}_S\, M \rightarrow M \equiv (\mathsf{mk\text{-}cmp}\, f\, M)$$

This theorem states that if f preserves wellformedness and is sound then mk-cmp f is also sound.

Corollary 1. mk-cmp *preserves soundness using the CBPV equational theory*

$$\forall f\, S\, M : (\forall S'\, N : \mathsf{wf}_{S'}\, N \to \mathsf{wf}_{S'}\, (f\, N)) \to$$
$$(\forall S'\, N : \mathsf{wf}_{S'}\, N \to N \Leftrightarrow_{S'} (f\, N)) \to$$
$$\mathsf{wf}_S\, M \;\to\; M \equiv (\mathsf{mk\text{-}cmp}\, f\, M)$$

This corollary states that if f preserves wellformedness and is included in the CBPV equational theory, then mk-cmp f is sound. This corollary follows directly from Theorem 13 along with the fact that the equational theory is sound (Theorem 11).

The proof structure for the optimizations mk-cmp f, where f is any of the eight optimizations described in Fig. 10, proceeds as follows:
First, we prove that f preserves wellformedness:

$$\forall S\, M : \mathsf{wf}_S\, M \to \mathsf{wf}_S\, (f\, M).$$

Then, we prove that the equational theory includes f:

$$\forall S\, M : \mathsf{wf}_S\, M \to M \Leftrightarrow_S (f\, M).$$

Finally, using the mk-cmp corollary (Corollary 1) along with these results we can conclude that the optimization is correct:

$$\forall S\, M : \mathsf{wf}_S\, M \to M \equiv (\mathsf{mk\text{-}cmp}\, f\, M).$$

Our CBPV equational theory made it simple to prove all these optimizations correct in Coq with a minimal amount of effort. This demonstrates the power of our approach in reducing the cost of verifying compiler optimizations. The same approach will similarly facilitate reasoning about more complex optimizations.

5   Conclusion and Future Work

We developed a sound equational theory for a variant of Levy's low-level CBPV language and showed how it makes verifying several typical optimizations trivial. For the sake of explaining and comparing our proof method, we also applied it to the pure untyped CBV λ-calculus. Similar to prior work on proving equivalences using $\mathcal{N}$ we did not target completeness (w.r.t contextual equivalence). We rather focused on sound reasoning techniques that are ideal for verified compilers.

The adequacy of reduction, that we rely on in our soundness proof, can alternatively be derived from the Standardization Theorem for λ-calculus [5, 18], which states that there is a "standard" reduction sequence for any multi-step reduction [3]. Takahashi gave a simple proof of the Standardization Theorem for call-by-name [18] and Crary adapted and formalized her proof for call-by-value [5]. We plan to investigate whether adapting and scaling this proof method to our CBPV language, which allows mutual recursion, provides a simpler proof than using $\mathcal{N}$.

A   Appendix: Definition of $\mathcal{P}$ for CBPV

$$\frac{P\,M \quad P\,N \quad R\,M\,N}{\mathcal{P}\,R\,P\,M\,N} \qquad \frac{P\,(\mathrm{prd}\,V) \quad P\,(\mathrm{prd}\,V) \quad \mathcal{P}\,R\,P\,V\,V'}{\mathcal{P}\,R\,P\,(\mathrm{prd}\,V)\,(\mathrm{prd}\,V')}$$

$$\frac{P\,(\mathrm{force}\,V) \quad P\,(\mathrm{force}\,V') \quad \mathcal{P}\,R\,P\,V\,V'}{\mathcal{P}\,R\,P\,(\mathrm{force}\,V)\,(\mathrm{force}\,V')} \qquad \frac{P\,(\lambda x.M) \quad P\,(\lambda x.N) \quad \mathcal{P}\,R\,P\,M\,N}{\mathcal{P}\,R\,P\,(\lambda x.M)\,(\lambda x.N)}$$

$$\frac{P\,(\mathrm{thunk}\,M) \quad P\,(\mathrm{thunk}\,N) \quad \mathcal{P}\,R\,P\,M\,N}{\mathcal{P}\,R\,P\,(\mathrm{thunk}\,M)\,(\mathrm{thunk}\,N)}$$

$$\frac{P\,(M\,\mathrm{to}\,x\,\mathrm{in}\,N) \quad P\,(M'\,\mathrm{to}\,x\,\mathrm{in}\,N) \quad \mathcal{P}\,R\,P\,M\,M'}{\mathcal{P}\,R\,P\,(M\,\mathrm{to}\,x\,\mathrm{in}\,N)\,(M'\,\mathrm{to}\,x\,\mathrm{in}\,N)}$$

$$\frac{P\,(M\,\mathrm{to}\,x\,\mathrm{in}\,N) \quad P\,(M\,\mathrm{to}\,x\,\mathrm{in}\,N') \quad \mathcal{P}\,R\,P\,N\,N'}{\mathcal{P}\,R\,P\,(M\,\mathrm{to}\,x\,\mathrm{in}\,N)\,(M\,\mathrm{to}\,x\,\mathrm{in}\,N')}$$

$$\frac{P\,(V\cdot M) \quad P\,(V'\cdot M) \quad \mathcal{P}\,R\,P\,V\,V'}{\mathcal{P}\,R\,P\,(V\cdot M)\,(V'\cdot M)} \qquad \frac{P\,(V\cdot M) \quad P\,(V\cdot N) \quad \mathcal{P}\,R\,P\,M\,N}{\mathcal{P}\,R\,P\,(V\cdot M)\,(V\cdot N)}$$

$$\frac{P\,(V_1 \oplus V_2) \quad P\,(V_1' \oplus V_2) \quad \mathcal{P}\,R\,P\,V_1\,V_1'}{\mathcal{P}\,R\,P\,(V_1 \oplus V_2)\,(V_1' \oplus V_2)} \qquad \frac{P\,(V_1 \oplus V_2) \quad P\,(V_1 \oplus V_2') \quad \mathcal{P}\,R\,P\,V_2\,V_2'}{\mathcal{P}\,R\,P\,(V_1 \oplus V_2)\,(V_1 \oplus V_2')}$$

$$\frac{P\,(\mathrm{if0}\ V\ M_1\ M_2) \quad P\,(\mathrm{if0}\ V'\ M_1\ M_2) \quad \mathcal{P}\,R\,P\,V\,V'}{\mathcal{P}\,R\,P\,(\mathrm{if0}\ V\ M_1\ M_2)\,(\mathrm{if0}\ V'\ M_1\ M_2)}$$

$$\frac{P\,(\mathrm{if0}\ V\ M_1\ M_2) \quad P\,(\mathrm{if0}\ V\ M_1'\ M_2) \quad \mathcal{P}\,R\,P\,M_1\,M_1'}{\mathcal{P}\,R\,P\,(\mathrm{if0}\ V\ M_1\ M_2)\,(\mathrm{if0}\ V\ M_1'\ M_2)}$$

$$\frac{P\,(\mathrm{if0}\ V\ M_1\ M_2) \quad P\,(\mathrm{if0}\ V\ M_1\ M_2') \quad \mathcal{P}\,R\,P\,M_2\,M_2'}{\mathcal{P}\,R\,P\,(\mathrm{if0}\ V\ M_1\ M_2)\,(\mathrm{if0}\ V\ M_1\ M_2')}$$

$$\frac{P\,(\mathrm{letrec}\,\overline{x_i = M_i}^{\,i}\,\mathrm{in}\,N) \quad P\,(\mathrm{letrec}\,\overline{x_i = M_i}^{\,i}\,\mathrm{in}\,N') \quad \mathcal{P}\,R\,P\,N\,N'}{\mathcal{P}\,R\,P\,(\mathrm{letrec}\,\overline{x_i = M_i}^{\,i}\,\mathrm{in}\,N)\,(\mathrm{letrec}\,\overline{x_i = M_i}^{\,i}\,\mathrm{in}\,N')}$$

$$\frac{P\,(\mathrm{letrec}\,\overline{x_i = M_i}^{\,i}\,\mathrm{in}\,N) \quad P\,(\mathrm{letrec}\,\overline{x_i = M_i'}^{\,i}\,\mathrm{in}\,N) \quad \mathcal{P}\,R\,P\,M_j\,M_j' \text{ where } j \text{ is one of the } i\text{'s}}{\mathcal{P}\,R\,P\,(\mathrm{letrec}\,\overline{x_i = M_i}^{\,i}\,\mathrm{in}\,N)\,(\mathrm{letrec}\,\overline{x_i = M_i'}^{\,i}\,\mathrm{in}\,N)}$$

Fig. 11. Rules defining conditional compatible closure $\mathcal{P}$ of a (binary) relation R and a (unary) predicate P.

References

1. Abramsky, S.: The lazy λ- calculus. In: Research Topics in Functional Programming, pp. 65–116. Addison Wesley (1990)
2. Abramsky, S., Ong, C.H.L.: Full abstraction in the lazy lambda calculus. Inf. Comput. **105**(2), 159–267 (1993)
3. Barendregt, H.: The Lambda Calculus: Its Syntax and Semantics. Studies in Logic and the Foundations of Mathematics. Elsevier Science, New York (2013)
4. Biernacki, D., Lenglet, S., Polesiuk, P.: Proving soundness of extensional normal-form bisimilarities. In: Mathematical Foundations of Program Semantics (MFPS). Electronic Notes in Theoretical Computer Science (2017)
5. Crary, K.: A simple proof of call-by-value standardization. Technical report CMU-CS-09-137, Carnegie Mellon University (2009). https://www.cs.cmu.edu/~crary/papers/2009/standard.pdf
6. Downen, P., Maurer, L., Ariola, Z.M., Jones, S.P.: Sequent calculus as a compiler intermediate language. In: Proceedings of the Sixth ACM SIGPLAN International Conference on Functional Programming (ICFP), pp. 74–88 (2016)
7. Garbuzov, D., Mansky, W., Rizkallah, C., Zdancewic, S.: Structural operational semantics for control flow graph machines. CoRR abs/1805.05400 (2018). http://arxiv.org/abs/1805.05400
8. Kumar, R., Myreen, M.O., Norrish, M., Owens, S.: Cakeml: a verified implementation of ML. In: Proceedings of the 41st ACM SIGPLAN-SIGACT Symposium on Principles of Programming Languages, POPL 2014, pp. 179–191. ACM, New York (2014). https://doi.org/10.1145/2535838.2535841, http://doi.acm.org/10.1145/2535838.2535841
9. Lassen, S.B.: Bisimulation in untyped lambda calculus: Böhm trees and bisimulation up to context. Electron. Notes Theor. Comput. Sci. **20**, 346–374 (1999)
10. Lassen, S.B.: Eager normal form bisimulation. In: 20th Annual IEEE Symposium on Logic in Computer Science (LICS 2005), pp. 345–354 (2005)
11. Leroy, X.: A formally verified compiler back-end. J. Autom. Reasoning **43**(4), 363–446 (2009). https://doi.org/10.1007/s10817-009-9155-4
12. Levy, P.B.: Call-by-push-value: a subsuming paradigm. In: Girard, J.-Y. (ed.) TLCA 1999. LNCS, vol. 1581, pp. 228–243. Springer, Heidelberg (1999). https://doi.org/10.1007/3-540-48959-2_17
13. Mason, I., Talcott, C.: Equivalence in functional languages with effects. J. Funct. Program. **1**(3), 287–327 (1991)
14. Maurer, L., Downen, P., Ariola, Z.M., Peyton Jones, S.: Compiling without continuations. In: Proceedings of the 38th ACM SIGPLAN Conference on Programming Language Design and Implementation, pp. 482–494. ACM (2017)
15. Morris, J.H.: Lambda calculus models of programming languages. Ph.D. thesis, Massachusets Institute of Technology (1968)
16. Rizkallah, C., Garbuzov, D., Zdancewic, S.: Accompanying Coq formalization (2018). http://www.cse.unsw.edu.au/~crizkallah/publications/equational_theory_cbpv.tar.gz
17. Sangiorgi, D., Kobayashi, N., Sumii, E.: Environmental bisimulations for higher-order languages. In: 22nd Annual IEEE Symposium on Logic in Computer Science, LICS 2007, pp. 293–302. IEEE (2007)
18. Takahashi, M.: Parallel reductions in lambda-calculus. Inf. Comput. **118**(1), 120–127 (1995). https://doi.org/10.1006/inco.1995.1057

Program Verification in the Presence of Cached Address Translation

Hira Taqdees Syeda[1,2]($\boxtimes$) and Gerwin Klein[1,2]($\boxtimes$)

[1] Data61, CSIRO, Eveleigh, Australia
{Hira.Syeda,Gerwin.Klein}@data61.csiro.au
[2] School of Computer Science and Engineering, UNSW, Sydney, Australia

Abstract. Operating system (OS) kernels achieve isolation between user-level processes using multi-level page tables and translation lookaside buffers (TLBs). Controlling the TLB correctly is a fundamental security property—yet all large-scale formal OS verification projects leave correct functionality of the TLB as an assumption. We present a logic for reasoning about low-level programs in the presence of TLB address translation. We extract invariants and necessary conditions for correct TLB operation that mirror the informal reasoning of OS engineers. Our program logic reduces to a standard logic for user-level reasoning, reduces to side-condition checks for kernel-level reasoning, and can handle typical OS kernel tasks such as context switching and page table manipulations.

1 Introduction

We present a program logic in the interactive proof assistant Isabelle/HOL [15] for verifying programs in the presence of an ARMv7-style memory management unit (MMU), consisting of multi-level page tables and a translation lookaside buffer (TLB) for caching page table walks. This logic builds on our earlier work [17], a machine model with a sound abstraction of the ARMv7-style TLB. While program logics for reasoning in the presence of address translation exist [11], reasoning in the presence of a TLB has so far remained hard, and is left as an assumption in all large-scale operating system (OS) kernel verification projects such as seL4 [7] and CertiKOS [6].

Page table data structures encode a mapping from virtual to physical memory addresses. The OS kernel manages these, e.g. by adding, removing, or changing mappings, by keeping a page table structure per user process, and by maintaining invariants, such as never giving a user access to kernel-private data structures, ensuring that certain mappings are always present, or ensuring non-overlapping mappings between different page tables if so desired.

Since the TLB caches address translation, each of these operations may leave the TLB out of date w.r.t. the page table in memory, and the OS kernel must flush (or *invalidate*) the TLB before that lack of synchronisation can affect program execution. Since flushing the TLB is expensive, OS kernel designers work hard to delay and minimise flushes and to make them as specific as possible,

© Springer International Publishing AG, part of Springer Nature 2018
J. Avigad and A. Mahboubi (Eds.): ITP 2018, LNCS 10895, pp. 542–559, 2018.
https://doi.org/10.1007/978-3-319-94821-8_32

using additional TLB features such as process-specific address space identifiers (ASIDs) to only invalidate specific sets of entries. If this management is done correctly, the TLB has no effect other than speeding up execution. If it is done incorrectly, machine execution will diverge from the semantics usual program logics assume, e.g. wrong memory contents will be read/written, or unexpected memory access faults might occur.

The main contribution of this paper apart from the logic itself and its soundness is to show that it can be used to reason effectively and efficiently about kernel code, that it reduces to simple side-condition checks on kernel code that does not modify page tables, and that the logic reduces to standard Hoare logic for user-level code. The development of the logic, and the case study presented in the paper have led us to significantly extend the TLB model of our previous work [17], which provided soundness for memory operations, but not for ASID maintenance and page table root switches. The previous model would have required TLB invalidation where current kernel code (correctly) does not perform any. The case study that flushed out this deficiency is inspired by the seL4 kernel [7], and systematically covers all significant interactions with the TLB. In fact, we chose to model the ARMv7 TLB, because we aim to eventually integrate this logic with the existing seL4 proofs on ARM.

The logic is generic and can easily be adapted to, for instance, the shallow embedding the seL4 specifications use, or the more deeply embedded C semantics of the same project. It should also transfer readily to other settings such as the lower levels of CertiKOS in Coq.

After related work in Sect. 2, we introduce the Isabelle/HOL notation we use in Sect. 3. Section 4 presents the syntax and semantic operations of a small example language, as well as the program logic. Section 5 shows the main reduction theorems that simplify reasoning, and Sect. 6 concludes with the case study examples. The corresponding Isabelle/HOL theories are available online [18].

2 Related Work

The TLB has the nice property that it has no effect on the execution of a program apart from making it faster, *if* it is used correctly. For this reason, all large-scale formal OS kernel verifications so far have left correct TLB management as an assumption. This includes the OS kernel verification work in seL4 [7,8] and CertiKOS [6], which both do reason about page table structures, but omit the TLB. Similarly, Daum *et al.* [4] reason about user-level programs on top of seL4, including page tables, but not about the TLB.

Kolanski *et al.* [9–11] develop an extension of separation logic to formally reason about page tables, virtual memory access, and shared memory in Isabelle/HOL. We build directly on the abstract interface to page table encodings Kolanski developed, which makes our work independent of the precise page table format the architecture uses. Kolanski's model does not include the TLB and does not address TLB caching, consistency and invalidation, which we add here.

Nemati *et al.* [14] show the design, implementation and verification of a direct paging mechanism in a virtualization platform for ARMv7-A in HOL4 [16]. Similarly to others, they model the state parameters of the MMU, such as page table walks, but not the TLB or its maintenance operations.

Kovalev [12] and Alkassar *et al.* [1] *do* provide a TLB model, in particular a model of the Intel x64 TLB including selected maintenance operations and partial walks. Kovalev [12] states a reduction theorem for page table walks in ASID 0 for a specific hypervisor setup, which is based on ideas similar to the ones presented here. However, while other parts of this development are mechanised, this reduction theorem is not. As we will see in Sect. 4, the restriction to one ASID makes the model too conservative for usual OS code.

Barthe *et al.* [3] present an abstract TLB model including TLB flushes and invariants for enforcing isolation between guest operating systems, but stop short of a program logic and a proof that the abstraction is sound.

We build directly on our earlier work [17] which provides a detailed operational TLB model based on the ARM architecture manual [2] integrated with the ARM instruction set architecture (ISA) semantics of Fox and Myreen [5]. Reasoning directly about this detailed model is hard, because the TLB introduces non-determinism, because global state changes even on memory reads, and because it introduces new failure modes that need to be avoided. Our earlier work provides a tower of abstractions from this model, including soundness proof. The final abstraction is similar to the ideas by Kovalev [12] and Kolanski [10], but the case study in this paper shows that making efficient use of ASIDs requires additional complexity. We have therefore extended the existing tower of abstractions and soundness proofs and arrive at the TLB model we will show in Sect. 4.

3   Notation

This section introduces Isabelle/HOL syntax used in this paper, where different from standard mathematical notation. Isabelle denotes the space of total functions by $\Rightarrow$, and type variables are written 'a, 'b, etc. The notation t::τ means that HOL term t has HOL type τ. The option type

   datatype 'a option = None | Some 'a

adjoins a new element None to a type 'a. We use 'a option to model partial functions, writing $\lfloor$a$\rfloor$ instead of Some a and 'a $\rightharpoonup$ 'b instead of 'a $\Rightarrow$ 'b option. Some has an underspecified inverse called the, satisfying the $\lfloor$x$\rfloor$ = x.

Isabelle's type system does not include dependent types, but can encode numerals and machine words of fixed length. The type 'n word represents a word with n bits; concrete types include e.g. 32 word and 64 word. Function update is written f(x := y) where f::'a $\Rightarrow$ 'b, x::'a and y::'b, f(x $\mapsto$ y) stands for f (x:= Some y). We model the program state as a record type state. For every record field, there is a *selector* function of the same name. For example, if s has type state then heap s denotes the value of the heap field of s, and s($\lfloor$heap := id$\rfloor$) will update heap of s to be the identity function id.

4   Logic

This section presents a program logic for reasoning in the presence of cached address translation. We define the syntax of a simple Turing-complete heap language with TLB management primitives, introduce the abstract TLB and memory model the language works on, and show the rules of program logic.

4.1   Syntax and Program State

Figure 1 shows the Isabelle data types for the abstract syntax of the language. Control structures are the standard SKIP, IF, WHILE and assignment, where assignment expects the left-hand side to evaluate to a heap address. In addition, we have specific privileged commands for flushing the TLB, updating the current page table root, the current ASID, and the processor mode. The Flush operation has a number of variants: invalidate all entries, invalidate by virtual address or by virtual address/ASID pair, and invalidate an entire ASID [2, Chap. B3].

For simplicity, there are no local variables in this language, only the global heap. We identify values and pointers and admit arbitrary HOL functions for comparison, binary, and unary arithmetic expressions.

```
datatype aexp =                        datatype com =
    Const val                              SKIP
  | UnOp (val ⇒ val) aexp                | aexp := aexp
  | BinOp (val ⇒ val ⇒ val) aexp aexp   | com ;; com
  | HeapLookup aexp                      | IF bexp THEN com ELSE com
                                         | WHILE bexp DO com
datatype bexp =                          | Flush flush_type
    BConst bool                          | UpdateRoot aexp
  | BComp (val ⇒ val ⇒ bool) aexp aexp  | UpdateASID asid
  | BBinOp (bool ⇒ bool ⇒ bool) bexp bexp | SetMode mode_t
  | BNot bexp
                                        type_synonym asid = 8 word
datatype mode_t = Kernel | User         type_synonym val = 32 word

datatype flush_type = flushTLB        | flushvarange (val set)
                      | flushASID asid | flushASIDvarange asid (val set)
```

Fig. 1. Syntax of the heap based WHILE language.

Figure 2 illustrates the program state. It consists of the heap (physical memory), the set of inconsistent virtual addresses, the active page table root, the active ASID, the last known page table state for all inactive ASIDs (page tables snapshot), and the processor mode.

The first of these is for traditional heap manipulation, the rest for keeping track of the TLB. This state model is similar to the TLB-relevant machine state in our previous work on the ISA level [17], but it is not the same. The main

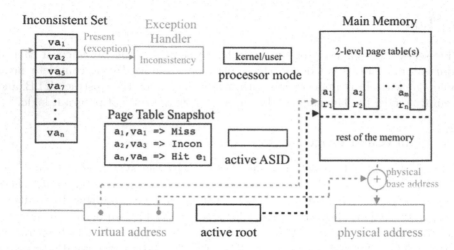

Fig. 2. Abstracted TLB memory model

idea of this previous work was that it is sufficient to keep track of the addresses on which the TLB and the in-memory page table may disagree. The logic presented in the present paper made case studies feasible and showed that this is not sufficient: to soundly model the effect of the `UpdateASID` command without requiring unnecessary flushes, this new model keeps track of a conservative estimate of what the TLB might remember from the time an ASID was last active. Essentially this is, for each ASID, a snapshot of the current page table state when that ASID was last active modulo all addresses that were inconsistent at that time. An ASID becomes active when the `UpdateASID` command updates the corresponding machine register. In addition, this model also relaxes which addresses become inconsistent when page tables are modified. Section 4.2 will provide more details.

Using the types `vaddr` and `paddr` for virtual and physical addresses from Kolanski's page table interface [11] we can declare the following:

datatype lookup_type = Miss | Incon | Hit tlb_entry
type_synonym iset = vaddr set
type_synonym ptable_snapshot = asid $\Rightarrow$ vaddr $\Rightarrow$ lookup_type

where it suffices for this paper to see `tlb_entry` as the result of a page table walk (see [17] for details).

With these, the record `state` has the components `heap :: paddr` $\rightharpoonup$ `val`, `iset :: iset`, `pt_snpshot :: ptable_snapshot`, `root :: paddr`, `asid :: asid`, and `mode :: mode_t`. Most of these are straightforward. The `iset` is the set of TLB-inconsistent addresses, and the snapshot is a function from ASID `a` to address `va` to `lookup_type`, where `Miss` encodes that the snapshot has no information for `va`, `Incon` encodes that `va` should not be used, and `Hit` is the result of the page table walk for `va` when `a` was last active.

4.2   Semantic Operations

This section presents the main semantic operations of the language. They describe the effects of memory accesses and the new TLB operations on the state.

We interpret the values `val` of the language as virtual addresses, which means memory read and write first undergo address translation. Both operations are sensitive to the current `mode` of the machine, since some mappings might be accessible in kernel mode only and lead to a page fault otherwise. To decode page tables, we reuse Kolanski's existing ARM page table formalisation [11], extended with this additional access control behaviour. Our interface to this formalisation is the function `pt_lookup`, which takes a heap, a page table root, and the current mode, and yields a partial function from virtual address to physical address. With this, we can formalise address translation, read, and write under a TLB.

Adding a TLB to address translation only adds a check that the virtual address is not part of the `iset`:

```
phy_ad :: iset ⇒ heap ⇒ root ⇒ mode_t ⇒ vaddr ⇀ paddr
phy_ad IS hp rt m va ≡ if va ∉ IS then pt_lookup hp rt m va else None
```

The memory read and write functions are then simply:

```
read :: iset ⇒ heap ⇒ root ⇒ mode_t ⇒ vaddr ⇀ val
read IS hp rt m va ≡ phy_ad IS hp rt m va ▷ load_value hp
```

```
write :: iset ⇒ heap ⇒ root ⇒ mode_t ⇒ vaddr ⇒ val ⇀ heap
write IS hp rt m va v ≡
case phy_ad IS hp rt m va of None ⇒ None | ⌊y⌋ ⇒ ⌊hp(y ↦ v)⌋
```

where $x ▷ g ≡$ case x of None $⇒$ None $| ⌊y⌋ ⇒ g\ y$. Both functions first perform address translation, then access the physical heap. Read returns `None` when the translation failed, write returns a new heap if successful and `None` otherwise.

The effect of a write operation extends further than the heap. If the operation modified the active page table, we may have to add new addresses to the TLB `iset`. For this, we compare the page table before and after:

```
pt_comp wlk wlk' =
{va | ¬ is_fault (wlk va) ∧ ¬ is_fault (wlk' va) ∧ wlk va ≠ wlk' va ∨
      ¬ is_fault (wlk va) ∧ is_fault (wlk' va)}
```

```
incon_comp a hp hp' rt rt' = pt_comp (pt_walk a hp rt) (pt_walk a hp' rt')
```

where `a` is the current ASID and `pt_walk` is a version of `pt_lookup` that returns more information, including whether the walk resulted in a page fault (missing mapping). We compare the results of page table walks in a heap `hp` from a root `rt` with walks in a different, updated heap `hp'` and potentially different root `rt'`. For heap writes, the root will be the same, and for root updates, the heaps will be the same. Two scenarios might add inconsistent entries: changing an existing mapping (first disjunct), or removing an existing mapping (second disjunct). Note that a single heap write can affect multiple mappings at once, for instance when it changes the pointer to an entire page table level. It is the effect of this comparison that OS engineers reason about informally when they compute which address need to be flushed from the TLB. We will show examples in Sect. 6.

The effect of a write is then

```
heap_iset_update_s (pp ↦ v) ≡
let hp = heap s; hp' = hp(pp ↦ v); rt = root s; a = asid s
in s(|heap := hp', iset := iset s ∪ incon_comp a hp hp' rt rt|)
```

and the effect of a page table root update is

```
root_iset_update_s rt' ≡
let rt = root s; hp = heap s; a = asid s
in s(|root := rt', iset := iset s ∪ incon_comp a hp hp rt rt'|)
```

For changing the current ASID, we will make use of the page table snapshots to determine which addresses become inconsistent, and we need to update the snapshot for the ASID we are switching away from.

```
lift_pt walk ≡ λva. if is_fault (walk va) then Miss else Hit (walk va)
to_incon V walk ≡ λva. if addr_val va ∈ V then Incon else walk va
snap_pt s = to_incon (IC s) (lift_pt (pt_walk (asid s) (heap s) (root s)))
new_snp s = (pt_snpshot s)(asid s := snap_pt s)
```

where IC s ≡ {vp | Addr vp ∈ iset s} and addr_val (Addr a) = a and Addr is the constructor for addresses.

Taking a snapshot is taking the pt_walks in the current state, marking all unmapped entries as Miss, and everything in the iset as Incon, and then storing that function under the current ASID in new_snp.

```
snp_comp a snp walk ≡ {va | snp a va ≠ Miss ∧ snp a va ≠ Hit (walk va)}
snp_incon a s ≡ snp_comp a (new_snp s) (pt_walk a (heap s) (root s))
```

Determining the iset for the new ASID a compares the Hit entries in the snapshot for a with the current pt_walk. We use new_snp s instead of pt_snpshot s, because a could also be the current ASID. The UpdateASID command then executes

```
asid_pt_snpshot_update_s a ≡
s(|asid := a, iset := snp_incon a s, pt_snpshot := new_snp s|)
```

The final set of semantic effects are flush operations. The functions

```
flush_iset :: flush_type ⇒ iset ⇒ asid ⇒ iset and
flush_snpshot :: flush_type ⇒ pt_snpshot ⇒ asid ⇒ pt_snpshot
```

simply remove the relevant entries from the iset, and set them to Miss in the pt_snpshot respectively. The flush instruction does both simultaneously:

```
iset_pt_snpshot_update_s f ≡
let is = iset s; snp = pt_snpshot s; a = asid s
in s(|iset := flush_iset f is a, pt_snpshot := flush_snpshot f snp a|)
```

4.3   Hoare Logic

With the syntax and the semantic operations of the previous sections it is straightforward to define an operational semantics for the language. We omit the details here and only briefly summarise the salient points before we focus on the rules of the program logic.

The semantics of arithmetic and Boolean expressions, $[A]$ s and $[B]_b$ s, are partial functions from program **state** to **val** and **bool**, respectively. While the rest is standard and omitted here, **HeapLookup** goes through virtual memory:

```
[HeapLookup vp] s =
(case [vp] s of None ⇒ None
| ⌊v⌋ ⇒ read (iset s) (heap s) (root s) (mode s) (Addr v))
```

For commands, we write $(c, s) \Rightarrow s'$ for *command* c *executed in state* s *terminates in state* s', where s' is of type **state option** with **None** indicating failure. More details about the semantics can be found at [18].

Our Hoare triples are partial for termination, but demand absence of failure.

$$\{P\} \ c \ \{Q\} \equiv \forall s \ s'. \ (c, s) \Rightarrow s' \land P \ s \longrightarrow (\exists r. \ s' = \lfloor r \rfloor \land Q \ r)$$

Figures 3 and 4 show the rules of the program logic. Their soundness derives directly from the operational semantics. Figure 3 summarises the rules for traditional commands such as SKIP, WHILE, etc. and Fig. 4 gives the rules for the commands that interact with the TLB. We note that the traditional rules are completely standard, as intended. We write $\langle\langle b \rangle\rangle$ s to denote that $[b]_b$ s $\neq$ **None**: the precondition in the IF and WHILE rules must be strong enough for failure free evaluation of b. The rules in Fig. 4 are in weakest-precondition form. They have a generic postcondition P and the weakest precondition that will establish P. We will now explain them.

$$\{P\} \ \text{SKIP} \ \{P\} \qquad \frac{\{P\} \ c \ \{Q\} \qquad P' \longrightarrow P}{\{P'\} \ c \ \{Q\}}$$

$$\frac{\{P \land \langle b \rangle\} \ c_1 \ \{Q\} \qquad \{P \land \neg\langle b \rangle\} \ c_2 \ \{Q\}}{\{P \land \langle\langle b \rangle\rangle\} \ \text{IF b THEN } c_1 \text{ ELSE } c_2 \ \{Q\}}$$

$$\frac{\{P \land \langle b \rangle\} \ c \ \{P\} \qquad P \longrightarrow \langle\langle b \rangle\rangle}{\{P\} \ \text{WHILE b DO } c \ \{P \land \neg\langle b \rangle\}} \qquad \frac{\{P\} \ c_1 \ \{Q\} \qquad \{Q\} \ c_2 \ \{R\}}{\{P\} \ c_1;; \ c_2 \ \{R\}}$$

Fig. 3. Hoare logic rules for standard commands.

$\{\lambda s.\ [\![1]\!]\ s = \lfloor vp \rfloor \wedge [\![r]\!]\ s = \lfloor v \rfloor \wedge vp \notin \mathcal{IC}\ s \wedge \mathtt{Addr}\ vp \hookrightarrow_s pp \wedge$
$\quad$ P (heap_iset_update$_s$ (pp $\mapsto$ v))$\}$
1 ::= r $\{\!| P |\!\}$

$\{\lambda s.\ \mathtt{mode}\ s = \mathtt{Kernel} \wedge [\![rte]\!]\ s = \lfloor rt \rfloor \wedge$ P (root_iset_update$_s$ Addr rt)$\}$
UpdateRoot rte $\{\!| P |\!\}$

$\{\lambda s.\ \mathtt{mode}\ s = \mathtt{Kernel} \wedge$ P (asid_pt_snpshot_update$_s$ a)$\}$ UpdateASID a $\{\!| P |\!\}$

$\{\lambda s.\ \mathtt{mode}\ s = \mathtt{Kernel} \wedge$ P (iset_pt_snpshot_update$_s$ f)$\}$ Flush f $\{\!| P |\!\}$

$\{\lambda s.\ \mathtt{mode}\ s = \mathtt{Kernel} \wedge$ P (s(|mode := flg|))$\}$ SetMode flg $\{\!| P |\!\}$

Fig. 4. Hoare logic rules for commands with TLB effects.

The assignment rule requires that the expressions 1 and r evaluate without failure. The assignment succeeds if the virtual address vp is *consistent* in the current state (vp $\notin \mathcal{IC}$ s) and vp is *mapped* (Addr vp $\hookrightarrow_s$ pp), where

vp $\hookrightarrow_s$ pp = (phy_ad (iset s) (heap s) (root s) (mode s) vp = $\lfloor pp \rfloor$)

The effect of the assignment is the heap and iset update heap_iset_update we described in Sect. 4.2.

The rule for the command UpdateRoot, only available in kernel mode, updates the current page table root to the value of the expression rte. The effect is modelled by root_iset_update defined in Sect. 4.2.

The UpdateASID command, also only available in kernel mode, sets the new ASID a, increases the iset using snp_incon, and records a page table snapshot for the old ASID using new_snp.

Finally, Flush is the instruction that the makes the iset smaller, and removes mappings in the snapshots of inactive ASIDs, using iset_asid_map_update from Sect. 4.2.

4.4 Discussion

Our previous work [17] motivated an abstract TLB model that we have extended and refined here. The program logic uses the (morally) same model, but there is still a break in logic: the TLB abstraction is on a machine-level ISA model; the program logic is for a higher-level language with explicit memory access, intended for languages such as C. The bridge between the two worlds would be a compiler correctness statement that takes the TLB into account. This may initially not sound straightforward: the high-level language makes fewer memory accesses visible than the low-level machine performs. In particular, a compiler will usually implement a stack for local variables, and memory areas for global variables, as well as for the code itself. These memory accesses are under address translation and might be relevant for TLB reasoning.

Sections 5 and 6 will show that we can ignore the TLB for kernel-level code, if we can assume that these memory areas (code, stack, globals) are statically known and that the compiler will not generate additional memory accesses outside these static areas. This is a reasonable assumption—otherwise kernel code could never be sure that privileged memory areas such as memory-mapped devices are not randomly overwritten by compiler-generated accesses. We will then have to prove that we never remove or change active mappings for these areas (adding new mappings for e.g. the stack would be fine). For user-level code, we will see that the issue becomes irrelevant.

The logic could be made slightly more precise by distinguishing between situations that must always be avoided, such as using inconsistent TLB entries, and page faults, which can be recoverable by executing a page fault handler. In kernel-level code, page faults are usually unwanted as modelled here, in user-level code they will usually be recoverable. We omit the distinction here for simplicity. Page fault handlers could for instance be modelled as exceptions in the logic.

In summary, we have so far provided a Hoare logic for reasoning about programs in the presence of cached address translation. The model as shown is specific to the ARMv7 architecture, but should generalise readily to similar architectures, since it uses an abstract interface for page table encoding. So far, reasoning is possible, and is at the right level of abstraction for code that manipulates page tables, but it is not yet convenient for code that does not interfere with virtual memory mappings or even runs in user mode.

5   Safe Set

This section introduces a reduction theorem that restricts and simplifies the assignment rule, which is the most frequent reasoning step in any usual program. The general assignment rule reasons about (a) consistency of the target address in the current state (b) valid address translation, and (c) potential update of the iset. The rule explicitly mentions page table walks, which means the proof engineer has to discharge page table obligations even if the memory write has nothing to do with page tables. This is not what systems programmers do. They instead establish invariants under which most of the code can be reasoned about without awareness of the TLB or page tables.

Given a TLB-consistent set of virtual addresses, this set can only become unsafe to write to when we change one of the page table mappings that translate the addresses in this set. If none of these are contained in the set, any write to the set is safe, even if it may change other mappings and increase the TLB iset. To formalise this notion, we re-use another function from Kolanski's page table interface [11]: ptable_trace. It takes a heap, a root, and a virtual address va, and returns the set of physical addresses visited in the page table walk for va. Memory writes outside the ptable_trace for va will not change the outcome of the walk for va. Generalising this notion to a set of virtual addresses, we define

ptrace_set V s = ⋃ptable_trace (heap s) (root s) ' V

where f ' V applies f to all elements of the set V, and ⋃ is the union of a set of sets. The ptrace_set V gives us the set of physical addresses that encode the translation for the virtual addresses in V. We can now define what a *safe set* is:

```
safe_set V s ≡ ∀va∈V. va ∈ C s ∧ (∃p. va ↪ₛ p ∧ p ∉ ptrace_set V s)
```

where C s ≡ {va | va ∉ iset s}. In words, a set V is a safe set in state s iff all addresses va ∈ V are consistent in the current state, if they map to a physical address p, and if that address is not part of the page table encoding for any of the addresses in V.

Our first observation is that once a set V is a safe set, assignments in V can no longer make it unsafe, and the safe set property will remain invariant:

Theorem 1. *Any write to the safe set will preserve the safe set. Formally:*

```
{⋋s. safe_set V s ∧
     (∃vp v. ⟦lval⟧ s = ⌊vp⌋ ∧ ⟦rval⟧ s = ⌊v⌋ ∧ Addr vp ∈ V)}
lval ::= rval {⋋s. safe_set V s}
```

Proof. See lemma `safe_set_preserved` in [18].

Our previous work [17] already contains a corresponding theorem for the concrete machine model. The following theorem is new. It develops the concept further into a simpler assignment rule where it is sufficient to check that the address is part of the safe set. We know with Theorem 1 that the safe set will remain invariant, so we could now ignore the iset completely, but since the proof engineer might want to keep track of it for other purposes, we still record it in the rule. However, in contrast to the general assignment rule, if the post condition does not mention the TLB, now neither will the precondition.

Theorem 2. *In the assignment rule, it is sufficient to check the static safe set instead of the dynamic inconsistency set* IC*.*

```
{⋋s. (∃vp v. ⟦lval⟧ s = ⌊vp⌋ ∧ ⟦rval⟧ s = ⌊v⌋ ∧ Addr vp ∈ V ∧
         Q (heap_iset_updateₛ (the_phy_ad vp s ↦ v))) ∧ safe_set V s}
lval ::= rval {Q}
```

where
```
the_phy_ad vp s ≡ the (pt_lookup (heap s) (root s) (mode s) (Addr vp))
```

Proof. See lemma `weak_pre_write` in [18].

For code that is not interested in TLB effects, i.e. outside context switching and page table manipulations, this rule enables proof engineers to reason as if no TLB was present. The majority of OS and user-level code satisfies this condition. The rule still mentions address translation, but the translation is now static within V, i.e. can be computed once. The reduction to checking a static set of addresses also give us justification that compilers do not introduce additional complexity into reasoning under the TLB, they merely add addresses that need to be part of this safe set, e.g. the area of virtual memory that contains code, stack, and global variables.

6   Case Studies

In this section, we apply the program logic and its reduction theorems to the main scenarios where TLB effects are relevant. These are: kernel-level code without TLB or page table manipulations, standard user-level code, context-switching, and page table manipulations. Of these, page table manipulations turn out to be the least interesting, so we only summarise them, while we present the rest in more detail.

The case study uses the seL4 microkernel as inspiration to distill out code sequences for a toy kernel that manages page tables and the TLB, and prevents users from accessing these, as well as other kernel data structures, directly. It maintains a set of page tables, typically one per user, potentially shared. This setting applies to all major protected-mode OS kernels, e.g. Linux, Windows, MacOS, as well microkernels. While simplified, the case study aims to be realistic in demonstrating popular techniques for avoiding TLB flushes, such as ASIDs, and uses a so-called kernel window to reduce page tables switches. The kernel window is a set of virtual addresses, unavailable to the user, backed by kernel mappings with permissions that make them available only in kernel mode.[1] It is the combination of ASID use, context switching, and flush avoidance that led us to adjust our previous model [17] for this case study.

As is customary, the mappings for this kernel window are constant, and each user-level page table that the kernel maintains has a number of known kernel mapping entries which reside at the same position in the page table encoding. This gives us a ready candidate for safe-set reasoning about kernel code: all addresses in the kernel window minus the addresses that are used to encode the kernel mappings in page table data structures.

Since the aim is to show reasoning principles, not to prove correctness of a particular kernel, the examples below use two-level ARMv7 page tables and with a simple concrete encoding, and a specific layout. The encoding and layout should generalise readily to larger settings. In addition to the page tables (one per user) that are stored in the kernel window, we assume the existence of one further kernel data structure: a map `root_map` from page table roots to the ASID for the user of this page table. A real OS kernel might maintain these as part of a larger data structure. We ignore the details here, and use them only to formulate basic invariants the kernel must maintain.

The main invariants we use in this example are (a) all kernel data structures reside in physical kernel memory, (b) they do not overlap, (c) the current ASID is associated correctly with the current page table root, (d) all page tables contain the kernel mappings, (e) no page table contains mappings that allows user mode to resolve to physical kernel memory, and (f) the mapping from page table roots to ASIDs is injective.

---

[1] This is the technique attacked by Meltdown [13]. Since hardware manufacturers are promising to fix this major flaw, we present the more interesting setting instead of the less complex and slower scenario with a separate kernel address space.

The following two properties are true for most of the execution of the system, but are invalidated temporarily: (g) The kernel window minus the entries that encode kernel mappings is a safe set. This property only holds in kernel mode. (h) The ASID snapshots agree with the page table for that ASID/user. This property is invalidated for a specific ASID between page table manipulations and flush instructions.

Formally:

```
mmu_layout s ≡
kernel_data_area s ⊆ kernel_phy_mem ∧ non_overlapping (kernel_data s) ∧
root_map s (root s) = ⌊asid ε⌋ ∧ kernel_mappings s ∧
user_mappings s ∧ partial_inj (root_map s)
```

where we define partial injectivity as
```
partial_inj f ≡ ∀x y. x ≠ y ⟶ f x ≠ f y ∨ f x = None ∧ f y = None
```
The restriction on user mappings is easily phrased with our previous address translation predicates, where `roots s = set (root_log s)`, `root_log` is a list of page table roots with `root_map s r ≠ None`, and `set` turns a list into a set.
```
user_mappings s ≡
∀rt∈roots s.
```

$\forall$ `va pa. pt_lookup (heap s) rt User va = ⌊pa⌋ ⟶ pa ∉ kernel_phy_mem`
The presence of kernel mappings is more technical. We spare the reader the details of the formal page table encoding, but note that it represents a constant offset translation, such that for all virtual addresses `va` in the kernel window, we get `Addr va ↪ₛ Addr (va - offset)` for a constant `offset`, i.e. the outcome of the translation is easily described statically. This is a simple yet realistic setup, similar to what e.g. seL4 uses.

The memory area of the kernel data structures is the union of the footprint of all static data structures plus the footprint of all page tables. The memory area of a page table starting at root `rt` is the set of all addresses that can be produced by a `ptable_trace`.

```
pt_area s rt ≡ ⋃ptable_trace (heap s) rt ' UNIV
kernel_data s ≡ map (pt_area s) (root_log s) @ [rt_map_area]
kernel_data_area s ≡ ⋃set (kernel_data s)
```

The definition of non-overlapping is:
```
non_overlapping [] = True
non_overlapping (x · xs) = (x ∩ ⋃set xs = ∅ ∧ non_overlapping xs)
```

To avoid flushing the TLB, we maintain for most of the execution the additional invariant that the TLB is fully consistent for all ASIDs that we might switch to, and that for each ASID the TLB snapshot agrees with the page table that we *would* switch to for that ASID. This means, if there were page table modifications for a user we are about to switch to, we assume that the corresponding flush has already happened. Since the property is not valid for all ASIDs between page table modifications and flush, we provide a set of ASIDs as argument to exclude. If this set is empty, we will omit the argument in the notation.

```
asids_consistent S s ≡
∀r a. root_map s r = ⌊a⌋ ∧ a ∉ S ∪ {asid s} ⟶
      (∀v. pt_snpshot s a v = Miss ∨
           pt_snpshot s a v = Hit (pt_walk a (heap s) r v))
```

This concludes the formalisation of the necessary kernel invariants.

6.1 User Execution

The simplest of the reduction theorems is user-level execution: when the kernel has switched to user mode, the iset should be empty for the current ASID, and since the user cannot perform any actions that adds addresses to this set, it will remain empty. Most actions that have any effect on the iset are explicitly privileged, i.e. unavailable in user mode. Only assignments could possibly have an adverse effect.

The following theorem shows that they do not, and that any arbitrary assignment in user mode will preserve not only this property of the iset, but, almost trivially, also all kernel invariants. In that sense it is a simple demonstration of the separation that virtual memory achieves between kernel and user processes.

Theorem 3. *When the kernel invariants hold, we are in user mode, and the* iset *is empty, then these three conditions are preserved, and the heap is updated as expected. We assume that the address the left-hand side resolves to is mapped.*

```
{|λs. mmu_layout s ∧ mode s = User ∧
     IC s = ∅ ∧ [[lval]] s = ⌊vp⌋ ∧ [[rval]] s = ⌊v⌋ ∧ Addr vp ↪s p|}
lval ::= rval
{|λs. mmu_layout s ∧ mode s = User ∧ IC s = ∅ ∧ heap s p = ⌊v⌋|}
```

Proof. See lemma `user_safe_assignment` in [18].

The essence of the rule above is the same as Kolanski's assignment rule [11] without TLB. The invariant part of the rule could be moved to the definition of validity and be hidden from the user completely. Like Kolanski, we still had to assume that the address vp is mapped, because we do not distinguish between recoverable page faults and program failure. In the settings we are interested in, we aim to avoid page faults. In a setting with dynamically mapped pages, e.g. by a page fault handler, the logic can be extended to take this conditional execution into account, for instance using an exception mechanism or a conditional jump. In that case, the condition that addresses are mapped can be dropped, and we arrive at a standard Hoare logic assignment rule.

6.2 Kernel Execution

User execution boils down to standard reasoning. We can show that kernel execution without virtual memory modifications do as well.

As mentioned in Sect. 5, the safe set for kernel execution is the entire kernel window, i.e. the virtual addresses that are mapped by the global mappings, minus the addresses of the page table entries that encode these global mappings. Since we will need to re-establish this set every time we switch to a different page table, and it is always safe to reduce the safe set, we not only remove the kernel window encoding in the *current* page table, but also that of of all *other* page tables the kernel might switch to and call this set `kernel_safe`.

Since we fixed the global mappings in `mmu_layout`, we can give a short, closed form of translation for addresses in `kernel_safe`: `k_phy_ad vp = Addr vp - offset`. With these, we can formulate a theorem for assignments in kernel mode that do not touch any of the virtual memory data structures, i.e. when the write does not take place in any of the addresses covered by `kernel_data`.

Theorem 4. *If the* `mmu_layout` *invariants hold, we are in kernel mode, and we are performing a write in the kernel safe set that does not touch any MMU-relevant data structures, then the* `mmu_layout` *invariants are preserved and the effect is a simple heap update with known constant address translation.*

```
{λs. mmu_layout s ∧ mode s = Kernel ∧ safe_set (kernel_safe s) s ∧
     asids_consistent s ∧ [lval] s = ⌊vp⌋ ∧ [rval] s = ⌊v⌋ ∧
     Addr vp ∈ kernel_safe s ∧ k_phy_ad vp ∉ kernel_data_area s}
lval ::= rval
{λs. mmu_layout s ∧ mode s = Kernel ∧ safe_set (kernel_safe s) s ∧
     asids_consistent s ∧ heap s (k_phy_ad vp) = ⌊v⌋}
```

Proof. See lemma `kernel_safe_assignemnt` in [18]. □

This lemma covers kernel code that is uninteresting for the purposes of the MMU and TLB, which is the majority of code in a normal kernel. The Isabelle theories [18] also contain examples for page table modifications. The main difference to this theorem is that, while the write still happens in the safe set, and the safe set is preserved, there are now inconsistent addresses that need to be flushed before we return to user mode. These could be for the active page table, but also for an inactive page table, where the need for flushing is observed in the `asids_consistent` invariant.

6.3   Context Switch

We have so far shown reduction theorems for simpler reasoning when nothing interesting happens to the TLB. This section is the opposite: context switching. There are many ways for the OS to implement context switching—our example shows one where we change to a new address space, i.e. a new page table and ASID, without flushing the TLB, establishing the conditions of Theorem 3 for user-level reasoning.

Switching page table roots without flushing is non-trivial, and the ARM architecture manual [2, Chap. B3.10] even gives a specific sequence of instructions to achieve this. The manual uses this sequence, because speculative execution might otherwise contaminate the new ASID with mappings from the old page table, i.e. the TLB might still contain entries from the previous user. Theorem 5 shows that our model is conservative for speculative execution, but precise enough so we can reason about this sequence and see why it is safe.

The recommended sequence switches to a new user-level page table and ASID by using a reserved ASID (in this case 0). It first switches to this reserved ASID, then sets the new page table root, then switches to the ASID for that root, before it switches to user mode. A real kernel would at this point also restore registers, which we omit.

Theorem 5. *The context switch sequence to a new ASID a and new page table root r preserves the* mmu_layout *and ASID snapshot consistency invariants and establishes the conditions for user-level reasoning, provided that the TLB has no inconsistent addresses at this point, that the reserved ASID 0 is not used for any user page table, and that that r is a known page table associated with ASID a.*

$\{\![\lambda s.\ $ mmu_layout $s \wedge$ asids_consistent $s \wedge$ mode s = Kernel $\wedge$
$\quad \mathcal{IC}\ s$ = $\emptyset \wedge 0 \notin$ ran (root_map s) $\wedge$ root_map s (Addr r) = $\lfloor a \rfloor\}\!]$
UpdateASID 0;; UpdateRoot (Const r);; UpdateASID a;; SetMode User
$\{\![\lambda s.\ $ mmu_layout $s \wedge \mathcal{IC}\ s$ = $\emptyset \wedge$ mode s = User $\wedge$ asids_consistent $s\}\!]$

Proof. See lemma context_switch_invariants in [18].

Our previous ISA-level model [17] without ASID page table snapshots was not strong enough to admit this theorem without flushing the TLB. In particular, the fact that the TLB does not contain entries for the ASID we are switching to that are inconsistent with the current page table at that point would either be lost (making it unsound) or over-approximated (requiring a flush).

For compiler correctness, we would additionally need to know that ASID 0 does not have inconsistent entries for the code and data areas of the kernel, which is maintained if ASID 0 is used only in the way above. To make this more explicit, we could add a static set to the program logic for code and data that must always be consistency, and the condition asids_consistent would maintain that at least the global kernel mappings are consistent in ASID 0.

This concludes the case study examples for our logic. We have seen that we can reason about user code, 'uninteresting' kernel code, and kernel code that manipulates paging structures, each at their appropriate level of abstraction.

7   Summary

We have presented a program logic for reasoning about low-level OS code in the presence of cached address translation.

The model and case study use the ARMv7 architecture, but our interface to page table encodings is generic and should apply to all architectures with conventional multi-level page tables. The details of TLB maintenance may differ between architectures, i.e. Intel x86 does not require an explicit TLB flush on context switch, but the ideas of the model should again transfer readily.

The model can also capture the effect of defective hardware, such as the recent Meltdown attack [13] which exploits the fact that permission bits of TLB entries are not checked during speculative execution on some platforms, and uses a cache side channel to thereby make kernel-only TLB mappings readable to user space. To conservatively formalise the effect of this attack, one could change the model to ignore read restrictions in TLB entries. A system that can be proved safe under that conservative model, should then be safe under Meltdown.

We currently do not treat global locked (pinned) TLB entries, and the TLB in this version of the logic does not cache partial page table walks (as in e.g. ARMv7-A). Our previous work does cover partial walks—the main influence on the model is that the update of the iset becomes slightly more conservative. Pinned TLB entries would have the effect of explicitly allowing inconsistency between the TLB and the page table, with the TLB taking preference.

Our logic does not address concurrency aspects—they are orthogonal. In a multi-core setting, each core has its own TLB which reads from global memory. Modifying a page table that is active on another core is almost never safe, unless the change merely adds new mappings or the change happens in the same safe set style presented here, where the execution on all cores must adhere to the intersection of all safe sets.

Weak memory and caches do have an interaction point with the TLB, because page table walks are subject to both and caches can be either virtually or physically indexed. We expect our safe set reasoning to transfer directly, requiring cache flushes and/or barrier instructions in addition to TLB flushes. We leave a cache formalisation for future work.

The strength of the model and logic is its simplicity, which took multiple iterations to achieve, finding a balance between abstraction soundness, not too complex reasoning, and not too much conservatism for allowing optimisations and idioms used in real OS code, resulting in a program logic that feels familiar to proof engineers.

The logic allows us to prove reduction theorems that mirror the informal reasoning OS engineers perform when they write kernel code. It also allows us to drop into a simpler setting when we reason about code that does not affect virtual memory mappings. In these cases, we only need to show that memory accesses are within a set of safe addresses. Our work shows that reasoning in the presence of a TLB does not need to be significantly more onerous than without.

References

1. Alkassar, E., Cohen, E., Kovalev, M., Paul, W.J.: Verification of TLB virtualization implemented in C. In: Joshi, R., Müller, P., Podelski, A. (eds.) VSTTE 2012. LNCS, vol. 7152, pp. 209–224. Springer, Heidelberg (2012). https://doi.org/10.1007/978-3-642-27705-4_17
2. ARM Ltd.: ARM Architecture Reference Manual, ARM v7-A and ARM v7-R, aRM DDI 0406B, April 2008
3. Barthe, G., Betarte, G., Campo, J.D., Luna, C.: Cache-leakage resilient OS isolation in an idealized model of virtualization. In: 25th CSF, pp. 186–197 (2012)

 4. Daum, M., Billing, N., Klein, G.: Concerned with the unprivileged: user programs in kernel refinement. Form. Aspects Comput. **26**(6), 1205–1229 (2014)
 5. Fox, A., Myreen, M.O.: A trustworthy monadic formalization of the ARMv7 instruction set architecture. In: Kaufmann, M., Paulson, L.C. (eds.) ITP 2010. LNCS, vol. 6172, pp. 243–258. Springer, Heidelberg (2010). https://doi.org/10. 1007/978-3-642-14052-5_18
 6. Gu, L., Vaynberg, A., Ford, B., Shao, Z., Costanzo, D.: CertiKOS: a certified kernel for secure cloud computing. In: 2nd APSys (2011)
 7. Klein, G., Andronick, J., Elphinstone, K., Murray, T., Sewell, T., Kolanski, R., Heiser, G.: Comprehensive formal verification of an OS microkernel. Trans. Comp. Syst. **32**(1), 2:1–2:70 (2014)
 8. Klein, G., Elphinstone, K., Heiser, G., Andronick, J., Cock, D., Derrin, P., Elkaduwe, D., Engelhardt, K., Kolanski, R., Norrish, M., Sewell, T., Tuch, H., Winwood, S.: seL4: formal verification of an OS kernel. In: SOSP, Big Sky, MT, USA, October 2009, pp. 207–220 (2009)
 9. Kolanski, R.: A logic for virtual memory. In: SSV, Sydney, Australia, July 2008, pp. 61–77 (2008)
10. Kolanski, R.: Verification of programs in virtual memory using separation logic. Ph.D. thesis, UNSW, Sydney, Australia, July 2011. http://ts.data61.csiro.au/
11. Kolanski, R., Klein, G.: Types, maps and separation logic. In: Berghofer, S., Nipkow, T., Urban, C., Wenzel, M. (eds.) TPHOLs 2009. LNCS, vol. 5674, pp. 276–292. Springer, Heidelberg (2009). https://doi.org/10.1007/978-3-642-03359-9_20
12. Kovalev, M.: TLB virtualization in the context of hypervisor verification. Ph.D. thesis, Saarland University, Saarbrücken, Germany (2013)
13. Lipp, M., Schwarz, M., Gruss, D., Prescher, T., Haas, W., Mangard, S., Kocher, P., Genkin, D., Yarom, Y., Hamburg, M.: Meltdown. ArXiv e-prints 1801.01207, January 2018
14. Nemati, H., Guanciale, R., Dam, M.: Trustworthy virtualization of the ARMv7 memory subsystem. In: Italiano, G.F., Margaria-Steffen, T., Pokorný, J., Quisquater, J.-J., Wattenhofer, R. (eds.) SOFSEM 2015. LNCS, vol. 8939, pp. 578–589. Springer, Heidelberg (2015). https://doi.org/10.1007/978-3-662-46078-8_48
15. Nipkow, T., Paulson, L., Wenzel, M.: Isabelle/HOL—A Proof Assistant for Higher-Order Logic. LNCS, vol. 2283. Springer, Heidelberg (2002). https://doi.org/10. 1007/3-540-45949-9
16. Slind, K., Norrish, M.: A brief overview of HOL4. In: Mohamed, O.A., Muñoz, C., Tahar, S. (eds.) TPHOLs 2008. LNCS, vol. 5170, pp. 28–32. Springer, Heidelberg (2008). https://doi.org/10.1007/978-3-540-71067-7_6
17. Syeda, H.T., Klein, G.: Reasoning about translation lookaside buffers. In: 21st LPAR. EPiC Series in Computing, vol. 46, pp. 490–508 (2017)
18. Syeda, H.T., Klein, G., Kolanski, R.: Isabelle/HOL Program Logic for Cached Address Translation, January 2018. https://github.com/SEL4PROJ/tlb/tree/ITP18. https://doi.org/10.5281/zenodo.1246933

Verified Tail Bounds for Randomized Programs

Joseph Tassarotti[✉] and Robert Harper

Carnegie Mellon University, Pittsburgh, USA
jtassaro@andrew.cmu.edu

Abstract. We mechanize a theorem by Karp, along with several extensions, that provide an easy to use "cookbook" method for verifying tail bounds of randomized algorithms, much like the traditional "Master Theorem" gives bounds for deterministic algorithms. We apply these results to several examples: the number of comparisons performed by Quick-Sort, the span of parallel QuickSort, the height of randomly generated binary search trees, and the number of rounds needed for a distributed leader election protocol. Because the constants involved in our symbolic bounds are concrete, we are able to use them to derive numerical probability bounds for various input sizes for these examples.

1 Introduction

Formal verification of randomized algorithms remains a challenging problem. In recent years, a number of specialized program logics [8,10,11,37,42] and automated techniques [6,19,20] have been developed to analyze these programs. In addition, a number of randomized algorithms have been verified directly in interactive theorem provers [26,27,52] without using intermediary program logics. Besides establishing correctness results, much of this work has focused on verifying the *expected* or *average* cost of randomized algorithms. Although expectation bounds are an important first step in cost analysis, there are other stronger properties that often hold. For many randomized algorithms, we can establish *tail bounds* which bound the probability that the algorithm takes more than a given amount of time.

For example, it is well known that randomized QuickSort performs $O(n \log n)$ comparisons on average when sorting a list of length n, and this fact has been verified in theorem provers before [27,52]. However, not only does it do $O(n \log n)$ comparisons *on average*, but the probability that it does more than $O(n \log n)$ comparisons is vanishingly small for sufficiently large lists. To be precise, let W_n be the number of comparisons when sorting a list of length n. Then, for any positive k there exists c_k such that $\Pr[W_n > c_k n \log n] < \frac{1}{n^k}$. When we say that such c_k exist, we mean so in a constructive and practical sense: we can

Electronic supplementary material The online version of this chapter (https://doi.org/10.1007/978-3-319-94821-8_33) contains supplementary material, which is available to authorized users.

J. Avigad and A. Mahboubi (Eds.): ITP 2018, LNCS 10895, pp. 560–578, 2018.
https://doi.org/10.1007/978-3-319-94821-8_33

actually determine them and they are not absurdly large, so that one can derive interesting concrete bounds. For instance, when n is 10 million, the probability that W_n is greater than $8n \log_2 n$ is less than 10^{-9}. These kinds of tail bounds hold for many other classical randomized algorithms and are often stronger than asymptotic expectation bounds.

Despite this, there is a good reason for the prior emphasis on expectation bounds rather than tail bounds in the field of formal methods: tail bounds on running time are usually quite difficult to derive. Common approaches for deriving these bounds involve the use of methods from analytic combinatorics [31] or the theory of concentration of measure [25]. Although these techniques are very effective, to be able to use them in a theorem prover one would first need to be able to mechanize the extensive body of results that they depend upon.

The Need for "Cookbook" Methods. Let us contrast the difficulty mentioned above with the (relative) ease of analyzing *deterministic* algorithms. For deterministic divide-and-conquer algorithms, the cost is often given by recurrences of the form

$$W(x) = a(x) + \sum_{i-1}^{n} W(h_i(x)) \tag{1}$$

where the "toll" function $a(x)$ represents the cost to process an input and divide it into subproblems of size $h_1(x), \ldots, h_n(x)$, which are then solved recursively. Every undergraduate algorithms course covers "cookbook" techniques such as the Master Theorem [13,23] that can be used to straightforwardly derive asymptotic bounds on these kinds of recurrences. Moreover, these results can also be used to easily analyze recurrences for other types of resource use, such as the maximum stack depth or the span of parallel divide-and-conquer algorithms [15]. Recurrences for these kinds of resources have the form:

$$S(x) = b(x) + \max_{i=1}^{n} S(h_i(x)) \tag{2}$$

We will call recurrences of the form in Eq. 1 "work recurrences" and those of the form in Eq. 2 "span recurrences". Although Eq. 2 does not fit the format of the Master Theorem directly, when S is monotone the recurrence simplifies to $S(x) = b(x) + S(\max_{i=1}^{n}(h_i(x)))$ and so can be analyzed using the Master Theorem.

What is nice about these methods is that they give a process for carrying out the analysis: find the toll function, bound the size of recursive problems, and then use the theorem. Even if the first two steps might require some ingenuity, the method at least suggests an approach to decomposing the problem.

Besides being easy to use, results like the Master Theorem do not have many mathematical prerequisites. This makes them ideal for use in interactive theorem provers. Indeed, Eberl [28] has recently mechanized the more advanced Akra and Bazzi [2] recurrence theorem in Isabelle and has used it to derive asymptotic bounds for a number of recurrence relations.

For randomized divide-and-conquer algorithms, the same recurrence relations arise, except the $h_i(x)$ are random variables because the algorithms use randomness to divide the input into subproblems. Because of the similarity between deterministic and probabilistic recurrences, textbook authors sometimes give the following heuristic argument before presenting a formal analysis [23, pp. 175–177]: In an algorithm like QuickSort, the size of the sublists generated by the partitioning step can be extremely unbalanced in the worst case, but this happens very rarely. In fact, each sublist is unlikely to be much more than $\frac{3}{4}$ the length of the original list. And, for a deterministic recurrence like $W(n) = n + W(\frac{3}{4}n) + W(\frac{3}{4}n)$, the master theorem says the result will be $O(n \log n)$. Thus, intuitively, we should expect the average running time of Quicksort to be something like $O(n \log n)$.

This raises a natural question: Is there a variant of the Master Theorem that can be used to justify this kind of heuristic argument? Moreover, because Eq. 2 does *not* simplify to a version of Eq. 1 in the randomized setting[1], we ideally want something that can be used to analyze recurrences of both forms.

For the case where there is only a *single* recursive call (so that $n = 1$ above), Karp [38] developed such a result. At a high-level, using Karp's theorem involves two steps. First, bound the average size of the recursive subproblem by finding a function m such that $\mathrm{E}[h_1(x)] \leq m(x)$. Next, find a solution u to the *deterministic* recurrence relation

$$u(x) \geq a(x) + u(m(x))$$

Then the theorem says that for all positive integers w,

$$\Pr[W(x) > u(x) + wa(x)] \leq \left(\frac{m(x)}{x}\right)^w$$

There are a few side conditions on the functions m and u which are usually easy to check. Although this method generally does not give the tightest possible bounds, they are often strong. Recently, Karp's technique has been extended [51] to the case for $n > 1$ for both span and work recurrences.

Our Contribution. In this paper, we present a mechanization of Karp's theorem and these extensions in Coq, and use it to develop verified tail bounds for (1) the number of comparisons in sequential QuickSort, (2) the span arising from comparisons in parallel QuickSort, (3) the height of a randomly generated binary search tree, and (4) the number of rounds needed in a distributed randomized leader election protocol. By using the Coq-Interval library [41] we are able to instantiate our bounds in Coq to establish numerical results such as the 10^{-9} probability bound for QuickSort quoted above. To our knowledge, this is the first time these kinds of bounds have been mechanized.

We start by outlining the mechanization of probability theory that our work is based on (Sect. 2). We then describe Karp's theorem and its extensions in more detail (Sect. 3). To demonstrate how Karp's result is used, we describe our verification of the examples mentioned above, with a focus on the sequential QuickSort analysis (Sect. 4). Of course, formalization often requires changing parts of a paper proof, and our experience with Karp's theorem was no different. We discuss the issues we encountered and what we had to change in Sect. 5.

---

[1] This is because in general $\mathrm{E}[\max(X_1, X_2)] \geq \max(\mathrm{E}[X_1], \mathrm{E}[X_2])$.

Finally, we compare our approach to related work (Sect. 6) and conclude by discussing possible extensions and improvements to our development (Sect. 7).

The Coq development described in this paper is available at https://github.com/jtassarotti/coq-probrec.

2 Probability Preliminaries

2.1 Discrete Probability

We first need a set of basic results and definitions about probabilities and expectations to be able to even state Karp's theorem. We had to decide whether to use a measure-theoretic formulation or restrict ourselves to discrete distributions. Although the Isabelle standard library has an extensive formalization of measure theoretic probability, we are not aware of a similarly complete set of results in Coq (we discuss existing libraries later in Sect. 6). Moreover, the applications we had in mind only involved discrete distributions, so we did not need the extra generality of the measure-theoretic approach. To keep things simple, we decided to develop a small library for discrete probability theory. Defining probability and expectation for discrete distributions still involves infinite series over countable sets, which can raise some subtle issues involving convergence. We use the Coquelicot real analysis library [16] to deal with infinite series.

The definition of probability distributions is given in Fig. 1. We represent them as a record type parameterized by a countable type. We use the `ssreflect` [32] library's definition of countable types (`countType`), which consists of a type A equipped with a surjection from `nat` to A.

The distribution record consists of three fields: (1) a probability mass function `pmf : A → R` that assigns a probability to each element of A, (2) a proof that `pmf a` is non-negative for all a, and (3) a proof that the countable series that sums `pmf a` over all a converges and is equal to 1.

Random variables on a distribution (`rvar`) are functions from the underlying countable space A to some other type B. The expected value of a real-valued random variable is defined in the usual way as the series $\sum_{r \in \mathsf{img}(X)} \Pr\left[X = r\right] \cdot r$. Because the underlying distribution is discrete, the image of the random variable is a countable set, so we can define such a series.

Of course, expectations of discrete random variables do not always exist, because the above series may not converge absolutely. Because of this, even with the restriction to discrete probability, dealing with infinite series and issues of convergence can often be tedious. In actuality, many randomized algorithms only involve *finite* distributions. For random variables defined on such distributions, the expectation always exists, because the series is actually just a finite sum. For our mechanization of Karp's theorem, we restrict to these finite distributions.

2.2 Monadic Encoding

We represent sequential and parallel randomized algorithms in Coq using a monadic encoding. Variants of this kind of representation have been used in many prior formalizations and domain specific languages [3, 9, 46, 48].

```
Record distrib (A: countType) := mkDistrib {
  pmf :> A -> R; pmf_pos : forall a, pmf a >= 0;
  pmf_sum1 : is_series (countable_sum pmf) 1
}.

Definition pr {A: countType} (O: distrib A) (P: A -> bool) :=
  Series (countable_sum (fun a => if P a then O a else 0)).

Record rvar {A} (O: distrib A) (B: eqType) := mkRvar {
  rvar_fun :> A -> B;
}.

Definition pr_eq {A} {B: eqType} (O: distrib A) (X: rvar O B) (b: B) :=
  pr O (fun a => X a == b).

Definition Ex {A} (O: distrib A) (X: rrvar O) : Rbar :=
  Series (countable_sum (fun r => (pr_eq X r * r))).
```

Fig. 1. Basic definitions for discrete probability distributions and random variables.

The type `ldist A` represents probabilistic computations that result in values of type `A`. Such computations are represented as a finite list of values of type `A` paired with the probabilities that these values occur. The bind operation `dist_bind l f` represents the process of performing the computation represented by `l` to obtain a random element of type `A` (*i.e.*, "sampling" from the distribution represented by `l`), and then passing this to `f`. The return operation (`dist_ret`) applied to `a` corresponds to the probabilistic computation that simply returns `a` with probability 1. We use Coq's notation mechanism to represent binding `m` in `e` by writing `x ← m; e`, and write `mret a` for returning `a`.

3   Karp's Theorem

Now that we have a formalization of the basic concepts of probability theory and a way to describe randomized algorithms in Coq, we can give a more careful explanation of Karp's theorem and its extensions.

3.1   Unary Recurrences

The setting for Karp's theorem is more general than the informal account we gave in the introduction. Specifically, he assumes that there is a set I of algorithm inputs, a function $size : I \to \mathbb{R}^{\geq 0}$ such that $size(z)$ is the "size" of input z, and a family of random variables $h(z)$ which correspond to the new problem that is passed to the recursive call of the algorithm. The random variable $W(z)$, which represents the cost of the algorithm when run on input z, is assumed to obey the following unary recurrence:

$$W(z) = a(size(z)) + W(h(z)) \tag{3}$$

Although the intent of this recurrence is clear, it requires some care to interpret: on the right hand side, $h(z)$ is a random variable, but it is given as an argument to W, which technically has I as a domain, not I-valued random variables. Instead, we should read this not as the composition $W \circ h$ applied to z, but rather as a specification for the process which first generates a random problem according to $h(z)$ and then passes it to W. In other words, this part of the recurrence is really describing a monadic process of the form:

$$z' \leftarrow h(z); W(z')$$

Already, Eq. 3 addresses a detail that is often glossed over in informal treatments of randomized algorithms. In informal accounts, one often speaks about a random variable $W(n)$, which is meant to correspond to the number of steps taken by an algorithm when processing an instance of size n. The issue is that usually, the exact distribution depends not just on the size of the problem but also the particular instance, so it is somewhat sloppy to regard $W(n)$ as a random variable (admittedly, we did so in Sect. 1). For instance, when randomized QuickSort is run on a list containing duplicate elements, a good implementation will generally perform *fewer* total comparisons. Even if one tries to avoid this issue by, say, restricting only to lists that do not contain duplicates, one would still need to *prove* that the distribution depends on the size of the list alone. This is mostly harmless in informal treatments, but it is a detail that would otherwise have to be dealt with in a theorem prover.

We assume there is some constant d that is the "cut-off" point for the recurrence: when the input's size drops below d no further recursive calls are made. The function $a : \mathbb{R} \to \mathbb{R}^{\geq 0}$ is required to be continuous and increasing[2] on (d, ∞), but equal to 0 on the interval $[0, d]$. In addition, it is required that $0 \leq size(h(z)) \leq size(z)$, *i.e.*, the size of the subproblem is not bigger than the original.

Then, assume there exists some continuous function $m : \mathbb{R} \to \mathbb{R}$ such that for all z, $\mathrm{E}\left[size(h(z))\right] \leq m(size(z))$ and $0 \leq m(size(z)) \leq size(z))$. Moreover, the function $m(x)/x$ must be non-decreasing. Karp then argues that if there exists a solution to the deterministic recurrence relation $\tau(x) = a(x) + \tau(m(x))$, there must be a continuous minimal solution $u : \mathbb{R} \to \mathbb{R}$. He assumes such a solution exists and derives the following tail bound for W in terms of u:

Theorem 1 ([38]). *For all z and integer w such that $size(z) > d$,*

$$\Pr\left[W(z) > u(size(z)) + w \cdot a(size(z))\right] \leq \left(\frac{m(size(z))}{size(z)}\right)^w$$

Because u is the minimal solution to the deterministic recurrence, we can replace u with any other solution t in the above bound: if $W(z)$ is greater than the version with t, then by minimality of u, it must be bigger than the version

---

[2] In fact, the assumptions in [38] are slightly stronger than this. But as we discuss in Sect. 5, we discovered that the weaker assumptions mentioned here are sufficient.

with u. This means we do not need to find a closed form for the minimal solution u, because any solution will give us a bound.

It is important to note that m, a and u are all functions from $\mathbb{R}$ to $\mathbb{R}$. This means that we do not have to deal with subtle rounding issues that sometimes come up when attempting to formalize solutions to recurrences for algorithms. Eberl [28], in his formalization of the Akra-Bazzi theorem, has pointed out how important this can be. The trade-off is that establishing that the recurrence holds everywhere on the domain $\mathbb{R}$ can be harder, especially at the boundaries where the recurrence terminates.

3.2   Extension to Binary Work and Span Recurrences

Although Theorem 1 makes it easier to get strong tail bounds, it cannot be used in many cases because it only applies to programs with a single recursive call.

Tassarotti [51] describes an extension to cover the general case of work and span recurrences with $n > 1$ recursive calls. In our mechanization, we only handle the case where there are two recursive calls (so that $n = 2$) because this is sufficient for many examples. In this setting, we now have two random variables h_1 and h_2 giving the recursive subproblems. These variables are generally not independent: for QuickSort, h_1 would be the lower partition of the list and h_2 would be the upper partition. However, it is assumed that there is some function $g_1 : \mathbb{R} \to \mathbb{R}$ such that for all $z \in I$ and (z_1, z_2) in the support of $(h_1(z), h_2(z))$:

$$g_1\left(size(z_1)\right) + g_1\left(size(z_2)\right) \leq g_1\left(size(z)\right)$$

Informally, we can think of this function g_1 as a kind of ranking function, and the above inequality is saying that the combined rank of the two subproblems is no bigger than that of the original problem. The function m is now required to bound the expected value of the maximum size of the two subproblems:

$$\mathrm{E}\left[\max\left(size(h_1(z)), size(h_2(z))\right)\right] \leq m(size(z))$$

For bounding span recurrences of the form:

$$S(z) \leq a(size(z)) + \max(S(h_1(z)), S(h_2(z))) \tag{4}$$

we assume once more that u is a solution to the recurrence $u(x) \geq a(x) + t(m(x))$. Then we have:

Theorem 2. *For all z and integer w such that $size(z) > d$ and $g_1(size(z)) > 1$,*

$$\Pr\left[S(z) > u(size(z)) + w \cdot a(size(z))\right] \leq g_1(size(z)) \cdot \left(\frac{m(size(z))}{size(z)}\right)^w$$

The difference between the bound above and the one in Theorem 1 is the additional factor $g_1(size(z))$. Generally speaking, $g_1(size(z))$ will be bounded by a polynomial, so that in comparison to $\left(\frac{m(size(z))}{size(z)}\right)^w$, which decreases exponentially with respect to w, the effect is negligible.

The bound for binary work recurrences is slightly different. Given the recurrence:

$$W(z) \leq a(size(z)) + W(h_1(z)) + W(h_2(z)) \tag{5}$$

we need a second "ranking" function g_2 with the same property that $g_2\left(size(z_1)\right) + g_2\left(size(z_2)\right) \leq g_2\left(size(z)\right)$ for all z_1 and z_2 in the support of the joint distribution $(h_1(z), h_2(z))$ when $size(z) > d$. In the proof by Tassarotti [51], this second ranking function is used to transform the work recurrence into a span recurrence which is then bounded by Theorem 2, and this bound is converted back to a bound on the original recurrence. From the perspective of the user of the theorem, we now need u to solve the deterministic recurrence $u(x) \geq \frac{a(x)}{g_2(x)} + u(m(x))$, and we obtain the following bound:

Theorem 3. *For all z and integer w such that $size(z) > d$ and $g_1(size(z)) > 1$,*

$$\Pr\left[W(z) > g_2(size(z)) \cdot u(size(z)) + w \cdot a(size(z))\right] \leq g_1(size(z)) \cdot \left(\frac{m(size(z))}{size(z)}\right)^w$$

Observe that on the left side of the bound, we re-scale u by a factor of $g_2(size(z))$ because it was the solution to a recurrence in which we normalized everything by g_2.

The above results let us fairly easily obtain tail bounds for a wide variety of probabilistic recurrences arising in the analysis of randomized divide-and-conquer algorithms. In the next section, we demonstrate their use by verifying a series of examples. After showing how they are used, we return to the discussion of the results themselves in Sect. 5, where we describe issues we encountered when trying to translate the paper proofs into Coq.

4   Examples

We now apply the results developed in the previous sections to several examples.

4.1   Sequential QuickSort

Our first example is bounding the number of comparisons performed by a sequential implementation of randomized QuickSort. To count the number of comparisons that the monadic implementation of the algorithm performs, we combine the probabilistic monad from Sect. 2.2 with a version of the writer monad that increments a counter every time a comparison is done. This cost monad is defined by:

```
Definition cost A := (nat * A).
Definition cost_bind {A B} (f: A -> cost B) x :=
  (x.1 + (f (x.2)).1, (f (x.2)).2).
Definition cost_ret {A} (x: A) := (0, x).
```

A computation of type cost A is just a pair of a nat, representing the count of the number of comparisons, and an underlying value of type A. The bind operation sums costs in the obvious way. We can then define a version of comparison in this monad:

```
Definition compare (x y: nat) :=
  (1, ltngtP x y).
```

where `ltngtP` is a function from the `ssreflect` library that returns whether x < y, x = y, or x > y.

The code[3] for QuickSort is given in Fig. 2. This is the standard randomized functional version of QuickSort: For empty and singleton lists, `qs` simply returns the input. Otherwise, it selects an element uniformly at random from the list using `draw_pivot`. It then uses `partition` to split the list into three parts: elements smaller than the pivot, elements equal to the pivot, and elements larger than the pivot. Elements smaller and larger than the pivot are recursively sorted and then the results are joined together. Partition uses the `compare` operator defined above, which implicitly counts the comparisons it performs.

```
Fixpoint qs l : ldist (cost (list nat)) :=
  match l as l' return with
  | [::] => mret ([::])
  | [::a] => mret ([::a])
  | (a :: b :: l') =>
      p <- draw_pivot (a :: b :: l');
      '(lower, middle, upper) <- partition p l;
      ls <- qs (lower);
      us <- qs (upper);
      mret (ls ++ middle ++ us)
  end
```

Fig. 2. Simplified version of code for sequential QuickSort. In `ssreflect`, we write `[::]` for the empty list and [:: a] for a list containing the single element a. Because randomized QuickSort is not structurally recursive, the actual definition in our development defines it by well-founded recursion on the size of the input.

What is the probabilistic recurrence for this algorithm? In each round of the recursion, the algorithm performs n comparisons to partition a list of length n. So, taking the size function to be the length of the list, we have the toll function $a(x) = x$. There are two recursive calls, and we have to sum the comparisons performed by each to get the total, so we need to use Theorem 3.

The h_1 and h_2 functions giving the recursive subproblems correspond to the `lower` and `upper` sublists returned by `partition`. We now need to bound the expected value of the maximum of the sizes of these two lists. We first show:

$$E\left[\max\left(size(h_1(l)), size(h_2(l))\right)\right] \leq \frac{1}{size(l)} \sum_{i=0}^{size(l)-1} \max(i, size(l) - i - 1)$$

---

[3] The definition in our development is actually defined by well-founded induction on the size of the input, because the Coq termination checker cannot determine that this definition always terminates.

To get some intuition for this inequality, imagine the input list l was already sorted. In this situation, if the pivot we draw is in position i, then the sublist of elements less than i only contains elements to the left of i in l and the sublist of elements larger than i contains only elements to the right of i in l. The size of each sublist is therefore at most i and $size(l) - i - 1$, respectively, which corresponds to the ith term in the sum above. The factor of $\frac{1}{size(l)}$ is the probability of selecting each pivot index, because they are all equally likely. Of course, the input list is not actually sorted, but when we select pivot position i, we can consider where its position would be in the final sorted list, and the result is just a re-ordering of the terms in the sum.

Next we show by induction on n that:

$$\sum_{i=0}^{n-1} \max(i, n - i - 1) = \binom{n}{2} + \left\lfloor \frac{n}{2} \right\rfloor \cdot \left\lceil \frac{n}{2} \right\rceil \leq \frac{3n^2}{4}$$

We combine the two inequalities to conclude:

$$\mathrm{E}\left[\max\left(size(h_1(l)), size(h_2(l))\right)\right] \leq \frac{3}{4} \cdot size(l)$$

The above bound is for the case when the list has at least 2 elements; otherwise the recursion is over so that the sublists have length 0. Hence we can define m to be $m(x) = 0$ for $x < 4/3$ and $m(x) = \frac{3x}{4}$ otherwise. We use $4/3$ as the cut-off point rather than 2 because it makes the recurrence easier to solve.

To use Theorem 3, we need to come up with two "ranking" functions g_1 and g_2 such that $g_i(size(h_1(z))) + g_i(size(h_2(z))) \leq g_i(size(z))$ for each i. Ideally, we want g_1 to be as small as possible, because it scales the final bound we derive, whereas for g_2 we want to pick something that makes it easy to solve the recurrence $t(x) \geq a(x)/g_2(x) + t(m(x))$. Like the derivation of the bound m, these parts of the proof are not automatic and require some experimentation. We define the following choices for the parameters of Theorem 3:

$$g_1(x) = x \qquad g_2(x) = \begin{cases} \frac{1}{2} & x \leq 1 \\ \frac{x}{x-1} & 1 < x < 2 \\ x & x \geq 2 \end{cases} \qquad t(x) = \begin{cases} 1 & x \leq 1 \\ \log_{\frac{4}{3}} x + 1 & x > 1 \end{cases}$$

We can check g_1 and g_2 satisfy the necessary conditions, and that t is a solution to the resulting deterministic recurrence relation.

Writing $T(x)$ for the total number of comparisons performed on input x, Theorem 3 now gives us:

$$\Pr\left[T(x) > size(x) \cdot \log_{4/3}(size(x)) + 1 + w \cdot size(x)\right] \leq size(x) \cdot \left(\frac{3}{4}\right)^w$$

for l such that $size(x) > 1$. More concisely, if we set $n = size(x)$, then this becomes:

$$\Pr\left[T(x) > n \log_{4/3} n + 1 + wn\right] \leq n \cdot \left(\frac{3}{4}\right)^w$$

In Coq, this is rendered as:

```
Theorem bound x w:
    rsize x > 1 ->
    pr_gt (T x) (rsize x * (k * ln (rsize x) + 1) + INR w * rsize x)
       <= (rsize x) * (3/4)^w.
```

where $k = \frac{1}{\ln 4/3}$, `rsize` returns the length of a list as a real number, and INR : nat $\to$ R coerces its input into a real number.

To understand the significance of these bounds, consider the case when $w = \lfloor c \cdot \log_{4/3} n \rfloor$ for some constant c. Then, using the above we get:

$$\Pr\left[T(x) > (c+1)n\log_{4/3} n + 1\right] \leq \Pr\left[T(x) > n\log_{4/3} n + 1 + wn\right] \quad (6)$$

$$\leq n \cdot \left(\frac{3}{4}\right)^w \leq n \cdot \left(\frac{3}{4}\right)^{c\log_{4/3} n - 1} \quad (7)$$

$$= \frac{4}{3} \cdot \frac{1}{n^{c-1}} \quad (8)$$

so that when $c > 2$, the probability goes very quickly to 0 for lists of even moderate size.

We can now use the Coq-Interval library, which provides tactics for establishing numerical inequalities, to compute the value of this bound for particular choices of n. In particular, we can establish the claim from the introduction: when sorting a list with 10 million elements, the probability that QuickSort performs more than $8n\log_2 n$ comparisons is less than 10^{-9}.

```
Remark concrete2:
  forall l, rsize l = 10 ^ 7 ->
      pr_gt (T l) (10^7 * (8 * 1/(ln 2) * ln (10^7))) <= 1/(10^9).
```

4.2   Other Examples

We have mechanized the analysis of three other examples using Karp's theorem. A discussion of these examples is given in the appendix of the full version of this paper available as supplementary material. Here we give a brief description of the examples:

1. Parallel QuickSort: using Theorem 2 we show that the longest chain of sequential dependencies from comparisons in a parallel version of QuickSort is $O(\log(n))$ with high probability.
2. Binary search tree: we analyze the height of a binary search tree which is generated by inserting a set of elements under a random permutation. We show the height is $O(\log(n))$ with high probability using Theorem 2.
3. Randomized leader election: we consider a protocol for distributed leader election that has been analyzed by several authors [30,47]. The protocol consists of stages called "rounds". At the beginning of a round, each active node generates a random bit. If the bit is 1, the node remains "active" and sends a message to all the other nodes; otherwise, if the bit is 0 it becomes inactive and stops trying to become the leader. If every active node generates

a 0 within a round, no messages are sent and instead of becoming inactive, those nodes try again in the next round. When there is only one active node remaining, it is deemed the leader. We use Theorem 1 to show that with high probability at most $O(\log n)$ rounds are needed.

5 Changes Needed for Mechanization

Anyone who has mechanized something based on a paper proof has probably encountered issues that make it harder than just "translating" the steps of the proof into the formal system. Even when the paper proof is correct, there are inevitably parts of the argument that are more difficult to mechanize than they appear on paper, and this can require changing the strategy of the proof.

Our experience mechanizing Karp's theorem and its extensions was no different. In this section we describe obstacles that arose in our attempt to mechanize the proof.

5.1 Overview of Proof

To put the following discussion in context, we need to give a sketch of the paper proof. Recall that Theorem 1 says that if we have a probabilistic recurrence W with a corresponding deterministic recurrence solved by u, then for all z and integer w,

$$\Pr\left[W(z) > u(size(z)) + w \cdot a(size(z))\right] \leq \left(\frac{m(size(z))}{size(z)}\right)^w$$

The first thing one would naturally try to prove this is to proceed by induction on the size of z. However, immediately one realizes that the induction hypothesis needs to be strengthened: the bound above is only shown at each integer w, so there are "gaps" in between where we do not have an appropriately tight intermediate bound. To address this, Karp defines a function D_r which "interpolates" the bound $\left(\frac{m(size(z))}{size(z)}\right)^w$ to fill in these gaps. This function D_r is somewhat complicated, and is defined in a piecewise manner as follows:

1. If $r \leq 0$ and $x > 0$, $D_r(x) = 1$
2. If $r > 0$:
   (a) If $x \leq d$ then $D_r(x) = 0$
   (b) If $x > d$ and $u(x) \geq r$ then $D_r(x) = 1$
   (c) If $x > d$ and $u(x) < r$ then

$$D_r(x) = \left(\frac{m(x)}{x}\right)^{\left\lceil \frac{r-u(x)}{a(x)} \right\rceil} \frac{x}{u^{-1}\left(r - a(x)\left\lceil \frac{r-u(x)}{a(x)} \right\rceil\right)}$$

This definition is intricate, especially the last case. However, if we set $r = u(size(z)) + w \cdot a(size(z))$, then $D_r(size(z))$ simplifies to $\left(\frac{m(size(z))}{size(z)}\right)^w$, confirming the intuition that this is some kind of interpolation.

Next, define $K_r(z) = \Pr[W(z) > r]$. Then, the result follows by showing that

$$K_r(z) \leq D_r(size(z))$$

The probabilistic recurrence relation for W implies that:

$$K_r(z) \leq \mathrm{E}\left[K_{r-a(size(z))}(h(z))\right] \tag{9}$$

when $size(z) > d$. Karp's idea is to recursively define a sequence of functions K_r^i for $i \in \mathbb{N}$ which approximate K_r. These are defined by:

$$K_r^0(z) = \begin{cases} 1 & \text{if } r < u(d) \\ 0 & \text{otherwise} \end{cases}$$

$$K_r^{i+1}(z) = \mathrm{E}\left[K_{r-a(size(z))}^i(h(z))\right]$$

Note the similarity between the recursive case and the property in (9). For all i, $K_r^i(z) \leq 1$, so $\sup_i K_r^i(z)$ exists. Karp says then that $K_r(z) \leq \sup_i K_r^i(z)$, so it suffices to show that for all i, $K_r^i(z) \leq D_r(size(z))$.

The proof is by induction on i. The base case is straightforward. For the inductive case, the definition of K_r^{i+1} and the induction hypothesis give us:

$$\begin{aligned} K_r^{i+1}(z) &= \mathrm{E}\left[K_{r-a(size(z))}^i(h(z))\right] \\ &\leq \mathrm{E}\left[D_{r-a(size(z))}(h(z))\right] \end{aligned}$$

So we just need to show that this final expected value is $\leq D_r(size(z))$. The key is the following simple lemma, which lets us bound the expected value of suitable functions of random variables:

Lemma 1 ([38, **Lemma 3.1**]). *Let X be a random variable with values in the range $[0, x]$. Suppose $f : \mathbb{R} \to \mathbb{R}$ is a non-negative function such that $f(0) = 0$, and there exists some constant c such that for all $y \geq c$, $f(y) = 1$ and $f(y)/y$ is non-decreasing on the interval $(0, c]$. Then:*

$$\mathrm{E}[f(X)] \leq \frac{\mathrm{E}[X]\, f(\min(x, c))}{\min(x, c)}$$

Applying this with $X = size(h(z))$, $f = D_{r-a(size(z))}$, and suitable choice of c gives us the desired result. Of course, we need to check that this choice of f satisfies the conditions of the lemma. In particular, showing that $f(y)/y$ is non-decreasing is somewhat involved, and it is here that the various continuity assumptions on parameters like a are used.

Once the inductive proof is finished, we set $r = u(size(z)) + w \cdot a(size(z))$, to get the form of the bound in the statement of the theorem.

5.2   Changes

Termination Assumption. The first problem we had was that we were unable to prove that $K_r(z) \leq \sup_i K_r^i(z)$. In the original paper proof, this inequality is simply stated without further justification. Young [53] has suggested that in fact one may need stronger assumptions on W or h to be able to conclude this and suggests two alternatives. Either W can be assumed to be a minimal solution to the probabilistic recurrence, or one can assume that the recurrence terminates with probability 1, that is $\Pr[h^n(z) > d] \to 0$ as $n \to \infty$. In the end, we chose to make the latter assumption, because it is easy to show for most examples.

Existence of a Minimal Solution. Karp argues that if there is a solution to the deterministic recurrence relation, there must be a minimal solution u. The results in the theorem are then stated in terms of u. It seemed to us more efficient to simply state the results in terms of any continuous and invertible solution t to the recurrence relation. In this way, we avoid the need to prove the existence, continuity, and invertibility of the minimal solution. In fact, rather than assuming t is invertible on its full domain, we merely assume that there exists a function t' which is an inverse to t on the subdomain (d, ∞), that is: $t'(t(x)) = x$ for $x > d$ and $t(t'(x)) = x$ for $x > t(d)$. The definition of D is then changed to replace occurrences of u with t.

Division by Zero. The original piecewise definition of D above involves division by $u^{-1}(r - a(x)\lceil \frac{r-u(x)}{a(x)} \rceil)$. However, it is not clear that this is always non-zero on the domain considered, and this is not explicitly discussed in the paper proof. Since we replace the u^{-1} function with a user supplied function t', we found it easier to simply require an explicit assumption that t' is non-zero everywhere.

Unneeded Assumptions. In the original paper proof, the toll function a is assumed to be everywhere continuous and strictly increasing on $[d, \infty)$. This rules out recurrences like $W(z) = 1 + W(h(x))$ which show up in examples such as the leader election protocol. For that reason, there is actually an additional result in Karp [38] for the particular case where $a(x) = 0$ for $x \leq d$ and 1 otherwise.

However, after finishing the mechanization of Theorem 1, we suspected that the assumptions on a could be weakened, avoiding the need for the additional lemma. We changed the assumptions to only require that a was monotone and continuous on the interval (d, ∞). In turn, we require the function t which solves the deterministic recurrence to be strictly increasing on the interval (d, ∞). Our prior proof script worked mostly unchanged: most of the changes actually ended up deleting helper lemmas we had needed under the original assumptions. This is not because our proof scripts were highly automated or robust, but because the original proof really was not exploiting these stronger assumptions. Checking this carefully with respect to the original paper proof would have been rather tedious, but was straightforward with a mechanized version.

Extending to the Binary Case. In a technical report, Karpinski and Zimmermann [39] claimed to extend Karp's result to work and span recurrences with multiple recursive calls, so we initially tried to verify their result. The argument is fundamentally like Karp's original proof, so many steps were described briefly because they were intended to be similar to the corresponding parts of the proof of Theorem 1. However, we were unable to prove that their analogue of the D_r function satisfied the assumptions of Lemma 1, and so we were stuck at the corresponding step of the induction argument. It was at this point that we mechanized the results from Tassarotti [51] instead.

6 Related Work

6.1 Verification of Randomized Algorithms and Mechanized Probability Theory

Audebaud and Paulin-Mohring [3] developed a different monadic encoding for reasoning about randomized algorithms in Coq that can represent randomized algorithms that do not necessarily terminate. It would be interesting to try to generalize our version of Karp's theorem and apply them to programs expressed using this monad.

Barthe et al. [9] develop a probabilistic variant of Benton's relational Hoare logic [14] called pRHL to do relational reasoning about pairs of randomized programs. Extensions to and applications of pRHL for reasoning about probabilistic programs have been developed in a series of papers [7,10,11], and this kind of relational reasoning has been implemented in the EasyCrypt tool [5]. There are many other formal logics for reasoning about probabilistic programs (*e.g.*, [8,40,44,49]). Kaminski et al. [37] presented a weakest-precondition logic that can be used to establish expected running time. As an example, they proved a bound on the expected number of comparisons used by QuickSort. The soundness of their logic was later mechanized by Hölzl [34] in Isabelle.

Van der Weegen and McKinna [52] mechanized a proof of the average number of comparisons performed by QuickSort in Coq, and used monad transformers to elegantly separate reasoning about correctness and cost while still being able to extract efficient code. Eberl [27] has recently mechanized a similar result, as well as bounds on the expected depth and height of binary search trees [26]. Haslbeck et al. [33] have verified expected height bounds for treaps, which requires measure theoretic probability because of the way that treap algorithms sample from continuous distributions. See the overview by Eberl et al. [29] for a description of the mechanizations from [26,27,33]. Eberl [28] also mechanized the Akra-Bazzi theorem, a generalization of the Master Theorem for reasoning about deterministic divide and conquer recurrences.

More generally, multiple large developments of probability theory have been carried out in several theorem provers, including large amounts of measure theory [35,36], the Central Limit Theorem [4], Lévy and Hoeffding's inequalities [24], and information theory [1], to name just some of these results.

6.2   Techniques for Bounds on Randomized Algorithms

There are a vast number of tools and results that have been developed for analyzing properties of randomized algorithms; see [25,31,43,45] for expository accounts of both simple and more advanced techniques. Different "cookbook" methods like Karp's also exist: Bazzi and Mitter [12] developed a variant of the Akra-Bazzi master theorem for deriving asymptotic expectation bounds for work recurrences. Roura [50] presented a master theorem that also applies to recurrences like that of the expected work for QuickSort.

Chaudhuri and Dubhashi [22] extended the results of Karp [38] for unary probabilistic recurrence relations by weakening some of the assumptions of Theorem 1. Their proof used only "standard" techniques from probability theory like Markov's inequality and Chernoff bounds, so they argued that it is easier to understand. Of course, this approach may be less beneficial for mechanization if we do not have a pre-existing library of results.

7   Conclusion

We have described our mechanization of theorems by Karp [38] and Tassarotti [51] that make it easier to obtain tail bounds for various probabilistic recurrence relations arising in the study of randomized algorithms. To demonstrate the use of these results, we have explained our verification of four example applications. Moreover, we have shown that these results can be used to obtain concrete numerical bounds, fully checked in Coq, for input sizes of practical significance. To our knowledge, this is the first mechanization of these kinds of tail bounds in a theorem prover.

In future work, it would be interesting to try to automate the inference of the a, g_1, and g_2 functions used when applying Karp's theorem. The resulting deterministic recurrence could also probably be solved automatically, since more complex recurrences have been analyzed automatically in related work (*e.g.,* [21]). If these analyses are done as part of external tools, it would be useful to be able to produce proof certificates that could be checked using the Coq development we describe here, as in some other resource analysis tools [17,18].

It should also be possible to extend the applicability of our mechanization by handling arbitrary probability distributions instead of finite ones. Moreover, it may be possible to use tools like the probabilistic relational Hoare logic of Barthe et al. [9] to prove suitable refinements between imperative randomized algorithms and the functional versions we have analyzed here. This would allow one to derive corresponding tail bounds on the imperative versions.

Acknowledgments. The authors thank Jean-Baptiste Tristan, Jan Hoffmann, Justin Hsu, Guy Blelloch, Carlo Angiuli, Daniel Gratzer, Manuel Eberl, and the anonymous reviewers of this work for their feedback. This research was conducted with U.S. Government support under and awarded by DoD, Air Force Office of Scientific Research, National Defense Science and Engineering Graduate (NDSEG) Fellowship, 32 CFR 168a. This work was also supported by a gift from Oracle Labs. Any opinions, findings and conclusions or recommendations expressed in this material are those of the authors and do not necessarily reflect the views of these organizations.

References

1. Affeldt, R., Hagiwara, M.: Formalization of Shannon's theorems in SSReflect-Coq. In: ITP, pp. 233–249 (2012)
2. Akra, M., Bazzi, L.: On the solution of linear recurrence equations. Comp. Opt. Appl. **10**(2), 195–210 (1998)
3. Audebaud, P., Paulin-Mohring, C.: Proofs of randomized algorithms in Coq. Sci. Comput. Program. **74**(8), 568–589 (2009)
4. Avigad, J., Hölzl, J., Serafin, L.: A formally verified proof of the Central Limit Theorem. CoRR abs/1405.7012 (2014). http://arxiv.org/abs/1405.7012
5. Barthe, G., Crespo, J.M., Grégoire, B., Kunz, C., Béguelin, S.Z.: Computer-aided cryptographic proofs. In: ITP, pp. 11–27 (2012)
6. Barthe, G., Espitau, T., Ferrer Fioriti, L.M., Hsu, J.: Synthesizing probabilistic invariants via Doob's decomposition. In: Chaudhuri, S., Farzan, A. (eds.) CAV 2016. LNCS, vol. 9779, pp. 43–61. Springer, Cham (2016). https://doi.org/10.1007/978-3-319-41528-4_3
7. Barthe, G., Espitau, T., Grégoire, B., Hsu, J., Strub, P.: Proving uniformity and independence by self-composition and coupling. In: LPAR (2017)
8. Barthe, G., Gaboardi, M., Grégoire, B., Hsu, J., Strub, P.: A program logic for union bounds. In: ICALP, pp. 107:1–107:15 (2016)
9. Barthe, G., Grégoire, B., Béguelin, S.Z.: Formal certification of code-based cryptographic proofs. In: POPL, pp. 90–101 (2009)
10. Barthe, G., Grégoire, B., Béguelin, S.Z.: Probabilistic relational hoare logics for computer-aided security proofs. In: MPC, pp. 1–6 (2012)
11. Barthe, G., Grégoire, B., Hsu, J., Strub, P.: Coupling proofs are probabilistic product programs. In: POPL, pp. 161–174 (2017)
12. Bazzi, L., Mitter, S.K.: The solution of linear probabilistic recurrence relations. Algorithmica **36**(1), 41–57 (2003)
13. Bentley, J.L., Haken, D., Saxe, J.B.: A general method for solving divide-and-conquer recurrences. SIGACT News **12**(3), 36–44 (1980)
14. Benton, N.: Simple relational correctness proofs for static analyses and program transformations. In: POPL (2004)
15. Blelloch, G., Greiner, J.: Parallelism in sequential functional languages. In: Proceedings of the 7th International Conference on Functional Programming Languages and Computer Architecture, pp. 226–237 (1995)
16. Boldo, S., Lelay, C., Melquiond, G.: Coquelicot: a user-friendly library of real analysis for Coq. Math. Comput. Sci. **9**(1), 41–62 (2015)
17. Carbonneaux, Q., Hoffmann, J., Reps, T., Shao, Z.: Automated resource analysis with Coq proof objects. In: Majumdar, R., Kunčak, V. (eds.) CAV 2017. LNCS, vol. 10427, pp. 64–85. Springer, Cham (2017). https://doi.org/10.1007/978-3-319-63390-9_4
18. Carbonneaux, Q., Hoffmann, J., Shao, Z.: Compositional certified resource bounds. In: POPL, pp. 467–478 (2015)
19. Chakarov, A., Sankaranarayanan, S.: Probabilistic program analysis with martingales. In: Sharygina, N., Veith, H. (eds.) CAV 2013. LNCS, vol. 8044, pp. 511–526. Springer, Heidelberg (2013). https://doi.org/10.1007/978-3-642-39799-8_34

20. Chatterjee, K., Fu, H., Murhekar, A.: Automated recurrence analysis for almost-linear expected-runtime bounds. In: Majumdar, R., Kunčak, V. (eds.) CAV 2017. LNCS, vol. 10426, pp. 118–139. Springer, Cham (2017). https://doi.org/10.1007/978-3-319-63387-9_6

21. Chatterjee, K., Novotný, P., Zikelic, D.: Stochastic invariants for probabilistic termination. In: POPL, pp. 145–160 (2017)

22. Chaudhuri, S., Dubhashi, D.P.: Probabilistic recurrence relations revisited. Theor. Comput. Sci. **181**(1), 45–56 (1997)

23. Cormen, T.H., Leiserson, C.E., Rivest, R.L., Stein, C.: Introduction to Algorithms, 3rd edn. MIT Press (2009). http://mitpress.mit.edu/books/introduction-algorithms

24. Daumas, M., Lester, D., Martin-Dorel, É., Truffert, A.: Improved bound for stochastic formal correctness of numerical algorithms. Innovations Syst. Softw. Eng. **6**(3), 173–179 (2010)

25. Dubhashi, D.P., Panconesi, A.: Concentration of Measure for the Analysis of Randomized Algorithms. Cambridge University Press (2009). http://www.cambridge.org/gb/knowledge/isbn/item2327542/

26. Eberl, M.: Expected shape of random binary search trees. Archive of Formal Proofs 2017 (2017). https://www.isa-afp.org/entries/Random_BSTs.shtml

27. Eberl, M.: The number of comparisons in quicksort. Archive of Formal Proofs 2017 (2017). https://www.isa-afp.org/entries/Quick_Sort_Cost.shtml

28. Eberl, M.: Proving divide and conquer complexities in Isabelle/HOL. J. Autom. Reasoning **58**(4), 483–508 (2017)

29. Eberl, M., Haslbeck, M.W., Nipkow, T.: Verified analysis of random trees. In: ITP (2018)

30. Fill, J.A., Mahmoud, H.M., Szpankowski, W.: On the distribution for the duration of a randomized leader election algorithm. Ann. Appl. Probab. **6**(4), 1260–1283 (1996)

31. Flajolet, P., Sedgewick, R.: Analytic Combinatorics. Cambridge University Press (2009)

32. Gonthier, G., Mahboubi, A., Tassi, E.: A Small Scale Reflection Extension for the Coq system. Research Report RR-6455, Inria Saclay Ile de France (2016). https://hal.inria.fr/inria-00258384

33. Haslbeck, M.W., Eberl, M., Nipkow, T.: Treaps. Archive of Formal Proofs (2018). https://isa-afp.org/entries/Treaps.html

34. Hölzl, J.: Formalising semantics for expected running time of probabilistic programs. In: ITP, pp. 475–482 (2016)

35. Hölzl, J., Heller, A.: Three chapters of measure theory in Isabelle/HOL. In: ITP, pp. 135–151 (2011)

36. Hurd, J.: Formal Verification of Probabilistic Algorithms. Ph.D. thesis. Cambridge University, May 2003

37. Kaminski, B.L., Katoen, J.-P., Matheja, C., Olmedo, F.: Weakest precondition reasoning for expected run–times of probabilistic programs. In: Thiemann, P. (ed.) ESOP 2016. LNCS, vol. 9632, pp. 364–389. Springer, Heidelberg (2016). https://doi.org/10.1007/978-3-662-49498-1_15

38. Karp, R.M.: Probabilistic recurrence relations. J. ACM **41**(6), 1136–1150 (1994)

39. Karpinski, M., Zimmermann, W.: Probabilistic recurrence relations for parallel divide-and-conquer algorithms. Technical report TR-91-067, International Computer Science Institute (ICSI) (1991). https://www.icsi.berkeley.edu/ftp/global/pub/techreports/1991/tr-91-067.pdf

40. Kozen, D.: A probabilistic PDL. In: STOC, pp. 291–297 (1983)
41. Martin-Dorel, É., Melquiond, G.: Proving tight bounds on univariate expressions with elementary functions in Coq. J. Autom. Reason. **57**(3), 187–217 (2016)
42. McIver, A., Morgan, C., Kaminski, B.L., Katoen, J.: A new proof rule for almost-sure termination. PACMPL **2**(POPL), 33:1–33:28 (2018). http://doi.acm.org/10.1145/3158121
43. Mitzenmacher, M., Upfal, E.: Probability and Computing - Randomized Algorithms and Probabilistic Analysis. Cambridge University Press (2005)
44. Morgan, C., McIver, A., Seidel, K.: Probabilistic predicate transformers. ACM Trans. Program. Lang. Syst. **18**(3), 325–353 (1996)
45. Motwani, R., Raghavan, P.: Randomized Algorithms. Cambridge University Press (1995)
46. Petcher, A., Morrisett, G.: The foundational cryptography framework. In: Focardi, R., Myers, A. (eds.) POST 2015. LNCS, vol. 9036, pp. 53–72. Springer, Heidelberg (2015). https://doi.org/10.1007/978-3-662-46666-7_4
47. Prodinger, H.: How to select a loser. Disc. Math. **120**(1), 149–159 (1993)
48. Ramsey, N., Pfeffer, A.: Stochastic lambda calculus and monads of probability distributions. In: POPL, pp. 154–165 (2002)
49. Ramshaw, L.H.: Formalizing the Analysis of Algorithms. Ph.D. thesis. Stanford University (1979)
50. Roura, S.: Improved master theorems for divide-and-conquer recurrences. J. ACM **48**(2), 170–205 (2001)
51. Tassarotti, J.: Probabilistic recurrence relations for work and span of parallel algorithms. CoRR abs/1704.02061 (2017). http://arxiv.org/abs/1704.02061
52. van der Weegen, E., McKinna, J.: A machine-checked proof of the average-case complexity of quicksort in Coq. In: Berardi, S., Damiani, F., de'Liguoro, U. (eds.) TYPES 2008. LNCS, vol. 5497, pp. 256–271. Springer, Heidelberg (2009). https://doi.org/10.1007/978-3-642-02444-3_16
53. Young, N.: Answer to: Understanding proof of theorem 3.3 in Karp's probabilistic recurrence relations. Theoretical Computer Science Stack Exchange (2016). http://cstheory.stackexchange.com/q/37144

Verified Memoization and Dynamic Programming

Simon Wimmer$^{(\boxtimes)}$ ⓘ, Shuwei Hu ⓘ, and Tobias Nipkow ⓘ

Fakultät für Informatik, Technische Universität München, Munich, Germany
wimmers@in.tum.de

Abstract. We present a lightweight framework in Isabelle/HOL for the automatic verified (functional or imperative) memoization of recursive functions. Our tool constructs a memoized version of the recursive function and proves a correspondence theorem between the two functions. A number of simple techniques allow us to achieve bottom-up computation and space-efficient memoization. The framework's utility is demonstrated on a number of representative dynamic programming problems.

1 Introduction

Verification of functional properties of programs is most easily performed on functional programs. Performance, however, is more easily achieved with imperative programs. One method of improving performance of functional algorithms automatically is memoization. In particular dynamic programming is based on memoization. This paper presents a framework and a tool [24] (for Isabelle/HOL [16,17]) that memoizes pure functions automatically and proves that the memoized function is correct w.r.t. the original function. Memoization is parameterized by the underlying memory implementation which can be purely functional or imperative. We verify a collection of representative dynamic programming algorithms at the functional level and derive efficient implementations with the help of our tool. This appears to be the first tool that can memoize recursive functions (including dynamic programming algorithms) and prove a correctness theorem for the memoized version.

1.1 Related Work

Manual memoization has been used in specific projects before, e.g. [3,21], but this did not result in an automatic tool. One of the few examples of dynamic programming in the theorem proving literature is a formalization of the CYK algorithm where the memoizing version (using HOL functions as tables) is defined and verified by hand [2]. In total it requires 1000 lines of Isabelle text. Our version in Sect. 3.4 is a mere 120 lines and yields efficient imperative code.

ⓒ Springer International Publishing AG, part of Springer Nature 2018
J. Avigad and A. Mahboubi (Eds.): ITP 2018, LNCS 10895, pp. 579–596, 2018.
https://doi.org/10.1007/978-3-319-94821-8_34

Superficially very similar is the work by Itzhaky *et al.* [9] who present a system for developing optimized DP algorithms by interactive, stepwise refinement focusing on optimizations like parallelization. Their system contains an ad-hoc logical infrastructure powered by an SMT solver that checks the applicability conditions of each refinement step. However, no overall correctness theorem is generated and the equivalent of our memoization step is performed by their backend, a compiler to C++ which is part of the trusted core.

We build on existing infrastructure in Isabelle/HOL for generating executable code in functional and imperative languages automatically [1, 6, 7].

2   Memoization

2.1   Overview

The workhorse of our framework is a tool that automatically memoizes [15] recursive functions defined with Isabelle's function definition command [11]. More precisely, to memoize a function f, the idea is to pass on a memory between invocations of f and to check whether the value of f for x can already be found in the memory whenever $f\,x$ is to be computed. If $f\,x$ is not already present in the memory, we compute $f\,x$ using f's recursive definition and store the resulting value in the memory. The memory is threaded through with the help of a *state monad*. Starting from the defining equations of f, our algorithm produces a version f'_m that is defined in the state monad. The only place where the program actually interacts with the state is on recursive invocations of f'_m. Each defining equation of f of the form

$$f\,x = t$$

is re-written to

$$f'_m\,x =_m t_m$$

where t_m is a version of t defined in the state monad. The operator $=_m$ encapsulates the interaction with the state monad. Given x, it checks whether the state already contains a memoized value for $f\,x$ and returns that or runs the computation t_m and adds the computed value to the state. Termination proofs for f'_m are replayed from the termination proofs of f. To prove that f still describes the same function as f'_m, we use relational parametricity combined with induction. The following subsections will explain in further detail each of the steps involved in this process: *monadification* (i.e. defining f'_m in the state monad), replaying the termination proof, proving the correspondence of f and f'_m via relational parametricity, and implementing the memory. Moreover, we will demonstrate how this method can be adopted to also obtain a version f_h that is defined in the *heap monad* of Imperative HOL [4] and allows one to use imperative implementations of the memory.

2.2   Monadification

We define the state monad with memory of type $'m$ and pure (result) type $'a$:

datatype $('m, 'a)$ $state = State$ $(run\_state : 'm \to 'a \times 'm)$.

That is, given an initial state of type $'m$, a computation in the state monad produces a pair of a computation result of type $'a$ and a result state (of type $'m$). To make type $('m, 'a)$ $state$ a monad, we need to define the operators $return$ $(\langle \text{-} \rangle)$ and $bind$ $(\ggg)$:

$$return :: 'a \to ('m, 'a) \ state$$
$$bind :: ('m, 'a) \ state \to ('a \to ('m, 'b) \ state) \to ('m, 'b) \ state$$

$$\langle a \rangle \quad = \ State \ (\lambda M. \ (a, M))$$
$$s \ggg f = \ State \ (\lambda M. \ \textsf{case} \ run\_state \ s \ M \ \textsf{of} \ (a, M') \Rightarrow run\_state \ (f \ a) \ M')$$

The definition of $\ggg$ describes how states are threaded through the program.

There are a number of different styles of turning a purely functional program into a corresponding *monadified* version (see e.g. [5]). We opt for the call-by-value monadification style from [8]. This style of monadification has two distinct features: it fixes a strict call-by-value monadification order and generalizes suitably to higher-order functions. The type $M(\tau)$ of the monadified version of a computation of type τ can be described recursively as follows:

$$M(\tau) = ('m, M'(\tau)) \ state$$
$$M'(\tau_1 \to \tau_2) = M'(\tau_1) \to M(\tau_2)$$
$$M'(\tau_1 \oplus \tau_2) = M'(\tau_1) \oplus M'(\tau_2) \qquad \text{where } \oplus \in \{+, \times\}$$
$$M'(\tau) = \tau \qquad\qquad\qquad\qquad \text{otherwise}$$

As a running example, consider the *map* function on lists. Its type is

$$('a \to 'b) \to ('a \ list \to 'b \ list)$$

and its monadified version map_m has type

$$('m, ('a \to ('m, 'b) \ state) \to ('m, 'a \ list \to ('m, 'b \ list) \ state) \ state) \ state.$$

The definitions of map_m and map'_m are

$$map_m \ = \ \langle \lambda f'_m. \ \langle \lambda xs. \ map'_m \ f'_m \ xs \rangle \rangle$$
$$map'_m \ f'_m \ [] \ = \ \langle [] \rangle$$
$$map'_m \ f'_m \ (Cons \ x \ xs) \ = \ Cons_m \ \bullet \ (\langle f'_m \rangle \ \bullet \ \langle x \rangle) \ \bullet \ (map'_m \ \bullet \ \langle f'_m \rangle \ \bullet \ \langle xs \rangle)$$
$$Cons_m \ = \ \langle \lambda \ x. \ \langle \lambda \ xs. \ \langle Cons \ x \ xs \rangle \rangle \rangle$$

compared to *map*:

$$map\ f\ [] \ = \ []$$
$$map\ f\ (Cons\ x\ xs) \ = \ Cons\ (f\ x)\ (map\ f\ xs).$$

As can be seen in this definition, the idea of the translation is to wrap up all bound variables in a *return*, to replace function application with the "$\bullet$"-operator (see below), and to replace constants by their corresponding monadified versions. The combinator map'_{m} follows the recursive structure of *map* and combinator map_{m} lifts map'_{m} from type $M'(\tau)$ to $M(\tau)$ where τ is the type of *map*.

The lifted function application operator "$\bullet$" simply uses $\ggg$ to pass along the state:

$$f_{\mathsf{m}} \bullet x_{\mathsf{m}} = f_{\mathsf{m}} \ggg (\lambda f.\ x_{\mathsf{m}} \ggg f).$$

Our monadification algorithm can be seen as a set of rewrite rules, which are applied in a bottom-up manner. The algorithm will maintain a mapping Γ from terms to their corresponding monadified versions. If we monadify an equation of the form $f\ x = t$ into $f'_{\mathsf{m}}\ x'_{\mathsf{m}} = t_{\mathsf{m}}$[1], then the initial mapping Γ_0 includes the bindings $f \mapsto \langle f'_{\mathsf{m}} \rangle$ and $x \mapsto \langle x'_{\mathsf{m}} \rangle$. Let $\Gamma \vdash t \rightsquigarrow t_{\mathsf{m}}$ denote that t is monadified to t_{m} given Γ. Moreover, we say that a term is Γ-*pure* if none of its subterms are in the domain of Γ. Our monadification rules are the following:

$$\frac{e :: \tau_0 \to \tau_1 \to \ldots \to \tau_n \qquad e \text{ is } \Gamma\text{-pure} \qquad \forall i.\ M'(\tau_i) = \tau_i}{\Gamma \vdash e \rightsquigarrow \langle \lambda t_0.\ \langle \lambda t_1.\ \cdots \langle \lambda t_{n-1}.\ e\ t_0\ t_1 \cdots t_{n-1} \rangle \cdots \rangle \rangle} \ \text{PURE}$$

$$\frac{\Gamma[x \mapsto \langle x'_{\mathsf{m}} \rangle] \vdash t \rightsquigarrow t_{\mathsf{m}}}{\Gamma \vdash (\lambda x :: \tau.\ t) \rightsquigarrow \langle \lambda x'_{\mathsf{m}} :: M'(\tau).\ t_{\mathsf{m}} \rangle} \ \lambda \qquad \frac{\Gamma \vdash e \rightsquigarrow e_{\mathsf{m}} \qquad \Gamma \vdash x \rightsquigarrow x_{\mathsf{m}}}{\Gamma \vdash (e\ x) \rightsquigarrow (e_{\mathsf{m}} \bullet x_{\mathsf{m}})} \ \text{APP}$$

$$\frac{g \in \mathsf{dom}\ \Gamma}{\Gamma \vdash g \rightsquigarrow \Gamma(g)} \ \Gamma$$

The rules are ordered from highest to lowest priority.

An additional rule specifically treats case-expressions:

$$\frac{\begin{array}{c} g \text{ is a case-combinator with } n \text{ branches} \\ \Gamma \vdash t_1 \rightsquigarrow_\eta t'_1 \qquad \cdots \qquad \Gamma \vdash t_n \rightsquigarrow_\eta t'_n \end{array}}{\Gamma \vdash g\ t_1\ \ldots\ t_n \rightsquigarrow \langle g\ t'_1\ \ldots\ t'_n \rangle} \ \text{COMB}$$

Here $\Gamma \vdash t \rightsquigarrow_\eta t'$ denotes that t is fully η-expanded to $\lambda x_1 \ldots x_k.\ s$ first, and if $\Gamma[x_1 \mapsto \langle x'_1 \rangle, \ldots, x_k \mapsto \langle x'_k \rangle] \vdash s \rightsquigarrow s'$, then $\Gamma \vdash t \rightsquigarrow_\eta \lambda x'_1 \ldots x'_k.\ s'$. The value that is subject to the case analysis is another argument t_{n+1}, and the arguments to $t_1, \ldots, t_n$ are produced by the case-combinator.

As an example, consider the following unusual definition of the Fibonacci sequence:

$$fib\ n = 1 + sum\ ((\lambda f.\ map\ f\ [0..n-2])\ fib).$$

---

[1] If f has type $\tau_1 \to \cdots \to \tau_n$ and x has type τ, the variables f'_{m} and x_{m} are assumed to have types $M'(\tau_1) \to \cdots \to M'(\tau_n)$ and $M(\tau)$, respectively. For a term $x :: \tau$ that satisfies $M'(\tau) = \tau$, x and x'_{m} are used interchangeably.

Its right-hand side can be transformed via the following derivation:

$$
\cfrac{
\cfrac{
\cfrac{map \in \mathsf{dom}\,\Gamma'}{\Gamma' \vdash map \rightsquigarrow map_{\mathsf{m}}}\,\Gamma \quad
\cfrac{f \in \mathsf{dom}\,\Gamma'}{\Gamma' \vdash f \rightsquigarrow \langle f'_{\mathsf{m}}\rangle}\,\Gamma \quad
\cfrac{[0..n \quad 2] \rightsquigarrow \ldots}{}
}{\ldots \qquad \langle \lambda f'_{\mathsf{m}}.\ map_{\mathsf{m}} \bullet \langle f'_{\mathsf{m}}\rangle \bullet \langle \diamond\rangle\rangle}\ \lambda \quad
\cfrac{\diamond}{\cfrac{fib}{\langle fib'_{\mathsf{m}}\rangle}\,\Gamma}
}{
\Gamma \vdash fib\ n \rightsquigarrow \langle \lambda x.\ \langle 1 + x\rangle\rangle \bullet (\langle \lambda xs.\ \langle sum\ xs\rangle\rangle \bullet (\langle \lambda f'_{\mathsf{m}}.\ map_{\mathsf{m}} \bullet \langle f'_{\mathsf{m}}\rangle \bullet (\ldots)\rangle \bullet \langle fib'_{\mathsf{m}}\rangle)))
}
$$

$$
\cfrac{
\cfrac{([..])\,0}{\langle \lambda r.\ \langle [0..r]\rangle\rangle}\,\textsc{Pure} \quad
\cfrac{(-)}{\langle \lambda x.\ \langle \lambda y.\ \langle x - y\rangle\rangle\rangle}\,\textsc{Pure} \quad
\cfrac{n}{\langle n'_{\mathsf{m}}\rangle}\,\Gamma \quad
\cfrac{2}{\langle 2\rangle}\,\textsc{Pure}
}{
[0..n-2] \rightsquigarrow \langle \lambda r.\ \langle [0..r]\rangle\rangle \bullet (\langle \lambda x.\ \langle \lambda y.\ \langle x - y\rangle\rangle\rangle \bullet \langle n'_{\mathsf{m}}\rangle \bullet \langle 2\rangle)
} =: \diamond
$$

where $\Gamma' = \Gamma[f \mapsto \langle f'_{\mathsf{m}}\rangle]$ and $\Gamma = [fib \mapsto \langle fib'_{\mathsf{m}}\rangle, n \mapsto \langle n'_{\mathsf{m}}\rangle]$. Parts of the derivation tree have been elided, and the left-hand side and Γ have been left out where they are clear from the context. A double line represents multiple applications of the App-rule. Note that the term $\lambda f.\ map\ f\ [0..n-2]$ falls through the Pure-rule before being processed by the λ-rule. This is because the bound variable f has the function type $\tau = int \rightarrow int$, which means $M'(\tau) \neq \tau$.

2.3   Reasoning with Parametricity

We want to use *relational parametricity* [19,22] to prove the correspondence between a program and its memoized version. A memory m is said to be *consistent* with a pure function $f :: 'a \rightarrow 'r$, if it only memoizes actual values of f: if m maps a to r, then $r = f\ a$. We will use a relation $\Downarrow_R v\ s$ to assert that, given a consistent state m, $run\_state\ s\ m$ will produce a consistent state and a computation result v' with $R\ v\ v'$. Formally, $\Downarrow$ is defined with respect to the function f that we want to memoize, and an invariant inv_{m} on states:

$$
\Downarrow_R v\ s = \forall m.\ cmem\ m \wedge inv_{\mathsf{m}}\ m \longrightarrow
$$
$$
(\mathsf{case}\ run\_state\ s\ m\ \mathsf{of}\ (v',\ m') \Rightarrow R\ v\ v' \wedge cmem\ m' \wedge inv_{\mathsf{m}}\ m')
$$

where $cmem\ m$ expresses that m is consistent with f and $inv_{\mathsf{m}}\ m$ expresses that m correctly represents a memory. Using the function relator

$$
R \dashrightarrow S = \lambda f\ g.\ \forall x\ y.\ R\ x\ y \longrightarrow S\ (f\ x)\ (g\ y),
$$

one can state the parametricity theorems for our fundamental monad combinators as (the relations can be understood as types):

$$
(R \dashrightarrow \Downarrow_R)\ (\lambda x.\ x)\ return
$$
$$
(\Downarrow_R \dashrightarrow (R \dashrightarrow \Downarrow_S) \dashrightarrow \Downarrow_S)\ (\lambda v\ g.\ g\ v)\ (\ggg\!=)
$$
$$
(\Downarrow_{(R \dashrightarrow \Downarrow_S)} \dashrightarrow \Downarrow_R \dashrightarrow \Downarrow_S)\ (\lambda\ g\ x.\ g\ x)\ (\bullet).
$$

To prove the parametricity theorem, e.g. for map and map'_{m}, one needs only the first and third property, and the parametricity theorems for all previously monadified constants that appear in map and map'_{m}.

To prove the correspondence theorem for a monadification result, we use induction (following the recursion structure of the monadified function) together with parametricity reasoning.

Automating this induction proof is non-trivial. The reason is that the function definition command uses an elaborate extraction procedure based on congruence rules [11,20] to extract recursive function calls and a surrounding context, which is used to prove termination of the function and to prove the induction theorem. This may lead to complex (and necessary) assumptions in the induction hypotheses that specify for which sets of function arguments the hypotheses are valid. The challenge is to integrate the specific format of these assumptions with parametricity reasoning.

To combat this problem, we opt for a more specialized approach that performs an induction proof by exploiting parametricity but that goes beyond the infrastructure that is provided by Isabelle's integrated parametricity reasoning facility [12]. The main difference is that we use special variants of parametricity theorems that resemble the structure of the congruence theorems used by the function definition command. These produce the right pre-conditions to discharge the induction hypotheses.

Consider the fib function defined at the end of Sect. 2.2. Both its termination and its generated induction rule are based on the information that $\mathit{fib}\ n$ can call $\mathit{fib}\ x$ only if x is between 0 and $n - 2$. This information is extracted with the help of this congruence rule:

$$\frac{xs = ys \qquad \forall x.\ x \in set\ ys \longrightarrow f\ x = g\ x}{(map\ f\ xs) = (map\ g\ ys)}\ map\_cong$$

Similarly, for the monadified version fib'_m, its recursive calls are extracted by a pre-registered congruence rule:

$$\frac{xs = ys \qquad \forall x.\ x \in set\ ys \longrightarrow f'_m\ x = g'_m\ x}{(map_m \bullet \langle f'_m \rangle \bullet \langle xs \rangle) = (map_m \bullet \langle g'_m \rangle \bullet \langle ys \rangle)}\ map_m\text{-}cong$$

After initiating the induction proof of the correspondence theorem with our tool, we are left with a goal that grants us the induction hypothesis

$$\forall x.\ x \in set\ [0, \ldots, n - 2] \longrightarrow \Downarrow_= (\mathit{fib}\ x)\ (\mathit{fib}'_m\ x)$$

and asks us to prove

$$\Downarrow_{list\_all2\ (=)}\ (map\ \mathit{fib}\ [0, \ldots, n - 2])\ (map_m \bullet \langle \mathit{fib}'_m \rangle \bullet \langle [0, \ldots, n - 2] \rangle)$$

where $list\_all2\ S$ compares two lists of equal length elementwise by relation S. To solve this goal, our tool will apply another pre-registered parametricity theorem for map and map_m (which is derived from map_m_cong following a canonical pattern):

$$\frac{xs = ys \qquad \forall x.\ x \in set\ ys \longrightarrow \Downarrow_S\ (f\ x)\ (f'_m\ x)}{\Downarrow_{list\_all2}\ S\ (map\ f\ xs)\ (map_m \bullet \langle f'_m \rangle \bullet \langle ys \rangle)}\ map\_map_m$$

It generates the same pre-condition as the congruence rule and yields a goal that exatcly matches the aforementioned induction hypothesis.

2.4   Termination

When monadifying a recursive function f that was defined with the function definition command [11], we need to replay the termination proof of f to define f'_m. The termination proof—whether automatic or manual—relies on exhibiting a well-founded relation that is consistent with the recursive function calls of f. To capture the latter notion, the function definition command defines a relation f_rel between values in the domain of f. The idea for replaying the termination proof is that f and f'_m share the same domain and the same structure of recursive function calls. Thus, one tries to prove $f\_rel = f'_m\_rel$, and if this succeeds, the termination relation for f is also compatible with the one for f'_m, yielding termination of f'_m.

However, the structure of f_rel is sometimes too dissimilar to f'_m_rel, and thus an automated proof of the equality fails. The main reason for that is that monadification can reorder the control flow and thus can alter the order in which the function definition commands encounters the recursive function calls when analyzing f'_m_rel. Moreover, sometimes a congruence rule is unnecessarily used while defining f, causing our tool to fail if a corresponding parametric version has not been registered with our tool. In such cases, we try to fall back to the automated termination prover that is provided by the function definition command.

2.5   Technical Limitations

While the monadification procedure that was presented in the previous sections is designed to run automatically, it is not universally applicable to any Isabelle/HOL function without previous setup. This encompasses the following limiations:

- As outlined above, higher-order combinators such as *map* generally need to be pre-registered together with their corresponding congruence and parametricity theorems.
- Just like Isabelle's function definition command, our tool relies on a context analysis for recursive calls. If we define (note the *id*)

$$fib\ n = 1 + sum\ (id\ (\lambda f.\ map\ f\ [0..n-2])\ fib)$$

it becomes impossible to prove termination with the function definition command because the information that recursive calls happen only on values between 0 and $n-2$ is lost, and similarly our parametricity reasoner fails.
- Currently, our parametricity reasoner can only prove goals of the form

$$(R \dashrightarrow S)(\lambda x.\ f\ x)(\lambda y.\ g\ y)$$

if Isabelle's built-in parametricity reasoner can automatically show $R = (=)$. We plan to relax this limitation in the future.

Nevertheless, our tool works fully automatically for our case studies consisting of functions on lists and numbers that involve different higher-order combinators and non-trivial termination proofs.

2.6   Memoization

Compared to *monadification*, *memoization* of a program simply differs by replacing $=$ in each defining equation by $=_\mathsf{m}$ of type

$$('a \to ('m, 'r) \ state) \times 'a \to ('m, 'r) \ state \to bool.$$

The memoized version of a function of type $'a \to 'r$ then is of type $'a \to M'('r)$ where $'a$ should not contain function types. This seems to work only for functions with exactly one argument but our tool will automatically uncurry the function subject to memoization whenever necessary.

Concerning the memory type $'m$, we merely assume that it comes with two functions with the obvious intended meaning:

$$lookup :: 'a \to ('m, 'r \ option) \ state$$
$$update :: 'a \to 'r \to ('m, unit) \ state$$

We use a memoizing operation *retrieve_or_run* to define $=_\mathsf{m}$:

$$\big((f'_\mathsf{m}, x) =_\mathsf{m} t\big) \; = \; \big(f'_\mathsf{m} \ x = retrieve\_or\_run \ x \ (\lambda\_. \ t)\big)$$

$$retrieve\_or\_run \ x \ t' \; = \; lookup \ x \ggg \big(\lambda r. \ \textsf{case} \ r \ \textsf{of}$$
$$Some \ v \Rightarrow \langle v \rangle$$
$$\mid \ None \ \ \Rightarrow t' \ () \ggg (\lambda v. \ update \ x \ v \ggg \lambda\_. \ \langle v \rangle)\big).$$

Note that it is vital to wrap the additional λ-abstraction around t: otherwise call-by-value evaluation would build up a monadic expression that eagerly follows the full recursive branching of the original computation before any memoization is applied.

In order to specify the behavior of *lookup* and *update* we define an abstraction function $map\_of :: 'm \to 'a \to 'r \ option$ that turns a memory into a function:

$$map\_of \ heap \ k \; = \; fst \ (run\_state \ (lookup \ k) \ heap).$$

To guarantee that *retrieve_or_run* always produces a consistent memory, $lookup \ k$ should never add to the mapping, and $update \ k \ v$ should add at most the mapping $k \mapsto v$. (We will exploit the permissiveness of this specification in Sect. 2.9.) Formally, for all m with $inv_\mathsf{m} \ m$:

$$map\_of \ (snd \ (run\_state \ (lookup \ k) \ m)) \subseteq_m map\_of \ m$$
$$map\_of \ (snd \ (run\_state \ (update \ k \ v) \ m)) \subseteq_m (map\_of \ m)(k \mapsto v)$$

where $(m_1 \subseteq_m m_2) \longleftrightarrow (\forall a \in dom \ m_1. \ m_1 \ a = m_2 \ a)$. Additionally, inv_m is required to be invariant under *lookup* and *update*. This allows us to prove correctness of *retrieve_or_run*:

$$\Downarrow_= (f \ x) \ s \longrightarrow \Downarrow_= (f \ x) \ (retrieve\_or\_run \ x \ s).$$

Given that this is not a parametricity theorem, our method to inductively prove parametricity theorems for memoized functions needs to treat equations defined via $=_m$ specially before parametric reasoning can be initiated.

From the correctness of $retrieve\_or\_run$ and the correspondence theorem for f'_m we can derive correctness of f'_m:

$$\Downarrow_= (f\ x)\ (f'_m\ x).$$

As a corollary, we obtain:

$$run\_state\ (f'_m\ x)\ empty = (v,\ m) \longrightarrow f\ x = v \wedge cmem\ m.$$

A simple instantiation of our memory interface can be given with the help of the standard implementation of mappings via red-black trees in Isabelle/HOL.

2.7   Imperative Memoization

This section outlines how our approach to monadification and memoization can be extended from a purely functional to an imperative memory implementation. Imperative HOL [4] is a framework for specifying and reasoning about imperative programs in Isabelle/HOL. It provides a *heap monad*

datatype $'a\ Heap = Heap\ (execute : heap \to ('a \times heap)\ option)$

in which imperative programs can be expressed. The definition shows that the heap monad merely encapsulates a state monad (specialized to heaps) in an *option* monad to indicate failure. Our approach is simple: assuming that none of the operations in the memoized program fail (failures could only arise from *lookup* or *update*), the heap monad is equivalent to a state monad. This can be stated formally, where inv_h is a heap invariant:

$$\Downarrow^h_R\ f_m\ f_h\ =\ \forall\ heap.\ inv_h\ heap \longrightarrow$$
$$(\textsf{case}\ run\_state\ f_m\ heap\ \textsf{of}\ (v_1, heap_1) \Rightarrow \textsf{case}\ execute\ f_h\ heap\ \textsf{of}$$
$$Some\ (v_2, heap_2) \Rightarrow R\ v_1\ v_2 \wedge heap_1 = heap_2 \wedge inv_h\ heap_2$$
$$|\ None \Rightarrow False)$$

One could now be tempted to combine $\Downarrow$ and $\Downarrow^h$ into a relation between pure values and the heap monad by defining $\Downarrow'$ as a composition of the two:

$$\Downarrow'_R\ =\ \Downarrow_R \circ\circ \Downarrow^h_=$$

where $\circ\circ$ is the composition of binary relations. However, this would prohibit proving the analogue of the parametricity theorem for $\ggg=$. The reason is that $\Downarrow'$ would demand too strong a notion of non-failure: computations are never allowed to fail, no matter whether we start the computation with a consistent state or not. Instead we use a weaker notion (analogous to $\Downarrow$)

$$\Downarrow'_R\ v\ f_h\ =\ \forall heap.\ inv_m\ heap \wedge inv_h\ heap \wedge cmem\ heap \longrightarrow$$
$$(\textsf{case}\ execute\ f_h\ heap\ \textsf{of}\ None \Rightarrow False$$
$$|\ Some\ (v',\ heap') \Rightarrow inv_m\ heap' \wedge inv_h\ heap' \wedge R\ v\ v' \wedge cmem\ heap')$$

where inv_{m} and inv_{h} correspond to $\Downarrow$ and $\Downarrow^h$, respectively. The advantage is that one can prove

$$(\Downarrow_R \circ\circ \Downarrow^h_=)\ v\ f_{\mathsf{h}} \implies \Downarrow'_R\ v\ f_{\mathsf{h}},$$

to exploit compositionality where necessary, while still obtaining the analogous theorems for the elementary monad combinators (though not through reasoning via compositionality for $\ggg$). Using $\Downarrow'_R$ instead of $\Downarrow_R$, one can now use the same infrastructure for monadification and parametricity proofs to achieve imperative memoization.

2.8 Bottom-up Computation

In a classic imperative setting, dynamic programming algorithms are usually not expressed as recursive programs with memoization but rather as a computation that incrementally fills a table of memoized values according to some iteration strategy (typically in a bottom-up manner), using the recurrences to compute new values. The increased control over the computation order allows one to reduce the size of the memory drastically for some dynamic programming algorithms—examples of these can be found below. We propose a combination of two simple techniques to accomplish a similar behaviour and memory efficiency within our framework. The first, which is described in this section, is a notion of iterators for computations in the state monad that allows one to freely specify the computation order of a dynamic program. The second is to exploit our liberal interface for memories to use implementations that store only part of the previously seen computation results (to be exemplified in the next section).

Our interface for iterators consists of two functions $cnt :: {'a} \to bool$ and $nxt :: {'a} \to {'a}$ that indicate whether the iterator can produce any more elements and yield the next element, respectively. We can use these to iterate a computation in the state monad:

$$
\begin{aligned}
iter\_state\ f\ =\ &wfrec\ \{(nxt\ x,\ x)\ |\ cnt\ x\} \\
&(\lambda rec\ x.\ \text{if}\ cnt\ x\ \text{then}\ f\ x \ggg (\lambda\_.\ rec\ (nxt\ x))\ \text{else}\ \langle(\,)\rangle)
\end{aligned}
$$

where $wfrec$ takes the well-founded termination relation as its first argument. Given a size function on the iterator value, we can prove termination if

$$finite\ \{x\ |\ cnt\ x\}\quad \text{and}\quad \forall\, x.\ cnt\ x \longrightarrow size\ x < size\ (nxt\ x).$$

Provided that a given iteration strategy terminates in this sense, we can use it to compute the value of a memoized function:

$$(=\,\dashrightarrow\Downarrow_R)\ g\ f\ \longrightarrow\ \Downarrow_R\ (g\ x)\ (iter\_state\ cnt\ nxt\ f\ x \ggg (\lambda\_.\ f\ x)).$$

As an example, a terminating iterator that builds up a table of n rows and m columns in a row-by-row, left-to-right order can be specified as:

$$
\begin{aligned}
size\ (x,y) &= x * (m+1) + y \\
cnt\ (x,y) &= x \le n \wedge y \le m \\
nxt\ (x,y) &= \text{if}\ y < m\ \text{then}\ (x, y+1)\ \text{else}\ (x+1, 0)
\end{aligned}
$$

If the recursion pattern of f is consistent with nxt, the stack depth of the iterative version is at most one because every recursive call is already memoized.

2.9 Memory Implementations

To achieve a space-efficient implementation for the Minimum-Edit Distance problem or the Bellman-Ford algorithm, one needs to complement the bottom-up computation strategy from the last section with a memory that stores only the last two rows. We will showcase how such a memory can be implemented generically within our framework, and how to exploit compositionality to get an analogous imperative implementation without repeating the correctness proof.

Abstractly, we will implement a mapping $'k \rightarrow 'v\ option$ and split up $'k$ using two key functions $key_1 :: 'k \rightarrow 'k_2$ and $key_2 :: 'k \rightarrow 'k_1$. We demand that together, they are injective:

$$\forall k\ k'.\ key_1\ k = key_1\ k' \wedge key_2\ k = key_2\ k' \longrightarrow k = k'.$$

For a rectangular memory, for instance, key_1 and key_2 could map a key to its row and column index. We use two pairs of lookup and update functions, (l_1, u_1) and (l_2, u_2) to implement the memory for the two rows. We also store the row keys k_1 and k_2 that the currently stored values correspond to in the memory.

For the verification it is crucial that we have previously introduced a memory invariant. The invariant states that k_1 and k_2 are different, and that the first and second row only store key-value pairs that correspond to k_1 and k_2, respectively. The main additional insight that is used in the correctness proof for this memory implementation is the following monotonicity lemma, where $\cup_m$ denotes map union:

$$(m_1 \cup_m m_2) \subseteq_m (m'_1 \cup_m m'_2)$$
$$\text{if } m_1 \subseteq_m m'_1,\ m_2 \subseteq_m m'_2, \text{ and } dom\ m_1 \cap dom\ m'_2 = \{\}.$$

We now extend this formalization towards an imperative implementation that stores the two rows as arrays. To this end, assume we are also given a function $idx\_of :: 'k_2 \rightarrow nat$ with

$mem\_update\ k\ v = (\text{let } i = idx\_of\ f\ k \text{ in}$

$\quad$ if $i < size$ then $(Array.upd\ i\ (Some\ v)\ mem \ggg (\lambda\ \_.\ return\ ()))$

$\quad$ else $return\ ()$

To verify this implementation, we wrap lookup, update, and move in

$$state\_of\ s\ =\ State\ (\lambda\ heap.\ the\ (execute\ s\ heap))$$

where $the\ (Some\ x) = x$, and prove that these correctly implement the interface for the previous implementation in the state monad. As the second step, one relates—via parametricity reasoning—this implementation with an implementation in the heap monad, where lookup, update and move are used without the $state\_of$ wrapper: we can prove $\Downarrow^h_= (state\_of\ m)\ m$ if m never fails and preserves the memory invariant.

3   Examples

This section presents five representative but quite different examples of dynamic programming. We have also applied the tool to further examples that are not explained here, for instance the optimal binary search tree problem [18] and the Viterbi algorithm [23]. For the first example, Bellman-Ford, we start with a recursive function, prove its correctness and refine it to an imperative memoized algorithm with the help of the automation described above. Because the refinement steps are automatic and the same for all examples, they are not shown for the other examples.

The examples below employ lists: $x \cdot xs$ is the list with head x and tail xs; $xs @ ys$ is the concatenation of the lists xs and ys; $xs \,!\, i$ is the ith element of xs; $[i..j]$ is the list of integers from i to j, and similarly for the set $\{i..j\}$; *slice xs i j* is the sublist of xs from index i (starting with 0) to (but excluding) index j.

For the verification of the Knapsack problem and the Bellman-Ford algorithm, we followed Kleinberg and Tardos [10]. In both cases, the crucial part of the correctness argument involves a recurrence of the form

$$OPT \ (Suc \ n) \ t_1 \ \ldots \ t_k = \Pi\{u_1, \ldots, u_m\}$$

where each of the u_i involve terms of the form $OPT \ n$ and $\Pi \in \{Min, Max\}$. We prove this equality by proving two inequalities ($\leq$, $\geq$). The easier direction is the one where we just need to show that the left-hand side covers all the solutions that are covered by the right-hand side. This direction is not explicitly covered in the proof by Kleinberg and Tardos. For the other direction, we first prove that the unique minimum or maximum exist and then analyze the solution that computes the minimum or maximum, directly following the same kind of argument as Kleinberg and Tardos.

3.1   Bellman-Ford Algorithm

The Bellman-Ford Algorithm solves the *single-destination shortest path problem* (and the single-source shortest path problem by reversing the edges): given nodes $1, \ldots, n$, a sink $t \in \{1, \ldots, n\}$, and edge weights $W :: nat \rightarrow nat \rightarrow int$, we have to compute for each source $v \in \{1, \ldots, n\}$ the minimum weight of any path from v to t. The main idea of the algorithm is to consider paths in the order of increasing *path length*. Thus we define $OPT \ i \ v$ as the weight of the shortest path leading from v to t, and using at most i edges:

$$OPT \ i \ v = Min \ (\{\text{if } t = v \text{ then } 0 \text{ else } \infty\} \cup$$
$$\{weight \ (v \cdot xs) \mid length \ xs + 1 \leq i \wedge set \ xs \subseteq \{0..n\}\}).$$

If $OPT \ (n+1) \ s = OPT \ n \ s$ for all $s \in \{1, \ldots, n\}$, then there is no cycle of negative weight (from which t can be reached), and $OPT \ n$ represents shortest path lengths. Otherwise, we know that there is a cycle of negative weight.

Following Kleinberg and Tardos, we prove

$$OPT \ (Suc \ i) \ v = min \ (OPT \ i \ v) \ (Min \ \{OPT \ i \ w + W \ v \ w \mid w. \ w \leq n\}),$$

yielding a recursive solution (replacing sets by lists):

$BF\ 0\ j$ $\quad = ($if $t = j$ then 0 else $\infty)$
$BF\ (Suc\ k)\ j = min\_list\ (BF\ k,\ j \cdot [W\ j\ i + BF\ k,\ i\ .\ i \leftarrow [0..n]]).$

Applying our tool for memoization, we get:

$BF_m'\ 0\ j$ $\quad =_m if_m\ \langle t = j \rangle\ \langle 0 \rangle\ \langle \infty \rangle$
$BF_m'\ (Suc\ k)\ j =_m \langle \lambda xs.\ \langle min\_list\ xs \rangle \rangle \bullet (\langle \lambda x.\ \langle \lambda xs.\ \langle x \cdot xs \rangle \rangle \rangle \bullet BF_m'\ k\ j \bullet$
$\quad (map_m \bullet \langle \lambda i.\ \langle \lambda x.\ \langle W\ j\ i\ +\ x \rangle \rangle \bullet BF_m'\ k\ i \rangle \bullet \langle [0..n] \rangle)).$

Using the technique described in Sect. 2.8, we fill the table in the order
$(0,0),(0,1),\ldots,(n,0),\ldots,(n,n)$. The pairwise memory implementation from
Sect. 2.9 is used to only store two rows corresponding to the first part of the
pair, which are in turn indexed by the second one. Together, this yields a program that can compute the length of the shortest path in $O(n)$ space. The final
correctness theorem for this implementation is (with explicit context parameters
n and W):

$BF\ n\ W\ t\ i\ j = fst\ (run\_state$
$\quad (iter\_BF\ n\ W\ t\ (i,\ j) \ggg (\lambda\ \_.\ BF_m'\ n\ W\ t\ i\ j))\ Mapping.empty).$

Isabelle can be instructed to use this equation when generating code for BF.
Thus the efficient implementation becomes completely transparent for the user.

Lastly, we can choose how to implement the parameter for the edge weights
W. A common graph representation are adjacency lists of type $(nat \times int)\ list\ list$
that contain for each node v an association list of pairs of a neighbouring node
and the corresponding edge weight. To obtain an efficient implementation, the
outer list can be realized with Isabelle's immutable arrays. They come with a
function $IArray$ that maps $'a\ list$ to an immutable array and with the infix $!!$
array subscript function. Thus we can transform a list into an immutable array
first and then run the Bellman-Ford algorithm:

$BF\_ia\ n\ W\ t\ i\ j = ($let $W' = graph\_of\ (IArray\ W)$ in $fst\ (run\_state$
$\quad\quad (iter\_BF\ n\ W'\ t\ (i,\ j) \ggg (\lambda\ \_.\ BF_m'\ n\ W'\ t\ i\ j))$
$\quad\quad Mapping.empty))$
$graph_of\ a\ i\ j\quad = $ case $find\ (\lambda p.\ fst\ p = j)\ (a\ !!\ i)$ of
$\quad\quad None \Rightarrow \infty\ |\ Some\ x \Rightarrow snd\ x.$

Note that the defining equation for BF_h' looks exactly the same as for BF_m'
but for different underlying constants for the heap monad. For imperative memoization, the final theorems for BF or BF_ia would just differ in that run_state
would be replaced by $execute$ and the initial memory would be replaced by a
correctly initialized empty heap memory.

3.2   Knapsack Problem

In the Knapsack Problem, we are given n items $1, \ldots, n$, a weight assignment
$w :: nat \to nat$, and a value assignment $v :: nat \to nat$. Given a Knapsack, which

can carry at most weight W, the task is to compute a selection of items that fits in the Knapsack and maximizes the total value. Thus we define:

$$OPT \; n \; W = Max \left\{ \sum_{i \in S} v \; i \; \middle| \; S \subseteq \{1..n\} \wedge \sum_{i \in S} w \; i \; \leq \; W \right\}.$$

Again following Kleinberg and Tardos, we prove:

> $OPT \; (Suc \; i) \; W = ($if $W < w \; (Suc \; i)$ then $OPT \; i \; W$
> else $max \; (v \; (Suc \; i) + OPT \; i \; (W - w \; (Suc \; i))) \; (OPT \; i \; W)).$

This directly yields the following recursive solution:

> $knapsack \; 0 \; W \; = \; 0$
> $knapsack \; (Suc \; i) \; W \; = \; ($if $W < w \; (Suc \; i)$ then $knapsack \; i \; W$
> else $max \; (knapsack \; i \; W) \; (v \; (Suc \; i) + knapsack \; i \; (W - w \; (Suc \; i))))).$

Like Bellman-Ford, the algorithm can be memoized using a bottom-up computation and a memory, which stores only the last two rows. However, the algorithm's running time and space consumption are still exponential in the input size, assuming a binary encoding of W.

3.3   A Counting Problem

A variant of Project Euler problem #114[2] was posed in the 2018 edition of the "VerifyThis" competition[3] [14]. We consider a row consisting of n tiles, which can be either red or black, and we impose the condition that red tiles only come in blocks of three consecutive tiles. We are asked to compute $count(n)$, the number of valid rows of size n. This is an example of counting problems that can be solved with memoization.

Besides the base cases $count(0) = count(1) = count(2) = 1$, and $count(3) = 2$, one gets the following recursion:

$$count(n) = count(n - 1) + 1 + \sum_{i=3}^{n-1} count(n - i - 1) \;\; \text{if} \;\; n > 3.$$

These equations directly yield a recursive functional solution, which can be memoized as described for the examples above. The reasoning to prove the main recursion, however, is different. We define

$$count(n) = card \; \{l \mid length \; l = n \wedge valid \; l\}$$

where $valid$ is an inductively defined predicate describing a well-defined row. The reasoning trick is to prove the following case analysis on the validity of a single row

> $valid \; l \longleftrightarrow l = [] \vee (l \; ! \; 0 = B \wedge valid \; (tl \; l)) \vee$
> $length \; l \geq 3 \wedge (\forall i < length \; l. \; l \; ! \; i = R) \vee \ldots.$

---

[2] https://projecteuler.net/problem=114.
[3] http://www.pm.inf.ethz.ch/research/verifythis.html.

that is then used to split the defining set of $count(n)$ into *disjoint* subsets that correspond to the individual terms on the right-hand side of the recursion.

3.4   The Cocke-Younger-Kasami Algorithm

Given a grammar in Chomsky normal form, the CYK algorithm computes the set of nonterminals that produce (yield) some input string. We model productions in Chomsky normal form as pairs (A, r) of a nonterminal A and a r.h.s. r that is either of the form $T\ a$, where a is a terminal (of type $'t$), or $NN\ B\ C$, where B and C are nonterminals (of type $'n$). Below, $P :: ('n, 't)\ prods$ is a list of productions. The *yield* of a nonterminal is defined inductively as a relation:

$$\frac{(A,\ T\ a) \in set\ P}{yield\ P\ A\ [a]} \qquad \frac{(A,\ NN\ B\ C) \in set\ P \qquad yield\ P\ B\ u \qquad yield\ P\ C\ v}{yield\ P\ A\ (u\ @\ v)}$$

A functional programmer will start out with an implementation $CYK ::$ $('n,\ 't)\ prods \to\ 't\ list \to\ 'n\ list$ of the CYK algorithm defined by recursion on lists and prove its correctness: $set\ (CYK\ P\ w) = \{N\ |\ yield\ P\ N\ w\}$. However, memoizing the list argument leads to an inefficient implementation. An efficient implementation can be obtained from a version of the CYK algorithm that indexes into the (constant) list and memoizes the index arguments. Our starting point is the following function CYK_ix where w is not of type $'a\ list$ but an indexing function of type $nat \to\ 't$. Isabelle supports list comprehension syntax:

```
CYK_ix :: ('n, 't) prods → (nat → 't) → nat → nat → 'n list
CYK_ix P w i 0 = []
CYK_ix P w i (Suc 0) = [A . (A, T a) ← P, a = w i]
CYK_ix P w i n =
  [A. k ← [1..n−1], B ← CYK_ix P w i k, C ← CYK_ix P w (i+k) (n−k),
     (A, NN B' C') ← P, B' = B, C' = C]
```

The correctness theorem (proved by induction) explains the meaning of the arguments i and n:

$$set\ (CYK\_ix\ P\ w\ i\ n) = \{N\ |\ yield\ P\ N\ (slice\ w\ i\ (i + n))\}$$

As for Bellman-Ford, we obtain an imperative memoized version $CYK\_ix'_m$ and a correctness theorem that relates it to CYK_ix and parameter w is realized by an immutable array.

3.5   Minimum Edit Distance

The minimum edit distance between two lists xs and ys of type $'a\ list$ is the minimum cost of converting xs to ys by means of a sequence of the edit operations copy, replace, insert and delete:

datatype $'a\ ed\ =\ Copy\ |\ Repl\ 'a\ |\ Ins\ 'a\ |\ Del$

The cost of *Copy* is 0, all other operations have cost 1. Function *edit* defines how an $'a$ *ed list* transform one $'a$ *list* into another:

$$edit \ (Copy \cdot es) \ (x \cdot xs) = x \cdot edit \ es \ xs$$
$$edit \ (Repl \ a \cdot es) \ (\_ \cdot xs) = a \cdot edit \ es \ xs$$
$$edit \ (Ins \ a \cdot es) \ xs = a \cdot edit \ es \ xs$$
$$edit \ (Del \cdot es) \ (\_ \cdot xs) = edit \ es \ xs$$
$$edit \ [] \ xs = xs$$

We have omitted the cases where the second list becomes empty before the first. This time we start from two functions defined by recursion on lists:

$$min\_ed :: \ 'a \ list \rightarrow \ 'a \ list \rightarrow nat$$
$$min\_eds :: \ 'a \ list \rightarrow \ 'a \ list \rightarrow \ 'a \ ed \ list$$

Function *min_ed* computes the minimum edit distance and *min_eds* :: $'a$ *list* → $'a$ list → $'a$ ed list computes a list of edits with minimum cost. We omit their definitions. The relationship between them is trivial to prove: $min\_ed \ xs \ ys = cost \ (min\_eds \ xs \ ys)$. Therefore the following easy correctness and minimality theorems about *min_eds* also imply correctness and minimality of *min_ed*:

$$edit \ (min\_eds \ xs \ ys) \ xs = ys \qquad cost \ (min\_eds \ xs \ (edit \ es \ xs)) \leq cost \ es$$

As for CYK, we define a function by recursion on indices

$$min\_ed\_ix :: (nat \rightarrow \ 'a) \rightarrow (nat \rightarrow \ 'a) \rightarrow nat \rightarrow nat \rightarrow nat \rightarrow nat \rightarrow nat$$
$min\_ed\_ix \ xs \ ys \ m \ n \ i \ j =$
(if $m \leq i$ then if $n \leq j$ then 0 else $n - j$
else if $n \leq j$ then $m - i$
  else min_list
    $[1 + min\_ed\_ix \ xs \ ys \ m \ n \ i \ (j + 1),$
    $1 + min\_ed\_ix \ xs \ ys \ m \ n \ (i + 1) \ j,$
    (if $xs \ i = ys \ j$ then 0 else 1) +
    $min\_ed\_ix \ xs \ ys \ m \ n \ (i + 1) \ (j + 1)])$

and prove that it correctly refines *min_ed*: $min\_ed\_ix \ xs \ ys \ m \ n \ i \ j = min\_ed$ (*slice xs i m*) (*slice ys j n*). Although one can prove correctness of this indexed version directly, the route via the recursive functions on lists is simpler.

As before we obtain an imperative memoized version $min\_ed\_ix'_m$ and a correctness theorem that relates it to *min_ed_ix*.

4 Future Work

We plan to expand our work in two major directions in the future. Firstly, we want to use our memoization tool to allow for other monads than the state and the heap monad. The main task here is to find monads that play well with our style of parametric reasoning. In simple monads such as reader or writer monads, the monadic operations do not interfere with the original computation, so they fit well in this framework. For the state monad, we can give correspondence proofs because we thread an invariant—values stored in the state are consistent with the memoized functions—through our relations. For other monads, such as an IO monad, it is less clear what these invariants would look like. Moreover, our tool currently only adds monadic effects at recursive invocations of a function—for other monads one would certainly want to insert these in other places, too. This added flexibility would also allow us to save recursive function invocations in memoized functions: instead of performing the memoization at the equality sign, we could wrap memoization around each recursive invocation of the function. Furthermore, this would allow one to memoize repeated applications of non-recursive functions in the context of an enclosing function.

Our second goal is to integrate the memoization process with the Imperative Refinement Framework [13]. It allows stepwise refinement of functional programs and to replace functional by imperative data structures in a final refinement step. The main obstacle here is that the framework already comes with its own nondeterminism monad to facilitate refinement reasoning. This means that high-level programs are already stated in terms of this monad. We have started work to allow automated monadification of these programs by adding the state via a state transformer monad.

Acknowledgments. Tobias Nipkow is supported by DFG Koselleck grant NI 491/16-1. The authors would like to thank Andreas Lochbihler for a fruitful discussion on monadification.

References

1. Berghofer, S., Nipkow, T.: Executing higher order logic. In: Callaghan, P., Luo, Z., McKinna, J., Pollack, R., Pollack, R. (eds.) TYPES 2000. LNCS, vol. 2277, pp. 24–40. Springer, Heidelberg (2002). https://doi.org/10.1007/3-540-45842-5_2
2. Bortin, M.: A formalisation of the Cocke-Younger-Kasami algorithm. Archive of Formal Proofs (2016). http://isa-afp.org/entries/CYK.html, Formal proof development
3. Braibant, T., Jourdan, J., Monniaux, D.: Implementing and reasoning about hash-consed data structures in Coq. J. Autom. Reasoning **53**(3), 271–304 (2014). https://doi.org/10.1007/s10817-014-9306-0
4. Bulwahn, L., Krauss, A., Haftmann, F., Erkök, L., Matthews, J.: Imperative functional programming with Isabelle/HOL. In: Mohamed, O.A., Muñoz, C., Tahar, S. (eds.) TPHOLs 2008. LNCS, vol. 5170, pp. 134–149. Springer, Heidelberg (2008). https://doi.org/10.1007/978-3-540-71067-7_14
5. Erwig, M., Ren, D.: Monadification of functional programs. Sci. Comput. Program. **52**(1), 101–129 (2004). http://www.sciencedirect.com/science/article/pii/S0167642304000486, special Issue on Program Transformation
6. Haftmann, F., Krauss, A., Kunčar, O., Nipkow, T.: Data refinement in Isabelle/HOL. In: Blazy, S., Paulin-Mohring, C., Pichardie, D. (eds.) ITP 2013. LNCS, vol. 7998, pp. 100–115. Springer, Heidelberg (2013). https://doi.org/10.1007/978-3-642-39634-2_10

7. Haftmann, F., Nipkow, T.: Code generation via higher-order rewrite systems. In: Blume, M., Kobayashi, N., Vidal, G. (eds.) FLOPS 2010. LNCS, vol. 6009, pp. 103–117. Springer, Heidelberg (2010). https://doi.org/10.1007/978-3-642-12251-4_9

8. Hatcliff, J., Danvy, O.: A generic account of continuation-passing styles. In: Conf. Record of POPL'94: 21st ACM SIGPLAN-SIGACT Symposium on Principles of Programming Languages. pp. 458–471 (1994), http://doi.acm.org/10.1145/174675.178053

9. Itzhaky, S., Singh, R., Solar-Lezama, A., Yessenov, K., Lu, Y., Leiserson, C., Chowdhury, R.: Deriving divide-and-conquer dynamic programming algorithms using solver-aided transformations. In: Proceedings of the 2016 ACM SIGPLAN International Conference on Object-Oriented Programming, Systems, Languages, and Applications, OOPSLA 2016, pp. 145–164. ACM (2016). http://doi.acm.org/10.1145/2983990.2983993

10. Kleinberg, J.M., Tardos, É.: Algorithm Design. Addison-Wesley (2006)

11. Krauss, A.: Automating recursive definitions and termination proofs in higher-order logic. Ph.D. thesis, Technical University Munich (2009). http://mediatum2.ub.tum.de/doc/681651/document.pdf

12. Kuncar, O.: Types, abstraction and parametric polymorphism in higher-order logic. Ph.D. thesis, Technical University Munich, Germany (2016). http://nbn-resolving.de/urn:nbn:de:bvb:91-diss-20160408-1285267-1-5

13. Lammich, P.: Refinement to Imperative/HOL. In: Urban, C., Zhang, X. (eds.) ITP 2015. LNCS, vol. 9236, pp. 253–269. Springer, Cham (2015). https://doi.org/10.1007/978-3-319-22102-1_17

14. Lammich, P., Wimmer, S.: VerifyThis 2018 – Polished Isabelle solutions. Archive of Formal Proofs, April 2018. http://isa-afp.org/entries/VerifyThis2018.html, Formal proof development

15. Michie, D.: Memo functions and machine learning. Nature **218**, 19–22 (1968)

16. Nipkow, T., Klein, G.: Concrete Semantics with Isabelle/HOL. Springer, Cham (2014). https://doi.org/10.1007/978-3-319-10542-0. http://concrete-semantics.org

17. Nipkow, T., Wenzel, M., Paulson, L.C. (eds.): Isabelle/HOL. LNCS, vol. 2283. Springer, Heidelberg (2002). https://doi.org/10.1007/3-540-45949-9

18. Nipkow, T., Somogyi, D.: Optimal binary search tree. Archive of Formal Proofs (2018). http://isa-afp.org/entries/Optimal_BST.html, Formal proof development

19. Reynolds, J.C.: Types, abstraction and parametric polymorphism. In: IFIP Congress, pp. 513–523 (1983)

20. Slind, K.: Reasoning about terminating functional programs. Ph.D. thesis, Technical University Munich, Germany (1999). https://mediatum.ub.tum.de/node?id=601660

21. Verma, K.N., Goubault-Larrecq, J., Prasad, S., Arun-Kumar, S.: Reflecting BDDs in Coq. In: Jifeng, H., Sato, M. (eds.) ASIAN 2000. LNCS, vol. 1961, pp. 162–181. Springer, Heidelberg (2000). https://doi.org/10.1007/3-540-44464-5_13

22. Wadler, P.: Theorems for free! In: Proceedings of the Fourth International Conference on Functional Programming Languages and Computer Architecture, FPCA 1989, pp. 347–359. ACM (1989). http://doi.acm.org.eaccess.ub.tum.de/10.1145/99370.99404

23. Wimmer, S.: Hidden Markov models. Archive of Formal Proofs (2018). http://isa-afp.org/entries/Hidden_Markov_Models.html, Formal proof development

24. Wimmer, S., Hu, S., Nipkow, T.: Monadification, memoization and dynamic programming. Archive of Formal Proofs (2018). http://isa-afp.org/entries/Monad_Memo_DP.html, Formal proof development

MDP + TA = PTA: Probabilistic Timed Automata, Formalized (Short Paper)

Simon Wimmer[1]([⊠])([iD]) and Johannes Hölzl[2]([iD])

[1] TU München, Munich, Germany
wimmers@in.tum.de
[2] VU Amsterdam, Amsterdam, Netherlands
jhl890@vu.nl

Abstract. We present a formalization of probabilistic timed automata (PTA) in which we try to follow the formula "MDP + TA = PTA" as far as possible: our work starts from existing formalizations of Markov decision processes (MDP) and timed automata (TA) and combines them modularly. We prove the fundamental result for probabilistic timed automata: the region construction that is known from timed automata carries over to the probabilistic setting. In particular, this allows us to prove that minimum and maximum reachability probabilities can be computed via a reduction to MDP model checking, including the case where one wants to disregard unrealizable behavior.

1 Introduction

Timed automata (TA) [1] are a widely used formalism for modeling nondeterministic real-time systems. Markov decision processes (MDPs) with discrete time are popular for modeling probabilistic systems with nondeterminism. Probabilistic timed automata (PTA) fuse the concepts of TA and MDPs and allow probabilistic modeling of nondeterministic real-time systems. PRISM [3] implements model checking functionality for MDPs and PTA and has successfully been applied to a number of case studies [6].

We have previously formalized MDPs [2] and TA [8] in Isabelle/HOL. This paper presents an Isabelle/HOL formalization of PTA, which follows the formula "MDP + TA = PTA" as far as possible by combining our existing formalizations modularly. We prove the fundamental result for PTA: the region construction that is known from TA carries over to the probabilistic setting. In particular, we prove that minimum and maximum reachability probabilities (with respect to possbile resolutions of nondeterminism) can be computed via a reduction to MDP model checking, including the case where one wants to disregard unrealizable behavior. This work is a necessary first step towards our long-term goal of certifying the computation results of PRISM's backward reachability algorithm [4] for reducing PTA to MDP model checking. The formalization can be found in the Archive of Formal Proofs [9].

© Springer International Publishing AG, part of Springer Nature 2018
J. Avigad and A. Mahboubi (Eds.): ITP 2018, LNCS 10895, pp. 597–603, 2018.
https://doi.org/10.1007/978-3-319-94821-8_35

2   Preliminaries

Markov Chains. A probability mass function (PMF, or discrete distribution) $\mu :: \sigma\,pmf$ is a function $\sigma \Rightarrow \mathbb{R}_{\geq 0}$ with countable support $\{x \mid \mu\,x \neq 0\}$ whose range sums to 1. Any PMF forms a monad, thus we have $(map_{pmf}\,f\,\mu)\,y = \mu\,\{x \mid f\,x = y\}$ and $(ret_{pmf}\,x)\,x = 1$. A *Markov chain (MC)* is represented by the transition system $K :: \sigma \Rightarrow \sigma\,pmf$ (its kernel, which is commonly represented by a transition matrix $\mathbb{R}^{|\sigma| \times |\sigma|}$), mapping each state to a distribution of next states. The trace space $T_K\,s$ is the probability measure with the property $T_K\,s\,(x_0 \cdots x_n) = K\,s\,x_0 * \cdots * K\,x_{n-1}\,x_n$ (where $(x_0 \cdots x_n)$ is the set of state traces starting with $x_0, \cdots, x_n$). A *probabilistic coupling* with respect to a relation R exists between two PMFs μ and μ' (written $rel_{pmf}\,R\,\mu\,\mu'$) if there exists a distribution ν on the product type, such that the support of ν is a subset of R and the marginal distributions of ν are $\mu = map_{pmf}\,\pi_1\,\nu$ and $\mu' = map_{pmf}\,\pi_2\,\nu$. Probabilistic couplings allow us to relate two Markov chains.

Markov Decision Processes. MDPs are automata allowing probabilistic and non-deterministic choice. An MDP is represented by the transition system $K :: \sigma \Rightarrow \sigma\,pmf\,set$, where σ is the type of states, and the probability distributions over the next states of type $\sigma\,pmf$ are called *actions*. Each MDP gives rise to a set of MCs, each showing one possible behaviour. We introduce, coinductively, *configurations* $\sigma\,cfg$, where each $c :: \sigma\,cfg$ consists of a state σ, an action $\sigma\,pmf$, and a continuation $\sigma \Rightarrow \sigma\,cfg$. The configurations give rise to a Markov chain $K_c :: \sigma\,cfg \Rightarrow \sigma\,cfg\,pmf$, by mapping the continuations over the actions. Each $c :: \sigma\,cfg$ whose actions are closed under K and which is in state s induces a MC showing a possible behavior of the MDP K starting in s. To simplify the theory, we assume that $K\,x \neq \emptyset$. See [2] for details.

Timed Automata. Compared to standard finite automata, TA introduce a notion of clocks. Clocks are indexed by natural numbers and do not have any structure. A *clock valuation* u is a function of type $\mathbb{N} \Rightarrow \mathbb{R}$. Locations and transitions are guarded by *clock constraints*, which have to be fulfilled to stay in a location or to take a transition. Clock constraints are conjunctions of constraints of the form $c \sim d$ for a clock c, an integer d, and $\sim\,\in\,\{<, \leq, =, \geq, >\}$. We write $u \vdash cc$ if the clock constraint cc holds for the clock valuation u. We define a timed automaton A as a pair $(\mathcal{T}, \mathcal{I})$ where $\mathcal{I}$ is an assignment of clock constraints to locations (also named invariants) and $\mathcal{T}$ is a set of transitions written as $A \vdash l \longrightarrow^{g,r} l'$ where l and l' are the start and successor location, g is the guard of the transition, and r is a set of clocks that will be reset to zero when the transition is taken. States of TA are pairs of a location and a clock valuation. The operational semantics define two kinds of steps:

Delay: $(l, u) \rightarrow (l, u \oplus d)$ if $d \geq 0$ and $u \oplus d \vdash \mathcal{I}\,l$;
Action: $(l, u) \rightarrow (l', [r \rightarrow 0]u)$ if $A \vdash l \longrightarrow^{g,r} l'$, $u \vdash g$, and $[r \rightarrow 0]u \vdash \mathcal{I}\,l'$;

where $(u \oplus d)\,c = u\,c + d$ and $([r \rightarrow 0]u)\,c = if\,c \in r\,then\,0\,else\,u\,c$.

Regions. The initial decidability result [1] partitioned the set of clock valuations into a quotient of sets of clock valuations, the so-called regions, and showed that these yield a sound and complete abstraction[1]. Our formalization [8] proves this fundamental result and decidability of reachability properties for ordinary TA.

3 Probabilistic Timed Automata

PTA fuse the concepts of TA and MDPs: discrete transitions are replaced by probability distributions over pairs of a set of clocks to be reset and a successor location. An example of a PTA is depicted in the left part of Fig. 1.

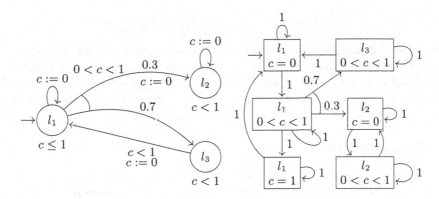

Fig. 1. Example of a PTA with one clock and its region graph

Definition. Consequently, the syntactic definition of PTA is very similar to TA. The only difference is that now transitions are of the form $A \vdash l \longrightarrow^g \mu$ for μ of type $(\mathbb{N} \, set \times \sigma) \, pmf$ (the clocks to reset and σ for the type of locations).

Typical presentations define the semantics of PTA based on the notion of so-called *probabilistic timed structures*, which are just a special type of MDP. We omit this detour and directly formalize PTA in terms of MDPs. Consequently, to formalize the semantics of a PTA, we define its kernel K as the smallest set that is compatible with

$$\frac{(l, u) \in S \qquad t \geq 0 \qquad u \oplus t \vdash \mathcal{I} \, l}{ret_{pmf} \, (l, u \oplus t) \in K \, (l, u)} \text{ DELAY}$$

$$\frac{(l, u) \in S \qquad A \vdash l \longrightarrow^g \mu \qquad u \vdash g}{map_{pmf} \, (\lambda(r, l). \, (l, [r \to 0]u)) \, \mu \in K \, (l, u)} \text{ ACTION}$$

where S is the set of valid states. A state (l, u) is valid if l belongs to A and if $u \vdash \mathcal{I} \, l$. For technical reasons there is a third rule to add self loops for non-valid

---

[1] We use the same notions as in [8]. Soundness: for every abstract run, there is a concrete instantiation. Completeness: every concrete run can be abstracted.

states. These do not change the semantics as they are not reachable from valid
states. The MDP K is uncountably infinite as S is generally infinite.

Region Graph. We want to reduce the computation of reachability probabilities
for A to a computation on a finite MDP. Analogous to TA, this reduction can
be obtained through the region quotient. More precisely, we will partition S into
a finite set of states $\mathcal{S}$ of the form (l, R), for l a location of A, and R a region
of A such that $\forall u \in R.\, u \vdash \mathcal{I}\, l$. With this notion, the finite MDP, coined *region
graph* in [5], is defined through its kernel $\mathcal{K}$:

$$\frac{(l, R) \in \mathcal{S} \qquad R' \in Succ\ R \qquad \forall u \in R'.\, u \vdash \mathcal{I}\, l}{ret_{pmf}\, (l, R') \in \mathcal{K}\ (l, R)} \ \text{DELAY}_R$$

$$\frac{(l, R) \in \mathcal{S} \qquad A \vdash l \longrightarrow^g \mu \qquad \forall u \in R.\, u \vdash g}{map_{pmf}\, (\lambda(r, l).\, (l, \{[r \to 0]u \mid u \in R\})\, \mu \in \mathcal{K}\ (l, R)} \ \text{ACTION}_R$$

Here $Succ\ R$ denotes the set of regions that can be reached from R by delaying
for an arbitrary amount of time. Again, for technical reasons there is a third rule
to add self loops for non-valid states. The *maximum* probability (under all valid
initial configurations) to reach the state in the upper right of the region graph
depicted in Fig. 1 is 0.7, while the *minimum* probability is 0.

4   Bisimulation

We relate the infinite MDP that defines the PTA with the finite region graph in
a way that directly allows us to prove correctness of the reduction for maximum
and minimum reachability probabilities in one go. Concretely, our agenda is to
first define abstraction and representation functions that map between configurations of the infinite MDPs and the finite region graph, and vice versa. We
then prove a more general bisimulation theorem on MDPs which states that the
path measure assigned to related paths is the same for related configurations.

Representation and Abstraction. We will use the overloaded notations α and
rep to denote the abstraction and representation functions for states, actions,
and configurations. The main difficulty of our formalization effort was to define
these such that one obtains the desired properties. What are these properties?
Chiefly, for a valid configuration c, the probability distributions of the successors
of c and $\alpha\, c$ should expose a *probabilistic coupling* w.r.t. the relation $\lambda c\, a.\, \alpha\, c = a$.
Moreover, the abstraction of a representative should yield the original object:
$\alpha\, (rep\ x) = x$. Finally, validity should be preserved, i.e. $\alpha\, (l, u) \in \mathcal{S} \leftrightarrow (l, u) \in S$.

The elementary abstraction functions are easy to define: $\alpha\, (l, u) = (l, [u]_{\mathcal{R}})$
for $[u]_{\mathcal{R}}$ the unique region with $u \in [u]_{\mathcal{R}}$, and $\alpha\, t = map_{pmf}\, \alpha\, t$ for an action t.
For a configuration c, $\alpha\, c$ is defined co-recursively in terms of c: the concrete
configuration c is maintained as the internal state of $\alpha\, c$ and states and actions
are simply mapped with α; the internal successor configuration however is determined by the continuation of c for the *unique* successor state s of c such that $\alpha\, s$
is the successor state of $\alpha\, c$. The definition of *rep* is more involved and omitted
for brevity.

Bisimulation Theorem. At the core of our argument lies the following bisimulation theorem on Markov chains:

$$T_K \, x \, A = T_L \, y \, B \quad \text{if } R\,x\,y \text{ and } \forall \omega\,\omega'. \, rel_{stream} \, R\,\omega\,\omega' \longrightarrow (\omega \in \Lambda \leftrightarrow \omega' \in B)$$
$$\text{and } \forall x\,y. \, R\,x\,y \longrightarrow rel_{pmf} \, R\,(K\,x)\,(L\,y)$$

where $T_{\{K,L\}}$ denotes the trace space induced by Markov chains K and L, respectively; x and y are states of K and L; $rel_{stream} \, R$ compares two traces pointwise by R; and A and B are sets of infinite traces of K and L. Finally, R has to be of the form $R\,s\,t = (s \in S \wedge f\,s = t)$ for some S and f.

For a configuration c with state s, we can instantiate this theorem by taking $K = K_c$, $L = \mathcal{K}_c$, $x = s$, $y = \alpha\,s$, $f = \alpha$, and S as the set of valid configurations of the PTA. The coupling property of K and L follows because

$$\mathcal{K}_c\,(\alpha\,c) = map_{pmf} \, \alpha \, (K_c\,c) \, .$$

For this instantiation, $rel_{stream} \, R\,x\,y$ essentially means that y is the pointwise abstraction of x. We consider reachability properties on state traces of the form $\varphi\,U\,\psi$ (where φ and ψ can be a mixture of predicates on location and clock), so

$$A = \{\omega \mid \varphi\,U\,\psi\,(smap, \, \omega)\} \text{ and } B = \{\omega \mid (\varphi \circ rep)\,U\,(\psi \circ rep)\,(smap\,\omega)\}$$

where $smap \, \omega$ maps traces of configurations to traces of MDP states. Consequently, the premise on A and B is easily satisfied if

$$\forall s\,t. \, \alpha\,s = \alpha\,t \longrightarrow \varphi\,s \leftrightarrow \varphi\,t \wedge \psi\,s \leftrightarrow \psi\,t \, ,$$

which matches exactly the property that is delivered by the region construction.

5   Taking Zenoness into Account

So far, we have considered bisimulation properties between trace spaces of pairs of related configurations. Minimum and maximum reachability probabilities, however, are considered in relation to a set of configurations C. To compute these probabilities, one can consider the set of configurations C_α on the finite MDP such that $\forall c \in C. \, \alpha\,c \in C_\alpha$ and $\forall c \in C_\alpha. \, rep\,c \in C$. If C is the set of valid configurations, for instance, then C_α is easily proved to be the set of valid configurations of the region graph.

Often one wants to restrict C such that unrealizable behaviors are excluded: a configuration should not be able to keep time from passing beyond a fixed deadline. A configuration is *zeno* if it admits such behaviours. In the example, a zeno configuration could continuously take the loop transition on l_1 without letting any time pass. In [5] a *computable* description of C_α is given for the case that C is restricted to configurations that only yield non-zeno behaviors with probability 1.

The critical component of our proof for the correctness of C_α (in the sense outlined above) is that *rep* chooses the successor states always such that at least half of the amount of time that could possibly elapse does elapse.

Interestingly, the proof of $\forall c \in C_\alpha.\, rep\, c \in C$ was much harder to formalize than the other direction, although roles seem to flipped in the argument of [5]. For the harder direction, we illustrate the structure of our proof on the part that is concerned with the single region R_∞ in which all clocks have elapsed beyond the *maximal clock constant of the automaton* (the region $c > 1$ in the example): any run on the region graph that stays in R_∞ forever is classified as non-zeno.

Our argument establishes that for a transition $(l, R_\infty) \to (l', R_\infty)$ of the region graph, the representing transition $(l, u) \to (l', u')$ will incur a time delay of 0.5 if $u \neq u'$. An informal argument can get away by claming that the transition can always be chosen such that the latter condition is satisfied. Unfortunately, this is not immediately true for the semantics given above as the abstract transition could always be a reset transition and thus time would never be allowed to elapse. A possible remedy is to fuse delay and action transitions into a single step. We rather want to keep them separate and instead employ a probabilistic argument: assuming that a transition with $l = l'$ occurs infinitely often, with probability 1 a step with $u \neq u'$ has to occur infinitely often.

6   Conclusion

Discussion. Our bisimulation argument neatly separates discrete, TA-related reasoning from probabilistic, MDP-related reasoning. In fact, most of the proofs take place on the discrete side, because none of the arguments to satisfy the bisimulation theorem are predominantly of probabilistic nature. As seen above, only the reasoning on zenoness needs to break with this style.

We found it crucial to carry out each argument on the right level of abstraction. There are three main levels to consider here (from low to high): Markov chains, MDPs and configuration traces, and states and state traces of the PTA. The theorem is usually stated on the highest level possible, and often we can move easily from a higher to a lower level by applying a number of rewrite rules. For the divergence argument, we even introduce another level of abstraction: since we are only concerned with time, the location part can be dropped, and thus we consider traces of clock valuations. The probabilistic argument described in the last section manifests a rare case where one needs to put in some upfront work on a lower level to hold the argument on the higher level together.

It is not yet clear to us whether it is necessary or advantageous to work with *rep*. In the current formalization it still plays an important rule by providing the diverging concrete witness for a diverging configuration of the region graph. The bisimulation argument in Sect. 4, however, can be made relying only on α.

Lastly, our simple definition of the PTA semantics and of the region graph shows that derived concepts can be surprisingly easy to define —even compared to an informal definition— if the necessary foundations have already been laid.

Related Work. We are not aware of any previous proof-assistant formalizations of PTA. There is, however, another formalization of TA and the region construction using PVS [11]. A formalization in Coq [7] is aimed at modeling a subclass of TA and proving properties of concrete automata.

Future Work. We conjecture that many further results for PTA can be formalized by following the formula "MDP + TA = PTA" in the style that we outlined above. In particular, most of the more practical *zone* based (as opposed to region based) exploration methods for the reduction to a finite MDP should lie within the scope of this technique. The backward reachability algorithm of PRISM [4] is an instance. This also means that verified or certified model checkers for PTA can be devised from a modular combination of verified tools for MDPs and TA. Work in this direction already exists for the latter [10] but not the former formalism.

Acknowledgments. We want to thank David Parker and Gethin Norman for clarifying our understanding of PTA model checking w.r.t. divergence. This project has received funding from the European Research Council (ERC) under the European Union's Horizon 2020 research and innovation program (grant agreement No 713999 - Matryoshka).

References

1. Alur, R., Dill, D.L.: A theory of timed automata. Th. Comp. Sci. **126**(2), 183–235 (1994). https://doi.org/10.1016/0304-3975(94)90010-8
2. Hölzl, J.: Markov chains and Markov decision processes in Isabelle/HOL. J. Autom. Reasoning **59**(3), 345–387 (2017). https://doi.org/10.1007/s10817-016-9401-5
3. Kwiatkowska, M., Norman, G., Parker, D.: PRISM 4.0: verification of probabilistic real-time systems. In: Gopalakrishnan, G., Qadeer, S. (eds.) CAV 2011. LNCS, vol. 6806, pp. 585–591. Springer, Heidelberg (2011). https://doi.org/10.1007/978-3-642-22110-1_47
4. Kwiatkowska, M., Norman, G., Sproston, J., Wang, F.: Symbolic model checking for probabilistic timed automata. Inf. Comput. **205**(7), 1027–1077 (2007)
5. Kwiatkowska, M.Z., Norman, G., Segala, R., Sproston, J.: Automatic verification of real-time systems with discrete probability distributions. Th. Comp. Sci. **282**(1)
6. Norman, G., Parker, D., Sproston, J.: Model checking for probabilistic timed automata. Formal Methods Syst. Des. **43**(2), 164–190 (2013)
7. Paulin-Mohring, C.: Modelisation of timed automata in Coq. In: Kobayashi, N., Pierce, B.C. (eds.) TACS 2001. LNCS, vol. 2215, pp. 298–315. Springer, Heidelberg (2001). https://doi.org/10.1007/3-540-45500-0_15
8. Wimmer, S.: Formalized timed automata. In: Blanchette, J.C., Merz, S. (eds.) ITP 2016. LNCS, vol. 9807, pp. 425–440. Springer, Cham (2016). https://doi.org/10.1007/978-3-319-43144-4_26
9. Wimmer, S., Hölzl, J.: Probabilistic timed automata. Archive of Formal Proofs (2018). Formal proof development. http://isa-afp.org/entries/Probabilistic_Timed_Automata.html
10. Wimmer, S., Lammich, P.: Verified model checking of timed automata. In: Beyer, D., Huisman, M. (eds.) TACAS 2018. LNCS, vol. 10805, pp. 61–78. Springer, Cham (2018). https://doi.org/10.1007/978-3-319-89960-2_4
11. Xu, Q., Miao, H.: Formal verification framework for safety of real-time system based on timed automata model in PVS. In: Proceedings of IASTED 2006, pp. 107–112 (2006)

Formalization of a Polymorphic Subtyping Algorithm

Jinxu Zhao[1(✉)], Bruno C. d. S. Oliveira[1], and Tom Schrijvers[2]

[1] The University of Hong Kong, Pokfulam, Hong Kong
{jxzhao,bruno}@cs.hku.hk
[2] KU Leuven, Leuven, Belgium
tom.schrijvers@cs.kuleuven.be

Abstract. Modern functional programming languages such as Haskell support sophisticated forms of type-inference, even in the presence of *higher-order polymorphism*. Central to such advanced forms of type-inference is an algorithm for polymorphic subtyping. This paper formalizes an algorithmic specification for polymorphic subtyping in the Abella theorem prover. The algorithmic specification is shown to be *decidable*, and *sound* and *complete* with respect to Odersky and Läufer's well-known declarative formulation of polymorphic subtyping.

While the meta-theoretical results are not new, as far as we know our work is the first to mechanically formalize them. Moreover, our algorithm differs from those currently in the literature by using a novel approach based on *worklist judgements*. Worklist judgements simplify the propagation of information required by the unification process during subtyping. Furthermore they enable a simple formulation of the meta-theoretical properties, which can be easily encoded in theorem provers.

1 Introduction

Most statically typed functional languages support a form of *(implicit) parametric polymorphism* [28]. Traditionally, functional languages have employed variants of the Hindley-Milner [5,14,23] type system, which supports full type-inference without any type annotations. However the Hindley-Milner type system only supports *first-order polymorphism*, where all universal quantifiers only occur at the top-level of a type. Modern functional programming languages such as Haskell go beyond Hindley-Milner and support *higher-order polymorphism*. With higher-order polymorphism there is no restriction on where universal quantifiers can occur. This enables more code reuse and more expressions to type-check, and has numerous applications [12,15,17,18].

Unfortunately, with higher-order polymorphism full type-inference becomes undecidable [35]. To recover decidability some type annotations on polymorphic arguments are necessary. A canonical example that requires higher-order polymorphism in Haskell is:

```
hpoly = (\f :: forall a. a -> a) -> (f 1, f 'c')
```

© Springer International Publishing AG, part of Springer Nature 2018
J. Avigad and A. Mahboubi (Eds.): ITP 2018, LNCS 10895, pp. 604–622, 2018.
https://doi.org/10.1007/978-3-319-94821-8_36

The function hpoly cannot be type-checked in Hindley-Milner. The type of hpoly is (forall a. a -> a) -> (Int, Char). The single universal quantifier does not appear at the top-level. Instead it is used to quantify a type variable a used in the first argument of the function. Notably hpoly requires a type annotation for the first argument (forall a. a -> a). Despite these additional annotations, the type-inference algorithm employed by GHC Haskell [16] preserves many of the desirable properties of Hindley-Milner. Like in Hindley-Milner type instantiation is *implicit*. That is, calling a polymorphic function never requires the programmer to provide the instantiations of the type parameters.

Central to type-inference with *higher-order polymorphism* is an algorithm for polymorphic subtyping. This algorithm allows us to check whether one type is more general than another, which is essential to detect valid instantiations of a polymorphic type. For example, the type forall a. a -> a is more general than Int -> Int. A simple declarative specification for polymorphic subtyping was proposed by Odersky and Läufer [26]. Since then several algorithms have been proposed that implement it. Most notably, the algorithm proposed by Peyton Jones et al. [16] forms the basis for the implementation of type inference in the GHC compiler. Dunfield and Krishnaswami [9] provided a very elegant formalization of another sound and complete algorithm, which has also inspired implementations of type-inference in some polymorphic programming languages (such as PureScript [30] or DDC [6]).

Unfortunately, while many aspects of programming languages and type systems have been mechanically formalized in theorem provers, there is little work on formalizing algorithms related to type-inference. The main exceptions to the rule are mechanical formalizations of algorithm $\mathcal{W}$ and other aspects of traditional Hindler-Milner type-inference [7,8,11,24,33]. However, as far as we know, there is no mechanisation of algorithms used by modern functional languages like Haskell, and polymorphic subtyping included is no exception. This is a shame because recently there has been a lot of effort in promoting the use of theorem provers to check the meta-theory of programming languages, e.g., through well-known examples like the POPLMARK challenge [3] and the CompCert project [21]. Mechanical formalizations are especially valuable for proving the correctness of the semantics and type systems of programming languages. Type-inference algorithms are arguably among the most non-trivial aspects of the implementations of programming languages. In particular the information discovery process required by many algorithms (through unification-like or constraint-based approaches), is quite subtle and tricky to get right. Moreover, extending type-inference algorithms with new programming language features is often quite delicate. Studying the meta-theory for such extensions would be greatly aided by the existence of a mechanical formalization of the base language, which could then be extended by the language designer.

Handling variable binding is particularly challenging in type inference, because the algorithms typically do not rely simply on local environments, but instead propagate information across judgements. Yet, there is little work on how to deal with these complex forms of binding in theorem provers. We believe

Type variables	a, b	
Types	$A, B, C ::=$	$1 \mid a \mid \forall a.A \mid A \to B$
Monotypes	$\tau \quad ::=$	$1 \mid a \mid \tau_1 \to \tau_2$
Contexts	$\Psi \quad ::=$	$\cdot \mid \Psi, a$

Fig. 1. Syntax of declarative system

that this is the primary reason why theorem provers have still not been widely adopted for formalizing type-inference algorithms.

This paper advances the state-of-the-art by formalizing an algorithm for polymorphic subtyping in the Abella theorem prover. We hope that this work encourages other researchers to use theorem provers for formalizing type-inference algorithms. In particular, we show that the problem we have identified above can be overcome by means of *worklist judgments*. These are a form of judgement that turns the complicated global propagation of unifications into a simple local substitution. Moreover, we exploit several ideas in the recent inductive formulation of a type-inference algorithm by Dunfield and Krishnaswami [9], which turn out to be useful for mechanisation in a theorem prover.

Building on these ideas we develop a complete formalization of polymorphic subtyping in the Abella theorem prover. Moreover, we show that the algorithm is *sound, complete* and *decidable* with respect to the well-known declarative formulation of polymorphic subtyping by Odersky and Läufer. While these meta-theoretical results are not new, as far as we know our work is the first to mechanically formalize them.

In summary the contributions of this paper are:

- **A mechanical formalization of a polymorphic subtyping algorithm.** We show that the algorithm is *sound, complete* and *decidable* in the Abella theorem prover, and make the Abella formalization available online[1].
- **Information propagation using worklist judgements:** we employ worklists judgements in our algorithmic specification of polymorphic subtyping to propagate information across judgements.

2   Overview: Polymorphic Subtyping

This section introduces Odersky and Läufer declarative subtyping rules, and discusses the challenges in formalizing a corresponding algorithmic version. Then the key ideas of our approach that address those challenges are introduced.

2.1   Declarative Polymorphic Subtyping

In implicitly polymorphic type systems, the subtyping relation compares the degree of polymorphism of types. In short, if a polymorphic type A can always

---

[1] https://github.com/JimmyZJX/Abella-subtyping-algorithm.

$$\boxed{\Psi \vdash A}$$

$$\frac{}{\Psi \vdash 1} \; \text{wf}_d\text{unit} \qquad \frac{a \in \Psi}{\Psi \vdash a} \; \text{wf}_d\text{var} \qquad \frac{\Psi \vdash A \quad \Psi \vdash B}{\Psi \vdash A \to B} \; \text{wf}_d\to \qquad \frac{\Psi, a \vdash A}{\Psi \vdash \forall a.A} \; \text{wf}_d\forall$$

$$\boxed{\Psi \vdash A \leq B}$$

$$\frac{a \in \Psi}{\Psi \vdash a \leq a} \; \leq\text{Var} \qquad \frac{}{\Psi \vdash 1 \leq 1} \; \leq\text{Unit} \qquad \frac{\Psi \vdash B_1 \leq A_1 \quad \Psi \vdash A_2 \leq B_2}{\Psi \vdash A_1 \to A_2 \leq B_1 \to B_2} \; \leq\to$$

$$\frac{\Psi \vdash \tau \quad \Psi \vdash [\tau/a]A \leq B}{\Psi \vdash \forall a.A \leq B} \; \leq\forall\text{L} \qquad \frac{\Psi, a \vdash A \leq B}{\Psi \vdash A \leq \forall a.B} \; \leq\forall\text{R}$$

Fig. 2. Well-formedness of declarative types and declarative subtyping

be instantiated to any instantiation of B, then A is "at least as polymorphic as" B, or we just say that A is "more polymorphic than" B, or $A \leq B$.

There is a very simple declarative formulation of polymorphic subtyping due to Odersky and Laüfer [26]. The syntax of this declarative system is shown in Fig. 1. Types, represented by A, B, C, are the unit type 1, type variables a, b, universal quantification $\forall a.A$ and function type $A \to B$. We allow nested universal quantifiers to appear in types, but not in monotypes. Contexts Ψ collect a list of type variables.

In Fig. 2, we give the well-formedness and subtyping relation for the declarative system. The cases without universal quantifiers are handled by Rules $\leq$Var, $\leq$Unit and $\leq\to$. The subtyping rule for function types ($\leq\to$) is standard, being contravariant on the argument types. Rule $\leq\forall$R says that if A is a subtype of B under the context extended with a, where a is fresh in A, then $A \leq \forall a.B$. Intuitively, if A is more general than the universally quantified type $\forall a.B$, then A must instantiate to $[\tau/a]B$ for every τ.

Finally, the most interesting rule is $\leq\forall$L, which instantiates $\forall a.A$ to $[\tau/a]A$, and concludes the subtyping $\forall a.A \leq B$ if the instantiation is a subtype of B. Notice that τ is *guessed*, and the algorithmic system should provide the means to compute this guess. Furthermore, the guess is a *monotype*, which rules out the possibility of polymorphic (or impredicative) instantiation. The restriction to monotypes and predicative instantiation is used by both Peyton Jones et al. [16] and Dunfield and Krishnaswami's [9] algorithms, which are adopted by several practical implementations of programming languages.

2.2 Finding Solutions for Variable Instantiation

The declarative system specifies the behavior of subtyping relations, but is not directly implementable: the rule $\leq\forall$L requires guessing a monotype τ. The core problem that an algorithm for polymorphic subtyping needs to solve is to find an algorithmic way to compute the monotypes, instead of guessing them. An additional challenge is that the declarative rule $\leq\to$ splits one judgment into two, and the (partial) solutions found for existential variables when processing the first judgment should be transfered to the second judgement.

Dunfield and Krishnaswami's Approach. An elegant algorithmic solution to computing the monotypes is presented by Dunfield and Krishnaswami [9]. Their algorithmic subtyping judgement has the form:

$$\Psi \vdash A \leq B \dashv \Phi$$

A notable difference to the declarative judgement is the presence of a so-called *output context* Φ, which refines the *input context* Ψ with solutions for existential variables found while processing the two types being compared for subtyping. Both Ψ and Φ are *ordered contexts* with the same structure. Ordered contexts are particularly useful to keep track of the correct scoping for variables, and are a notable different to older type-inference algorithms [5] that use global unification variables or constraints collected in a set.

Output contexts are useful to transfer information across judgements in Dunfield and Krishnaswami's approach. For example, the algorithmic rule corresponding to $\leq\rightarrow$ in their approach is:

$$\frac{\Psi \vdash B_1 <: A_1 \dashv \Phi \quad \Phi \vdash [\Phi]A_2 <: [\Phi]B_2 \dashv \Phi'}{\Psi \vdash A_1 \rightarrow A_2 <: B_1 \rightarrow B_2 \dashv \Phi'} <:\rightarrow$$

The information gathered by the output context when comparing the input types of the functions for subtyping is transfered to the second judgement by becoming the new input context, while any solution derived from the first judgment is applied to the types of the second judgment.

Example. If we want to show that $\forall a.a \rightarrow a$ is a subtype of $1 \rightarrow 1$, the declarative system will guess the proper $\tau = 1$ for Rule $\leq\forall L$:

$$\frac{\cdot \vdash 1 \quad \cdot \vdash 1 \rightarrow 1 \leq 1 \rightarrow 1}{\cdot \vdash \forall a.a \rightarrow a \leq 1 \rightarrow 1} \leq\forall L$$

Dunfield and Krishnaswami introduce an *existential variable*—denoted with α, β—whenever a monotype τ needs to be guessed. Below is a sample derivation of their algorithm; we omit the full set of algorithmic rules due to lack of space:

$$\frac{\dfrac{\overline{\quad\quad\quad\quad\quad\quad} \text{ InstRSolve}}{\alpha \vdash 1 \leq \alpha \dashv \alpha = 1} \quad \dfrac{\overline{\quad\quad\quad\quad\quad\quad\quad} <:\text{Unit}}{\alpha = 1 \vdash 1 \leq 1 \vdash \alpha = 1}}{\dfrac{\alpha \vdash \alpha \rightarrow \alpha \leq 1 \rightarrow 1 \dashv \alpha = 1}{\cdot \vdash \forall a.a \rightarrow a \leq 1 \rightarrow 1 \dashv \cdot} <:\forall L} <:\rightarrow$$

The first step applies Rule $<:\forall$L, which introduces a fresh existential variable, α, and opens the left-hand-side $\forall$-quantifier with it. Next, Rule $<:\rightarrow$ splits the judgment in two. For the first branch, Rule `InstRSolve` satisfies $1 \leq \alpha$ by solving α to 1, and stores the solution in its output context. The output context of the first branch is used as the input context of the second branch, and the judgment is updated according to current solutions. Finally, the second branch becomes a base case, and Rule $<:$`Unit` finishes the derivation, makes no change to the input context and propagates the output context back.

Dunfield and Krishnaswami's algorithmic specification is elegant and contains several useful ideas for a mechanical formalization of polymorphic subtyping. For example *ordered contexts* and *existential variables* enable a purely inductive formulation of polymorphic subtyping. However the binding/scoping structure of their algorithmic judgement is still fairly complicated and poses challenges when porting their approach to a theorem prover.

2.3   The Worklist Approach

We inherit Dunfield and Krishnaswami's ideas of ordered contexts, existential variables and the idea of solving those variables, but drop output contexts. Instead our algorithmic rule has the form:

$$\Gamma \vdash \Omega$$

where Ω is a list of judgments $A \leq B$ instead of a single one. This judgement form, which we call *worklist judgement*, simplifies two aspects of Dunfield and Krishnaswami's approach.

Firstly, as already stated, there are no output contexts. Secondly the form of the ordered contexts become simpler. The transfer of information across judgements is simplified because all judgements share the input context. Moreover the order of the judgements in the list allows information discovered when processing the earlier judgements to be easily transfered to the later judgements. In the worklist approach the rule for function types is:

$$\frac{\Gamma \vdash B_1 \leq A_1; A_2 \leq B_2; \Omega}{\Gamma \vdash A_1 \rightarrow A_2 \leq B_1 \rightarrow B_2; \Omega} \leq_a \rightarrow$$

The derivation of the previous example with the worklist approach is:

$$
\cfrac{
\cfrac{
\cfrac{
\cfrac{
\cfrac{}{\cdot \vdash \cdot} \text{a\_nil}
}{\cdot \vdash 1 \leq 1; \cdot} \leq_a\text{unit}
}{\alpha \vdash 1 \leq \alpha; \alpha \leq 1; \cdot} \leq_a\text{solve\_ex}
}{\alpha \vdash \alpha \rightarrow \alpha \leq 1 \rightarrow 1; \cdot} \leq_a\rightarrow
}{\cdot \vdash \forall a.a \rightarrow a \leq 1 \rightarrow 1; \cdot} \leq_a\forall\text{L}
$$

Type variables	a, b
Existential variables	α, β

Algorithmic types	A, B, C	$::=$	$1 \mid a \mid \alpha \mid \forall a.A \mid A \to B$
Algorithmic context	Γ	$::=$	$\cdot \mid \Gamma, a \mid \Gamma, \alpha$
Algorithmic judgments	Ω	$::=$	$\cdot \mid A \leq B; \Omega$

$$\boxed{\Gamma \vdash A}$$

$$\frac{}{\Gamma \vdash 1} \; \text{wf}_a\text{unit} \qquad \frac{a \in \Gamma}{\Gamma \vdash a} \; \text{wf}_a\text{var} \qquad \frac{\alpha \in \Gamma}{\Gamma \vdash \alpha} \; \text{wf}_a\text{exvar}$$

$$\frac{\Gamma \vdash A \quad \Gamma \vdash B}{\Gamma \vdash A \to B} \; \text{wf}_a{\to} \qquad \frac{\Gamma, a \vdash A}{\Gamma \vdash \forall a.A} \; \text{wf}_a\forall$$

Fig. 3. Syntax and well-formedness judgement for the algorithmic system.

To derive $\cdot \vdash \forall a.a \to a \leq 1 \to 1$ with the worklist approach, we first introduce an existential variable and change the judgement to $\alpha \vdash \alpha \to \alpha \leq 1 \to 1; \cdot$. Then, we split the judgment in two for the function types and the derivation comes to $\alpha \vdash 1 \leq \alpha; \alpha \leq 1; \cdot$. When the first judgment is solved with $\alpha = 1$, we immediately remove α from the context, while propagating the solution as a substitution to the rest of the judgment list, resulting in $\cdot \vdash 1 \leq 1; \cdot$, which finishes the derivation in two trivial steps.

With this form of eager propagation, solutions no longer need to be recorded in contexts, simplifying the encoding and reasoning in a proof assistant.

Key Results. Both the declarative and algorithmic systems are formalized in Abella. We have proven 3 important properties for this algorithm: *decidability*, ensuring that the algorithm always terminates; and *soundness* and *completeness*, showing the equivalence of the declarative and algorithmic systems.

3   A Worklist Algorithm for Polymorphic Subtyping

This section presents our algorithm for polymorphic subtyping. A novel aspect of our algorithm is the use of worklist judgments: a form of judgement that facilitates the propagation of information.

3.1   Syntax and Well-Formedness of the Algorithmic System

Figure 3 shows the syntax and the well-formedness judgement.

Existential Variables. In order to solve the unknown types τ, the algorithmic system extends the declarative syntax of types with *existential variables* α. They behave like unification variables, but are not globally defined. Instead, the ordered *algorithmic context*, inspired by Dunfield and Krishnaswami [9], defines their scope. Thus the type τ represented by the corresponding existential variable is always bound in the corresponding declarative context Ψ.

$$\boxed{\Gamma \vdash \Omega}$$

$$\frac{}{\Gamma \vdash \cdot} \ \texttt{a\_nil}$$

$$\frac{\Gamma \vdash \Omega}{\Gamma \vdash 1 \leq 1; \Omega} \leq_\texttt{a}\texttt{unit} \qquad \frac{a \in \Gamma \quad \Gamma \vdash \Omega}{\Gamma \vdash a \leq a; \Omega} \leq_\texttt{a}\texttt{var} \qquad \frac{\alpha \in \Gamma \quad \Gamma \vdash \Omega}{\Gamma \vdash \alpha \leq \alpha; \Omega} \leq_\texttt{a}\texttt{exvar}$$

$$\frac{\Gamma \vdash B_1 \leq A_1; A_2 \leq B_2; \Omega}{\Gamma \vdash A_1 \to A_2 \leq B_1 \to B_2; \Omega} \leq_\texttt{a}\!\!\to$$

$$\frac{\alpha \ \text{fresh} \quad \Gamma, \alpha \vdash [\alpha/a]A < B; \Omega}{\Gamma \vdash \forall a.A \leq B; \Omega} \leq_\texttt{a}\forall\text{L} \qquad \frac{b \ \text{fresh} \quad \Gamma, b \vdash A \leq B; \Omega}{\Gamma \vdash A \leq \forall b.B; \Omega} \leq_\texttt{a}\forall\text{R}$$

$$\frac{\alpha \notin FV(A) \cup FV(B) \quad \Gamma[\alpha_1, \alpha_2] \vdash \alpha_1 \to \alpha_2 \leq A \to B; [\alpha_1 \to \alpha_2/\alpha]\Omega}{\Gamma[\alpha] \vdash \alpha \leq A \to B; \Omega} \leq_\texttt{a}\texttt{instL}$$

$$\frac{\alpha \notin FV(A) \cup FV(B) \quad \Gamma[\alpha_1, \alpha_2] \vdash A \to B \leq \alpha_1 \to \alpha_2; [\alpha_1 \to \alpha_2/\alpha]\Omega}{\Gamma[\alpha] \vdash A \to B \leq \alpha; \Omega} \leq_\texttt{a}\texttt{instR}$$

$$\frac{\Gamma[\alpha][] \vdash [\alpha/\beta]\Omega}{\Gamma[\alpha][\beta] \vdash \alpha \leq \beta; \Omega} \leq_\texttt{a}\texttt{solve\_ex} \qquad \frac{\Gamma[\alpha][] \vdash [\alpha/\beta]\Omega}{\Gamma[\alpha][\beta] \vdash \beta \leq \alpha; \Omega} \leq_\texttt{a}\texttt{solve\_ex}'$$

$$\frac{\Gamma[a][] \vdash [a/\beta]\Omega}{\Gamma[a][\beta] \vdash a \leq \beta; \Omega} \leq_\texttt{a}\texttt{solve\_var} \qquad \frac{\Gamma[a][] \vdash [a/\beta]\Omega}{\Gamma[a][\beta] \vdash \beta \leq a; \Omega} \leq_\texttt{a}\texttt{solve\_var}'$$

$$\frac{\Gamma[] \vdash [1/\alpha]\Omega}{\Gamma[\alpha] \vdash \alpha \leq 1; \Omega} \leq_\texttt{a}\texttt{solve\_unit} \qquad \frac{\Gamma[] \vdash [1/\alpha]\Omega}{\Gamma[\alpha] \vdash 1 \leq \alpha; \Omega} \leq_\texttt{a}\texttt{solve\_unit}'$$

Fig. 4. Algorithmic subtyping

Worklist Judgements. The form of our algorithmic judgements is non-standard. Our algorithm keeps track of an explicit list of outstanding work: the list Ω of (reified) *algorithmic judgements* of the form $A \leq B$, to which a substitution can be applied once and for all to propagate the solution of an existential variable.

Hole Notation. To facilitate context manipulation, we use the syntax $\Gamma[\Gamma_M]$ to denote a context of the form $\Gamma_L, \Gamma_M, \Gamma_R$ where Γ is the context $\Gamma_L, \bullet, \Gamma_R$ with a hole ($\bullet$). Hole notations with the same name implicitly share the same Γ_L and Γ_R. A multi-hole notation like $\Gamma[\alpha][\beta]$ means $\Gamma_1, \alpha, \Gamma_2, \beta, \Gamma_3$.

3.2   Algorithmic Subtyping

The algorithmic subtyping judgement, defined in Fig. 4, has the form $\Gamma \vdash \Omega$, where Ω collects multiple subtyping judgments $A \leq B$. The algorithm treats Ω as a worklist. In every step it takes one task from the worklist for processing, possibly pushes some new tasks on the worklist, and repeats this process until

$$
\cfrac{
  \cfrac{
    \cfrac{
      \cfrac{
        \cfrac{
          \cfrac{
            \cfrac{
              \cfrac{
                \cfrac{\rule{2.5em}{0.4pt}}{\alpha_1 \vdash \cdot}\ \texttt{a\_nil}
              }{\alpha_1 \vdash 1 \leq 1; \cdot}\ \leq_a\texttt{unit}
            }{\alpha_1, \alpha_2 \vdash \alpha_1 \leq \alpha_2; 1 \leq 1; \cdot}\ \leq_a\texttt{solve\_ex}
          }{\alpha_1, \alpha_2, \beta \vdash \alpha_1 \leq \beta; \beta \leq \alpha_2; 1 \leq 1; \cdot}\ \leq_a\texttt{solve\_ex}
        }{\alpha_1, \alpha_2, \beta \vdash \beta \to \beta \leq \alpha_1 \to \alpha_2; 1 \leq 1; \cdot}\ \leq_a{\to}
      }{\alpha, \beta \vdash \beta \to \beta \leq \alpha; 1 \leq 1; \cdot}\ \leq_a\texttt{instR}
    }{\alpha \vdash \forall a.\, a \to a \leq \alpha; 1 \leq 1; \cdot}\ \leq_a\forall L
  }{\alpha \vdash \alpha \to 1 \leq (\forall a.\, a \to a) \to 1; \cdot}\ \leq_a{\to}
}{\cdot \vdash \forall a.\, a \to 1 \leq (\forall a.\, a \to a) \to 1; \cdot}\ \leq_a\forall L
$$

$$
\cfrac{
  \cfrac{
    \cfrac{
      \cfrac{
        \cfrac{\textit{stuck}}{\alpha, b \vdash \alpha \leq b; \cdot}\ ?
      }{\alpha \vdash \alpha \leq \forall b.\, b; \cdot}\ \leq_a\forall R
    }{\alpha \vdash 1 \leq 1; \alpha \leq \forall b.\, b; \cdot}\ \leq_a\texttt{unit}
  }{\alpha \vdash 1 \to \alpha \leq 1 \to \forall b.\, b; \cdot}\ \leq_a{\to}
}{\cdot \vdash \forall a.\, 1 \to a \leq 1 \to \forall b.\, b; \cdot}\ \leq_a\forall L
$$

Fig. 5. Successful and failing derivations for the algorithmic subtyping relation

the list is empty. This last and single base case is handled by Rule a_nil. The remaining rules all deal with the first task in the worklist. Logically we can discern 3 groups of rules.

Firstly, we have five rules that are similar to those in the declarative system, mostly just adapted to the worklist style. For instance, Rule $\leq_a{\to}$ consumes one judgment and pushes two to the worklist. A notable difference with the declarative Rule $\leq\forall L$ is that Rule $\leq_a\forall L$ requires no guessing of a type τ to instantiate the polymorphic type $\forall a.A$, but instead introduces an existential variable α to the context and to A. In accordance with the declarative system, where the monotype τ should be bound in the context Ψ, here α should only be solved to a monotype bound in Γ. More generally, for any algorithmic context $\Gamma[\alpha]$, the algorithmic variable α can only be solved to a monotype that is well-formed with respect to Γ_L.

Secondly, Rules $\leq_a\texttt{instL}$ and $\leq_a\texttt{instR}$ partially instantiate existential types α, to function types. The domain and range of the new function type are undetermined: they are set to two fresh existential variables α_1 and α_2. To make sure that $\alpha_1 \to \alpha_2$ has the same scope as α, the new variables α_1 and α_2 are inserted in the same position in the context where the old variable α was. To propagate the instantiation to the remainder of the worklist, α is substituted for $\alpha_1 \to \alpha_2$ in Ω. The *occurs-check* side-condition is necessary to prevent a diverging infinite instantiation. For example $1 \to \alpha \leq \alpha$ would diverge with no such check.

Thirdly, in the remaining six rules an existential variable can be immediately solved. Each of the six similar rules removes an existential variable from the context, performs a substitution on the remainder of the worklist and continues.

The algorithm on judgment list is designed to share the context across all judgments. However, the declarative system does not share a single context in its derivation. This gap is filled by strengthening and weakening lemmas of both systems, where most of them are straightforward to prove, except for the strengthening lemma of the declarative system, which is a little trickier.

Example. We illustrate the subtyping rules through a sample derivation in the left of Fig. 5, which shows that $\forall a.\, a \to 1 \leq (\forall a.\, a \to a) \to 1$. Thus the derivation starts with an empty context and a judgment list with only one element.

In step 1, we have only one judgment, and that one has a top-level $\forall$ on the left hand side. So the only choice is rule $\leq_a\forall L$, which opens the universally quantified type with an unknown existential variable α. Variable α will be solved later to some monotype that is well-formed within the context before α. That is, the empty context $\cdot$ in this case. In step 2, rule $\leq_a\to$ is applied to the worklist, splitting the first judgment into two. Step 3 is similar to step 1, where the left-hand-side $\forall$ of the first judgment is opened according to rule $\leq_a\forall L$ with a fresh existential variable. In step 4, the first judgment has an arrow on the left hand side, but the right-hand-side type is an existential variable. It is obvious that α should be solved to a monotype of the form $\sigma \to \tau$. Rule instR implements this, but avoids guessing σ and τ by "splitting" α into two existential variables, α_1 and α_2, which will be solved to some σ and τ later. Step 5 applies Rule $\leq_a\to$ again. Notice that after the split, β appears in two judgments. When the first β is solved during any step of derivation, the next β will be substituted by that solution. This propagation mechanism ensures the consistent solution of the variables, while keeping the context as simple as possible. Steps 6 and 7 solve existential variables. The existential variable that is right-most in the context is always solved in terms of the other. Therefore in step 6, β is solved in terms of α_1, and in step 7, α_2 is solved in terms of α_1. Additionally, in step 6, when β is solved, the substitution $[\alpha_1/\beta]$ is propagated to the rest of the judgment list, and thus the second judgment becomes $\alpha_1 \leq \alpha_2$. Steps 8 and 9 trivially finish the derivation. Notice that α_1 is not instantiated at the end. This means that any well-scoped instantiation is fine.

A Failing Derivation. We illustrate the role of ordered contexts through another example: $\forall a.\ 1 \to a \leq 1 \to \forall b.\ b$. From the declarative perspective, a should be instantiated to some τ first, then b is introduced to the context, so that $b \notin FV(\tau)$. As a result, we cannot find τ such that $\tau \leq b$. The right of Fig. 5 shows the algorithmic derivation, which also fails due to the scoping—α is introduced earlier than b, thus it cannot be solved to b.

4   Metatheory

This section presents the 3 main meta-theoretical results that we have proved in Abella. The first two are soundness and completeness of our algorithm with respect to Odersky and Läufer's declarative subtyping. The third result is our algorithm's decidability.

4.1   Transfer to the Declarative System

To state the correctness of the algorithmic subtyping rules, Fig. 6 introduces two *transfer* judgements to relate the declarative and the algorithmic system. The first judgement, transfer of contexts $\Gamma \to \Psi$, removes existential variables from the algorithmic context Γ to obtain a declarative context Ψ. The second judgement, transfer of the judgement list $\Gamma \mid \Omega \rightsquigarrow \Omega'$, replaces all occurrences of

$$\boxed{\Gamma \to \Psi}$$

$$\frac{}{\cdot \to \cdot} \to \cdot \qquad \frac{\Gamma \to \Psi}{\Gamma, a \to \Psi, a} \to\mathbf{var} \qquad \frac{\Gamma \to \Psi}{\Gamma, \alpha \to \Psi} \to\mathbf{exvar}$$

$$\boxed{\Gamma \mid \Omega \rightsquigarrow \Omega'}$$

$$\frac{}{\cdot \mid \Omega \rightsquigarrow \Omega} \rightsquigarrow\cdot \qquad \frac{\Gamma \mid \Omega \rightsquigarrow \Omega'}{\Gamma, a \mid \Omega \rightsquigarrow \Omega'} \rightsquigarrow\mathbf{var} \qquad \frac{\Gamma \to \Psi \quad \Psi \vdash \tau \quad \Gamma \mid [\tau/\alpha]\Omega \rightsquigarrow \Omega'}{\Gamma, \alpha \mid \Omega \rightsquigarrow \Omega'} \rightsquigarrow\mathbf{exvar}$$

Fig. 6. Transfer rules

existential variables in Ω by well-scoped mono-types. Notice that this judgment is not decidable, i.e. a pair of Γ and Ω may be related with multiple Ω'. However, if there exists some substitution that transforms Ω to Ω', and each subtyping judgment in Ω' holds, we know that Ω is potentially satisfiable.

The following two lemmas generalize Rule $\rightsquigarrow$**exvar** from substituting the first existential variable to substituting any existential variable.

Lemma 1 (Insert). *If* $\Gamma \to \Psi$ *and* $\Psi \vdash \tau$ *and* $\Gamma, \Gamma_1 \mid [\tau/\alpha]\Omega \rightsquigarrow \Omega'$, *then* $\Gamma, \alpha, \Gamma_1 \mid \Omega \rightsquigarrow \Omega'$.

Lemma 2 (Extract). *If* $\Gamma, \alpha, \Gamma_1 \mid \Omega \rightsquigarrow \Omega'$, *then* $\exists \tau$ *s.t.* $\Gamma \to \Psi, \Psi \vdash \tau$ *and* $\Gamma, \Gamma_1 \mid [\tau/\alpha]\Omega \rightsquigarrow \Omega'$.

In order to match the shape of algorithmic subtyping relation for the following proofs, we define a relation $\Psi \vdash \Omega$ for the declarative system, meaning that all the declarative judgments hold under context Ψ.

Definition 1 (Declarative Subtyping Worklist)

$$\Psi \vdash \Omega := \forall (A \le B) \in \Omega, \Psi \vdash A \le B$$

4.2   Soundness

Our algorithm is sound with respect to the declarative specification. For any derivation of a list of algorithmic judgments $\Gamma \vdash \Omega$, we can find a valid transfer $\Gamma \mid \Omega \rightsquigarrow \Omega'$ such that all judgments in Ω' hold in Ψ, with $\Gamma \to \Psi$.

Theorem 1 (Soundness). *If* $\Gamma \vdash \Omega$ *and* $\Gamma \to \Psi$, *then there exists* Ω', *s.t.* $\Gamma \mid \Omega \rightsquigarrow \Omega'$ *and* $\Psi \vdash \Omega'$.

The proof proceeds by induction on the derivation of $\Gamma \vdash \Omega$, finished off by appropriate applications of the insertion and extraction lemmas.

4.3   Completeness

Completeness of the algorithm means that any declarative derivation has an algorithmic counterpart.

Theorem 2 (Completeness). *If $\Psi \vdash \Omega'$ and $\Gamma \to \Psi$ and $\Gamma \mid \Omega \rightsquigarrow \Omega'$, then $\Gamma \vdash \Omega$.*

The proof proceeds by induction on the derivation of $\Psi \vdash \Omega'$. As the declarative system does not involve information propagation across judgments, the induction can focus on the subtyping derivation of the first judgment without affecting other judgments. The difficult cases correspond to the $\leq_a$instL and $\leq_a$instR rules. When the proof by induction on $\Psi \vdash \Omega'$ reaches the $\leq\to$ case, the first declarative judgment has a shape like $A_1 \to A_2 \leq B_1 \to B_2$. One of the possibile cases for the first corresponding algorithmic judgement is $\alpha \leq A \to B$. However, the case analysis does not indicate that α is fresh in A and B. Thus we cannot apply Rule $\leq_a$instL and make use of the induction hypothesis. The following lemma helps us out in those cases: it rules out subtypings with infinite types as solutions (e.g. $\alpha \leq 1 \to \alpha$) and guarantees that α is free in A and B.

Lemma 3 (Prune Transfer for Instantiation). *If $\Psi \vdash A_1 \to A_2 \leq B_1 \to B_2; \Omega'$ and $\Gamma \to \Psi$ and $\Gamma \mid (\alpha \leq A \to B; \Omega) \rightsquigarrow (A_1 \to A_2 \leq B_1 \to B_2; \Omega')$, then $\alpha \notin FV(A) \cup FV(B)$.*

A similar lemma holds for the symmetric case $(A \to B \leq \alpha; \Omega)$.

4.4   Decidability

The third key result for our algorithm is decidability.

Theorem 3 (Decidability). *Given any well-formed judgment list Ω under Γ, it is decidable whether $\Gamma \vdash \Omega$ or not.*

We have proven this theorem by means of a lexicographic group of induction measurements $\langle |\Omega|_\forall, |\Gamma|_\alpha, |\Omega|_\to \rangle$ on the worklist Ω and algorithmic context Γ. The worklist measures $|\cdot|_\forall$ and $|\cdot|_\to$ count the number of universal quantifiers and function types respectively.

Definition 2 (Worklist Measures)

$$
\begin{aligned}
|1|_\forall = |a|_\forall = |\alpha|_\forall = 0 && |1|_\to = |a|_\to = |\alpha|_\to = 0 \\
|A \to B|_\forall = |A|_\forall + |B|_\forall && |A \to B|_\to = |A|_\to + |B|_\to + 1 \\
|\forall x.A|_\forall = |A|_\forall + 1 && |\forall x.A|_\to = |A|_\to \\
|\Omega|_\forall = \textstyle\sum_{A \leq B \in \Omega} |A|_\forall + |B|_\forall && |\Omega|_\to = \textstyle\sum_{A \leq B \in \Omega} |A|_\to + |B|_\to
\end{aligned}
$$

The context measure $|\cdot|_\alpha$ counts the number of unsolved existential variables.

Definition 3 (Context Measure)

$$
|\cdot|_\alpha = 0 \qquad |\Gamma, a|_\alpha = |\Gamma|_\alpha \qquad |\Gamma, \alpha|_\alpha = |\Gamma|_\alpha + 1
$$

It is not difficult to see that all but two of the algorithm's rules decrease one of the three measures. The two exceptions are the Rules $\leq_a$instL and $\leq_a$instR; both increment the number of existential variables and the number of function types without affecting the number of universal quantifiers. To handle these rules, we handle a special class of judgements, which we call *instantiation judgements* Ω_i, separately. They take the form:

Definition 4 (Ω_i)

$$\Omega_i := \cdot \mid \alpha \leq A; \Omega_i' \mid A \leq \alpha; \Omega_i' \quad \text{where } \alpha \notin FV(A) \cup FV(\Omega_i')$$

These instantiation judgements are these ones consumed and produced by the Rules $\leq_a$instL and $\leq_a$instR. The following lemma handles their decidability.

Lemma 4 (Instantiation Decidability). *For any context Γ and judgment list Ω_i, Ω, it is decidable whether $\Gamma \vdash \Omega_i, \Omega$ if both of the conditions hold*

(1) $\forall \Gamma', \Omega'$ *s.t.* $|\Omega'|_\forall < |\Omega_i, \Omega|_\forall$, *it is decidable whether* $\Gamma' \vdash \Omega'$.
(2) $\forall \Gamma', \Omega'$ *s.t.* $|\Omega'|_\forall = |\Omega_i, \Omega|_\forall$ *and* $|\Gamma'|_\alpha = |\Gamma|_\alpha - |\Omega_i|$, *it is decidable whether* $\Gamma' \vdash \Omega'$.

In other words, for any instantiation judgment prefix Ω_i, the algorithm either reduces the number of $\forall$'s or solves one existential variable per instantiation judgment. The proof of this lemma is by induction on the measure $2*|\Omega_i|_\rightarrow + |\Omega_i|$ of the instantiation judgment list.

In summary, the decidability theorem can be shown through a lexicographic group of induction measurements $\langle |\Omega|_\forall, |\Omega|_\alpha, |\Omega|_\rightarrow \rangle$. The critical case is that, whenever we encounter an instantiation judgment at the front of the worklist, we refer to Lemma 4, which reduces the number of unsolved variables by consuming that instantiation judgment, or reduces the number of $\forall$-quantifiers. Other cases are relatively straightforward.

5   The Choice of Abella

We have chosen the Abella (v2.0.5) proof assistant [10] to develop our formalization. Our development is only based on the reasoning logic of Abella, and does not make use of its specification logic. Abella is particularly helpful due to its built-in support for variable bindings, and its λ-tree syntax [22] is a form of HOAS, which helps with the encoding and reasoning about substitutions. For instance, the type $\forall x.x \rightarrow a$ is encoded as `all (x\ arrow x a)`, where `x\ arrow x a` is a lambda abstraction in Abella. An opening $[b/x](x \rightarrow a)$ is encoded as an application `all (x\ arrow x a)`, which can be simplified(evaluated) to `arrow b a`. Name supply and freshness conditions are controlled by the ∇-quantifier. The expression `nabla x, F` means that `x` is a unique variable in `F`, i.e. it is different from any other names occurring elsewhere. Such variables are called nominal constants. They can be of any type, in other words, every type may contain unlimited number of such atomic nominal constants.

Encoding of the Declarative System. As a concrete example, our declarative context and well-formedness rules are encoded as follows.

```
Kind ty      type.
Type i       ty.                  % the unit type
Type all     (ty → ty) → ty.      % forall-quantifier
Type arrow   ty → ty → ty.        % function type
Type bound   ty → o.              % variable collection in contexts


Define env : olist → prop by
   env nil;
   nabla x, env (bound x :: E) := env E.

Define wft : olist → ty → prop by
   wft E i;
   nabla x, wft (E x) x := nabla x, member (bound x) (E x);
   wft E (arrow A B) := wft E A ∧ wft E B;
   wft E (all A) := nabla x, wft (bound x :: E) (A x).
```

We use the type olist just as normal list of o with two constructors, namely nil : olist and (::) : o → olist → olist, where o purely means "the element type of olist". The member : o → olist → prop relation is also pre-defined. The second case of the relation wft states rule wf$_d$var. The encoding (E x) basically means that the context *may* contain x. If we write (E x) as E, then the context should not contain x, and both wft E x and member (bound x) E make no sense. Instead, we treat E : ty → olist as an *abstract structure* of a context, such as x\ bound x :: bound a :: nil For the fourth case of the relation wft, the type $\forall x.A$ in our target language is expressed as (all A), and its opening A, (A x).

Encoding of the Algorithmic System. In terms of the algorithmic system, notably, Abella handles the $\leq_a$instL and $\leq_a$instR rules in a nice way:

```
% sub_alg_list : enva → [subty_judgment] → prop
Define subal : olist → olist → prop by
   subal E nil;
   subal E (subt i i :: Exp) := subal E Exp;
   % some cases omitted ...
   % <: instL
   nabla x, subal (E x) (subt x (arrow A B) :: Exp x) :=
       exists E1 E2 F, nabla x y z, append E1 (exvar x :: E2) (E x) ∧
          append E1 (exvar y :: exvar z :: E2) (F y z) ∧
             subal (F y z) (subt (arrow y z) (arrow A B) :: Exp (arrow y z));
   % <: instR is symmetric to <: instL, omitted here
   % other cases omitted ...
```

Thanks to the way Abella deals with nominal constants, the pattern subt x (arrow A B) implicitly states that $x \notin FV(A) \land x \notin FV(B)$. If the condition were not required, we would have encoded the pattern as subt x (arrow (A x) (B x)) instead.

File(s)	SLOC	# of Theorems	Description
olist.thm, nat.thm	303	55	Basic data structures
higher.thm, order.thm	164	15	Declarative system
higher_alg.thm	618	44	Algorithmic system
trans.thm	411	46	Transfer
sound.thm	166	2	Soundness theorem
depth.thm	143	12	Definition of depth
complete.thm	626	28	Lemmas and Completeness theorem
decidable.thm	1077	53	Lemmas and Decidability theorem
Total	3627	267	(33 definitions in total)

Fig. 7. Statistics for the proof scripts

5.1   Statistics and Discussion

Some basic statistics on our proof script are shown in Fig. 7. The proof consists of 3627 lines of code with a total of 33 definitions and 267 theorems. We have to mention that Abella provides few built-in tactics and does not support user-defined ones, and we would reduce significant lines of code if Abella provided more handy tactics. Moreover, the definition of natural numbers, the plus operation and less-than relation are defined within our proof due to Abella's lack of packages. However, the way Abella deals with name bindings is very helpful for type system formalizations and substitution-intensive formalizations, such as this one.

6   Related Work

Type Inference for Polymorphic Subtyping. Higher-order polymorphism is a practical and important programming language feature. Due to the undecidability of type-inference for System F [35], different decidable partial type-inference approaches were developed. The subtyping relation of this paper, originally proposed by Odersky and Laüfer [26], is *predicative* ($\forall$'s only instantiate to monotypes), which is considered a reasonable and practical trade-off. There is also work on partial impredicative type-inference algorithms [19,20,34]. However, unlike the predicative subtyping relation for System F, the subtyping for impredicative System F is undecidable [31]. Therefore such algorithms have to navigate through the design space to impose restrictions that allow for a decidable algorithm. As a result such algorithms tend to be more complex, and are less adopted in practice.

Gundry et al. [13] revisited the Hindley-Milner type system. They make use of ordered contexts on the unification during type inference, and their algorithm works differently from algorithm $\mathcal{W}$. Dunfield and Krishnaswami [9] adopted a similar idea on ordered contexts and presented an algorithmic approach for predicative polymorphic subtyping that tracks the (partial) solutions of existential variables in the algorithmic context—this denotes a delayed substitution that is incrementally applied to outstanding work as it is encountered. Their algorithm comes with 40 pages of manual proofs on the soundness, completeness and decidability. We have tried to mechanize these proofs directly, but have not been successful yet because most proof assistants do not naturally support output contexts and their more complex ordered contexts. Their theorems have statements that are more complex than those in the worklist approach. One of the reasons for the added complexity is that, when the constraints are not strict enough, the algorithm may not instantiate all existential variables. However in order to match the declarative judgement, all the unsolved variables should be properly assigned. For example, their generalized completeness theorem is:

Theorem 4 (Generalized Completeness of Subtyping [9]). *If* $\Psi \longrightarrow \Phi$ *and* $\Psi \vdash A$ *and* $\Psi \vdash B$ *and* $[\Phi]\Psi \vdash [\Phi]A \leq [\Phi]B$ *then there exist* Δ *and* Φ' *such that* $\Delta \longrightarrow \Phi'$ *and* $\Phi \longrightarrow \Phi'$ *and* $\Psi \vdash [\Psi]A <: [\Psi]B \dashv \Delta$.

Here, the auxiliary relation $\Psi \longrightarrow \Psi'$ extends a context Ψ to a context Ψ'. This is used to extend the algorithm's input and output contexts Ψ and Δ, with possibly unassigned existential variables, to a complete (i.e., fully-assigned) contexts Φ and Φ' suitable for the declarative specification.

While we are faced with a similar gap between algorithm and specification, which we tackle with our transfer relations $\Gamma \to \Psi$, our completeness statement is much shorter because our algorithm does not return an output context which needs to be transferred. Moreover, we have cleanly encapsulated any substitutions to the worklist in the worklist transfer judgement $\Gamma \mid \Omega \rightsquigarrow \Omega'$.

Peyton Jones et al. [16] developed a higher-rank predicative bidirectional type system. They enriched their subtyping relations with deep skolemisation, while other relations remain similar to ours. Their algorithm is unification-based with a structure similar to algorithm $\mathcal{W}$'s.

Unification Algorithms. Our algorithm works similarly to some unification algorithms that use a set of unification constraints and single-step simplification transitions. Some work [1, 27] adopts this idea in dependently typed inference and reconstruction. These approaches collect a set of constraints and nondeterministically process one of them at a time. Those approaches consider various forms of constraints, including term unification, context unification and solution for metavariables. In contrast, our algorithm is presented in a simpler form, using ordered (worklist) judgements, which is sufficient for the subtyping problem.

Formalizations of Type-Inference Algorithms in Theorem Provers. The well-known POPLMARK challenge [3] has encouraged the development of new proof assistant features for facilitating the development and verification of type systems. As a result, many theorem provers and packages now provide methods for dealing with variable binding [2, 4, 32], and more and more type system designers choose to formalize their proofs with these tools. Yet, difficulties with mechanising algorithmic aspects, like unification and constraint solving, have received very little attention. Moreover, while most type system judgements only feature local (input) contexts, which have a simple binding/scoping structure, many traditional type-inference algorithms require more complex binding structures with output contexts.

Naraschewski and Nipkow [24] published the first formal verification of algorithm $\mathcal{W}$ in Isabelle/HOL [25]. The treatment of new variables is a little tricky in their formalization, while most other parts follow the structure of Damas's manual proof closely. Following Naraschewski and Nipkow other researchers [7,8] prove a similar result in Coq [29]. Nominal techniques [32] in Isabelle/HOL have also been used for a similar verification [33]. Moreover, Garrigue [11] mechanized a type inference algorithm for Core ML extended with structural polymorphism and recursion.

7 Conclusion and Future Work

In this paper we have shown how to mechanise an algorithmic subtyping relation for higher-order polymorphism, together with its proofs of soundness, completeness and decidability, in the Abella proof assistant. In ongoing work we are extending our mechanisation with a bidirectional type inference algorithm. The main difficulty there is communicating the instantiations of existential variables from the subtyping algorithm to the type inference. To make this possible we are exploring a continuation passing style formulation, which generalises the worklist approach. Another possible extension is to have the algorithm return an explicit witness for the subtyping as part of type-directed elaboration into System F.

Acknowledgement. We sincerely thank the anonymous reviewers for their insightful comments. This work has been sponsored by the Hong Kong Research Grant Council projects number 17210617 and 17258816, and by the Research Foundation - Flanders.

References

1. Abel, A., Pientka, B.: Higher-order dynamic pattern unification for dependent types and records. In: Ong, L. (ed.) TLCA 2011. LNCS, vol. 6690, pp. 10–26. Springer, Heidelberg (2011). https://doi.org/10.1007/978-3-642-21691-6_5
2. Aydemir, B., Charguéraud, A., Pierce, B.C., Pollack, R., Weirich, S.: Engineering formal metatheory. In: Proceedings of the 35th Annual ACM SIGPLAN-SIGACT Symposium on Principles of Programming Languages, POPL 2008 (2008)
3. Aydemir, B.E., Bohannon, A., Fairbairn, M., Foster, J.N., Pierce, B.C., Sewell, P., Vytiniotis, D., Washburn, G., Weirich, S., Zdancewic, S.: Mechanized metatheory for the masses: The POPLmark challenge. In: The 18th International Conference on Theorem Proving in Higher Order Logics (2005)
4. Chlipala, A.: Parametric higher-order abstract syntax for mechanized semantics. In: Proceedings of the 13th ACM SIGPLAN International Conference on Functional Programming, ICFP 2008 (2008)
5. Damas, L., Milner, R.: Principal type-schemes for functional programs. In: Proceedings of the 9th ACM SIGPLAN-SIGACT Symposium on Principles of Programming Languages, POPL 1982 (1982)

6. Disciple Development Team: The Disciplined Disciple Compiler (2017). http:// disciple.ouroborus.net/
7. Dubois, C.: Proving ML type soundness within Coq. In: Aagaard, M., Harrison, J. (eds.) TPHOLs 2000. LNCS, vol. 1869, pp. 126–144. Springer, Heidelberg (2000). https://doi.org/10.1007/3-540-44659-1_9
8. Dubois, C., Menissier-Morain, V.: Certification of a type inference tool for ML: Damas-Milner within Coq. J. Autom. Reasoning **23**(3), 319–346 (1999)
9. Dunfield, J., Krishnaswami, N.R.: Complete and easy bidirectional typechecking for higher-rank polymorphism. In: Proceedings of the 18th ACM SIGPLAN International Conference on Functional Programming, ICFP 2013 (2013)
10. Gacek, A.: The abella interactive theorem prover (system description). In: Armando, A., Baumgartner, P., Dowek, G. (eds.) IJCAR 2008. LNCS (LNAI), vol. 5195, pp. 154–161. Springer, Heidelberg (2008). https://doi.org/10.1007/978-3-540-71070-7_13
11. Garrigue, J.: A certified implementation of ML with structural polymorphism and recursive types. Mathe. Struct. Comput. Sci. **25**(4), 867–891 (2015)
12. Gill, A., Launchbury, J., Peyton Jones, S.L.: A short cut to deforestation. In: Proceedings of the Conference on Functional Programming Languages and Computer Architecture, FPCA 1993 (1993)
13. Gundry, A., McBride, C., McKinna, J.: Type inference in context. In: Proceedings of the Third ACM SIGPLAN Workshop on Mathematically Structured Functional Programming, MSFP 2010 (2010)
14. Hindley, R.: The principal type-scheme of an object in combinatory logic. Trans. Am. Mathe. Soc. **146**, 29–60 (1969)
15. Jones, M.P.: Functional programming with overloading and higher-order polymorphism. In: Jeuring, J., Meijer, E. (eds.) AFP 1995. LNCS, vol. 925, pp. 97–136. Springer, Heidelberg (1995). https://doi.org/10.1007/3-540-59451-5_4
16. Peyton Jones, S., Vytiniotis, D., Weirich, S., Shields, M.: Practical type inference for arbitrary-rank types. J. Funct. Program. **17**(1), 1–82 (2007)
17. Lämmel, R., Jones, S.P.: Scrap your boilerplate: a practical design pattern for generic programming. In: Proceedings of the 2003 ACM SIGPLAN International Workshop on Types in Languages Design and Implementation, TLDI 2003 (2003)
18. Launchbury, J., Peyton Jones, S.L.: State in Haskell. LISP Symbolic Comput. **8**(4), 293–341 (1995)
19. Le Botlan, D., Rémy, D.: MLF: raising ML to the power of system F. In: Proceedings of the Eighth ACM SIGPLAN International Conference on Functional Programming, ICFP 2003 (2003)
20. Leijen, D.: HMF: Simple type inference for first-class polymorphism. In: Proceedings of the 13th ACM SIGPLAN International Conference on Functional Programming, ICFP 2008 (2008)
21. Leroy, X., et al.: The CompCert verified compiler. Documentation and user's manual, INRIA Paris-Rocquencourt (2012)
22. Miller, D.: Abstract syntax for variable binders: an overview. In: Lloyd, J., et al. (eds.) CL 2000. LNCS (LNAI), vol. 1861, pp. 239–253. Springer, Heidelberg (2000). https://doi.org/10.1007/3-540-44957-4_16
23. Milner, R.: A theory of type polymorphism in programming. J. Comput. Syst. Sci. **17**(3), 348–375 (1978)
24. Naraschewski, W., Nipkow, T.: Type inference verified: algorithm W in Isabelle/HOL. J. Autom. Reason. **23**(3), 299–318 (1999)

25. Nipkow, T., Paulson, L.C., Wenzel, M.: Isabelle/HOL: A Proof Assistant for High-erorderlogic, vol. 2283. Springer Science & Business Media, Heidelberg (2002). https://doi.org/10.1007/3-540-45949-9
26. Odersky, M., Läufer, K.: Putting type annotations to work. In: Proceedings of the 23rd ACM SIGPLAN-SIGACT Symposium on Principles of Programming Languages, POPL 1996 (1996)
27. Reed, J.: Higher-order constraint simplification in dependent type theory. In: Proceedings of the Fourth International Workshop on Logical Frameworks and Meta-Languages: Theory and Practice, LFMTP 2009 (2009)
28. Reynolds, J.C.: Types, abstraction and parametric polymorphism. In: Information Processing, pp. 513–523 (1983)
29. The Coq development team: The Coq proof assistant (2017). https://coq.inria.fr/
30. The PureScript development team: PureScript (2017). http://www.purescript.org/
31. Tiuryn, J., Urzyczyn, P.: The subtyping problem for second-order types is undecidable. In: Proceedings 11th Annual IEEE Symposium on Logic in Computer Science (1996)
32. Urban, C.: Nominal techniques in Isabelle/HOL. J. Autom. Reason. **40**(4), 327–356 (2008)
33. Urban, C., Nipkow, T.: Nominal verification of algorithm W. from semantics to computer science. In: Essays in Honour of Gilles Kahn, pp. 363–382 (2008)
34. Vytiniotis, D., Weirich, S., Peyton Jones, S.: FPH: first-class polymorphism for Haskell. In: Proceedings of the 13th ACM SIGPLAN International Conference on Functional Programming, ICFP 2008 (2008)
35. Wells, J.B.: Typability and type checking in system F are equivalent and undecidable. Ann. Pure Appl. Logic **98**(1–3), 111–156 (1999)

An Agda Formalization
of Üresin & Dubois' Asynchronous
Fixed-Point Theory

Ran Zmigrod, Matthew L. Daggitt[✉], and Timothy G. Griffin

Computer Laboratory, University of Cambridge, Cambridge, UK
mld46@cam.ac.uk

Abstract. In this paper we describe an Agda-based formalization of results from Üresin & Dubois' "Parallel Asynchronous Algorithms for Discrete Data." That paper investigates a large class of iterative algorithms that can be transformed into asynchronous processes. In their model each node asynchronously performs partial computations and communicates results to other nodes using unreliable channels. Üresin & Dubois provide sufficient conditions on iterative algorithms that guarantee convergence to unique fixed points for the associated asynchronous iterations. Proving such sufficient conditions for an iterative algorithm is often dramatically simpler than reasoning directly about an asynchronous implementation. These results are used extensively in the literature of distributed computation, making formal verification worthwhile.

Our Agda library provides users with a collection of sufficient conditions, some of which mildly relax assumptions made in the original paper. Our primary application has been in reasoning about the correctness of network routing protocols. To do so we have derived a new sufficient condition based on the ultrametric theory of Alexander Gurney. This was needed to model the complex policy-rich routing protocol that maintains global connectivity in the internet. Additionally we highlight and discuss two propositions from Üresin & Dubois, which during the course of the formalisation, turned out to be false.

1 Introduction

Many applications work with an iterative algorithm $\mathbf{F}$ and an initial state $\mathbf{x}(0)$ where successive states are computed as

$$\mathbf{x}(t+1) = \mathbf{F}(\mathbf{x}(t))$$

until a fixed point ξ is reached at some time t' when $\mathbf{x}(t') = \xi = \mathbf{F}(\xi)$. Here we assume that $\mathbf{x}(t)$ represents an n-dimensional vector in some state space. If we rewrite $\mathbf{F}$ as

$$\mathbf{F}(\mathbf{x}) = (\mathbf{F}_1(\mathbf{x}), \ldots, \mathbf{F}_n(\mathbf{x})),$$

The original version of this chapter was revised: Two proofs were corrected. For detailed information please see the erratum. The erratum to this chapter is available at https://doi.org/10.1007/978-3-319-94821-8_38

© The Author(s) 2018
J. Avigad and A. Mahboubi (Eds.): ITP 2018, LNCS 10895, pp. 623–639, 2018.
https://doi.org/10.1007/978-3-319-94821-8_37

then we can imagine that it may be possible to assign the computation of each $\mathbf{F}_i$ to a distinct processor. This might be performed in parallel with shared memory or in a completely distributed manner. However, enforcing correctness using global synchronization mechanisms may incur performance penalties that negate the gains from the parallelization. Furthermore, global synchronization is infeasible for applications such as network routing.

This leads to the question: When can we use the $\mathbf{F}_i$ to correctly implement an *asynchronous* version of $\mathbf{F}$-iteration? There are many answers to this question that depend on properties of the state space and the function $\mathbf{F}$ – see the survey paper by Frommer & Syzld [9].

Many of the approaches discussed in [9] rely on the rich structure of vector spaces over continuous domains. However, our motivation arises from network routing protocols where the state space is comprised of discrete data. Happily, Üresin and Dubois [21] have developed a theory of asynchronous iterations over discrete state spaces. They prove that if $\mathbf{F}$ is an Asynchronously Contracting Operator (ACO, see Sect. 3), then the associated asynchronous iteration will always converge to the correct fixed point. Their proof uses very weak assumptions about inter-process communication (indeed, in the case that the state space is finite they show that ACO is a necessary condition as well). These weak assumptions are a good model for the case of distributed routing protocols where messages can be delayed, lost, duplicated or reordered. Henceforth we will refer to Üresin and Dubois [21] as **UD**.

Proving that a given $\mathbf{F}$ is an ACO can be dramatically simpler than reasoning directly about an asynchronous implementation. However, in many cases it still remains non-trivial and so **UD** also derive several sufficient conditions that imply the ACO condition. These conditions are typically easier to prove for many common iterative algorithms. For example, they provide sufficient conditions for special cases where the state space is a partial order and $\mathbf{F}$ is order preserving.

In this paper we describe an Agda [3] formalization of the sufficient conditions and associated proofs from **UD**. This represents one part of a larger project in which we are developing formalized proofs of the asynchronous convergence for policy-rich distributed Bellman-Ford routing protocols (see [5]). This work required formalizing a new sufficient condition not found in **UD**, based on the ultrametric theory of Gurney [11]. During formalization it also became apparent that two of the other sufficient conditions in the original paper are incorrect. We provide a counter-example. In addition we suggest how to strengthen one of the sufficient conditions so that correctness is still guaranteed.

Many other applications of the results of **UD** can be found in the literature (for example, [4,6,7,16]). The proofs in **UD** are mathematically rigorous in the traditional sense, but their definitions are somewhat informal and they occasionally claim the existence of objects without providing an explicit construction. In our opinion a formal verification of the results is therefore a useful exercise.

There have been other efforts to formalize asynchronous computation such as Meseguer and Ölveczky [17] for real-time systems and Henrio, Khan, and Kammüller [13,14] for distributed languages. However, as far as we know our work is the first attempt to formalize the results of **UD**.

Our Agda development can be found on Github [1]. We hope that this will be a valuable resource for others interested in asynchronous iterations.

2 Preliminaries

In this section we introduce the components of the model of asynchronous computation that underpin **UD**'s results together with their Agda formalizations. Naturally, when formalizing mathematical proofs, there are concerns over steps that are considered trivial in the informal proof. We therefore highlight key features in the proof which are in practice significantly more complex than perhaps implied by the original reasoning.

Definition 1. *An* iterative algorithm *consists of an initial state $x(0)$ and an operator F such that $\forall t \in \mathbb{N}$, $x(t+1) = F(x(t))$.*

We begin by formalizing the product state space $S = S_1 \times \cdots \times S_n$. This is encoded by a Fin n-indexed family of Setoids. The type S is a function that takes i and returns the Carrier type of the i-th setoid. We can now formalize the iterative algorithm as follows:

```
sync-iter : S → ℕ → S
sync-iter x₀ zero    = x₀
sync-iter x₀ (suc K) = F (sync-iter x₀ K)
```

Routing Example. We briefly outline how this work can be applied to reasoning about convergence of a very general class of internet routing protocols. Full details can be found in Daggitt et al. [5].

Routing problems can be formalized as a tuple $(R, \oplus, E, \bar{0}, \infty)$, where:

- R is the set of routes,
- $\oplus : R \to R \to R$ is the choice operator, returning the preferred route,
- E is a set of functions of the form $R \to R$ representing generalized edge weights,
- $\bar{0}$ is the trivial route from a node to itself,
- ∞ is the invalid route.

A network configuration is represented as an $n \times n$ adjacency matrix $\mathbf{A}$ over E. The state space is made up of $n \times n$ matrices $\mathbf{X}$ over R. Matrix addition, $\mathbf{X} \oplus \mathbf{X}'$, is just the pointwise application of $\oplus$. The application of $\mathbf{A}$ to state $\mathbf{X}$ is defined as

$$(\mathbf{A}(\mathbf{X}))_{ij} = \left(\bigoplus_k \mathbf{A}_{ik}(\mathbf{X}_{kj}) \right).$$

That is, each node i choose the best extensions of the routes to j advertised by its neighbors. Finally, the iterative algorithm $\mathbf{F}$ is defined as

$$\mathbf{F}(\mathbf{X}) = \mathbf{A}(\mathbf{X}) \oplus \mathbf{I}, \tag{1}$$

where $\mathbf{I}$ is the matrix defined as $\mathbf{I}_{ii} = \bar{0}$, and $\mathbf{I}_{ij} = \infty$ for $i \neq j$. As explained in [5], an asynchronous version of $\mathbf{F}$ provides a good model of Distributed Bellman-Ford (DBF) routing protocols. At each asynchronous iteration in the distributed setting, each node i will compute only the i-th row of $\mathbf{F}(\mathbf{X})$ from the rows communicated by its adjacent neighbors.

Shortest paths routing is probably the simplest example where $\oplus = \min$ and E is the set of all f_w with $f_w(r) = w + r$.

2.1   Schedules

Schedules determine the asynchronous behaviour; they dictate when nodes release new information and the timing of that information propagating to other nodes. Let I be the set of nodes participating in the asynchronous process.

Definition 2. *A schedule ζ is a pair of functions $\alpha : \mathbb{N} \to \mathcal{P}(I)$ and $\beta : \mathbb{N} \to I \to I \to \mathbb{N}$ which satisfy the following properties:*

A1 $: \forall t \in \mathbb{N}, i, j \in I. \ \beta(t+1, i, j) \leq t$
A2 $: \forall t \in \mathbb{N}, i \ \ \in I. \ \exists t'. \ t < t' \wedge i \in \alpha(t')$
A3 $: \forall t \in \mathbb{N}, i, j \in I. \ \exists t'. \ \forall t''. \ t' < t'' \Rightarrow \beta(t'', i, j) \neq t$

The activation function α takes a time t and returns a subset of I containing the nodes that updated their value at time t. The data flow function β takes a time t and two nodes i and j and returns the time at which the data used by i at time t was generated by j.

Assumption A1 captures the notion of causality by ensuring that data can only be used after it was generated. A2 says that each node continues to activate indefinitely. Lastly, A3 says that the data generated at time t will only be used for a finite number of future updates.

Generalization 1. UD use a shared-memory model with all nodes communicating via shared memory, and so their definition of β takes only a single node i. However this model does not capture processes in which nodes communicate in a pairwise fashion without shared memory (e.g. internet routing). We have therefore augmented our definition of β to take two nodes, a source and destination. Their original definition can be recovered by providing a β function that is constant in its third argument.

Generalization 2. UD assumed that all nodes are active initially (i.e. $\alpha(0) = I$), which is unlikely to be true in a distributed context. Fortunately this assumption turns out to be unnecessary.

We formalize schedules in Agda as a dependent record. The number of nodes in the computation is passed as a parameter and the nodes themselves are represented by the Fin n type. The three properties are named causality, nonstarvation, and finite respectively.

```
record Schedule (n : ℕ) : Set where
  field
    α               : (t : 𝕋) → Subset n
    β               : (t : 𝕋)(i j : Fin n) → 𝕋
    causality       : ∀ t i j → β (suc t) i j ≤ t
    nonstarvation   : ∀ t i → ∃ λ k → i ∈ α (t + suc k)
    finite          : ∀ t i j → ∃ λ k → ∀ l → β (k + l) i j ≢ t
```

In the definition we use $\mathbb{T}$ as an alias for $\mathbb{N}$ to help semantically differentiate between times and other natural numbers. It would also be possible to implicitly capture causality by changing the return type of β to Fin t instead of $\mathbb{T}$. However, it turns out that in practice when using β we nearly always want a regular time, and therefore each call to β would require a conversion to $\mathbb{T}$. We thus decide to keep causality as an explicit field of Schedule.

Another choice made when designing the formalisation of nonstarvation and finite was to replace the conditions such as $\forall y.\ x \leq y \implies P(y)$ with $\forall y.\ P(x+y)$. This removes the need to pass around proof terms, and consequently often makes using these properties easier to use. This same technique is used throughout the rest of our library.

An asynchronously iteration can be constructed by combining an iterative algorithm with a schedule.

Definition 3. *An* asynchronous iteration *over a schedule* $\mathscr{S} = (\alpha,\ \beta)$, *an initial state* $\boldsymbol{x}(0)$, *and an operator* $\boldsymbol{F}$, *is denoted as* $(\boldsymbol{F},\ \boldsymbol{x}(0),\ \mathscr{S})$ *such that* $\forall t \in \mathbb{N}, i \in I$

$$
x_i(t+1) = \begin{cases} \boldsymbol{x}_i(t) & \text{if } i \notin \alpha(t+1) \\ F_i(\boldsymbol{x}_0(\beta(t+1,i,0)),\ldots,\boldsymbol{x}_{n-1}(\beta(t+1,i,n-1))) & \text{otherwise} \end{cases}
$$

We formalize this in Agda as follows:

```
async-Iter' : Schedule n → S → ∀ {t} → Acc _<_ t → S
async-Iter' 𝒮 x₀ {zero}    _          i = x₀ i
async-Iter' 𝒮 x₀ {suc t} (acc rec) i with i ∈? α 𝒮 (suc t)
... | yes _ = F (λ j → async-Iter' 𝒮 x₀
                      (rec (β 𝒮 (suc t) i j) (s≤s (causality 𝒮 t i j))) j) i
... | no _  = async-Iter' 𝒮 x[0] (rec t ≤-refl) i
```

Those unfamiliar with Agda may wonder why the Acc argument is necessary. While we can see that this function will terminate as each recursive call goes from time t to time $\beta(t,\ i,\ j)$ which is strictly smaller due to causality, the Agda termination checker cannot detect this without help. Acc is a data-type found in the Agda standard library that helps the termination checker by providing an argument to the function that always becomes structurally smaller with each recursive call. Using the proof that the natural numbers are well-founded, this complexity is hidden from the user in the main function:

async-iter : Schedule $n \to S \to \mathbb{T} \to S$
async-iter $\mathscr{S}$ x_0 t = async-iter' $\mathscr{S}$ x_0 ($<$-wellFounded t)

3   Convergence Theorem

UD define a class of **F**s called Asynchronously Contracting Operators (ACOs). They then prove that if an operator is an ACO, then it will converge to the correct fixed point for all possible schedules.

Definition 4. *An operator F is an* asynchronously contracting operator (ACO) *on a subset $D(0)$ of the state space $S = S_0 \times S_1 \times \cdots \times S_{n-1}$ iff there exists a sequence of sets $D(K)$ such that*

(i) $\forall K \in \mathbb{N}.\ D(K) = D_0(K) \times D_1(K) \times \cdots \times D_{n-1}(K)$
(ii) $\exists \xi \in S.\ \exists T \in \mathbb{N}.\ \forall K \in \mathbb{N}.$

$$K < T \Rightarrow D(K+1) \subseteq D(K)$$
$$K \geq T \Rightarrow D(K) = \{\xi\}$$

(iii) $\forall K \in \mathbb{N}.\ x \in D(K) \Rightarrow F(x) \in D(K+1)$

The sequence $D(K)$ can be seen as a form of approximation for the process with each iteration providing a higher accuracy. Each set contains the possible states at a moment in time. $D(0)$ contains many possible states as the algorithm has just begun, and each set in the sequence removes some incorrect states. This occurs until $D(T) = \{\xi\}$ when the converged state has been found.

Generalization 3. The definition of ACO in **UD** used the clause $K < T \Rightarrow D(K+1) \subset D(K)$, where we have relaxed this to $K < T \Rightarrow D(K+1) \subseteq D(K)$. This relaxation is also found in the survey by Frommer and Szyld [9].

The definition of an ACO is captured in the following record type:

record **ACO** p : Set ___ where
  field
       D   : $\mathbb{N} \to \forall\ i \to S_i\ i \to$ Set p
       D-decreasing : $\forall\ K \to$ D (suc K) $\subseteq$ D K
       D-finish   : $\exists_2\ \lambda\ T\ \xi \to \forall\ K \to$ IsSingleton ξ (D ($T + K$))
       F-monotonic : $\forall\ K\ \{t\} \to t \in$ D $K \to$ F $t \in$ D (suc K)

The variable p represents the universe level of the family of sets D, while the universe level of ACO is inferred automatically (Set ___). The sets themselves are implemented as a double-indexed family of predicates over $S_i\ i$.

The following theorem is the main sufficient condition proved in **UD**.

Theorem 1. *If* **F** *is an ACO on a set* $D(0)$, *then for all schedules* $\mathscr{S}$, *any asynchronous iteration* $\mathbf{x}(k) = (\mathbf{F}, \mathbf{x}(0), \mathscr{S})$ *with* $\mathbf{x}(0) \in D(0)$, *converges to the unique fixed point* ξ *of* **F** *in* $D(0)$.

In order to prove this theorem, **UD** consider the concept of a *pseudo-periodic schedule*. It is then proved that every schedule (Definition 2) is in fact pseudo-periodic, which greatly simplifies reasoning about schedules. This is perhaps the least rigorous aspect of the work of **UD**, as they state this without proof.

Definition 5. *A schedule* $\mathscr{S} = (\alpha, \beta)$ *is* pseudo-periodic *if there exists an increasing function* $\varphi : \mathbb{N} \to \mathbb{N}$ *such that:*

(i) $\varphi(0) = 0$
(ii) $\forall K \in \mathbb{N}, i \in I.\ \exists t \in \mathbb{N}.\ i \in \alpha(t) \wedge \varphi(K) \leq t < \varphi(K+1)$
(iii) $\forall K, t \in \mathbb{N}, i, j \in I.\ t \geq \varphi(K+1) \implies \beta(t, i, j) \geq \tau_i(K) \geq \varphi(K)$

where $\tau_i(K)$ *is the earliest time after* $\varphi(K)$ *that element* i *is updated.*

The intuition behind φ is that by time $\varphi(K+1)$ every node is guaranteed to be using data generated at least as recently as $\varphi(K)$. Hence the interval $(\varphi(K), \varphi(K+1)]$ is known as the kth pseudo-period.

We formalize the pseudo-periodic property in Agda as follows:

```
record IsPseudoperiodic {n : ℕ} (𝒮 : Schedule n) : Set where
  open Schedule 𝒮
  field
    φ : ℕ → 𝕋
    τ : ℕ → Fin n → 𝕋

    φ-increasing : ∀ K → K ≤ φ K
    τ-active     : ∀ K i → i ∈ α (τ K i)
    τ-after-φ    : ∀ K i → φ K ≤ τ K i
    τ-expired    : ∀ K t i j → τ K j ≤ β (φ (suc K) + t) i j
```

Note that this represents a simplification of **UD**'s definition. We worked backwards from the proof of Theorem 1 and identified only those properties required. This simplification may have to change if we extend our library to include **UD**'s proof that the ACO condition is also necessary (in the case of finite state spaces).

UD assert that for any schedule there exist an infinite number of possible functions φ, but they do not provide any explicit constructions. This is one area where we had initial concerns when planning our proof strategy in Agda.

We start by defining nextActive, which takes a time t and a node index i and returns the first time after t for which that i is active.

```
nextActive' : (t k : T) {i : Fin n} → i ∈ α (t + suc k) → Acc _<_ k → T
nextActive' t zero {i} _ _ = suc t
nextActive' t (suc k) {i} i∈a[t+1+K] (acc rs) with i ∈? α t
... | yes i∈a = t
... | no i∉a rewrite +-suc t (suc k) = nextActive' (suc t) k i∈a[t+1+K] _

nextActive : T → Fin n → T
nextActive t i with nonstarvation t i
... | (K , i∈a[t+1+K]) = nextActive' t K i∈a[t+1+K] (<-wellFounded K)
```

We then define allActive, which returns the first time after t such that all nodes have activated since t.

```
allActive : T → T
allActive t = max t (nextActive t)
```

We then need to define three auxiliary functions: $pointExpiry_{ij}$ returns a time after which i does not use the data generated by j at time t.

```
pointExpiry_{ij} : Fin n → Fin n → T → T
pointExpiry_{ij} i j t = proj₁ (finite t i j)
```

$expiry_{ij}$ returns a time after which i only uses data generated by j after time t.

```
expiry_{ij} : T → Fin n → Fin n → T
expiry_{ij} t i j = List.max t (applyUpTo (pointExpiry_{ij} i j) (suc t))
```

$expiry_i$ returns a time after which i only uses data generated after time t.

```
expiry_i : T → Fin n → T
expiry_i t i = max t (expiry_{ij} t i)
```

Using these we can define the function expiry that returns a time after which all nodes only use data generated after time t.

```
expiry : T → T
expiry t = max t (expiry_i t)
```

Finally, we construct φ as follows:

```
φ : N → T
φ zero    = zero
φ (suc K) = suc (expiry (allActive (φ K)))
```

Therefore we find a time t such that all nodes have been activated after $\varphi(K)$ and then $\varphi(K+1)$ is defined as the time after which all data used was generated after t. The function τ (as defined in property (iii) of pseudo-periodic schedules) is simply a special call to nextActive.

$\tau : \mathbb{N} \to \text{Fin } n \to \mathbb{T}$
$\tau\ K\ i = \text{nextActive } (\varphi\ K)\ i$

We now prove that φ and τ satisfy the properties required to be pseudo-periodic as given in Definition 5. The property φ-increasing is relatively simple, given that proofs that the various functions are increasing:

φ-increasing : $\forall\ K \to K \leq \varphi\ K$
φ-increasing zero   = z≤n
φ-increasing (suc K) = s≤s (begin
  K $\leq\langle$ φ-increasing K $\rangle$
  $\varphi\ K$ $\leq\langle$ allActive-increasing $(\varphi\ K)$ $\rangle$
  allActive $(\varphi\ K)$ $\leq\langle$ expiry-increasing (allActive $(\varphi\ K)$) $\rangle$
  expiry (allActive $(\varphi\ K)$) ∎)

The second property says that τ is always active and it can be satisfied by using properties of nextActive:

τ-active : $\forall\ K\ i \to i \in \alpha\ (\tau\ K\ i)$
τ-active $K = $ nextActive-active $(\varphi\ K)$

The third property can be easily proved using the fact that nextActive is increasing:

τ-after-φ : $\forall\ K\ i \to \varphi\ K \leq \tau\ K\ i$
τ-after-φ zero $i = $ z≤n
τ-after-φ (suc K) $i = $ nextActive-increasing $(\varphi$ (suc K)) i

The final property states that at all points during a pseudo-period, no nodes use information generated in a previous pseudo-period. This is the most complex of the four properties to prove.

τ-expired : $\forall\ K\ t\ i\ j \to \tau\ K\ j \leq \beta\ (\varphi$ (suc K) $+ t)\ i\ j$
τ-expired $K\ t\ i\ j = $ expiry-expired (begin
  expiry (nextActive $\_$ j) $\leq\langle$ expiry-monotone (nextActive≤allActive $\_$ j) $\rangle$
  expiry (allActive $(\varphi\ K)$) $\leq\langle$ n≤1+n (expiry (allActive $(\varphi\ K)$)) $\rangle$
  φ (suc K) $\leq\langle$ m≤m+n $(\varphi$ (suc K)) t $\rangle$
  φ (suc K) $+ t$ ∎) $i\ j$

As previously mentioned the construction of φ is not discussed in **UD**. Nevertheless, filling this gap required significant effort in our Agda development.

The proof of Theorem 1 requires an additional fact about the functions τ_i: for each K, once all i have been updated after some time t, then $\mathbf{x}(t) \in D(K)$.

Lemma 1. $\forall t,\ K \in \mathbb{N},\ i \in I.\ \tau_i(K) \le t \implies x_i(t) \in D_i(K).$

In **UD** Lemma 1 is proved by a fairly easy induction on K. However, in Agda the construction, called τ-stability, turned out to be more difficult. Several smaller lemmas were required, the biggest of which is that the asynchronous iteration remains within $D(0)$, the proof of which is called async[t]'$\in D_0$.

async[t]'$\in D_0$: $\forall$ {t} (acc_t : Acc _<_ t) → async-Iter' $\mathscr{S}$ x_0 acc_t ∈ D 0
async[t]'$\in D_0$ {zero} _ $i = x_0 \in D_0$ i
async[t]'$\in D_0$ {suc t} $\overline{(\text{acc } rec)}$ i with i ∈? α (suc t)
... | yes $i \in a$ = D-decreasing 0 (F-monotonic 0 (λ j →
    async[t]'$\in D_0$ (rec (β (suc t) i j) ($s{\le}s$ (causality t i j))) j)) i
... | no $i \notin a$ = async[t]'$\in D_0$ (rec t ($s{\le}s$ $\le$-refl)) i

τ-stability' : $\forall$ {t K i} (acc_t : Acc _<_ t) → τ K $i \le t$ →
            async-Iter' $\mathscr{S}$ x_0 acc_t $i \in_u$ D K i
τ-stability' {_}   {zero} {i} acc_t _   = async[t]'$\in D_0$ acc_t i
τ-stability' {zero} {suc K} {i} _ $\tau{\le}0$ =
    contradiction $\tau{\le}0$ ($<{\Rightarrow}\not\ge$ $0{<}\tau[1{+}K]$)
τ-stability' {suc t} {suc K} {i} (acc rec) $\tau{\le}1{+}t$ with i ∈? α (suc t)
... | yes _ = F-monotonic K (λ j → τ-stability' _ ($\tau[1{+}K]$-expired $\tau{\le}1{+}t$)) i
... | no $i \notin a$ with τ (suc K) $i \overset{?}{=}$ suc t
... | no $\tau{\not\equiv}1{+}t$ = τ-stability' _ ($<{\Leftrightarrow}\le$pred ($\le{+}\not\equiv{\Rightarrow}<$ $\tau{\le}1{+}t$ $\tau{\not\equiv}1{+}t$))
... | yes $\tau{\equiv}1{+}t$ =
      contradiction (subst ($i \in_s$_) (cong α $\tau{\equiv}1{+}t$) (τ-active (suc K) i)) $i \notin a$

τ-stability : $\forall$ {t K i} → τ K $i \le t$ → asyncIter $\mathscr{S}$ x_0 t $i \in_u$ D K i
τ-stability {t} = τ-stability' ($<$-wellFounded t)

We now construct the final proof of convergence. To do this we must construct a time after which the result of the asynchronous iteration is always equal to the fixed point. **UD** prove that $\varphi(T + 1)$, where T is from the ACO, is the convergence time. This is because each pseudo-period, every node is updated at least once and a total of T updates must occur before convergence. In the Agda, we first extract T and ξ from D-Finish. We then prove Theorem 1 as follows.

T : 𝕋
T = proj₁ D-finish

ξ : S
ξ = proj₁ (proj₂ D-finish)

tᶜ : 𝕋
tᶜ = φ (suc T)

async[tᶜ]∈D[T] : ∀ t → asyncIter 𝒮 x₀ (tᶜ + t) ∈ D T
async[tᶜ]∈D[T] t j = τ-stability (begin
    τ T j   ≤⟨ τ-expired T 0 j j ⟩
    β (tᶜ + 0) j j ≡⟨ cong (λ v → β v j j) (+-identityʳ tᶜ) ⟩
    β tᶜ j j   ≤⟨ β-decreasing j j 1≤tᶜ ⟩
    tᶜ   ≤⟨ m≤m+n tᶜ t ⟩
    tᶜ + t   ∎)
    where open ≤-Reasoning

async-converge : ∀ K → asyncIter 𝒮 x₀ (tᶜ + K) ≈ ξ
async-converge K = D[T]≈{ξ} (async[tᶜ]∈D[T] K)

4   The Library

UD show that being an ACO is a sufficient (and sometimes a necessary) condition for convergence. However in practice, constructing the sets $D(K)$ can still be a non-trivial exercise. Therefore, an extensive array of sufficient (but often not necessary) conditions have been constructed that in practice can be simpler and more intuitive to apply. These conditions are nearly always a reduction back to ACOs.

In this section we discuss three different proposed sufficient conditions. The first two are from **UD** and the third is a modified version of a new sufficient condition found in a recent paper by Gurney [11] (which was essential for the results described in Daggitt et al. [5]). During the formalization process, we discovered counterexamples to the two conditions from **UD**.

4.1   Synchronous Iteration Conditions

The first set of sufficient conditions makes use of the synchronous iteration of the algorithm, which **UD** refer to as $\mathbf{y}(t)$, as opposed to the asynchronous iteration $\mathbf{x}(t)$. The conditions involve the existence of partial orderings, $\preceq_i$, over each S_i, which are lifted to the partial order $\preceq$ over S in the usual point-wise manner. **UD** then make the following claim (Proposition 3 in **UD**):

Claim 1. An operator F has a fixed-point ξ to which every asynchronous iteration converges for every starting state $y(0) \in D(0)$ if:

(i) $\forall a \in D(0).\ F(a) \in D(0)$
(ii) $\forall a, b \in D(0).\ a \preceq b \implies F(a) \preceq F(b)$
(iii) $\forall K \in \mathbb{N}.\ y(K+1) \preceq y(K)$
(iv) The sequence $\{y(K)\}$ converges.

They attempt to prove this by first showing a reduction from these conditions to an ACO and then using Theorem 1 to obtain the required result.

However this claim is not true. While the asynchronous iteration does converge from every starting state in $D(0)$, it does not necessarily converges to the *same* fixed point. The flaw in the original proof is that **UD** tacitly assume that the set $D(0)$ for the ACO they construct is the same as the original $D(0)$ specified in the conditions above. However the only elements that are provably in the ACO's $D(0)$ is the set $\{y(t) \mid t \in \mathbb{N}\}$. We now present a counter-example to the claim.

Consider the degenerate asynchronous environment that contains only a single node (i.e. $I = \{0\}$) and let F be the identity function (i.e. $F(a) = a$). Let $D(0) = \{x, y\}$ where the only relationships in the partial order are $x \preceq x$ and $y \preceq y$. Clearly (i), (ii), (iii) and (iv) all trivially hold as F is the identity function. However x and y are both fixed points, and which fixed point is reached depends on whether the iteration starts at x or y. Hence Claim 1 cannot be true.

We can strengthen the conditions by changing requirement (iv) to "There exists a ξ such that for all $y(0)$ the sequence $\{y(K)\}$ converges to ξ". The library formalises these conditions in Agda as:

```
record SynchronousConditions p o : Set (lsuc (a ⊔ ℓ ⊔ p ⊔ o)) where

    field
        D₀            : Pred Sᵢ p
        D₀-cong       : ∀ {x y} → x ∈ D₀ → x ≈ y → y ∈ D₀
        D₀-closed     : ∀ {x} → x ∈ D₀ → F x ∈ D₀

        _≤_ : Rel Sᵢ o
        ≤-isPartialOrder : IsIndexedPartialOrder S _≈_ _≤_

        F-monotone    : ∀ {x y} → x ∈ D₀ → y ∈ D₀ → x ≤ y → F x ≤ F y
        F-cong        : ∀ {x y} → x ≈ y → F x ≈ F y
        iter-decreasing : ∀ {x} → x ∈ D₀ → ∀ K → synclter x (suc K) ≤ synclter x K

        ξ             : S
        ξ-fixed       : F ξ ≈ ξ
        iter-converge : ∀ {x} → x ∈ D₀ → ∃ λ T → synclter x T ≈ ξ
```

The reduction of these conditions to an ACO runs as follows. The sequence of sets D required by the definition of an ACO are defined as follows:

$$D(K) = \{\mathbf{x} \mid \xi \preceq \mathbf{x} \preceq \mathbf{y}(K) \wedge \mathbf{x} \in D_0\}$$

which is directly translated in Agda as:

D : ℕ → ∀ i → Mᵢ i → Set p
D K i = (λ x → (ξ i ⪯ x) × (x ⪯ sync-iter x₀ K i)) ∩ D₀ i

The field D-decreasing can be proven using iter-decreasing and D-finish is a consequence of iter-converge and ξ-fixed. F-monotonic is the same for both ACO and SynchronousConditions.

Routing Example. Classical routing theory [2] assumes that distributivity holds:

$$\forall e \in E : x, y \in S : e(x \oplus y) = e(x) \oplus e(y) \tag{2}$$

and under this assumption one can prove that every entry of every routing table improves monotonically with each iteration when the protocol starts from the initial state $\mathbf{I}$. Therefore for classical routing problems such as shortest-paths, it is relatively easy to construct an instance of SynchronousConditions.

4.2 Finite Conditions

Another set of sufficient conditions proposed by **UD** are applicable when the initial set $D(0)$ is finite. Like Proposition 1, it requires that $\mathbf{F}$ is monotonic and $D(0)$ be closed over $\mathbf{F}$. Instead of reasoning about the synchronous iteration of the operator, it adds an additional requirement that $\mathbf{F}$ is non-expansive over $D(0)$.

Claim 2. An operator $\mathbf{F}$ has a fixed-point ξ to which every asynchronous iteration converges for every starting state $\mathbf{y}(0) \in D(0)$ if:

  (i) $D(0)$ is finite
  (ii) $\forall \mathbf{a} \in D(0). \ \mathbf{F}(\mathbf{a}) \in D(0)$
  (iii) $\forall \mathbf{a} \in D(0). \ \mathbf{F}(\mathbf{a}) \preceq \mathbf{a}$
  (iv) $\forall \mathbf{a}, \mathbf{b} \in D(0). \ \mathbf{a} \preceq \mathbf{b} \implies \mathbf{F}(\mathbf{a}) \preceq \mathbf{F}(\mathbf{b})$

UD's attempted proof for Claim 2 is a reduction to the conditions for Claim 1. Like Claim 1, the conditions therefore guarantee convergence but not to a unique solution. Similarly the counterexample for Claim 1 is also a counterexample for Claim 2.

Unlike Claim 1, we do not have a proposed strengthening of Claim 2 which would guarantee the uniqueness of the fixed point. The reason is that the finiteness condition, while guaranteeing the existence of a fixed point when combined with the other conditions, does not help to prove uniqueness. Instead much stronger conditions would be required, for example the assumption of the existence of a metric space over the computation as discussed in the next subsection. Any such stronger conditions tend to make finiteness superfluous.

4.3   Ultrametrics

The notion of convergence has an intuitive interpretation in metric spaces. In such spaces, convergence is equivalent to every application of the operator $\mathbf{F}$ moving you closer (in discrete steps) to the fixed point ξ.

There already exist results of this type. For instance El Tarazi [8] shows that if there is a normed linear space over each the values at each node i, then convergence occurs if there exists a fixed point $\mathbf{x}^*$ and a $\gamma \in (0, 1]$ such that:

$$||\mathbf{F}(\mathbf{x}) - \mathbf{x}^*|| \leqslant \gamma ||\mathbf{x} - \mathbf{x}^*||$$

However in many ways this is a very strong sufficient condition as the existence of a norm requires the existence of an additive operator on the space. For many processes, including our example of network routing, this may not be true.

Instead there is a more general result by Gurney [11] based on ultrametrics. An ultrametric [19] is a metric where the standard triangle inequality has been replaced by the strong triangle inequality. As far as we are aware, this result seems to have appeared only in [11]. The work proves that the ultrametric conditions not only imply the existence of an ACO but are actually equivalent to the existence of an ACO and therefore equivalent to saying the process converges. As with the theorems of **UD** we are primarily concerned with the usability of the theorems and therefore only prove the forwards direction.

Definition 6. *An* ultrametric space (S, Γ, d) *is a set* S, *a totally ordered set* Γ *with a least element* 0, *and a function* $d : S \to S \to \Gamma$ *such that:*

M1 : $d(x, y) = 0 \Leftrightarrow x = y$
M2 : $d(x, y) = d(y, x)$
M3 : $d(x, z) \leqslant \max(d(x, y), d(y, z))$

Definition 7. *A function* $f : S \to S$ *is strictly contracting on orbits in an ultrametric space* (S, Γ, d) *if:*

$$x \neq f(x) \implies d(x, f(x)) > d(f(x), f(f(x)))$$

i.e. the distance between iterations strictly decreases.

Definition 8. *An operator* $f : S \to S$ *is strictly contracting on a fixed point* x^* *in an ultrametric space* (S, Γ, d) *if:*

$$x \neq x^* \implies d(x^*, x) > d(x^*, f(x))$$

Theorem 2 (Gurney [11]). *If there exists* (S_i, Γ, d_i), *and we take* $S = \prod_i S_i$ *and* $d(\boldsymbol{x}, \boldsymbol{y}) = \max_i d_i(\boldsymbol{x}_i, \boldsymbol{y}_i)$ *then* $\boldsymbol{F}$ *is an ACO if:*

1. *Γ is finite*
2. *$\boldsymbol{F}$ is strictly contracting on orbits over (S, Γ, d)*
3. *$\boldsymbol{F}$ is strictly contracting on a fixed point over (S, Γ, d)*
4. *S is non-empty*

These conditions are constructed in Agda as:

```
record UltrametricConditions : Set (a ⊔ ℓ) where
  field
    dᵢ                        : ∀ {i} → Sᵢ i → Sᵢ i → ℕ

  d : S → S → ℕ
  d x y = max 0 (λ i → dᵢ (x i) (y i))

  field
    dᵢ-isUltrametric     : ∀ {i} → IsUltrametric (Sᵢ i) dᵢ
    F-strContrOnOrbits : F StrContrOnOrbitsOver d
    F-strContrOnFP      : F StrContrOnFixedPointOver d
    d-bounded             : Bounded d

    element                 : S
    _=?_                     : Decidable _≈_
    F-cong                  : F Preserves _≈_ ⟶ _≈_
```

Note that in our formalisation we currently assume $\Gamma = $ Fin n for some n in order to simplify the theory. We plan to generalize this at some point.

Our Agda proof is very similar to the original proof by Gurney [11]. One of the key differences is that Gurney assumes that F is contracting where as we assume that F is strictly contracting on a fixed point. This is because in our use-case it is not possible to construct a contracting metric. The relationship between the two properties is not entirely clear, but the resulting proofs are very similar.

Routing Example. The Border Gateway Protocol [18] is used by all Internet Service Providers (ISPs) to maintain connectivity in the global internet. As explained in [5], distributivity (Eq. 2) cannot be guaranteed in this setting primarily because of the competing interests of service providers and the very expressive policy languages needed to implement these interests in routing.

Consequently, a great deal of research has been directed at finding sufficient conditions that guarantee convergence for policy-rich protocols such as BGP (see for example [10,20]). One reasonable condition is that the algebra be *strictly increasing*:

$$\forall e \in E : x \in S : x = x \oplus e(x) \neq e(x) \tag{3}$$

This says that a route x must be strictly more preferred than any extension $e(x)$.

However, now individual routing table entries are no longer guaranteed to improve monotonically, and so there is no natural ordering on the state space. Assuming Eq. 3, [5] show how to construct suitable ultrametrics d_i over the routing tables in such a way that they fulfill the properties required by Theorem 2. It is based on the observation that the *worst* routing table entry in the state will always improve after each iteration.

5   Conclusion

In this paper we have taken the mathematically rigorous yet informal proof of Üresin and Dubois' theory regarding the convergence of asynchronous iterations [21] and formalized it constructively in Agda. After explicitly constructing the previously unspecified pseudo-periodic sequences and mildly weakening some assumptions, we have succeeded in formalizing the core theorem of the paper. However some of the auxiliary propositions proposed in the original paper turned out to be false. In our opinion this alone justifies the formalization process.

Furthermore, we have described our library of proofs and sufficient conditions for asynchronous convergence, including a recent, new ultrametric condition. We hope that the library of sufficient conditions will be a valuable resource for those wanting to formally verify the convergence of a wide range of asynchronous iterations. The library is available on Github [1].

We are primarily interested in proving convergence and therefore we have thus far only formalized the sufficient conditions from Üresin and Dubois and not their proof that the ACO condition is also necessary in the case of finite state spaces. This would be an interesting extension to our development. In addition it would be interesting to see if other related work such as [15,22,23], using different models, could be integrated into our formalization.

References

1. Agda routing library. https://github.com/MatthewDaggitt/agda-routing/tree/itp2018
2. Baras, J.S., Theodorakopoulos, G.: Path problems in networks. Synth. Lect. Commun. Netw. **3**(1), 1–77 (2010)
3. Bove, A., Dybjer, P., Norell, U.: A brief overview of Agda – a functional language with dependent types. In: Berghofer, S., Nipkow, T., Urban, C., Wenzel, M. (eds.) TPHOLs 2009. LNCS, vol. 5674, pp. 73–78. Springer, Heidelberg (2009). https://doi.org/10.1007/978-3-642-03359-9_6
4. Chau, C.K.: Policy-based routing with non-strict preferences. SIGCOMM Comput. Commun. Rev. **36**(4), 387–398 (2006)
5. Daggitt, M.L., Gurney, A.J.T., Griffin, T.G.: Asynchronous convergence of policy-rich distributed bellman-ford routing protocols. In: SIGCOMM Proceedings. ACM (2018, to appear)
6. Ducourthial, B., Tixeuil, S.: Self-stabilization with path algebra. Theor. Comput. Sci. **293**(1), 219–236 (2003). Max-Plus Algebras
7. Edwards, S.A., Lee, E.A.: The semantics and execution of a synchronous block-diagram language. Sci. Comput. Program. **48**(1), 21–42 (2003)
8. El Tarazi, M.N.: Some convergence results for asynchronous algorithms. Numer. Math. **39**(3), 325–340 (1982)
9. Frommer, A., Szyld, D.B.: On asynchronous iterations. J. Comput. Appl. Math. **123**(1), 201–216 (2000)
10. Griffin, T.G., Shepherd, F.B., Wilfong, G.: The stable paths problem and interdomain routing. IEEE/ACM Trans. Network. **10**(2), 232–243 (2002)
11. Gurney, A.J.T.: Asynchronous iterations in ultrametric spaces. Technical report (2017). https://arxiv.org/abs/1701.07434

12. Hendrick, C.: Routing information protocol (RIP), RFC 1058 (1988)
13. Henrio, L., Kammüller, F.: Functional active objects: typing and formalisation. Electron. Notes Theor. Comput. Sci. **255**, 83–101 (2009). FOCLASA
14. Henrio, L., Khan, M.U.: Asynchronous components with futures: semantics and proofs in Isabelle/HOL. Electron. Notes Theor. Comput. Sci. **264**(1), 35–53 (2010)
15. Lee, H., Welch, J.L.: Applications of probabilistic quorums to iterative algorithms. In: Proceedings 21st International Conference on Distributed Computing Systems, pp. 21–28, April 2001
16. Lee, H., Welch, J.L.: Randomized registers and iterative algorithms. Distrib. Comput. **17**(3), 209–221 (2005)
17. Meseguer, J., Ölveczky, P.C.: Formalization and correctness of the PALS architectural pattern for distributed real-time systems. In: ICFEM, pp. 303–320 (2010)
18. Rekhter, Y., Li, T.: A Border Gateway Protocol (BGP) (1995)
19. Schörner, E.: Ultrametric fixed point theorems and applications. Valuat. Theory Appl. **2**, 353–359 (2003)
20. Sobrinho, J.L.: An algebraic theory of dynamic network routing. IEEE/ACM Trans. Netw. **13**(5), 1160–1173 (2005)
21. Üresin, A., Dubois, M.: Parallel asynchronous algorithms for discrete data. J. ACM **37**(3), 588–606 (1990)
22. Üresin, A., Dubois, M.: Effects of asynchronism on the convergence rate of iterative algorithms. J. Parallel Distrib. Comput. **34**(1), 66–81 (1996)
23. Wei, J.: Parallel asynchronous iterations of least fixed points. Parallel Comput. **19**(8), 887–895 (1993)

Erratum to: Interactive Theorem Proving

Jeremy Avigad and Assia Mahboubi

Erratum to:
J. Avigad and A. Mahboubi (Eds.):
Interactive Theorem Proving, **LNCS 10895,**
https://doi.org/10.1007/978-3-319-94821-8

Erratum to "Backwards and Forwards with Separation Logic":

In the original version the following typo was introduced on page 80, line 30: "In [22] fHobor". In the updated version this typo was corrected to "In [22] Hobor".

Erratum to "An Agda Formalization of Üresin and Dubois' Asynchronous Fixed-Point Theory":

Our work formalised several proofs by Uresin & Dubois. In the original version of our paper two of the subsidiary proofs, while being technically correct, didn't actually prove what Uresin & Dubois say they prove. This was not made clear in our paper, which repeats their incorrect claim. The updated version lays out the corrected proofs and outlines why our original proofs were wrong.

The updated online version of these chapters can be found at
https://doi.org/10.1007/978-3-319-94821-8_5
https://doi.org/10.1007/978-3-319-94821-8_37

Author Index

Printed in the United States
By Bookmasters